W9-AZP-824

Pathways to College Mathematics

A Brief Guide to Getting
the Most from This Book

Feature	Description	Benefit	Page
Section-Opening Scenarios	Every section opens with a scenario presenting a unique application of mathematics in your life outside the classroom.	Realizing that mathematics is everywhere will help motivate your learning.	112
Learning Objectives	Every section begins with a list of objectives. Each objective is restated in the margin where the objective is covered.	The objectives focus your reading by emphasizing what is most important and where to find it.	2
Detailed Worked-Out Examples	Examples are clearly written and provide step-by-step solutions. No steps are omitted, and each step is thoroughly explained to the right of the mathematics.	The blue annotations will help you understand the solutions by providing the reason why every mathematics step is true and necessary.	103
Applications Using Real-World Data	Interesting applications from nearly every discipline, supported by up-to-date real world data, are included in every section.	Ever wonder how you'll use mathematics? These applications will show you how mathematics can solve real problems.	183
Explanatory Voice Balloons	Voice Balloons help to demystify mathematics. They translate mathematical language into plain English, clarify problem-solving procedures, and present alternative ways of understanding concepts.	Does math ever look foreign to you? This feature often translates math into everyday English. If you understand what you're reading, it makes problem solving a lot easier.	104
Great Question!	Answers to students' questions offer suggestions for problem solving, point out common errors to avoid, and provide informal hints and suggestions.	By seeing common mistakes, you'll be able to avoid them. This feature should help you not to feel anxious or threatened when asking questions in class.	116
Achieving Success	Achieving Success boxes offer strategies for success in learning math, as well as suggestions for future college coursework.	Follow these suggestions to help achieve your full academic potential in mathematics and beyond.	202
Brief Reviews	Brief Reviews cover skills you already learned but may have forgotten.	Having these refresher boxes easily accessible will help ease anxiety about skills you may have forgotten.	32

2 Work the Problems

Feature	Description	Benefit	Page
Check Point Examples	Each example is followed by a similar matched problem, called a Check Point, that offers you the opportunity to work a similar exercise. The answers to the Check Points are provided in the answer section.	You learn best by doing. You'll solidify your understanding of worked examples if you try a similar problem right away to be sure you understand what you've just read.	423
Extensive and Varied Exercise Sets	An abundant collection of exercises is included in an Exercise Set at the end of each section. Exercises are organized within categories. The exercises in the first category, Practice Exercises, follow the same order as the section's worked examples.	The parallel order of the Practice Exercises lets you refer to the worked examples and use them as models for solving these problems.	520

3 Review for Quizzes and Tests

Feature	Description	Benefit	Page
Detailed Chapter Summary	Each chapter contains a detailed review chart that summarizes the definitions and concepts in every section of the chapter. Examples that illustrate these key concepts are also included in the review.	Study each detailed summary and you'll know the most important material in the chapter!	316
Chapter Review Exercises	After the detailed summary, a comprehensive collection of review exercises for each of the chapter's sections provides the opportunity to practice the skills you just reviewed.	Practice makes perfect. These exercises contain the most significant problems for each of the chapter's sections.	97
Chapter Test	Each chapter contains a practice test with approximately 25 problems that cover the important concepts in the chapter.	You can use the chapter test to determine whether you have mastered the material covered in the chapter. Try taking the test under test conditions: Time yourself and don't peek at the answers until you're done.	492
Video Lecture Series	Available to students within MyMathLab, the Video Lecture Series highlights key examples from every chapter of the textbook.	These videos let you review each objective from the textbook where you need extra help.	509

Annotated Instructor's Edition

Pathways to College Mathematics

Robert Blitzer

Miami Dade College

PEARSON

Boston Columbus Hoboken Indianapolis New York San Francisco
Amsterdam Cape Town Dubai London Madrid Milan Munich Paris Montréal Toronto
Delhi Mexico City São Paulo Sydney Hong Kong Seoul Singapore Taipei Tokyo

Editorial Director: *Chris Hoag*
Senior Acquisitions Editor: *Dawn Giovanniello*
Editorial Assistant: *Megan Tripp*
Program Manager: *Beth Kaufman*
Project Manager: *Kathleen A. Manley*
Program Management Team Lead: *Karen Wernholm*
Project Management Team Lead: *Christina Lepre*
Media Producer: *Vicki Dreyfus*
TestGen Content Manager: *Marty Wright*
MathXL Content Developer: *Rebecca Williams*
Marketing Manager: *Alicia Frankel*
Marketing Assistant: *Kelly Cross*
Senior Author Support/Technology Specialist: *Joe Vetere*
Rights and Permissions Project Manager: *Diahanne Lucas Dowridge*
Senior Procurement Specialist: *Carol Melville*
Associate Director of Design: *Andrea Nix*
Program Design Lead: *Beth Paquin*
Cover Design: *Studio Montage*
Production Coordination and Composition: *codeMantra*
Illustrations: *Scientific Illustrators and codeMantra*
Cover Image Frisbee: Daniel R. Burch/Getty Images; Cute turtle holding a blank white board: Oleg Belov/Shutterstock;
Turtle jumps and catches the frisbee: Oleg Belov/Shutterstock

This work is solely for the use of instructors and administrators for the purpose of teaching courses and assessing student learning. Unauthorized dissemination, publication or sale of the work, in whole or in part (including posting on the internet) will destroy the integrity of the work and is strictly prohibited.

Copyright © 2016 by Pearson Education, Inc. All Rights Reserved. Printed in the United States of America. This publication is protected by copyright, and permission should be obtained from the publisher prior to any prohibited reproduction, storage in a retrieval system, or transmission in any form or by any means, electronic, mechanical, photocopying, recording, or otherwise. For information regarding permissions, request forms and the appropriate contacts within the Pearson Education Global Rights & Permissions department, please visit www.pearsoned.com/permissions/.

Acknowledgments of third-party content appear on page C1–C2, which constitutes an extension of this copyright page.

PEARSON, ALWAYS LEARNING, and MYMATHLAB are exclusive trademarks in the U.S. and/or other countries owned by Pearson Education, Inc. or its affiliates.

Unless otherwise indicated herein, any third-party trademarks that may appear in this work are the property of their respective owners and any references to third-party trademarks, logos or other trade dress are for demonstrative or descriptive purposes only. Such references are not intended to imply any sponsorship, endorsement, authorization, or promotion of Pearson's products by the owners of such marks, or any relationship between the owner and Pearson Education, Inc. or its affiliates, authors, licensees or distributors.

Library of Congress Cataloging-in-Publication Data

Blitzer, Robert.

   Pathways to college mathematics/Robert F. Blitzer, Miami-Dade College. —1st edition.
     pages cm
   ISBN 978-0-13-410716-5
   1. Mathematics—Textbooks.   I. Title.
   QA39.3.B586 2016
   510—dc23
                      2015002451

2   17

www.pearsonhighered.com

ISBN-13: 978-0-13-410716-5
ISBN-10: 0-13-410716-0
AIE ISBN-13: 978-0-13-417575-1
ISBN-10: 0-13-417575-1

Contents

Preface

Pathways to College Mathematics provides a general survey of topics to prepare students for success in a variety of college math courses, including college algebra, statistics, liberal arts mathematics, quantitative reasoning, finite mathematics, and mathematics for education majors. The prerequisite is basic math or prealgebra.

The book has four major goals:

1. To provide skills necessary for STEM and non-STEM students that can be applied in future college math courses.
2. To show students how mathematics can solve authentic problems that apply to their lives.
3. To enable students to develop problem-solving skills, while fostering critical thinking, within an interesting setting.
4. To provide students with learning strategies that lead to persistence and success in mathematics.

One major obstacle in the way of achieving these goals is the fact that many students do not read their textbook. This has been a source of frustration for me and my colleagues in the classroom. Anecdotal evidence gathered over years highlights two basic reasons why students do not take advantage of their textbook:

"I'll never use this information."
"I can't follow the explanations."

I've written every page of *Pathways to College Mathematics* with the intent of eliminating these two objections. The ideas and tools I've used to do so are described for the student in "A Brief Guide to Getting the Most from This Book," which appears at the front of this book and are described in the following features.

Pedagogical Features

- *Learning Objectives.* Learning objectives, framed in the context of a student question (What am I supposed to learn?), are clearly stated at the beginning of each section. These objectives help students recognize and focus on the section's most important ideas. The objectives are restated in the margin at their point of use.

- *Chapter-Opening and Section-Opening Scenarios.* Every chapter and every section open with a scenario presenting a unique application of mathematics in students' lives outside the classroom. These scenarios are revisited in the course of the chapter or section in an example, discussion, or exercise.

- *Innovative Applications.* A wide variety of interesting applications, supported by up-to-date real world data, are included in every section.

- *Detailed Worked-Out Examples.* Each example is titled, making the purpose of the example clear. Examples are clearly written and provide students with detailed step-by-step solutions. No steps are omitted and each step is thoroughly explained to the right of the mathematics.

- *Explanatory Voice Balloons.* Voice balloons are used in a variety of ways to demystify mathematics. They translate mathematical language into everyday English, help clarify problem-solving procedures, present alternative ways of understanding concepts, and connect problem solving to concepts students have already learned.

- *Check Point Examples.* Each example is followed by a similar matched problem, called a Check Point, offering students the opportunity to test for conceptual understanding by working a similar exercise. The answers to the Check Points are provided in the answer section in the back of the book.

- *Concept and Vocabulary Checks.* This feature offers short-answer exercises, mainly fill-in-the-blank and true/false items, that assess students' understanding of the definitions and concepts presented in each section. The exercises can be used for classroom discussion in order to engage participation in the learning process. The Concept and Vocabulary Checks appear as separate features preceding the Exercise Sets.

- *Extensive and Varied Exercise Sets.* An abundant collection of exercises is included in an Exercise Set at the end of each section. Exercises are organized into six categories, a format that makes it easy to create well-rounded homework assignments.

  - Practice Exercises follow the same order as the section's worked examples. This parallel order enables students to refer to the titled examples and their detailed explanations to successfully achieve each section's objectives.

  - Practice Plus Exercises contain more challenging practice exercises that often require students to combine several skills or concepts. These exercises provide the option of creating assignments that take practice exercises to a more challenging level.

  - Application Exercises give the option of assigning realistic, relevant, and unique applications consistent with your students' needs and interests.

  - Critical Thinking Exercises require students to employ analytic skills that go beyond applying each section's basic objectives. The exercises ask students to make sense of complex problems and persevere in solving them.

  - Technology Exercises enable students to use technological tools to explore and deepen their understanding of concepts.

  - Group Exercises contain projects and collaborative activities that give students the opportunity to work cooperatively as they think and talk about mathematics.

- *Brief Reviews.* The Brief Review boxes summarize mathematical skills that students should have learned previously, but which many students still need to review. This feature appears whenever a particular skill is first needed and eliminates the need for you to reteach that skill.

- *Great Question!* This feature presents a variety of study tips in the context of students' questions. Answers to questions offer suggestions for problem solving, point out common errors to avoid, and provide informal hints and suggestions. As a secondary benefit, this feature should help students not to feel anxious or threatened when asking questions in class.

- *Achieving Success.* The Achieving Success boxes at the end of most sections offer strategies for persistence and success in college mathematics courses.

- *Blitzer Bonuses.* These enrichment essays provide historical, interdisciplinary, and otherwise interesting connections to the mathematics under study, showing students that math is an interesting and dynamic discipline.

- ***Detailed Chapter Review Summary Charts.*** Each chapter contains a review chart that summarizes the definitions and concepts in every section of the chapter. Examples that illustrate these key concepts are also included in the chart. For further review, the chart refers students to similar worked-out examples, by page number, from the chapter.

- ***End-of-Chapter Materials.*** A comprehensive collection of review exercises for each of the chapter's sections follows the review chart. This is followed by a chapter test that enables students to test their understanding of the material covered in the chapter.

I hope that my love for learning, as well as my respect for the diversity of students I have taught and learned from over the years, is apparent throughout *Pathways to College Mathematics*. By connecting mathematics to the whole spectrum of learning, it is my intent to show students that their world is profoundly mathematical, and indeed, π is in the sky.

Robert Blitzer

Supplements List

Student Resources

NEW Learning Guide: Activities-Based

Organized by the textbook's learning objectives, this new Learning Guide helps students learn how to make the most of their textbook, while also providing additional practice for each section and guidance for test preparation. New activities also give students an opportunity to discover and reinforce the concepts in an active learning environment, and are ideal for groupwork in class. Published in an unbound, binder-ready format, the Learning Guide can serve as the foundation for a course notebook for students.

Student's Solutions Manual

Includes fully worked solutions to the odd-numbered section exercises plus all Check Points, Chapter Reviews, and Chapter Tests.

Video Lecture Series

These videos, available in MyMathLab, provide students with extra help for each section of the textbook.

Instructor Resources

Annotated Instructor's Edition

Answers to exercises are printed on the same text page with longer answers at the back of the text.

Instructor's Solutions Manual (download only)

Includes fully worked solutions to every exercise in the text. Available in MyMathLab® and on the Instructor's Resource Center.

TestGen® (download only)

TestGen® enables teachers to build, edit, and print tests using a computerized test bank of questions developed to cover all the objectives of the text. TestGen® is algorithmically based, allowing teachers to create multiple but equivalent versions of the same question or test with the click of the button. Teachers can also modify test bank questions or add new questions. The software and testbank are available for download on the Instructor's Resource Center.

PowerPoint Lecture Slides (download only)

Available in MyMathLab® and on the Instructor's Resource Center, these fully editable slides include definitions, key concepts, and examples for use in a lecture setting and are available for each section of the text.

Acknowledgments

I would like to thank the following people for reviewing *Pathways to College Mathematics*.

Applebee, Jennifer *Middlesex County College*

Calandrino, Constance *Hudson County Community College*

Crockett, Suzonne *Lamar State College Orange*

Drent, Gigi *Kaua'i Community College*

Ghosh, Indranil *Austin Peay State University*

Girard, Ryan *Kaua'i Community College*

Hildebrand, Jeff *Georgia Gwinnett College*

Maurer, Vikki *Linn-Benton Community College*

Miceli, Patricia *Wright College*

Millard, Erin *Kaua'i Community College*

Pain, Karen *Palm Beach State College*

Peace, Matthew *Florida Gateway College*

Riola, Christopher *Moraine Valley Community College*

Roemer, Cynthia *Union County College*

San Nicolas, Kristine *Central Texas College*

Sehr, Barbara *Indiana University Kokomo*

Slocum, Craig *Moraine Valley Community College*

Vanderlaan, Cynthia *Purdue University at Fort Wayne*

Veneziale, Diane *Burlington County College*

Venter, Alexsis *Arapahoe Community College*

Additional acknowledgments are extended to Brad Davis, for preparing the answer section and annotated answers, and serving as accuracy checker; Dan Miller and Kelly Barber, for preparing the solutions manuals; the codeMantra formatting team for the book's brilliant paging; Brian Morris and Kevin Morris at Scientific Illustrators, for superbly illustrating the book; and Rebecca Dunn, project manager, and Kathleen Manley, production editor, whose collective talents kept every aspect of this project moving through its many stages.

I would like to thank my editor at Pearson, Dawn Giovanniello, and editorial assistant, Megan Tripp, who guided and coordinated the book from manuscript through production. Thanks to Beth Paquin and Studio Montage for the quirky cover and interior design. Finally, thanks to marketing manager Alicia Frankel for your innovative marketing efforts, and to the entire Pearson sales force, for your confidence and enthusiasm about the book.

Robert Blitzer

To the Student

The bar graph shows some of the qualities that students say make a great teacher.

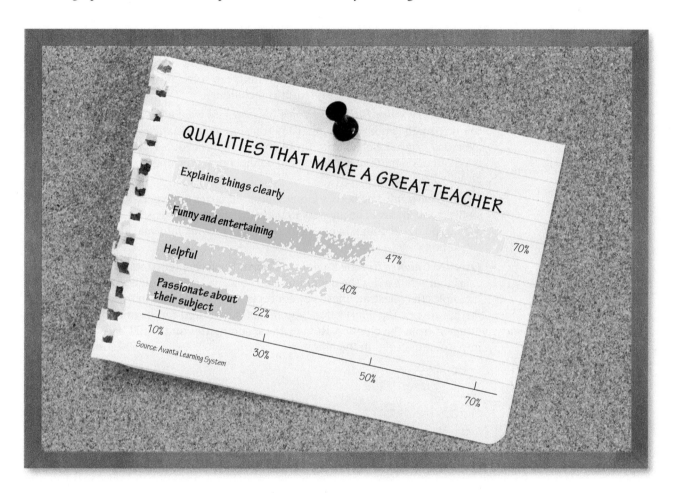

QUALITIES THAT MAKE A GREAT TEACHER

Explains things clearly — 70%

Funny and entertaining — 47%

Helpful — 40%

Passionate about their subject — 22%

10%   30%   50%   70%

Source: Avanta Learning System

It was my goal to incorporate each of these qualities throughout the pages of this book to help you gain control over the part of your life that involves numbers and mathematical ideas, providing the skills needed for success in future college math courses.

Explains Things Clearly

I understand that your primary purpose in reading *Pathways to College Mathematics* is to acquire a solid understanding of the required topics in your pathways math course. In order to achieve this goal, I've carefully explained each topic. Important definitions and procedures are set off in boxes, and worked-out examples that present solutions in a step-by-step manner appear in every section. Each example is followed by a similar matched problem, called a Check Point, for you to try so that you can actively participate in the learning process as you read the book. (Answers to all Check Points appear in the back of the book.)

Funny/Entertaining

Who says that a math textbook can't be entertaining? From our quirky cover to the photos in the chapter and section openers, prepare to expect the unexpected. I hope some of the book's enrichment essays, called Blitzer Bonuses, will put a smile on your face from time to time.

Helpful

I designed the book's features to help you acquire knowledge of fundamental mathematics, as well as to show you how math can solve authentic problems that apply to your life. These helpful features include:

- *Explanatory Voice Balloons*: Voice balloons are used in a variety of ways to make math less intimidating. They translate mathematical language into everyday English, help clarify problem-solving procedures, present alternative ways of understanding concepts, and connect new concepts to concepts you have already learned.
- *Great Question!*: The book's Great Question! boxes are based on questions students ask in class. The answers to these questions give suggestions for problem solving, point out common errors to avoid, and provide informal hints and suggestions.
- *Detailed Chapter Review Summary Charts*: Each chapter contains a review chart that summarizes the definitions and concepts in every section of the chapter. Examples that illustrate these key concepts are also included in the chart. For further review, the chart refers you to similar worked-out examples, by page number, from the chapter. Review these summaries and you'll know the most important material in the chapter!

Passionate about Their Subject

I passionately believe that no other discipline comes close to math in offering a more extensive set of tools for application and development of your mind. I wrote the book in Point Reyes National Seashore, 40 miles north of San Francisco. The park consists of 75,000 acres with miles of pristine surf-washed beaches, forested ridges, and bays bordered by white cliffs. It was my hope to convey the beauty and excitement of mathematics using nature's unspoiled beauty as a source of inspiration and creativity. Enjoy the pages that follow as you empower yourself with the mathematics needed to succeed in college, your career, and in your life.

Regards,

Bob

Robert Blitzer

About the Author

Bob Blitzer is a native of Manhattan and received a Bachelor of Arts degree with dual majors in mathematics and psychology (minor: English literature) from the City College of New York. His unusual combination of academic interests led him toward a Master of Arts in mathematics from the University of Miami and a doctorate in behavioral sciences from Nova University. Bob's love for teaching mathematics was nourished for nearly 30 years at Miami Dade College, where he received numerous teaching awards, including Innovator of the Year from the League for Innovations in the Community College and an endowed chair based on excellence in the classroom. In addition to *Pathways to College Mathematics*, Bob has written textbooks covering developmental mathematics, introductory algebra, intermediate algebra, college algebra, algebra and trigonometry, precalculus, and liberal arts mathematics, all published by Pearson. When not secluded in his Northern California writer's cabin, Bob can be found hiking the beaches and trails of Point Reyes National Seashore, and tending to the chores required by his beloved entourage of horses, chickens, and irritable roosters.

Numerical Pathways

Surfing the web, you hear politicians discussing the problem of the national debt that exceeds $15 trillion. They state that the interest on the debt equals government spending on veterans, homeland security, education, and transportation combined. They make it seem like the national debt is a real problem, but later you realize that you don't really know what a number like 15 trillion means. If the national debt were evenly divided among all citizens of the country, how much would every man, woman, and child have to pay? Is economic doomsday about to arrive?

Here's where you'll find this application:

Literacy with numbers, called numeracy, is a prerequisite for functioning in a meaningful way personally, professionally, and as a citizen. In this chapter, our focus is on understanding numbers, their properties, and their applications.

- The problem of placing a national debt that exceeds $15 trillion in perspective appears as Example 9 in Section 1.6.
- Confronting a national debt in excess of $15 trillion starts with grasping just how colossal $1 trillion actually is. The Blitzer Bonus on page 72 should help provide insight into this mind-boggling number.

1.1 Number Theory: Prime and Composite Numbers

After you have read this section, you should be able to:

1. Determine divisibility.

2. Write the prime factorization of a composite number.

3. Find the greatest common divisor of two numbers.

4. Solve problems using the greatest common divisor.

5. Find the least common multiple of two numbers.

6. Solve problems using the least common multiple.

Number Theory and Divisibility

You are organizing an intramural league at your college. You need to divide 40 men and 24 women into all-male and all-female teams so that each team has the same number of people. The men's teams should have the same number of players as the women's teams. What is the largest number of people that can be placed on a team?

This problem can be solved using a branch of mathematics called **number theory**. Number theory is primarily concerned with the properties of numbers used for counting, namely $1, 2, 3, 4, 5$, and so on. The set of counting numbers is also called the set of **natural numbers**. We represent this set by the letter **N**.

The Set of Natural Numbers

$$\mathbf{N} = \{1, 2, 3, 4, 5, 6, 7, 8, 9, 10, 11, \ldots\}$$

The three dots after 11 indicate that there is no final number and that the list goes on forever.

A Brief Review • Sets

• A **set** is a collection of objects whose contents can be clearly determined. The objects in a set are called the **elements**, or members, of the set.

• Capital letters are generally used to name sets.

• One method used to designate a set involves listing the elements inside a pair of braces, { }. Commas are used to separate the elements of the set.

We can solve the intramural league problem. However, to do so we must understand the concept of divisibility. For example, there are a number of different ways to divide the 24 women into teams, including

1 team with all 24 women:	$1 \times 24 = 24$
2 teams with 12 women per team:	$2 \times 12 = 24$
3 teams with 8 women per team:	$3 \times 8 = 24$
4 teams with 6 women per team:	$4 \times 6 = 24$
6 teams with 4 women per team:	$6 \times 4 = 24$
8 teams with 3 women per team:	$8 \times 3 = 24$
12 teams with 2 women per team:	$12 \times 2 = 24$
24 teams with 1 woman per team:	$24 \times 1 = 24.$

The natural numbers that are multiplied together resulting in a product of 24 are called *factors* of 24. Any natural number can be expressed as a product of two or more natural numbers. The natural numbers that are multiplied are called the **factors** of the product. Notice that a natural number may have many factors.

$$2 \times 12 = 24 \qquad 3 \times 8 = 24 \qquad 6 \times 4 = 24$$

Factors of 24   Factors of 24   Factors of 24

Great Question!

What's the difference between a factor and a divisor?

There is no difference. The words *factor* and *divisor* mean the same thing. Thus, 8 is a factor and a divisor of 24.

Great Question!

What's the difference between $b \mid a$ and b/a?

It's easy to confuse these notations. The symbol $b \mid a$ means b divides a. The symbol b/a means b divided by a (that is, $b \div a$, the quotient of b and a). For example, $5 \mid 35$ means 5 divides 35, whereas $5/35$ means 5 divided by 35, which is equivalent to the fraction $\frac{1}{7}$.

1 Determine divisibility.

The numbers 1, 2, 3, 4, 6, 8, 12, and 24 are all factors of 24. Each of these numbers divides 24 without a remainder.

In general, let a and b represent natural numbers. We say that a is **divisible** by b if the operation of dividing a by b leaves a remainder of 0.

A natural number is divisible by all of its factors. Thus, 24 is divisible by 1, 2, 3, 4, 6, 8, 12, and 24. Using the factor 8, we can express this divisibility in a number of ways:

24 is **divisible** by 8.

8 is a **divisor** of 24.

8 **divides** 24.

Mathematicians use a special notation to indicate divisibility.

> **Divisibility**
>
> If a and b are natural numbers, a is **divisible** by b if the operation of dividing a by b leaves a remainder of 0. This is the same as saying that b is a **divisor** of a, or b **divides** a. All three statements are symbolized by writing
>
> $$b \mid a.$$

Using this new notation, we can write

$$12 \mid 24.$$

Twelve divides 24 because 24 divided by 12 leaves a remainder of 0. By contrast, 13 does not divide 24 because 24 divided by 13 does not leave a remainder of 0. The notation

$$13 \nmid 24$$

means that 13 does not divide 24.

Table 1.1 shows some common rules for divisibility. Divisibility rules for 7 and 11 are difficult to remember and are not included in the table.

Table 1.1 Rules of Divisibility

Divisible By	Test	Example
2	The last digit is 0, 2, 4, 6, or 8.	5,892,796 is divisible by 2 because the last digit is 6.
3	The sum of the digits is divisible by 3.	52,341 is divisible by 3 because the sum of the digits is $5 + 2 + 3 + 4 + 1 = 15$, and 15 is divisible by 3.
4	The last two digits form a number divisible by 4.	3,947,136 is divisible by 4 because 36 is divisible by 4.
5	The number ends in 0 or 5.	28,160 and 72,805 end in 0 and 5, respectively. Both are divisible by 5.
6	The number is divisible by both 2 and 3. (In other words, the number is even and the sum of its digits is divisible by 3.)	954 is divisible by 2 because it ends in 4. 954 is also divisible by 3 because the digit sum is 18, which is divisible by 3. Because 954 is divisible by both 2 and 3, it is divisible by 6.
8	The last three digits form a number that is divisible by 8.	593,777,832 is divisible by 8 because 832 is divisible by 8.
9	The sum of the digits is divisible by 9.	5346 is divisible by 9 because the sum of the digits, 18, is divisible by 9.
10	The last digit is 0.	998,746,250 is divisible by 10 because the number ends in 0.
12	The number is divisible by both 3 and 4. (In other words, the sum of the digits is divisible by 3 and the last two digits form a number divisible by 4.)	614,608,176 is divisible by 3 because the digit sum is 39, which is divisible by 3. It is also divisible by 4 because the last two digits form 76, which is divisible by 4. Because 614,608,176 is divisible by both 3 and 4, it is divisible by 12.

Technology

Calculators and Divisibility

You can use a calculator to verify each result in Example 1. Consider part (a):

$$4 \mid 3,754,086.$$

Divide 3,754,086 by 4 using the following keystrokes:

$$3754086 \boxed{\div} 4.$$

Press $\boxed{=}$ or $\boxed{\text{ENTER}}$. The number displayed is 938521.5. This is not a natural number. The 0.5 shows that the division leaves a nonzero remainder. Thus, 4 does not divide 3,754,086. The given statement is false.

Now consider part (b):

$$9 \nmid 4,119,706,413.$$

Use your calculator to divide the number on the right by 9:

$$4119706413 \boxed{\div} 9.$$

Press $\boxed{=}$ or $\boxed{\text{ENTER}}$. The display is 457745157. This is a natural number. The remainder of the division is 0, so 9 does divide 4,119,706,413. The given statement is false.

Example 1   Using the Rules of Divisibility

Which one of the following statements is true?

a. $4 \mid 3,754,086$ **b.** $9 \nmid 4,119,706,413$ **c.** $8 \mid 677,840$

Solution

a. $4 \mid 3,754,086$ states that 4 divides 3,754,086. **Table 1.1** indicates that for 4 to divide a number, the last two digits must form a number that is divisible by 4. Because 86 is not divisible by 4, the given statement is false.

b. $9 \nmid 4,119,706,413$ states that 9 does *not* divide 4,119,706,413. Based on **Table 1.1**, if the sum of the digits is divisible by 9, then 9 does indeed divide this number. The sum of the digits is $4 + 1 + 1 + 9 + 7 + 0 + 6 + 4 + 1 + 3 = 36$, which is divisible by 9. Because 4,119,706,413 is divisible by 9, the given statement is false.

c. $8 \mid 677,840$ states that 8 divides 677,840. **Table 1.1** indicates that for 8 to divide a number, the last three digits must form a number that is divisible by 8. Because 840 is divisible by 8, then 8 divides 677,840, and the given statement is true.

The statement given in part (c) is the only true statement.

 **Check Point 1**   Which one of the following statements is true?   b

a. $8 \mid 48,324$ **b.** $6 \mid 48,324$ **c.** $4 \nmid 48,324$

Great Question!

Why is it so important to work each of the book's Check Points?

You learn best by doing. Do not simply look at the worked examples and conclude that you know how to solve them. To be sure you understand the worked examples, try each Check Point. Check your answer in the answer section before continuing your reading. Expect to read this book with pencil and paper handy to work the Check Points.

Prime Factorization

By developing some other ideas of number theory, we will be able to solve the school athletics program problem. We begin with the definition of a prime number.

Prime Numbers

A **prime number** is a natural number greater than 1 that has only itself and 1 as factors.

Using this definition, we see that the number 7 is a prime number because it has only 1 and 7 as factors. Said in another way, 7 is prime because it is divisible by only 1 and 7. The first ten prime numbers are 2, 3, 5, 7, 11, 13, 17, 19, 23, and 29. Each number in this list has exactly two divisors, itself and 1. By contrast, 9 is not a prime number; in addition to being divisible by 1 and 9, it is also divisible by 3. The number 9 is an example of a *composite number*.

Great Question!

Can a prime number be even?

The number 2 is the only even prime number. Every other even number has at least three factors: 1, 2, and the number itself.

Composite Numbers

A **composite number** is a natural number greater than 1 that is divisible by a number other than itself and 1.

2 Write the prime factorization of a composite number.

Using this definition, the first ten composite numbers are 4, 6, 8, 9, 10, 12, 14, 15, 16, and 18. Each number in this list has at least three distinct divisors.

By the definitions on the previous page, both prime numbers and composite numbers must be natural numbers *greater than* 1, so **the natural number 1 is neither prime nor composite**.

Every composite number can be expressed as the product of prime numbers. For example, the composite number 45 can be expressed as

$$45 = 3 \times 3 \times 5.$$

Note that 3 and 5 are prime numbers. Expressing a composite number as the product of prime numbers is called **prime factorization**. The prime factorization of 45 is $3 \times 3 \times 5$. The order in which we write these factors does not matter. This means that

$$45 = 3 \times 3 \times 5$$
$$\text{or} \quad 45 = 5 \times 3 \times 3$$
$$\text{or} \quad 45 = 3 \times 5 \times 3.$$

The ancient Greeks proved that if the order of the factors is disregarded, there is only one prime factorization possible for any given composite number. This statement is called the **Fundamental Theorem of Arithmetic**.

The Fundamental Theorem of Arithmetic

Every composite number can be expressed as a product of prime numbers in one and only one way (if the order of the factors is disregarded).

One method used to find the prime factorization of a composite number is called a **factor tree**. To use this method, begin by selecting any two numbers, other than 1, whose product is the number to be factored. One or both of the factors may not be prime numbers. Continue to factor composite numbers. Stop when all numbers are prime.

Great Question!

In Example 2, do I have to start the factor tree for 700 with 7 · 100?

No. It does not matter how you begin a factor tree. For example, in Example 2 you can factor 700 by starting with 5 and 140 ($5 \times 140 = 700$).

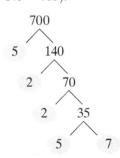

The prime factorization of 700 is

$$700 = 5 \times 2 \times 2 \times 5 \times 7$$
$$= 2^2 \times 5^2 \times 7.$$

This is the same prime factorization we obtained in Example 2.

Example 2   Prime Factorization Using a Factor Tree

Find the prime factorization of 700.

Solution

Start with any two numbers, other than 1, whose product is 700, such as 7 and 100. This forms the first branch of the tree. Continue factoring the composite number or numbers that result (in this case 100), branching until each branch ends with a prime number.

The prime factors are shown on light blue ovals. Thus, the prime factorization of 700 is

$$700 = 7 \times 2 \times 2 \times 5 \times 5.$$

We can use exponents to show the repeated prime factors:

$$700 = 7 \times 2^2 \times 5^2.$$

Using a dot to indicate multiplication and arranging the factors from least to greatest, we can write

$$700 = 2^2 \cdot 5^2 \cdot 7.$$

 **Check Point 2**   Find the prime factorization of 120. $2^3 \cdot 3 \cdot 5$

A Brief Review • Exponents

- If n is a natural number,

$$\underset{\text{Base}}{b^{\overset{\text{Exponent or Power}}{n}}} = \underbrace{b \cdot b \cdot b \cdot \cdots \cdot b.}_{\substack{b \text{ appears as a} \\ \text{factor } n \text{ times.}}}$$

- b^n is read "the nth power of b" or "b to the nth power." Thus, the nth power of b is defined as the product of n factors of b. The expression b^n is called an **exponential expression**. Furthermore, $b^1 = b$.

Exponential Expression	Read	Evaluation
8^1	8 to the first power	$8^1 = 8$
5^2	5 to the second power or 5 squared	$5^2 = 5 \cdot 5 = 25$
6^3	6 to the third power or 6 cubed	$6^3 = 6 \cdot 6 \cdot 6 = 216$
10^4	10 to the fourth power	$10^4 = 10 \cdot 10 \cdot 10 \cdot 10 = 10,000$
2^5	2 to the fifth power	$2^5 = 2 \cdot 2 \cdot 2 \cdot 2 \cdot 2 = 32$

Greatest Common Divisor

3 Find the greatest common divisor of two numbers.

The greatest common divisor (GCD) of two or more natural numbers is the largest number that is a divisor (or factor) of all the numbers. For example, 8 is the greatest common divisor of 32 and 40 because it is the largest natural number that divides both 32 and 40. Some pairs of numbers have 1 as their greatest common divisor. Such number pairs are said to be **relatively prime**. For example, the greatest common divisor of 5 and 26 is 1. Thus, 5 and 26 are relatively prime.

The greatest common divisor can be found using prime factorizations.

Finding the Greatest Common Divisor Using Prime Factorizations

To find the greatest common divisor of two or more numbers,

1. Write the prime factorization of each number.
2. Select each prime factor with the smallest exponent that is common to each of the prime factorizations.
3. Form the product of the numbers from step 2. The greatest common divisor is the product of these factors.

Example 3   Finding the Greatest Common Divisor

Find the greatest common divisor of 216 and 234.

Solution

Step 1   Write the prime factorization of each number. Begin by writing the prime factorizations of 216 and 234.

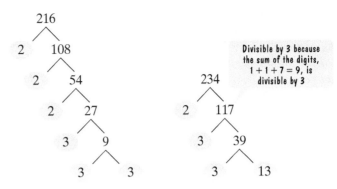

The factor tree on the left indicates that
$$216 = 2^3 \times 3^3.$$

The factor tree on the right indicates that
$$234 = 2 \times 3^2 \times 13.$$

Step 2 Select each prime factor with the smaller exponent that is common to each of the prime factorizations. Look at the factorizations of 216 and 234 from step 1. Can you see that 2 is a prime number common to the factorizations of 216 and 234? Likewise, 3 is also a prime number common to the two factorizations. By contrast, 13 is a prime number that is not common to both factorizations.

$$216 = 2^3 \times 3^3$$
$$234 = 2 \times 3^2 \times 13$$

2 is a prime number common to both factorizations. 3 is a prime number common to both factorizations.

Now we need to use these prime factorizations to determine which exponent is appropriate for 2 and which exponent is appropriate for 3. The appropriate exponent is the smaller exponent associated with the prime number in the factorizations. The exponents associated with 2 in the factorizations are 1 and 3, so we select 1. Therefore, one factor for the greatest common divisor is 2^1, or 2. The exponents associated with 3 in the factorizations are 2 and 3, so we select 2. Therefore, another factor for the greatest common divisor is 3^2.

$$216 = 2^3 \times 3^3$$

The smaller exponent on 2 is 1.   The smaller exponent on 3 is 2.

$$234 = 2^1 \times 3^2 \times 13$$

Step 3 Form the product of the numbers from step 2. The greatest common divisor is the product of these factors.

$$\text{Greatest common divisor} = 2 \times 3^2 = 2 \times 9 = 18$$

The greatest common divisor of 216 and 234 is 18.

Check Point 3 Find the greatest common divisor of 225 and 825. 75

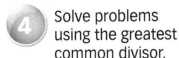

Solve problems using the greatest common divisor.

Example 4 Solving a Problem Using the Greatest Common Divisor

For an intramural league, you need to divide 40 men and 24 women into all-male and all-female teams so that each team has the same number of people. What is the largest number of people that can be placed on a team?

"For thousands of years, people have loved numbers and found patterns and structures among them. The allure of numbers is not limited to or driven by a desire to change the world in a practical way. When we observe how numbers are connected to one another, we are seeing the inner workings of a fundamental concept."

Edward B. Burger and Michael Starbird, *Coincidences, Chaos, and All That Math Jazz,* W. W. Norton and Company, 2005

Solution

Because 40 men are to be divided into teams, the number of men on each team must be a divisor of 40. Because 24 women are to be divided into teams, the number of women placed on a team must be a divisor of 24. Although the teams are all-male and all-female, the same number of people must be placed on each team. The largest number of people that can be placed on a team is the largest number that will divide into 40 and 24 without a remainder. This is the greatest common divisor of 40 and 24.

To find the greatest common divisor of 40 and 24, begin with their prime factorizations.

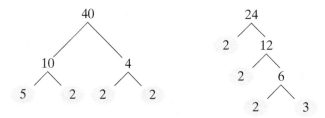

The factor trees indicate that

$$40 = 2^3 \times 5 \qquad \text{and} \qquad 24 = 2^3 \times 3.$$

We see that 2 is a prime number common to both factorizations. The exponents associated with 2 in the factorizations are 3 and 3, so we select 3.

$$\text{Greatest common divisor} = 2^3 = 2 \times 2 \times 2 = 8$$

The largest number of people that can be placed on a team is 8. Thus, the 40 men can form five teams with 8 men per team. The 24 women can form three teams with 8 women per team.

 Check Point 4   A chorus teacher needs to divide 192 men and 288 women into all-male and all-female singing groups so that each group has the same number of people. What is the largest number of people that can be placed in each singing group?   96

Least Common Multiple

5 Find the least common multiple of two numbers.

The **least common multiple** (LCM) of two or more natural numbers is the smallest natural number that is divisible by all of the numbers. One way to find the least common multiple is to make a list of the numbers that are divisible by each number. This list represents the **multiples** of each number. For example, if we wish to find the least common multiple of 15 and 20, we can list the sets of multiples of 15 and multiples of 20.

$\begin{cases} \text{Numbers Divisible by 15:} \\ \quad \text{Multiples of 15:} \end{cases}$ $\{15, 30, 45, 60, 75, 90, 105, 120, \dots\}$

$\begin{cases} \text{Numbers Divisible by 20:} \\ \quad \text{Multiples of 20:} \end{cases}$ $\{20, 40, 60, 80, 100, 120, 140, 160, \dots\}$

Two common multiples of 15 and 20 are 60 and 120. The least common multiple is 60. Equivalently, 60 is the smallest number that is divisible by both 15 and 20.

Sometimes a partial list of the multiples for each of two numbers does not reveal the smallest number that is divisible by both given numbers. A more efficient method for finding the least common multiple is to use prime factorizations.

> **Finding the Least Common Multiple Using Prime Factorizations**
>
> To find the least common multiple of two or more numbers,
>
> **1.** Write the prime factorization of each number.
>
> **2.** Select every prime factor that occurs, raised to the greatest power to which it occurs, in these factorizations.
>
> **3.** Form the product of the numbers from step 2. The least common multiple is the product of these factors.

Example 5   Finding the Least Common Multiple

Find the least common multiple of 144 and 300.

Solution

Step 1   Write the prime factorization of each number. Write the prime factorizations of 144 and 300.

$$144 = 2^4 \times 3^2$$
$$300 = 2^2 \times 3 \times 5^2$$

Step 2   Select every prime factor that occurs, raised to the greater power to which it occurs, in these factorizations. The prime factors that occur are 2, 3, and 5. The greater exponent that appears on 2 is 4, so we select 2^4. The greater exponent that appears on 3 is 2, so we select 3^2. The only exponent that occurs on 5 is 2, so we select 5^2. Thus, we have selected 2^4, 3^2, and 5^2.

Step 3   Form the product of the numbers from step 2. The least common multiple is the product of these factors.

$$\text{Least common multiple} = 2^4 \times 3^2 \times 5^2 = 16 \times 9 \times 25 = 3600$$

The least common multiple of 144 and 300 is 3600. The smallest natural number divisible by both 144 and 300 is 3600.

 **Check Point 5**   Find the least common multiple of 18 and 30.   90

 Solve problems using the least common multiple.

Example 6   Solving a Problem Using the Least Common Multiple

A movie theater runs its films continuously. One movie runs for 80 minutes and a second runs for 120 minutes. Both movies begin at 4:00 P.M. When will the movies begin again at the same time?

Solution

The shorter movie lasts 80 minutes, or 1 hour, 20 minutes. It begins at 4:00, so it will be shown again at 5:20. The longer movie lasts 120 minutes, or 2 hours. It begins at 4:00, so it will be shown again at 6:00. We are asked to find when the movies will begin again at the same time. Therefore, we are looking for the least common multiple of 80 and 120. Find the least common multiple and then add this number of minutes to 4:00 P.M.

Begin with the prime factorizations of 80 and 120:

$$80 = 2^4 \times 5$$
$$120 = 2^3 \times 3 \times 5.$$

Great Question!

Can I solve Example 6 by making a partial list of starting times for each movie?

Yes. Here's how it's done:

Shorter Movie (Runs 1 hour, 20 minutes):

   4:00, 5:20, 6:40, 8:00, . . .

Longer Movie (Runs 2 hours):

   4:00, 6:00, 8:00, . . .

The list reveals that both movies start together again at 8:00 P.M.

Using $80 = 2^4 \times 5$ and $120 = 2^3 \times 3 \times 5$, we select each prime factor, with the greater exponent from each factorization.

$$\text{Least common multiple} = 2^4 \times 3 \times 5 = 16 \times 3 \times 5 = 240$$

Therefore, it will take 240 minutes, or 4 hours, for the movies to begin again at the same time. By adding 4 hours to 4:00 P.M., they will start together again at 8:00 P.M.

Check Point 6   A movie theater runs two documentary films continuously. One documentary runs for 40 minutes and a second documentary runs for 60 minutes. Both movies begin at 3:00 P.M. When will the movies begin again at the same time?   5:00 P.M.

Achieving Success

Practice! Practice! Practice!

The way to learn mathematics is by seeing solutions to examples and **doing exercises**. This means working the Check Points and the assigned exercises in the Exercise Sets. There are no alternatives. It's easy to read a solution, or watch your professor solve an example, and believe you know what to do. However, learning mathematics requires that you actually **perform solutions by yourself**. Get in the habit of working exercises every day. The more time you spend solving exercises, the easier the process becomes.

Concept and Vocabulary Check

Exercises in the Concept and Vocabulary Check are intended for group and class discussions.

In Exercises 1–4, fill in each blank so that the resulting statement is true.

1. A natural number greater than 1 that has only itself and 1 as factors is called a/an ___prime___ number.

2. A natural number greater than 1 that is divisible by a number other than itself and 1 is called a/an ___composite___ number.

3. The largest number that is a factor of two or more natural numbers is called their greatest common divisor

4. The smallest number that is divisible by two or more natural numbers is called their least common multiple

In Exercises 5–8, determine whether each statement is true or false. If the statement is false, make the necessary change(s) to produce a true statement.   Changes to false statements will vary.

5. The notation $b \mid a$ means that b is divisible by a.   false

6. $b \nmid a$ means that b does not divide a.   true

7. The words *factor* and *divisor* have opposite meanings.   false

8. A number can only be divisible by exactly one number.   false

Respond to Exercises 9–18 using verbal or written explanations.

9. If a is a factor of c, what does this mean?   9–18. Answers will vary.

10. How do you know that 45 is divisible by 5?

11. What does "a is divisible by b" mean?

12. Describe the difference between a prime number and a composite number.

13. What does the Fundamental Theorem of Arithmetic state?

14. What is the greatest common divisor of two or more natural numbers?

15. Describe how to find the greatest common divisor of two natural numbers.

16. What is the least common multiple of two or more natural numbers?

17. Describe how to find the least common multiple of two natural numbers.

18. The process of finding the greatest common divisor of two natural numbers is similar to the process of finding the least common multiple of the numbers. Describe how the two processes differ.

Exercise Set 1.1

Practice Exercises

Use rules of divisibility to determine whether each number given in Exercises 1–10 is divisible by The numbers that divide the given number are shown.

  a. 2   **b.** 3   **c.** 4   **d.** 5   **e.** 6
  f. 8   **g.** 9   **h.** 10   **i.** 12.

1. 6944 2, 4, 8
2. 7245 3, 5, 9
3. 21,408 2, 3, 4, 6, 8, 12
4. 25,025 5
5. 26,428 2, 4
6. 89,001 3, 9
7. 374,832 2, 3, 4, 6, 8, 9, 12
8. 347,712 2, 3, 4, 6, 8, 12
9. 6,126,120 2, 3, 4, 5, 6, 8, 9, 10, 12 **10.** 5,941,221 3

In Exercises 11–24, use a calculator to determine whether each statement is true or false. If the statement is true, explain why this is so using one of the rules of divisibility in **Table 1.1** *on page 3.*

11. $3 \mid 5958$ true*
12. $3 \mid 8142$ true*
13. $4 \mid 10{,}612$ true*
14. $4 \mid 15{,}984$ true*
15. $5 \mid 38{,}814$ false
16. $5 \mid 48{,}659$ false
17. $6 \mid 104{,}538$ true*
18. $6 \mid 163{,}944$ true*
19. $8 \mid 20{,}104$ true*
20. $8 \mid 28{,}096$ true*
21. $9 \mid 11{,}378$ false
22. $9 \mid 23{,}772$ false
23. $12 \mid 517{,}872$ true*
24. $12 \mid 785{,}172$ true*

In Exercises 25–44, find the prime factorization of each composite number.

25. 75 $3 \cdot 5^2$
26. 45 $3^2 \cdot 5$
27. 56 $2^3 \cdot 7$
28. 48 $2^4 \cdot 3$
29. 105 $3 \cdot 5 \cdot 7$
30. 180 $2^2 \cdot 3^2 \cdot 5$
31. 500 $2^2 \cdot 5^3$
32. 360 $2^3 \cdot 3^2 \cdot 5$
33. 663 $3 \cdot 13 \cdot 17$
34. 510 $2 \cdot 3 \cdot 5 \cdot 17$
35. 885 $3 \cdot 5 \cdot 59$
36. 999 $3^3 \cdot 37$
37. 1440 $2^5 \cdot 3^2 \cdot 5$
38. 1280 $2^8 \cdot 5$
39. 1996 $2^2 \cdot 499$
40. 1575 $3^2 \cdot 5^2 \cdot 7$
41. 3675 $3 \cdot 5^2 \cdot 7^2$
42. 8316 $2^2 \cdot 3^3 \cdot 7 \cdot 11$
43. 85,800
44. 30,600 $2^3 \cdot 3^2 \cdot 5^2 \cdot 17$ **43.** $2^3 \cdot 3 \cdot 5^2 \cdot 11 \cdot 13$

In Exercises 45–56, find the greatest common divisor of the numbers.

45. 42 and 56 14
46. 25 and 70 5
47. 16 and 42 2
48. 66 and 90 6
49. 60 and 108 12
50. 96 and 212 4
51. 72 and 120 24
52. 220 and 400 20
53. 342 and 380 38
54. 224 and 430 2
55. 240 and 285 15
56. 150 and 480 30

In Exercises 57–68, find the least common multiple of the numbers.

57. 42 and 56 168
58. 25 and 70 350
59. 16 and 42 336
60. 66 and 90 990
61. 60 and 108 540
62. 96 and 212 5088
63. 72 and 120 360
64. 220 and 400
65. 342 and 380
66. 224 and 430
67. 240 and 285
68. 150 and 480
64. 4400 **65.** 3420 **66.** 48,160 **67.** 4560 **68.** 2400

Practice Plus

In Exercises 69–74, determine all values of d that make each statement true.

69. $9 \mid 12{,}34d$ 8
70. $9 \mid 23{,}42d$ 7
71. $8 \mid 76{,}523{,}45d$ 6
72. $8 \mid 88{,}888{,}82d$ 4
73. $4 \mid 963{,}23d$ 2, 6
74. $4 \mid 752{,}67d$ 2, 6

A **perfect number** *is a natural number that is equal to the sum of its factors, excluding the number itself. In Exercises 75–78, determine whether or not each number is perfect.*

75. 28 yes
76. 6 yes
77. 20 no
78. 50 no

A prime number is an **emirp** *("prime" spelled backward) if it becomes a different prime number when its digits are reversed. In Exercises 79–82, determine whether or not each prime number is an emirp.*

79. 41 no
80. 43 no
81. 107 yes
82. 113 yes

A prime number p such that 2p + 1 is also a prime number is called a **Germain prime**, *named after the French mathematician Sophie Germain (1776–1831), who made major contributions to number theory. In Exercises 83–86, determine whether or not each prime number is a Germain prime.*

83. 13 no
84. 11 yes
85. 241 no
86. 97 no

87. Find the product of the greatest common divisor of 24 and 27 and the least common multiple of 24 and 27. Compare this result to the product of 24 and 27. Write a conjecture based on your observation.

88. Find the product of the greatest common divisor of 48 and 72 and the least common multiple of 48 and 72. Compare this result to the product of 48 and 72. Write a conjecture based on your observation.   3456; 3456; Answers will vary. An example is: The product of the greatest common divisor and least common multiple of two numbers equals the product of the two numbers.

Application Exercises

89. In Carl Sagan's novel *Contact,* Ellie Arroway, the book's heroine, has been working at SETI, the Search for Extraterrestrial Intelligence, listening to the crackle of the cosmos. One night, as the radio telescopes are turned toward Vega, they suddenly pick up strange pulses through the background noise. Two pulses are followed by a pause, then three pulses, five, seven,

$$11, \ 13, \ 17, \ 19, \ 23, \ 29, \ 31, \ldots$$

continuing through 97. Then it starts all over again. Ellie is convinced that only intelligent life could generate the structure in the sequence of pulses. "It's hard to imagine some radiating plasma sending out a regular set of mathematical signals like this." What is it about the structure of the pulses that the book's heroine recognizes as the sign of intelligent life? Asked in another way, what is significant about the numbers of pulses? The numbers are the prime numbers between 2 and 97.

*See Answers to Selected Exercises.   **87.** 648; 648; Answers will vary. An example is: The product of the greatest common divisor and least common multiple of two numbers equals the product of the two numbers.

90. a. Multiples of 18: 18, 36, 54, 72, 90, 108, 126, 144, 162, 180, 198, 216; Multiples of 12: 12, 24, 36, 48, 60, 72, 84, 96, 108, 120, 132, 144, 156, 168, 180, 192, 204, 216; 6 times
c. Answers will vary; an example is: When each species has a prime number of years as the length of its life cycle, the two species do not have to share the forest as often.

90. There are two species of insects, *Magicicada septendecim* and *Magicicada tredecim,* that live in the same environment. They have a life cycle of exactly 17 and 13 years, respectively. For all but their last year, they remain in the ground feeding on the sap of tree roots. Then, in their last year, they emerge en masse from the ground as fully formed cricketlike insects, taking over the forest in a single night. They chirp loudly, mate, eat, lay eggs, then die six weeks later.

(*Source:* Marcus du Sautoy, *The Music of the Primes,* HarperCollins, 2003.)

   a. Suppose that the two species have life cycles that are not prime, say 18 and 12 years, respectively. List the set of multiples of 18 that are less than or equal to 216. List the set of multiples of 12 that are less than or equal to 216. Over a 216-year period, how many times will the two species emerge in the same year and compete to share the forest?

   b. Recall that both species have evolved prime-number life cycles, 17 and 13 years, respectively. Find the least common multiple of 17 and 13. How often will the two species have to share the forest? 221; once every 221 years

   c. Compare your answers to parts (a) and (b). What explanation can you offer for each species having a prime number of years as the length of its life cycle?

91. A relief worker needs to divide 300 bottles of water and 144 cans of food into groups that each contain the same number of items. Also, each group must have the same type of item (bottled water or canned food). What is the largest number of relief supplies that can be put in each group? 12

92. A chorus teacher needs to divide 180 men and 144 women into all-male and all-female singing groups so that each group has the same number of people. What is the largest number of people that can be placed in each singing group? 36

93. You have in front of you 310 five-dollar bills and 460 ten-dollar bills. Your problem: Place the five-dollar bills and the ten-dollar bills in stacks so that each stack has the same number of bills, and each stack contains only one kind of bill (five-dollar or ten-dollar). What is the largest number of bills that you can place in each stack? 10

94. Harley collects sports cards. He has 360 football cards and 432 baseball cards. Harley plans to arrange his cards in stacks so that each stack has the same number of cards. Also, each stack must have the same type of card (football or baseball). Every card in Harley's collection is to be placed in one of the stacks. What is the largest number of cards that can be placed in each stack? 72

95. You and your brother both work the 4:00 P.M. to 8:00 P.M. shift. You have every sixth night off. Your brother has every tenth night off. Both of you were off on June 1. Your brother would like to see a bargain 5:00 movie with you. When will the two of you have the same night off again? July 1

96. A movie theater runs its films continuously. One movie is a short documentary that runs for 40 minutes. The other movie is a full-length feature that runs for 100 minutes. Each film is shown in a separate theater. Both movies begin at noon. When will the movies begin again at the same time? 3:20 P.M.

97. Two people are jogging around the sidewalk surrounding the campus in the same direction. One person can run completely around the track in 15 minutes. The second person takes 18 minutes. If they both start running in the same place at the same time, how long will it take them to be together at this place again if they continue to run? 90 min or $1\frac{1}{2}$ hr

98. Two people are in a bicycle race around a circular track. One rider can race completely around the track in 40 seconds. The other rider takes 45 seconds. If they both begin the race at a designated starting point, how long will it take them to be together at this starting point again if they continue to race around the track? 360 sec or 6 min

Critical Thinking Exercises

99. Write a four-digit natural number that is divisible by 4 and not by 8. Answers will vary; an example is 1020.

100. Find the greatest common divisor and the least common multiple of $2^{17} \cdot 3^{25} \cdot 5^{31}$ and $2^{14} \cdot 3^{37} \cdot 5^{30}$. Express answers in the same form as the numbers given.

101. A middle-aged man observed that his present age was a prime number. He also noticed that the number of years in which his age would again be prime was equal to the number of years ago in which his age was prime. How old is the man? 53

102. A movie theater runs its films continuously. One movie runs for 85 minutes and a second runs for 100 minutes. The theater has a 15-minute intermission after each movie, at which point the movie is shown again. If both movies start at noon, when will the two movies start again at the same time? 2:20 A.M. on the third day

103. The difference between consecutive prime numbers is always an even number, except for two particular prime numbers. What are those numbers? 2 and 3

100. GCD: $2^{14} \cdot 3^{25} \cdot 5^{30}$; LCM: $2^{17} \cdot 3^{37} \cdot 5^{31}$

Technology Exercises

Use the divisibility rules listed in **Table 1.1** *on page 3 to answer the questions in Exercises 104–106. Then, using a calculator, perform the actual division to determine whether your answer is correct.*

104. Is 67,234,096 divisible by 4? yes

105. Is 12,541,750 divisible by 3? no

106. Is 48,201,651 divisible by 9? yes

Group Exercises 107–112. Answers will vary.

The following topics from number theory are appropriate for either individual or group research projects. A report should be given to the class on the researched topic. Useful references include books about numbers and number theory, books whose purpose is to excite the reader about mathematics, history of mathematics books, encyclopedias, and the Internet.

107. Euclid and Number Theory

108. An Unsolved Problem from Number Theory

109. Perfect Numbers

110. Deficient and Abundant Numbers

111. Formulas That Yield Primes

112. The Sieve of Eratosthenes

1.2 The Integers; Order of Operations

What am I supposed to learn?

After you have read this section, you should be able to:

1. Define the integers.

2. Graph integers on a number line.

3. Use the symbols $<$ and $>$.

4. Find the absolute value of an integer.

5. Perform operations with integers.

6. Use the order of operations agreement.

Can you cheat death? Life expectancy for the average American man is 75.2 years; for a woman, it's 80.4. But what's in your hands if you want to eke out a few more birthday candles? In this section, we use operations on a set of numbers called the *integers* to indicate factors within your control that can stretch your probable life span. Start by flossing. (See Example 5 on page 18.)

Mirror II (1963), George Tooker. Addison Gallery, Phillips Academy, MA. © George Tooker.

Defining the Integers

In Section 1.1, we applied some ideas of number theory to the set of natural, or counting, numbers:

$$\text{Natural numbers} = \{1, 2, 3, 4, 5, \ldots\}.$$

When we combine the number 0 with the natural numbers, we obtain the set of **whole numbers**:

$$\text{Whole numbers} = \{0, 1, 2, 3, 4, 5, \ldots\}.$$

The whole numbers do not allow us to describe certain everyday situations. For example, if the balance in your checking account is $30 and you write a check for $35, your checking account is overdrawn by $5. We can write this as -5, read *negative* 5. The set consisting of the natural numbers, 0, and the negatives of the natural numbers is called the set of **integers**.

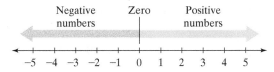

$$\text{Integers} = \{\ldots, -4, -3, -2, -1, 0, 1, 2, 3, 4, \ldots\}$$

Negative integers   Positive integers

Notice that the term *positive integers* is another name for the natural numbers. The positive integers can be written in two ways:

1. Use a "$+$" sign. For example, $+4$ is "positive four."

2. Do not write any sign. For example, 4 is assumed to be "positive four."

The Number Line; The Symbols $<$ and $>$

The **number line** is a graph we use to visualize the set of integers, as well as other sets of numbers. The number line is shown in **Figure 1.1**.

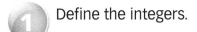

Negative numbers   Zero   Positive numbers

$$-5 \quad -4 \quad -3 \quad -2 \quad -1 \quad 0 \quad 1 \quad 2 \quad 3 \quad 4 \quad 5$$

Figure 1.1 The number line

The number line extends indefinitely in both directions, shown by the arrows on the left and the right. Zero separates the positive numbers from the negative

1 **Define the integers.**

numbers on the number line. The positive integers are located to the right of 0 and the negative integers are located to the left of 0. **Zero is neither positive nor negative.** For every positive integer on a number line, there is a corresponding negative integer on the opposite side of 0.

 Graph integers on a number line.

Integers are graphed on a number line by placing a dot at the correct location for each number.

 **Graphing Integers on a Number Line**

Graph:

a. −3   **b.** 4   **c.** 0.

Solution

Place a dot at the correct location for each integer.

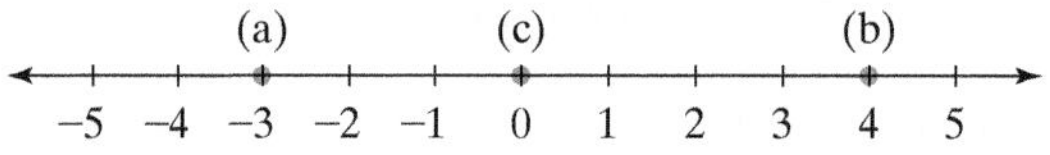

✔ **Check Point 1**   Graph:

▶ **a.** −4   **b.** 0   **c.** 3.   *

Use the symbols < and >.

We will use the following symbols for comparing two integers:

< means "is less than."

> means "is greater than."

On the number line, the integers increase from left to right. The *lesser* of two integers is the one farther to the *left* on a number line. The *greater* of two integers is the one farther to the *right* on a number line.

Look at the number line in **Figure 1.2**. The integers −4 and −1 are graphed.

Figure 1.2

Observe that −4 is to the left of −1 on the number line. This means that −4 is less than −1.

$$-4 < -1$$ −4 is less than −1 because −4 is to the **left** of −1 on the number line.

In **Figure 1.2**, we can also observe that −1 is to the right of −4 on the number line. This means that −1 is greater than −4.

$$-1 > -4$$ −1 is greater than −4 because −1 is to the **right** of −4 on the number line.

The symbols < and > are called **inequality symbols**. These symbols always point to the lesser of the two integers when the inequality statement is true.

−4 is less than −1. $$-4 < -1$$ The symbol points to −4, the lesser number.

−1 is greater than −4. $$-1 > -4$$ The symbol still points to −4, the lesser number.

*See Answers to Selected Exercises.

Example 2   **Using the Symbols** $<$ **and** $>$

Insert either $<$ or $>$ in the shaded area between the integers to make each statement true:

a. $-4\ \blacksquare\ 3$ **b.** $-1\ \blacksquare\ -5$ **c.** $-5\ \blacksquare\ -2$ **d.** $0\ \blacksquare\ -3$.

Solution

The solution is illustrated by the number line in **Figure 1.3**.

$$\xleftarrow{\qquad\begin{array}{ccccccccccc} \bullet & \bullet & \bullet & \bullet & \bullet & | & | & | & \bullet & | & | \\ -5 & -4 & -3 & -2 & -1 & 0 & 1 & 2 & 3 & 4 & 5 \end{array}\qquad}\rightarrow$$ **Figure 1.3**

a. $-4 < 3$ (negative 4 is less than 3) because -4 is to the left of 3 on the number line.

b. $-1 > -5$ (negative 1 is greater than negative 5) because -1 is to the right of -5 on the number line.

c. $-5 < -2$ (negative 5 is less than negative 2) because -5 is to the left of -2 on the number line.

d. $0 > -3$ (zero is greater than negative 3) because 0 is to the right of -3 on the number line.

✓ **Check Point 2**   Insert either $<$ or $>$ in the shaded area between the integers to make each statement true:

a. $6\ \blacksquare\ -7$ $>$ **b.** $-8\ \blacksquare\ -1$ $<$
c. $-25\ \blacksquare\ -2$ $<$ **d.** $-14\ \blacksquare\ 0$. $<$

The symbols $<$ and $>$ may be combined with an equal sign, as shown in the following table:

	Symbols	Meaning	Examples	Explanation
This inequality is true if either the $<$ part or the $=$ part is true.	$a \le b$	a is less than or equal to b.	$2 \le 9$ $9 \le 9$	Because $2 < 9$ Because $9 = 9$
This inequality is true if either the $>$ part or the $=$ part is true.	$b \ge a$	b is greater than or equal to a.	$9 \ge 2$ $2 \ge 2$	Because $9 > 2$ Because $2 = 2$

Absolute Value

④ Find the absolute value of an integer.

Absolute value describes distance from 0 on a number line. If a represents an integer, the symbol $|a|$ represents its absolute value, read "the absolute value of a." For example,

$$|-5| = 5.$$

The absolute value of -5 is 5 because -5 is 5 units from 0 on a number line.

> **Absolute Value**
>
> The **absolute value** of an integer a, denoted by $|a|$, is the distance from 0 to a on the number line. Because absolute value describes a distance, it is never negative.

Example 3   **Finding Absolute Value**

Find the absolute value:

a. $|-3|$ **b.** $|5|$ **c.** $|0|$.

Great Question!

Other than using a number line, is there another way to remember that -1 is greater than -5?

Yes. Think of negative integers as amounts of money that you *owe*. It's better to owe less, so

$$-1 > -5.$$

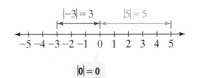

Figure 1.4 Absolute value describes distance from 0 on a number line.

Solution

The solution is illustrated in **Figure 1.4**.

a. $|-3| = 3$ The absolute value of -3 is 3 because -3 is 3 units from 0.

b. $|5| = 5$ 5 is 5 units from 0.

c. $|0| = 0$ 0 is 0 units from itself.

Example 3 illustrates that the absolute value of a positive integer or 0 is the number itself. The absolute value of a negative integer, such as -3, is the number without the negative sign. Zero is the only real number whose absolute value is 0: $|0| = 0$. **The absolute value of any integer other than 0 is always positive.**

Great Question!

What's the difference between $|-3|$ and $-|3|$?

They're easy to confuse.

$$|-3| = 3$$

-3 is 3 units from 0.

$$-|3| = -3$$

The negative is not inside the absolute value bars and is not affected by the absolute value.

Check Point 3 Find the absolute value:

a. $|-8|$ 8   **b.** $|6|$ 6   **c.** $-|8|$.   -8

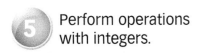

Addition of Integers

It has not been a good day! First, you lost your wallet with $50 in it. Then, you borrowed $10 to get through the day, which you somehow misplaced. Your loss of $50 followed by a loss of $10 is an overall loss of $60. This can be written

$$-50 + (-10) = -60.$$

The result of adding two or more numbers is called the **sum** of the numbers. The sum of -50 and -10 is -60.

You can think of gains and losses of money to find sums. For example, to find $17 + (-13)$, think of a gain of $17 followed by a loss of $13. There is an overall gain of $4. Thus, $17 + (-13) = 4$. In the same way, to find $-17 + 13$, think of a loss of $17 followed by a gain of $13. There is an overall loss of $4, so $-17 + 13 = -4$.

Using gains and losses, we can develop the following rules for adding integers:

Rules for Addition of Integers

Rule	Examples
If the integers have the same sign, **1.** Add their absolute values. **2.** The sign of the sum is the same as the sign of the two numbers.	$-11 + (-15) = -26$ Add absolute values: $11 + 15 = 26.$ Use the common sign.
If the integers have different signs, **1.** Subtract the smaller absolute value from the larger absolute value. **2.** The sign of the sum is the same as the sign of the number with the larger absolute value.	$-13 + 4 = -9$ Subtract absolute values: $13 - 4 = 9.$ Use the sign of the number with the greater absolute value. $13 + (-6) = 7$ Subtract absolute values: $13 - 6 = 7.$ Use the sign of the number with the greater absolute value.

5 Perform operations with integers.

Technology

Calculators and Adding Integers

You can use a calculator to add integers. Here are the keystrokes for finding $-11 + (-15)$:

Scientific Calculator

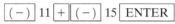

Graphing Calculator

$(-)$ 11 $+$ $(-)$ 15 ENTER

Here are the keystrokes for finding $-13 + 4$:

Scientific Calculator

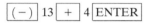

Graphing Calculator

$(-)$ 13 $+$ 4 ENTER

Great Question!

Other than gains and losses of money, is there another good analogy for adding integers?

Yes. Think of temperatures above and below zero on a thermometer. Picture the thermometer as a number line standing straight up. For example,

$$-11 + (-15) = -26$$

> If it's 11 below zero and the temperature falls 15 degrees, it will then be 26 below zero.

$$-13 + 4 = -9$$

> If it's 13 below zero and the temperature rises 4 degrees, the new temperature will be 9 below zero.

$$13 + (-6) = 7.$$

> If it's 13 above zero and the temperature falls 6 degrees, it will then be 7 above zero.

Using the analogies of gains and losses of money or temperatures can make the formal rules for addition of integers easy to use.

Can you guess what number is displayed if you use a calculator to find a sum such as $18 + (-18)$? If you gain 18 and then lose 18, there is neither an overall gain nor an overall loss. Thus,

$$18 + (-18) = 0.$$

We call 18 and -18 **additive inverses**. Additive inverses have the same absolute value, but lie on opposite sides of zero on the number line. Thus, -7 is the additive inverse of 7, and 5 is the additive inverse of -5. In general, the sum of any integer and its additive inverse is 0:

$$a + (-a) = 0.$$

Subtraction of Integers

Suppose that a computer that normally sells for $1500 has a price reduction of $600. The computer's reduced price, $900, can be expressed in two ways:

$$1500 - 600 = 900 \quad \text{or} \quad 1500 + (-600) = 900.$$

This means that

$$1500 - 600 = 1500 + (-600).$$

To subtract 600 from 1500, we add 1500 and the additive inverse of 600. Generalizing from this situation, we define subtraction as follows:

Definition of Subtraction

For all integers a and b,

$$a - b = a + (-b).$$

In words, to subtract b from a, add the additive inverse of b to a. The result of subtraction is called the **difference**.

Technology

Calculators and Subtracting Integers

You can use a calculator to subtract integers. Here are the keystrokes for finding $17 - (-11)$:

Scientific Calculator

$17 \boxed{-} 11 \boxed{+/_-} \boxed{=}$

Graphing Calculator

$17 \boxed{-} \boxed{(-)} 11 \boxed{ENTER}$

Here are the keystrokes for finding $-18 - (-5)$:

Scientific Calculator

$18 \boxed{+/_-} \boxed{-} 5 \boxed{+/_-} \boxed{=}$

Graphing Calculator

$\boxed{(-)} 18 \boxed{-} \boxed{(-)} 5 \boxed{ENTER}$

Don't confuse the subtraction key on a graphing calculator, $\boxed{-}$, with the sign change or additive inverse key, $\boxed{(-)}$. What happens if you do?

Example 4   Subtracting Integers

Subtract:

a. $17 - (-11)$

b. $-18 - (-5)$

c. $-18 - 5$.

Solution

a. $17 - (-11) = 17 + 11 = 28$

> Change the subtraction to addition.   Replace −11 with its additive inverse.

b. $-18 - (-5) = -18 + 5 = -13$

> Change the subtraction to addition.   Replace −5 with its additive inverse.

c. $-18 - 5 = -18 + (-5) = -23$

> Change the subtraction to addition.   Replace 5 with its additive inverse.

✓ **Check Point 4**   Subtract:

a. $30 - (-7)$ 37

b. $-14 - (-10)$ −4

▶ **c.** $-14 - 10$.   −24

Great Question!

Is there a practical way to think about what it means to subtract a negative integer?

Yes. Think of taking away a debit. Let's apply this analogy to $17 - (-11)$. Your checking account balance is $17 after an erroneous $11 charge was made against your account. When you bring this error to the bank's attention, they will take away the $11 debit and your balance will go up to $28:

$$17 - (-11) = 28.$$

Subtraction is used to solve problems in which the word *difference* appears. The difference between integers a and b is expressed as $a - b$.

Example 5   An Application of Subtraction Using the Word *Difference*

Life expectancy for the average American man is 75.2 years; for a woman, it's 80.4 years. The number line in **Figure 1.5**, with points representing eight integers, indicates factors, many within our control, that can stretch or shrink one's probable life span.

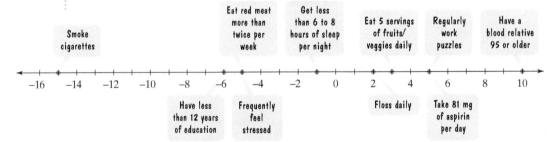

Stretching or Shrinking One's Life Span

Years of Life Gained or Lost

Figure 1.5
Source: Newsweek

a. What is the difference in the life span between a person who regularly works puzzles and a person who eats red meat more than twice per week?

b. What is the difference in the life span between a person with less than 12 years of education and a person who smokes cigarettes?

Solution

a. We begin with the difference in the life span between a person who regularly works puzzles and a person who eats red meat more than twice per week. Refer to **Figure 1.5** to determine years of life gained or lost.

The difference	is	the change in life span for a person who regularly works puzzles	minus	the change in life span for a person who eats red meat more than twice per week.
=		5	−	(−5)

$$= 5 + 5 = 10$$

The difference in the life span is 10 years.

b. Now we consider the difference in the life span between a person with less than 12 years of education and a person who smokes cigarettes.

The difference	is	the change in life span for a person with less than 12 years of education	minus	the change in life span for a person who smokes cigarettes.
=		−6	−	(−15)

$$= -6 + 15 = 9$$

The difference in the life span is 9 years.

Check Point 5 Use the number line in **Figure 1.5** to answer the following questions:

a. What is the difference in the life span between a person who eats five servings of fruits/veggies daily and a person who frequently feels stressed? 8 years

b. What is the difference in the life span between a person who gets less than 6 to 8 hours of sleep per night and a person who smokes cigarettes? 14 years

Multiplication of Integers

The result of multiplying two or more numbers is called the **product** of the numbers. You can think of multiplication as repeated addition or subtraction that starts at 0. For example,

$$3(-4) = 0 + (-4) + (-4) + (-4) = -12$$

The numbers have different signs and the product is negative.

and

$$(-3)(-4) = 0 - (-4) - (-4) - (-4) = 0 + 4 + 4 + 4 = 12.$$

The numbers have the same sign and the product is positive.

These observations give us the following rules for multiplying integers:

Rules for Multiplying Integers

Rule	**Examples**
1. The product of two integers with different signs is found by multiplying their absolute values. The product is negative.	• $7(-5) = -35$
2. The product of two integers with the same sign is found by multiplying their absolute values. The product is positive.	• $(-6)(-11) = 66$
3. The product of 0 and any integer is 0: $a \cdot 0 = 0$ and $0 \cdot a = 0$.	• $-17(0) = 0$
4. If no number is 0, a product with an odd number of negative factors is found by multiplying absolute values. The product is negative.	• $-2(-3)(-5) = -30$ Three (odd) negative factors
5. If no number is 0, a product with an even number of negative factors is found by multiplying absolute values. The product is positive.	• $-2(3)(-5) = 30$ Two (even) negative factors

Exponential Notation

Because exponents indicate repeated multiplication, rules for multiplying real numbers can be used to evaluate exponential expressions.

Example 6   Evaluating Exponential Expressions

Evaluate:

a. $(-6)^2$ **b.** -6^2 **c.** $(-5)^3$ **d.** $(-2)^4$.

Solution

a. $(-6)^2 = (-6)(-6) = 36$

Base is −6.   Same signs give positive product.

b. $-6^2 = -(6 \cdot 6) = -36$

Base is 6. The negative is not inside parentheses and is not taken to the second power.

c. $(-5)^3 = (-5)(-5)(-5) = -125$

An odd number of negative factors gives a negative product.

d. $(-2)^4 = (-2)(-2)(-2)(-2) = 16$

An even number of negative factors gives a positive product.

✔ **Check Point 6**   Evaluate:

a. $(-5)^2$ 25   **b.** -5^2 −25   **c.** $(-4)^3$ −64   **d.** $(-3)^4$. 81

Blitzer Bonus
●●●●●●●●●●●●●●●

Integers, Karma, and Exponents

On Friday the 13th, are you a bit more careful crossing the street even if you don't consider yourself superstitious? Numerology, the belief that certain integers have greater significance and can be lucky or unlucky, is widespread in many cultures.

Integer	Connotation	Culture	Origin	Example
4	Negative	Chinese	The word for the number 4 sounds like the word for death.	Many buildings in China have floor-numbering systems that skip 40–49.
7	Positive	United States	In dice games, this prime number is the most frequently rolled number with two dice.	There was a spike in the number of couples getting married on 7/7/07.
8	Positive	Chinese	It's considered a sign of prosperity.	The Beijing Olympics began at 8 P.M. on 8/8/08.
13	Negative	Various	Various reasons, including the number of people at the Last Supper	Many buildings around the world do not label any floor "13."
18	Positive	Jewish	The Hebrew letters spelling *chai*, or living, are the 8th and 10th in the alphabet, adding up to 18.	Monetary gifts for celebrations are often given in multiples of 18.
666	Negative	Christian	The New Testament's Book of Revelation identifies 666 as the "number of the beast," which some say refers to Satan.	In 2008, Reeves, Louisiana, eliminated 666 as the prefix of its phone numbers.

Source: The New York Times

Although your author is not a numerologist, he is intrigued by curious exponential representations for 666:

$$666 = 6 + 6 + 6 + 6^3 + 6^3 + 6^3$$
$$666 = 1^3 + 2^3 + 3^3 + 4^3 + 5^3 + 6^3 + 5^3 + 4^3 + 3^3 + 2^3 + 1^3$$
$$666 = 2^2 + 3^2 + 5^2 + 7^2 + 11^2 + 13^2 + 17^2$$

Sum of the squares of the first seven prime numbers

$$666 = 1^6 - 2^6 + 3^6.$$

Division of Integers

The result of dividing the integer a by the nonzero integer b is called the **quotient** of the numbers. We can write this quotient as $a \div b$ or $\frac{a}{b}$.

A relationship exists between multiplication and division. For example,

$$\frac{-12}{4} = -3 \text{ means that } 4(-3) = -12.$$

$$\frac{-12}{-4} = 3 \text{ means that } -4(3) = -12.$$

Technology

Multiplying and Dividing on a Calculator

Example: $(-173)(-256)$

Scientific Calculator

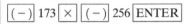

Graphing Calculator

The number 44288 should be displayed.
Division is performed in the same manner, using $\div$ instead of $\times$. What happens when you divide by 0? Try entering

$$8 \div 0$$

and pressing $=$ or ENTER.

Because of the relationship between multiplication and division, the rules for obtaining the sign of a quotient are the same as those for obtaining the sign of a product.

Rules for Dividing Integers

Rule	Examples
1. The quotient of two integers with different signs is found by dividing their absolute values. The quotient is negative.	• $\dfrac{80}{-4} = -20$ • $\dfrac{-15}{5} = -3$
2. The quotient of two integers with the same sign is found by dividing their absolute values. The quotient is positive.	• $\dfrac{27}{9} = 3$ • $\dfrac{-45}{-3} = 15$
3. Zero divided by any nonzero integer is zero.	• $\dfrac{0}{-5} = 0$ (because $-5(0) = 0$)
4. Division by 0 is undefined.	• $\dfrac{-8}{0}$ is undefined (because 0 cannot be multiplied by an integer to obtain -8).

⑥ Use the order of operations agreement.

Order of Operations

Suppose that you want to find the value of $3 + 7 \cdot 5$. Which procedure shown below is correct?

$$3 + 7 \cdot 5 = 3 + 35 = 38 \qquad \text{or} \qquad 3 + 7 \cdot 5 = 10 \cdot 5 = 50$$

If you know the answer, you probably know certain rules, called the **order of operations**, that make sure there is only one correct answer. One of these rules states that if a problem contains no parentheses, perform multiplication before addition. Thus, the procedure on the left is correct because the multiplication of 7 and 5 is first. Then the addition is performed. The correct answer is 38.

Here are the rules for determining the order in which operations should be performed:

Great Question!

How can I remember the order of operations?

This sentence may help: Please excuse my dear Aunt Sally.

Please	Parentheses
Excuse	Exponents
{ My	{ Multiplication
{ Dear	{ Division
{ Aunt	{ Addition
{ Sally	{ Subtraction

Order of Operations

1. Perform all operations within grouping symbols.
2. Evaluate all exponential expressions.
3. Do all multiplications and divisions in the order in which they occur, working from left to right.
4. Finally, do all additions and subtractions in the order in which they occur, working from left to right.

In the third step, be sure to do all multiplications and divisions *as they occur* from left to right. For example,

$$8 \div 4 \cdot 2 = 2 \cdot 2 = 4$$ Do the division first because it occurs first.

$$8 \cdot 4 \div 2 = 32 \div 2 = 16.$$ Do the multiplication first because it occurs first.

Example 7   Using the Order of Operations

Simplify: $6^2 - 24 \div 2^2 \cdot 3 + 1.$

Solution

There are no grouping symbols. Thus, we begin by evaluating exponential expressions. Then we multiply or divide. Finally, we add or subtract.

$$6^2 - 24 \div 2^2 \cdot 3 + 1$$

$$= 36 - 24 \div 4 \cdot 3 + 1$$ Evaluate exponential expressions: $6^2 = 6 \cdot 6 = 36$ and $2^2 = 2 \cdot 2 = 4.$

$$= 36 - 6 \cdot 3 + 1$$ Perform the multiplications and divisions from left to right. Start with $24 \div 4 = 6.$

$$= 36 - 18 + 1$$ Now do the multiplication: $6 \cdot 3 = 18.$

$$= 18 + 1$$ Finally, perform the additions and subtractions from left to right. Subtract: $36 - 18 = 18.$

$$= 19$$ Add: $18 + 1 = 19.$

▶ ✔ **Check Point 7**   Simplify: $7^2 - 48 \div 4^2 \cdot 5 + 2.$ 36

Example 8   Using the Order of Operations

Simplify: $(-6)^2 - (5 - 7)^2(-3).$

Solution

Because grouping symbols appear, we perform the operation within parentheses first.

$$(-6)^2 - (5 - 7)^2(-3)$$

$$= (-6)^2 - (-2)^2(-3)$$ Work inside parentheses first: $5 - 7 = 5 + (-7) = -2.$

$$= 36 - 4(-3)$$ Evaluate exponential expressions: $(-6)^2 = (-6)(-6) = 36$ and $(-2)^2 = (-2)(-2) = 4.$

$$= 36 - (-12)$$ Multiply: $4(-3) = -12.$

$$= 48$$ Subtract: $36 - (-12) = 36 + 12 = 48.$

▶ ✔ **Check Point 8**   Simplify: $(-8)^2 - (10 - 13)^2(-2).$ 82

Achieving Success

Learn from your mistakes. Being human means making mistakes. By finding and understanding your errors, you will become a better math student.

Source of Error	Remedy
Not Understanding a Concept	Review the concept by finding a similar example in your textbook or class notes. Ask your teacher questions to help clarify the concept.
Skipping Steps	Show clear step-by-step solutions. Detailed solution procedures help organize your thoughts and enhance understanding. Doing too many steps mentally often results in preventable mistakes.
Carelessness	Write neatly. Not being able to read your own math writing leads to errors. Avoid writing in pen so you won't have to put huge marks through incorrect work.

"You can achieve your goal if you persistently pursue it."

—Cha Sa-Soon, a 68-year-old South Korean woman who passed her country's written driver's-license exam on her 950th try (*Source: Newsweek*)

Concept and Vocabulary Check

Exercises in the Concept and Vocabulary Check are intended for group and class discussions.

In Exercises 1–4, fill in each blank so that the resulting statement is true.

1. The integers are defined by the set $\{\ldots, -3, -2, -1, 0, 1, 2, 3, \ldots\}$

2. If $a < b$, then a is located to the _____left_____ of b on a number line.

3. On a number line, the absolute value of a, denoted by $|a|$, represents _the distance from 0 to_ a

4. Two integers that have the same absolute value, but lie on opposite sides of zero on a number line, are called __additive inverses__.

In Exercises 5–8, determine whether each statement is true or false. If the statement is false, make the necessary change(s) to produce a true statement.   Changes to false statements will vary.

5. The sum of a positive integer and a negative integer is always a positive integer.   false

6. The difference between 0 and a negative integer is always a positive integer.   true

7. The product of a positive integer and a negative integer is never a positive integer.   true

8. The quotient of 0 and a negative integer is undefined.   false

Respond to Exercises 9–16 using verbal or written explanations.

9. How does the set of integers differ from the set of whole numbers?   9–16. Answers will vary.

10. Explain how to graph an integer on a number line.

11. Explain how to add integers.

12. Explain how to subtract integers.

13. Explain how to multiply integers.

14. Explain how to divide integers.

15. Describe what it means to raise a number to a power. In your description, include a discussion of the difference between -5^2 and $(-5)^2$.

16. Why is $\frac{0}{4}$ equal to 0, but $\frac{4}{0}$ undefined?

Exercise Set 1.2

Practice Exercises

In Exercises 1–4, start by drawing a number line that shows integers from −5 to 5. Then graph each of the following integers on your number line.

1. 3 *   **2.** 5 *   **3.** −4 *   **4.** −2 *

In Exercises 5–12, insert either < or > in the shaded area between the integers to make the statement true.

5. −2 ■ 7   <   **6.** −1 ■ 13   <

7. −13 ■ −2   <   **8.** −1 ■ −13   >

9. 8 ■ −50   >   **10.** 7 ■ −9   >

11. −100 ■ 0   <   **12.** 0 ■ −300   >

In Exercises 13–18, find the absolute value.

13. $|-14|$ 14   **14.** $|-16|$ 16

15. $|14|$ 14   **16.** $|16|$ 16

17. $|-300,000|$ 300,000   **18.** $|-1,000,000|$ 1,000,000

In Exercises 19–30, find each sum.

19. $-7 + (-5)$ −12   **20.** $-3 + (-4)$ −7

21. $12 + (-8)$ 4   **22.** $13 + (-5)$ 8

23. $6 + (-9)$ −3   **24.** $3 + (-11)$ −8

25. $-9 + (+4)$ −5   **26.** $-7 + (+3)$ −4

27. $-9 + (-9)$ −18   **28.** $-13 + (-13)$ −26

29. $9 + (-9)$ 0   **30.** $13 + (-13)$ 0

In Exercises 31–42, perform the indicated subtraction.

31. $13 - 8$ 5

32. $14 - 3$ 11

33. $8 - 15$ −7

34. $9 - 20$ −11

35. $4 - (-10)$ 14

36. $3 - (-17)$ 20

37. $-6 - (-17)$ 11

38. $-4 - (-19)$ 15

39. $-12 - (-3)$ −9

40. $-19 - (-2)$ −17

41. $-11 - 17$ −28

42. $-19 - 21$ −40

In Exercises 43–52, find each product.

43. $6(-9)$ −54

44. $5(-7)$ −35

45. $(-7)(-3)$ 21

46. $(-8)(-5)$ 40

47. $(-2)(6)$ −12

48. $(-3)(10)$ −30

49. $(-13)(-1)$ 13

50. $(-17)(-1)$ 17

51. $0(-5)$ 0

52. $0(-8)$ 0

In Exercises 53–66, evaluate each exponential expression.

53. 5^2 25

54. 6^2 36

55. $(-5)^2$ 25

56. $(-6)^2$ 36

57. 4^3 64

58. 2^3 8

59. $(-5)^3$ −125

60. $(-4)^3$ −64

61. $(-5)^4$ 625

62. $(-4)^4$ 256

63. -3^4 −81

64. -1^4 −1

65. $(-3)^4$ 81

66. $(-1)^4$ 1

In Exercises 67–80, find each quotient, or, if applicable, state that the expression is undefined.

67. $\frac{-12}{4}$ −3

68. $\frac{-40}{5}$ −8

69. $\frac{21}{-3}$ −7

70. $\frac{60}{-6}$ −10

71. $\frac{-90}{-3}$ 30

72. $\frac{-66}{-6}$ 11

73. $\frac{0}{-7}$ 0

74. $\frac{0}{-8}$ 0

75. $\frac{-7}{0}$ undefined

76. $\frac{0}{0}$ undefined

77. $(-480) \div 24$ −20

78. $(-300) \div 12$ −25

79. $(465) \div (-15)$ −31

80. $(-594) \div (-18)$ 33

In Exercises 81–100, use the order of operations to find the value of each expression.

81. $7 + 6 \cdot 3$ 25

82. $-5 + (-3) \cdot 8$ −29

83. $(-5) - 6(-3)$ 13

84. $-8(-3) - 5(-6)$ 54

85. $6 - 4(-3) - 5$ 13

86. $3 - 7(-1) - 6$ 4

87. $3 - 5(-4 - 2)$ 33

88. $3 - 9(-1 - 6)$ 66

89. $(2 - 6)(-3 - 5)$ 32

90. $9 - 5(6 - 4) - 10$ −11

91. $3(-2)^2 - 4(-3)^2$ −24

92. $5(-3)^2 - 2(-2)^3$ 61

93. $(2 - 6)^2 - (3 - 7)^2$ 0

94. $(4 - 6)^2 - (5 - 9)^3$ 68

95. $6(3 - 5)^3 - 2(1 - 3)^3$ −32

96. $-3(-6 + 8)^3 - 5(-3 + 5)^3$ −64

97. $8^2 - 16 \div 2^2 \cdot 4 - 3$ 45

98. $10^2 - 100 \div 5^2 \cdot 2 - (-3)$ 95

99. $24 \div [3^2 \div (8 - 5)] - (-6)$ 14

100. $30 \div [5^2 \div (7 - 12)] - (-9)$ 3

Practice Plus

In Exercises 101–110, use the order of operations to find the value of each expression.

101. $8 - 3[-2(2 - 5) - 4(8 - 6)]$ 14

102. $8 - 3[-2(5 - 7) - 5(4 - 2)]$ 26

103. $-2^2 + 4[16 \div (3 - 5)]$ −36

104. $-3^2 + 2[20 \div (7 - 11)]$ −19

105. $4|10 - (8 - 20)|$ 88

106. $-5|7 - (20 - 8)|$ −25

107. $[-5^2 + (6 - 8)^3 - (-4)] - [|-2|^3 + 1 - 3^2]$ −29

108. $[-4^2 + (7 - 10)^3 - (-27)] - [|-2|^5 + 1 - 5^2]$ −24

109. $\dfrac{12 \div 3 \cdot 5|2^2 + 3^2|}{7 + 3 - 6^2}$ −10

110. $\dfrac{-3 \cdot 5^2 + 89}{(5 - 6)^2 - 2|3 - 7|}$ −2

In Exercises 111–114, express each sentence as a single numerical expression. Then use the order of operations to simplify the expression. **111.** $-10 - (-2)^3; -2$ **112.** $-100 - (-5)^3; 25$

111. Cube −2. Subtract this exponential expression from −10.

112. Cube −5. Subtract this exponential expression from −100.

113. Subtract 10 from 7. Multiply this difference by 2. Square this product. $[2(7 - 10)]^2; 36$

114. Subtract 11 from 9. Multiply this difference by 2. Raise this product to the fourth power. $[2(9 - 11)]^4; 256$

Application Exercises

115. The peak of Mount McKinley, the highest point in the United States, is 20,320 feet above sea level. Death Valley, the lowest point in the United States, is 282 feet below sea level. What is the difference in elevation between the peak of Mount McKinley and Death Valley? 20,602 ft

116. The peak of Mount Kilimanjaro, the highest point in Africa, is 19,321 feet above sea level. Qattara Depression, Egypt, the lowest point in Africa, is 436 feet below sea level. What is the difference in elevation between the peak of Mount Kilimanjaro and the Qattara Depression? 19,757 ft

In Exercises 117–126, we return to the number line that shows factors that can stretch or shrink one's probable life span.

Stretching or Shrinking One's Life Span

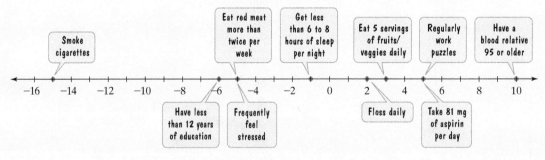

Years of Life Gained or Lost

Source: Newsweek

117. If you have a blood relative 95 or older and you smoke cigarettes, do you stretch or shrink your life span? By how many years? shrink by 5 years

118. If you floss daily and eat red meat more than twice per week, do you stretch or shrink your life span? By how many years? shrink by 3 years

119. If you frequently feel stressed and have less than 12 years of education, do you stretch or shrink your life span? By how many years? shrink by 11 years

120. If you get less than 6 to 8 hours of sleep per night and smoke cigarettes, do you stretch or shrink your life span? By how many years? shrink by 16 years

121. What happens to the life span for a person who takes 81 mg of aspirin per day and eats red meat more than twice per week? no change

122. What happens to the life span for a person who regularly works puzzles and a person who frequently feels stressed? no change

123. What is the difference in the life span between a person who has a blood relative 95 or older and a person who smokes cigarettes? 25 years

124. What is the difference in the life span between a person who has a blood relative 95 or older and a person who has less than 12 years of education? 16 years

125. What is the difference in the life span between a person who frequently feels stressed and a person who has less than 12 years of education? 1 year

126. What is the difference in the life span between a person who gets less than 6 to 8 hours of sleep per night and a person who frequently feels stressed? 4 years

The accompanying bar graph shows the amount of money, in billions of dollars, collected and spent by the U.S. government in selected years from 2001 through 2011. Use the information from the graph to solve Exercises 127–130. Express answers in billions of dollars.

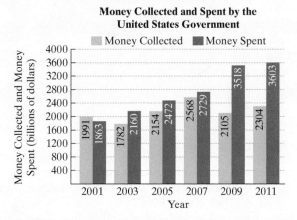

Money Collected and Spent by the United States Government

Source: Budget of the U.S. Government

127. a. In 2001, what was the difference between the amount of money collected and the amount spent? Was there a budget surplus or deficit in 2001? $128 billion; surplus

b. In 2011, what was the difference between the amount of money collected and the amount spent? Was there a budget surplus or deficit in 2011? −$1299 billion; deficit

c. What is the difference between the 2001 surplus and the 2011 deficit? $1427 billion

128. a. In 2001, what was the difference between the amount of money collected and the amount spent? Was there a budget surplus or deficit in 2001? $128 billion; surplus

b. In 2009, what was the difference between the amount of money collected and the amount spent? Was there a budget surplus or deficit in 2009? −$1413 billion; deficit

c. What is the difference between the 2001 surplus and the 2009 deficit? $1541 billion

129. What is the difference between the 2007 deficit and the 2011 deficit? $1138 billion

130. What is the difference between the 2007 deficit and the 2009 deficit? $1252 billion

Critical Thinking Exercises

In Exercises 131–132, insert one pair of parentheses to make each calculation correct.

131. $8 - 2 \cdot 3 - 4 = 10$ $8 - 2 \cdot (3 - 4) = 10$

132. $8 - 2 \cdot 3 - 4 = 14$ $(8 - 2) \cdot 3 - 4 = 14$

Technology Exercises

Scientific calculators that have parentheses keys allow for the entry and computation of relatively complicated expressions in a single step. For example, the expression $15 + (10 - 7)^2$ can be evaluated by entering the following keystrokes:

$$15 \boxed{+} \boxed{(}\, 10 \boxed{-} 7 \boxed{)}\, \boxed{y^x}\, 2 \boxed{=}\,.$$

Find the value of each expression in Exercises 133–135 in a single step on your scientific calculator.

133. $8 - 2 \cdot 3 - 9$ −7   **134.** $(8 - 2) \cdot (3 - 9)$ −36

135. $5^3 + 4 \cdot 9 - (8 + 9 \div 3)$ 150

1.3 : The Rational Numbers

What am I supposed to learn?

After you have read this section, you should be able to:

① Define the rational numbers.

② Reduce rational numbers.

③ Convert between mixed numbers and improper fractions.

④ Express rational numbers as decimals.

⑤ Express decimals in the form $\frac{a}{b}$.

⑥ Multiply and divide rational numbers.

⑦ Add and subtract rational numbers.

⑧ Use the order of operations agreement with rational numbers.

⑨ Apply the density property of rational numbers.

⑩ Solve problems involving rational numbers.

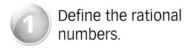

You are making eight dozen chocolate chip cookies for a large neighborhood block party. The recipe lists the ingredients needed to prepare five dozen cookies, such as $\frac{3}{4}$ cup sugar. How do you adjust the amount of sugar, as well as the amounts of each of the other ingredients, given in the recipe?

Adapting a recipe to suit a different number of portions usually involves working with numbers that are not integers. For example, the number describing the amount of sugar, $\frac{3}{4}$ (cup), is not an integer, although it consists of the quotient of two integers, 3 and 4. Before returning to the problem of changing the size of a recipe, we study a new set of numbers consisting of the quotients of integers.

Defining the Rational Numbers

If two integers are added, subtracted, or multiplied, the result is always another integer. This, however, is not always the case with division. For example, 10 divided by 5 is the integer 2. By contrast, 5 divided by 10 is $\frac{1}{2}$, and $\frac{1}{2}$ is not an integer. To permit divisions such as $\frac{5}{10}$, we enlarge the set of integers, calling the new collection the *rational numbers*. The set of **rational numbers** consists of all the numbers that can be expressed as a quotient of two integers, with the denominator not 0.

> **The Rational Numbers**
>
> The set of **rational numbers** is the set of all numbers which can be expressed in the form $\frac{a}{b}$, where a and b are integers and b is not equal to 0. The integer a is called the **numerator**, and the integer b is called the **denominator**.

The following numbers are examples of rational numbers:

$$\tfrac{1}{2}, \ \tfrac{-3}{4}, \ 5, \ 0.$$

The integer 5 is a rational number because it can be expressed as the quotient of integers: $5 = \frac{5}{1}$. Similarly, 0 can be written as $\frac{0}{1}$.

In general, every integer a is a rational number because it can be expressed in the form $\frac{a}{1}$.

Reducing Rational Numbers

A rational number is **reduced to its lowest terms**, or **simplified**, when the numerator and denominator have no common divisors other than 1. Reducing rational numbers to lowest terms is done using the **Fundamental Principle of Rational Numbers**.

① Define the rational numbers.

Great Question!

Is the rational number $\frac{-3}{4}$ the same as $-\frac{3}{4}$?

We know that the quotient of two numbers with different signs is a negative number. Thus,

$$\frac{-3}{4} = -\frac{3}{4} \quad \text{and} \quad \frac{3}{-4} = -\frac{3}{4}.$$

② Reduce rational numbers.
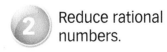

> **The Fundamental Principle of Rational Numbers**
>
> If $\frac{a}{b}$ is a rational number and c is any number other than 0,
>
> $$\frac{a \cdot c}{b \cdot c} = \frac{a}{b}.$$
>
> The rational numbers $\frac{a}{b}$ and $\frac{a \cdot c}{b \cdot c}$ are called **equivalent fractions**.

When using the Fundamental Principle to reduce a rational number, the simplification can be done in one step by finding the greatest common divisor of the numerator and the denominator, and using it for c. Thus, **to reduce a rational number to its lowest terms, divide both the numerator and the denominator by their greatest common divisor**.

For example, consider the rational number $\frac{12}{100}$. The greatest common divisor of 12 and 100 is 4. We reduce to lowest terms as follows:

$$\frac{12}{100} = \frac{3 \cdot \cancel{4}}{25 \cdot \cancel{4}} = \frac{3}{25} \quad \text{or} \quad \frac{12}{100} = \frac{12 \div 4}{100 \div 4} = \frac{3}{25}.$$

Example 1   Reducing a Rational Number

Reduce $\frac{130}{455}$ to lowest terms.

Solution

Begin by finding the greatest common divisor of 130 and 455.

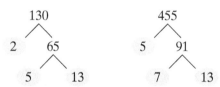

Thus, $130 = 2 \cdot 5 \cdot 13$ and $455 = 5 \cdot 7 \cdot 13$. The greatest common divisor is $5 \cdot 13$, or 65. Divide the numerator and the denominator of the given rational number by $5 \cdot 13$ or by 65.

$$\frac{130}{455} = \frac{2 \cdot \cancel{5} \cdot \cancel{13}}{\cancel{5} \cdot 7 \cdot \cancel{13}} = \frac{2}{7} \quad \text{or} \quad \frac{130}{455} = \frac{130 \div 65}{455 \div 65} = \frac{2}{7}$$

There are no common divisors of 2 and 7 other than 1. Thus, the rational number $\frac{2}{7}$ is in its lowest terms.

✓ **Check Point 1**   Reduce $\frac{72}{90}$ to lowest terms. $\frac{4}{5}$

---

③ Convert between mixed numbers and improper fractions.

Mixed Numbers and Improper Fractions

A **mixed number** consists of the sum of an integer and a rational number, expressed without the use of an addition sign. Here is an example of a mixed number:

$3\frac{4}{5}.$ The integer is 3 and the rational number is $\frac{4}{5}$. $3\frac{4}{5}$ means $3 + \frac{4}{5}$.

The mixed number $3\frac{4}{5}$ is read "three and four-fifths."

An **improper fraction** is a rational number for which the numerator is greater than the denominator. An example of an improper fraction is $\frac{19}{5}$.

The mixed number $3\frac{4}{5}$ can be converted to the improper fraction $\frac{19}{5}$ using the boxed procedure at the top of the next page.

> **Converting a Positive Mixed Number to an Improper Fraction**
>
> **1.** Multiply the denominator of the rational number by the integer and add the numerator to this product.
> **2.** Place the sum in step 1 over the denominator in the mixed number.

Example 2   **Converting from a Mixed Number to an Improper Fraction**

Convert $3\frac{4}{5}$ to an improper fraction.

Solution

$$3\frac{4}{5} = \frac{5 \cdot 3 + 4}{5}$$

Multiply the denominator by the integer and add the numerator.

Place the sum over the mixed number's denominator.

$$= \frac{15 + 4}{5} = \frac{19}{5}$$

✔ **Check Point 2**   Convert $2\frac{5}{8}$ to an improper fraction. $\frac{21}{8}$

When converting a negative mixed number to an improper fraction, copy the negative sign and then follow the previous procedure. For example,

$$-2\frac{3}{4} = -\frac{4 \cdot 2 + 3}{4} = -\frac{8 + 3}{4} = -\frac{11}{4}.$$

Copy the negative sign from step to step and convert $2\frac{3}{4}$ to an improper fraction.

A positive improper fraction can be converted to a mixed number using the following procedure:

> **Converting a Positive Improper Fraction to a Mixed Number**
>
> **1.** Divide the denominator into the numerator. Record the quotient and the remainder.
> **2.** Write the mixed number using the following form:
>
> $$\text{quotient } \frac{\text{remainder}}{\text{original denominator}}.$$
>
> *integer part*   *rational number part*

Example 3   **Converting from an Improper Fraction to a Mixed Number**

Convert $\frac{42}{5}$ to a mixed number.

Solution

Step 1   **Divide the denominator into the numerator.**

$$\begin{array}{r} 8 \\ 5{\overline{\smash{\big)}\,42}} \\ \underline{40} \\ 2 \end{array}$$

quotient

remainder

Great Question!

Does $-2\frac{3}{4}$ mean that I need to add $\frac{3}{4}$ to -2?

No. $-2\frac{3}{4}$ means

$$-\left(2\frac{3}{4}\right) \quad \text{or} \quad -\left(2 + \frac{3}{4}\right).$$

$-2\frac{3}{4}$ does not mean

$$-2 + \frac{3}{4}.$$

Math illiteracy affects 42 out of every 5 people

Step 2   Write the mixed number using quotient $\dfrac{\text{remainder}}{\text{original denominator}}$. We use

$$5)\overline{42} \begin{array}{l} 8 \\ \underline{40} \\ 2 \end{array}$$, and obtain $\dfrac{42}{5} = 8\dfrac{2}{5}$.

> **Check Point 3**   Convert $\dfrac{5}{3}$ to a mixed number. $1\frac{2}{3}$

When converting a negative improper fraction to a mixed number, copy the negative sign and then follow the previous procedure. For example,

$$-\dfrac{29}{8} = -3\dfrac{5}{8}.$$

Convert $\dfrac{29}{8}$ to a mixed number.

$$8)\overline{29} \begin{array}{l} 3 \leftarrow \text{quotient} \\ \underline{24} \\ 5 \leftarrow \text{remainder} \end{array}$$

Copy the negative sign.

Great Question!

When should I use mixed numbers and when are improper fractions preferable?

In applied problems, answers are usually expressed as mixed numbers, which many people find more meaningful than improper fractions. However, improper fractions are often easier to work with when performing operations with fractions.

④   Express rational numbers as decimals.

Rational Numbers and Decimals

We have seen that a rational number is the quotient of integers. Rational numbers can also be expressed as decimals. As shown in the place-value chart in the margin, it is convenient to represent rational numbers with denominators of 10, 100, 1000, and so on as decimals. For example,

$$\dfrac{7}{10} = 0.7, \quad \dfrac{3}{100} = 0.03, \quad \text{and} \quad \dfrac{8}{1000} = 0.008.$$

Any rational number $\dfrac{a}{b}$ can be expressed as a decimal by dividing the denominator, b, into the numerator, a.

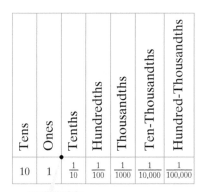

Tens	Ones	Tenths	Hundredths	Thousandths	Ten-Thousandths	Hundred-Thousandths
10	1	$\frac{1}{10}$	$\frac{1}{100}$	$\frac{1}{1000}$	$\frac{1}{10,000}$	$\frac{1}{100,000}$

decimal point

Example 4   **Expressing Rational Numbers as Decimals**

Express each rational number as a decimal:

a. $\dfrac{5}{8}$ **b.** $\dfrac{7}{11}$.

Solution

In each case, divide the denominator into the numerator.

a.

$$8)\overline{5.000} \begin{array}{l} 0.625 \\ \underline{4\,8} \\ 20 \\ \underline{16} \\ 40 \\ \underline{40} \\ 0 \end{array} \qquad \dfrac{5}{8} = 0.625$$

b.

$$11)\overline{7.0000\ldots} \begin{array}{l} 0.6363\ldots \\ \underline{6\,6} \\ 40 \\ \underline{33} \\ 70 \\ \underline{66} \\ 40 \\ \underline{33} \\ 70 \\ \vdots \end{array} \qquad \dfrac{7}{11} = 0.6363\ldots$$

In Example 4, the decimal for $\frac{5}{8}$, namely 0.625, stops and is called a **terminating decimal**. Other examples of terminating decimals are

$$\tfrac{1}{4} = 0.25, \quad \tfrac{2}{5} = 0.4, \quad \text{and} \quad \tfrac{7}{8} = 0.875.$$

By contrast, the division process for $\frac{7}{11}$ results in 0.6363..., with the digits 63 repeating over and over indefinitely. To indicate this, write a bar over the digits that repeat. Thus,

$$\tfrac{7}{11} = 0.\overline{63}.$$

The decimal for $\frac{7}{11}, 0.\overline{63}$, is called a **repeating decimal**. Other examples of repeating decimals are

$$\tfrac{1}{3} = 0.333\ldots = 0.\overline{3} \quad \text{and} \quad \tfrac{2}{3} = 0.666\ldots = 0.\overline{6}.$$

Rational Numbers and Decimals

Any rational number can be expressed as a decimal. The resulting decimal will either terminate (stop), or it will have a digit that repeats or a block of digits that repeats.

✓ **Check Point 4**   Express each rational number as a decimal:

▶ **a.** $\frac{3}{8}$ 0.375   **b.** $\frac{5}{11}$. $0.\overline{45}$

A Brief Review • Percents

Numerical information involving rational numbers and decimals is often presented using *percents*.

- **Percents** are the result of expressing numbers as part of 100. The word *percent* means *per hundred*. For example, a Harris Interactive poll found that 55 out of every 100 college students preferred print textbooks. Thus, $\frac{55}{100} = 55\%$, indicating that 55% of college students preferred print textbooks. The percent sign, %, is used to indicate the number of parts out of 100 parts.

- To express a decimal as a percent, move the decimal point two places to the right and attach a percent sign.

  Example

  Move decimal point two places right.

  $0.25 = 025.\%$ ← Attach a percent sign.

Thus, 0.25 = 25%.

- To express a percent as a decimal, move the decimal point two places to the left and remove the percent sign.

  Example

  $55\% = 55.\% = 0.55\%$ ← The percent sign is removed.

  The decimal point starts at the far right.

  The decimal point is moved two places to the left.

Thus, 55% = 0.55.

A more thorough review of percents can be found in Appendix A.

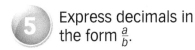

Express decimals in the form $\frac{a}{b}$.

Tens	Ones	Tenths	Hundredths	Thousandths	Ten-Thousandths	Hundred-Thousandths
10	1	$\frac{1}{10}$	$\frac{1}{100}$	$\frac{1}{1000}$	$\frac{1}{10,000}$	$\frac{1}{100,000}$

decimal point

Reversing Directions: Expressing Decimals as Quotients of Two Integers

Terminating decimals can be expressed with denominators of 10, 100, 1000, 10,000, and so on. Use the place-value chart shown in the margin. The digits to the right of the decimal point are the numerator of the rational number. To find the denominator, observe the last digit to the right of the decimal point. The place value of this digit will indicate the denominator.

Example 5   Expressing Terminating Decimals in $\frac{a}{b}$ Form

Express each terminating decimal as a quotient of integers:

a. 0.7   **b.** 0.49   **c.** 0.048.

Solution

a. $0.7 = \frac{7}{10}$ because the 7 is in the tenths position.

b. $0.49 = \frac{49}{100}$ because the last digit on the right, 9, is in the hundredths position.

c. $0.048 = \frac{48}{1000}$ because the last digit on the right, 8, is in the thousandths position. Reducing to lowest terms, $\frac{48}{1000} = \frac{48 \div 8}{1000 \div 8} = \frac{6}{125}$.

✓ **Check Point 5**   Express each terminating decimal as a quotient of integers, reduced to lowest terms:

▶ **a.** 0.9 $\frac{9}{10}$ **b.** 0.86 $\frac{43}{50}$ **c.** 0.053. $\frac{53}{1000}$

A Brief Review • Solving One-Step Equations

- Solving an equation involves determining all values that result in a true statement when substituted into the equation. Such values are solutions of the equation.

  Example
  The solution of $x - 4 = 10$ is 14 because $14 - 4 = 10$ is a true statement.

- Two basic rules can be used to solve equations:

  1. We can add or subtract the same quantity on both sides of an equation.

  2. We can multiply or divide both sides of an equation by the same quantity, as long as we do not multiply or divide by zero.

Examples of Equations That Can Be Solved in One Step

Equation	How to Solve	Solving the Equation	The Equation's Solution
$x - 4 = 10$	Add 4 to both sides.	$x - 4 + 4 = 10 + 4$ $x = 14$	14
$y + 12 = 17$	Subtract 12 from both sides.	$y + 12 - 12 = 17 - 12$ $y = 5$	5
$99n = 53$	Divide both sides by 99.	$\frac{99n}{99} = \frac{53}{99}$ $n = \frac{53}{99}$	$\frac{53}{99}$
$\frac{z}{5} = 9$	Multiply both sides by 5.	$5 \cdot \frac{z}{5} = 5 \cdot 9$ $z = 45$	45

Equations whose solutions require more than one step are discussed in Chapter 2.

Why have we provided this brief review of equations that can be solved in one step? If you are given a rational number as a repeating decimal, there is a technique for expressing the number as a quotient of integers that requires solving a one-step equation. We begin by illustrating the technique with an example. Then we will summarize the steps in the procedure and apply them to another example.

Example 6 Expressing a Repeating Decimal in $\frac{a}{b}$ Form

Express $0.\overline{6}$ as a quotient of integers.

Solution

Step 1 Let n equal the repeating decimal. Let $n = 0.\overline{6}$, so that $n = 0.66666\ldots$.

Step 2 If there is one repeating digit, multiply both sides of the equation in step 1 by 10.

$$n = 0.66666\ldots \qquad \text{This is the equation from step 1.}$$
$$10n = 10(0.66666\ldots) \qquad \text{Multiply both sides by 10.}$$
$$10n = 6.66666\ldots \qquad \text{Multiplying by 10 moves the decimal point one place to the right.}$$

Step 3 Subtract the equation in step 1 from the equation in step 2. Be sure to line up the decimal points before subtracting.

Remember from algebra that n means $1n$. Thus, $10n - 1n = 9n$.

$$
\begin{aligned}
10n &= 6.66666\ldots &\text{This is the equation from step 2.}\\
- \quad n &= 0.66666\ldots &\text{This is the equation from step 1.}\\
\hline
9n &= 6
\end{aligned}
$$

Step 4 Divide both sides of the equation in step 3 by the number in front of n and solve for n. We solve $9n = 6$ for n by dividing both sides by 9.

$$9n = 6 \qquad \text{This is the equation from step 3.}$$
$$\frac{9n}{9} = \frac{6}{9} \qquad \text{Divide both sides by 9.}$$
$$n = \frac{6}{9} = \frac{2}{3} \qquad \text{Reduce } \tfrac{6}{9} \text{ to lowest terms:}$$
$$\frac{6}{9} = \frac{2 \cdot \cancel{3}}{3 \cdot \cancel{3}} = \frac{2}{3}.$$

We began the solution process with $n = 0.\overline{6}$, and now we have $n = \frac{2}{3}$. Therefore,

$$0.\overline{6} = \frac{2}{3}.$$

Here are the steps for expressing a repeating decimal as a quotient of integers. Assume that the repeating digit or digits begin directly to the right of the decimal point.

Expressing a Repeating Decimal as a Quotient of Integers

Step 1 Let n equal the repeating decimal.

Step 2 Multiply both sides of the equation in step 1 by 10 if one digit repeats, by 100 if two digits repeat, by 1000 if three digits repeat, and so on.

Step 3 Subtract the equation in step 1 from the equation in step 2.

Step 4 Divide both sides of the equation in step 3 by the number in front of n and solve for n.

 **Check Point 6**   Express $0.\overline{2}$ as a quotient of integers. $\frac{2}{9}$

Example 7   Expressing a Repeating Decimal in $\frac{a}{b}$ Form

Express $0.\overline{53}$ as a quotient of integers.

Solution

Step 1   Let n equal the repeating decimal.   Let $n = 0.\overline{53}$, so that $n = 0.535353 \ldots$.

Step 2   If there are two repeating digits, multiply both sides of the equation in step 1 by 100.

$$n = 0.535353 \ldots$$ This is the equation from step 1.

$$100n = 100(0.535353 \ldots)$$ Multiply both sides by 100.

$$100n = 53.535353 \ldots$$ Multiplying by 100 moves the decimal point two places to the right.

Step 3   Subtract the equation in step 1 from the equation in step 2.

$$100n = 53.535353 \ldots$$ This is the equation from step 2.

$$- \quad n = 0.535353 \ldots$$ This is the equation from step 1.

$$99n = 53$$

Step 4   Divide both sides of the equation in step 3 by the number in front of n and solve for n.   We solve $99n = 53$ for n by dividing both sides by 99.

$$99n = 53$$ This is the equation from step 3.

$$\frac{99n}{99} = \frac{53}{99}$$ Divide both sides by 99.

$$n = \frac{53}{99}$$

Because n equals $0.\overline{53}$ and n equals $\frac{53}{99}$,

$$0.\overline{53} = \frac{53}{99}.$$

Check Point 7   Express $0.\overline{79}$ as a quotient of integers. $\frac{79}{99}$

Multiplying and Dividing Rational Numbers

6 Multiply and divide rational numbers.

The product of two rational numbers is found as follows:

Multiplying Rational Numbers

The product of two rational numbers is the product of their numerators divided by the product of their denominators.

If $\frac{a}{b}$ and $\frac{c}{d}$ are rational numbers, then $\frac{a}{b} \cdot \frac{c}{d} = \frac{a \cdot c}{b \cdot d}$.

Example 8   Multiplying Rational Numbers

Multiply. If possible, reduce the product to its lowest terms:

a. $\frac{3}{8} \cdot \frac{5}{11}$

b. $\left(-\frac{2}{3} \right) \left(-\frac{9}{4} \right)$

c. $\left(3\frac{2}{3} \right) \left(1\frac{1}{4} \right)$.

Great Question!

Is it OK if I divide by common factors before I multiply?

Yes. You can divide numerators and denominators by common factors *before* performing multiplication. Then multiply the remaining factors in the numerators and multiply the remaining factors in the denominators. For example,

$$\frac{7}{15} \cdot \frac{20}{21} = \frac{7}{\underset{3}{\cancel{15}}} \cdot \frac{\overset{4}{\cancel{20}}}{\underset{3}{\cancel{21}}} = \frac{1 \cdot 4}{3 \cdot 3} = \frac{4}{9}.$$

Solution

a. $\frac{3}{8} \cdot \frac{5}{11} = \frac{3 \cdot 5}{8 \cdot 11} = \frac{15}{88}$

b. $\left(-\frac{2}{3} \right) \left(-\frac{9}{4} \right) = \frac{(-2)(-9)}{3 \cdot 4} = \frac{18}{12} = \frac{3 \cdot \cancel{6}}{2 \cdot \cancel{6}} = \frac{3}{2}$ or $1\frac{1}{2}$

c. $\left(3\frac{2}{3} \right) \left(1\frac{1}{4} \right) = \frac{11}{3} \cdot \frac{5}{4} = \frac{11 \cdot 5}{3 \cdot 4} = \frac{55}{12}$ or $4\frac{7}{12}$

✓ **Check Point 8**   Multiply. If possible, reduce the product to its lowest terms:

a. $\frac{4}{11} \cdot \frac{2}{3}$ $\frac{8}{33}$

b. $\left(-\frac{3}{7} \right) \left(-\frac{14}{4} \right)$ $\frac{3}{2}$ or $1\frac{1}{2}$

▶ **c.** $\left(3\frac{2}{5} \right) \left(1\frac{1}{2} \right)$. $\frac{51}{10}$ or $5\frac{1}{10}$

Two numbers whose product is 1 are called **reciprocals**, or **multiplicative inverses**, of each other. Thus, the reciprocal of 2 is $\frac{1}{2}$ and the reciprocal of $\frac{1}{2}$ is 2 because $2 \cdot \frac{1}{2} = 1$. In general, if $\frac{c}{d}$ is a nonzero rational number, its reciprocal is $\frac{d}{c}$ because $\frac{c}{d} \cdot \frac{d}{c} = 1$.

Reciprocals are used to find the quotient of two rational numbers.

Dividing Rational Numbers

The quotient of two rational numbers is the product of the first number and the reciprocal of the second number.

If $\frac{a}{b}$ and $\frac{c}{d}$ are rational numbers and $\frac{c}{d}$ is not 0, then $\frac{a}{b} \div \frac{c}{d} = \frac{a}{b} \cdot \frac{d}{c} = \frac{a \cdot d}{b \cdot c}$.

Example 9   Dividing Rational Numbers

Divide. If possible, reduce the quotient to its lowest terms:

a. $\frac{4}{5} \div \frac{1}{10}$

b. $-\frac{3}{5} \div \frac{7}{11}$

c. $4\frac{3}{4} \div 1\frac{1}{2}$.

Solution

a. $\frac{4}{5} \div \frac{1}{10} = \frac{4}{5} \cdot \frac{10}{1} = \frac{4 \cdot 10}{5 \cdot 1} = \frac{40}{5} = 8$

b. $-\frac{3}{5} \div \frac{7}{11} = -\frac{3}{5} \cdot \frac{11}{7} = \frac{-3(11)}{5 \cdot 7} = -\frac{33}{35}$

c. $4\frac{3}{4} \div 1\frac{1}{2} = \frac{19}{4} \div \frac{3}{2} = \frac{19}{4} \cdot \frac{2}{3} = \frac{19 \cdot 2}{4 \cdot 3} = \frac{38}{12} = \frac{19 \cdot \cancel{2}}{6 \cdot \cancel{2}} = \frac{19}{6}$ or $3\frac{1}{6}$

 **Check Point 9** Divide. If possible, reduce the quotient to its lowest terms:

a. $\frac{9}{11} \div \frac{5}{4}$ $\frac{36}{55}$

b. $-\frac{8}{15} \div \frac{2}{5}$ $-\frac{4}{3}$ or $-1\frac{1}{3}$

▶ **c.** $3\frac{3}{8} \div 2\frac{1}{4}$. $\frac{3}{2}$ or $1\frac{1}{2}$

Adding and Subtracting Rational Numbers

⑦ Add and subtract rational numbers.

Rational numbers with identical denominators are added and subtracted using the following rules:

> **Adding and Subtracting Rational Numbers with Identical Denominators**
>
> The sum or difference of two rational numbers with identical denominators is the sum or difference of their numerators over the common denominator.
> If $\frac{a}{b}$ and $\frac{c}{b}$ are rational numbers, then $\frac{a}{b} + \frac{c}{b} = \frac{a+c}{b}$ and $\frac{a}{b} - \frac{c}{b} = \frac{a-c}{b}$.

Example 10 **Adding and Subtracting Rational Numbers with Identical Denominators**

Perform the indicated operations:

a. $\frac{3}{7} + \frac{2}{7}$

b. $\frac{11}{12} - \frac{5}{12}$

c. $-5\frac{1}{4} - \left(-2\frac{3}{4}\right)$.

Solution

a. $\frac{3}{7} + \frac{2}{7} = \frac{3+2}{7} = \frac{5}{7}$

b. $\frac{11}{12} - \frac{5}{12} = \frac{11-5}{12} = \frac{6}{12} = \frac{1 \cdot 6}{2 \cdot 6} = \frac{1}{2}$

c. $-5\frac{1}{4} - \left(-2\frac{3}{4}\right) = -\frac{21}{4} - \left(-\frac{11}{4}\right) = -\frac{21}{4} + \frac{11}{4} = \frac{-21+11}{4} = \frac{-10}{4} = -\frac{5}{2}$ or $-2\frac{1}{2}$

 **Check Point 10** Perform the indicated operations:

a. $\frac{5}{12} + \frac{3}{12}$ $\frac{2}{3}$

b. $\frac{7}{4} - \frac{1}{4}$ $\frac{3}{2}$ or $1\frac{1}{2}$

▶ **c.** $-3\frac{3}{8} - \left(-1\frac{1}{8}\right)$. $-\frac{9}{4}$ or $-2\frac{1}{4}$

If the rational numbers to be added or subtracted have different denominators, we use the least common multiple of their denominators to rewrite the rational numbers. The least common multiple of the denominators is called the **least common denominator (LCD)**.

Rewriting rational numbers with a least common denominator is done using the Fundamental Principle of Rational Numbers, discussed at the beginning of this section. Recall that if $\frac{a}{b}$ is a rational number and c is a nonzero number, then

$$\frac{a}{b} = \frac{a}{b} \cdot \frac{c}{c} = \frac{a \cdot c}{b \cdot c}.$$

Multiplying the numerator and the denominator of a rational number by the same nonzero number is equivalent to multiplying by 1, resulting in an equivalent fraction.

Example 11   Adding Rational Numbers with Unlike Denominators

Find the sum: $\frac{3}{4} + \frac{1}{6}$.

Solution

The smallest number divisible by both 4 and 6 is 12. Therefore, 12 is the least common multiple of 4 and 6, and will serve as the least common denominator. To obtain a denominator of 12, multiply the denominator and the numerator of the first rational number, $\frac{3}{4}$, by 3. To obtain a denominator of 12, multiply the denominator and the numerator of the second rational number, $\frac{1}{6}$, by 2.

$$\frac{3}{4} + \frac{1}{6} = \frac{3}{4} \cdot \frac{3}{3} + \frac{1}{6} \cdot \frac{2}{2}$$

Rewrite each rational number as an equivalent fraction with a denominator of 12; $\frac{3}{3} = 1$ and $\frac{2}{2} = 1$, and multiplying by 1 does not change a number's value.

$$= \frac{9}{12} + \frac{2}{12}$$

Multiply.

$$= \frac{11}{12}$$

Add numerators and put this sum over the least common denominator.

✓ **Check Point 11**   Find the sum: $\frac{1}{5} + \frac{3}{4}$. $\frac{19}{20}$

If the least common denominator cannot be found by inspection, use prime factorizations of the denominators and the method for finding their least common multiple, discussed in Section 1.1.

Example 12   Subtracting Rational Numbers with Unlike Denominators

Perform the indicated operation: $\frac{1}{15} - \frac{7}{24}$.

Solution

We need to first find the least common denominator, which is the least common multiple of 15 and 24. What is the smallest number divisible by both 15 and 24? The answer is not obvious, so we begin with the prime factorization of each number.

$$15 = 5 \cdot 3$$
$$24 = 8 \cdot 3 = 2^3 \cdot 3$$

The different factors are 5, 3, and 2. Using the greater number of times each factor appears in either factorization, the least common multiple is $5 \cdot 3 \cdot 2^3 = 5 \cdot 3 \cdot 8 = 120$. We will now express each rational number with a denominator of 120, which is the

Technology

Here is a possible keystroke sequence on a graphing calculator for the subtraction problem in Example 12:

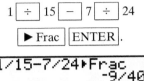

$$1 \boxed{\div} 15 \boxed{-} 7 \boxed{\div} 24$$

$$\boxed{\blacktriangleright \text{Frac}} \boxed{\text{ENTER}}.$$

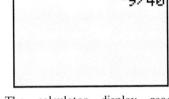

```
1/15-7/24▶Frac
             -9/40
```

The calculator display reads −9/40, serving as a check for our answer in Example 12.

least common denominator. For the first rational number, $\frac{1}{15}$, 120 divided by 15 is 8. Thus, we will multiply the numerator and the denominator by 8. For the second rational number, $\frac{7}{24}$, 120 divided by 24 is 5. Thus, we will multiply the numerator and the denominator by 5.

$$\frac{1}{15} - \frac{7}{24} = \frac{1}{15} \cdot \frac{8}{8} - \frac{7}{24} \cdot \frac{5}{5}$$ *Rewrite each rational number as an equivalent fraction with a denominator of 120.*

$$= \frac{8}{120} - \frac{35}{120}$$ *Multiply.*

$$= \frac{8 - 35}{120}$$ *Subtract the numerators and put this difference over the least common denominator.*

$$= \frac{-27}{120}$$ *Perform the subtraction.*

$$= \frac{-9 \cdot \cancel{3}}{40 \cdot \cancel{3}}$$ *Reduce to lowest terms.*

$$= -\frac{9}{40}$$

 Check Point 12 Perform the indicated operation: $\frac{3}{10} - \frac{7}{12}$. $-\frac{17}{60}$

Order of Operations with Rational Numbers

In the previous section, we presented rules for determining the order in which operations should be performed: operations in grouping symbols; exponential expressions; multiplication/division (left to right); addition/subtraction (left to right). In our next example, we apply the order of operations to an expression with rational numbers.

8 Use the order of operations agreement with rational numbers.

Example 13   **Using the Order of Operations**

Simplify: $\left(\frac{1}{2}\right)^3 - \left(\frac{1}{2} - \frac{3}{4}\right)^2 (-4)$.

Solution

Because grouping symbols appear, we perform the operation within parentheses first.

$$\left(\frac{1}{2}\right)^3 - \left(\frac{1}{2} - \frac{3}{4}\right)^2 (-4)$$

$$= \left(\frac{1}{2}\right)^3 - \left(-\frac{1}{4}\right)^2 (-4)$$ *Work inside parentheses first:* $\frac{1}{2} - \frac{3}{4} = \frac{2}{4} - \frac{3}{4} = \frac{2}{4} + \left(-\frac{3}{4}\right) = -\frac{1}{4}$.

$$= \frac{1}{8} - \frac{1}{16}(-4)$$ *Evaluate exponential expressions:* $\left(\frac{1}{2}\right)^3 = \frac{1}{2} \cdot \frac{1}{2} \cdot \frac{1}{2} = \frac{1}{8}$ *and* $\left(-\frac{1}{4}\right)^2 = \left(-\frac{1}{4}\right)\left(-\frac{1}{4}\right) = \frac{1}{16}$.

$$= \frac{1}{8} - \left(-\frac{1}{4}\right)$$ *Multiply:* $\frac{1}{16} \cdot \left(\frac{-4}{1}\right) = -\frac{4}{16} = -\frac{1}{4}$.

$$= \frac{3}{8}$$ *Subtract:* $\frac{1}{8} - \left(-\frac{1}{4}\right) = \frac{1}{8} + \frac{1}{4} = \frac{1}{8} + \frac{2}{8} = \frac{3}{8}$.

 Check Point 13 Simplify: $\left(-\frac{1}{2}\right)^2 - \left(\frac{7}{10} - \frac{8}{15}\right)^2 (-18)$. $\frac{3}{4}$

9 Apply the density property of rational numbers.

Density of Rational Numbers

It is always possible to find a rational number between any two distinct rational numbers. Mathematicians express this idea by saying that the set of rational numbers is **dense**.

Density of the Rational Numbers

If r and t represent rational numbers, with $r < t$, then there is a rational number s such that s is between r and t:

$$r < s < t.$$

One way to find a rational number between two given rational numbers is to find the rational number halfway between them. Add the given rational numbers and divide their sum by 2, thereby finding the average of the numbers.

Example 14   Illustrating the Density Property

Find the rational number halfway between $\frac{1}{2}$ and $\frac{3}{4}$.

Solution

First, add $\frac{1}{2}$ and $\frac{3}{4}$.

$$\frac{1}{2} + \frac{3}{4} = \frac{2}{4} + \frac{3}{4} = \frac{5}{4}$$

Next, divide this sum by 2.

$$\frac{5}{4} \div \frac{2}{1} = \frac{5}{4} \cdot \frac{1}{2} = \frac{5}{8}$$

The number $\frac{5}{8}$ is halfway between $\frac{1}{2}$ and $\frac{3}{4}$. Thus,

$$\frac{1}{2} < \frac{5}{8} < \frac{3}{4}.$$

Great Question!

Why does $\frac{1}{2} < \frac{5}{8} < \frac{3}{4}$ seem meaningless to me?

The inequality $\frac{1}{2} < \frac{5}{8} < \frac{3}{4}$ is more obvious if all denominators are changed to 8:

$$\frac{4}{8} < \frac{5}{8} < \frac{6}{8}.$$

We can repeat the procedure of Example 14 and find the rational number halfway between $\frac{1}{2}$ and $\frac{5}{8}$. Repeated application of this procedure implies the following surprising result.

Between any two distinct rational numbers are *infinitely many* rational numbers.

 **Check Point 14** Find the rational number halfway between $\frac{1}{3}$ and $\frac{1}{2}$. $\frac{5}{12}$

Problem Solving with Rational Numbers

10 Solve problems involving rational numbers.

A common application of rational numbers involves preparing food for a different number of servings than what the recipe gives. The amount of each ingredient can be found as follows:

Amount of ingredient needed

$$= \frac{\text{desired serving size}}{\text{recipe serving size}} \times \text{ingredient amount in the recipe.}$$

Example 15   Adjusting the Size of a Recipe

A chocolate-chip cookie recipe for five dozen cookies requires $\frac{3}{4}$ cup sugar. If you want to make eight dozen cookies, how much sugar is needed?

Solution

Amount of sugar needed

$$= \frac{\text{desired serving size}}{\text{recipe serving size}} \times \text{sugar amount in recipe}$$

$$= \frac{8 \text{ dozen}}{5 \text{ dozen}} \times \frac{3}{4} \text{ cup}$$

The amount of sugar needed, in cups, is determined by multiplying the rational numbers:

$$\frac{8}{5} \times \frac{3}{4} = \frac{8 \cdot 3}{5 \cdot 4} = \frac{24}{20} = \frac{6 \cdot \cancel{4}}{5 \cdot \cancel{4}} = 1\frac{1}{5}.$$

Thus, $1\frac{1}{5}$ cups of sugar is needed. (Depending on the measuring cup you are using, you may need to round the sugar amount to $1\frac{1}{4}$ cups.)

✔ **Check Point 15**   A chocolate-chip cookie recipe for five dozen cookies requires two eggs. If you want to make seven dozen cookies, exactly how many eggs are needed? Now round your answer to a realistic number that does not involve a fractional part of an egg. $2\frac{4}{5}$ eggs; 3 eggs

---

Blitzer Bonus

NUMB3RS: Solving Crime with Mathematics

NUMB3RS was a prime-time TV crime series. The show's hero, Charlie Eppes, is a brilliant mathematician who uses his powerful skills to help the FBI identify and catch criminals. The episodes are entertaining and the basic premise shows how math is a powerful weapon in the never-ending fight against crime. *NUMB3RS* is significant because it was the first popular weekly drama that revolved around mathematics. A team of mathematician advisors ensured that the equations seen in the scripts were real and relevant to the episodes. The mathematical content of the show included many pathway topics from this book, ranging from prime numbers to probability theory to basic geometry.

Episodes of *NUMB3RS* began with a spoken tribute about the importance of mathematics:

> "We all use math everywhere. To tell time, to predict the weather, to handle money . . . Math is more than formulas and equations. Math is more than numbers. It is logic. It is rationality. It is using your mind to solve the biggest mysteries we know."

Achieving Success

Read the textbook. There is a big difference between high school math classes and college courses. In high school, teachers generally cover all material for tests in class through lectures and/or activities. In college, students are responsible for information in their textbook, whether it is covered in class or not. The ability to read textbooks and learn from their pages is an essential skill for achieving success in college mathematics. Begin practicing this skill now. Read this book even if your teacher covers everything in class. Not only will you reinforce what you've learned in class, but you'll also be developing a skill that is a key component for success in higher education.

Concept and Vocabulary Check

Exercises in the Concept and Vocabulary Check are intended for group and class discussions.

In Exercises 1–4, fill in each blank so that the resulting statement is true.   4. reciprocal/multiplicative inverse

1. The set of __rational numbers__ is the set of all numbers which can be expressed in the form $\frac{a}{b}$, where a and b are _____integers_____ and b is not equal to _____zero_____.

2. The number $\frac{17}{5}$ is an example of _an improper fraction_ because _the numerator is greater than the denominator_

3. Numbers in the form $\frac{a}{b}$ (see Exercise 1) can be expressed as decimals. The decimals either ___terminate/stop___ or _have repeating digits_.

4. The quotient of two fractions is the product of the first number and the _____ of the second number.

In Exercises 5–8, determine whether each statement is true or false. If the statement is false, make the necessary change(s) to produce a true statement.   Changes to false statements will vary.

5. $\frac{1}{2} + \frac{1}{5} = \frac{2}{7}$ false

6. $\frac{1}{2} \div 4 = 2$ false

7. Every fraction has infinitely many equivalent fractions.   true

8. $\dfrac{3+7}{30} = \dfrac{\overset{1}{3}+7}{\underset{10}{30}} = \dfrac{8}{10} = \dfrac{4}{5}$ false

Respond to Exercises 9–18 using verbal or written explanations.

9. What is a rational number?   9–18. Answers will vary.

10. Explain how to reduce a rational number to its lowest terms.

11. Explain how to convert from a mixed number to an improper fraction. Use $7\frac{2}{3}$ as an example.

12. Explain how to convert from an improper fraction to a mixed number. Use $\frac{47}{5}$ as an example.

13. Explain how to write a rational number as a decimal.

14. Explain how to write $0.\overline{9}$ as a quotient of integers.

15. Explain how to multiply rational numbers. Use $\frac{5}{6} \cdot \frac{1}{2}$ as an example.

16. Explain how to divide rational numbers. Use $\frac{5}{6} \div \frac{1}{2}$ as an example.

17. Explain how to add rational numbers with different denominators. Use $\frac{5}{6} + \frac{1}{2}$ as an example.

18. What does it mean when we say that the set of rational numbers is dense?

Exercise Set 1.3

Practice Exercises

In Exercises 1–12, reduce each rational number to its lowest terms.

1. $\frac{10}{15}$ $\frac{2}{3}$ **2.** $\frac{18}{45}$ $\frac{2}{5}$ **3.** $\frac{15}{18}$ $\frac{5}{6}$

4. $\frac{16}{64}$ $\frac{1}{4}$ **5.** $\frac{24}{42}$ $\frac{4}{7}$ **6.** $\frac{32}{80}$ $\frac{2}{5}$

7. $\frac{60}{108}$ $\frac{5}{9}$ **8.** $\frac{112}{128}$ $\frac{7}{8}$ **9.** $\frac{342}{380}$ $\frac{9}{10}$

10. $\frac{210}{252}$ $\frac{5}{6}$ **11.** $\frac{308}{418}$ $\frac{14}{19}$ **12.** $\frac{144}{300}$ $\frac{12}{25}$

In Exercises 13–18, convert each mixed number to an improper fraction.

13. $2\frac{3}{8}$ $\frac{19}{8}$ **14.** $2\frac{7}{9}$ $\frac{25}{9}$ **15.** $-7\frac{3}{5}$ $-\frac{38}{5}$

16. $-6\frac{2}{5}$ $-\frac{32}{5}$ **17.** $12\frac{7}{16}$ $\frac{199}{16}$ **18.** $11\frac{5}{16}$ $\frac{181}{16}$

In Exercises 19–24, convert each improper fraction to a mixed number.

19. $\frac{23}{5}$ $4\frac{3}{5}$ **20.** $\frac{47}{8}$ $5\frac{7}{8}$ **21.** $-\frac{76}{9}$ $-8\frac{4}{9}$

22. $-\frac{59}{9}$ $-6\frac{5}{9}$ **23.** $\frac{711}{20}$ $35\frac{11}{20}$ **24.** $\frac{788}{25}$ $31\frac{13}{25}$

In Exercises 25–36, express each rational number as a decimal.

25. $\frac{3}{4}$ 0.75   **26.** $\frac{3}{5}$ 0.6   **27.** $\frac{7}{20}$ 0.35   **28.** $\frac{3}{20}$ 0.15

29. $\frac{7}{8}$ 0.875   **30.** $\frac{5}{16}$ 0.3125   **31.** $\frac{9}{11}$ $0.\overline{81}$ **32.** $\frac{3}{11}$ $0.\overline{27}$

33. $\frac{22}{7}$ $3.\overline{142857}$ **34.** $\frac{20}{3}$ $6.\overline{6}$ **35.** $\frac{2}{7}$ $0.\overline{285714}$ **36.** $\frac{5}{7}$ $0.\overline{714285}$

In Exercises 37–48, express each terminating decimal as a quotient of integers. If possible, reduce to lowest terms.

37. 0.3 $\frac{3}{10}$ **38.** 0.9 $\frac{9}{10}$ **39.** 0.4 $\frac{2}{5}$

40. 0.6 $\frac{3}{5}$ **41.** 0.39 $\frac{39}{100}$ **42.** 0.59 $\frac{59}{100}$

43. 0.82 $\frac{41}{50}$ **44.** 0.64 $\frac{16}{25}$ **45.** 0.725 $\frac{29}{40}$

46. 0.625 $\frac{5}{8}$ **47.** 0.5399 $\frac{5399}{10,000}$ **48.** 0.7006 $\frac{3503}{5000}$

In Exercises 49–56, express each repeating decimal as a quotient of integers. If possible, reduce to lowest terms.

49. $0.\overline{7}$ $\frac{7}{9}$ **50.** $0.\overline{1}$ $\frac{1}{9}$ **51.** $0.\overline{9}$ 1   **52.** $0.\overline{3}$ $\frac{1}{3}$

53. $0.\overline{36}$ $\frac{4}{11}$ **54.** $0.\overline{81}$ $\frac{9}{11}$ **55.** $0.\overline{257}$ $\frac{257}{999}$ **56.** $0.\overline{529}$ $\frac{529}{999}$

In Exercises 57–104, perform the indicated operations. If possible, reduce the answer to its lowest terms.

57. $\frac{3}{8} \cdot \frac{7}{11}$ $\frac{21}{88}$

58. $\frac{5}{8} \cdot \frac{3}{11}$ $\frac{15}{88}$

59. $\left(-\frac{1}{10}\right)\left(\frac{7}{12}\right)$ $-\frac{7}{120}$

60. $\left(-\frac{1}{8}\right)\left(\frac{5}{9}\right)$ $-\frac{5}{72}$

61. $\left(-\frac{2}{3}\right)\left(-\frac{9}{4}\right)$ $\frac{3}{2}$

62. $\left(-\frac{5}{4}\right)\left(-\frac{6}{7}\right)$ $\frac{15}{14}$

63. $\left(3\frac{3}{4}\right)\left(1\frac{3}{5}\right)$ 6

64. $\left(2\frac{4}{5}\right)\left(1\frac{1}{4}\right)$ $3\frac{1}{2}$

65. $\frac{5}{4} \div \frac{3}{8}$ $\frac{10}{3}$

66. $\frac{5}{8} \div \frac{4}{3}$ $\frac{15}{32}$

67. $-\frac{7}{8} \div \frac{15}{16}$ $-\frac{14}{15}$

68. $-\frac{13}{20} \div \frac{4}{5}$ $-\frac{13}{16}$

69. $6\frac{3}{5} \div 1\frac{1}{10}$ 6

70. $1\frac{3}{4} \div 2\frac{5}{8}$ $\frac{2}{3}$

71. $\frac{2}{11} + \frac{3}{11}$ $\frac{5}{11}$

72. $\frac{5}{13} + \frac{2}{13}$ $\frac{7}{13}$

73. $\frac{5}{6} - \frac{1}{6}$ $\frac{2}{3}$

74. $\frac{7}{12} - \frac{5}{12}$ $\frac{1}{6}$

75. $\frac{7}{12} - \left(-\frac{1}{12}\right)$ $\frac{2}{3}$

76. $\frac{5}{16} - \left(-\frac{5}{16}\right)$ $\frac{5}{8}$

77. $\frac{1}{2} + \frac{1}{5}$ $\frac{7}{10}$

78. $\frac{1}{3} + \frac{1}{5}$ $\frac{8}{15}$

79. $\frac{3}{4} + \frac{3}{20}$ $\frac{9}{10}$

80. $\frac{2}{5} + \frac{2}{15}$ $\frac{8}{15}$

81. $\frac{5}{24} + \frac{7}{30}$ $\frac{53}{120}$

82. $\frac{7}{108} + \frac{55}{144}$ $\frac{193}{432}$

83. $\frac{13}{18} - \frac{2}{9}$ $\frac{1}{2}$

84. $\frac{13}{15} - \frac{2}{45}$ $\frac{37}{45}$

85. $\frac{4}{3} - \frac{3}{4}$ $\frac{7}{12}$

86. $\frac{3}{2} - \frac{2}{3}$ $\frac{5}{6}$

87. $\frac{1}{15} - \frac{27}{50}$ $-\frac{71}{150}$

88. $\frac{4}{15} - \frac{1}{6}$ $\frac{1}{10}$

89. $2\frac{2}{3} + 1\frac{3}{4}$ $\frac{53}{12}$ or $4\frac{5}{12}$

90. $2\frac{1}{8} + 3\frac{3}{4}$ $\frac{47}{8}$ or $5\frac{7}{8}$

91. $3\frac{2}{3} - 2\frac{1}{2}$ $\frac{7}{6}$ or $1\frac{1}{6}$

92. $3\frac{3}{4} - 2\frac{1}{3}$ $\frac{17}{12}$ or $1\frac{5}{12}$

93. $-5\frac{2}{3} + 3\frac{1}{6}$ $-\frac{5}{2}$ or $-2\frac{1}{2}$

94. $-2\frac{1}{2} + 1\frac{3}{4}$ $-\frac{3}{4}$

95. $-1\frac{4}{7} - \left(-2\frac{5}{14}\right)$ $\frac{11}{14}$

96. $-1\frac{4}{9} - \left(-2\frac{5}{18}\right)$ $\frac{5}{6}$

97. $\left(\frac{1}{2} - \frac{1}{3}\right) \div \frac{5}{8}$ $\frac{4}{15}$

98. $\left(\frac{1}{2} + \frac{1}{4}\right) \div \left(\frac{1}{2} + \frac{1}{3}\right)$ $\frac{9}{10}$

99. $-\frac{9}{4}\left(\frac{1}{2}\right) + \frac{3}{4} \div \frac{5}{6}$ $-\frac{9}{40}$

100. $\left[-\frac{4}{7} - \left(-\frac{2}{5}\right)\right]\left[-\frac{3}{8} + \left(-\frac{1}{9}\right)\right]$ $\frac{1}{12}$

101. $\dfrac{\frac{7}{9} - 3}{\frac{5}{6}} \div \frac{3}{2} + \frac{3}{4}$ $-1\frac{1}{36}$

102. $\dfrac{\frac{17}{25}}{\frac{3}{5} - 4} \div \frac{1}{5} + \frac{1}{2}$ $-\frac{1}{2}$

103. $\frac{1}{4} - 6(2 + 8) \div \left(-\frac{1}{3}\right)\left(-\frac{1}{9}\right)$ $-19\frac{3}{4}$

104. $\frac{3}{4} - 4(2 + 7) \div \left(-\frac{1}{2}\right)\left(-\frac{1}{6}\right)$ $-11\frac{1}{4}$

In Exercises 105–110, find the rational number halfway between the two numbers in each pair.

105. $\frac{1}{4}$ and $\frac{1}{3}$ $\frac{7}{24}$

106. $\frac{2}{3}$ and $\frac{5}{6}$ $\frac{3}{4}$

107. $\frac{1}{2}$ and $\frac{2}{3}$ $\frac{7}{12}$

108. $\frac{3}{5}$ and $\frac{2}{3}$ $\frac{19}{30}$

109. $-\frac{2}{3}$ and $-\frac{5}{6}$ $-\frac{3}{4}$

110. -4 and $-\frac{7}{2}$ $-\frac{15}{4}$

Different operations with the same rational numbers usually result in different answers. Exercises 111–112 illustrate some curious exceptions. **111.** Both are equal to $\frac{169}{36}$. **112.** Both are equal to $\frac{13}{2}$.

111. Show that $\frac{13}{4} + \frac{13}{9}$ and $\frac{13}{4} \times \frac{13}{9}$ give the same answer.

112. Show that $\frac{169}{30} + \frac{13}{15}$ and $\frac{169}{30} \div \frac{13}{15}$ give the same answer.

Practice Plus

In Exercises 113–116, perform the indicated operations. Leave denominators in prime factorization form.

113. $\dfrac{5}{2^2 \cdot 3^2} - \dfrac{1}{2 \cdot 3^2}$ $\dfrac{1}{2^2 \cdot 3}$

114. $\dfrac{7}{3^2 \cdot 5^2} - \dfrac{1}{3 \cdot 5^3}$ $\dfrac{32}{3^2 \cdot 5^3}$

115. $\dfrac{1}{2^4 \cdot 5^3 \cdot 7} + \dfrac{1}{2 \cdot 5^4} - \dfrac{1}{2^3 \cdot 5^2}$ $-\dfrac{289}{2^4 \cdot 5^4 \cdot 7}$

116. $\dfrac{1}{2^3 \cdot 17^8} + \dfrac{1}{2 \cdot 17^9} - \dfrac{1}{2^2 \cdot 3 \cdot 17^8}$ $\dfrac{29}{2^3 \cdot 3 \cdot 17^9}$

Application Exercises

A study of teens under the age of 18 revealed how they dealt with stress. The circle graphs below show the breakdown of the number of men and women who responded to stress in four different ways. Use this information to solve Exercises 117–118.

How Teens Deal with Stress

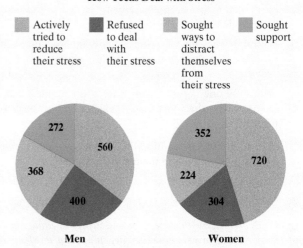

■ Actively tried to reduce their stress   ■ Refused to deal with their stress   ■ Sought ways to distract themselves from their stress   ■ Sought support

Men   **Women**

Source: Johns Hopkins Bloomberg School of Public Health

117. a. What fractional part of the men actively tried to reduce their stress? Reduce this fraction to its lowest terms.

b. Express the rational number in part (a) as a decimal. What percentage of teen males actively tried to reduce their stress? $0.35 = 35\%$

c. What fractional part of the women actively tried to reduce their stress? Reduce this fraction to its lowest terms.

d. Express the rational number in part (c) as a decimal. What percentage of teen females actively tried to reduce their stress? $0.45 = 45\%$

e. What is the difference between the percentage of teen females and teen males who actively tried to reduce their stress? 10%

118. a. What fractional part of the men refused to deal with their stress? Reduce this fraction to its lowest terms.

b. Express the rational number in part (a) as a decimal. What percentage of teen males refused to deal with their stress? $0.25 = 25\%$

117. a. $\frac{560}{1600} = \frac{7}{20}$ **c.** $\frac{720}{1600} = \frac{9}{20}$ **118. a.** $\frac{400}{1600} = \frac{1}{4}$

c. What fractional part of the women refused to deal with their stress? Reduce this fraction to its lowest terms.

d. Express the rational number in part (c) as a decimal. What percentage of teen females refused to deal with their stress? $0.19 = 19\%$

e. What is the difference between the percentage of teen males and teen females who refused to deal with their stress?   6%

Use the following list of ingredients for chocolate brownies to solve Exercises 119–124.

Ingredients for 16 Brownies

$\frac{2}{3}$ *cup butter, 5 ounces unsweetened chocolate, $1\frac{1}{2}$ cups sugar, 2 teaspoons vanilla, 2 eggs, 1 cup flour*

119. How much of each ingredient is needed to make 8 brownies?   *

120. How much of each ingredient is needed to make 12 brownies?   *

121. How much of each ingredient is needed to make 20 brownies?   *

122. How much of each ingredient is needed to make 24 brownies?   *

123. With only one cup of butter, what is the greatest number of brownies that you can make? (Ignore part of a brownie.)

124. With only one cup of sugar, what is the greatest number of brownies that you can make? (Ignore part of a brownie.)

A mix for eight servings of instant potatoes requires $2\frac{2}{3}$ cups of water. Use this information to solve Exercises 125–126.

125. If you want to make 11 servings, how much water is needed?

126. If you want to make six servings, how much water is needed?

The sounds created by plucked or bowed strings of equal diameter and tension produce various notes depending on the lengths of the strings. If a string is half as long as another, its note will be an octave higher than the longer string. Using a length of 1 unit to represent middle C, the diagram shows different fractions of the length of this unit string needed to produce the notes D, E, F, G, A, B, and c one octave higher than middle C.

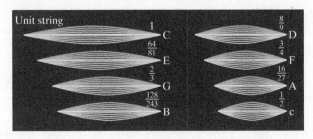

For many of the strings, the length is $\frac{8}{9}$ of the length of the previous string. For example, the A string is $\frac{8}{9}$ of the length of the string needed to produce the note for G: $\frac{8}{9} \cdot \frac{2}{3} = \frac{16}{27}$.
Use this information to solve Exercises 127–128.

127. **a.** Which strings from D through c are $\frac{8}{9}$ of the length of the preceding string?   D, E, G, A, B

b. How is your answer to part (a) shown on this one-octave span of the piano keyboard?

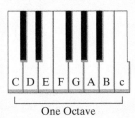

There are black keys to the left of the keys for the notes D, E, G, A, and B.

One Octave

128. **a.** Which strings from D through c are not $\frac{8}{9}$ of the length of the preceding string?   F, c

b. How is your answer to part (a) shown on the one-octave span on the piano keyboard in Exercise 127(b)?

129. A board $7\frac{1}{2}$ inches long is cut from a board that is 2 feet long. If the width of the saw cut is $\frac{1}{16}$ inch, what is the length of the remaining piece? $16\frac{7}{16}$ in.

130. A board that is $7\frac{1}{4}$ inches long is cut from a board that is 3 feet long. If the width of the saw cut is $\frac{1}{16}$ inch, what is the length of the remaining piece? $28\frac{11}{16}$ in.

131. A franchise is owned by three people. The first owns $\frac{5}{12}$ of the business and the second owns $\frac{1}{4}$ of the business. What fractional part of the business is owned by the third person? $\frac{1}{3}$

132. At a workshop on enhancing creativity, $\frac{1}{4}$ of the participants are musicians, $\frac{2}{5}$ are artists, $\frac{1}{10}$ are actors, and the remaining participants are writers. What fraction of the people attending the workshop are writers? $\frac{1}{4}$

133. If you walk $\frac{3}{4}$ mile and then jog $\frac{2}{5}$ mile, what is the total distance covered? How much farther did you walk than jog?

134. Many companies pay people extra when they work more than a regular 40-hour work week. The overtime pay is often $1\frac{1}{2}$ times the regular hourly rate. This is called time and a half. A summer job for students pays \$12 an hour and offers time and a half for the hours worked over 40. If a student works 46 hours during one week, what is the student's total pay before taxes?   \$588

135. A will states that $\frac{3}{5}$ of the estate is to be divided among relatives. Of the remaining estate, $\frac{1}{4}$ goes to charity. What fraction of the estate goes to charity? $\frac{1}{10}$

136. The legend of a map indicates that 1 inch = 16 miles. If the distance on the map between two cities is $2\frac{3}{8}$ inches, how far apart are the cities?   38 mi

Critical Thinking Exercises

137. Shown below is a short excerpt from "The Star-Spangled Banner." The time is $\frac{3}{4}$, which means that each measure must contain notes that add up to $\frac{3}{4}$. The values of the different notes tell musicians how long to hold each note.

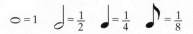

Use vertical lines to divide this line of "The Star-Spangled Banner" into measures.   *

*See Answers to Selected Exercises.

138. Use patterns to predict the addition problem and the sum that will appear in the fourth row. Then perform the arithmetic to verify your conjecture.

$\frac{1}{1\cdot2}+\frac{1}{2\cdot3}+\frac{1}{3\cdot4}+\frac{1}{4\cdot5}+\frac{1}{5\cdot6}=\frac{5}{6}$ $\dfrac{1}{1\cdot2}+\dfrac{1}{2\cdot3}=\dfrac{2}{3}$

$$\dfrac{1}{1\cdot2}+\dfrac{1}{2\cdot3}+\dfrac{1}{3\cdot4}=\dfrac{3}{4}$$

$$\dfrac{1}{1\cdot2}+\dfrac{1}{2\cdot3}+\dfrac{1}{3\cdot4}+\dfrac{1}{4\cdot5}=\dfrac{4}{5}$$

Technology Exercises

139. Use a calculator to express the following rational numbers as decimals.

a. $\dfrac{197}{800}$ 0.24625   **b.** $\dfrac{4539}{3125}$ 1.45248   **c.** $\dfrac{7}{6250}$ 0.00112

140. Some calculators have a fraction feature. This feature allows you to perform operations with fractions and displays the answer as a fraction reduced to its lowest terms. If your calculator has this feature, use it to verify any five of the answers that you obtained in Exercises 57–104.   Answers will vary.

Group Exercise   Answers will vary.

141. Each member of the group should present an application of rational numbers. The application can be based on research or on how the group member uses rational numbers in his or her life. If you are not sure where to begin, ask yourself how your life would be different if fractions and decimals were concepts unknown to our civilization.

1.4 : The Irrational Numbers

What am I supposed to learn?

After you have read this section, you should be able to:

1. Define the irrational numbers.

2. Simplify square roots.

3. Perform operations with square roots.

4. Rationalize denominators.

For the followers of the Greek mathematician Pythagoras in the sixth century B.C., numbers took on a life-and-death importance. The "Pythagorean Brotherhood" was a secret group whose members were convinced that properties of whole numbers were the key to understanding the universe. Members of the Brotherhood (which admitted women) thought that all numbers that were not whole numbers could be represented as the ratio of whole numbers. A crisis occurred for the Pythagoreans when they discovered the existence of a number that was not rational. Because the

Pythagoras

Shown here is Renaissance artist Raphael Sanzio's (1483–1520) image of Pythagoras from *The School of Athens* mural. Detail of left side.
*Stanza della Segnatura, Vatican Palace, Vatican State.
Scala/Art Resource, NY.*

Pythagoreans viewed numbers with reverence and awe, the punishment for speaking about this number was death. However, a member of the Brotherhood revealed the secret of the number's existence. When he later died in a shipwreck, his death was viewed as punishment from the gods.

The triangle in **Figure 1.6** led the Pythagoreans to the discovery of a number that could not be expressed as the quotient of integers. Based on their understanding of the relationship between the sides of this triangle, they knew that the length of the side shown in red had to be a number that, when squared, is equal to 2. The Pythagoreans discovered that this number seemed to be close to the rational numbers

$$\frac{14}{10}, \frac{141}{100}, \frac{1414}{1000}, \frac{14{,}141}{10{,}000}, \text{ and so on.}$$

However, they were shocked to find that there is no quotient of integers whose square is equal to 2.

Length = ?

Length: 1 unit

Length: 1 unit

Figure 1.6

The positive number whose square is equal to 2 is written $\sqrt{2}$. We read this "the square root of 2," or "radical 2." The symbol $\sqrt{}$ is called the **radical sign**. The number under the radical sign, in this case 2, is called the **radicand**. The entire symbol $\sqrt{2}$ is called a **radical**.

Mathematicians have proved that $\sqrt{2}$ cannot be represented as a quotient of integers. This means that there is no terminating or repeating decimal that can be multiplied by itself to give 2. We can, however, give a decimal approximation for $\sqrt{2}$. We use the symbol $\approx$, which means "is approximately equal to." Thus,

$$\sqrt{2} \approx 1.414214.$$

We can verify that this is only an approximation by multiplying 1.414214 by itself. The product is not exactly 2:

$$1.414214 \times 1.414214 = 2.000001237796.$$

A number like $\sqrt{2}$, whose decimal representation does not come to an end and does not have a block of repeating digits, is an example of an **irrational number**.

 Define the irrational numbers.

> ### The Irrational Numbers
>
> The set of **irrational numbers** is the set of numbers whose decimal representations are neither terminating nor repeating.

Perhaps the best known of all the irrational numbers is π (pi). This irrational number represents the distance around a circle (its circumference) divided by the diameter of the circle. In the *Star Trek* episode "Wolf in the Fold," Spock foils an evil computer by telling it to "compute the last digit in the value of π." Because π is an irrational number, there is no last digit in its decimal representation:

$$\pi = 3.14159265358979323846426433832795\ldots.$$

The nature of the irrational number π has fascinated mathematicians for centuries. Amateur and professional mathematicians have taken up the challenge of calculating π to more and more decimal places. Although such an exercise may seem pointless, it serves as the ultimate stress test for new high-speed computers and also as a test for the long-standing, but still unproven, conjecture that the distribution of digits in π is completely random.

Blitzer Bonus
• • • • • • • • • • • • • • • •

The Best and Worst of π

In 2011, Japanese mathematician Shigeru Kondo calculated π to ten trillion decimal places. The calculations used 8900 hours, or nearly 371 days, of computer time.

The most inaccurate version of π came from the 1897 General Assembly of Indiana. Bill No. 246 stated that "π was by law 4."

> ### Technology
>
> You can obtain decimal approximations for irrational numbers using a calculator. For example, to approximate $\sqrt{2}$, use the following keystrokes:
>
> **Scientific Calculator**   **Graphing Calculator**
>
> $2\ \boxed{\sqrt{}}$ or $2\ \boxed{\begin{smallmatrix}\text{2ND}\\\text{INV}\end{smallmatrix}}\ \boxed{x^2}$ $\boxed{\sqrt{}}\ 2\ \boxed{\text{ENTER}}$ or $\boxed{\begin{smallmatrix}\text{2ND}\\\text{INV}\end{smallmatrix}}\ \boxed{x^2}\ 2\ \boxed{\text{ENTER}}$
>
> > Some graphing calculators show an open parenthesis after displaying $\sqrt{}$. In this case, enter a closed parenthesis, $\boxed{)}$, after 2.
>
> The display may read 1.41421356237, although your calculator may show more or fewer digits. Between which two integers would you graph $\sqrt{2}$ on a number line?

The U.N. building is designed with three golden rectangles.

Square Roots

The United Nations Building in New York was designed to represent its mission of promoting world harmony. Viewed from the front, the building looks like three rectangles stacked upon each other. In each rectangle, the width divided by the height is $\sqrt{5} + 1$ to 2, approximately 1.618 to 1. The ancient Greeks believed that such a rectangle, called a **golden rectangle**, was the most pleasing of all rectangles. The comparison 1.618 to 1 is approximate because $\sqrt{5}$ is an irrational number.

The **principal square root** of a nonnegative number n, written $\sqrt{n}$, is the positive number that when multiplied by itself gives n. Thus,

$$\sqrt{36} = 6 \text{ because } 6 \cdot 6 = 36$$

and

$$\sqrt{81} = 9 \text{ because } 9 \cdot 9 = 81.$$

Notice that both $\sqrt{36}$ and $\sqrt{81}$ are rational numbers because 6 and 9 are terminating decimals. Thus, **not all square roots are irrational**.

Numbers such as 36 and 81 are called *perfect squares*. A **perfect square** is a number that is the square of a whole number. The first few perfect squares are listed below.

$0 = 0^2$	$16 = 4^2$	$64 = 8^2$	$144 = 12^2$
$1 = 1^2$	$25 = 5^2$	$81 = 9^2$	$169 = 13^2$
$4 = 2^2$	$36 = 6^2$	$100 = 10^2$	$196 = 14^2$
$9 = 3^2$	$49 = 7^2$	$121 = 11^2$	$225 = 15^2$

The principal square root of a perfect square is a whole number. For example,

$$\sqrt{0} = 0, \sqrt{1} = 1, \sqrt{4} = 2, \sqrt{9} = 3, \sqrt{16} = 4, \sqrt{25} = 5, \sqrt{36} = 6,$$

and so on.

Simplifying Square Roots

2 Simplify square roots.

A rule for simplifying square roots can be generalized by comparing $\sqrt{25 \cdot 4}$ and $\sqrt{25} \cdot \sqrt{4}$. Notice that

$$\sqrt{25 \cdot 4} = \sqrt{100} = 10 \quad \text{and} \quad \sqrt{25} \cdot \sqrt{4} = 5 \cdot 2 = 10.$$

Because we obtain 10 in both situations, the original radicals must be equal. That is,

$$\sqrt{25 \cdot 4} = \sqrt{25} \cdot \sqrt{4}.$$

This result is a particular case of the **product rule for square roots** that can be generalized as follows:

The Product Rule for Square Roots

If a and b represent nonnegative numbers, then

$$\sqrt{ab} = \sqrt{a} \cdot \sqrt{b} \quad \text{and} \quad \sqrt{a} \cdot \sqrt{b} = \sqrt{ab}.$$

The square root of a product is the product of the square roots.

Example 1 shows how the product rule is used to remove from the square root any perfect squares that occur as factors.

Great Question!

Is the square root of a sum the sum of the square roots?

No. There are no addition or subtraction rules for square roots:

$$\sqrt{a + b} \neq \sqrt{a} + \sqrt{b}$$
$$\sqrt{a - b} \neq \sqrt{a} - \sqrt{b}.$$

For example, if $a = 9$ and $b = 16$,

$$\sqrt{9 + 16} = \sqrt{25} = 5$$

and

$$\sqrt{9} + \sqrt{16} = 3 + 4 = 7.$$

Thus,

$$\sqrt{9 + 16} \neq \sqrt{9} + \sqrt{16}.$$

Example 1   Simplifying Square Roots

Simplify, if possible:

a. $\sqrt{75}$　　　　**b.** $\sqrt{500}$　　　　**c.** $\sqrt{17}$.

Solution

a. $\sqrt{75} = \sqrt{25 \cdot 3}$ *25 is the greatest perfect square that is a factor of 75.*

    $= \sqrt{25} \cdot \sqrt{3}$ $\sqrt{ab} = \sqrt{a} \cdot \sqrt{b}$

    $= 5\sqrt{3}$ *Write $\sqrt{25}$ as 5.*

b. $\sqrt{500} = \sqrt{100 \cdot 5}$ *100 is the greatest perfect square factor of 500.*

    $= \sqrt{100} \cdot \sqrt{5}$ $\sqrt{ab} = \sqrt{a} \cdot \sqrt{b}$

    $= 10\sqrt{5}$ *Write $\sqrt{100}$ as 10.*

c. Because 17 has no perfect square factors (other than 1), $\sqrt{17}$ cannot be simplified.

 **Check Point 1**   Simplify, if possible:

▶ a. $\sqrt{12}$ $2\sqrt{3}$ b. $\sqrt{60}$ $2\sqrt{15}$ c. $\sqrt{55}$.   cannot be simplified

Multiplying Square Roots

③ Perform operations with square roots.

If a and b are nonnegative, then we can use the product rule

$$\sqrt{a} \cdot \sqrt{b} = \sqrt{a \cdot b}$$

to multiply square roots. The product of the square roots is the square root of the product. Once the square roots are multiplied, simplify the square root of the product when possible.

Example 2	Multiplying Square Roots

Multiply:

a. $\sqrt{2} \cdot \sqrt{5}$ b. $\sqrt{7} \cdot \sqrt{7}$ c. $\sqrt{6} \cdot \sqrt{12}$.

Solution

> It is possible to multiply irrational numbers and obtain a rational number for the product.

a. $\sqrt{2} \cdot \sqrt{5} = \sqrt{2 \cdot 5} = \sqrt{10}$

b. $\sqrt{7} \cdot \sqrt{7} = \sqrt{7 \cdot 7} = \sqrt{49} = 7$

c. $\sqrt{6} \cdot \sqrt{12} = \sqrt{6 \cdot 12} = \sqrt{72} = \sqrt{36 \cdot 2} = \sqrt{36} \cdot \sqrt{2} = 6\sqrt{2}$

 **Check Point 2**   Multiply:

▶ a. $\sqrt{3} \cdot \sqrt{10}$ $\sqrt{30}$ b. $\sqrt{10} \cdot \sqrt{10}$ 10   c. $\sqrt{6} \cdot \sqrt{2}$. $2\sqrt{3}$

Dividing Square Roots

Another property for square roots involves division.

> **The Quotient Rule for Square Roots**
>
> If a and b represent nonnegative numbers and $b \neq 0$, then
>
> $$\frac{\sqrt{a}}{\sqrt{b}} = \sqrt{\frac{a}{b}} \quad \text{and} \quad \sqrt{\frac{a}{b}} = \frac{\sqrt{a}}{\sqrt{b}}.$$
>
> The quotient of two square roots is the square root of the quotient.

Once the square roots are divided, simplify the square root of the quotient when possible.

Example 3   Dividing Square Roots

Find the quotient:

a. $\dfrac{\sqrt{75}}{\sqrt{3}}$ b. $\dfrac{\sqrt{90}}{\sqrt{2}}$.

Solution

a. $\dfrac{\sqrt{75}}{\sqrt{3}} = \sqrt{\dfrac{75}{3}} = \sqrt{25} = 5$

b. $\dfrac{\sqrt{90}}{\sqrt{2}} = \sqrt{\dfrac{90}{2}} = \sqrt{45} = \sqrt{9 \cdot 5} = \sqrt{9} \cdot \sqrt{5} = 3\sqrt{5}$

Check Point 3   Find the quotient:

a. $\dfrac{\sqrt{80}}{\sqrt{5}}$ 4   b. $\dfrac{\sqrt{48}}{\sqrt{6}}$. $2\sqrt{2}$

Adding and Subtracting Square Roots

The number that multiplies a square root is called the square root's **coefficient**. For example, in $3\sqrt{5}$, 3 is the coefficient of the square root.

Square roots with the same radicand can be added or subtracted by adding or subtracting their coefficients:

$$a\sqrt{c} + b\sqrt{c} = (a + b)\sqrt{c} \qquad a\sqrt{c} - b\sqrt{c} = (a - b)\sqrt{c}.$$

Sum of coefficients times the common square root   Difference of coefficients times the common square root

Example 4   Adding and Subtracting Square Roots

Add or subtract as indicated:

a. $7\sqrt{2} + 5\sqrt{2}$ b. $2\sqrt{5} - 6\sqrt{5}$ c. $3\sqrt{7} + 9\sqrt{7} - \sqrt{7}$.

Solution

a. $7\sqrt{2} + 5\sqrt{2} = (7 + 5)\sqrt{2}$
$= 12\sqrt{2}$

b. $2\sqrt{5} - 6\sqrt{5} = (2 - 6)\sqrt{5}$
$= -4\sqrt{5}$

c. $3\sqrt{7} + 9\sqrt{7} - \sqrt{7} = 3\sqrt{7} + 9\sqrt{7} - 1\sqrt{7}$ Write $\sqrt{7}$ as $1\sqrt{7}$.
$= (3 + 9 - 1)\sqrt{7}$
$= 11\sqrt{7}$

Check Point 4   Add or subtract as indicated:

a. $8\sqrt{3} + 10\sqrt{3}$ $18\sqrt{3}$

b. $4\sqrt{13} - 9\sqrt{13}$ $-5\sqrt{13}$

c. $7\sqrt{10} + 2\sqrt{10} - \sqrt{10}$. $8\sqrt{10}$

In some situations, it is possible to add and subtract square roots that do not contain a common square root by first simplifying.

Great Question!

Can I combine $\sqrt{2} + \sqrt{7}$?

No. Sums or differences of square roots that cannot be simplified and that do not contain a common radicand cannot be combined into one term by adding or subtracting coefficients. Some examples:

- $5\sqrt{3} + 3\sqrt{5}$ cannot be combined by adding coefficients. The square roots, $\sqrt{3}$ and $\sqrt{5}$, are different.
- $28 + 7\sqrt{3}$, or $28\sqrt{1} + 7\sqrt{3}$, cannot be combined by adding coefficients. The square roots, $\sqrt{1}$ and $\sqrt{3}$, are different.

Example 5 **Adding and Subtracting Square Roots by First Simplifying**

Add or subtract as indicated:

a. $\sqrt{2} + \sqrt{8}$

b. $4\sqrt{50} - 6\sqrt{32}$.

Solution

a. $\sqrt{2} + \sqrt{8}$

$= \sqrt{2} + \sqrt{4 \cdot 2}$ Split 8 into two factors such that one factor is a perfect square.

$= 1\sqrt{2} + 2\sqrt{2}$ $\sqrt{4 \cdot 2} = \sqrt{4} \cdot \sqrt{2} = 2\sqrt{2}$

$= (1 + 2)\sqrt{2}$ Add coefficients and retain the common square root.

$= 3\sqrt{2}$ Simplify.

b. $4\sqrt{50} - 6\sqrt{32}$

$= 4\sqrt{25 \cdot 2} - 6\sqrt{16 \cdot 2}$ 25 is the greatest perfect square factor of 50 and 16 is the greatest perfect square factor of 32.

$= 4 \cdot 5\sqrt{2} - 6 \cdot 4\sqrt{2}$ $\sqrt{25 \cdot 2} = \sqrt{25}\sqrt{2} = 5\sqrt{2}$ and $\sqrt{16 \cdot 2} = \sqrt{16}\sqrt{2} = 4\sqrt{2}$

$= 20\sqrt{2} - 24\sqrt{2}$ Multiply.

$= (20 - 24)\sqrt{2}$ Subtract coefficients and retain the common square root.

$= -4\sqrt{2}$ Simplify.

Check Point 5 Add or subtract as indicated:

a. $\sqrt{3} + \sqrt{12}$ $3\sqrt{3}$

b. $4\sqrt{8} - 7\sqrt{18}$. $-13\sqrt{2}$

Rationalizing Denominators

4 Rationalize denominators.

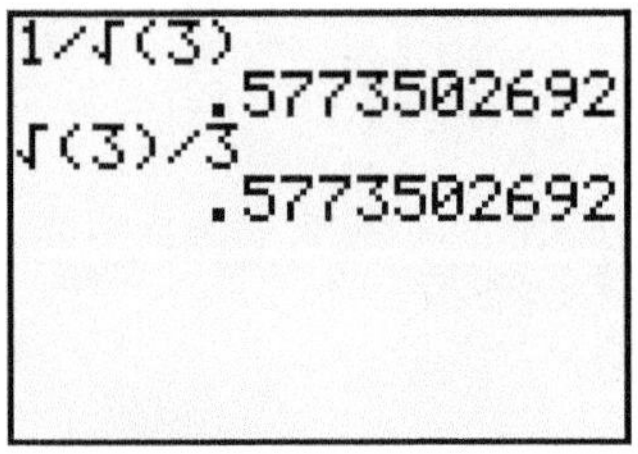

Figure 1.7 The calculator screen shows approximate values for $\frac{1}{\sqrt{3}}$ and $\frac{\sqrt{3}}{3}$.

The calculator screen in **Figure 1.7** shows approximate values for $\frac{1}{\sqrt{3}}$ and $\frac{\sqrt{3}}{3}$. The two approximations are the same. This is not a coincidence:

$$\frac{1}{\sqrt{3}} = \frac{1}{\sqrt{3}} \cdot \boxed{\frac{\sqrt{3}}{\sqrt{3}}} = \frac{\sqrt{3}}{\sqrt{9}} = \frac{\sqrt{3}}{3}$$

Any nonzero number divided by itself is 1. Multiplication by 1 does not change the value of $\frac{1}{\sqrt{3}}$.

This process involves rewriting a radical expression as an equivalent expression in which the denominator no longer contains any radicals. The process is called **rationalizing the denominator**. If the denominator contains the square root of a natural number that is not a perfect square, **multiply the numerator and the denominator by the smallest number that produces the square root of a perfect square in the denominator.**

Example 6 Rationalizing Denominators

Rationalize the denominator:

a. $\dfrac{15}{\sqrt{6}}$ **b.** $\sqrt{\dfrac{3}{5}}$ **c.** $\dfrac{12}{\sqrt{8}}$.

Solution

a. If we multiply the numerator and the denominator of $\dfrac{15}{\sqrt{6}}$ by $\sqrt{6}$, the denominator becomes $\sqrt{6} \cdot \sqrt{6} = \sqrt{36} = 6$. Therefore, we multiply by 1, choosing $\dfrac{\sqrt{6}}{\sqrt{6}}$ for 1.

$$\frac{15}{\sqrt{6}} = \frac{15}{\sqrt{6}} \cdot \frac{\sqrt{6}}{\sqrt{6}} = \frac{15\sqrt{6}}{\sqrt{36}} = \frac{15\sqrt{6}}{6} = \frac{5\sqrt{6}}{2}$$

Multiply by 1. Simplify: $\frac{15}{6} = \frac{5 \cdot \cancel{3}}{2 \cdot \cancel{3}} = \frac{5}{2}$.

b. $\sqrt{\dfrac{3}{5}} = \dfrac{\sqrt{3}}{\sqrt{5}} = \dfrac{\sqrt{3}}{\sqrt{5}} \cdot \dfrac{\sqrt{5}}{\sqrt{5}} = \dfrac{\sqrt{15}}{\sqrt{25}} = \dfrac{\sqrt{15}}{5}$

Multiply by 1.

c. The *smallest* number that will produce a perfect square in the denominator of $\dfrac{12}{\sqrt{8}}$ is $\sqrt{2}$, because $\sqrt{8} \cdot \sqrt{2} = \sqrt{16} = 4$. We multiply by 1, choosing $\dfrac{\sqrt{2}}{\sqrt{2}}$ for 1.

$$\frac{12}{\sqrt{8}} = \frac{12}{\sqrt{8}} \cdot \frac{\sqrt{2}}{\sqrt{2}} = \frac{12\sqrt{2}}{\sqrt{16}} = \frac{12\sqrt{2}}{4} = 3\sqrt{2}$$

✓ **Check Point 6** Rationalize the denominator:

a. $\dfrac{25}{\sqrt{10}}$ $\frac{5\sqrt{10}}{2}$ **b.** $\sqrt{\dfrac{2}{7}}$ $\frac{\sqrt{14}}{7}$ **c.** $\dfrac{5}{\sqrt{18}}$. $\frac{5\sqrt{2}}{6}$

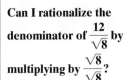

Great Question!

What exactly does rationalizing a denominator do to an irrational number in the denominator?

Rationalizing a numerical denominator makes that denominator a rational number.

Great Question!

Can I rationalize the denominator of $\dfrac{12}{\sqrt{8}}$ by multiplying by $\dfrac{\sqrt{8}}{\sqrt{8}}$?

Yes. However, it takes more work to simplify the result.

Blitzer Bonus

Golden Rectangles

The early Greeks believed that the most pleasing of all rectangles were **golden rectangles**, whose ratio of width to height is

$$\frac{w}{h} = \frac{\sqrt{5} + 1}{2}.$$

The Parthenon at Athens fits into a golden rectangle once the triangular pediment is reconstructed.

Irrational Numbers and Other Kinds of Roots

Irrational numbers appear in the form of roots other than square roots. The symbol $\sqrt[3]{}$ represents the **cube root** of a number. For example,

$$\sqrt[3]{8} = 2 \text{ because } 2 \cdot 2 \cdot 2 = 8 \quad \text{and} \quad \sqrt[3]{64} = 4 \text{ because } 4 \cdot 4 \cdot 4 = 64.$$

Although these cube roots are rational numbers, most cube roots are not. For example,

$$\sqrt[3]{217} \approx 6.0092 \text{ because } (6.0092)^3 \approx 216.995, \text{ not exactly } 217.$$

There is no end to the kinds of roots for numbers. For example, $\sqrt[4]{}$ represents the **fourth root** of a number. Thus, $\sqrt[4]{81} = 3$ because $3 \cdot 3 \cdot 3 \cdot 3 = 81$. Although the fourth root of 81 is rational, most fourth roots, fifth roots, and so on tend to be irrational.

Blitzer Bonus

A Radical Idea: Time Is Relative

What does travel in space have to do with square roots? Imagine that in the future we will be able to travel at velocities approaching the speed of light (approximately 186,000 miles per second). According to Einstein's theory of special relativity, time would pass more quickly on Earth than it would in the moving spaceship. The special-relativity equation

$$R_a = R_f \sqrt{1 - \left(\frac{v}{c}\right)^2}$$

gives the aging rate of an astronaut, R_a, relative to the aging rate of a friend, R_f, on Earth. In this formula, v is the astronaut's speed and c is the speed of light. As the astronaut's speed approaches the speed of light, we can substitute c for v.

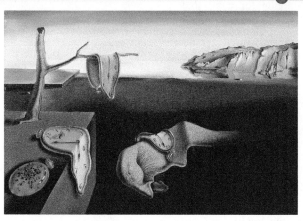

The Persistence of Memory (1931), Salvador Dali.
© 2011 MoMA/ARS

$$R_a = R_f \sqrt{1 - \left(\frac{v}{c}\right)^2} \qquad \text{Einstein's equation gives the aging rate of an astronaut, } R_a, \text{ relative to the aging rate of a friend, } R_f \text{ on Earth.}$$

$$R_a = R_f \sqrt{1 - \left(\frac{c}{c}\right)^2} \qquad \text{The velocity, } v, \text{ is approaching the speed of light, } c, \text{ so let } v = c.$$

$$= R_f \sqrt{1 - 1} \qquad \left(\frac{c}{c}\right)^2 = 1^2 = 1 \cdot 1 = 1$$

$$= R_f \sqrt{0} \qquad \text{Simplify the radicand: } 1 - 1 = 0.$$

$$= R_f \cdot 0 \qquad \sqrt{0} = 0$$

$$= 0 \qquad \text{Multiply: } R_f \cdot 0 = 0.$$

Close to the speed of light, the astronaut's aging rate, R_a, relative to that of a friend, R_f, on Earth is nearly 0. What does this mean? As we age here on Earth, the space traveler would barely get older. The space traveler would return to an unknown futuristic world in which friends and loved ones would be long gone.

Achieving Success

Don't Panic at the Length of the Exercise Sets.

You are not expected to work all, or even most, of the problems. Your professor will provide guidance on which exercises to work by assigning those problems that are consistent with the goals and objectives of your course.

Concept and Vocabulary Check

Exercises in the Concept and Vocabulary Check are intended for group and class discussions.

In Exercises 1–10, fill in each blank so that the resulting statement is true.

1. The set of irrational numbers is the set of numbers whose decimal representations are neither ____terminating____ nor ____repeating____.

2. The irrational number _____π_____ represents the circumference of a circle divided by the diameter of the circle.

3. The square root of *n*, represented by _____$\sqrt{n}$_____, is the nonnegative number that when multiplied by itself gives _____*n*_____.

4. $\sqrt{49 \cdot 6} = \sqrt{49} \cdot \sqrt{6} = 7\sqrt{6}$

5. The number that multiplies a square root is called the square root's ____coefficient____.

6. $8\sqrt{3} + 10\sqrt{3} = (\underline{8} + \underline{10})\sqrt{3} = \underline{18\sqrt{3}}$

7. $\sqrt{50} + \sqrt{32} = \sqrt{25 \cdot 2} + \sqrt{16 \cdot 2} = \sqrt{25} \cdot \sqrt{2} + \sqrt{16} \cdot \sqrt{2} = \underline{5}\sqrt{2} + \underline{4}\sqrt{2} = \underline{9\sqrt{2}}$

8. The process of rewriting a radical expression as an equivalent expression in which the denominator no longer contains any radicals is called __rationalizing the denominator__

9. The number $\sqrt{\dfrac{2}{7}}$ can be rewritten without a radical in the denominator by multiplying the numerator and denominator by _____$\sqrt{7}$_____.

10. The number $\dfrac{5}{\sqrt{12}}$ can be rewritten without a radical in the denominator by multiplying the numerator and denominator by _____$\sqrt{3}$_____, which is the smallest number that will produce a perfect square in the denominator.

In Exercises 11–14, determine whether each statement is true or false. If the statement is false, make the necessary change(s) to produce a true statement. Changes to false statements will vary.

11. The product of any two irrational numbers is always an irrational number. false

12. $\sqrt{9} + \sqrt{16} = \sqrt{25}$ false

13. $\sqrt{\sqrt{16}} = 2$ true

14. $\dfrac{\sqrt{64}}{2} = \sqrt{32}$ false

Respond to Exercises 15–20 using verbal or written explanations.

15. Describe the difference between a rational number and an irrational number. **15–20.** Answers will vary.

16. Using $\sqrt{50}$, explain how to simplify a square root.

17. Describe how to multiply square roots.

18. Explain how to add square roots with the same radicand.

19. Explain how to add $\sqrt{3} + \sqrt{12}$.

20. Describe what it means to rationalize a denominator. Use $\dfrac{2}{\sqrt{5}}$ in your explanation.

Exercise Set 1.4

Practice Exercises

Evaluate each expression in Exercises 1–10.

1. $\sqrt{9}$ 3
2. $\sqrt{16}$ 4
3. $\sqrt{25}$ 5
4. $\sqrt{49}$ 7
5. $\sqrt{64}$ 8
6. $\sqrt{100}$ 10
7. $\sqrt{121}$ 11
8. $\sqrt{144}$ 12
9. $\sqrt{169}$ 13
10. $\sqrt{225}$ 15

*In Exercises 11–16, use a calculator with a square root key to find a decimal approximation for each square root. Round the number displayed to the nearest **a.** tenth, **b.** hundredth, **c.** thousandth.*

11. $\sqrt{173}$ **a.** 13.2 **b.** 13.15 **c.** 13.153
12. $\sqrt{3176}$ **a.** 56.4 **b.** 56.36 **c.** 56.356
13. $\sqrt{17{,}761}$ **a.** 133.3 **b.** 133.27 **c.** 133.270
14. $\sqrt{779{,}264}$ **a.** 882.8 **b.** 882.76 **c.** 882.759
15. $\sqrt{\pi}$ **a.** 1.8 **b.** 1.77 **c.** 1.772
16. $\sqrt{2\pi}$ **a.** 2.5 **b.** 2.51 **c.** 2.507

In Exercises 17–24, simplify the square root.

17. $\sqrt{20}$ $\quad 2\sqrt{5}$ **18.** $\sqrt{50}$ $\quad 5\sqrt{2}$

19. $\sqrt{80}$ $\quad 4\sqrt{5}$ **20.** $\sqrt{12}$ $\quad 2\sqrt{3}$

21. $\sqrt{250}$ $\quad 5\sqrt{10}$ **22.** $\sqrt{192}$ $\quad 8\sqrt{3}$

23. $7\sqrt{28}$ $\quad 14\sqrt{7}$ **24.** $3\sqrt{52}$ $\quad 6\sqrt{13}$

In Exercises 25–56, perform the indicated operation. Simplify the answer when possible.

25. $\sqrt{7}\cdot\sqrt{6}$ $\quad \sqrt{42}$ **26.** $\sqrt{19}\cdot\sqrt{3}$ $\quad \sqrt{57}$

27. $\sqrt{6}\cdot\sqrt{6}$ $\quad 6$ **28.** $\sqrt{5}\cdot\sqrt{5}$ $\quad 5$

29. $\sqrt{3}\cdot\sqrt{6}$ $\quad 3\sqrt{2}$ **30.** $\sqrt{12}\cdot\sqrt{2}$ $\quad 2\sqrt{6}$

31. $\sqrt{2}\cdot\sqrt{26}$ $\quad 2\sqrt{13}$ **32.** $\sqrt{5}\cdot\sqrt{50}$ $\quad 5\sqrt{10}$

33. $\dfrac{\sqrt{54}}{\sqrt{6}}$ $\quad 3$ **34.** $\dfrac{\sqrt{75}}{\sqrt{3}}$ $\quad 5$

35. $\dfrac{\sqrt{90}}{\sqrt{2}}$ $\quad 3\sqrt{5}$ **36.** $\dfrac{\sqrt{60}}{\sqrt{3}}$ $\quad 2\sqrt{5}$

37. $\dfrac{-\sqrt{96}}{\sqrt{2}}$ $\quad -4\sqrt{3}$ **38.** $\dfrac{-\sqrt{150}}{\sqrt{3}}$ $\quad -5\sqrt{2}$

39. $7\sqrt{3}+6\sqrt{3}$ $\quad 13\sqrt{3}$ **40.** $8\sqrt{5}+11\sqrt{5}$ $\quad 19\sqrt{5}$

41. $4\sqrt{13}-6\sqrt{13}$ $\quad -2\sqrt{13}$ **42.** $6\sqrt{17}-8\sqrt{17}$ $\quad -2\sqrt{17}$

43. $\sqrt{5}+\sqrt{5}$ $\quad 2\sqrt{5}$ **44.** $\sqrt{3}+\sqrt{3}$ $\quad 2\sqrt{3}$

45. $4\sqrt{2}-5\sqrt{2}+8\sqrt{2}$ $\quad 7\sqrt{2}$

46. $6\sqrt{3}+8\sqrt{3}-16\sqrt{3}$ $\quad -2\sqrt{3}$

47. $\sqrt{5}+\sqrt{20}$ $\quad 3\sqrt{5}$

48. $\sqrt{3}+\sqrt{27}$ $\quad 4\sqrt{3}$

49. $\sqrt{50}-\sqrt{18}$ $\quad 2\sqrt{2}$

50. $\sqrt{63}-\sqrt{28}$ $\quad \sqrt{7}$

51. $3\sqrt{18}+5\sqrt{50}$ $\quad 34\sqrt{2}$

52. $4\sqrt{12}+2\sqrt{75}$ $\quad 18\sqrt{3}$

53. $\dfrac{1}{4}\sqrt{12}-\dfrac{1}{2}\sqrt{48}$ $\quad -\dfrac{3}{2}\sqrt{3}$

54. $\dfrac{1}{5}\sqrt{300}-\dfrac{2}{3}\sqrt{27}$ $\quad 0$

55. $3\sqrt{75}+2\sqrt{12}-2\sqrt{48}$ $\quad 11\sqrt{3}$

56. $2\sqrt{72}+3\sqrt{50}-\sqrt{128}$ $\quad 19\sqrt{2}$

In Exercises 57–66, rationalize the denominator.

57. $\dfrac{5}{\sqrt{3}}$ $\quad \dfrac{5\sqrt{3}}{3}$ **58.** $\dfrac{12}{\sqrt{5}}$ $\quad \dfrac{12\sqrt{5}}{5}$

59. $\dfrac{21}{\sqrt{7}}$ $\quad 3\sqrt{7}$ **60.** $\dfrac{30}{\sqrt{5}}$ $\quad 6\sqrt{5}$

61. $\dfrac{12}{\sqrt{30}}$ $\quad \dfrac{2\sqrt{30}}{5}$ **62.** $\dfrac{15}{\sqrt{50}}$ $\quad \dfrac{3\sqrt{2}}{2}$

63. $\dfrac{15}{\sqrt{12}}$ $\quad \dfrac{5\sqrt{3}}{2}$ **64.** $\dfrac{13}{\sqrt{40}}$ $\quad \dfrac{13\sqrt{10}}{20}$

65. $\sqrt{\dfrac{2}{5}}$ $\quad \dfrac{\sqrt{10}}{5}$ **66.** $\sqrt{\dfrac{5}{7}}$ $\quad \dfrac{\sqrt{35}}{7}$

Practice Plus

In Exercises 67–74, perform the indicated operations. Simplify the answer when possible.

67. $3\sqrt{8}-\sqrt{32}+3\sqrt{72}-\sqrt{75}$ $\quad 20\sqrt{2}-5\sqrt{3}$

68. $3\sqrt{54}-2\sqrt{24}-\sqrt{96}+4\sqrt{63}$ $\quad \sqrt{6}+12\sqrt{7}$

69. $3\sqrt{7}-5\sqrt{14}\cdot\sqrt{2}$ $\quad -7\sqrt{7}$

70. $4\sqrt{2}-8\sqrt{10}\cdot\sqrt{5}$ $\quad -36\sqrt{2}$

71. $\dfrac{\sqrt{32}}{5}+\dfrac{\sqrt{18}}{7}$ $\quad \dfrac{43\sqrt{2}}{35}$

72. $\dfrac{\sqrt{27}}{2}+\dfrac{\sqrt{75}}{7}$ $\quad \dfrac{31\sqrt{3}}{14}$

73. $\dfrac{\sqrt{2}}{\sqrt{3}}+\dfrac{\sqrt{3}}{\sqrt{2}}$ $\quad \dfrac{5\sqrt{6}}{6}$

74. $\dfrac{\sqrt{2}}{\sqrt{7}}+\dfrac{\sqrt{7}}{\sqrt{2}}$ $\quad \dfrac{9\sqrt{14}}{14}$

Application Exercises

The formula

$$d=\sqrt{\dfrac{3h}{2}}$$

models the distance, d, in miles, that a person h feet high can see to the horizon. Use this formula to solve Exercises 75–76.

75. The pool deck on a cruise ship is 72 feet above the water. How far can passengers on the pool deck see? Write the answer in simplified radical form. Then use the simplified radical form and a calculator to express the answer to the nearest tenth of a mile. $\quad 6\sqrt{3}$ mi; 10.4 mi

76. The captain of a cruise ship is on the star deck, which is 120 feet above the water. How far can the captain see? Write the answer in simplified radical form. Then use the simplified radical form and a calculator to express the answer to the nearest tenth of a mile. $\quad 6\sqrt{5}$ mi; 13.4 mi

Police use the formula $v=2\sqrt{5L}$ to estimate the speed of a car, v, in miles per hour, based on the length, L, in feet, of its skid marks upon sudden braking on a dry asphalt road. Use the formula to solve Exercises 77–78.

77. A motorist is involved in an accident. A police officer measures the car's skid marks to be 245 feet long. Estimate the speed at which the motorist was traveling before braking. If the posted speed limit is 50 miles per hour and the motorist tells the officer he was not speeding, should the officer believe him? Explain. $\quad$ 70 mph; He was speeding.

78. A motorist is involved in an accident. A police officer measures the car's skid marks to be 45 feet long. Estimate the speed at which the motorist was traveling before braking. If the posted speed limit is 35 miles per hour and the motorist tells the officer she was not speeding, should the officer believe her? Explain. $\quad$ 30 mph; She was not speeding.

79. The graph shows the median heights for boys of various ages in the United States from birth through 60 months, or five years old.

Boys' Heights

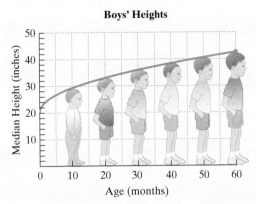

Source: Laura Walther Nathanson, *The Portable Pediatrician for Parents*

a. Use the graph to estimate the median height, to the nearest inch, of boys who are 50 months old. 41 in.

b. The formula $h = 2.9\sqrt{x} + 20.1$ models the median height, *h*, in inches, of boys who are *x* months of age. According to the formula, what is the median height of boys who are 50 months old? Use a calculator and round to the nearest tenth of an inch. How well does your estimate from part (a) describe the median height obtained from the formula? 40.6 in.; quite well

80. The graph shows the median heights for girls of various ages in the United States from birth through 60 months, or five years old.

Girls' Heights

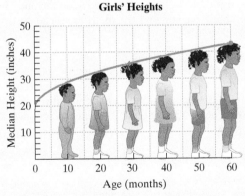

Source: Laura Walther Nathanson, *The Portable Pediatrician for Parents*

a. Use the graph to estimate the median height, to the nearest inch, of girls who are 50 months old. 41 in.

b. The formula $h = 3.1\sqrt{x} + 19$ models the median height, *h*, in inches, of girls who are *x* months of age. According to the formula, what is the median height of girls who are 50 months old? Use a calculator and round to the nearest tenth of an inch. How well does your estimate from part (a) describe the median height obtained from the formula? 40.9 in.; quite well

Autism is a neurological disorder that impedes language and derails social and emotional development. New findings suggest that the condition is not a sudden calamity that strikes children at the age of 2 or 3, but a developmental problem linked to abnormally rapid brain growth during infancy. The graphs show that the heads of severely autistic children start out smaller than average and then go through a period of explosive growth. Exercises 81–82 involve mathematical models for the data shown by the graphs.

Developmental Differences between Healthy Children and Severe Autistics

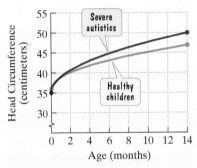

Source: *The Journal of the American Medical Association*

81. The data for one of the two groups shown by the graphs can be modeled by

$$H = 2.9\sqrt{x} + 36,$$

where *H* is the head circumference, in centimeters, at age *x* months, $0 \le x \le 14$.

a. According to the model, what is the head circumference at birth? 36 cm

b. According to the model, what is the head circumference at 9 months? 44.7 cm

c. According to the model, what is the head circumference at 14 months? Use a calculator and round to the nearest tenth of a centimeter. 46.9 cm

d. Use the values that you obtained in parts (a) through (c) and the graphs shown above to determine if the given model describes healthy children or severe autistics. healthy children

82. The data for one of the two groups shown by the graphs can be modeled by

$$H = 4\sqrt{x} + 35,$$

where *H* is the head circumference, in centimeters, at age *x* months, $0 \le x \le 14$.

a. According to the model, what is the head circumference at birth? 35 cm

b. According to the model, what is the head circumference at 9 months? 47 cm

c. According to the model, what is the head circumference at 14 months? Use a calculator and round to the nearest centimeter. 50 cm

d. Use the values that you obtained in parts (a) through (c) and the graphs shown above to determine if the given model describes healthy children or severe autistics. severe autistics

83. The popular comic strip *FoxTrot* follows the off-the-wall lives of the Fox family. Youngest son Jason is forever obsessed by his love of math. In the math-themed strip shown below, Jason shares his opinion in a coded message about the mathematical abilities of his sister Paige.

A = 11, B = 8, C = 1, D = 17,
E = 10, F = 2, G = 5, H = 19,
I = 13, J = 7, K = 26, L = 6,
M = 22, N = 20, O = 15, P = 16,
Q = 24, R = 9, S = 23, T = 12,
U = 3, V = 25, W = 4, X = 18,
Y = 21, Z = 14

Foxtrot © 2009 Bill Amend. Reprinted with permission of Universal Uclick. All rights reserved.

Solve problems A through Z in the left pane. Then decode Jason Fox's message involving his opinion about the mathematical abilities of his sister Paige shown on the first line. <small>Paige Fox is bad at math.</small>

Hints: Here is the solution for problem C and partial solutions for problems Q and U.

These are from trigonometry.

$$C = \sin\frac{\pi}{2} = \sin 90° = 1$$

$$Q = \int_0^2 9x^2 dx = 3x^3 \Big|_0^2 = 3 \cdot 2^3 - 3 \cdot 0^3 = \underline{\quad}$$

This is from calculus.

$$U = -3 \cos \pi = -3 \cos 180° = -3(-1) = \underline{\quad}$$

Note: The comic strip *FoxTrot* is now printed in more than one thousand newspapers. What made cartoonist Bill Amend, a college physics major, put math in the comic? "I always try to use math in the strip to make the joke accessible to anyone," he said, "But if you understand math, hopefully you'll like it that much more!" We highly recommend the math humor in Amend's *FoxTrot* collection *Math, Science, and Unix Underpants* (Andrews McMeel Publishing 2009).

The Blitzer Bonus on page 52 gives Einstein's special-relativity equation

$$R_a = R_f \sqrt{1 - \left(\frac{v}{c}\right)^2}$$

for the aging rate of an astronaut, R_a, relative to the aging rate of a friend on Earth, R_f, where v is the astronaut's speed and c is the speed of light. Take a few minutes to read the essay and then solve Exercises 84–85.

84. You are moving at 80% of the speed of light. Substitute $0.8c$ in the equation shown above. What is your aging rate relative to a friend on Earth? If 100 weeks have passed for your friend, how long were you gone? <small>0.6 R_f; 60 weeks</small>

85. You are moving at 90% of the speed of light. Substitute $0.9c$ in the equation above. What is your aging rate, correct to two decimal places, relative to a friend on Earth? If 100 weeks have passed for your friend, how long, to the nearest week, were you gone? <small>0.44 R_f; 44 weeks</small>

86. Read the Blitzer Bonus on page 52. The future is now: You have the opportunity to explore the cosmos in a starship traveling near the speed of light. The experience will enable you to understand the mysteries of the universe in deeply personal ways, transporting you to unimagined levels of knowing and being. The downside: You return from your two-year journey to a futuristic world in which friends and loved ones are long gone. Do you explore space or stay here on Earth? What are the reasons for your choice?

Critical Thinking Exercises

In Exercises 87–89, insert either < or > in the shaded area between the numbers to make each statement true.

87. $\sqrt{2}$ ▪ 1.5   <

88. $-\pi$ ▪ -3.5 >

89. $-\dfrac{3.14}{2}$ ▪ $-\dfrac{\pi}{2}$ >

90. How does doubling a number affect its square root?

91. Between which two consecutive integers is $-\sqrt{47}$? <small>−7 and −6</small>

92. Simplify: $\sqrt{2} + \sqrt{\dfrac{1}{2}}$. <small>$\frac{3\sqrt{2}}{2}$</small>

93. Provide an example to show that the following statement is false: The difference between two distinct irrational numbers is always an irrational number.

86. Answers will vary.   **90.** The square root is multiplied by $\sqrt{2}$.   **93.** Answers will vary; an example is: $(3 + \sqrt{2}) - (1 + \sqrt{2}) = 2$.

Group Exercises

The following topics related to irrational numbers are appropriate for either individual or group research projects. A report should be given to the class on the researched topic.   94–98. Answers will vary.

94. A History of How Irrational Numbers Developed

95. Pi: Its History, Applications, and Curiosities

96. Proving That $\sqrt{2}$ Is Irrational

97. Imaginary Numbers: Their History, Applications, and Curiosities

98. The Golden Rectangle in Art and Architecture

1.5 | Real Numbers and Their Properties

What am I supposed to learn?

After you have read this section, you should be able to:

 Recognize subsets of the real numbers.

 Recognize properties of real numbers.

 Recognize subsets of the real numbers.

The Set of Real Numbers

The vampire legend is death as seducer; he/she sucks our blood to take us to a perverse immortality. The vampire resembles us, but appears hidden among mortals. In this section, you will find vampires in the world of numbers. Mathematicians even use the labels *vampire* and *weird* to describe sets of numbers. However, the label that appears most frequently is *real*. The union of the rational numbers and the irrational numbers is the set of **real numbers**.

The sets that make up the real numbers are summarized in **Table 1.2**. We refer to these sets as **subsets** of the real numbers, meaning that all elements in each subset are also elements in the set of real numbers.

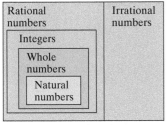

This diagram shows that every real number is rational or irrational.

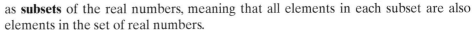
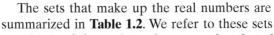

Table 1.2 Important Subsets of the Real Numbers		
Name	**Description**	**Examples**
Natural numbers	$\{1, 2, 3, 4, 5, \ldots\}$ These are the numbers that we use for counting.	$2, 3, 5, 17$
Whole numbers	$\{0, 1, 2, 3, 4, 5, \ldots\}$ The set of whole numbers includes 0 and the natural numbers.	$0, 2, 3, 5, 17$
Integers	$\{\ldots, -5, -4, -3, -2, -1, 0, 1, 2, 3, 4, 5, \ldots\}$ The set of integers includes the whole numbers and the negatives of the natural numbers.	$-17, -5, -3, -2, 0, 2, 3, 5, 17$
Rational numbers	The set of rational numbers is the set of all numbers that can be expressed in the form $\frac{a}{b}$, where a and b are integers, $b \neq 0$. Rational numbers can be expressed as terminating or repeating decimals.	$-17 = \frac{-17}{1}, -5 = \frac{-5}{1}, -3, -2,$ $0, 2, 3, 5, 17,$ $\frac{2}{5} = 0.4,$ $\frac{-2}{3} = -0.6666\ldots = -0.\overline{6}$
Irrational numbers	The set of irrational numbers is the set of all numbers whose decimal representations are neither terminating nor repeating. Irrational numbers cannot be expressed as a quotient of integers.	$\sqrt{2} \approx 1.414214$ $-\sqrt{3} \approx -1.73205$ $\pi \approx 3.142$ $-\frac{\pi}{2} \approx -1.571$

Blitzer Bonus

Weird Numbers

Mathematicians use the label **weird** to describe a number if

1. The sum of its factors, excluding the number itself, is greater than the number.

2. No partial collection of the factors adds up to the number.

The number 70 is weird. Its factors are 1, 2, 5, 7, 10, 14, and 35. The sum of these factors is 74, which is greater than 70. Two or more numbers in the list of factors cannot be added to obtain 70.

   Weird numbers are rare. Below 10,000, the weird numbers are 70, 836, 4030, 5830, 7192, 7912, and 9272. It is not known whether an odd weird number exists.

Example 1   Classifying Real Numbers

Consider the following set of numbers:

$$\left\{ -7, -\frac{3}{4}, 0, 0.\overline{6}, \sqrt{5}, \pi, 7.3, \sqrt{81} \right\}.$$

List the numbers in the set that are

a. natural numbers.   **b.** whole numbers.   **c.** integers.

d. rational numbers.   **e.** irrational numbers.   **f.** real numbers.

Solution

a. Natural numbers: The natural numbers are the numbers used for counting. The only natural number in the set is $\sqrt{81}$ because $\sqrt{81} = 9$. (9 multiplied by itself, or 9^2, is 81.)

b. Whole numbers: The whole numbers consist of the natural numbers and 0. The elements of the set that are whole numbers are 0 and $\sqrt{81}$.

c. Integers: The integers consist of the natural numbers, 0, and the negatives of the natural numbers. The elements of the set that are integers are $\sqrt{81}$, 0, and -7.

d. Rational numbers: All numbers in the set that can be expressed as the quotient of integers are rational numbers. These include $-7 \left(-7 = \frac{-7}{1} \right)$, $-\frac{3}{4}$, $0 \left(0 = \frac{0}{1} \right)$, and $\sqrt{81} \left(\sqrt{81} = \frac{9}{1} \right)$. Furthermore, all numbers in the set that are terminating or repeating decimals are also rational numbers. These include $0.\overline{6}$ and 7.3.

e. Irrational numbers: The irrational numbers in the set are $\sqrt{5} \, (\sqrt{5} \approx 2.236)$ and $\pi \, (\pi \approx 3.14)$. Both $\sqrt{5}$ and π are only approximately equal to 2.236 and 3.14, respectively. In decimal form, $\sqrt{5}$ and π neither terminate nor have blocks of repeating digits.

f. Real numbers: All the numbers in the given set are real numbers.

✓ **Check Point 1**   Consider the following set of numbers:

$$\left\{ -9, -1.3, 0, 0.\overline{3}, \frac{\pi}{2}, \sqrt{9}, \sqrt{10} \right\}.$$

List the numbers in the set that are

a. natural numbers. $\sqrt{9}$ **b.** whole numbers. $0, \sqrt{9}$

c. integers. $-9, 0, \sqrt{9}$ **d.** rational numbers. $-9, -1.3, 0, 0.\overline{3}, \sqrt{9}$

e. irrational numbers. $\frac{\pi}{2}, \sqrt{10}$ **f.** real numbers. $-9, -1.3, 0, 0.\overline{3}, \frac{\pi}{2}, \sqrt{9}, \sqrt{10}$

Blitzer Bonus

Vampire Numbers

Like legendary vampires that lie concealed among humans, vampire numbers lie hidden within the set of real numbers, mostly undetected. By definition, vampire numbers have an even number of digits. Furthermore, they are the product of two numbers whose digits all survive, in scrambled form, in the vampire. For example, 1260, 1435, and 2187 are vampire numbers.

$$21 \times 60 = 1260 \qquad 35 \times 41 = 1435 \qquad 27 \times 81 = 2187$$

| The digits 2, 1, 6, and 0 lie scrambled in the vampire number. | The digits 3, 5, 4, and 1 lurk within the vampire number. | The digits 2, 7, 8, and 1 survive in the vampire number. |

As the real numbers grow increasingly larger, is it necessary to pull out a wooden stake with greater frequency? How often can you expect to find vampires hidden among the giants? And is it possible to find a weird vampire?

   On the right of the equal sign is a 40-digit vampire number that was discovered using a Pascal program on a personal computer:

$$98,765,432,198,765,432,198 \times 98,765,432,198,830,604,534 = 9,754,610,597,415,368,368,844,499,268,390,128,385,732.$$

Source: Clifford Pickover, *Wonders of Numbers,* Oxford University Press, 2001.

Properties of the Real Numbers

 2 Recognize properties of real numbers.

When you use your calculator to add two real numbers, you can enter them in either order. The fact that two real numbers can be added in either order is called the **commutative property of addition**. You probably use this property, as well as other properties of the real numbers listed in **Table 1.3**, without giving it much thought. The properties of the real numbers are especially useful in algebra, as we shall see in Chapter 2.

Blitzer Bonus
●●●●●●●●●●●●●●●●●

The Associative Property and the English Language

In the English language, phrases can take on different meanings depending on the way the words are associated with commas.

Here are four examples.

- Woman, without her man, is nothing.
  Woman, without her, man is nothing.
- Mr. Rogers, the secretary is two hours late.
  Mr. Rogers, the secretary, is two hours late.
- Do not break your bread or roll in your soup.
  Do not break your bread, or roll in your soup.
- In the parade will be several hundred students carrying flags, and many teachers.
  In the parade will be several hundred students, carrying flags and many teachers.

Table 1.3	Properties of the Real Numbers	
Name	**Meaning**	**Examples**
Closure Property of Addition	The sum of any two real numbers is a real number.	$4\sqrt{2}$ is a real number and $5\sqrt{2}$ is a real number, so $4\sqrt{2} + 5\sqrt{2}$, or $9\sqrt{2}$, is a real number.
Closure Property of Multiplication	The product of any two real numbers is a real number.	10 is a real number and $\frac{1}{2}$ is a real number, so $10 \cdot \frac{1}{2}$, or 5, is a real number.
Commutative Property of Addition	Changing order when adding does not affect the sum. $a + b = b + a$	• $13 + 7 = 7 + 13$ • $\sqrt{2} + \sqrt{5} = \sqrt{5} + \sqrt{2}$
Commutative Property of Multiplication	Changing order when multiplying does not affect the product. $ab = ba$	• $13 \cdot 7 = 7 \cdot 13$ • $\sqrt{2} \cdot \sqrt{5} = \sqrt{5} \cdot \sqrt{2}$
Associative Property of Addition	Changing grouping when adding does not affect the sum. $(a + b) + c = a + (b + c)$	$(7 + 2) + 5 = 7 + (2 + 5)$ $9 + 5 = 7 + 7$ $14 = 14$
Associative Property of Multiplication	Changing grouping when multiplying does not affect the product. $(ab)c = a(bc)$	$(7 \cdot 2) \cdot 5 = 7 \cdot (2 \cdot 5)$ $14 \cdot 5 = 7 \cdot 10$ $70 = 70$
Distributive Property of Multiplication over Addition	Multiplication distributes over addition. $a \cdot (b + c) = a \cdot b + a \cdot c$	$7(4 + \sqrt{3}) = 7 \cdot 4 + 7 \cdot \sqrt{3}$ $= 28 + 7\sqrt{3}$
Identity Property of Addition	Zero can be deleted from a sum. $a + 0 = a$ $0 + a = a$	• $\sqrt{3} + 0 = \sqrt{3}$ • $0 + \pi = \pi$
Identity Property of Multiplication	One can be deleted from a product. $a \cdot 1 = a$ $1 \cdot a = a$	• $\sqrt{3} \cdot 1 = \sqrt{3}$ • $1 \cdot \pi = \pi$
Inverse Property of Addition	The sum of a real number and its additive inverse gives 0, the additive identity. $a + (-a) = 0$ $(-a) + a = 0$	• $\sqrt{3} + (-\sqrt{3}) = 0$ • $-\pi + \pi = 0$
Inverse Property of Multiplication	The product of a nonzero real number and its multiplicative inverse gives 1, the multiplicative identity. $a \cdot \dfrac{1}{a} = 1, a \neq 0$ $\dfrac{1}{a} \cdot a = 1, a \neq 0$	• $\sqrt{3} \cdot \dfrac{1}{\sqrt{3}} = 1$ • $\dfrac{1}{\pi} \cdot \pi = 1$

Great Question!

Is there an easy way to distinguish between the commutative and associative properties?

Commutative: Changes *order*.
Associative: Changes *grouping*.

Example 2 Identifying Properties of the Real Numbers

Name the property illustrated:

a. $\sqrt{3} \cdot 7 = 7 \cdot \sqrt{3}$ **b.** $(4 + 7) + 6 = 4 + (7 + 6)$

c. $2(3 + \sqrt{5}) = 6 + 2\sqrt{5}$ **d.** $\sqrt{2} + (\sqrt{3} + \sqrt{7}) = \sqrt{2} + (\sqrt{7} + \sqrt{3})$

e. $17 + (-17) = 0$ **f.** $\sqrt{2} \cdot 1 = \sqrt{2}.$

Solution

a. $\sqrt{3} \cdot 7 = 7 \cdot \sqrt{3}$ *Commutative property of multiplication*

b. $(4 + 7) + 6 = 4 + (7 + 6)$ *Associative property of addition*

c. $2(3 + \sqrt{5}) = 6 + 2\sqrt{5}$ *Distributive property of multiplication over addition*

d. $\sqrt{2} + (\sqrt{3} + \sqrt{7}) = \sqrt{2} + (\sqrt{7} + \sqrt{3})$ *The only change between the left and the right sides is in the order that $\sqrt{3}$ and $\sqrt{7}$ are added. The order is changed from $\sqrt{3} + \sqrt{7}$ to $\sqrt{7} + \sqrt{3}$ using the commutative property of addition.*

e. $17 + (-17) = 0$ *Inverse property of addition*

f. $\sqrt{2} \cdot 1 = \sqrt{2}$ *Identity property of multiplication*

✓ **Check Point 2**   Name the property illustrated:

a. $(4 \cdot 7) \cdot 3 = 4 \cdot (7 \cdot 3)$ associative property of multiplication

b. $3(\sqrt{5} + 4) = 3(4 + \sqrt{5})$ commutative property of addition

c. $3(\sqrt{5} + 4) = 3\sqrt{5} + 12$ distributive property of multiplication over addition

d. $2(\sqrt{3} + \sqrt{7}) = (\sqrt{3} + \sqrt{7})2$ commutative property of multiplication

e. $1 + 0 = 1$ identity property of addition

f. $-4\left(-\dfrac{1}{4}\right) = 1.$ inverse property of multiplication

Although the entire set of real numbers is closed with respect to addition and multiplication, some of the subsets of the real numbers do not satisfy the closure property for a given operation. If an operation on a set results in just one number that is not in that set, then the set is not closed for that operation.

Example 3 Verifying Closure

a. Are the integers closed with respect to multiplication?

b. Are the irrational numbers closed with respect to multiplication?

c. Are the natural numbers closed with respect to division?

Solution

a. Consider some examples of the multiplication of integers:

$$3 \cdot 2 = 6 \quad 3(-2) = -6 \quad -3(-2) = 6 \quad -3 \cdot 0 = 0.$$

The product of any two integers is always a positive integer, a negative integer, or zero, which is an integer. Thus, the integers are closed under the operation of multiplication.

b. If we multiply two irrational numbers, must the product always be an irrational number? The answer is no. On the right is an example.

$$\sqrt{7} \cdot \sqrt{7} = \sqrt{49} = 7$$

Both irrational   Not an irrational number

This means that the irrational numbers are not closed under the operation of multiplication.

c. If we divide any two natural numbers, must the quotient always be a natural number? The answer is no. Here is an example:

$$4 \div 8 = \tfrac{1}{2}.$$

Both natural numbers   Not a natural number

Thus, the natural numbers are not closed under the operation of division.

✓ **Check Point 3**

a. Are the natural numbers closed with respect to multiplication?   yes

▶ **b.** Are the integers closed with respect to division?   no

The commutative property involves a change in order with no change in the final result. However, changing the order in which we subtract and divide real numbers can produce different answers. For example,

$$7 - 4 \neq 4 - 7 \quad \text{and} \quad 6 \div 2 \neq 2 \div 6.$$

Because the real numbers are not commutative with respect to subtraction and division, it is important that you enter numbers in the correct order when using a calculator to perform these operations.

The associative property does not hold for the operations of subtraction and division. The examples below show that if we change groupings when subtracting or dividing three numbers, the answer may change.

$$(6 - 1) - 3 \neq 6 - (1 - 3) \qquad (8 \div 4) \div 2 \neq 8 \div (4 \div 2)$$
$$5 - 3 \neq 6 - (-2) \qquad\qquad 2 \div 2 \neq 8 \div 2$$
$$2 \neq 8 \qquad\qquad\qquad 1 \neq 4$$

Blitzer Bonus

• • • • • • • • • • • • • • • • •

Beyond the Real Numbers

Only real numbers greater than or equal to zero have real number square roots. The square root of -1, $\sqrt{-1}$, is not a real number. This is because there is no real number that can be multiplied by itself that results in -1. Multiplying any real number by itself can never give a negative product. In the sixteenth century, mathematician Girolamo Cardano (1501–1576) wrote that square roots of negative numbers would cause "mental tortures." In spite of these "tortures," mathematicians invented a new number, called *i,* to represent $\sqrt{-1}$. The number *i* is not a real number; it is called an **imaginary number**. Thus, $\sqrt{9} = 3$, $-\sqrt{9} = -3$, but $\sqrt{-9}$ is not a real number. However, $\sqrt{-9}$ is an imaginary number, represented by $3i$. The adjective *real* as a way of describing what we now call the real numbers was first used by the French mathematician and philosopher René Descartes (1596–1650) in response to the concept of imaginary numbers.

Source: © 2000 Roz Chast from Cartoonbank.com. All rights reserved.

Achieving Success

Ask! Ask! Ask!

Do not be afraid to ask questions in class. Your professor may not realize that a concept is unclear until you raise your hand and ask a question. Other students who have problems asking questions in class will be appreciative that you have spoken up. Be polite and professional, but ask as many questions as required.

Concept and Vocabulary Check

Exercises in the Concept and Vocabulary Check are intended for group and class discussions.

In Exercises 1–8, fill in each blank so that the resulting statement is true.

1. Every real number is either _____rational_____ or _____irrational_____.

2. The _____closure_____ property of addition states that the sum of any two real numbers is a real number.

3. If a and b are real numbers, the commutative property of multiplication states that _____$ab = ba$_____.

4. If a, b, and c are real numbers, the associative property of addition states that $(a + b) + c = a + (b + c)$

5. If a, b, and c are real numbers, the distributive property states that $a(b + c) = ab + ac$.

6. The _____identity_____ property of addition states that zero can be deleted from a sum.

7. The _____identity_____ property of multiplication states that _____1_____ can be deleted from a product.

8. The product of a nonzero real number and its _multiplicative inverse_ gives 1, the _multiplicative identity_.

In Exercises 9–16, determine whether each statement is true or false. If the statement is false, make the necessary change(s) to produce a true statement. Changes to false statements will vary.

9. Every rational number is an integer. false

10. Some whole numbers are not integers. false

11. Some rational numbers are not positive. true

12. Irrational numbers cannot be negative. false

13. Subtraction is a commutative operation. false

14. $(24 \div 6) \div 2 = 24 \div (6 \div 2)$ false

15. $7 \cdot a + 3 \cdot a = a \cdot (7 + 3)$ true

16. $2 \cdot a + 5 = 5 \cdot a + 2$ false

Respond to Exercises 17–24 using verbal or written explanations.

17. What does it mean when we say that the rational numbers are a subset of the real numbers? 17–24. Answers will vary.

18. What does it mean if we say that a set is closed under a given operation?

19. State the commutative property of addition and give an example.

20. State the commutative property of multiplication and give an example.

21. State the associative property of addition and give an example.

22. State the associative property of multiplication and give an example.

23. State the distributive property of multiplication over addition and give an example.

24. Does $7 \cdot (4 \cdot 3) = 7 \cdot (3 \cdot 4)$ illustrate the commutative property or the associative property? Explain your answer.

Exercise Set 1.5

Practice Exercises

In Exercises 1–4, list all numbers from the given set that are

    a. *natural numbers.*   **b.** *whole numbers.*

    c. *integers.*   **d.** *rational numbers.*

    e. *irrational numbers.*   **f.** *real numbers.*

1. $\{-9, -\frac{4}{5}, 0, 0.25, \sqrt{3}, 9.2, \sqrt{100}\}$ *

2. $\{-7, -0.\overline{6}, 0, \sqrt{49}, \sqrt{50}\}$ *

3. $\{-11, -\frac{5}{6}, 0, 0.75, \sqrt{5}, \pi, \sqrt{64}\}$ *

4. $\{-5, -0.\overline{3}, 0, \sqrt{2}, \sqrt{4}\}$ *

5. Give an example of a whole number that is not a natural number.   0

6. Give an example of an integer that is not a whole number.   −1

7. Give an example of a rational number that is not an integer. $\frac{1}{2}$

8. Give an example of a rational number that is not a natural number. $\frac{1}{2}$

9. Give an example of a number that is an integer, a whole number, and a natural number.   1

10. Give an example of a number that is a rational number, an integer, and a real number.   1

11. Give an example of a number that is an irrational number and a real number. $\sqrt{2}$

12. Give an example of a number that is a real number, but not an irrational number.   1   **6–12.** Answers will vary. Examples are given.

Complete each statement in Exercises 13–15 to illustrate the commutative property.

13. $3 + (4 + 5) = 3 + (5 + \underline{\quad})$ 4

14. $\sqrt{5} \cdot 4 = 4 \cdot \underline{\quad}$ $\sqrt{5}$

15. $9 \cdot (6 + 2) = 9 \cdot (2 + \underline{\quad})$ 6

Complete each statement in Exercises 16–17 to illustrate the associative property.

16. $(3 + 7) + 9 = \underline{\quad} + (7 + \underline{\quad})$ 3; 9

17. $(4 \cdot 5) \cdot 3 = \underline{\quad} \cdot (5 \cdot \underline{\quad})$ 4; 3

Complete each statement in Exercises 18–20 to illustrate the distributive property.

18. $3 \cdot (6 + 4) = 3 \cdot 6 + 3 \cdot \underline{\quad}$ 4

19. $\underline{\quad} \cdot (4 + 5) = 7 \cdot 4 + 7 \cdot 5$ 7

20. $2 \cdot (\underline{\quad} + 3) = 2 \cdot 7 + 2 \cdot 3$ 7

Use the distributive property to simplify the radical expressions in Exercises 21–28.

21. $5(6 + \sqrt{2})$ $30 + 5\sqrt{2}$ **22.** $4(3 + \sqrt{5})$ $12 + 4\sqrt{5}$

23. $\sqrt{7}(3 + \sqrt{2})$ $3\sqrt{7} + \sqrt{14}$ **24.** $\sqrt{6}(7 + \sqrt{5})$ $7\sqrt{6} + \sqrt{30}$

25. $\sqrt{3}(5 + \sqrt{3})$ $5\sqrt{3} + 3$

26. $\sqrt{7}(9 + \sqrt{7})$ $9\sqrt{7} + 7$

27. $\sqrt{6}(\sqrt{2} + \sqrt{6})$ $2\sqrt{3} + 6$

28. $\sqrt{10}(\sqrt{2} + \sqrt{10})$ $2\sqrt{5} + 10$

*See Answers to Selected Exercises.

In Exercises 29–44, state the name of the property illustrated.

29. $6 + (-4) = (-4) + 6$ commutative property of addition

30. $11 \cdot (7 + 4) = 11 \cdot 7 + 11 \cdot 4$ distributive property

31. $6 + (2 + 7) = (6 + 2) + 7$ associative property of addition

32. $6 \cdot (2 \cdot 3) = 6 \cdot (3 \cdot 2)$ commutative property of multiplication

33. $(2 + 3) + (4 + 5) = (4 + 5) + (2 + 3)$ commutative property of addition

34. $7 \cdot (11 \cdot 8) = (11 \cdot 8) \cdot 7$ commutative property of multiplication

35. $2(-8 + 6) = -16 + 12$ distributive property

36. $-8(3 + 11) = -24 + (-88)$ distributive property

37. $(2\sqrt{3}) \cdot \sqrt{5} = 2(\sqrt{3} \cdot \sqrt{5})$ associative property of multiplication

38. $\sqrt{2}\pi = \pi\sqrt{2}$ commutative property of multiplication

39. $\sqrt{17} \cdot 1 = \sqrt{17}$ identity property of multiplication

40. $\sqrt{17} + 0 = \sqrt{17}$ identity property of addition

41. $\sqrt{17} + (-\sqrt{17}) = 0$ inverse property of addition

42. $\sqrt{17} \cdot \dfrac{1}{\sqrt{17}} = 1$ inverse property of multiplication

43. $\dfrac{1}{\sqrt{2} + \sqrt{7}}(\sqrt{2} + \sqrt{7}) = 1$ inverse property of multiplication

44. $(\sqrt{2} + \sqrt{7}) + -(\sqrt{2} + \sqrt{7}) = 0$ inverse property of addition

In Exercises 45–49, use two numbers to show that

45. the natural numbers are not closed with respect to subtraction. $1 - 2 = -1$

46. the natural numbers are not closed with respect to division.

47. the integers are not closed with respect to division. $4 \div 8 = \frac{1}{2}$

48. the irrational numbers are not closed with respect to subtraction. $\sqrt{2} - \sqrt{2} = 0$

49. the irrational numbers are not closed with respect to multiplication. $\sqrt{2} \cdot \sqrt{2} = 2$

46. $4 \div 8 = \frac{1}{2}$ **45–49.** Answers will vary. Examples are given.

Practice Plus

In Exercises 50–53, determine if each statement is true or false. Do not use a calculator.

50. $468(787 + 289) = 787 + 289(468)$ false

51. $468(787 + 289) = 787(468) + 289(468)$ true

52. $58 \cdot 9 + 32 \cdot 9 = (58 + 32) \cdot 9$ true

53. $58 \cdot 9 \cdot 32 \cdot 9 = (58 \cdot 32) \cdot 9$ false

Application Exercises

In Exercises 54–57, use the definition of vampire numbers from the Blitzer Bonus on page 58 to determine which products are vampires.

54. $15 \times 93 = 1395$ vampire

55. $80 \times 86 = 6880$ vampire

56. $20 \times 51 = 1020$ not a vampire

57. $146 \times 938 = 136{,}948$ vampire

A **narcissistic number** *is an n-digit number equal to the sum of each of its digits raised to the nth power. Here's an example:*

$$153 = 1^3 + 5^3 + 3^3.$$

Three digits, so exponents are 3

In Exercises 58–61, determine which real numbers are narcissistic.

58. 370   narcissistic

59. 371   narcissistic

60. 372   not narcissistic

61. 9474   narcissistic

62. The expressions

$$\frac{D(A + 1)}{24} \quad \text{and} \quad \frac{DA + D}{24}$$

describe the drug dosage for children between the ages of 2 and 13. In each expression, D stands for an adult dose and A represents the child's age.

a. Name the property that explains why these expressions are equal for all values of D and A.   distributive property

b. If an adult dose of ibuprofen is 200 milligrams, what is the proper dose for a 12-year-old child? Use both forms of the expressions to answer the question. Which form is easier to use?   approx. 108 mg; Answers will vary.

65. Answers will vary. An example is: In a group of at least 367 people, at least two people will have the same birthday, but a subset of the group containing 50 people would not necessarily have two people with the same birthday.

Critical Thinking Exercises

In Exercises 63–64, name the property used to go from step to step each time that "(why?)" occurs.

63. $7 + 2(x + 9)$
$= 7 + (2x + 18)$ (why?)   dist. prop.
$= 7 + (18 + 2x)$ (why?)   comm. prop. of add.
$= (7 + 18) + 2x$ (why?)   assoc. prop. of add.
$= 25 + 2x$
$= 2x + 25$ (why?)   comm. prop. of add.

64. $5(x + 4) + 3x$
$= (5x + 20) + 3x$ (why?)   dist. prop.
$= (20 + 5x) + 3x$ (why?)   comm. prop. of add.
$= 20 + (5x + 3x)$ (why?)   assoc. prop. of add.
$= 20 + (5 + 3)x$ (why?)   dist. prop.
$= 20 + 8x$
$= 8x + 20$ (why?)   comm. prop. of add.

65. Closure illustrates that a characteristic of a set is not necessarily a characteristic of all of its subsets. The real numbers are closed with respect to multiplication, but the irrational numbers, a subset of the real numbers, are not. Give an example of a set that is not mathematical that has a particular characteristic, but which has a subset without this characteristic.

1.6 : Exponents and Scientific Notation

What am I supposed to learn?

After you have read this section, you should be able to:

① Use properties of exponents.

② Convert from scientific notation to decimal notation.

③ Convert from decimal notation to scientific notation.

④ Perform computations using scientific notation.

⑤ Solve applied problems using scientific notation.

Bigger than the biggest thing ever and then some. Much bigger than that in fact, really amazingly immense, a totally stunning size, real 'wow, that's big', time . . . Gigantic multiplied by colossal multiplied by staggeringly huge is the sort of concept we're trying to get across here.

Douglas Adams, *The Restaurant at the End of the Universe*

Although Adams's description may not quite apply to this $15.2 trillion national debt, exponents can be used to explore the meaning of this "staggeringly huge" number. In this section, you will learn to use exponents to provide a way of putting large and small numbers in perspective.

Properties of Exponents

We have seen that exponents are used to indicate repeated multiplication. Now consider the multiplication of two exponential expressions, such as $b^4 \cdot b^3$. We are multiplying 4 factors of b and 3 factors of b. We have a total of 7 factors of b:

4 factors of b 3 factors of b

$$b^4 \cdot b^3 = (b \cdot b \cdot b \cdot b)(b \cdot b \cdot b) = b^7.$$

Total: 7 factors of b

① Use properties of exponents.

The product is exactly the same if we add the exponents:

$$b^4 \cdot b^3 = b^{4+3} = b^7.$$

Properties of exponents allow us to perform operations with exponential expressions without having to write out long strings of factors. Three such properties are given in **Table 1.4**.

Table 1.4 Properties of Exponents

Property	Meaning	Examples
The Product Rule $b^m \cdot b^n = b^{m+n}$	When multiplying exponential expressions with the same base, add the exponents. Use this sum as the exponent of the common base.	$9^6 \cdot 9^{12} = 9^{6+12} = 9^{18}$
The Power Rule $(b^m)^n = b^{m \cdot n}$	When an exponential expression is raised to a power, multiply the exponents. Place the product of the exponents on the base and remove the parentheses.	$(3^4)^5 = 3^{4 \cdot 5} = 3^{20}$ $(5^3)^8 = 5^{3 \cdot 8} = 5^{24}$
The Quotient Rule $\dfrac{b^m}{b^n} = b^{m-n}$	When dividing exponential expressions with the same base, subtract the exponent in the denominator from the exponent in the numerator. Use this difference as the exponent of the common base.	$\dfrac{5^{12}}{5^4} = 5^{12-4} = 5^8$ $\dfrac{9^{40}}{9^5} = 9^{40-5} = 9^{35}$

The third property in **Table 1.4**, $\dfrac{b^m}{b^n} = b^{m-n}$, called the quotient rule, can lead to a zero exponent when subtracting exponents. Here is an example:

$$\frac{4^3}{4^3} = 4^{3-3} = 4^0.$$

We can see what this zero exponent means by evaluating 4^3 in the numerator and the denominator:

$$\frac{4^3}{4^3} = \frac{4 \cdot 4 \cdot 4}{4 \cdot 4 \cdot 4} = \frac{64}{64} = 1.$$

This means that 4^0 must equal 1. This example illustrates the zero exponent rule.

The Zero Exponent Rule

If b is any real number other than 0,

$$b^0 = 1.$$

Example 1   Using the Zero Exponent Rule

Use the zero exponent rule to simplify:

a. 7^0 b. π^0 c. $(-5)^0$ d. -5^0.

Solution

a. $7^0 = 1$ b. $\pi^0 = 1$ c. $(-5)^0 = 1$ d. $-5^0 = -1$.

Only 5 is raised to the 0 power.

Check Point 1   Use the zero exponent rule to simplify:

a. 19^0 1   b. $(3\pi)^0$ 1   c. $(-14)^0$ 1   d. -14^0. −1

Great Question!

What's the difference between $\dfrac{4^3}{4^5}$ and $\dfrac{4^5}{4^3}$?

$\dfrac{4^3}{4^5}$ and $\dfrac{4^5}{4^3}$ represent different numbers:

$$\frac{4^3}{4^5} = 4^{3-5} = 4^{-2} = \frac{1}{4^2} = \frac{1}{16}$$

$$\frac{4^5}{4^3} = 4^{5-3} = 4^2 = 16.$$

The quotient rule can result in a negative exponent. Consider, for example, $4^3 \div 4^5$:

$$\frac{4^3}{4^5} = 4^{3-5} = 4^{-2}.$$

We can see what this negative exponent means by evaluating the numerator and the denominator:

$$\frac{4^3}{4^5} = \frac{\cancel{4} \cdot \cancel{4} \cdot \cancel{4}}{\cancel{4} \cdot \cancel{4} \cdot \cancel{4} \cdot 4 \cdot 4} = \frac{1}{4^2}.$$

Notice that $\dfrac{4^3}{4^5}$ equals both 4^{-2} and $\dfrac{1}{4^2}$. This means that 4^{-2} must equal $\dfrac{1}{4^2}$. This example is a particular case of the negative exponent rule.

The Negative Exponent Rule

If b is any real number other than 0 and m is a natural number,

$$b^{-m} = \frac{1}{b^m}.$$

Example 2   Using the Negative Exponent Rule

Use the negative exponent rule to simplify:

a. 8^{-2} **b.** 5^{-3} **c.** 7^{-1}.

Solution

a. $8^{-2} = \dfrac{1}{8^2} = \dfrac{1}{8 \cdot 8} = \dfrac{1}{64}$ **b.** $5^{-3} = \dfrac{1}{5^3} = \dfrac{1}{5 \cdot 5 \cdot 5} = \dfrac{1}{125}$ **c.** $7^{-1} = \dfrac{1}{7^1} = \dfrac{1}{7}$

✓ **Check Point 2**   Use the negative exponent rule to simplify:

a. 9^{-2} $\dfrac{1}{81}$ **b.** 6^{-3} $\dfrac{1}{216}$ **c.** 12^{-1}. $\dfrac{1}{12}$

Powers of Ten

Exponents and their properties allow us to represent and compute with numbers that are large or small. For example, one billion, or 1,000,000,000, can be written as 10^9. In terms of exponents, 10^9 might not look very large, but consider this: If you can count to 200 in one minute and decide to count for 12 hours a day at this rate, it would take you in the region of 19 years, 9 days, 5 hours, and 20 minutes to count to 10^9!

Powers of ten follow two basic rules:

1. **A positive exponent tells how many 0s follow the 1.** For example, 10^9 (one billion) is a 1 followed by nine zeros: 1,000,000,000. A googol, 10^{100}, is a 1 followed by one hundred zeros. (A googol far exceeds the number of protons, neutrons, and electrons in the universe.) A googol is a veritable pipsqueak compared to the googolplex, 10 raised to the googol power, or $10^{10^{100}}$; that's a 1 followed by a googol zeros. (If each zero in a googolplex were no larger than a grain of sand, there would not be enough room in the universe to represent the number.)

2. **A negative exponent tells how many places there are to the right of the decimal point.** For example, 10^{-9} (one billionth) has nine places to the right of the decimal point. The nine places include eight 0s and the 1.

$$10^{-9} = 0.\underbrace{000000001}_{\text{nine places}}$$

Earthquakes and Powers of Ten

The earthquake that ripped through northern California on October 17, 1989, measured 7.1 on the Richter scale, killed more than 60 people, and injured more than 2400. Shown here is San Francisco's Marina district, where shock waves tossed houses off their foundations and into the street.

The Richter scale is misleading because it is not actually a 1 to 8, but rather a 1 to 10 million scale. Each level indicates a tenfold increase in magnitude from the previous level, making a 7.0 earthquake a million times greater than a 1.0 quake. The following is a translation of the Richter scale:

Richter Number (R)	Magnitude (10^{R-1})
1	$10^{1-1} = 10^0 = 1$
2	$10^{2-1} = 10^1 = 10$
3	$10^{3-1} = 10^2 = 100$
4	$10^{4-1} = 10^3 = 1000$
5	$10^{5-1} = 10^4 = 10{,}000$
6	$10^{6-1} = 10^5 = 100{,}000$
7	$10^{7-1} = 10^6 = 1{,}000{,}000$
8	$10^{8-1} = 10^7 = 10{,}000{,}000$

Table 1.5 Names of Large Numbers	
10^2	hundred
10^3	thousand
10^6	million
10^9	billion
10^{12}	trillion
10^{15}	quadrillion
10^{18}	quintillion
10^{21}	sextillion
10^{24}	septillion
10^{27}	octillion
10^{30}	nonillion
10^{100}	googol
10^{googol}	googolplex

Scientific Notation

As of December 2011, the national debt of the United States was about $15.2 trillion. This is the amount of money the government has had to borrow over the years, mostly by selling bonds, because it has spent more than it has collected in taxes. A stack of $1 bills equaling the national debt would measure more than 950,000 miles. That's more than two round trips from Earth to the moon. Because a trillion is 10^{12} (see **Table 1.5**), the national debt can be expressed as

$$15.2 \times 10^{12}.$$

Because $15.2 = 1.52 \times 10$, the national debt can be expressed as

$$15.2 \times 10^{12} = (1.52 \times 10) \times 10^{12} = 1.52 \times (10 \times 10^{12})$$
$$= 1.52 \times 10^{1+12} = 1.52 \times 10^{13}$$

The number 1.52×10^{13} is written in a form called *scientific notation*.

Scientific Notation

A positive number is written in **scientific notation** when it is expressed in the form

$$a \times 10^n,$$

where a is a number greater than or equal to 1 and less than 10 ($1 \le a < 10$) and n is an integer.

It is customary to use the multiplication symbol, $\times$, rather than a dot, when writing a number in scientific notation.

Here are three examples of numbers in scientific notation:

- The universe is 1.375×10^{10} years old.
- In 2010, humankind generated 1.2 zettabytes, or 1.2×10^{21} bytes, of digital information. (A byte consists of eight binary digits, or bits, 0 or 1.)
- The length of the AIDS virus is 1.1×10^{-4} millimeter.

Convert from scientific notation to decimal notation.

We can use n, the exponent on the 10 in $a \times 10^n$, to change a number in scientific notation to decimal notation. If n is **positive**, move the decimal point in a to the **right** n places. If n is **negative**, move the decimal point in a to the **left** $|n|$ places.

Example 3   Converting from Scientific to Decimal Notation

Write each number in decimal notation:

a. 1.375×10^{10}

b. 1.1×10^{-4}.

Solution

In each case, we use the exponent on the 10 to move the decimal point. In part (a), the exponent is positive, so we move the decimal point to the right. In part (b), the exponent is negative, so we move the decimal point to the left.

a. $1.375 \times 10^{10} = 13{,}750{,}000{,}000$

$n = 10$

Move the decimal point 10 places to the right.

b. $1.1 \times 10^{-4} = 0.00011$

$n = -4$

Move the decimal point $|-4|$ places, or 4 places, to the left.

 **Check Point 3**   Write each number in decimal notation:

a. 7.4×10^9 7,400,000,000

b. 3.017×10^{-6}.   0.000003017

Convert from decimal notation to scientific notation.

To convert a positive number from decimal notation to scientific notation, we reverse the procedure of Example 3.

Converting from Decimal to Scientific Notation

Write the number in the form $a \times 10^n$.

- Determine a, the numerical factor. Move the decimal point in the given number to obtain a number greater than or equal to 1 and less than 10.
- Determine n, the exponent on 10^n. The absolute value of n is the number of places the decimal point was moved. The exponent n is positive if the given number is greater than 10 and negative if the given number is between 0 and 1.

Example 4   Converting from Decimal Notation to Scientific Notation

Write each number in scientific notation:

a. 4,600,000

b. 0.000023.

Technology

You can use your calculator's EE (enter exponent) or EXP key to convert from decimal to scientific notation. Here is how it's done for 0.000023:

Many Scientific Calculators

Keystrokes	Display
.000023 EE =	2.3 − 05

Many Graphing Calculators

Use the mode setting for scientific notation.

Keystrokes	Display
.000023 ENTER	2.3E − 5

Solution

a. $4{,}600{,}000 = 4.6 \times 10^6$

This number is greater than 10, so n is positive in $a \times 10^n$.

Move the decimal point in 4,600,000 to get $1 \le a < 10$.

The decimal point moved 6 places from 4,600,000 to 4.6.

b. $0.000023 = 2.3 \times 10^{-5}$

This number is less than 1, so n is negative in $a \times 10^n$.

Move the decimal point in 0.000023 to get $1 \le a < 10$.

The decimal point moved 5 places from 0.000023 to 2.3.

Check Point 4 Write each number in scientific notation:

a. $7{,}410{,}000{,}000$ 7.41×10^9 **b.** 0.000000092. 9.2×10^{-8}

Example 5 Expressing the U.S. Population in Scientific Notation

As of December 2011, the population of the United States was approximately 312 million. Express the population in scientific notation.

Solution

Because a million is 10^6, the 2011 population can be expressed as

$$312 \times 10^6.$$

This factor is not between 1 and 10, so the number is not in scientific notation.

The voice balloon indicates that we need to convert 312 to scientific notation.

$$312 \times 10^6 = (3.12 \times 10^2) \times 10^6 = 3.12 \times 10^{2+6} = 3.12 \times 10^8$$

$312 = 3.12 \times 10^2$

In scientific notation, the population is 3.12×10^8.

Great Question!

I read that the U.S. population exceeds $\frac{3}{10}$ of a billion. Yet you described it as 312 million. Which description is correct?

Both descriptions are correct. We can use exponential properties to express 312 million in billions.

$$312 \text{ million} = 312 \times 10^6 = (0.312 \times 10^3) \times 10^6 = 0.312 \times 10^{3+6} = 0.312 \times 10^9$$

Because 10^9 is a billion, U.S. population exceeds $\frac{3}{10}$ of a billion.

Check Point 5 Express 410×10^7 in scientific notation. 4.1×10^9

4   Perform computations using scientific notation.

Computations with Scientific Notation

We use the product rule for exponents to multiply numbers in scientific notation:

$$(a \times 10^n) \times (b \times 10^m) = (a \times b) \times 10^{n+m}.$$

Add the exponents on 10 and multiply the other parts of the numbers separately.

Example 6   Multiplying Numbers in Scientific Notation

Multiply: $(3.4 \times 10^9)(2 \times 10^{-5})$. Write the product in decimal notation.

Solution

$$
\begin{aligned}
(3.4 \times 10^9)(2 \times 10^{-5}) &= (3.4 \times 2) \times (10^9 \times 10^{-5}) && \text{Regroup factors.}\\
&= 6.8 \times 10^{9+(-5)} && \text{Add the exponents on 10}\\
&&& \text{and multiply the other parts.}\\
&= 6.8 \times 10^4 && \text{Simplify.}\\
&= 68{,}000 && \text{Write the product in decimal}\\
&&& \text{notation.}
\end{aligned}
$$

✔ **Check Point 6**   Multiply: $(1.3 \times 10^7)(4 \times 10^{-2})$. Write the product in decimal notation.   520,000

We use the quotient rule for exponents to divide numbers in scientific notation:

$$\frac{a \times 10^n}{b \times 10^m} = \left(\frac{a}{b}\right) \times 10^{n-m}.$$

Subtract the exponents on 10 and divide the other parts of the numbers separately.

Example 7   Dividing Numbers in Scientific Notation

Divide: $\dfrac{8.4 \times 10^{-7}}{4 \times 10^{-4}}$. Write the quotient in decimal notation.

Solution

$$
\begin{aligned}
\frac{8.4 \times 10^{-7}}{4 \times 10^{-4}} &= \left(\frac{8.4}{4}\right) \times \left(\frac{10^{-7}}{10^{-4}}\right) && \text{Regroup factors.}\\
&= 2.1 \times 10^{-7-(-4)} && \text{Subtract the exponents on 10 and divide}\\
&&& \text{the other parts.}\\
&= 2.1 \times 10^{-3} && \text{Simplify: } -7-(-4) = -7+4 = -3.\\
&= 0.0021 && \text{Write the quotient in decimal notation.}
\end{aligned}
$$

✔ **Check Point 7**   Divide: $\dfrac{6.9 \times 10^{-8}}{3 \times 10^{-2}}$. Write the quotient in decimal notation.   0.0000023

Multiplication and division involving very large or very small numbers can be performed by first converting each number to scientific notation.

Example 8   Using Scientific Notation to Multiply

Multiply: $0.00064 \times 9{,}400{,}000{,}000$. Express the product in **a.** scientific notation and **b.** decimal notation.

Technology

$(3.4 \times 10^9)(2 \times 10^{-5})$
on a Calculator

Many Scientific Calculators

3.4 [EE] 9 [×] 2 [EE] 5 [+/−] [=]

Display: 6.8 04

Many Graphing Calculators

3.4 [EE] 9 [×] 2 [EE] [(−)] 5 [ENTER]

Display: 6.8 E 4

Solution

a. $0.00064 \times 9,400,000,000$

$\qquad = 6.4 \times 10^{-4} \times 9.4 \times 10^{9}$ *Write each number in scientific notation.*

$\qquad = (6.4 \times 9.4) \times (10^{-4} \times 10^{9})$ *Regroup factors.*

$\qquad = 60.16 \times 10^{-4+9}$ *Add the exponents on 10 and multiply the other parts.*

$\qquad = 60.16 \times 10^{5}$ *Simplify.*

$\qquad = (6.016 \times 10) \times 10^{5}$ *Express 60.16 in scientific notation.*

$\qquad = 6.016 \times 10^{6}$ *Add exponents on 10:*
$10^{1} \times 10^{5} = 10^{1+5} = 10^{6}.$

b. The answer in decimal notation is obtained by moving the decimal point in 6.016 six places to the right. The product is 6,016,000.

✓ **Check Point 8**   Multiply: $0.0036 \times 5,200,000$. Express the product in **a.** scientific notation and **b.** decimal notation.   a. 1.872×10^{4} b. $18,720$

Applications: Putting Numbers in Perspective

⑤ Solve applied problems using scientific notation.

Due to tax cuts and spending increases, the United States began accumulating large deficits in the 1980s. To finance the deficit, the government had borrowed $15.2 trillion as of December 2011. The graph in **Figure 1.8** shows the national debt increasing over time.

The National Debt

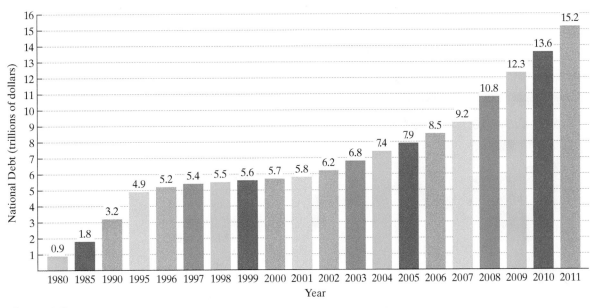

Figure 1.8
Source: Office of Management and Budget

Example 9 shows how we can use scientific notation to comprehend the meaning of a number such as 15.2 trillion.

Example 9   The National Debt

As of December 2011, the national debt was $15.2 trillion, or 15.2×10^{12} dollars. At that time, the U.S. population was approximately 312,000,000 (312 million), or 3.12×10^{8}. If the national debt was evenly divided among every individual in the United States, how much would each citizen have to pay?

Technology

Here is the keystroke sequence for solving Example 9 using a calculator:

15.2 $\boxed{\text{EE}}$ 12 $\boxed{\div}$ 3.12 $\boxed{\text{EE}}$ 8.

The quotient is displayed by pressing $\boxed{=}$ on a scientific calculator or $\boxed{\text{ENTER}}$ on a graphing calculator. The answer can be displayed in scientific or decimal notation. Consult your manual.

Solution

The amount each citizen must pay is the total debt, 15.2×10^{12} dollars, divided by the number of citizens, 3.12×10^8.

$$\frac{15.2 \times 10^{12}}{3.12 \times 10^8} = \left(\frac{15.2}{3.12}\right) \times \left(\frac{10^{12}}{10^8}\right)$$
$$\approx 4.87 \times 10^{12-8}$$
$$= 4.87 \times 10^4$$
$$= 48{,}700$$

Every U.S. citizen would have to pay approximately $48,700 to the federal government to pay off the national debt.

If a number is written in the form $a \times 10^n$, the digits in a are called **significant digits**.

National Debt: 15.2×10^{12} U.S. Population: 3.12×10^8

Three significant digits   Three significant digits

Because these were the given numbers in Example 9, we rounded the answer, 4.87×10^4, to three significant digits. When multiplying or dividing in scientific notation where rounding is necessary and rounding instructions are not given, **round the scientific notation answer to the least number of significant digits found in any of the given numbers**.

Check Point 9   As of December 2011, the United States had spent $2.6 trillion for the wars in Iraq and Afghanistan. (Source: costsofwar.org) At that time, the U.S. population was approximately 312 million (3.12×10^8). If the cost of these wars was evenly divided among every individual in the United States, how much would each citizen have to pay? $8300

Blitzer Bonus

Seven Ways to Spend $1 Trillion

Confronting a national debt of $15.2 trillion starts with grasping just how colossal $1 trillion (1×10^{12}) actually is. To help you wrap your head around this mind-boggling number and to put the national debt in further perspective, consider what $1 trillion will buy:

- 40,816,326 new cars based on an average sticker price of $24,500 each
- 5,574,136 homes based on the national median price of $179,400 for existing single-family homes
- one year's salary for 14.7 million teachers based on the average teacher salary of $68,000 in California
- the annual salaries of all 535 members of Congress for the next 10,742 years based on current salaries of $174,000 per year
- the salary of basketball superstar LeBron James for 50,000 years based on an annual salary of $20 million
- annual base pay for 59.5 million U.S. privates (that's 100 times the total number of active-duty soldiers in the Army) based on basic pay of $16,794 per year
- salaries to hire all 2.8 million residents of the state of Kansas in full-time minimum-wage jobs for the next 23 years based on the federal minimum wage of $7.25 per hour

Source: Kiplinger.com

Concept and Vocabulary Check

Exercises in the Concept and Vocabulary Check are intended for group and class discussions.

In Exercises 1–5, fill in each blank so that the resulting statement is true.

1. When multiplying expressions with the same base, _____add_____ the exponents.

2. When an exponential expression is raised to a power, _____multiply_____ the exponents.

3. When dividing exponential expressions with the same base, _____subtract_____ the exponents.

4. Any nonzero real number raised to the zero power is equal to _____one_____.

5. A positive number is written in scientific notation when the first factor is _a number greater than_ or equal to 1 and less than 10 and the second factor is _10 to an integer power_.

In Exercises 6–10, determine whether each statement is true or false. If the statement is false, make the necessary change(s) to produce a true statement. Changes to false statements will vary.

6. $2^3 \cdot 2^5 = 4^8$ false

7. $\dfrac{10^8}{10^4} = 10^2$ false

8. $5^{-2} = -5^2$ false

9. A trillion is one followed by 12 zeros.   true

10. According to *Mother Jones* magazine, sending all 2009 U.S. high school graduates to private colleges would cost $347 billion. Because a billion is 10^9, the cost in scientific notation is 347×10^9 dollars.   false

Respond to Exercises 11–20 using verbal or written explanations.

11. Explain the product rule for exponents. Use $2^3 \cdot 2^5$ in your explanation.   **11–20.** Answers will vary.

12. Explain the power rule for exponents. Use $(3^2)^4$ in your explanation.

13. Explain the quotient rule for exponents. Use $\dfrac{5^8}{5^2}$ in your explanation.

14. Explain the zero exponent rule and give an example.

15. Explain the negative exponent rule and give an example.

16. How do you know if a number is written in scientific notation?

17. Explain how to convert from scientific to decimal notation and give an example.

18. Explain how to convert from decimal to scientific notation and give an example.

19. Suppose you are looking at a number in scientific notation. Describe the size of the number you are looking at if the exponent on ten is **a.** positive, **b.** negative, **c.** zero.

20. Describe one advantage of expressing a number in scientific notation over decimal notation.

Exercise Set 1.6

Practice Exercises

In Exercises 1–12, use properties of exponents to simplify each expression. First express the answer in exponential form. Then evaluate the expression.

1. $2^2 \cdot 2^3$ $2^5 = 32$

2. $3^3 \cdot 3^2$ $3^5 = 243$

3. $4 \cdot 4^2$ $4^3 = 64$

4. $5 \cdot 5^2$ $5^3 = 125$

5. $(2^2)^3$ $2^6 = 64$

6. $(3^3)^2$ $3^6 = 729$

7. $(1^4)^5$ $1^{20} = 1$

8. $(1^3)^7$ $1^{21} = 1$

9. $\dfrac{4^7}{4^5}$ $4^2 = 16$

10. $\dfrac{6^7}{6^5}$ $6^2 = 36$

11. $\dfrac{2^8}{2^4}$ $2^4 = 16$

12. $\dfrac{3^8}{3^4}$ $3^4 = 81$

In Exercises 13–24, use the zero and negative exponent rules to simplify each expression.

13. 3^0 1

14. 9^0 1

15. $(-3)^0$ 1

16. $(-9)^0$ 1

17. -3^0 −1

18. -9^0 −1

19. 2^{-2} $\frac{1}{4}$

20. 3^{-2} $\frac{1}{9}$

21. 4^{-3} $\frac{1}{64}$

22. 2^{-3} $\frac{1}{8}$

23. 2^{-5} $\frac{1}{32}$

24. 2^{-6} $\frac{1}{64}$

In Exercises 25–30, use properties of exponents to simplify each expression. First express the answer in exponential form. Then evaluate the expression.

25. $3^4 \cdot 3^{-2}$ $3^2 = 9$

26. $2^5 \cdot 2^{-2}$ $2^3 = 8$

27. $3^{-3} \cdot 3$ $3^{-2} = \frac{1}{9}$

28. $2^{-3} \cdot 2$ $2^{-2} = \frac{1}{4}$

29. $\dfrac{2^3}{2^7}$ $2^{-4} = \frac{1}{16}$

30. $\dfrac{3^4}{3^7}$ $3^{-3} = \frac{1}{27}$

In Exercises 31–42, use properties of exponents to simplify each expression. Express answers in exponential form with positive exponents only. Assume that any variables in denominators are not equal to zero.

31. $(x^5 \cdot x^3)^{-2}$ $\frac{1}{x^{16}}$

32. $(x^2 \cdot x^4)^{-3}$ $\frac{1}{x^{18}}$

33. $\frac{(x^3)^4}{(x^2)^7}$ $\frac{1}{x^2}$

34. $\frac{(x^2)^5}{(x^3)^4}$ $\frac{1}{x^2}$

35. $\left(\frac{x^5}{x^2}\right)^{-4}$ $\frac{1}{x^{12}}$

36. $\left(\frac{x^7}{x^2}\right)^{-3}$ $\frac{1}{x^{15}}$

37. $\frac{2x^5 \cdot 3x}{15x^6}$ $\frac{2}{5}$

38. $\frac{4x^7 \cdot 5x}{10x^8}$ 2

39. $(-2x^3y^{-4})(3x^{-1}y)$ $-\frac{6x^2}{y^3}$

40. $(-5x^4y^{-3})(4x^{-1}y)$ $-\frac{20x^3}{y^2}$

41. $\frac{30x^2y^5}{-6x^8y^{-3}}$ $-\frac{5y^8}{x^6}$

42. $\frac{24x^2y^{13}}{-8x^5y^{-2}}$ $-\frac{3y^{15}}{x^3}$

In Exercises 43–58, express each number in decimal notation.

43. 2.7×10^2 270

44. 4.7×10^3 4700

45. 9.12×10^5 912,000

46. 8.14×10^4 81,400

47. 8×10^7 80,000,000

48. 7×10^6 7,000,000

49. 1×10^5 100,000

50. 1×10^8 100,000,000

51. 7.9×10^{-1} 0.79

52. 8.6×10^{-1} 0.86

53. 2.15×10^{-2} 0.0215

54. 3.14×10^{-2} 0.0314

55. 7.86×10^{-4} 0.000786

56. 4.63×10^{-5} 0.0000463

57. 3.18×10^{-6} 0.00000318

58. 5.84×10^{-7} 0.000000584

In Exercises 59–78, express each number in scientific notation.

59. 370 3.7×10^2

60. 530 5.3×10^2

61. 3600 3.6×10^3

62. 2700 2.7×10^3

63. 32,000 3.2×10^4

64. 64,000 6.4×10^4

65. 220,000,000 2.2×10^8

66. 370,000,000,000 3.7×10^{11}

67. 0.027 2.7×10^{-2}

68. 0.014 1.4×10^{-2}

69. 0.0037 3.7×10^{-3}

70. 0.00083 8.3×10^{-4}

71. 0.00000293 2.93×10^{-6}

72. 0.000000647 6.47×10^{-7}

73. 820×10^5 8.2×10^7

74. 630×10^8 6.3×10^{10}

75. 0.41×10^6 4.1×10^5

76. 0.57×10^9 5.7×10^8

77. 2100×10^{-9} 2.1×10^{-6}

78. $97,000 \times 10^{-11}$ 9.7×10^{-7}

In Exercises 79–92, perform the indicated operation and express each answer in decimal notation.

79. $(2 \times 10^3)(3 \times 10^2)$

80. $(5 \times 10^2)(4 \times 10^4)$

81. $(2 \times 10^9)(3 \times 10^{-5})$

82. $(4 \times 10^8)(2 \times 10^{-4})$

83. $(4.1 \times 10^2)(3 \times 10^{-4})$ 0.123

84. $(1.2 \times 10^3)(2 \times 10^{-5})$ 0.024

85. $\frac{12 \times 10^6}{4 \times 10^2}$ 30,000

86. $\frac{20 \times 10^{20}}{10 \times 10^{15}}$ 200,000

87. $\frac{15 \times 10^4}{5 \times 10^{-2}}$ 3,000,000

88. $\frac{18 \times 10^2}{9 \times 10^{-3}}$ 200,000

89. $\frac{6 \times 10^3}{2 \times 10^5}$ 0.03

90. $\frac{8 \times 10^4}{2 \times 10^7}$ 0.004

91. $\frac{6.3 \times 10^{-6}}{3 \times 10^{-3}}$ 0.0021

92. $\frac{9.6 \times 10^{-7}}{3 \times 10^{-3}}$ 0.00032

In Exercises 93–102, perform the indicated operation by first expressing each number in scientific notation. Write the answer in scientific notation.

93. $(82,000,000)(3,000,000,000)$ 2.46×10^{17} *

94. $(94,000,000)(6,000,000,000)$ 5.64×10^{17} *

95. $(0.0005)(6,000,000)$ $(5.0 \times 10^{-4})(6.0 \times 10^6) = 3 \times 10^3$

96. $(0.000015)(0.004)$ $(1.5 \times 10^{-5})(4.0 \times 10^{-3}) = 6 \times 10^{-8}$

97. $\frac{9,500,000}{500}$ $\frac{9.5 \times 10^6}{5 \times 10^2} = 1.9 \times 10^4$

98. $\frac{30,000}{0.0005}$ $\frac{3 \times 10^4}{5 \times 10^{-4}} = 6 \times 10^7$

99. $\frac{0.00008}{200}$ $\frac{8 \times 10^{-5}}{2 \times 10^2} = 4 \times 10^{-7}$

100. $\frac{0.0018}{0.0000006}$ $\frac{1.8 \times 10^{-3}}{6 \times 10^{-7}} = 3 \times 10^3$

101. $\frac{480,000,000,000}{0.00012}$ $\frac{4.8 \times 10^{11}}{1.2 \times 10^{-4}} = 4 \times 10^{15}$

102. $\frac{0.000000096}{16,000}$ $\frac{9.6 \times 10^{-8}}{1.6 \times 10^4} = 6 \times 10^{-12}$

Practice Plus

In Exercises 103–106, perform the indicated operations. Express each answer as a fraction reduced to its lowest terms.

103. $\frac{2^4}{2^5} + \frac{3^3}{3^5}$ $\frac{11}{18}$

104. $\frac{3^5}{3^6} + \frac{2^3}{2^6}$ $\frac{11}{24}$

105. $\frac{2^6}{2^4} - \frac{5^4}{5^6}$ $\frac{99}{25} = 3\frac{24}{25}$

106. $\frac{5^6}{5^4} - \frac{2^4}{2^6}$ $\frac{99}{4} = 24\frac{3}{4}$

In Exercises 107–110, perform the indicated computations. Express answers in scientific notation.

107. $(5 \times 10^3)(1.2 \times 10^{-4}) \div (2.4 \times 10^2)$ 2.5×10^{-3}

108. $(2 \times 10^2)(2.6 \times 10^{-3}) \div (4 \times 10^3)$ 1.3×10^{-4}

109. $\frac{(1.6 \times 10^4)(7.2 \times 10^{-3})}{(3.6 \times 10^8)(4 \times 10^{-3})}$ 8×10^{-5}

110. $\frac{(1.2 \times 10^6)(8.7 \times 10^{-2})}{(2.9 \times 10^6)(3 \times 10^{-3})}$ 1.2×10^1

79. 600,000   **80.** 20,000,000   **81.** 60,000   **82.** 80,000
*See Answers to Selected Exercises.

Application Exercises

The bar graph shows the total amount Americans paid in federal taxes, in trillions of dollars, and the U.S. population, in millions, from 2007 through 2010. Exercises 111–112 are based on the numbers displayed by the graph.

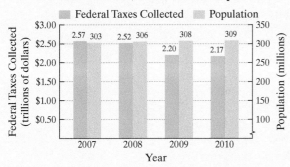

Federal Taxes and the United States Population

Sources: Internal Revenue Service and U.S. Census Bureau

111. a. In 2010, the United States government collected $2.17 trillion in taxes. Express this number in scientific notation. 2.17×10^{12}

b. In 2010, the population of the United States was approximately 309 million. Express this number in scientific notation. 3.09×10^8

c. Use your scientific notation answers from parts (a) and (b) to answer this question: If the total 2010 tax collections were evenly divided among all Americans, how much would each citizen pay? Express the answer in scientific and decimal notations. 7.02×10^3; $7020

112. a. In 2009, the United States government collected $2.20 trillion in taxes. Express this number in scientific notation. 2.20×10^{12}

b. In 2009, the population of the United States was approximately 308 million. Express this number in scientific notation. 3.08×10^8

c. Use your scientific notation answers from parts (a) and (b) to answer this question: If the total 2009 tax collections were evenly divided among all Americans, how much would each citizen pay? Express the answer in scientific and decimal notations. 7.14×10^3; $7140

The bar graph quantifies our love for movies by showing the number of tickets sold, in millions, and the average price per ticket for five selected years. Exercises 113–114 are based on the numbers displayed by the graph. Do not round the answers.

United States Film Admissions and Admission Charges

Source: Motion Picture Association of America

113. Use scientific notation to compute the amount of money that the motion picture industry made from box-office receipts in 2010. Express the answer in scientific notation. 1.0586×10^{10}

114. Use scientific notation to compute the amount of money that the motion picture industry made from box-office receipts in 2005. Express the answer in scientific notation. 8.832×10^9

115. The mass of one oxygen molecule is 5.3×10^{-23} gram. Find the mass of 20,000 molecules of oxygen. Express the answer in scientific notation. 1.06×10^{-18} g

116. The mass of one hydrogen atom is 1.67×10^{-24} gram. Find the mass of 80,000 hydrogen atoms. Express the answer in scientific notation. 1.336×10^{-19} g

Critical Thinking Exercises

In Exercises 117–123, determine whether each statement is true or false. If the statement is false, make the necessary change(s) to produce a true statement. Changes to false statements will vary.

117. $4^{-2} < 4^{-3}$ false **118.** $5^{-2} > 2^{-5}$ true

119. $5^2 \cdot 5^{-2} > 2^5 \cdot 2^{-5}$ false **120.** $534.7 = 5.347 \times 10^3$ false

121. $\dfrac{8 \times 10^{30}}{4 \times 10^{-5}} = 2 \times 10^{25}$ false

122. $(7 \times 10^5) + (2 \times 10^{-3}) = 9 \times 10^2$ false

123. $(4 \times 10^3) + (3 \times 10^2) = 43 \times 10^2$ true

124. Give an example of a number for which there is no advantage to using scientific notation instead of decimal notation. Explain why this is the case. Answers will vary.

125. The mad Dr. Frankenstein has gathered enough bits and pieces (so to speak) for $2^{-1} + 2^{-2}$ of his creature-to-be. Write a fraction that represents the amount of his creature that must still be obtained. $\frac{1}{4}$

Technology Exercises 126–129. Answers will vary.

126. Use a calculator in fraction mode to check your answers in Exercises 19–24.

127. Use a calculator to check any three of your answers in Exercises 43–58.

128. Use a calculator to check any three of your answers in Exercises 59–78.

129. Use a calculator with an $\boxed{\text{EE}}$ or $\boxed{\text{EXP}}$ key to check any four of your computations in Exercises 79–102. Display the result of the computation in scientific notation and in decimal notation.

Group Exercises 130–131. Answers will vary.

130. **Putting Numbers into Perspective.** A large number can be put into perspective by comparing it with another number. For example, we put the $15.2 trillion national debt (Example 9) into perspective by comparing this number to the number of U.S. citizens.

For this project, each group member should consult an almanac, a newspaper, or the Internet to find a number greater than one million. Explain to other members of the group the context in which the large number is used. Express the number in scientific notation. Then put the number into perspective by comparing it with another number.

131. Refer to the Blitzer Bonus on page 72. Group members should use scientific notation to verify any three of the bulleted items on ways to spend $1 trillion.

1.7 Arithmetic and Geometric Sequences

What am I supposed to learn?

After you have read this section, you should be able to:

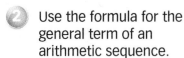 **1** Write terms of an arithmetic sequence.

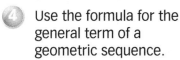 **2** Use the formula for the general term of an arithmetic sequence.

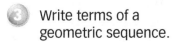 **3** Write terms of a geometric sequence.

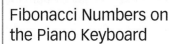 **4** Use the formula for the general term of a geometric sequence.

Sequences

Many creations in nature involve intricate mathematical designs, including a variety of spirals. For example, the arrangement of the individual florets in the head of a sunflower forms spirals. In some species, there are 21 spirals in the clockwise direction and 34 in the counterclockwise direction. The precise numbers depend on the species of sunflower: 21 and 34, or 34 and 55, or 55 and 89, or even 89 and 144.

This observation becomes even more interesting when we consider a sequence of numbers investigated by Leonardo of Pisa, also known as Fibonacci, an Italian mathematician of the thirteenth century. The **Fibonacci sequence** of numbers is an infinite sequence that begins as follows:

$$1, 1, 2, 3, 5, 8, 13, 21, 34, 55, 89, 144, 233, \ldots .$$

The first two terms are 1. Every term thereafter is the sum of the two preceding terms. For example, the third term, 2, is the sum of the first and second terms: $1 + 1 = 2$. The fourth term, 3, is the sum of the second and third terms: $1 + 2 = 3$, and so on. Did you know that the number of spirals in a daisy or a sunflower, 21 and 34, are two Fibonacci numbers? The number of spirals in a pinecone, 8 and 13, and a pineapple, 8 and 13, are also Fibonacci numbers.

We can think of a **sequence** as a list of numbers that are related to each other by a rule. The numbers in a sequence are called its **terms**. The letter a with a subscript is used to represent the terms of a sequence. Thus, a_1 represents the first term of the sequence, a_2 represents the second term, a_3 the third term, and so on. This notation is shown for the first six terms of the Fibonacci sequence:

$$1, \quad 1, \quad 2, \quad 3, \quad 5, \quad 8.$$

$a_1 = 1 \quad a_2 = 1 \quad a_3 = 2 \quad a_4 = 3 \quad a_5 = 5 \quad a_6 = 8$

Blitzer Bonus

Fibonacci Numbers on the Piano Keyboard

One Octave

Numbers in the Fibonacci sequence can be found in an octave on the piano keyboard. The octave contains 2 black keys in one cluster, 3 black keys in another cluster, a total of 5 black keys, 8 white keys, and a total of 13 keys altogether. The numbers 2, 3, 5, 8, and 13 are the third through seventh terms of the Fibonacci sequence.

Arithmetic Sequences

The bar graph in **Figure 1.9** shows how much Americans spent on their pets, rounded to the nearest billion dollars, each year from 2001 through 2010.

The graph illustrates that each year spending increased by $2 billion. The sequence of annual spending

$$29, 31, 33, 35, 37, 39, 41, \ldots$$

shows that each term after the first, 29, differs from the preceding term by a constant amount, namely 2. This sequence is an example of an *arithmetic sequence*.

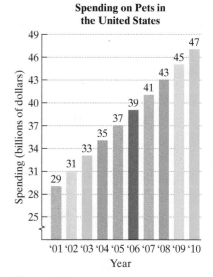

Figure 1.9
Source: American Pet Products Manufacturers Association

> **Definition of an Arithmetic Sequence**
>
> An **arithmetic sequence** is a sequence in which each term after the first differs from the preceding term by a constant amount. The difference between consecutive terms is called the **common difference** of the sequence.

The common difference, d, is found by subtracting any term from the term that directly follows it. In the following examples, the common difference is found by subtracting the first term from the second term: $a_2 - a_1$.

Arithmetic Sequence	**Common Difference**
$29, 31, 33, 35, 37, \ldots$	$d = 31 - 29 = 2$
$-5, -2, 1, 4, 7, \ldots$	$d = -2 - (-5) = -2 + 5 = 3$
$8, 3, -2, -7, -12, \ldots$	$d = 3 - 8 = -5$

If the first term of an arithmetic sequence is a_1, each term after the first is obtained by adding d, the common difference, to the previous term.

 Write terms of an arithmetic sequence.

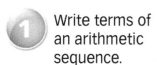 **Writing the Terms of an Arithmetic Sequence**

Write the first six terms of the arithmetic sequence with first term 6 and common difference 4.

Solution

The first term is 6. The second term is $6 + 4$, or 10. The third term is $10 + 4$, or 14, and so on. The first six terms are

$$6, 10, 14, 18, 22, \text{ and } 26.$$

✓ **Check Point 1** Write the first six terms of the arithmetic sequence with first term 100 and common difference 20. $100, 120, 140, 160, 180, \text{ and } 200$

Example 2 **Writing the Terms of an Arithmetic Sequence**

Write the first six terms of the arithmetic sequence with $a_1 = 5$ and $d = -2$.

Solution

The first term, a_1, is 5. The common difference, d, is -2. To find the second term, we add -2 to 5, giving 3. For the next term, we add -2 to 3, and so on. The first six terms are

$$5, 3, 1, -1, -3, \text{ and } -5.$$

✓ **Check Point 2** Write the first six terms of the arithmetic sequence with $a_1 = 8$ and $d = -3$. $8, 5, 2, -1, -4, \text{ and } -7$

The General Term of an Arithmetic Sequence

Use the formula for the general term of an arithmetic sequence.

Consider an arithmetic sequence whose first term is a_1 and whose common difference is d. We are looking for a formula for the general term, a_n. Let's begin by writing the first six terms. The first term is a_1. The second term is $a_1 + d$. The third term is $a_1 + d + d$, or $a_1 + 2d$. Thus, we start with a_1 and add d to each successive term. The first six terms are

$$a_1, \qquad a_1 + d, \qquad a_1 + 2d, \qquad a_1 + 3d, \qquad a_1 + 4d, \qquad a_1 + 5d.$$

a_1, first term	a_2, second term	a_3, third term	a_4, fourth term	a_5, fifth term	a_6, sixth term

Applying the pattern of the terms results in the following formula for the general term, or the nth term, of an arithmetic sequence:

General Term of an Arithmetic Sequence

The nth term (the general term) of an arithmetic sequence with first term a_1 and common difference d is

$$a_n = a_1 + (n-1)d.$$

Example 3   Using the Formula for the General Term of an Arithmetic Sequence

Find the eighth term of the arithmetic sequence whose first term is 4 and whose common difference is -7.

Solution

To find the eighth term, a_8, we replace n in the formula with 8, a_1 with 4, and d with -7.

$$a_n = a_1 + (n-1)d$$
$$a_8 = 4 + (8-1)(-7) = 4 + 7(-7) = 4 + (-49) = -45$$

The eighth term is -45. We can check this result by writing the first eight terms of the sequence:

$$4, -3, -10, -17, -24, -31, -38, -45.$$

✓ **Check Point 3**   Find the ninth term of the arithmetic sequence whose first term is 6 and whose common difference is -5. -34

The process of finding formulas to describe real-world phenomena is called **mathematical modeling**. Such formulas, together with the meaning assigned to the variables, are called **mathematical models**. Example 4 illustrates how the formula for the general term of an arithmetic sequence can be used to develop a mathematical model.

Example 4   Using an Arithmetic Sequence to Model Changes in the U.S. Population

The graph in **Figure 1.10** shows the percentage of the U.S. population by race/ethnicity for 2010, with projections by the U.S. Census Bureau for 2050.

The data show that in 2010, 64% of the U.S. population was white. On average, this is projected to decrease by approximately 0.45% per year.

a. Write a formula for the nth term of the arithmetic sequence that describes the percentage of the U.S. population that will be white n years after 2009.

b. What percentage of the U.S. population is projected to be white in 2030?

U.S. Population by Race/Ethnicity

■ 2010 Census   ■ 2050 Projections

White: 64, 46
Latino: 16, 30
African American: 12, 15
Asian: 9, 5

Figure 1.10
Source: U.S. Census Bureau

Solution

a. With a yearly decrease of 0.45%, we can express the percentage of the white population by the following arithmetic sequence:

$$64, \quad 64 - 0.45 = 63.55, \quad 63.55 - 0.45 = 63.10, \ \ldots .$$

a_1: percentage of whites in the population in 2010, 1 year after 2009

a_2: percentage of whites in the population in 2011, 2 years after 2009

a_3: percentage of whites in the population in 2012, 3 years after 2009

In this sequence, $64, 63.55, 63.10, \ldots$, the first term, a_1, represents the percentage of the population that was white in 2010. Each subsequent year this amount decreases by 0.45%, so $d = -0.45$. We use the formula for the general term of an arithmetic sequence to write the nth term of the sequence that describes the percentage of whites in the population n years after 2009.

$$a_n = a_1 + (n - 1)d \qquad \text{This is the formula for the general term of an arithmetic sequence.}$$

$$a_n = 64 + (n - 1)(-0.45) \qquad a_1 = 64 \text{ and } d = -0.45.$$

$$a_n = 64 - 0.45n + 0.45 \qquad \text{Distribute } -0.45 \text{ to each term in parentheses.}$$

$$a_n = -0.45n + 64.45 \qquad \text{Simplify.}$$

Thus, the percentage of the U.S. population that will be white n years after 2009 can be described by

$$a_n = -0.45n + 64.45.$$

b. Now we need to project the percentage of the population that will be white in 2030. The year 2030 is 21 years after 2009. Thus, $n = 21$. We substitute 21 for n in $a_n = -0.45n + 64.45$.

$$a_{21} = -0.45(21) + 64.45 = 55$$

The 21st term of the sequence is 55. Thus, 55% of the U.S. population is projected to be white in 2030.

Check Point 4   The data in **Figure 1.10** show that in 2010, 16% of the U.S. population was Latino. On average, this is projected to increase by approximately 0.35% per year.

a. Write a formula for the nth term of the arithmetic sequence that describes the percentage of the U.S. population that will be Latino n years after 2009. $a_n = 0.35n + 15.65$

b. What percentage of the U.S. population is projected to be Latino in 2030?   23%

Geometric Sequences

Figure 1.11 at the top of the next page shows a sequence in which the number of squares is increasing. From left to right, the number of squares is $1, 5, 25, 125$, and 625. In this sequence, each term after the first, 1, is obtained by multiplying the preceding term by a constant amount, namely 5. This sequence of increasing numbers of squares is an example of a *geometric sequence*.

Figure 1.11 A geometric sequence of squares

> ### Definition of a Geometric Sequence
> A **geometric sequence** is a sequence in which each term after the first is obtained by multiplying the preceding term by a fixed nonzero constant. The amount by which we multiply each time is called the **common ratio** of the sequence.

The common ratio, r, is found by dividing any term after the first term by the term that directly precedes it. In the examples below, the common ratio is found by dividing the second term by the first term: $\dfrac{a_2}{a_1}$.

<table>
<tr><th>Geometric Sequence</th><th>Common Ratio</th></tr>
<tr><td>$1, 5, 25, 125, 625, \ldots$</td><td>$r = \frac{5}{1} = 5$</td></tr>
<tr><td>$4, 8, 16, 32, 64, \ldots$</td><td>$r = \frac{8}{4} = 2$</td></tr>
<tr><td>$6, -12, 24, -48, 96, \ldots$</td><td>$r = \frac{-12}{6} = -2$</td></tr>
<tr><td>$9, -3, 1, -\frac{1}{3}, \frac{1}{9}, \ldots$</td><td>$r = \frac{-3}{9} = -\frac{1}{3}$</td></tr>
</table>

Great Question!

What happens to the terms of a geometric sequence when the common ratio is negative?

When the common ratio of a geometric sequence is negative, the signs of the terms alternate.

 3   Write terms of a geometric sequence.

How do we write out the terms of a geometric sequence when the first term and the common ratio are known? We multiply the first term by the common ratio to get the second term, multiply the second term by the common ratio to get the third term, and so on.

Example 5   Writing the Terms of a Geometric Sequence

Write the first six terms of the geometric sequence with first term 6 and common ratio $\frac{1}{3}$.

Solution

The first term is 6. The second term is $6 \cdot \frac{1}{3}$, or 2. The third term is $2 \cdot \frac{1}{3}$, or $\frac{2}{3}$. The fourth term is $\frac{2}{3} \cdot \frac{1}{3}$, or $\frac{2}{9}$, and so on. The first six terms are

$$6, 2, \frac{2}{3}, \frac{2}{9}, \frac{2}{27}, \text{ and } \frac{2}{81}.$$

✓ **Check Point 5**   Write the first six terms of the geometric sequence with first term 12 and common ratio $-\frac{1}{2}$. $12, -6, 3, -\frac{3}{2}, \frac{3}{4}, \text{ and } -\frac{3}{8}$

The General Term of a Geometric Sequence

4   Use the formula for the general term of a geometric sequence.

Consider a geometric sequence whose first term is a_1 and whose common ratio is r. We are looking for a formula for the general term, a_n. Let's begin by writing the first six terms. The first term is a_1. The second term is a_1r. The third term is $a_1r \cdot r$, or a_1r^2.

The fourth term is $a_1 r^2 \cdot r$, or $a_1 r^3$, and so on. Starting with a_1 and multiplying each successive term by r, the first six terms are

$$a_1, \qquad a_1 r, \qquad a_1 r^2, \qquad a_1 r^3, \qquad a_1 r^4, \qquad a_1 r^5.$$

| a_1, first term | a_2, second term | a_3, third term | a_4, fourth term | a_5, fifth term | a_6, sixth term |

Applying the pattern of the terms results in the following formula for the general term, or the nth term, of a geometric sequence:

> **General Term of a Geometric Sequence**
>
> The nth term (the general term) of a geometric sequence with first term a_1 and common ratio r is
>
> $$a_n = a_1 r^{n-1}.$$

Great Question!

When using $a_1 r^{n-1}$ to find the nth term of a geometric sequence, what should I do first?

Be careful with the order of operations when evaluating

$$a_1 r^{n-1}.$$

First, subtract 1 in the exponent and then raise r to that power. Finally, multiply the result by a_1.

Example 6   Using the Formula for the General Term of a Geometric Sequence

Find the eighth term of the geometric sequence whose first term is -4 and whose common ratio is -2.

Solution

To find the eighth term, a_8, we replace n in the formula with 8, a_1 with -4, and r with -2.

$$a_n = a_1 r^{n-1}$$
$$a_8 = -4(-2)^{8-1} = -4(-2)^7 = -4(-128) = 512$$

The eighth term is 512. We can check this result by writing the first eight terms of the sequence: $-4, 8, -16, 32, -64, 128, -256, 512$.

Check Point 6   Find the seventh term of the geometric sequence whose first term is 5 and whose common ratio is -3.   3645

Example 7   Geometric Population Growth

The table shows the population of the United States in 2000 and 2010, with estimates given by the Census Bureau for 2001 through 2009.

Year	2000	2001	2002	2003	2004	2005	2006	2007	2008	2009	2010
Population (millions)	281.4	284.0	286.6	289.3	292.0	294.7	297.4	300.2	303.0	305.8	308.7

a. Show that the population is increasing geometrically.

b. Write the general term for the geometric sequence modeling the population of the United States, in millions, n years after 1999.

c. Project the U.S. population, in millions, for the year 2020.

Solution

a. First, we use the sequence of population growth, 281.4, 284.0, 286.6, 289.3, and so on, to divide the population for each year by the population in the preceding year.

$$\frac{284.0}{281.4} \approx 1.009, \qquad \frac{286.6}{284.0} \approx 1.009, \qquad \frac{289.3}{286.6} \approx 1.009$$

Blitzer Bonus

Geometric Population Growth

Economist Thomas Malthus (1766–1834) predicted that population growth would increase as a geometric sequence and food production would increase as an arithmetic sequence. He concluded that eventually population would exceed food production. If two sequences, one geometric and one arithmetic, are increasing, the geometric sequence will eventually overtake the arithmetic sequence, regardless of any head start that the arithmetic sequence might initially have.

Continuing in this manner, we will keep getting approximately 1.009. This means that the population is increasing geometrically with $r \approx 1.009$. The population of the United States in any year shown in the sequence is approximately 1.009 times the population the year before.

b. The sequence of the U.S. population growth is

$$281.4, 284.0, 286.6, 289.3, 292.0, 294.7, \ldots.$$

Because the population is increasing geometrically, we can find the general term of this sequence using

$$a_n = a_1 r^{n-1}.$$

In this sequence, $a_1 = 281.4$ and [from part (a)] $r \approx 1.009$. We substitute these values into the formula for the general term. This gives the general term for the geometric sequence modeling the U.S. population, in millions, n years after 1999.

$$a_n = 281.4(1.009)^{n-1}$$

c. We can use the formula for the general term, a_n, in part (b) to project the U.S. population for the year 2020. The year 2020 is 21 years after 1999—that is, $2020 - 1999 = 21$. Thus, $n = 21$. We substitute 21 for n in $a_n = 281.4(1.009)^{n-1}$.

$$a_{21} = 281.4(1.009)^{21-1} = 281.4(1.009)^{20} \approx 336.6$$

The model projects that the United States will have a population of approximately 336.6 million in the year 2020.

Check Point 7 Write the general term for the geometric sequence

$$3, 6, 12, 24, 48, \ldots.$$

Then use the formula for the general term to find the eighth term. $a_n = 3(2)^{n-1}; 384$

Achieving Success

Do not wait until the last minute to study for an exam. Cramming is a high-stress activity that forces your brain to make a lot of weak connections. No wonder crammers tend to forget everything they learned minutes after taking a test.

Preparing for Chapter Tests Using the Book

• Study the chapter review chart. The chart contains definitions, concepts, procedures, and examples. If there is a topic that you are not sure of, read and review the examples from the chapter that are referenced in the chart.

• Work the assigned exercises from the Review Exercises. The Review Exercises contain the most significant problems for each of the chapter's sections.

• Find a quiet place to take the Chapter Test. Do not use notes, index cards, or any other resources. Check your answers and ask your teacher to review any exercises you missed.

> ## Concept and Vocabulary Check
>
> Exercises in the Concept and Vocabulary Check are intended for group and class discussions.

In Exercises 1–4, fill in each blank so that the resulting statement is true.

1. A sequence in which each term after the first differs from the preceding term by a constant amount is called a/an _____arithmetic_____ sequence. The difference between consecutive terms is called the ___common difference___ of the sequence.

2. The nth term of the sequence described in Exercise 1 is given by the formula ___$a_n = a_1 + (n - 1)d$___, where a_1 is ____the first term____ and d is _the common difference_.

3. A sequence in which each term after the first is obtained by multiplying the preceding term by a fixed nonzero number is called a/an ____geometric____ sequence. The amount by which we multiply each time is called the ____common ratio____ of the sequence.

4. The nth term of the sequence described in Exercise 3 is given by the formula _____$a_n = a_1 r^{n-1}$_____, where a_1 is ____the first term____ and r is ___the common ratio___.

In Exercises 5–11, determine whether each statement is true or false. If the statement is false, make the necessary change(s) to produce a true statement.   Changes to false statements will vary.

5. The common difference for the arithmetic sequence given by $1, -1, -3, -5, \ldots$ is 2.   false

6. The sequence $1, 4, 8, 13, 19, 26, \ldots$ is an arithmetic sequence.   false

7. The nth term of an arithmetic sequence whose first term is a_1 and whose common difference is d is $a_n = a_1 + nd$.   false

8. The sequence $2, 6, 24, 120, \ldots$ is an example of a geometric sequence.   false

9. Adjacent terms in a geometric sequence have a common difference.   false

10. A sequence that is not arithmetic must be geometric.   false

11. If a sequence is geometric, we can write as many terms as we want by repeatedly multiplying by the common ratio.   true

Respond to Exercises 12–18 using verbal or written explanations.   12–18. Answers will vary.

12. What is a sequence? Give an example with your description.

13. What is an arithmetic sequence? Give an example with your description.

14. What is the common difference in an arithmetic sequence?

15. What is a geometric sequence? Give an example with your description.

16. What is the common ratio in a geometric sequence?

17. If you are given a sequence that is arithmetic or geometric, how can you determine which type of sequence it is?

18. For the first 30 days of a flu outbreak, the number of students at your school who become ill is increasing. Which is worse: The number of students with the flu is increasing arithmetically or is increasing geometrically? Explain your answer.

Exercise Set 1.7

Practice Exercises

In Exercises 1–20, write the first six terms of the arithmetic sequence with the first term, a_1, and common difference, d.

1. $a_1 = 8, d = 2$ 8, 10, 12, 14, 16, and 18

2. $a_1 = 5, d = 3$ 5, 8, 11, 14, 17, and 20

3. $a_1 = 200, d = 20$ 200, 220, 240, 260, 280, and 300

4. $a_1 = 300, d = 50$ 300, 350, 400, 450, 500, and 550

5. $a_1 = -7, d = 4$ $-7, -3, 1, 5, 9,$ and 13

6. $a_1 = -8, d = 5$ $-8, -3, 2, 7, 12,$ and 17

7. $a_1 = -400, d = 300$ $-400, -100, 200, 500, 800,$ and 1100

8. $a_1 = -500, d = 400$ $-500, -100, 300, 700, 1100,$ and 1500

9. $a_1 = 7, d = -3$ $7, 4, 1, -2, -5,$ and -8

10. $a_1 = 9, d = -5$ $9, 4, -1, -6, -11,$ and -16

11. $a_1 = 200, d = -60$ $200, 140, 80, 20, -40,$ and -100

12. $a_1 = 300, d = -90$ $300, 210, 120, 30, -60,$ and -150

13. $a_1 = \frac{5}{2}, d = \frac{1}{2}$ $\frac{5}{2}, 3, \frac{7}{2}, 4, \frac{9}{2},$ and 5

14. $a_1 = \frac{3}{4}, d = \frac{1}{4}$ $\frac{3}{4}, 1, \frac{5}{4}, \frac{3}{2}, \frac{7}{4},$ and 2

15. $a_1 = \frac{3}{2}, d = \frac{1}{4}$ $\frac{3}{2}, \frac{7}{4}, 2, \frac{9}{4}, \frac{5}{2},$ and $\frac{11}{4}$

16. $a_1 = \frac{3}{2}, d = -\frac{1}{4}$ $\frac{3}{2}, \frac{5}{4}, 1, \frac{3}{4}, \frac{1}{2},$ and $\frac{1}{4}$

17. $a_1 = 4.25, d = 0.3$ 4.25, 4.55, 4.85, 5.15, 5.45, and 5.75

18. $a_1 = 6.3, d = 0.25$ 6.3, 6.55, 6.8, 7.05, 7.3, and 7.55

19. $a_1 = 4.5, d = -0.75$ 4.5, 3.75, 3, 2.25, 1.5, and 0.75

20. $a_1 = 3.5, d = -1.75$ 3.5, 1.75, 0, $-1.75, -3.5,$ and -5.25

In Exercises 21–40, find the indicated term for the arithmetic sequence with first term, a_1, and common difference, d.

21. Find a_6, when $a_1 = 13$, $d = 4$.　33

22. Find a_{16}, when $a_1 = 9$, $d = 2$.　39

23. Find a_{50}, when $a_1 = 7$, $d = 5$.　252

24. Find a_{60}, when $a_1 = 8$, $d = 6$.　362

25. Find a_9, when $a_1 = -5$, $d = 9$.　67

26. Find a_{10}, when $a_1 = -8$, $d = 10$.　82

27. Find a_{200}, when $a_1 = -40$, $d = 5$.　955

28. Find a_{150}, when $a_1 = -60$, $d = 5$.　685

29. Find a_{10}, when $a_1 = 8$, $d = -10$.　-82

30. Find a_{11}, when $a_1 = 10$, $d = -6$.　-50

31. Find a_{60}, when $a_1 = 35$, $d = -3$.　-142

32. Find a_{70}, when $a_1 = -32$, $d = 4$.　244

33. Find a_{12}, when $a_1 = 12$, $d = -5$.　-43

34. Find a_{20}, when $a_1 = -20$, $d = -4$.　-96

35. Find a_{90}, when $a_1 = -70$, $d = -2$.　-248

36. Find a_{80}, when $a_1 = 106$, $d = -12$.　-842

37. Find a_{12}, when $a_1 = 6$, $d = \frac{1}{2}$.　$\frac{23}{2}$

38. Find a_{14}, when $a_1 = 8$, $d = \frac{1}{4}$.　$\frac{45}{4}$

39. Find a_{50}, when $a_1 = 14$, $d = -0.25$.　1.75

40. Find a_{110}, when $a_1 = -12$, $d = -0.5$.　-66.5

In Exercises 41–48, write a formula for the general term (the nth term) of each arithmetic sequence. Then use the formula for a_n to find a_{20}, the 20th term of the sequence.

41. $1, 5, 9, 13, \ldots$　$a_n = 1 + (n-1)4; 77$

42. $2, 7, 12, 17, \ldots$　$a_n = 2 + (n-1)5; 97$

43. $7, 3, -1, -5, \ldots$　$a_n = 7 + (n-1)(-4); -69$

44. $6, 1, -4, -9, \ldots$　$a_n = 6 + (n-1)(-5); -89$

45. $a_1 = 9$, $d = 2$　$a_n = 9 + (n-1)2; 47$

46. $a_1 = 6$, $d = 3$　$a_n = 6 + (n-1)3; 63$

47. $a_1 = -20$, $d = -4$　$a_n = -20 + (n-1)(-4); -96$

48. $a_1 = -70$, $d = -5$　$a_n = -70 + (n-1)(-5); -165$

In Exercises 49–70, write the first six terms of the geometric sequence with the first term, a_1, and common ratio, r.

49. $a_1 = 4$, $r = 2$　4, 8, 16, 32, 64, and 128

50. $a_1 = 2$, $r = 3$　2, 6, 18, 54, 162, and 486

51. $a_1 = 1000$, $r = 1$　1000, 1000, 1000, 1000, 1000, and 1000

52. $a_1 = 5000$, $r = 1$　5000, 5000, 5000, 5000, 5000, and 5000

53. $a_1 = 3$, $r = -2$　3, -6, 12, -24, 48, and -96

54. $a_1 = 2$, $r = -3$　2, -6, 18, -54, 162, and -486

55. $a_1 = 10$, $r = -4$　10, -40, 160, -640, 2560, and -10,240

56. $a_1 = 20$, $r = -4$　20, -80, 320, -1280, 5120, and -20,480

57. $a_1 = 2000$, $r = -1$　2000, -2000, 2000, -2000, 2000, and -2000

58. $a_1 = 3000$, $r = -1$　3000, -3000, 3000, -3000, 3000, and -3000

59. $a_1 = -2$, $r = -3$　-2, 6, -18, 54, -162, and 486

60. $a_1 = -4$, $r = -2$　-4, 8, -16, 32, -64, and 128

61. $a_1 = -6$, $r = -5$　-6, 30, -150, 750, -3750, and 18,750

62. $a_1 = -8$, $r = -5$　-8, 40, -200, 1000, -5000, and 25,000

63. $a_1 = \frac{1}{4}$, $r = 2$　$\frac{1}{4}, \frac{1}{2}, 1, 2, 4$, and 8

64. $a_1 = \frac{1}{2}$, $r = 2$　$\frac{1}{2}, 1, 2, 4, 8$, and 16

65. $a_1 = \frac{1}{4}$, $r = \frac{1}{2}$　$\frac{1}{4}, \frac{1}{8}, \frac{1}{16}, \frac{1}{32}, \frac{1}{64}$, and $\frac{1}{128}$

66. $a_1 = \frac{1}{5}$, $r = \frac{1}{2}$　$\frac{1}{5}, \frac{1}{10}, \frac{1}{20}, \frac{1}{40}, \frac{1}{80}$, and $\frac{1}{160}$

67. $a_1 = -\frac{1}{16}$, $r = -4$　$-\frac{1}{16}, \frac{1}{4}, -1, 4, -16$, and 64

68. $a_1 = -\frac{1}{8}$, $r = -2$　$-\frac{1}{8}, \frac{1}{4}, -\frac{1}{2}, 1, -2$, and 4

69. $a_1 = 2$, $r = 0.1$　2, 0.2, 0.02, 0.002, 0.0002, and 0.00002

70. $a_1 = -1000$, $r = 0.1$　-1000, -100, -10, -1, -0.1, and -0.01

In Exercises 71–90, find the indicated term for the geometric sequence with first term, a_1, and common ratio, r.

71. Find a_7, when $a_1 = 4$, $r = 2$.　256

72. Find a_5, when $a_1 = 4$, $r = 3$.　324

73. Find a_{20}, when $a_1 = 2$, $r = 3$.　$2,324,522,934 \approx 2.32 \times 10^9$

74. Find a_{20}, when $a_1 = 2$, $r = 2$.　1,048,576

75. Find a_{100}, when $a_1 = 50$, $r = 1$.　50

76. Find a_{200}, when $a_1 = 60$, $r = 1$.　60

77. Find a_7, when $a_1 = 5$, $r = -2$.　320

78. Find a_4, when $a_1 = 4$, $r = -3$.　-108

79. Find a_{30}, when $a_1 = 2$, $r = -1$.　-2

80. Find a_{40}, when $a_1 = 6$, $r = -1$.　-6

81. Find a_6, when $a_1 = -2$, $r = -3$.　486

82. Find a_5, when $a_1 = -5$, $r = -2$.　-80

83. Find a_8, when $a_1 = 6$, $r = \frac{1}{2}$.　$\frac{3}{64}$

84. Find a_8, when $a_1 = 12$, $r = \frac{1}{2}$.　$\frac{3}{32}$

85. Find a_6, when $a_1 = 18$, $r = -\frac{1}{3}$.　$-\frac{2}{27}$

86. Find a_4, when $a_1 = 9$, $r = -\frac{1}{3}$.　$-\frac{1}{3}$

87. Find a_{40}, when $a_1 = 1000$, $r = -\frac{1}{2}$.　$\approx -1.82 \times 10^{-9}$

88. Find a_{30}, when $a_1 = 8000$, $r = -\frac{1}{2}$.　≈ 0.000014901

89. Find a_8, when $a_1 = 1,000,000$, $r = 0.1$.　0.1

90. Find a_8, when $a_1 = 40,000$, $r = 0.1$.　0.004

In Exercises 91–98, write a formula for the general term (the nth term) of each geometric sequence. Then use the formula for a_n to find a_7, the seventh term of the sequence.

91. $3, 12, 48, 192, \ldots$　$a_n = 3(4)^{n-1}; 12,288$

92. $3, 15, 75, 375, \ldots$　$a_n = 3(5)^{n-1}; 46,875$

93. $18, 6, 2, \frac{2}{3}, \ldots$　$a_n = 18\left(\frac{1}{3}\right)^{n-1}; \frac{2}{81}$

94. $12, 6, 3, \frac{3}{2}, \ldots$　$a_n = 12\left(\frac{1}{2}\right)^{n-1}; \frac{3}{16}$

95. $1.5, -3, 6, -12, \ldots$　$a_n = 1.5(-2)^{n-1}; 96$

96. $5, -1, \frac{1}{5}, -\frac{1}{25}, \ldots$　$a_n = 5\left(-\frac{1}{5}\right)^{n-1}; \frac{1}{3125}$

97. $0.0004, -0.004, 0.04, -0.4, \ldots$　$a_n = 0.0004(-10)^{n-1}; 400$

98. $0.0007, -0.007, 0.07, -0.7, \ldots$　$a_n = 0.0007(-10)^{n-1}; 700$

Determine whether each sequence in Exercises 99–114 is arithmetic or geometric. Then find the next two terms.

99. 2, 6, 10, 14, . . . arithmetic; 18 and 22

100. 3, 8, 13, 18, . . . arithmetic; 23 and 28

101. 5, 15, 45, 135, . . . geometric; 405 and 1215

102. 15, 30, 60, 120, . . . geometric; 240 and 480

103. −7, −2, 3, 8, . . . arithmetic; 13 and 18

104. −9, −5, −1, 3, . . . arithmetic; 7 and 11

105. $3, \frac{3}{2}, \frac{3}{4}, \frac{3}{8}, \ldots$ geometric; $\frac{3}{16}$ and $\frac{3}{32}$

106. $6, 3, \frac{3}{2}, \frac{3}{4}, \ldots$ geometric; $\frac{3}{8}$ and $\frac{3}{16}$

107. $\frac{1}{2}, 1, \frac{3}{2}, 2, \ldots$ arithmetic; $\frac{5}{2}$ and 3

108. $\frac{2}{3}, 1, \frac{4}{3}, \frac{5}{3}, \ldots$ arithmetic; 2 and $\frac{7}{3}$

109. 7, −7, 7, −7, . . . geometric; 7 and −7

110. 6, −6, 6, −6, . . . geometric; 6 and −6

111. 7, −7, −21, −35, . . . arithmetic; −49 and −63

112. 6, −6, −18, −30, . . . arithmetic; −42 and −54

113. $\sqrt{5}, 5, 5\sqrt{5}, 25, \ldots$ geometric; $25\sqrt{5}$ and 125

114. $\sqrt{3}, 3, 3\sqrt{3}, 9, \ldots$ geometric; $9\sqrt{3}$ and 27

Practice Plus

The sum, S_n, of the first n terms of an arithmetic sequence is given by

$$S_n = \frac{n}{2}(a_1 + a_n),$$

in which a_1 is the first term and a_n is the nth term. The sum, S_n, of the first n terms of a geometric sequence is given by

$$S_n = \frac{a_1(1 - r^n)}{1 - r},$$

in which a_1 is the first term and r is the common ratio $(r \neq 1)$. In Exercises 115–122, determine whether each sequence is arithmetic or geometric. Then use the appropriate formula to find S_{10}, the sum of the first ten terms.

115. 4, 10, 16, 22, . . . arithmetic; 310

116. 7, 19, 31, 43, . . . arithmetic; 610

117. 2, 6, 18, 54, . . . geometric; 59,048

118. 3, 6, 12, 24, . . . geometric; 3069

119. 3, −6, 12, −24, . . . geometric; −1023

120. 4, −12, 36, −108, . . . geometric; −59,048

121. −10, −6, −2, 2, . . . arithmetic; 80

122. −15, −9, −3, 3, . . . arithmetic; 120

123. Use the appropriate formula shown above to find 1 + 2 + 3 + 4 + · · · + 100, the sum of the first 100 natural numbers. 5050

124. Use the appropriate formula shown above to find 2 + 4 + 6 + 8 + · · · + 200, the sum of the first 100 positive even integers. 10,100

Application Exercises

The bar graph shows changes in the percentage of college graduates for Americans ages 25 and older from 1990 to 2010. Exercises 125–126 involve developing arithmetic sequences that model the data.

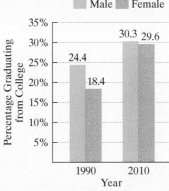

Percentage of College Graduates for Americans Ages 25 and Older

Source: U.S Census Bureau

125. In 1990, 18.4% of American women ages 25 and older had graduated from college. On average, this percentage has increased by approximately 0.6 each year.

   a. Write a formula for the *n*th term of the arithmetic sequence that models the percentage of American women ages 25 and older who had graduated from college *n* years after 1989. $a_n = 0.6n + 17.8$

   b. Use the model from part (a) to project the percentage of American women ages 25 and older who will be college graduates by 2019. 35.8%

126. In 1990, 24.4% of American men ages 25 and older had graduated from college. On average, this percentage has increased by approximately 0.3 each year.

   a. Write a formula for the *n*th term of the arithmetic sequence that models the percentage of American men ages 25 and older who had graduated from college *n* years after 1989. $a_n = 0.3n + 24.1$

   b. Use the model from part (a) to project the percentage of American men ages 25 and older who will be college graduates by 2019. 33.1%

127. Company A pays $24,000 yearly with raises of $1600 per year. Company B pays $28,000 yearly with raises of $1000 per year. Which company will pay more in year 10? How much more? Company A; $1400

128. Company A pays $23,000 yearly with raises of $1200 per year. Company B pays $26,000 yearly with raises of $800 per year. Which company will pay more in year 10? How much more? Company A; $600

In Exercises 129–130, suppose you save $1 the first day of a month, $2 the second day, $4 the third day, and so on. That is, each day you save twice as much as you did the day before.

129. What will you put aside for savings on the fifteenth day of the month? $16,384

130. What will you put aside for savings on the thirtieth day of the month? $536,870,912

131. A professional baseball player signs a contract with a beginning salary of $3,000,000 for the first year and an annual increase of 4% per year beginning in the second year. That is, beginning in year 2, the athlete's salary will be 1.04 times what it was in the previous year. What is the athlete's salary for year 7 of the contract? Round to the nearest dollar. $3,795,957

132. You are offered a job that pays $30,000 for the first year with an annual increase of 5% per year beginning in the second year. That is, beginning in year 2, your salary will be 1.05 times what it was in the previous year. What can you expect to earn in your sixth year on the job? Round to the nearest dollar. $38,288

In Exercises 133–134, you will develop geometric sequences that model the population growth for California and Texas, the two most-populated U.S. states.

133. The table shows the population of California for 2000 and 2010, with estimates given by the U.S. Census Bureau for 2001 through 2009.

Year	2000	2001	2002	2003	2004	2005
Population in millions	33.87	34.21	34.55	34.90	35.25	35.60

Year	2006	2007	2008	2009	2010
Population in millions	36.00	36.36	36.72	37.09	37.25

a. Divide the population for each year by the population in the preceding year. Round to two decimal places and show that California has a population increase that is approximately geometric. ≈ 1.01 for all but one division

b. Write the general term of the geometric sequence modeling California's population, in millions, n years after 1999. $a_n = 33.87(1.01)^{n-1}$

c. Use your model from part (b) to project California's population, in millions, for the year 2020. Round to two decimal places. 41.33 million

134. The table shows the population of Texas for 2000 and 2010, with estimates given by the U.S. Census Bureau for 2001 through 2009.

Year	2000	2001	2002	2003	2004	2005
Population in millions	20.85	21.27	21.70	22.13	22.57	23.02

Year	2006	2007	2008	2009	2010
Population in millions	23.48	23.95	24.43	24.92	25.15

a. Divide the population for each year by the population in the preceding year. Round to two decimal places and show that Texas has a population increase that is approximately geometric. ≈ 1.02 for all but one division

b. Write the general term of the geometric sequence modeling Texas's population, in millions, n years after 1999. $a_n = 20.85(1.02)^{n-1}$

c. Use your model from part (b) to project Texas's population, in millions, for the year 2020. Round to two decimal places. 30.98 million

Critical Thinking Exercises

135. A person is investigating two employment opportunities. They both have a beginning salary of $20,000 per year. Company A offers an increase of $1000 per year. Company B offers 5% more than during the preceding year. Which company will pay more in the sixth year? Company B

136. Would you rather have $10,000,000 and a brand new BMW, or would you rather have what you would receive on days 28, 29, and 30 combined if we give you 1¢ today, 2¢ tomorrow, 4¢ on day 3, 8¢ on day 4, 16¢ on day 5, and so on? Explain. $10,000,000 and a BMW; the other option yields less than $10,000,000.

Group Exercise

137. Enough curiosities involving the Fibonacci sequence exist to warrant a flourishing Fibonacci Association. It publishes a quarterly journal. Do some research on the Fibonacci sequence in the library or on the Internet, and find one property that interests you. After doing this research, get together with your group to share these intriguing properties. Answers will vary.

Chapter 1 Summary

Definitions and Concepts	**Examples**

Section 1.1 Number Theory: Prime and Composite Numbers

The set of natural numbers is $\{1, 2, 3, 4, 5, \ldots\}$.
$b \mid a$ (b divides a: a is divisible by b) for natural numbers a and b if the operation of dividing a by b leaves a remainder of 0.

Rules of Divisibility

Divisible By	Test
2	The last digit is 0, 2, 4, 6, or 8.
3	The sum of the digits is divisible by 3.
4	The last two digits form a number divisible by 4.
5	The number ends in 0 or 5.
6	The number is divisible by both 2 and 3. (In other words, the number is even and the sum of its digits is divisible by 3.)
8	The last three digits form a number that is divisible by 8.
9	The sum of the digits is divisible by 9.
10	The last digit is 0.
12	The number is divisible by both 3 and 4. (In other words, the sum of the digits is divisible by 3 and the last two digits form a number divisible by 4.)

- Which of the numbers 2, 3, 4, 5, 6, 8, 9, 10, and 12 divide 614,608,176?

  $2 \mid 614{,}608{,}176$ because the last digit is 6.

  $3 \mid 614{,}608{,}176$ because the sum of the digits is $6 + 1 + 4 + 6 + 0 + 8 + 1 + 7 + 6 = 39$, and 39 is divisible by 3.

  $4 \mid 614{,}608{,}176$ because the last two digits form the number 76, which is divisible by 4.

  $5 \nmid 614{,}608{,}176$ because the number does not end in 0 or 5.

  $6 \mid 614{,}608{,}176$ because the number is divisible by both 2 and 3.

  $8 \mid 614{,}608{,}176$ because the last three digits form the number 176, which is divisible by 8.

  $9 \nmid 614{,}608{,}176$ because the sum of the digits, 39, is not divisible by 9.

  $10 \nmid 614{,}608{,}176$ because the last digit is not 0.

  $12 \mid 614{,}608{,}176$ because the number is divisible by both 3 and 4.

Additional Example to Review

Example 1, page 4

A prime number is a natural number greater than 1 that has only itself and 1 as factors. A composite number is a natural number greater than 1 that is divisible by a number other than itself and 1.

The Fundamental Theorem of Arithmetic: Every composite number can be expressed as a product of prime numbers in one and only one way (if the order of the factors is disregarded).

- Find the prime factorization of 300.

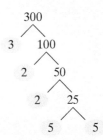

$$300 = 3 \cdot 2^2 \cdot 5^2 = 2^2 \cdot 3 \cdot 5^2$$

Additional Example to Review

Example 2, page 5

The greatest common divisor of two or more natural numbers is the largest number that is a divisor (or factor) of all the numbers. To find the greatest common divisor of two or more numbers,

1. Write the prime factorization of each number.
2. Select each prime factor with the smallest exponent that is common to each of the prime factorizations.
3. Form the product of the numbers from step 2. The greatest common divisor is the product of these factors.

• Find the greatest common divisor of 168 and 180.

$168 = 2 \cdot 2 \cdot 2 \cdot 3 \cdot 7 = 2^3 \cdot 3 \cdot 7$
$180 = 2 \cdot 2 \cdot 3 \cdot 3 \cdot 5 = 2^2 \cdot 3^2 \cdot 5$

Greatest common divisor $= 2^2 \cdot 3 = 4 \cdot 3 = 12$

Additional Examples to Review

Example 3, page 6; Example 4, page 7

The least common multiple of two or more natural numbers is the smallest natural number that is divisible by all of the numbers. To find the least common multiple of two or more numbers,

1. Write the prime factorization of each number.
2. Select every prime factor that occurs, raised to the greatest power to which it occurs, in these factorizations.
3. Form the product of the numbers from step 2. The least common multiple is the product of these factors.

• Find the least common multiple of 168 and 180.
$168 = 2^3 \cdot 3 \cdot 7$ and $180 = 2^2 \cdot 3^2 \cdot 5$
Least common multiple $= 2^3 \cdot 3^2 \cdot 5 \cdot 7 = 8 \cdot 9 \cdot 5 \cdot 7 = 2520$
The smallest natural number divisible by both 168 and 180 is 2520.

Additional Examples to Review

Example 5, page 9; Example 6, page 9

Section 1.2 The Integers; Order of Operations

The set of whole numbers is $\{0, 1, 2, 3, 4, 5, \ldots\}$. The set of integers is $\{\ldots, -3, -2, -1, 0, 1, 2, 3, \ldots\}$. Integers are graphed on a number line by placing a dot at the correct location for each number.

$a < b$ (a is less than b) means a is to the left of b on a number line.

$a > b$ (a is greater than b) means a is to the right of b on a number line.

• $-2 < 5$ and $5 > -2$.

Additional Examples to Review

Example 1, page 14; Example 2, page 15

$|a|$, the absolute value of a, is the distance of a from 0 on a number line. The absolute value of a positive number is the number itself. The absolute value of 0 is 0: $|0| = 0$. The absolute value of a negative number is the number without the negative sign. For example, $|-8| = 8$.

• $|-1000| = 1000$
• $|1| = 1$
• $|0| = 0$

Additional Example to Review

Example 3, page 15

Addition of Integers

If the integers have the same sign,

1. Add their absolute values.
2. The sign of the sum is the same as the sign of the two numbers.

If the integers have different signs,

1. Subtract the smaller absolute value from the larger absolute value.
2. The sign of the sum is the same as the sign of the number with the larger absolute value.

The result of addition is called the sum.

• $12 + 17 = 29$
• $-12 + (-17) = -29$
• $12 + (-17) = -5$
• $-12 + 17 = 5$

Additional Examples to Review

See the box at the bottom of page 16.

Subtraction of Integers

Additive inverses have the same absolute value, but lie on opposite sides of zero on a number line.

For all integers a and b,

$$a - b = a + (-b).$$

In words, to subtract b from a, add the additive inverse of b to a. The result of subtraction is called the difference.

- $25 - (-13) = 25 + 13 = 38$
- $-25 - (-13) = -25 + 13 = -12$
- $-25 - 13 = -25 + (-13) = -38$

Additional Examples to Review

Example 4, page 18; Example 5, page 18

Multiplication of Integers

1. The product of two integers with different signs is found by multiplying their absolute values. The product is negative.
2. The product of two integers with the same sign is found by multiplying their absolute values. The product is positive.
3. The product of 0 and any integer is 0: $a \cdot 0 = 0$ and $0 \cdot a = 0$.
4. If no number is 0, a product with an odd number of negative factors is found by multiplying absolute values. The product is negative.
5. If no number is 0, a product with an even number of negative factors is found by multiplying absolute values. The product is positive.

- $8(-10) = -80$
- $-8(10) = -80$
- $8(10) = 80$
- $(-8)(-10) = 80$
- $(-8)(-10)(-2) = -160$

  Three (odd) negative factors

- $(-2)^3 = (-2)(-2)(-2) = -8$
- $-8(10)(-2) = 160$

  Two (even) negative factors

- $(-10)^2 = (-10)(-10) = 100$

Additional Examples to Review

See the box on page 20; Example 6, page 20

Division of Integers

1. The quotient of two integers with different signs is found by dividing their absolute values. The quotient is negative.
2. The quotient of two integers with the same sign is found by dividing their absolute values. The quotient is positive.
3. Zero divided by any nonzero integer is zero.
4. Division by 0 is undefined.

- $\dfrac{-20}{5} = -4$
- $\dfrac{20}{5} = 4$
- $\dfrac{0}{-5} = 0$
- $\dfrac{20}{-5} = -4$
- $\dfrac{-20}{-5} = 4$
- $\dfrac{-5}{0}$ is undefined.

Additional Examples to Review

See the box near the top of page 22.

Order of Operations

1. Perform all operations within grouping symbols.
2. Evaluate all exponential expressions.
3. Do all multiplications and divisions from left to right.
4. Do all additions and subtractions from left to right.

- $-10 - (8 - 14) = -10 - (-6) = -10 + 6 = -4$
- $30 \div (10 - 5^2) = 30 \div (10 - 25) = 30 \div (-15) = -2$
- $(9 - 12)^3(3 - 7)^2 = (-3)^3(-4)^2 = -27(16) = -432$

Additional Examples to Review

Example 7, page 23; Example 8, page 23

Section 1.3 The Rational Numbers

The set of rational numbers is the set of all numbers which can be expressed in the form $\frac{a}{b}$, where a and b are integers and b is not equal to 0.

A rational number is reduced to its lowest terms, or simplified, by dividing both the numerator and the denominator by their greatest common divisor.

- $\dfrac{375}{1000} = \dfrac{5 \cdot 75}{5 \cdot 200} = \dfrac{\cancel{5} \cdot 25 \cdot 3}{\cancel{5} \cdot 25 \cdot 8} = \dfrac{3}{8}$

Additional Example to Review

Example 1, page 29

A mixed number consists of the sum of an integer and a rational number, expressed without the use of an addition sign. An improper fraction is a rational number whose numerator is greater than its denominator.

Converting a Positive Mixed Number to an Improper Fraction

1. Multiply the denominator of the rational number by the integer and add the numerator to this product.
2. Place the sum in step 1 over the denominator in the mixed number.

Converting a Positive Improper Fraction to a Mixed Number

1. Divide the denominator into the numerator. Record the quotient and the remainder.
2. Write the mixed number using the following form:

$$\text{quotient } \frac{\text{remainder}}{\text{original denominator}}.$$

- Convert $4\frac{2}{3}$ to an improper fraction.

$$4\frac{2}{3} = \frac{3 \cdot 4 + 2}{3} = \frac{12 + 2}{3} = \frac{14}{3}$$

- Convert $\frac{37}{5}$ to a mixed number.

$$\begin{array}{r} 7 \\ 5\overline{)37} \\ 35 \\ \hline 2 \end{array} \qquad \frac{37}{5} = 7\frac{2}{5}$$

Additional Examples to Review

Example 2, page 30; Example 3, page 30

Any rational number can be expressed as a decimal. The resulting decimal will either terminate (stop), or it will have a digit that repeats or a block of digits that repeat. The rational number $\frac{a}{b}$ is expressed as a decimal by dividing b into a.

- $\dfrac{3}{8} = 0.375$

$$\begin{array}{r} 0.375 \\ 8\overline{)3.000} \end{array}$$

- $\dfrac{7}{12} = 0.58\overline{3}$

$$\begin{array}{r} 0.58333\cdots \\ 12\overline{)7.00000\cdots} \end{array}$$

Additional Example to Review

Example 4, page 31

To express a terminating decimal as a quotient of integers, the digits to the right of the decimal point are the numerator. The place value of the last digit to the right of the decimal point determines the denominator.

- $0.3 = \dfrac{3}{10}$ • $0.37 = \dfrac{37}{100}$ • $0.019 = \dfrac{19}{1000}$

Additional Example to Review

Example 5, page 33

To express a repeating decimal as a quotient of integers, use the following procedure:

Step 1 Let n equal the repeating decimal.

Step 2 Multiply both sides of the equation in step 1 by 10 if one digit repeats, by 100 if two digits repeat, by 1000 if three digits repeat, and so on.

Step 3 Subtract the equation in step 1 from the equation in step 2.

Step 4 Divide both sides of the equation in step 3 by the number in front of n and solve for n.

- Express $0.\overline{24}$ as a quotient of integers.

1. $n = 0.\overline{24} = 0.242424\ldots$
2. $100n = 24.242424\ldots$
3. $\begin{aligned} 100n &= 24.242424\ldots \\ -n &= 0.242424\ldots \\ \hline 99n &= 24 \end{aligned}$
4. $\dfrac{99n}{99} = \dfrac{24}{99}$

$$n = \frac{24}{99} \qquad \text{Conclusion: } 0.\overline{24} = \frac{24}{99}$$

Additional Examples to Review

Example 6, page 34; Example 7, page 35

The product of two rational numbers is the product of their numerators divided by the product of their denominators.

- $\left(-\dfrac{3}{5}\right)\left(\dfrac{2}{7}\right) = \dfrac{-3(2)}{5 \cdot 7} = \dfrac{-6}{35} = -\dfrac{6}{35}$

- $\dfrac{3}{20}\left(6\dfrac{1}{4}\right) = \dfrac{3}{20} \cdot \dfrac{25}{4} = \dfrac{3}{\overset{}{\underset{4}{20}}} \cdot \dfrac{\overset{5}{25}}{4} = \dfrac{3 \cdot 5}{4 \cdot 4} = \dfrac{15}{16}$

Additional Example to Review

Example 8, page 36

Two numbers whose product is 1 are called reciprocals, or multiplicative inverses, of each other. The quotient of two rational numbers is the product of the first number and the reciprocal of the second number.

- $\dfrac{3}{5} \div 2\dfrac{1}{5} = \dfrac{3}{5} \div \dfrac{11}{5} = \dfrac{3}{5} \cdot \dfrac{5}{11} = \dfrac{3}{\underset{1}{\cancel{5}}} \cdot \dfrac{\overset{1}{\cancel{5}}}{11} = \dfrac{3 \cdot 1}{1 \cdot 11} = \dfrac{3}{11}$

Additional Example to Review

Example 9, page 36

The sum or difference of two rational numbers with identical denominators is the sum or difference of their numerators over the common denominator.

- $\dfrac{3}{5} + \dfrac{1}{5} = \dfrac{3 + 1}{5} = \dfrac{4}{5}$

- $\dfrac{11}{12} - \dfrac{1}{12} = \dfrac{11 - 1}{12} = \dfrac{10}{12} = \dfrac{2 \cdot 5}{2 \cdot 6} = \dfrac{5}{6}$

- $-3\dfrac{1}{4} - 5\dfrac{3}{4} = -\dfrac{13}{4} - \dfrac{23}{4} = \dfrac{-13 - 23}{4} = \dfrac{-36}{4} = -9$

Additional Example to Review

Example 10, page 37

Add or subtract rational numbers with unlike denominators by first expressing each rational number with the least common denominator and then following the procedure above.

- $\dfrac{3}{10} - \dfrac{7}{12}$

  $10 = 2 \cdot 5$ and $12 = 2^2 \cdot 3$, so the LCD is $2^2 \cdot 3 \cdot 5 = 60$.

  $\dfrac{3}{10} - \dfrac{7}{12} = \dfrac{3}{10} \cdot \dfrac{6}{6} - \dfrac{7}{12} \cdot \dfrac{5}{5} = \dfrac{18}{60} - \dfrac{35}{60} = \dfrac{18 - 35}{60}$

  $= \dfrac{-17}{60} = -\dfrac{17}{60}$

Additional Examples to Review

Example 11, page 38; Example 12, page 38

The order of operations can be applied to an expression with rational numbers.

- $\dfrac{5}{6} - 3\left(\dfrac{1}{2} + \dfrac{2}{3}\right) = \dfrac{5}{6} - 3\left(\dfrac{1}{2} \cdot \dfrac{3}{3} + \dfrac{2}{3} \cdot \dfrac{2}{2}\right) = \dfrac{5}{6} - 3\left(\dfrac{3}{6} + \dfrac{4}{6}\right)$

  > The least common denominator is 6.

  $= \dfrac{5}{6} - 3\left(\dfrac{3 + 4}{6}\right) = \dfrac{5}{6} - \dfrac{3}{1} \cdot \dfrac{7}{6} = \dfrac{5}{6} - \dfrac{3 \cdot 7}{6} = \dfrac{5}{6} - \dfrac{21}{6}$

  $= \dfrac{5 - 21}{6} = \dfrac{-16}{6} = \dfrac{-2 \cdot 8}{2 \cdot 3} = \dfrac{-8}{3} = -\dfrac{8}{3} \text{ or } -2\dfrac{2}{3}$

Additional Example to Review

Example 13, page 39

Density of the Rational Numbers

Given any two distinct rational numbers, there is always a rational number between them. To find the rational number halfway between two rational numbers, add the rational numbers and divide their sum by 2.

- Find the rational number halfway between $\frac{1}{5}$ and $\frac{1}{3}$.

$$\left(\frac{1}{5} + \frac{1}{3}\right) \div 2 = \left(\frac{1}{5} \cdot \frac{3}{3} + \frac{1}{3} \cdot \frac{5}{5}\right) \div 2 = \left(\frac{3}{15} + \frac{5}{15}\right) \div 2$$

The least common denominator is 15.

$$= \frac{8}{15} \div \frac{2}{1} = \frac{\overset{4}{8}}{15} \cdot \frac{1}{\underset{1}{2}} = \frac{4}{15}$$

Thus, $\frac{1}{5} < \frac{4}{15} < \frac{1}{3}$.

Additional Example to Review

Example 14, page 40

Section 1.4 The Irrational Numbers

The set of irrational numbers is the set of numbers whose decimal representations are neither terminating nor repeating. Examples of irrational numbers are $\sqrt{2} \approx 1.414$ and $\pi \approx 3.142$.

Simplifying square roots: Use the product rule, $\sqrt{ab} = \sqrt{a} \cdot \sqrt{b}$, to remove from the square root any perfect squares that occur as factors.

- $\sqrt{50} = \sqrt{25 \cdot 2} = \sqrt{25} \cdot \sqrt{2} = 5\sqrt{2}$

Additional Example to Review

Example 1, page 47

Multiplying square roots: $\sqrt{a} \cdot \sqrt{b} = \sqrt{ab}$. The product of square roots is the square root of the product.

- $\sqrt{10} \cdot \sqrt{8} = \sqrt{10 \cdot 8} = \sqrt{80} = \sqrt{16 \cdot 5}$
  $$= \sqrt{16} \cdot \sqrt{5} = 4\sqrt{5}$$

Additional Example to Review

Example 2, page 48

Dividing square roots: $\frac{\sqrt{a}}{\sqrt{b}} = \sqrt{\frac{a}{b}}$. The quotient of square roots is the square root of the quotient.

- $\frac{\sqrt{500}}{\sqrt{2}} = \sqrt{\frac{500}{2}} = \sqrt{250} = \sqrt{25 \cdot 10}$
  $$= \sqrt{25} \cdot \sqrt{10} = 5\sqrt{10}$$

Additional Example to Review

Example 3, page 49

Adding and subtracting square roots: If the radicals have the same radicand, add or subtract their coefficients. The answer is the sum or difference of the coefficients times the common square root. Addition or subtraction is sometimes possible by first simplifying the square roots.

- $2\sqrt{13} - 9\sqrt{13} = (2 - 9)\sqrt{13} = -7\sqrt{13}$
- $2\sqrt{32} + 4\sqrt{50} = 2\sqrt{16 \cdot 2} + 4\sqrt{25 \cdot 2}$
  $$= 2 \cdot 4\sqrt{2} + 4 \cdot 5\sqrt{2} = 8\sqrt{2} + 20\sqrt{2}$$
  $$= (8 + 20)\sqrt{2} = 28\sqrt{2}$$

Additional Examples to Review

Example 4, page 49; Example 5, page 50

Rationalizing denominators: Multiply the numerator and the denominator by the smallest number that produces a perfect square radicand in the denominator.

- $\frac{15}{\sqrt{3}} = \frac{15}{\sqrt{3}} \cdot \frac{\sqrt{3}}{\sqrt{3}} = \frac{15\sqrt{3}}{\sqrt{9}} = \frac{15\sqrt{3}}{3} = \frac{\overset{5}{15}\sqrt{3}}{\underset{1}{3}} = 5\sqrt{3}$

Additional Example to Review

Example 6, page 51

Section 1.5 Real Numbers and Their Properties

The set of real numbers is obtained by combining the rational numbers with the irrational numbers.

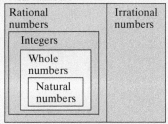

Real numbers

Rational numbers	Irrational numbers
Integers	
Whole numbers	
Natural numbers	

This diagram shows that every real number is rational or irrational.

Subsets of the Real Numbers

Name	Description
Natural numbers	$\{1, 2, 3, 4, 5, \ldots\}$ These are the numbers that we use for counting.
Whole numbers	$\{0, 1, 2, 3, 4, 5, \ldots\}$ The set of whole numbers includes 0 and the natural numbers.
Integers	$\{\ldots, -5, -4, -3, -2, -1, 0, 1, 2, 3, 4, 5, \ldots\}$ The set of integers includes the whole numbers and the negatives of the natural numbers.
Rational numbers	The set of rational numbers is the set of all numbers that can be expressed in the form $\dfrac{a}{b}$, where a and b are integers, $b \neq 0$. Rational numbers can be expressed as terminating or repeating decimals.
Irrational numbers	The set of irrational numbers is the set of all numbers whose decimal representations are neither terminating nor repeating. Irrational numbers cannot be expressed as a quotient of integers.

- Consider the set $\left\{ -6, -\frac{2}{7}, 0, 0.\overline{3}, \sqrt{5}, \sqrt{16}, 5\pi \right\}$

  Natural numbers: $\left\{ \sqrt{16} \text{ (or 4)} \right\}$

  Whole numbers: $\left\{ 0, \sqrt{16} \right\}$

  Integers: $\left\{ -6, 0, \sqrt{16} \right\}$

  Rational numbers: $\left\{ -6, -\frac{2}{7}, 0, 0.\overline{3} \left(\text{or } \frac{1}{3} \right), \sqrt{16} \right\}$

  Irrational numbers: $\left\{ \sqrt{5}, 5\pi \right\}$

  Real numbers: $\left\{ -6, -\frac{2}{7}, 0, 0.\overline{3}, \sqrt{5}, \sqrt{16}, 5\pi \right\}$

Additional Example to Review

Example 1, page 58

Properties of the Real Numbers

Name	Meaning
Closure Property of Addition	The sum of any two real numbers is a real number.
Closure Property of Multiplication	The product of any two real numbers is a real number.
Commutative Property of Addition	Changing order when adding does not affect the sum. $a + b = b + a$
Commutative Property of Multiplication	Changing order when multiplying does not affect the product. $ab = ba$
Associative Property of Addition	Changing grouping when adding does not affect the sum. $(a + b) + c = a + (b + c)$
Associative Property of Multiplication	Changing grouping when multiplying does not affect the product. $(ab)c = a(bc)$
Distributive Property of Multiplication over Addition	Multiplication distributes over addition. $a \cdot (b + c) = a \cdot b + a \cdot c$
Identity Property of Addition	Zero can be deleted from a sum. $a + 0 = a$ $0 + a = a$
Identity Property of Multiplication	One can be deleted from a product. $a \cdot 1 = a$ $1 \cdot a = a$
Inverse Property of Addition	The sum of a real number and its additive inverse gives 0, the additive identity. $a + (-a) = 0$ $(-a) + a = 0$
Inverse Property of Multiplication	The product of a nonzero real number and its multiplicative inverse gives 1, the multiplicative identity. $a \cdot \dfrac{1}{a} = 1, a \neq 0$ $\dfrac{1}{a} \cdot a = 1, a \neq 0$

- $7(4 \cdot 3) = 7(3 \cdot 4)$
  Commutative property of multiplication
- $7(4 + 3) = 7 \cdot 4 + 7 \cdot 3$ Distributive property
- $(7 + 4) + 3 = 7 + (4 + 3)$
  Associative property of addition
- $7 \cdot \frac{1}{7} = 1$ Inverse property of multiplication

Additional Examples to Review

Example 2, page 60; Example 3, page 60

Section 1.6 Exponents and Scientific Notation

Properties of Exponents

- Product rule: $b^m \cdot b^n = b^{m+n}$
- Power rule: $\left(b^m\right)^n = b^{m \cdot n}$
- Quotient rule: $\dfrac{b^m}{b^n} = b^{m-n}, b \neq 0$
- Zero exponent rule: $b^0 = 1, b \neq 0$
- Negative exponent rule: $b^{-m} = \dfrac{1}{b^m}, b \neq 0$

- $2^3 \cdot 2^5 = 2^{3+5} = 2^8 = 2 \cdot 2 \cdot 2 \cdot 2 \cdot 2 \cdot 2 \cdot 2 \cdot 2 = 256$
- $\dfrac{5^6}{5^3} = 5^{6-3} = 5^3 = 5 \cdot 5 \cdot 5 = 125$
- $9^{-2} = \dfrac{1}{9^2} = \dfrac{1}{9 \cdot 9} = \dfrac{1}{81}$
- $4^0 + 4^{-3} = 1 + \dfrac{1}{4^3} = 1 + \dfrac{1}{4 \cdot 4 \cdot 4}$

$$= 1 + \dfrac{1}{64} = \dfrac{64}{64} + \dfrac{1}{64} = \dfrac{65}{64}$$

Additional Examples to Review

Table 1.4, page 65; Example 1, page 65; Example 2, page 66

A positive number in scientific notation is expressed as $a \times 10^n$, where $1 \leq a < 10$ and n is an integer.

Changing from Scientific to Decimal Notation: If n is positive, move the decimal point in a to the right n places. If n is negative, move the decimal point in a to the left $|n|$ places.

- $3.4 \times 10^5 = 340{,}000$

  $n = 5$

  Move the decimal point 5 places to the right.

- $6.1 \times 10^{-3} = 0.0061$

  $n = -3$

  Move the decimal point $|-3| = 3$ places to the left.

Additional Example to Review

Example 3, page 68

Changing from Decimal to Scientific Notation: Move the decimal point in the given number to obtain a, where $1 \leq a < 10$. The number of places the decimal point moves gives the absolute value of n in $a \times 10^n$; n is positive if the number is greater than 10 and negative if the number is less than 1.

- $2800 = 2.8 \times 10^3$ The decimal point moved 3 places from 2800 to 2.8.
- $0.000019 = 1.9 \times 10^{-5}$ The decimal point moved 5 places from 0.000019 to 1.9.

Additional Examples to Review

Example 4, page 68; Example 5, page 69

The product and quotient rules for exponents are used to multiply and divide numbers in scientific notation. If a number is written in scientific notation, $a \times 10^n$, the digits in a are called significant digits. If rounding is necessary, round the scientific notation answer to the least number of significant digits found in any of the given numbers.

- Write the product in scientific notation and decimal notation.

$$(8 \times 10^9)(6 \times 10^{-4}) = (8 \times 6) \times (10^9 \times 10^{-4}) = 48 \times 10^{9+(-4)} = 48 \times 10^5$$

$$= (4.8 \times 10) \times 10^5 = 4.8 \times 10^6 = 4{,}800{,}000 \qquad \boxed{\text{Not in scientific notation}}$$

  Scientific notation   Decimal notation

- Use scientific notation to divide and write the answer in scientific notation.

$$\frac{50{,}000{,}000}{0.0025} = \frac{5 \times 10^7}{2.5 \times 10^{-3}} = \left(\frac{5}{2.5}\right) \times \left(\frac{10^7}{10^{-3}}\right) = 2 \times 10^{7-(-3)} = 2 \times 10^{7+3} = 2 \times 10^{10}$$

Additional Examples to Review

Example 6, page 70; Example 7, page 70; Example 8, page 70; Example 9, page 71

Section 1.7 Arithmetic and Geometric Sequences

In an arithmetic sequence, each term after the first differs from the preceding term by a constant, the common difference. Subtract any term from the term that directly follows it to find the common difference.

- Write the first six terms of the arithmetic sequence with first term 6 and common difference -2.

$$6, 6 + (-2) = 4, 4 + (-2) = 2, 2 + (-2) = 0,$$
$$0 + (-2) = -2, -2 + (-2) = -4$$

The first six terms are $6, 4, 2, 0, -2,$ and -4.

Additional Examples to Review

Example 1, page 77; Example 2, page 77

The general term, or the nth term, of an arithmetic sequence is

$$a_n = a_1 + (n - 1)d,$$

where a_1 is the first term and d is the common difference.

- Find the eighth term of the arithmetic sequence whose first term is 5 and whose common difference is -4.
  Given: $a_1 = 5, d = -4$; Find: a_8.
  Use $a_n = a_1 + (n - 1)d$, with $n = 8$.

$$\begin{aligned} a_8 &= 5 + (8 - 1)(-4) \\ &= 5 + 7(-4) \\ &= 5 + (-28) \\ &= -23 \end{aligned}$$

Additional Examples to Review

Example 3, page 78; Example 4, page 78

In a geometric sequence, each term after the first is obtained by multiplying the preceding term by a nonzero constant, the common ratio. Divide any term after the first by the term that directly precedes it to find the common ratio.

- Write the first five terms of the geometric sequence with first term 20 and common ratio $-\frac{1}{2}$.

$$20, \quad 20\left(-\frac{1}{2}\right) = -10, \quad -10\left(-\frac{1}{2}\right) = 5,$$

$$5\left(-\frac{1}{2}\right) = -\frac{5}{2}, \quad -\frac{5}{2}\left(-\frac{1}{2}\right) = \frac{5}{4}$$

The first five terms are $20, -10, 5, -\frac{5}{2},$ and $\frac{5}{4}$.

Additional Example to Review

Example 5, page 80

The general term, or the nth term, of a geometric sequence is

$$a_n = a_1 r^{n-1},$$

where a_1 is the first term and r is the common ratio.

- Find the sixth term of the geometric sequence whose first term is -10 and whose common ratio is 2.
  Given: $a_1 = -10, r = 2$; Find: a_6.
  Use $a_n = a_1 r^{n-1}$ with $n = 6$.

$$a_6 = -10 \cdot 2^{6-1} = -10 \cdot 2^5 = -10(32) = -320$$

Additional Examples to Review

Example 6, page 81; Example 7, page 81

Review Exercises

Section 1.1 Number Theory: Prime and Composite Numbers

In Exercises 1 and 2, determine whether the number is divisible by each of the following numbers: 2, 3, 4, 5, 6, 8, 9, 10, and 12. If you are using a calculator, explain the divisibility shown by your calculator using one of the rules of divisibility.

1. 238,632   divisible by 2, 3, 4, 6, 8, and 12
2. 421,153,470   divisible by 2, 3, 5, 6, 9, and 10

In Exercises 3–5, find the prime factorization of each composite number.

3. 705 $3 \cdot 5 \cdot 47$
4. 960 $2^6 \cdot 3 \cdot 5$
5. 6825 $3 \cdot 5^2 \cdot 7 \cdot 13$

In Exercises 6–8, find the greatest common divisor and the least common multiple of the numbers.

6. 30 and 48   GCD: 6; LCM: 240
7. 36 and 150   GCD: 6; LCM: 900
8. 216 and 254   GCD: 2; LCM: 27,432

9. For an intramural league, you need to divide 24 men and 60 women into all-male and all-female teams so that each team has the same number of people. What is the largest number of people that can be placed on a team?   12

10. The library at a college runs videotapes of two study skill programs continuously. One videotape runs for 42 minutes and a second runs for 56 minutes. Both videotapes begin at 9:00 A.M. When will the videos of the two programs begin again at the same time?   11:48 A.M.

Section 1.2 The Integers; Order of Operations

In Exercises 11–12, insert either $<$ or $>$ in the shaded area between the integers to make the statement true.

11. $-93 \blacksquare 17$ $<$
12. $-2 \blacksquare -200$ $>$

In Exercises 13–15, find the absolute value.

13. $|-860|$ 860
14. $|53|$ 53
15. $|0|$ 0

Perform the indicated operations in Exercises 16–28.

16. $8 + (-11)$ -3
17. $-6 + (-5)$ -11
18. $-7 - 8$ -15
19. $-7 - (-8)$ 1
20. $(-9)(-11)$ 99
21. $5(-3)$ -15
22. $\frac{-36}{-4}$ 9
23. $\frac{20}{-5}$ -4
24. $-40 \div 5 \cdot 2$ -16
25. $-6 + (-2) \cdot 5$ -16
26. $6 - 4(-3 + 2)$ 10
27. $28 \div (2 - 4^2)$ -2
28. $36 - 24 \div 4 \cdot 3 - 1$ 17

29. For the year 2015, the Congressional Budget Office projects a budget deficit of −$57 billion. For the same year, the Brookings Institution forecasts a budget deficit of −$715 billion. What is the difference between the CBO projection and the Brookings projection? $658 billion

Section 1.3 The Rational Numbers

In Exercises 30–32, reduce each rational number to its lowest terms.

30. $\frac{40}{75}$ $\frac{8}{15}$
31. $\frac{36}{150}$ $\frac{6}{25}$
32. $\frac{165}{180}$ $\frac{11}{12}$

In Exercises 33–34, convert each mixed number to an improper fraction.

33. $5\frac{9}{11}$ $\frac{64}{11}$
34. $-3\frac{2}{7}$ $-\frac{23}{7}$

In Exercises 35–36, convert each improper fraction to a mixed number.

35. $\frac{27}{5}$ $5\frac{2}{5}$
36. $-\frac{17}{9}$ $-1\frac{8}{9}$

In Exercises 37–40, express each rational number as a decimal.

37. $\frac{4}{5}$ 0.8
38. $\frac{3}{7}$ $0.\overline{428571}$
39. $\frac{5}{8}$ 0.625
40. $\frac{9}{16}$ 0.5625

In Exercises 41–44, express each terminating decimal as a quotient of integers in lowest terms.

41. 0.6 $\frac{3}{5}$
42. 0.68 $\frac{17}{25}$
43. 0.588 $\frac{147}{250}$
44. 0.0084 $\frac{21}{2500}$

In Exercises 45–47, express each repeating decimal as a quotient of integers in lowest terms.

45. $0.\overline{5}$ $\frac{5}{9}$
46. $0.\overline{34}$ $\frac{34}{99}$
47. $0.\overline{113}$ $\frac{113}{999}$

In Exercises 48–58, perform the indicated operations. Where possible, reduce the answer to lowest terms.

48. $\frac{3}{5} \cdot \frac{7}{10}$ $\frac{21}{50}$
49. $\left(3\frac{1}{3}\right)\left(1\frac{3}{4}\right)$ $\frac{35}{6}$
50. $\frac{4}{5} \div \frac{3}{10}$ $\frac{8}{3}$
51. $-1\frac{2}{3} \div 6\frac{2}{3}$ $-\frac{1}{4}$
52. $\frac{2}{9} + \frac{4}{9}$ $\frac{2}{3}$
53. $\frac{7}{9} + \frac{5}{12}$ $\frac{43}{36}$
54. $\frac{3}{4} - \frac{2}{15}$ $\frac{37}{60}$
55. $\frac{1}{3} + \frac{1}{2} \cdot \frac{4}{5}$ $\frac{11}{15}$
56. $\frac{3}{8}\left(\frac{1}{2} + \frac{1}{3}\right)$ $\frac{5}{16}$
57. $\frac{1}{2} - \frac{2}{3} \div \frac{5}{9} + \frac{3}{10}$ $-\frac{2}{5}$
58. $\left(\frac{1}{2} + \frac{1}{3}\right) \div \left(\frac{1}{4} - \frac{3}{8}\right)$ $-\frac{20}{3}$ or $-6\frac{2}{3}$

In Exercises 59–60, find the rational number halfway between the two numbers in each pair.

59. $\frac{1}{7}$ and $\frac{1}{8}$ $\frac{15}{112}$
60. $\frac{3}{4}$ and $\frac{3}{5}$ $\frac{27}{40}$

61. A recipe for chicken enchiladas is meant for six people and requires $4\frac{1}{2}$ pounds of chicken. If you want to serve 15 people, how much chicken is needed? $11\frac{1}{4}$ or about 11 pounds

62. The gas tank of a car is filled to its capacity. The first day, $\frac{1}{4}$ of the tank's gas is used for travel. The second day, $\frac{1}{3}$ of the tank's original amount of gas is used for travel. What fraction of the tank is filled with gas at the end of the second day? $\frac{5}{12}$ of the tank

Section 1.4 The Irrational Numbers

In Exercises 63–66, simplify the square root.

63. $\sqrt{28}$ $2\sqrt{7}$
64. $\sqrt{72}$ $6\sqrt{2}$
65. $\sqrt{150}$ $5\sqrt{6}$
66. $\sqrt{300}$ $10\sqrt{3}$

In Exercises 67–75, perform the indicated operation. Simplify the answer when possible.

67. $\sqrt{6} \cdot \sqrt{8}$ $4\sqrt{3}$
68. $\sqrt{10} \cdot \sqrt{5}$ $5\sqrt{2}$
69. $\frac{\sqrt{24}}{\sqrt{2}}$ $2\sqrt{3}$
70. $\frac{\sqrt{27}}{\sqrt{3}}$ 3

71. $\sqrt{5} + 4\sqrt{5}$ $5\sqrt{5}$

72. $7\sqrt{11} - 13\sqrt{11}$ $-6\sqrt{11}$

73. $\sqrt{50} + \sqrt{8}$ $7\sqrt{2}$

74. $\sqrt{3} - 6\sqrt{27}$ $-17\sqrt{3}$

75. $2\sqrt{18} + 3\sqrt{8}$ $12\sqrt{2}$

In Exercises 76–77, rationalize the denominator.

76. $\dfrac{30}{\sqrt{5}}$ $6\sqrt{5}$

77. $\sqrt{\dfrac{2}{3}}$ $\frac{\sqrt{6}}{3}$

78. Paleontologists use the mathematical model $W = 4\sqrt{2x}$ to estimate the walking speed of a dinosaur, W, in feet per second, where x is the length, in feet, of the dinosaur's leg. What is the walking speed of a dinosaur whose leg length is 6 feet? Express the answer in simplified radical form. Then use your calculator to estimate the walking speed to the nearest tenth of a foot per second. $8\sqrt{3} \approx 13.9$ ft/sec

Section 1.5 Real Numbers and Their Properties

79. Consider the set
$$\left\{-17, -\tfrac{9}{13}, 0, 0.75, \sqrt{2}, \pi, \sqrt{81}\right\}.$$
List all numbers from the set that are **a.** natural numbers, **b.** whole numbers, **c.** integers, **d.** rational numbers, **e.** irrational numbers, **f.** real numbers.   *

80. Give an example of an integer that is not a natural number.   0

81. Give an example of a rational number that is not an integer. $\frac{1}{2}$

82. Give an example of a real number that is not a rational number. $\sqrt{2}$

In Exercises 83–90, state the name of the property illustrated.

83. $3 + 17 = 17 + 3$ commutative property of addition

84. $(6 \cdot 3) \cdot 9 = 6 \cdot (3 \cdot 9)$ associative property of multiplication

85. $\sqrt{3}(\sqrt{5} + \sqrt{3}) = \sqrt{15} + 3$ distributive property

86. $(6 \cdot 9) \cdot 2 = 2 \cdot (6 \cdot 9)$ commutative property of multiplication

87. $\sqrt{3}(\sqrt{5} + \sqrt{3}) = (\sqrt{5} + \sqrt{3})\sqrt{3}$

88. $(3 \cdot 7) + (4 \cdot 7) = (4 \cdot 7) + (3 \cdot 7)$

89. $-3\left(-\dfrac{1}{3}\right) = 1$ inverse property of multiplication

90. $\sqrt{7} \cdot 1 = \sqrt{7}$ identity property of multiplication

In Exercises 91–92, give an example to show that

91. The natural numbers are not closed with respect to division. $2 \div 6 = \frac{1}{3}$

92. The whole numbers are not closed with respect to subtraction. $0 - 2 = -2$ **91–92.** Answers will vary. Examples are given.

Section 1.6 Exponents and Scientific Notation

In Exercises 93–103, evaluate each expression.

93. $6 \cdot 6^2$ 216

94. $2^3 \cdot 2^3$ 64

95. $\left(2^2\right)^2$ 16

96. $\left(3^3\right)^2$ 729

97. $\dfrac{5^6}{5^4}$ 25

98. 7^0 1

99. $(-7)^0$ 1

100. 6^{-3} $\frac{1}{216}$

101. 2^{-4} $\frac{1}{16}$

102. $\dfrac{7^4}{7^6}$ $\frac{1}{49}$

103. $3^5 \cdot 3^{-2}$ 27

In Exercises 104–107, express each number in decimal notation.

104. 4.6×10^2 460

105. 3.74×10^4 37,400

106. 2.55×10^{-3} 0.00255

107. 7.45×10^{-5} 0.0000745

In Exercises 108–113, express each number in scientific notation.

108. 7520 7.52×10^3

109. 3,590,000 3.59×10^6

110. 0.00725 7.25×10^{-3}

111. 0.000000409 4.09×10^{-7}

112. 420×10^{11} 4.2×10^{13}

113. 0.97×10^{-4} 9.7×10^{-5}

In Exercises 114–117, perform the indicated operation and express each answer in decimal notation.

114. $(3 \times 10^7)(1.3 \times 10^{-5})$ 390

115. $(5 \times 10^3)(2.3 \times 10^2)$ 1,150,000

116. $\dfrac{6.9 \times 10^3}{3 \times 10^5}$ 0.023

117. $\dfrac{2.4 \times 10^{-4}}{6 \times 10^{-6}}$ 40

In Exercises 118–121, perform the indicated operation by first expressing each number in scientific notation. Write the answer in scientific notation.

118. $(60,000)(540,000)$ $(6 \times 10^4)(5.4 \times 10^5) = 3.24 \times 10^{10}$

119. $(91,000)(0.0004)$ $(9.1 \times 10^4)(4 \times 10^{-4}) = 3.64 \times 10^1$

120. $\dfrac{8,400,000}{4000}$ $\dfrac{8.4 \times 10^6}{4 \times 10^3} = 2.1 \times 10^3$

121. $\dfrac{0.000003}{0.00000006}$ $\dfrac{3 \times 10^{-6}}{6 \times 10^{-8}} = 5 \times 10^1$

In 2011, the United States government spent more than it had collected in taxes, resulting in a budget deficit of \$1.3 trillion. In Exercises 122–124, you will use scientific notation to put a number like 1.3 trillion in perspective. Use 10^{12} for 1 trillion.

122. Express 1.3 trillion in scientific notation. 1.3×10^{12}

123. There are approximately 32,000,000 seconds in a year. Express this number in scientific notation. 3.2×10^7

124. Use your scientific notation answers from Exercises 122 and 123 to answer this question: How many years is 1.3 trillion seconds? (*Note*: 1.3 trillion seconds would take us back in time to a period when Neanderthals were using stones to make tools.)   40,625 years

125. The human body contains approximately 3.2×10^4 microliters of blood for every pound of body weight. Each microliter of blood contains approximately 5×10^6 red blood cells. Express in scientific notation the approximate number of red blood cells in the body of a 180-pound person. 2.88×10^{13}

*See Answers to Selected Exercises.   **80–82.** Answers will vary. Examples are given.   **87.** commutative property of multiplication   **88.** commutative property of addition

Section 1.7 Arithmetic and Geometric Sequences

In Exercises 126–128, write the first six terms of the arithmetic sequence with the first term, a_1, and common difference, d.

126. $a_1 = 7, d = 4$ $\quad$ 7, 11, 15, 19, 23, and 27

127. $a_1 = -4, d = -5$ $\quad$ $-4, -9, -14, -19, -24,$ and -29

128. $a_1 = \frac{3}{2}, d = -\frac{1}{2}$ $\quad$ $\frac{3}{2}, 1, \frac{1}{2}, 0, -\frac{1}{2},$ and -1

In Exercises 129–131, find the indicated term for the arithmetic sequence with first term, a_1, and common difference, d.

129. Find a_6, when $a_1 = 5, d = 3$. $\quad$ 20

130. Find a_{12}, when $a_1 = -8, d = -2$. $\quad$ -30

131. Find a_{14}, when $a_1 = 14, d = -4$. $\quad$ -38

In Exercises 132–133, write a formula for the general term (the nth term) of each arithmetic sequence. Then use the formula for a_n to find a_{20}, the 20th term of the sequence.

132. $-7, -3, 1, 5, \ldots$ $\quad$ $a_n = -7 + (n-1)4; 69$

133. $a_1 = 200, d = -20$ $\quad$ $a_n = 200 + (n-1)(-20); -180$

In Exercises 134–136, write the first six terms of the geometric sequence with the first term, a_1, and common ratio, r.

134. $a_1 = 3, r = 2$ $\quad$ 3, 6, 12, 24, 48, and 96

135. $a_1 = \frac{1}{2}, r = \frac{1}{2}$ $\quad$ $\frac{1}{2}, \frac{1}{4}, \frac{1}{8}, \frac{1}{16}, \frac{1}{32},$ and $\frac{1}{64}$

136. $a_1 = 16, r = -\frac{1}{2}$ $\quad$ $16, -8, 4, -2, 1,$ and $-\frac{1}{2}$

In Exercises 137–139, find the indicated term for the geometric sequence with first term, a_1, and common ratio, r.

137. Find a_4, when $a_1 = 2, r = 3$. $\quad$ 54

138. Find a_6, when $a_1 = 16, r = \frac{1}{2}$. $\quad$ $\frac{1}{2}$

139. Find a_5, when $a_1 = -3, r = 2$. $\quad$ -48

In Exercises 140–141, write a formula for the general term (the nth term) of each geometric sequence. Then use the formula for a_n to find a_8, the eighth term of the sequence.

140. $1, 2, 4, 8, \ldots$ $\quad$ $a_n = 1(2)^{n-1}; 128$

141. $100, 10, 1, \frac{1}{10}, \ldots$ $\quad$ $a_n = 100\left(\frac{1}{10}\right)^{n-1}; \frac{1}{100,000}$

Determine whether each sequence in Exercises 142–145 is arithmetic or geometric. Then find the next two terms.

142. $4, 9, 14, 19, \ldots$ $\quad$ arithmetic; 24 and 29

143. $2, 6, 18, 54, \ldots$ $\quad$ geometric; 162 and 486

144. $1, \frac{1}{4}, \frac{1}{16}, \frac{1}{64}, \ldots$ $\quad$ geometric; $\frac{1}{256}$ and $\frac{1}{1024}$

145. $0, -7, -14, -21, \ldots$ $\quad$ arithmetic; -28 and -35

146. The bar graph shows the number of hours per week devoted to housework by wives and husbands in 1965 and 2010.

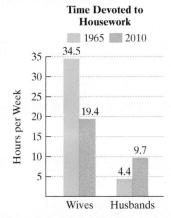

Source: James Henslin, *Sociology*, Eleventh Edition. Pearson, 2012.

In 1965, wives averaged 34.5 hours per week doing housework. On average, this has decreased by approximately 0.3 hour per year since then. **a.** $a_n = 34.8 - 0.3n$

a. Write a formula for the nth term of the arithmetic sequence that describes the number of hours per week devoted to housework by wives n years after 1964.

b. Use the model to project the number of hours per week wives will devote to housework in 2020. $\quad$ 18 hours per week

147. The table shows the population of Florida for 2000 and 2010, with estimates given by the U.S. Census Bureau for 2001 through 2009.

Year	2000	2001	2002	2003	2004	2005
Population in millions	15.98	16.24	16.50	16.76	17.03	17.30

Year	2006	2007	2008	2009	2010
Population in millions	17.58	17.86	18.15	18.44	18.80

a. Divide the population for each year by the population in the preceding year. Round to two decimal places and show that Florida has a population increase that is approximately geometric. $\quad$ ≈ 1.02 for each division

b. Write the general term of the geometric sequence modeling Florida's population, in millions, n years after 1999. $\quad$ $a_n = 15.98(1.02)^{n-1}$

c. Use your the model from part (b) to project Florida's population, in millions, for the year 2030. Round to two decimal places. $\quad$ 28.95 million

Chapter 1 Test

1. Which of the numbers 2, 3, 4, 5, 6, 8, 9, 10, and 12 divide 391,248? 2, 3, 4, 6, 8, 9, and 12

2. Find the prime factorization of 252. $2^2 \cdot 3^2 \cdot 7$

3. Find the greatest common divisor and the least common multiple of 48 and 72. GCD: 24; LCM: 144

Perform the indicated operations in Exercises 4–6.

4. $-6 - (5 - 12)$ 1

5. $(-3)(-4) \div (7 - 10)$ -4

6. $(6 - 8)^2(5 - 7)^3$ -32

7. Express $\frac{7}{12}$ as a decimal. $0.58\overline{3}$

8. Express $0.\overline{64}$ as a quotient of integers in lowest terms. $\frac{64}{99}$

In Exercises 9–11, perform the indicated operations. Where possible, reduce the answer to its lowest terms.

9. $\left(-\frac{3}{7}\right) \div \left(-2\frac{1}{7}\right)$ $\frac{1}{5}$

10. $\frac{19}{24} - \frac{7}{40}$ $\frac{37}{60}$

11. $\frac{1}{2} - 8\left(\frac{1}{4} + 1\right)$ $-\frac{19}{2}$

12. Find the rational number halfway between $\frac{1}{2}$ and $\frac{2}{3}$. $\frac{7}{12}$

13. Multiply and simplify: $\sqrt{10} \cdot \sqrt{5}$. $5\sqrt{2}$

14. Add: $\sqrt{50} + \sqrt{32}$. $9\sqrt{2}$

15. Rationalize the denominator: $\dfrac{6}{\sqrt{2}}$. $3\sqrt{2}$

16. List all the rational numbers in this set:
$$\left\{-7, -\frac{4}{5}, 0, 0.25, \sqrt{3}, \sqrt{4}, \frac{22}{7}, \pi\right\}.$$
$-7, -\frac{4}{5}, 0, 0.25, \sqrt{4},$ and $\frac{22}{7}$

In Exercises 17–18, state the name of the property illustrated.

17. $3(2 + 5) = 3(5 + 2)$ commutative property of addition

18. $6(7 + 4) = 6 \cdot 7 + 6 \cdot 4$ distributive property

In Exercises 19–21, evaluate each expression.

19. $3^3 \cdot 3^2$ 243

20. $\dfrac{4^6}{4^3}$ 64

21. 8^{-2} $\frac{1}{64}$

22. Multiply and express the answer in decimal notation.
$$(3 \times 10^8)(2.5 \times 10^{-5})$$ 7500

23. Divide by first expressing each number in scientific notation. Write the answer in scientific notation.
$$\frac{49,000}{0.007}$$ $\frac{4.9 \times 10^4}{7 \times 10^{-3}} = 7 \times 10^6$

In Exercises 24–26 use 10^6 for one million and 10^9 for one billion to rewrite the number in each statement in scientific notation.

24. The 2009 economic stimulus package allocated \$53.6 billion for grants to states for education. $\$5.36 \times 10^{10}$

25. The population of the United States at the time the economic stimulus package was voted into law was approximately 307 million. 3.07×10^8

26. Use your scientific notation answers from Exercises 24 and 25 to answer this question:

    If the cost for grants to states for education was evenly divided among every individual in the United States, how much would each citizen have to pay? $\approx \$175$

27. Write the first six terms of the arithmetic sequence with first term, a_1, and common difference, d.
$$a_1 = 1, d = -5$$ $1, -4, -9, -14, -19,$ and -24

28. Find a_9, the ninth term of the arithmetic sequence with the first term, a_1, and common difference, d.
$$a_1 = -2, d = 3$$ 22

29. Write the first six terms of the geometric sequence with first term, a_1, and common ratio, r.
$$a_1 = 16, r = \frac{1}{2}$$ $16, 8, 4, 2, 1,$ and $\frac{1}{2}$

30. Find a_7, the seventh term of the geometric sequence with the first term, a_1, and common ratio, r.
$$a_1 = 5, r = 2$$ 320

Algebraic Pathways: Equations and Inequalities

The belief that humor and laughter can have positive effects on our lives is not new.

The Bible tells us, "A merry heart doeth good like a medicine, but a broken spirit drieth the bones." (Proverbs 17:22)

Some random humor factoids:

• The average adult laughs 15 times each day. (Newhouse News Service)

• Forty-six percent of people who are telling a joke laugh more than the people they are telling it to. (*U.S. News and World Report*)

• Algebra can be used to model the influence that humor plays in our responses to negative life events. (Bob Blitzer, *Pathways to College Mathematics*)

That last tidbit that your author threw into the list is true. Based on our sense of humor, there is actually a formula that predicts how we will respond to difficult life events.

Formulas can be used to explain what is happening in the present and to make predictions about what might occur in the future. In this chapter, you will learn to use formulas and mathematical models in new ways that will help you to recognize patterns, logic, and order in a world that can appear chaotic to the untrained eye.

Here's where you'll find this application:

Humor opens Section 2.2, and the advantage of having a sense of humor becomes laughingly evident in the models in Example 6 on page 119.

101

2.1 | Algebraic Expressions and Formulas

What am I supposed to learn?

After you have read this section, you should be able to:

1. Evaluate algebraic expressions.

2. Use mathematical models.

3. Understand the vocabulary of algebraic expressions.

4. Simplify algebraic expressions.

Feeling attractive with a suntan that gives you a "healthy glow"? Think again. Direct sunlight is known to promote skin cancer. Although sunscreens protect you from burning, dermatologists are concerned with the long-term damage that results from the sun even without sunburn.

Algebraic Expressions

Let's see what this child's "healthy glow" has to do with algebra. The biggest difference between arithmetic and algebra is the use of *variables* in algebra. A **variable** is a letter that represents a variety of different numbers. For example, we can let x represent the number of minutes that a person can stay in the sun without burning with no sunscreen. With a number 6 sunscreen, exposure time without burning is six times as long, or 6 times x. This can be written $6 \cdot x$, but it is usually expressed as $6x$. Placing a number and a letter next to one another indicates multiplication.

Notice that $6x$ combines the number 6 and the variable x using the operation of multiplication. A combination of variables and numbers using the operations of addition, subtraction, multiplication, or division, as well as powers or roots, is called an **algebraic expression**. Here are some examples of algebraic expressions:

$$x + 6 \qquad x - 6 \qquad 6x \qquad \frac{x}{6} \qquad 3x + 5 \qquad \sqrt{x} + 7.$$

| The variable x increased by 6 | The variable x decreased by 6 | 6 times the variable x | The variable x divided by 6 | 5 more than 3 times the variable x | 7 more than the square root of the variable x |

Evaluating Algebraic Expressions

1 Evaluate algebraic expressions.

Evaluating an algebraic expression means finding the value of the expression for a given value of the variable. For example, we can evaluate $6x$ (from the sunscreen example) when $x = 15$. We substitute 15 for x. We obtain $6 \cdot 15$, or 90. This means that if you can stay in the sun for 15 minutes without burning when you don't put on any lotion, then with a number 6 lotion, you can "cook" for 90 minutes without burning.

Many algebraic expressions contain more than one operation. Evaluating an algebraic expression correctly involves carefully applying the order of operations agreement that we studied in Chapter 1.

The Order of Operations Agreement

1. Perform operations within the innermost parentheses and work outward. If the algebraic expression involves a fraction, treat the numerator and the denominator as if they were each enclosed in parentheses.

2. Evaluate all exponential expressions.

3. Perform multiplications and divisions as they occur, working from left to right.

4. Perform additions and subtractions as they occur, working from left to right.

Example 1   Evaluating an Algebraic Expression

Evaluate $7 + 5(x - 4)^3$ for $x = 6$.

Solution

$$
\begin{aligned}
7 + 5(x - 4)^3 &= 7 + 5(6 - 4)^3 && \text{Replace } x \text{ with 6.} \\
&= 7 + 5(2)^3 && \text{First work inside parentheses: } 6 - 4 = 2. \\
&= 7 + 5(8) && \text{Evaluate the exponential expression:} \\
&&& 2^3 = 2 \cdot 2 \cdot 2 = 8. \\
&= 7 + 40 && \text{Multiply: } 5(8) = 40. \\
&= 47 && \text{Add.}
\end{aligned}
$$

 **Check Point 1**   Evaluate $8 + 6(x - 3)^2$ for $x = 13$.   608

Example 2   Evaluating an Algebraic Expression

Evaluate $x^2 + 5x - 3$ for $x = -6$.

Solution

We substitute -6 for each of the two occurrences of x. Then we use the order of operations to evaluate the algebraic expression.

$$
\begin{aligned}
x^2 + 5x - 3 && \text{This is the given algebraic expression.} \\
= (-6)^2 + 5(-6) - 3 && \text{Substitute } -6 \text{ for each } x. \\
= 36 + 5(-6) - 3 && \text{Evaluate the exponential expression:} \\
&& (-6)^2 = (-6)(-6) = 36. \\
= 36 + (-30) - 3 && \text{Multiply: } 5(-6) = -30. \\
= 6 - 3 && \text{Add and subtract from left to right.} \\
&& \text{First add: } 36 + (-30) = 6. \\
= 3 && \text{Subtract.}
\end{aligned}
$$

Check Point 2   Evaluate $x^2 + 4x - 7$ for $x = -5$.   -2

Great Question!

Is there a difference between evaluating x^2 for $x = -6$ and evaluating $-x^2$ for $x = 6$?

Yes. Notice the difference between these evaluations:

- x^2 for $x = -6$

$$x^2 = (-6)^2$$
$$= (-6)(-6) = 36$$

- $-x^2$ for $x = 6$

$$-x^2 = -6^2 = -6 \cdot 6 = -36$$

The negative is not inside parentheses and is not taken to the second power.

Work carefully when evaluating algebraic expressions with exponents and negatives.

Example 3   Evaluating an Algebraic Expression

Evaluate $-2x^2 + 5xy - y^3$ for $x = 4$ and $y = -2$.

Solution

We substitute 4 for each x and -2 for each y. Then we use the order of operations to evaluate the algebraic expression.

$$
\begin{aligned}
-2x^2 + 5xy - y^3 && \text{This is the given algebraic expression.} \\
= -2 \cdot 4^2 + 5 \cdot 4(-2) - (-2)^3 && \text{Substitute 4 for } x \text{ and } -2 \text{ for } y. \\
= -2 \cdot 16 + 5 \cdot 4(-2) - (-8) && \text{Evaluate the exponential expressions:} \\
&& 4^2 = 4 \cdot 4 = 16 \text{ and} \\
&& (-2)^3 = (-2)(-2)(-2) = -8. \\
= -32 + (-40) - (-8) && \text{Multiply: } -2 \cdot 16 = -32 \text{ and} \\
&& 5(4)(-2) = 20(-2) = -40. \\
= -72 - (-8) && \text{Add and subtract from left to right. First add:} \\
&& -32 + (-40) = -72. \\
= -64 && \text{Subtract: } -72 - (-8) = -72 + 8 = -64.
\end{aligned}
$$

Check Point 3   Evaluate $-3x^2 + 4xy - y^3$ for $x = 5$ and $y = -1$.   -94

2 Use mathematical models.

Formulas and Mathematical Models

An **equation** is formed when an equal sign is placed between two algebraic expressions. One aim of algebra is to provide a compact, symbolic description of the world. These descriptions involve the use of *formulas*. A **formula** is an equation that uses variables to express a relationship between two or more quantities.

Here are two examples of formulas related to heart rate and exercise.

Couch-Potato Exercise

$$H = \frac{1}{5}(220 - a)$$

| Heart rate, in beats per minute, | is | $\frac{1}{5}$ of | the difference between 220 and your age. |

Working It

$$H = \frac{9}{10}(220 - a)$$

| Heart rate, in beats per minute, | is | $\frac{9}{10}$ of | the difference between 220 and your age. |

The process of finding formulas to describe real-world phenomena is called **mathematical modeling**. Such formulas, together with the meaning assigned to the variables, are called **mathematical models**. We often say that these formulas model, or describe, the relationships among the variables.

Example 4   **Modeling Caloric Needs**

The bar graph in **Figure 2.1** shows the estimated number of calories per day needed to maintain energy balance for various gender and age groups for moderately active lifestyles. (Moderately active means a lifestyle that includes physical activity equivalent to walking 1.5 to 3 miles per day at 3 to 4 miles per hour, in addition to the light physical activity associated with typical day-to-day life.)

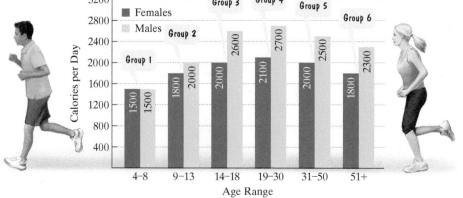

Calories Needed to Maintain Energy Balance for Moderately Active Lifestyles

- Females
- Males

Group 1: Females 1500, Males 1500
Group 2 (9–13): Females 1800, Males 2000
Group 3 (14–18): Females 2000, Males 2600
Group 4 (19–30): Females 2100, Males 2700
Group 5 (31–50): Females 2000, Males 2500
Group 6 (51+): Females 1800, Males 2300

Age Range: 4–8, 9–13, 14–18, 19–30, 31–50, 51+

Calories per Day

Figure 2.1
Source: U.S.D.A.

The mathematical model

$$F = -66x^2 + 526x + 1030$$

describes the number of calories needed per day, F, by females in age group x with moderately active lifestyles. According to the model, how many calories per day are needed by females between the ages of 19 and 30, inclusive, with this lifestyle? Does this underestimate or overestimate the number shown by the graph in **Figure 2.1**? By how much?

Solution

Because the 19–30 age range is designated as group 4, we substitute 4 for x in the given model. Then we use the order of operations to find F, the number of calories needed per day by females between the ages of 19 and 30.

$F = -66x^2 + 526x + 1030$	This is the given mathematical model.
$F = -66 \cdot 4^2 + 526 \cdot 4 + 1030$	Replace each occurrence of x with 4.
$F = -66 \cdot 16 + 526 \cdot 4 + 1030$	Evaluate the exponential expression: $4^2 = 4 \cdot 4 = 16$.
$F = -1056 + 2104 + 1030$	Multiply from left to right: $-66 \cdot 16 = -1056$ and $526 \cdot 4 = 2104$.
$F = 2078$	Add.

The formula indicates that females in the 19–30 age range with moderately active lifestyles need 2078 calories per day. **Figure 2.1** indicates that 2100 calories are needed. Thus, the mathematical model underestimates caloric needs by $2100 - 2078$ calories, or by 22 calories per day.

 **Check Point 4**   The mathematical model

$$M = -120x^2 + 998x + 590$$

describes the number of calories needed per day, M, by males in age group x with moderately active lifestyles. According to the model, how many calories per day are needed by males between the ages of 19 and 30, inclusive, with this lifestyle? Does this underestimate or overestimate the number shown by the graph in **Figure 2.1**? By how much?   2662 calories; underestimates by 38 calories

The Vocabulary of Algebraic Expressions

 Understand the vocabulary of algebraic expressions.

We have seen that an algebraic expression combines numbers and variables. Here is another example of an algebraic expression:

$$7x - 9y - 3.$$

The **terms** of an algebraic expression are those parts that are separated by addition. For example, we can rewrite $7x - 9y - 3$ as

$$7x + (-9y) + (-3).$$

This expression contains three terms, namely $7x$, $-9y$, and -3.

The numerical part of a term is called its **coefficient**. In the term $7x$, the 7 is the coefficient. In the term $-9y$, the -9 is the coefficient.

Coefficients of 1 and -1 are not written. Thus, the coefficient of x, meaning $1x$, is 1. Similarly, the coefficient of $-y$, meaning $-1y$, is -1.

A term that consists of just a number is called a **numerical term** or a **constant**. The numerical term of $7x - 9y - 3$ is -3.

The parts of each term that are multiplied are called the **factors** of the term. The factors of the term $7x$ are 7 and x.

Like terms are terms that have the same variable factors. For example, $3x$ and $7x$ are like terms.

4 Simplify algebraic expressions.

Simplifying Algebraic Expressions

The properties of real numbers that we discussed in Chapter 1 can be applied to algebraic expressions.

Properties of Real Numbers	
Property	**Example**
Commutative Property of Addition $a + b = b + a$	$13x^2 + 7x = 7x + 13x^2$
Commutative Property of Multiplication $ab = ba$	$x \cdot 6 = 6x$
Associative Property of Addition $(a + b) + c = a + (b + c)$	$3 + (8 + x) = (3 + 8) + x = 11 + x$
Associative Property of Multiplication $(ab)c = a(bc)$	$-2(3x) = (-2 \cdot 3)x = -6x$
Distributive Property $a(b + c) = ab + ac$	$5(3x + 7) = 5 \cdot 3x + 5 \cdot 7 = 15x + 35$
$a(b - c) = ab - ac$	$4(2x - 5) = 4 \cdot 2x - 4 \cdot 5 = 8x - 20$

The distributive property in the form
$$ba + ca = (b + c)a$$
enables us to add or subtract like terms. For example,
$$3x + 7x = (3 + 7)x = 10x$$
$$7y^2 - y^2 = 7y^2 - 1y^2 = (7 - 1)y^2 = 6y^2.$$
This process is called **combining like terms**.

An algebraic expression is **simplified** when parentheses have been removed and like terms have been combined.

Great Question!

Do I have to use the distributive property to combine like terms? Can't I just do it in my head?

Yes, you can combine like terms mentally. Add or subtract the coefficients of the terms. Use this result as the coefficient of the terms' variable factor(s).

Example 5   Simplifying an Algebraic Expression

Simplify: $5(3x - 7) - 6x$.

Solution

$5(3x - 7) - 6x$

$= 5 \cdot 3x - 5 \cdot 7 - 6x$ Use the distributive property to remove the parentheses.

$= 15x - 35 - 6x$ Multiply.

$= (15x - 6x) - 35$ Group like terms.

$= 9x - 35$ Combine like terms: $15x - 6x = (15 - 6)x = 9x$.

▶ **Check Point 5**   Simplify: $7(2x - 3) - 11x$. $3x - 21$

Example 6 Simplifying an Algebraic Expression

Simplify: $6(2x^2 + 4x) + 10(4x^2 + 3x)$.

Solution

52x^2 and 54x are not like terms. They contain different variable factors, x^2 and x, and cannot be combined.

$$6(2x^2 + 4x) + 10(4x^2 + 3x)$$

$$= 6 \cdot 2x^2 + 6 \cdot 4x + 10 \cdot 4x^2 + 10 \cdot 3x \quad \text{Use the distributive property to remove the parentheses.}$$

$$= 12x^2 + 24x + 40x^2 + 30x \quad \text{Multiply.}$$

$$= (12x^2 + 40x^2) + (24x + 30x) \quad \text{Group like terms.}$$

$$= 52x^2 + 54x \quad \text{Combine like terms:}$$
$$12x^2 + 40x^2 = (12 + 40)x^2 = 52x^2$$
$$\text{and } 24x + 30x = (24 + 30)x = 54x.$$

✓ **Check Point 6** Simplify: $7(4x^2 + 3x) + 2(5x^2 + x)$. $38x^2 + 23x$

It is not uncommon to see algebraic expressions with parentheses preceded by a negative sign or subtraction. An expression of the form $-(a + b)$ can be simplified as follows:

$$-(a + b) = -1(a + b) = (-1)a + (-1)b = -a + (-b) = -a - b.$$

Do you see a fast way to obtain the simplified expression on the right? **If a negative sign or a subtraction symbol appears outside parentheses, drop the parentheses and change the sign of every term within the parentheses.** For example,

$$-(3x^2 - 7x - 4) = -3x^2 + 7x + 4.$$

Example 7 Simplifying an Algebraic Expression

Simplify: $8x + 2[5 - (x - 3)]$.

Solution

$$8x + 2[5 - (x - 3)]$$

$$= 8x + 2[5 - x + 3] \quad \text{Drop parentheses and change the sign of each term in parentheses: } -(x - 3) = -x + 3.$$

$$= 8x + 2[8 - x] \quad \text{Simplify inside brackets: } 5 + 3 = 8.$$

$$= 8x + 16 - 2x \quad \text{Apply the distributive property:}$$

$$2[8 - x] = 2 \cdot 8 - 2x = 16 - 2x.$$

$$= (8x - 2x) + 16 \quad \text{Group like terms.}$$

$$= 6x + 16 \quad \text{Combine like terms: } 8x - 2x = (8 - 2)x = 6x.$$

✓ **Check Point 7** Simplify: $6x + 4[7 - (x - 2)]$. $2x + 36$

Achieving Success

Algebra is cumulative. This means that the topics build on one another. Understanding each topic depends on understanding the previous material. Do not let yourself fall behind.

Blitzer Bonus
● ● ● ● ● ● ● ● ● ● ● ● ● ●

Using Algebra to Measure Blood-Alcohol Concentration

The amount of alcohol in a person's blood is known as blood-alcohol concentration (BAC), measured in grams of alcohol per deciliter of blood. A BAC of 0.08, meaning 0.08%, indicates that a person has 8 parts alcohol per 10,000 parts blood. In every state in the United States, it is illegal to drive with a BAC of 0.08 or higher.

How Do I Measure My Blood-Alcohol Concentration?

Here's a formula that models BAC for a person who weighs w pounds and who has n drinks* per hour.

$$\text{BAC} = \frac{600n}{w(0.6n + 169)}$$

Number of drinks consumed in an hour

Blood-alcohol concentration

Body weight, in pounds

* A drink can be a 12-ounce can of beer, a 5-ounce glass of wine, or a 1.5-ounce shot of liquor. Each contains approximately 14 grams, or $\frac{1}{2}$ ounce, of alcohol.

Blood-alcohol concentration can be used to quantify the meaning of "tipsy."

BAC	Effects on Behavior
0.05	Feeling of well-being; mild release of inhibitions; absence of observable effects
0.08	Feeling of relaxation; mild sedation; exaggeration of emotions and behavior; slight impairment of motor skills; increase in reaction time
0.12	Muscle control and speech impaired; difficulty performing motor skills; uncoordinated behavior
0.15	Euphoria; major impairment of physical and mental functions; irresponsible behavior; some difficulty standing, walking, and talking
0.35	Surgical anesthesia; lethal dosage for a small percentage of people
0.40	Lethal dosage for 50% of people; severe circulatory and respiratory depression; alcohol poisoning/overdose

Source: National Clearinghouse for Alcohol and Drug Information

Keeping in mind the meaning of "tipsy," we can use our model to compare blood-alcohol concentrations of a 120-pound person and a 200-pound person for various numbers of drinks.

We determined each BAC using a calculator, rounding to three decimal places.

Blood-Alcohol Concentrations of a 120-Pound Person

$$\text{BAC} = \frac{600n}{120(0.6n + 169)}$$

n (number of drinks per hour)	1	2	3	4	5	6	7	8	9	10
BAC (blood-alcohol concentration)	0.029	0.059	0.088	0.117	0.145	0.174	0.202	0.230	0.258	0.286

Illegal to drive

Blood-Alcohol Concentrations of a 200-Pound Person

$$\text{BAC} = \frac{600n}{200(0.6n + 169)}$$

n (number of drinks per hour)	1	2	3	4	5	6	7	8	9	10
BAC (blood-alcohol concentration)	0.018	0.035	0.053	0.070	0.087	0.104	0.121	0.138	0.155	0.171

Illegal to drive

Like all mathematical models, the formula for BAC gives approximate rather than exact values. There are other variables that influence blood-alcohol concentration that are not contained in the model. These include the rate at which an individual's body processes alcohol, how quickly one drinks, sex, age, physical condition, and the amount of food eaten prior to drinking.

Concept and Vocabulary Check

Exercises in the Concept and Vocabulary Check are intended for group and class discussions.

In Exercises 1–6, fill in each blank so that the resulting statement is true.

1. Finding the value of an algebraic expression for a given value of the variable is called _____evaluating_____ the expression.

2. When an equal sign is placed between two algebraic expressions, an _____equation_____ is formed.

3. The parts of an algebraic expression that are separated by addition are called the _____terms_____ of the expression.

4. In the algebraic expression $7x$, 7 is called the _____coefficient_____ because it is the numerical part.

5. In the algebraic expression $7x$, 7 and x are called _____factors_____ because they are multiplied together.

6. The algebraic expressions $3x$ and $7x$ are called _____like terms_____ because they contain the same variable to the same power.

In Exercises 7–14, determine whether each statement is true or false. If the statement is false, make the necessary change(s) to produce a true statement.   Changes to false statements will vary.

7. The term x has no coefficient.   false

8. $5 + 3(x - 4) = 8(x - 4) = 8x - 32$ false

9. $-x - x = -x + (-x) = 0$ false

10. $x - 0.02(x + 200) = 0.98x - 4$ true

11. $3 + 7x = 10x$ false

12. $b \cdot b = 2b$ false

13. $(3y - 4) - (8y - 1) = -5y - 3$ true

14. $-4y + 4 = -4(y + 4)$ false

Respond to Exercises 15–20 using verbal or written explanations.

15. What is an algebraic expression? Provide an example with your description.   **15–20.** Answers will vary.

16. What does it mean to evaluate an algebraic expression? Provide an example with your description.

17. What is a term? Provide an example with your description.

18. What are like terms? Provide an example with your description.

19. Explain how to add like terms. Give an example.

20. What does it mean to simplify an algebraic expression?

Exercise Set 2.1

Practice Exercises

In Exercises 1–34, evaluate the algebraic expression for the given value or values of the variables.

1. $5x + 7$; $x = 4$ 27
2. $9x + 6$; $x = 5$ 51
3. $-7x - 5$; $x = -4$ 23
4. $-6x - 13$; $x = -3$ 5
5. $x^2 + 4$; $x = 5$ 29
6. $x^2 + 9$; $x = 3$ 18
7. $x^2 - 6$; $x = -2$ -2
8. $x^2 - 11$; $x = -3$ -2
9. $-x^2 + 4$; $x = 5$ -21
10. $-x^2 + 9$; $x = 3$ 0
11. $-x^2 - 6$; $x = -2$ -10
12. $-x^2 - 11$; $x = -3$ -20
13. $x^2 + 4x$; $x = 10$ 140
14. $x^2 + 6x$; $x = 9$ 135
15. $8x^2 + 17$; $x = 5$ 217
16. $7x^2 + 25$; $x = 3$ 88
17. $x^2 - 5x$; $x = -11$ 176
18. $x^2 - 8x$; $x = -5$ 65
19. $x^2 + 5x - 6$; $x = 4$ 30
20. $x^2 + 7x - 4$; $x = 6$ 74
21. $4 + 5(x - 7)^3$; $x = 9$ 44
22. $6 + 5(x - 6)^3$; $x = 8$ 46
23. $x^2 - 3(x - y)$; $x = 2, y = 8$ 22
24. $x^2 - 4(x - y)$; $x = 3, y = 8$ 29
25. $2x^2 - 5x - 6$; $x = -3$ 27
26. $3x^2 - 4x - 9$; $x = -5$ 86
27. $-5x^2 - 4x - 11$; $x = -1$ -12
28. $-6x^2 - 11x - 17$; $x = -2$ -19
29. $3x^2 + 2xy + 5y^2$; $x = 2, y = 3$ 69
30. $4x^2 + 3xy + 2y^2$; $x = 3, y = 2$ 62
31. $-x^2 - 4xy + 3y^3$; $x = -1, y = -2$ -33
32. $-x^2 - 3xy + 4y^3$; $x = -3, y = -1$ -22
33. $\dfrac{2x + 3y}{x + 1}$; $x = -2, y = 4$ -8
34. $\dfrac{2x + y}{xy - 2x}$; $x = -2, y = 4$ 0

The formula

$$C = \frac{5}{9}(F - 32)$$

expresses the relationship between Fahrenheit temperature, F, and Celsius temperature, C. In Exercises 35–36, use the formula to convert the given Fahrenheit temperature to its equivalent temperature on the Celsius scale.

35. 50°F 10°C

36. 86°F 30°C

A football was kicked vertically upward from a height of 4 feet with an initial speed of 60 feet per second. The formula

$$h = 4 + 60t - 16t^2$$

describes the ball's height above the ground, h, in feet, t seconds after it was kicked. Use this formula to solve Exercises 37–38.

37. What was the ball's height 2 seconds after it was kicked? 60 ft

38. What was the ball's height 3 seconds after it was kicked? 40 ft

In Exercises 39–60, simplify each algebraic expression.

39. $7x + 10x$ 17x

40. $5x + 13x$ 18x

41. $5x^2 - 8x^2$ $-3x^2$

42. $7x^2 - 10x^2$ $-3x^2$

43. $3(x + 5)$ 3x + 15

44. $4(x + 6)$ 4x + 24

45. $4(2x - 3)$ 8x − 12

46. $3(4x - 5)$ 12x − 15

47. $5(3x + 4) - 4$ 15x + 16

48. $2(5x + 4) - 3$ 10x + 5

49. $5(3x - 2) + 12x$ 27x − 10

50. $2(5x - 1) + 14x$ 24x − 2

51. $7(3y - 5) + 2(4y + 3)$ 29y − 29

52. $4(2y - 6) + 3(5y + 10)$ 23y + 6

53. $5(3y - 2) - (7y + 2)$ 8y − 12

54. $4(5y - 3) - (6y + 3)$ 14y − 15

55. $3(-4x^2 + 5x) - (5x - 4x^2)$ $-8x^2 + 10x$

56. $2(-5x^2 + 3x) - (3x - 5x^2)$ $-5x^2 + 3x$

57. $7 - 4[3 - (4y - 5)]$ 16y − 25

58. $6 - 5[8 - (2y - 4)]$ 10y − 54

59. $8x - 3[5 - (7 - 6x)]$ −10x + 6

60. $7x - 4[6 - (8 - 5x)]$ −13x + 8

Practice Plus

In Exercises 61–64, simplify each algebraic expression.

61. $18x^2 + 4 - [6(x^2 - 2) + 5]$ $12x^2 + 11$

62. $14x^2 + 5 - [7(x^2 - 2) + 4]$ $7x^2 + 15$

63. $2(3x^2 - 5) - [4(2x^2 - 1) + 3]$ $-2x^2 - 9$

64. $4(6x^2 - 3) - [2(5x^2 - 1) + 1]$ $14x^2 - 11$

Application Exercises

The maximum heart rate, in beats per minute, that you should achieve during exercise is 220 minus your age:

$$220 - a.$$

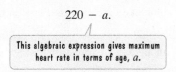

This algebraic expression gives maximum heart rate in terms of age, *a*.

The bar graph shows the target heart rate ranges for four types of exercise goals. The lower and upper limits of these ranges are fractions of the maximum heart rate, 220 − a. Exercises 65–66 are based on the information in the graph.

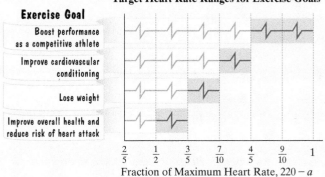

Target Heart Rate Ranges for Exercise Goals

Fraction of Maximum Heart Rate, 220 − *a*

65. If your exercise goal is to improve cardiovascular conditioning, the graph shows the following range for target heart rate, *H*, in beats per minute:

Lower limit of range $H = \frac{7}{10}(220 - a)$

Upper limit of range $H = \frac{4}{5}(220 - a).$

a. What is the lower limit of the heart rate range, in beats per minute, for a 20-year-old with this exercise goal? 140 beats per minute

b. What is the upper limit of the heart rate range, in beats per minute, for a 20-year-old with this exercise goal? 160 beats per minute

66. If your exercise goal is to improve overall health, the graph shows the following range for target heart rate, *H*, in beats per minute:

Lower limit of range $H = \frac{1}{2}(220 - a)$

Upper limit of range $H = \frac{3}{5}(220 - a).$

a. What is the lower limit of the heart rate range, in beats per minute, for a 30-year-old with this exercise goal? 95 beats per minute

b. What is the upper limit of the heart rate range, in beats per minute, for a 30-year-old with this exercise goal? 114 beats per minute

Hello, officer! The bar graph shows the percentage of Americans in various age groups who had contact with a police officer, for anything from an arrest to asking directions, in a recent year.

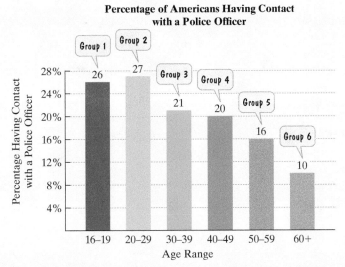

Percentage of Americans Having Contact with a Police Officer

Source: Bureau of Justice Statistics

The mathematical model

$$P = -0.5x^2 + 0.1x + 26.9$$

describes the percentage of Americans, P, in age group x who had contact with a police officer. Use this formula to solve Exercises 67–68.

67. According to the formula, what percentage of Americans ages 60 and older had contact with a police officer? Does this underestimate or overestimate the percentage shown by the graph? By how much?   9.5%; underestimates by 0.5

68. According to the formula, what percentage of Americans between the ages of 20 and 29, inclusive, had contact with a police officer? Does this underestimate or overestimate the percentage shown by the graph? By how much?   25.1%; underestimates by 1.9

***Salary after College.** In 2010, MonsterCollege surveyed 1250 U.S. college students expecting to graduate in the next several years. Respondents were asked the following question:*

> *What do you think your starting salary will be at your first job after college?*

The line graph at the top of the next column shows the percentage of college students who anticipated various starting salaries.

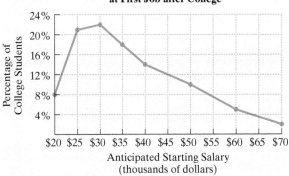

Anticipated Starting Salary at First Job after College

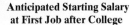
Source: MonsterCollege™

The mathematical model

$$p = -0.01s^2 + 0.8s + 3.7$$

describes the percentage of college students, p, who anticipated a starting salary, s, in thousands of dollars. Use this information to solve Exercises 69–70.   69. b. 18.7%; less than the estimate

69. a. Use the line graph to estimate the percentage of students who anticipated a starting salary of $30 thousand.   22%

   b. Use the formula to find the percentage of students who anticipated a starting salary of $30 thousand. How does this compare with your estimate in part (a)?

70. a. Use the line graph to estimate the percentage of students who anticipated a starting salary of $40 thousand.   14%

   b. Use the formula to find the percentage of students who anticipated a starting salary of $40 thousand. How does this compare with your estimate in part (a)?
   19.7%; exceeds the estimate

Critical Thinking Exercise

71. A business that manufactures small alarm clocks has weekly fixed costs of $5000. The average cost per clock for the business to manufacture x clocks is described by

$$\frac{0.5x + 5000}{x}.$$

a. $50.50, $5.50, and $1.00 per clock, respectively

   a. Find the average cost when $x = 100, 1000,$ and 10,000.

   b. Like all other businesses, the alarm clock manufacturer must make a profit. To do this, each clock must be sold for at least 50¢ more than what it costs to manufacture. Due to competition from a larger company, the clocks can be sold for $1.50 each and no more. Our small manufacturer can only produce 2000 clocks weekly. Does this business have much of a future? Explain.   no; Answers will vary.

2.2 Linear Equations in One Variable

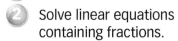

What am I supposed to learn?

After you have read this section, you should be able to:

1. Solve linear equations.

2. Solve linear equations containing fractions.

3. Identify equations with no solution or infinitely many solutions.

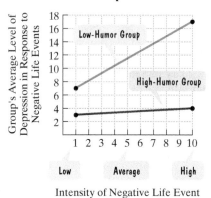

Sense of Humor and Depression

Group's Average Level of Depression in Response to Negative Life Events

Low-Humor Group

High-Humor Group

Intensity of Negative Life Event

Low   Average   High

Figure 2.2

Source: Steven Davis and Joseph Palladino, *Psychology*, Fifth Edition. Prentice Hall, 2007.

The belief that humor and laughter can have positive benefits on our lives is not new. The graphs in **Figure 2.2** indicate that persons with a low sense of humor have higher levels of depression in response to negative life events than those with a high sense of humor. These graphs can be modeled by the following formulas:

Low-Humor Group

$$D = \frac{10}{9}x + \frac{53}{9}$$

High-Humor Group

$$D = \frac{1}{9}x + \frac{26}{9}.$$

In each formula, x represents the intensity of a negative life event (from 1, low, to 10, high) and D is the level of depression in response to that event.

Suppose that the low-humor group averages a level of depression of 10 in response to a negative life event. We can determine the intensity of that event by substituting 10 for D in the low-humor model, $D = \frac{10}{9}x + \frac{53}{9}$:

$$10 = \frac{10}{9}x + \frac{53}{9}.$$

The two sides of an equation can be reversed. So, we can also express this equation as

$$\frac{10}{9}x + \frac{53}{9} = 10.$$

Notice that the highest exponent on the variable is 1. Such an equation is called a *linear equation in one variable*. In this section, we will study how to solve such equations. We return to the models for sense of humor and depression later in the section.

Solving Linear Equations in One Variable

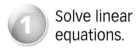

Solve linear equations.

We begin with the general definition of a linear equation in one variable.

Definition of a Linear Equation

A **linear equation in one variable** x is an equation that can be written in the form

$$ax + b = 0,$$

where a and b are real numbers, and $a \neq 0$.

An example of a linear equation in one variable is

$$4x + 12 = 0.$$

Solving an equation in x involves determining all values of x that result in a true statement when substituted into the equation. Such values are **solutions**, or **roots**, of the equation. For example, substitute -3 for x in $4x + 12 = 0$. We obtain

$$4(-3) + 12 = 0, \quad \text{or} \quad -12 + 12 = 0.$$

This simplifies to the true statement $0 = 0$. Thus, -3 is a solution of the equation $4x + 12 = 0$. We also say that -3 **satisfies** the equation $4x + 12 = 0$, because when we substitute -3 for x, a true statement results. The set of all such solutions is called the equation's **solution set**. For example, the solution set of the equation $4x + 12 = 0$ is $\{-3\}$.

Two or more equations that have the same solution set are called **equivalent equations**. For example, the equations

$$4x + 12 = 0 \quad \text{and} \quad 4x = -12 \quad \text{and} \quad x = -3$$

are equivalent equations because the solution set for each is $\{-3\}$. To solve a linear equation in x, we transform the equation into an equivalent equation one or more times. Our final equivalent equation should be of the form

$$x = \text{a number.}$$

The solution set of this equation is the set consisting of the number.

To generate equivalent equations, we will use the following properties:

The Addition and Multiplication Properties of Equality

The Addition Property of Equality

The same real number or algebraic expression may be added to both sides of an equation without changing the equation's solution set.

$$a = b \text{ and } a + c = b + c \text{ are equivalent equations.}$$

The Multiplication Property of Equality

The same nonzero real number may multiply both sides of an equation without changing the equation's solution set.

$$a = b \text{ and } ac = bc \text{ are equivalent equations as long as } c \neq 0.$$

Because subtraction is defined in terms of addition, the addition property also lets us subtract the same number from both sides of an equation without changing the equation's solution set. Similarly, because division is defined in terms of multiplication, the multiplication property of equality can be used to divide both sides of an equation by the same nonzero number to obtain an equivalent equation.

Table 2.1 illustrates how these properties are used to isolate x to obtain an equation of the form $x = $ a number.

Table 2.1 Using Properties of Equality to Solve Equations

	Equation	How to Isolate x	Solving the Equation	The Equation's Solution Set
These equations are solved using the Addition Property of Equality.	$x - 3 = 8$	Add 3 to both sides.	$x - 3 + 3 = 8 + 3$ $x = 11$	$\{11\}$
	$x + 7 = -15$	Subtract 7 from both sides.	$x + 7 - 7 = -15 - 7$ $x = -22$	$\{-22\}$
These equations are solved using the Multiplication Property of Equality.	$6x = 30$	Divide both sides by 6 (or multiply both sides by $\frac{1}{6}$).	$\dfrac{6x}{6} = \dfrac{30}{6}$ $x = 5$	$\{5\}$
	$\dfrac{x}{5} = 9$	Multiply both sides by 5.	$5 \cdot \dfrac{x}{5} = 5 \cdot 9$ $x = 45$	$\{45\}$

Example 1 — Using Properties of Equality to Solve an Equation

Solve and check: $2x + 3 = 17$.

Solution

Our goal is to obtain an equivalent equation with x isolated on one side and a number on the other side.

$2x + 3 = 17$	This is the given equation.
$2x + 3 - 3 = 17 - 3$	Subtract 3 from both sides.
$2x = 14$	Simplify.
$\dfrac{2x}{2} = \dfrac{14}{2}$	Divide both sides by 2.
$x = 7$	Simplify: $\dfrac{2x}{2} = 1x = x$ and $\dfrac{14}{2} = 7$.

Now we check the proposed solution, 7, by replacing x with 7 in the original equation.

$2x + 3 = 17$	This is the original equation.
$2 \cdot 7 + 3 \overset{?}{=} 17$	Substitute 7 for x. The question mark indicates that we do not yet know if the two sides are equal.
$14 + 3 \overset{?}{=} 17$	Multiply: $2 \cdot 7 = 14$.
$17 = 17$ This statement is true.	Add: $14 + 3 = 17$.

Because the check results in a true statement, we conclude that the solution set of the given equation is $\{7\}$.

▶ **Check Point 1**   Solve and check: $4x + 5 = 29$. $\{6\}$

Here is a step-by-step procedure for solving a linear equation in one variable. Not all of these steps are necessary to solve every equation.

> ### Solving a Linear Equation
>
> 1. Simplify the algebraic expression on each side by removing grouping symbols and combining like terms.
> 2. Collect all the variable terms on one side and all the constants, or numerical terms, on the other side.
> 3. Isolate the variable and solve.
> 4. Check the proposed solution in the original equation.

Example 2   Solving a Linear Equation

Solve and check: $2(x - 4) - 5x = -5$.

Solution

Step 1   Simplify the algebraic expression on each side.

$$2(x - 4) - 5x = -5 \qquad \text{This is the given equation.}$$
$$2x - 8 - 5x = -5 \qquad \text{Use the distributive property.}$$
$$-3x - 8 = -5 \qquad \text{Combine like terms: } 2x - 5x = -3x.$$

Step 2   Collect variable terms on one side and constants on the other side. The only variable term in $-3x - 8 = -5$ is $-3x$, and $-3x$ is already on the left side. We will collect constants on the right side by adding 8 to both sides.

$$-3x - 8 + 8 = -5 + 8 \qquad \text{Add 8 to both sides.}$$
$$-3x = 3 \qquad \text{Simplify.}$$

Step 3   Isolate the variable and solve. We isolate the variable, x, by dividing both sides of $-3x = 3$ by -3.

$$\frac{-3x}{-3} = \frac{3}{-3} \qquad \text{Divide both sides by } -3.$$
$$x = -1 \qquad \text{Simplify: } \frac{-3x}{-3} = 1x = x \text{ and } \frac{3}{-3} = -1.$$

Step 4   Check the proposed solution in the original equation. Substitute -1 for x in the original equation.

$$2(x - 4) - 5x = -5 \qquad \text{This is the original equation.}$$
$$2(-1 - 4) - 5(-1) \stackrel{?}{=} -5 \qquad \text{Substitute } -1 \text{ for } x.$$
$$2(-5) - 5(-1) \stackrel{?}{=} -5 \qquad \text{Simplify inside parentheses:}$$
$$\qquad\qquad -1 - 4 = -1 + (-4) = -5.$$
$$-10 - (-5) \stackrel{?}{=} -5 \qquad \text{Multiply: } 2(-5) = -10 \text{ and } 5(-1) = -5.$$
$$\underbrace{-5 = -5}_{\text{This statement is true.}} \qquad -10 - (-5) = -10 + 5 = -5$$

Because the check results in a true statement, we conclude that the solution set of the given equation is $\{-1\}$.

 **Check Point 2**   Solve and check: $6(x - 3) - 10x = -10$. $\{-2\}$

Great Question!

What are the differences between what I'm supposed to do with algebraic expressions and algebraic equations?

We simplify algebraic expressions. We solve algebraic equations. Although basic rules of algebra are used in both procedures, notice the differences between the procedures:

Simplifying an Algebraic Expression

Simplify: $3(x - 7) - (5x - 11)$.

> This is not an equation.
> There is no equal sign.

$$
\begin{aligned}
\text{Solution}\quad & 3(x - 7) - (5x - 11) \\
& = 3x - 21 - 5x + 11 \\
& = (3x - 5x) + (-21 + 11) \\
& = -2x + (-10) \\
& = -2x - 10
\end{aligned}
$$

> Stop! Further simplification is not possible. Avoid the common error of setting $-2x - 10$ equal to 0.

Solving an Algebraic Equation

Solve: $3(x - 7) - (5x - 11) = 14$.

> This is an equation.
> There is an equal sign.

$$
\begin{aligned}
\text{Solution}\quad & 3(x - 7) - (5x - 11) = 14 \\
& 3x - 21 - 5x + 11 = 14 \\
& -2x - 10 = 14
\end{aligned}
$$

Add 10 to both sides. $\quad -2x - 10 + 10 = 14 + 10$

$$-2x = 24$$

Divide both sides by -2. $\quad \dfrac{-2x}{-2} = \dfrac{24}{-2}$

$$x = -12$$

The solution set is $\{-12\}$.

Great Question!

Do I have to solve $5x - 12 = 8x + 24$ by collecting variable terms on the left and numbers on the right?

No. If you prefer, you can solve the equation by collecting variable terms on the right and numbers on the left. To collect variable terms on the right, subtract $5x$ from both sides:

$$5x - 12 - 5x = 8x + 24 - 5x$$
$$-12 = 3x + 24.$$

To collect numbers on the left, subtract 24 from both sides:

$$-12 - 24 = 3x + 24 - 24$$
$$-36 = 3x.$$

Now isolate x by dividing both sides by 3:

$$\frac{-36}{3} = \frac{3x}{3}$$
$$-12 = x.$$

This is the same equation that we obtained in Example 3.

Example 3   Solving a Linear Equation

Solve and check: $5x - 12 = 8x + 24$.

Solution

Step 1   Simplify the algebraic expression on each side. Neither side contains grouping symbols or like terms that can be combined. Therefore, we can skip this step.

Step 2   Collect variable terms on one side and constants on the other side. One way to do this is to collect variable terms on the left and constants on the right. This is accomplished by subtracting $8x$ from both sides and adding 12 to both sides.

$$5x - 12 = 8x + 24 \qquad \text{This is the given equation.}$$

$$5x - 12 - 8x = 8x + 24 - 8x \qquad \text{Subtract } 8x \text{ from both sides.}$$

$$-3x - 12 = 24 \qquad \text{Simplify: } 5x - 8x = -3x.$$

$$-3x - 12 + 12 = 24 + 12 \qquad \text{Add 12 to both sides and collect constants on the right side.}$$

$$-3x = 36 \qquad \text{Simplify.}$$

Step 3   Isolate the variable and solve. We isolate the variable, x, by dividing both sides of $-3x = 36$ by -3.

$$\frac{-3x}{-3} = \frac{36}{-3} \qquad \text{Divide both sides by } -3.$$

$$x = -12 \qquad \text{Simplify.}$$

Step 4 Check the proposed solution in the original equation. Because we isolated the variable and obtained $x = -12$, substitute -12 for x in the original equation.

$$5x - 12 = 8x + 24 \qquad \text{This is the original equation.}$$
$$5(-12) - 12 \overset{?}{=} 8(-12) + 24 \qquad \text{Substitute } -12 \text{ for } x.$$
$$-60 - 12 \overset{?}{=} -96 + 24 \qquad \text{Multiply: } 5(-12) = -60 \text{ and } 8(-12) = -96.$$

This statement is true. $\quad -72 = -72 \qquad$ Add: $-60 + (-12) = -72$ and $-96 + 24 = -72$.

Because the check results in a true statement, we conclude that the solution set of the given equation is $\{-12\}$.

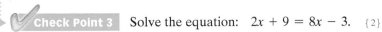 **Check Point 3** Solve the equation: $2x + 9 = 8x - 3.$ $\{2\}$

Example 4 Solving a Linear Equation

Solve and check: $2(x - 3) - 17 = 13 - 3(x + 2)$.

Solution

Step 1 Simplify the algebraic expression on each side.

> Do not begin with $13 - 3$. Multiplication (the distributive property) is applied before subtraction.

$$2(x - 3) - 17 = 13 - 3(x + 2) \qquad \text{This is the given equation.}$$
$$2x - 6 - 17 = 13 - 3x - 6 \qquad \text{Use the distributive property.}$$
$$2x - 23 = -3x + 7 \qquad \text{Combine like terms.}$$

Step 2 Collect variable terms on one side and constants on the other side. We will collect variable terms of $2x - 23 = -3x + 7$ on the left by adding $3x$ to both sides. We will collect the numbers on the right by adding 23 to both sides.

$$2x - 23 + 3x = -3x + 7 + 3x \qquad \text{Add } 3x \text{ to both sides.}$$
$$5x - 23 = 7 \qquad \text{Simplify: } 2x + 3x = 5x.$$
$$5x - 23 + 23 = 7 + 23 \qquad \text{Add 23 to both sides.}$$
$$5x = 30 \qquad \text{Simplify.}$$

Step 3 Isolate the variable and solve. We isolate the variable, x, by dividing both sides of $5x = 30$ by 5.

$$\frac{5x}{5} = \frac{30}{5} \qquad \text{Divide both sides by 5.}$$
$$x = 6 \qquad \text{Simplify.}$$

Step 4 Check the proposed solution in the original equation. Substitute 6 for x in the original equation.

$$2(x - 3) - 17 = 13 - 3(x + 2) \qquad \text{This is the original equation.}$$
$$2(6 - 3) - 17 \overset{?}{=} 13 - 3(6 + 2) \qquad \text{Substitute 6 for } x.$$
$$2(3) - 17 \overset{?}{=} 13 - 3(8) \qquad \text{Simplify inside parentheses.}$$
$$6 - 17 \overset{?}{=} 13 - 24 \qquad \text{Multiply.}$$
$$-11 = -11 \qquad \text{Subtract.}$$

The true statement $-11 = -11$ verifies that the solution set is $\{6\}$.

 **Check Point 4** Solve and check: $4(2x + 1) = 29 + 3(2x - 5).$ $\{5\}$

Solve linear equations containing fractions.

Linear Equations with Fractions

Equations are easier to solve when they do not contain fractions. How do we remove fractions from an equation? We begin by multiplying both sides of the equation by the least common denominator of any fractions in the equation. The least common denominator is the smallest number that all denominators will divide into. Multiplying every term on both sides of the equation by the least common denominator will eliminate the fractions in the equation. Example 5 shows how we "clear an equation of fractions."

Example 5   Solving a Linear Equation Involving Fractions

Solve and check: $\dfrac{3x}{2} = \dfrac{8x}{5} - 4$.

Solution

The denominators are 2 and 5. The smallest number that is divisible by both 2 and 5 is 10. We begin by multiplying both sides of the equation by 10, the least common denominator.

$$\frac{3x}{2} = \frac{8x}{5} - 4 \qquad \text{\small This is the given equation.}$$

$$10 \cdot \frac{3x}{2} = 10\left(\frac{8x}{5} - 4\right) \qquad \text{\small Multiply both sides by 10.}$$

$$10 \cdot \frac{3x}{2} = 10 \cdot \frac{8x}{5} - 10 \cdot 4 \qquad \text{\small Use the distributive property. Be sure to multiply all terms by 10.}$$

$$\overset{5}{10} \cdot \frac{3x}{\underset{1}{2}} = \overset{2}{10} \cdot \frac{8x}{\underset{1}{5}} - 40 \qquad \text{\small Divide out common factors in the multiplications.}$$

$$15x = 16x - 40 \qquad \text{\small Complete the multiplications: } 5 \cdot 3x = 15x \text{ and } 2 \cdot 8x = 16x. \text{ The fractions are now cleared.}$$

At this point, we have an equation similar to those we have previously solved. Collect the variable terms on one side and the constants on the other side.

$$15x - 16x = 16x - 40 - 16x \qquad \text{\small Subtract 16x from both sides to get the variable terms on the left.}$$

$$-x = -40 \qquad \text{\small Simplify.}$$

We're not finished. A negative sign should not precede x.

Isolate x by multiplying or dividing both sides of this equation by -1.

$$\frac{-x}{-1} = \frac{-40}{-1} \qquad \text{\small Divide both sides by } -1.$$

$$x = 40 \qquad \text{\small Simplify.}$$

Check the proposed solution. Substitute 40 for x in the original equation. You should obtain $60 = 60$. This true statement verifies that the solution set is $\{40\}$.

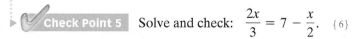

Check Point 5   Solve and check: $\dfrac{2x}{3} = 7 - \dfrac{x}{2}$. $\{6\}$

Example 6 ### An Application: Responding to Negative Life Events

In the section opener, we introduced line graphs, repeated in **Figure 2.2**, indicating that persons with a low sense of humor have higher levels of depression in response to negative life events than those with a high sense of humor. These graphs can be modeled by the following formulas:

Low-Humor Group	High-Humor Group
$D = \dfrac{10}{9}x + \dfrac{53}{9}$	$D = \dfrac{1}{9}x + \dfrac{26}{9}.$

In each formula, x represents the intensity of a negative life event (from 1, low, to 10, high) and D is the average level of depression in response to that event. If the high-humor group averages a level of depression of 3.5, or $\frac{7}{2}$, in response to a negative life event, what is the intensity of that event? How is the solution shown on the red line graph in **Figure 2.2**?

Solution

We are interested in the intensity of a negative life event with an average level of depression of $\frac{7}{2}$ for the high-humor group. We substitute $\frac{7}{2}$ for D in the high-humor model and solve for x, the intensity of the negative life event.

$$D = \frac{1}{9}x + \frac{26}{9}$$ This is the given formula for the high-humor group.

$$\frac{7}{2} = \frac{1}{9}x + \frac{26}{9}$$ Replace D with $\frac{7}{2}$.

$$18 \cdot \frac{7}{2} = 18\left(\frac{1}{9}x + \frac{26}{9}\right)$$ Multiply both sides by 18, the least common denominator.

$$18 \cdot \frac{7}{2} = 18 \cdot \frac{1}{9}x + 18 \cdot \frac{26}{9}$$ Use the distributive property.

$$\overset{9}{\cancel{18}} \cdot \frac{7}{\underset{1}{\cancel{2}}} = \overset{2}{\cancel{18}} \cdot \frac{1}{\underset{1}{\cancel{9}}}x + \overset{2}{\cancel{18}} \cdot \frac{26}{\underset{1}{\cancel{9}}}$$ Divide out common factors in the multiplications.

$$63 = 2x + 52$$ Complete the multiplications. The fractions are now cleared.

$$63 - 52 = 2x + 52 - 52$$ Subtract 52 from both sides to get constants on the left.

$$11 = 2x$$ Simplify.

$$\frac{11}{2} = \frac{2x}{2}$$ Divide both sides by 2.

$$\frac{11}{2} = x$$ Simplify.

The formula indicates that if the high-humor group averages a level of depression of 3.5 in response to a negative life event, the intensity of that event is $\frac{11}{2}$, or 5.5. This is illustrated on the line graph for the high-humor group in **Figure 2.3**.

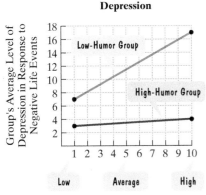

Sense of Humor and Depression

Group's Average Level of Depression in Response to Negative Life Events

Low   Average   High

Intensity of Negative Life Event

Figure 2.2 (repeated)

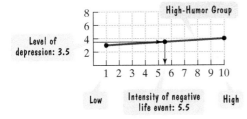

Level of depression: 3.5

High-Humor Group

Low   Intensity of negative life event: 5.5   High

Figure 2.3

$\frac{37}{10}$ or 3.7; If a horizontal line is drawn from 10 on the scale for level of depression until it touches the blue line graph for the low-humor group and then a vertical line is drawn from that point on the blue line graph to the scale for the intensity of the negative life event, the vertical line will touch the scale at 3.7.

Check Point 6 Use the model for the low-humor group given in Example 6 to solve this problem. If the low-humor group averages a level of depression of 10 in response to a negative life event, what is the intensity of that event? How is the solution shown on the blue line graph in **Figure 2.2**?

Identify equations with no solution or infinitely many solutions.

Equations with No Solution or Infinitely Many Solutions

Thus far, each equation that we have solved has had a single solution. However, some equations are not true for even one real number. By contrast, other equations are true for all real numbers.

If you attempt to solve an equation with no solution, you will eliminate the variable and obtain a false statement, such as $2 = 5$. If you attempt to solve an equation that is true for every real number, you will eliminate the variable and obtain a true statement, such as $4 = 4$.

Example 7   Attempting to Solve an Equation with No Solution

Solve: $2x + 6 = 2(x + 4)$.

Solution

$$2x + 6 = 2(x + 4)$$ This is the given equation.
$$2x + 6 = 2x + 8$$ Use the distributive property.
$$2x + 6 - 2x = 2x + 8 - 2x$$ Subtract $2x$ from both sides.
$$6 = 8$$ Simplify.

Keep reading. 6 = 8 is not the solution.

The original equation, $2x + 6 = 2(x + 4)$, is equivalent to the statement $6 = 8$, which is false for every value of x. The equation has no solution. The solution set is $\varnothing$, the empty set.

A Brief Review • The Empty Set

- The **empty set**, also called the **null set**, is the set that contains no elements.
- The empty set is represented by { } or $\varnothing$.

Check Point 7   Solve: $3x + 7 = 3(x + 1)$. $\varnothing$

Example 8   Solving an Equation for Which Every Real Number Is a Solution

Solve: $4x + 6 = 6(x + 1) - 2x$.

Solution

$$4x + 6 = 6(x + 1) - 2x$$ This is the given equation.
$$4x + 6 = 6x + 6 - 2x$$ Apply the distributive property on the right side.
$$4x + 6 = 4x + 6$$ Combine like terms on the right side: $6x - 2x = 4x$.

Can you see that the equation $4x + 6 = 4x + 6$ is true for every value of x? Let's continue solving the equation by subtracting $4x$ from both sides.

$$4x + 6 - 4x = 4x + 6 - 4x$$

Keep reading. 6 = 6 is not the solution. $$6 = 6$$

Great Question!

Do I have to use sets to write the solution of an equation?

Because of the fundamental role that sets play in mathematics, it's a good idea to use set notation to expess an equation's solution set. If an equation has no solution, its solution set is $\varnothing$, the empty set. If an equation with variable x is true for every real number, its solution set is $\{x \mid x \text{ is a real number}\}$.

The original equation, $4x + 6 = 6(x + 1) - 2x$, is equivalent to the statement $6 = 6$, which is true for every value of x. Thus, the solution set consists of the set of all real numbers, expressed in set-builder notation as $\{x \mid x \text{ is a real number}\}$. Try substituting any real number of your choice for x in the original equation. You will obtain a true statement.

Check Point 8 Solve: $7x + 9 = 9(x + 1) - 2x$. $\{x \mid x \text{ is a real number}\}$

Achieving Success

Some of the difficulties facing students who take a college math class for the first time are a result of the differences between a high school math class and a college math course. It is helpful to understand these differences throughout your college math courses.

High School Math Class	College Math Course
Attendance is required.	Attendance may be optional.
Teachers monitor progress and performance closely.	Students receive grades, but may not be informed by the professor if they are in trouble.
There are frequent tests, as well as make-up tests if grades are poor.	There are usually no more than three or four tests per semester. Make-up tests are rarely allowed.
Grades are often based on participation and effort.	Grades are usually based exclusively on test grades.
Students have contact with their instructor every day.	Students usually meet with their instructor two or three times per week.
Teachers cover all material for tests in class through lectures and activities.	Students are responsible for information whether or not it is covered in class.
A course is covered over the school year, usually ten months.	A course is covered in a semester, usually four months.
Extra credit is often available for struggling students.	Extra credit is almost never offered.

Source: BASS, ALAN, *MATH STUDY SKILLS*, 1st, © 2008. Printed and Electronically reproduced by permission of Pearson Education, Inc., Upper Saddle River, New Jersey.

Concept and Vocabulary Check

Exercises in the Concept and Vocabulary Check are intended for group and class discussions.

In Exercises 1–9, fill in each blank so that the resulting statement is true.

1. An equation in the form $ax + b = 0$, $a \neq 0$, such as $3x + 17 = 0$, is called a/an _____linear_____ equation in one variable.

2. Two or more equations that have the same solution set are called _____equivalent_____ equations.

3. The addition property of equality states that if $a = b$, then $a + c =$ _____$b + c$_____.

4. The multiplication property of equality states that if $a = b$ and $c \neq 0$, then $ac =$ _____bc_____.

5. The first step in solving $7 + 3(x - 2) = 2x + 10$ is to apply the distributive property/simplify the left side

6. The algebraic expression $7(x - 4) + 2x$ can be _____simplified_____, whereas the algebraic equation $7(x - 4) + 2x = 35$ can be _____solved_____.

7. The equation $\dfrac{x}{4} = 2 + \dfrac{x}{3}$ can be cleared of fractions by multiplying both sides by the least common denominator of $\dfrac{x}{4}$ and $\dfrac{x}{3}$, which is _____12_____.

8. In solving an equation, if you eliminate the variable and obtain a statement such as $2 = 3$, the equation has _____no_____ solution. The solution set can be expressed using the symbol _____$\varnothing$_____.

9. In solving an equation with variable x, if you eliminate the variable and obtain a statement such as $6 = 6$, the equation is _____true_____ for every value of x. The solution set can be expressed in set-builder notation as __$\{x \mid x \text{ is a real number}\}$__.

In Exercises 10–13, determine whether each statement is true or false. If the statement is false, make the necessary change(s) to produce a true statement. Changes to false statements will vary.

10. The equation $2x + 5 = 0$ is equivalent to $2x = 5$. false

11. The equation $x + \frac{1}{3} = \frac{1}{2}$ is equivalent to $x + 2 = 3$. false

12. The equation $3x = 2x$ has no solution.   false

13. The equation $3(x + 4) = 3(4 + x)$ has precisely one solution.   false

Respond to Exercises 14–19 using verbal or written explanations.

14. What is the solution set of an equation?   **14–19.** Answers will vary.

15. State the addition property of equality and give an example.

16. State the multiplication property of equality and give an example.

17. How do you know if an equation has one solution, no solution, or infinitely many solutions?

18. What is the difference between solving an equation such as $2(x - 4) + 5x = 34$ and simplifying an algebraic expression such as $2(x - 4) + 5x$? If there is a difference, which topic should be taught first? Why?

19. Suppose that you solve $\dfrac{x}{5} - \dfrac{x}{2} = 1$ by multiplying both sides by 20, rather than the least common denominator of 5 and 2 (namely, 10). Describe what happens. If you get the correct solution, why do you think we clear the equation of fractions by multiplying by the *least* common denominator?

Exercise Set 2.2

Practice Exercises

In Exercises 1–58, solve and check each equation.

1. $x - 7 = 3$ {10}

2. $x - 3 = -17$ {−14}

3. $x + 5 = -12$ {−17}

4. $x + 12 = -14$ {−26}

5. $\dfrac{x}{3} = 4$ {12}

6. $\dfrac{x}{5} = 3$ {15}

7. $5x = 45$ {9}

8. $6x = 18$ {3}

9. $8x = -24$ {−3}

10. $5x = -25$ {−5}

11. $-8x = 2$ $\left\{-\dfrac{1}{4}\right\}$

12. $-6x = 3$ $\left\{-\dfrac{1}{2}\right\}$

13. $5x + 3 = 18$ {3}

14. $3x + 8 = 50$ {14}

15. $6x - 3 = 63$ {11}

16. $5x - 8 = 72$ {16}

17. $4x - 14 = -82$ {−17}

18. $9x - 14 = -77$ {−7}

19. $14 - 5x = -41$ {11}

20. $25 - 6x = -83$ {18}

21. $9(5x - 2) = 45$ $\left\{\dfrac{7}{5}\right\}$

22. $10(3x + 2) = 70$ $\left\{\dfrac{5}{3}\right\}$

23. $5x - (2x - 10) = 35$

24. $11x - (6x - 5) = 40$ {7}

25. $3x + 5 = 2x + 13$ {8}

26. $2x - 7 = 6 + x$ {13}

27. $8x - 2 = 7x - 5$ {−3}

28. $13x + 14 = -5 + 12x$ {−19}

29. $7x + 4 = x + 16$ {2}

30. $8x + 1 = x + 43$ {6}

31. $8y - 3 = 11y + 9$ {−4}

32. $5y - 2 = 9y + 2$ {−1}

33. $2(4 - 3x) = 2(2x + 5)$

34. $3(5 - x) = 4(2x + 1)$ {1}

35. $8(y + 2) = 2(3y + 4)$

36. $3(3y - 1) = 4(3 + 3y)$ {−5}

37. $3(x + 1) = 7(x - 2) - 3$ {5}

38. $5x - 4(x + 9) = 2x - 3$ {−33}

$23.\ \left\{\dfrac{25}{3}\right\}$

39. $5(2x - 8) - 2 = 5(x - 3) + 3$ {6}

$33.\ \left\{-\dfrac{1}{5}\right\}$

40. $7(3x - 2) + 5 = 6(2x - 1) + 24$ {3}

$35.\ \{-4\}$

41. $6 = -4(1 - x) + 3(x + 1)$ {1}

42. $100 = -(x - 1) + 4(x - 6)$ {41}

43. $10(z + 4) - 4(z - 2) = 3(z - 1) + 2(z - 3)$ {−57}

44. $-2(z - 4) - (3z - 2) = -2 - (6z - 2)$ {−10}

45. $\dfrac{2x}{3} - 5 = 7$ {18}

46. $\dfrac{3x}{4} - 9 = -6$ {4}

47. $\dfrac{x}{3} + \dfrac{x}{2} = \dfrac{5}{6}$ {1}

48. $\dfrac{x}{4} - \dfrac{x}{5} = 1$ {20}

49. $20 - \dfrac{z}{3} = \dfrac{z}{2}$ {24}

50. $\dfrac{z}{5} - \dfrac{1}{2} = \dfrac{z}{6}$ {15}

51. $\dfrac{y}{3} + \dfrac{2}{5} = \dfrac{y}{5} - \dfrac{2}{5}$ {−6}

52. $\dfrac{y}{12} + \dfrac{1}{6} = \dfrac{y}{2} - \dfrac{1}{4}$ {1}

53. $\dfrac{3x}{4} - 3 = \dfrac{x}{2} + 2$ {20}

54. $\dfrac{3x}{5} - \dfrac{2}{5} = \dfrac{x}{3} + \dfrac{2}{5}$ {3}

55. $\dfrac{3x}{5} - x = \dfrac{x}{10} - \dfrac{5}{2}$ {5}

56. $2x - \dfrac{2x}{7} = \dfrac{x}{2} + \dfrac{17}{2}$ {7}

57. $\dfrac{x - 3}{5} - 1 = \dfrac{x - 5}{4}$ {−7}

58. $\dfrac{x - 2}{3} - 4 = \dfrac{x + 1}{4}$ {59}

In Exercises 59–78, solve each equation. Use set notation to express solution sets for equations with no solution or equations that are true for all real numbers.

59. $3x - 7 = 3(x + 1)$ ∅

60. $2(x - 5) = 2x + 10$ ∅

61. $2(x + 4) = 4x + 5 - 2x + 3$ {x | x is a real number}

62. $3(x - 1) = 8x + 6 - 5x - 9$ {x | x is a real number}

63. $7 + 2(3x - 5) = 8 - 3(2x + 1)$ $\left\{\dfrac{2}{3}\right\}$

64. $2 + 3(2x - 7) = 9 - 4(3x + 1)$ $\left\{\dfrac{4}{3}\right\}$

65. $4x + 1 - 5x = 5 - (x + 4)$ {x | x is a real number}

66. $5x - 5 = 3x - 7 + 2(x + 1)$ {x | x is a real number}

67. $4(x + 2) + 1 = 7x - 3(x - 2)$ ∅

68. $5x - 3(x + 1) = 2(x + 3) - 5$ ∅

69. $3 - x = 2x + 3$ {0}

70. $5 - x = 4x + 5$ {0}

71. $\dfrac{x}{3} + 2 = \dfrac{x}{3}$ ∅

72. $\dfrac{x}{4} + 3 = \dfrac{x}{4}$ ∅

73. $\dfrac{x}{3} = \dfrac{x}{2}$ {0}

74. $\dfrac{x}{4} = \dfrac{x}{3}$ {0}

75. $\dfrac{x - 2}{5} = \dfrac{3}{10}$ $\left\{\dfrac{7}{2}\right\}$

76. $\dfrac{x + 4}{8} = \dfrac{3}{16}$ $\left\{-\dfrac{5}{2}\right\}$

77. $\dfrac{x}{2} - \dfrac{x}{4} + 4 = x + 4$ {0}

78. $\dfrac{x}{2} + \dfrac{2x}{3} + 3 = x + 3$ {0}

Practice Plus

79. Evaluate $x^2 - x$ for the value of x satisfying $4(x-2) + 2 = 4x - 2(2-x)$.   2

80. Evaluate $x^2 - x$ for the value of x satisfying $2(x-6) = 3x + 2(2x-1)$.   6

81. Evaluate $x^2 - (xy - y)$ for x satisfying $\frac{x}{5} - 2 = \frac{x}{3}$ and y satisfying $-2y - 10 = 5y + 18$.   161

82. Evaluate $x^2 - (xy - y)$ for x satisfying $\frac{3x}{2} + \frac{3x}{4} = \frac{x}{4} - 4$ and y satisfying $5 - y = 7(y+4) + 1$.   −5

In Exercises 83–90, solve each equation.

83. $[(3+6)^2 \div 3] \cdot 4 = -54x$ {−2}

84. $2^3 - [4(5-3)^3] = -8x$ {3}

85. $5 - 12x = 8 - 7x - [6 \div 3(2 + 5^3) + 5x]$ ∅

86. $2(5x + 58) = 10x + 4(21 \div 3.5 - 11)$ ∅

87. $0.7x + 0.4(20) = 0.5(x + 20)$ {10}

88. $0.5(x + 2) = 0.1 + 3(0.1x + 0.3)$ {0}

89. $4x + 13 - \{2x - [4(x-3) - 5]\} = 2(x-6)$ {−2}

90. $-2\{7 - [4 - 2(1-x) + 3]\} = 10 - [4x - 2(x-3)]$ $\left\{\frac{4}{3}\right\}$

Application Exercises

The latest guidelines, which apply to both men and women, give healthy weight ranges, rather than specific weights, for your height. The further you are above the upper limit of your range, the greater are the risks of developing weight-related health problems. The bar graph shows these ranges for various heights for people between the ages of 19 and 34, inclusive.

Healthy Weight Ranges for Men and Women, Ages 19 to 34

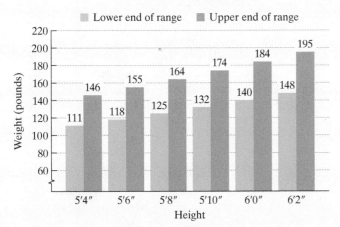

Source: U.S. Department of Health and Human Services

The mathematical model

$$\frac{W}{2} - 3H = 53$$

describes a weight, W, in pounds, that lies within the healthy weight range for a person whose height is H inches over 5 feet. Use this information to solve Exercises 91–92.

91. 142 pounds; 13 pounds   92. 178 pounds; 6 pounds

91. Use the formula to find a healthy weight for a person whose height is 5′6″. (*Hint:* $H = 6$ because this person's height is 6 inches over 5 feet.) How many pounds is this healthy weight below the upper end of the range shown by the bar graph?

92. Use the formula to find a healthy weight for a person whose height is 6′0″. (*Hint:* $H = 12$ because this person's height is 12 inches over 5 feet.) How many pounds is this healthy weight below the upper end of the range shown by the bar graph?

The formula

$$p = 15 + \frac{5d}{11}$$

describes the pressure of sea water, p, in pounds per square inch, at a depth of d feet below the surface. Use the formula to solve Exercises 93–94.

93. The record depth for breath-held diving, by Francisco Ferreras (Cuba) off Grand Bahama Island, on November 14, 1993, involved pressure of 201 pounds per square inch. To what depth did Ferreras descend on this ill-advised venture? (He was underwater for 2 minutes and 9 seconds!)   409.2 feet

94. At what depth is the pressure 20 pounds per square inch?   11 feet

Critical Thinking Exercises

95. Suppose you are an algebra teacher grading the following solution on an examination:

$$\text{Solve:}\quad -3(x - 6) = 2 - x.$$
$$\text{Solution:}\quad -3x - 18 = 2 - x$$
$$-2x - 18 = 2$$
$$-2x = -16$$
$$x = 8.$$

You should note that 8 checks, and the solution set is {8}. The student who worked the problem therefore wants full credit. Can you find any errors in the solution? If full credit is 10 points, how many points would you give the student? Justify your position.

96. Although the formulas in Example 6 on page 119 are correct, some people object to representing the variables with numbers, such as a 1-to-10 scale for the intensity of a negative life event. What might be their objection to quantifying the variables in this situation?   Answers will vary.

97. Write three equations whose solution set is {5}.

98. If x represents a number, write an English sentence about the number that results in an equation with no solution.

95. The student made two mistakes, not changing the sign of the second term when distributing −3 and subtracting 18 on the right when adding 18 on the left. Answers will vary.

97. Answers will vary; examples are: $x = 5$, $x + 3 = 8$, and $2x = 10$.

98. Answers will vary; an example is: A number plus 2 is equal to the number.

2.3 : Applications of Linear Equations

What am I supposed to learn?

After you have read this section, you should be able to:

1. Use linear equations to solve problems.

2. Solve a formula for a variable.

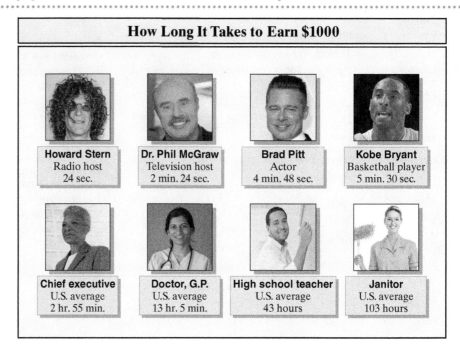

Source: *Time*

In this section, you'll see examples and exercises focused on how much money Americans earn. These situations illustrate a step-by-step strategy for solving problems. As you become familiar with this strategy, you will learn to solve a wide variety of problems.

Problem Solving with Linear Equations

1. **Use linear equations to solve problems.**

We have seen that a model is a mathematical representation of a real-world situation. In this section, we will be solving problems that are presented in English. This means that we must obtain models by translating from the ordinary language of English into the language of algebraic equations. To translate, however, we must understand the English prose and be familiar with the forms of algebraic language. Below are some general steps we will follow in solving word problems.

Strategy for Solving Word Problems

Step 1   Read the problem carefully several times until you can state in your own words what is given and what the problem is looking for. Let *x* (or any variable) represent one of the unknown quantities in the problem.

Step 2   If necessary, write expressions for any other unknown quantities in the problem in terms of *x*.

Step 3   Write an equation in *x* that models the verbal conditions of the problem.

Step 4   Solve the equation and answer the problem's question.

Step 5   Check the solution *in the original wording* of the problem, not in the equation obtained from the words.

The most difficult step in the strategy for solving word problems involves translating verbal conditions into an algebraic equation. Translations of some commonly used English phrases are listed in **Table 2.2**. We choose to use x to represent the variable, but we could use any letter.

Great Question!

Table 2.2 looks long and intimidating. What's the best way to get through the table?

Cover the right column with a sheet of paper and attempt to formulate the algebraic expression for the English phrase in the left column on your own. Then slide the paper down and check your answer. Work through the entire table in this manner.

Table 2.2 Algebraic Translations of English Phrases

English Phrase	Algebraic Expression
Addition	
The sum of a number and 7	$x + 7$
Five more than a number; a number plus 5	$x + 5$
A number increased by 6; 6 added to a number	$x + 6$
Subtraction	
A number minus 4	$x - 4$
A number decreased by 5	$x - 5$
A number subtracted from 8	$8 - x$
The difference between a number and 6	$x - 6$
The difference between 6 and a number	$6 - x$
Seven less than a number	$x - 7$
Seven minus a number	$7 - x$
Nine fewer than a number	$x - 9$
Multiplication	
Five times a number	$5x$
The product of 3 and a number	$3x$
Two-thirds of a number (used with fractions)	$\frac{2}{3}x$
Seventy-five percent of a number (used with decimals)	$0.75x$
Thirteen multiplied by a number	$13x$
A number multiplied by 13	$13x$
Twice a number	$2x$
Division	
A number divided by 3	$\frac{x}{3}$
The quotient of 7 and a number	$\frac{7}{x}$
The quotient of a number and 7	$\frac{x}{7}$
The reciprocal of a number	$\frac{1}{x}$
More than one operation	
The sum of twice a number and 7	$2x + 7$
Twice the sum of a number and 7	$2(x + 7)$
Three times the sum of 1 and twice a number	$3(1 + 2x)$
Nine subtracted from 8 times a number	$8x - 9$
Twenty-five percent of the sum of 3 times a number and 14	$0.25(3x + 14)$
Seven times a number, increased by 24	$7x + 24$
Seven times the sum of a number and 24	$7(x + 24)$

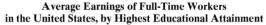
Example 1   Education Pays Off

The graph in **Figure 2.4** shows average yearly earnings in the United States by highest educational attainment.

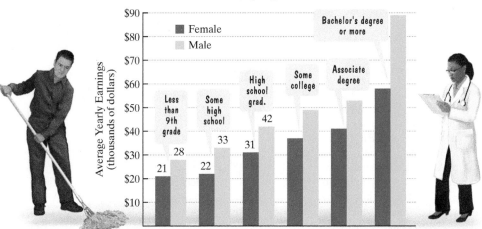

Average Earnings of Full-Time Workers in the United States, by Highest Educational Attainment

Figure 2.4
Source: U.S. Census Bureau

The average yearly salary of a man with an associate degree exceeds that of a man with some college by $4 thousand. The average yearly salary of a man with a bachelor's degree or more exceeds that of a man with some college by $40 thousand. Combined, three men with each of these educational attainments earn $191 thousand. Find the average yearly salary of men with each of these levels of education.

Solution

Step 1   Let *x* represent one of the unknown quantities. We know something about salaries of men with associate degrees and bachelor's degrees or more: They exceed the salary of a man with some college by $4 thousand and $40 thousand, respectively. We will let

x = the average yearly salary of a man with some college
(in thousands of dollars).

Step 2   Represent other unknown quantities in terms of *x*. Because a man with an associate degree earns $4 thousand more than a man with some college, let

$x + 4$ = the average yearly salary of a man with an associate degree.

Because a man with a bachelor's degree or more earns $40 thousand more than a man with some college, let

$x + 40$ = the average yearly salary of a man with a bachelor's degree or more.

Step 3   Write an equation in *x* that models the conditions. Combined, three men with each of these educational attainments earn $191 thousand.

Salary: some college	plus	salary: associate degree	plus	salary: bachelor's degree or more	equal	$191 thousand.
x	$+$	$(x + 4)$	$+$	$(x + 40)$	$=$	191

Step 4   Solve the equation and answer the question.

$$x + (x + 4) + (x + 40) = 191$$ This is the equation that models the problem's conditions.

$$3x + 44 = 191$$ Remove parentheses, regroup, and combine like terms.

$$3x = 147$$ Subtract 44 from both sides.

$$x = 49$$ Divide both sides by 3.

Great Question!

Example 1 involves using the word *exceeds* to represent two of the unknown quantities. Can you help me to write algebraic expressions for quantities described using *exceeds*?

Modeling with the word *exceeds* can be a bit tricky. It's helpful to identify the smaller quantity. Then add to this quantity to represent the larger quantity. For example, suppose that Tim's height exceeds Tom's height by *a* inches. Tom is the shorter person. If Tom's height is represented by *x*, then Tim's height is represented by $x + a$.

Because we isolated the variable in the model and obtained $x = 49$,

$$\text{average salary with some college} = x = 49$$
$$\text{average salary with an associate degree} = x + 4 = 49 + 4 = 53$$
$$\text{average salary with a bachelor's degree or more} = x + 40 = 49 + 40 = 89.$$

Men with some college average $49 thousand per year, men with associate degrees average $53 thousand per year, and men with bachelor's degrees or more average $89 thousand per year.

Step 5 Check the proposed solution in the original wording of the problem. The problem states that combined, three men with each of these educational attainments earn $191 thousand. Using the salaries we determined in step 4, the sum is

$$\text{\$49 thousand} + \text{\$53 thousand} + \text{\$89 thousand, or \$191 thousand,}$$

which satisfies the problem's conditions.

 **Check Point 1**   The average yearly salary of a woman with an associate degree exceeds that of a woman with some college by $4 thousand. The average yearly salary of a woman with a bachelor's degree or more exceeds that of a woman with some college by $21 thousand. Combined, three women with each of these educational attainments earn $136 thousand. Find the average yearly salary of women with each of these levels of education. (These salaries are illustrated by the bar graph on the ▶ previous page.)   some college: $37,000; associate degree: $41,000; bachelor's degree: $58,000

Your author teaching math in 1969

Example 2   Modeling Attitudes of College Freshmen

Researchers have surveyed college freshmen every year since 1969. **Figure 2.5** shows that attitudes about some life goals have changed dramatically over the years. In particular, the freshman class of 2010 was more interested in making money than the freshmen of 1969 had been. In 1969, 42% of first-year college students considered "being well-off financially" essential or very important. For the period from 1969 through 2010, this percentage increased by approximately 0.9 each year. If this trend continues, by which year will all college freshmen consider "being well-off financially" essential or very important?

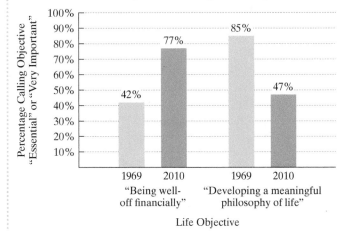

Life Objectives of College Freshmen, 1969–2010

Figure 2.5
Source: Higher Education Research Institute

Solution

Step 1 Let *x* represent one of the unknown quantities. We are interested in the year when all college freshmen, or 100% of the freshmen, will consider this life objective essential or very important. Let

x = the number of years after 1969 when all freshmen will consider "being well-off financially" essential or very important.

Step 2 **Represent other unknown quantities in terms of *x*.** There are no other unknown quantities to find, so we can skip this step.

Step 3 **Write an equation in *x* that models the conditions.**

The 1969 percentage	increased by	0.9 each year for x years	equals	100% of the freshmen.
42	+	0.9x	=	100

Step 4 **Solve the equation and answer the question.**

$$42 + 0.9x = 100 \qquad \text{This is the equation that models the problem's conditions.}$$

$$42 - 42 + 0.9x = 100 - 42 \qquad \text{Subtract 42 from both sides.}$$

$$0.9x = 58 \qquad \text{Simplify.}$$

$$\frac{0.9x}{0.9} = \frac{58}{0.9} \qquad \text{Divide both sides by 0.9.}$$

$$x = 64.\overline{4} \approx 64 \qquad \text{Simplify and round to the nearest whole number.}$$

Using current trends, by approximately 64 years after 1969, or in 2033, all freshmen will consider "being well-off financially" essential or very important.

Step 5 **Check the proposed solution in the original wording of the problem.** The problem states that all freshmen (100%, represented by 100 using the model) will consider the objective essential or very important. Does this approximately occur if we increase the 1969 percentage, 42%, by 0.9 each year for 64 years, our proposed solution?

$$42 + 0.9(64) = 42 + 57.6 = 99.6 \approx 100$$

This verifies that using trends shown in **Figure 2.5**, all first-year college students will consider the objective essential or very important approximately 64 years after 1969.

A Brief Review • Clearing an Equation of Decimals

- You can clear an equation of decimals by multiplying each side by a power of 10. The exponent on 10 will be equal to the greatest number of digits to the right of any decimal point in the equation.

- Multiplying a decimal number by 10^n has the effect of moving the decimal point n places to the right.

Example

$$42 + 0.9x = 100$$

The greatest number of digits to the right of any decimal point in the equation is 1. Multiply each side by 10^1, or 10.

$$10(42 + 0.9x) = 10(100)$$
$$10(42) + 10(0.9x) = 10(100)$$
$$420 + 9x = 1000$$
$$420 - 420 + 9x = 1000 - 420$$
$$9x = 580$$
$$\frac{9x}{9} = \frac{580}{9}$$
$$x = 64.\overline{4} \approx 64$$

It is not a requirement to clear decimals before solving an equation. Compare this solution to the one in step 4 of Example 2. Which method do you prefer?

✓ **Check Point 2** **Figure 2.5** on page 127 shows that the freshman class of 2010 was less interested in developing a philosophy of life than the freshmen of 1969 had been. In 1969, 85% of the freshmen considered this objective essential or very important. Since then, this percentage has decreased by approximately 0.9 each year. If this trend continues, by which year will only 25% of college freshmen ▶ consider "developing a meaningful philosophy of life" essential or very important?

by 67 years after 1969, or in 2036

Blitzer Bonus
●●●●●●●●●●●●●●●●●●

Forty-One Years of Change

Compared to 1969, college freshmen in 2010 had making money on their minds. Here are some other random factoids and statistics about life then versus life today.

	1969	2010
President	Richard Nixon	Barack Obama
U.S. Population	203 million	309 million
Life Expectancy	Male: 66.6 years Female: 73.1 years	Male: 76.2 years Female: 81.1 years
Cost per Gallon of Premium Gas	$0.36	$3.15
Average Cost of a New Car	$3270	$29,600
Average Income	$8550	$40,250
Unemployment	3.5%	9%
Minimum Voting Age	21	18
Technology	Teens listen to music on record players and stereos. People talk on phones through land lines.	Most teens listen to music digitally on iPods and mp3 players. Cellphones, smartphones, and laptops revolutionize global communication.
Percentage of Teens Enjoying Parents' Company	57%	64%
Percentage of Teens Stating the Importance of Patriotism	55%	31%

Sources: Channel One News, infoplease.com, *Scholastic Scope*

Great Question!
●●●●●●●●●●●●●●●●●

Why are algebraic word problems important?

There is great value in reasoning through the steps for solving a word problem. This value comes from the problem-solving skills that you will attain and is often more important than the specific problem or its solution.

Example 3   Selecting a Monthly Text Message Plan

You are choosing between two texting plans. Plan A has a monthly fee of $20.00 with a charge of $0.05 per text. Plan B has a monthly fee of $5.00 with a charge of $0.10 per text. Both plans include photo and video texts. For how many text messages will the costs for the two plans be the same?

Solution

Step 1   Let *x* represent one of the unknown quantities. Let

    x = the number of text messges for which the two plans cost the same.

Step 2   Represent other unknown quantities in terms of *x*. There are no other unknown quantities, so we can skip this step.

Step 3   Write an equation in x that models the conditions. The monthly cost for plan A is the monthly fee, $20.00, plus the per-text charge, $0.05, times the number of text messages, x. The monthly cost for plan B is the monthly fee, $5.00, plus the per-text charge, $0.10, times the number of text messages, x.

The monthly cost for plan A	must equal	the monthly cost for plan B.

$$20 + 0.05x = 5 + 0.10x$$

Step 4   Solve the equation and answer the question.

$$20 + 0.05x = 5 + 0.10x \quad \text{This is the equation that models the problem's conditions.}$$

$$20 = 5 + 0.05x \quad \text{Subtract } 0.05x \text{ from both sides.}$$

$$15 = 0.05x \quad \text{Subtract 5 from both sides.}$$

$$\frac{15}{0.05} = \frac{0.05x}{0.05} \quad \text{Divide both sides by } 0.05.$$

$$300 = x \quad \text{Simplify.}$$

Because x represents the number of text messages for which the two plans cost the same, the costs will be the same for 300 texts per month.

Step 5   Check the proposed solution in the original wording of the problem. The problem states that the costs for the two plans should be the same. Let's see if they are with 300 text messages:

$$\text{Cost for plan A} = \$20 + \$0.05(300) = \$20 + \$15 = \$35$$

Monthly fee	Per-text charge

$$\text{Cost for plan B} = \$5 + \$0.10(300) = \$5 + \$30 = \$35.$$

With 300 text messages, both plans cost $35 for the month. Thus, the proposed solution, 300 text messages, satisfies the problem's conditions.

Check Point 3   You are choosing between two texting plans. Plan A has a monthly fee of $15.00 with a charge of $0.08 per text. Plan B has a monthly fee of $3.00 with a charge of $0.12 per text. For how many text messages will the costs for the two plans be the same?   300 text messages

Example 4   A Price Reduction on a Digital Camera

Your local computer store is having a terrific sale on digital cameras. After a 40% price reduction, you purchase a digital camera for $276. What was the camera's price before the reduction?

Solution

Step 1   Let x represent one of the unknown quantities. We will let

$x = $ the original price of the digital camera prior to the reduction.

Step 2   Represent other unknown quantities in terms of x. There are no other unknown quantities to find, so we can skip this step.

Great Question!

Why is the 40% reduction written as 0.4x in Example 4?

- 40% is written 0.40 or 0.4.
- "Of" represents multiplication, so 40% of the original price is $0.4x$.

Notice that the orginal price, x, reduced by 40% is $x - 0.4x$ and *not* $x - 0.4$.

Step 3 Write an equation in *x* that models the conditions. The camera's original price minus the 40% reduction is the reduced price, $276.

Original price	minus	the reduction (40% of the original price)	is	the reduced price, $276.
x	$-$	$0.4x$	$=$	276

Step 4 Solve the equation and answer the question.

$$x - 0.4x = 276 \quad \text{This is the equation that models the problem's conditions.}$$

$$0.6x = 276 \quad \text{Combine like terms: } x - 0.4x = 1x - 0.4x = 0.6x.$$

$$\frac{0.6x}{0.6} = \frac{276}{0.6} \quad \text{Divide both sides by 0.6.}$$

$$x = 460 \quad \text{Simplify: } 0.6\overline{)276.0}$$

The digital camera's price before the reduction was $460.

Step 5 Check the proposed solution in the original wording of the problem. The price before the reduction, $460, minus the 40% reduction should equal the reduced price given in the original wording, $276:

$$460 - 40\% \text{ of } 460 = 460 - 0.4(460) = 460 - 184 = 276.$$

This verifies that the digital camera's price before the reduction was $460.

Check Point 4 After a 30% price reduction, you purchase a new computer for $840. What was the computer's price before the reduction? $1200

Solving a Formula for One of Its Variables

2 Solve a formula for a variable.

We know that solving an equation is the process of finding the number (or numbers) that make the equation a true statement. All of the equations we have solved contained only one letter, x.

By contrast, formulas contain two or more letters, representing two or more variables. An example is the formula for the perimeter of a rectangle:

$$P = 2l + 2w. \quad \text{A rectangle's perimeter is the sum of twice its length and twice its width.}$$

We say that this formula is solved for the variable P because P is alone on one side of the equation and the other side does not contain a P.

Solving a formula for a variable means rewriting the formula so that the variable is isolated on one side of the equation. It does not mean obtaining a numerical value for that variable.

To solve a formula for one of its variables, treat that variable as if it were the only variable in the equation. Think of the other variables as if they were numbers. Isolate all terms with the specified variable on one side of the equation and all terms without the specified variable on the other side. Then divide both sides by the same nonzero quantity to get the specified variable alone. The next two examples show how to do this.

Example 5 ▦ Solving a Formula for a Variable

Solve the formula $P = 2l + 2w$ for l.

Solution

First, isolate $2l$ on the right by subtracting $2w$ from both sides. Then solve for l by dividing both sides by 2.

> We need to isolate l.

$P = 2l + 2w$	This is the given formula.
$P - 2w = 2l + 2w - 2w$	Isolate $2l$ by subtracting $2w$ from both sides.
$P - 2w = 2l$	Simplify.
$\dfrac{P - 2w}{2} = \dfrac{2l}{2}$	Solve for l by dividing both sides by 2.
$\dfrac{P - 2w}{2} = l$	Simplify.

Equivalently, $l = \dfrac{P - 2w}{2}$.

✓ **Check Point 5** Solve the formula $P = 2l + 2w$ for w. $w = \dfrac{P - 2l}{2}$

Example 6 ▦ Solving a Formula for a Variable

The total price of an article purchased on a monthly deferred payment plan is described by the following formula:

$$T = D + pm.$$

In this formula, T is the total price, D is the down payment, p is the monthly payment, and m is the number of months one pays. Solve the formula for p.

Solution

First, isolate pm on the right by subtracting D from both sides. Then, isolate p from pm by dividing both sides of the formula by m.

> We need to isolate p.

$T = D + pm$	This is the given formula. We want p alone.
$T - D = D - D + pm$	Isolate pm by subtracting D from both sides.
$T - D = pm$	Simplify.
$\dfrac{T - D}{m} = \dfrac{pm}{m}$	Now isolate p by dividing both sides by m.
$\dfrac{T - D}{m} = p$	Simplify: $\dfrac{pm}{m} = \dfrac{p\cancel{m}}{\cancel{m}} = \dfrac{p}{1} = p$.

✓ **Check Point 6** Solve the formula $T = D + pm$ for m. $m = \dfrac{T - D}{p}$

Concept and Vocabulary Check

Exercises in the Concept and Vocabulary Check are intended for group and class discussions.

In Exercises 1–6, fill in each blank so that the resulting statement is true.

1. According to the U.S. Office of Management and Budget, the 2011 budget for defense exceeded the budget for education by $658.6 billion. If x represents the budget for education, in billions of dollars, the budget for defense can be represented by __$x + 658.6$__.

2. In 2000, 31% of U.S. adults viewed a college education as essential for success. For the period from 2000 through 2010, this percentage increased by approximately 2.4 each year. The percentage of U.S. adults who viewed a college education as essential for success x years after 2000 can be represented by __$31 + 2.4x$__.

3. A text message plan costs $4.00 per month plus $0.15 per text. The monthly cost for x text messages can be represented by __$4 + 0.15x$__.

4. I purchased a computer after a 15% price reduction. If x represents the computer's original price, the reduced price can be represented by __$x - 0.15x$ or $0.85x$__

5. Solving a formula for a variable means rewriting the formula so that the variable is __isolated on one side__

6. In order to solve $y = mx + b$ for x, we first __subtract b__ and then __divide by m__.

Respond to Exercises 7–8 using verbal or written explanations.

7. In your own words, describe a step-by-step approach for solving algebraic word problems.   7–8. Answers will vary.

8. Explain what it means to solve a formula for a variable.

Exercise Set 2.3

Practice Exercises

Use the five-step strategy for solving word problems to find the number or numbers described in Exercises 1–10.

1. When five times a number is decreased by 4, the result is 26. What is the number?   6

2. When two times a number is decreased by 3, the result is 11. What is the number?   7

3. When a number is decreased by 20% of itself, the result is 20. What is the number?   25

4. When a number is decreased by 30% of itself, the result is 28. What is the number?   40

5. When 60% of a number is added to the number, the result is 192. What is the number?   120

6. When 80% of a number is added to the number, the result is 252. What is the number?   140

7. 70% of what number is 224?   320

8. 70% of what number is 252?   360

9. One number exceeds another by 26. The sum of the numbers is 64. What are the numbers?   19 and 45

10. One number exceeds another by 24. The sum of the numbers is 58. What are the numbers?   17 and 41

Practice Plus

In Exercises 11–18, write each English phrase as an algebraic expression. Then simplify the expression. Let x represent the number.

11. A number decreased by the sum of the number and four

12. A number decreased by the difference between eight and the number $x - (8 - x); 2x - 8$

13. Six times the product of negative five and a number

14. Ten times the product of negative four and a number

15. The difference between the product of five and a number and twice the number $5x - 2x; 3x$

16. The difference between the product of six and a number and negative two times the number $6x - (-2x); 8x$

17. The difference between eight times a number and six more than three times the number $8x - (3x + 6); 5x - 6$

18. Eight decreased by three times the sum of a number and six $8 - 3(x + 6); -3x - 10$

Application Exercises

How will you spend your average life expectancy of 78 years? The bar graph shows the average number of years you will devote to each of your most time-consuming activities. Exercises 19–20 are based on the data displayed by the graph.

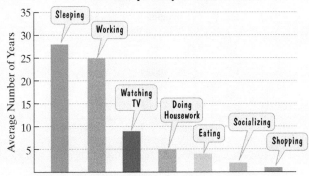

How You Will Spend Your Average Life Expectancy of 78 Years

Source: U.S. Bureau of Labor Statistics

19. According to the U.S. Bureau of Labor Statistics, you will devote 37 years to sleeping and watching TV. The number of years sleeping will exceed the number of years watching TV by 19. Over your lifetime, how many years will you spend on each of these activities?   TV: 9 years; sleeping: 28 years

11. $x - (x + 4); -4$ 13. $6(-5x); -30x$ 14. $10(-4x); -40x$

20. According to the U.S. Bureau of Labor Statistics, you will devote 32 years to sleeping and eating. The number of years sleeping will exceed the number of years eating by 24. Over your lifetime, how many years will you spend on each of these activities?   eating: 4 years; sleeping: 28 years

The bar graph shows average yearly earnings in the United States for people with a college education, by final degree earned. Exercises 21–22 are based on the data displayed by the graph.

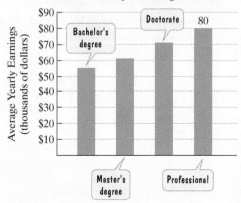

Average Earnings of Full-Time Workers in the U.S., by Final Degree Earned

Source: U.S. Census Bureau

21. The average yearly salary of an American whose final degree is a master's is $49 thousand less than twice that of an American whose final degree is a bachelor's. Combined, two people with each of these educational attainments earn $116 thousand. Find the average yearly salary of Americans with each of these final degrees.   bachelor's: $55 thousand; master's: $61 thousand

22. The average yearly salary of an American whose final degree is a doctorate is $39 thousand less than twice that of an American whose final degree is a bachelor's. Combined, two people with each of these educational attainments earn $126 thousand. Find the average yearly salary of Americans with each of these final degrees.   bachelor's: $55 thousand; doctorate: $71 thousand

Even as Americans increasingly view a college education as essential for success, many believe that a college education is becoming less available to qualified students. Exercises 23–24 are based on the data displayed by the graph.

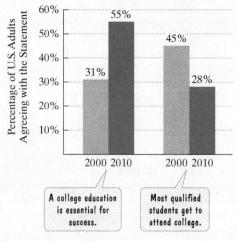

Changing Attitudes Toward College in the United States

Source: Public Agenda

23. In 2000, 31% of U.S. adults viewed a college education as essential for success. For the period 2000 through 2010, the percentage viewing a college education as essential for success increased on average by approximately 2.4 each year. If this trend continues, by which year will 67% of all American adults view a college education as essential for success?   by 2015

24. The data displayed by the graph at the bottom of the previous column indicate that in 2000, 45% of U.S. adults believed most qualified students get to attend college. For the period from 2000 through 2010, the percentage who believed that a college education is available to most qualified students decreased by approximately 1.7 each year. If this trend continues, by which year will only 11% of all American adults believe that most qualified students get to attend college?   by 2020

On average, every minute of every day, 158 babies are born. The bar graph represents the results of a single day of births, deaths, and population increase worldwide. Exercises 25–26 are based on the information displayed by the graph.

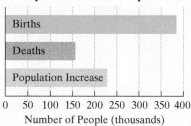

Daily Growth of World Population

Source: James Henslin, *Sociology,* Eleventh Edition, Pearson, 2012.

25. Each day, the number of births in the world is 84 thousand less than three times the number of deaths.
    a. If the population increase in a single day is 228 thousand, determine the number of births and deaths per day.
    b. If the population increase in a single day is 228 thousand, by how many millions of people does the worldwide population increase each year? Round to the nearest million.   83 million
    c. Based on your answer to part (b), approximately how many years does it take for the population of the world to increase by an amount greater than the entire U.S. population (315 million)?   approximately 4 years

26. Each day, the number of births in the world exceeds twice the number of deaths by 72 thousand.
    a. If the population increase in a single day is 228 thousand, determine the number of births and deaths per day.
    b. If the population increase in a single day is 228 thousand, by how many millions of people does the worldwide population increase each year? Round to the nearest million.   83 million
    c. Based on your answer to part (b), approximately how many years does it take for the population of the world to increase by an amount greater than the entire U.S. population (315 million)?   approximately 4 years

27. A new car worth $24,000 is depreciating in value by $3000 per year. After how many years will the car's value be $9000?

28. A new car worth $45,000 is depreciating in value by $5000 per year. After how many years will the car's value be $10,000?   after 7 years

25. a. births: 384,000; deaths: 156,000   26. a. births: 384,000; deaths: 156,000   27. after 5 years

29. You are choosing between two health clubs. Club A offers membership for a fee of $40 plus a monthly fee of $25. Club B offers membership for a fee of $15 plus a monthly fee of $30. After how many months will the total cost at each health club be the same? What will be the total cost for each club? after 5 months; $165

30. You need to rent a rug cleaner. Company A will rent the machine you need for $22 plus $6 per hour. Company B will rent the same machine for $28 plus $4 per hour. After how many hours of use will the total amount spent at each company be the same? What will be the total amount spent at each company? 3 hours; $40

31. The bus fare in a city is $1.25. People who use the bus have the option of purchasing a monthly discount pass for $15.00. With the discount pass, the fare is reduced to $0.75. Determine the number of times in a month the bus must be used so that the total monthly cost without the discount pass is the same as the total monthly cost with the discount pass. 30 times

32. A discount pass for a bridge costs $30 per month. The toll for the bridge is normally $5.00, but it is reduced to $3.50 for people who have purchased the discount pass. Determine the number of times in a month the bridge must be crossed so that the total monthly cost without the discount pass is the same as the total monthly cost with the discount pass. 20 times

33. You are choosing between two plans at a discount warehouse. Plan A offers an annual membership fee of $100 and you pay 80% of the manufacturer's recommended list price. Plan B offers an annual membership fee of $40 and you pay 90% of the manufacturer's recommended list price. How many dollars of merchandise would you have to purchase in a year to pay the same amount under both plans? What will be the cost for each plan? $600 of merchandise; $580

34. You are choosing between two plans at a discount warehouse. Plan A offers an annual membership fee of $300 and you pay 70% of the manufacturer's recommended list price. Plan B offers an annual membership fee of $40 and you pay 90% of the manufacturer's recommended list price. How many dollars of merchandise would you have to purchase in a year to pay the same amount under both plans? What will be the cost for each plan? $1300 of merchandise; $1210

35. In 2010, there were 13,300 students at college A, with a projected enrollment increase of 1000 students per year. In the same year, there were 26,800 students at college B, with a projected enrollment decline of 500 students per year. According to these projections, when will the colleges have the same enrollment? What will be the enrollment in each college at that time? 2019; 22,300 students

36. In 2000, the population of Greece was 10,600,000, with projections of a population decrease of 28,000 people per year. In the same year, the population of Belgium was 10,200,000, with projections of a population decrease of 12,000 people per year. (*Source:* United Nations) According to these projections, when will the two countries have the same population? What will be the population at that time? 2025; 9,900,000

37. After a 20% reduction, you purchase a television for $336. What was the television's price before the reduction? $420

38. After a 30% reduction, you purchase a dictionary for $30.80. What was the dictionary's price before the reduction? $44

39. Including 8% sales tax, an inn charges $162 per night. Find the inn's nightly cost before the tax is added. $150

40. Including 5% sales tax, an inn charges $252 per night. Find the inn's nightly cost before the tax is added. $240

In Exercises 41–58, solve each formula for the specified variable. Do you recognize the formula? If so, what does it describe?

41. $A = LW$ for L $\quad L = \dfrac{A}{W}$

42. $D = RT$ for R $\quad R = \dfrac{D}{T}$

43. $A = \frac{1}{2}bh$ for b $\quad b = \dfrac{2A}{h}$

44. $V = \frac{1}{3}Bh$ for B $\quad B = \dfrac{3V}{h}$

45. $I = Prt$ for P $\quad P = \dfrac{I}{rt}$

46. $C = 2\pi r$ for r $\quad r = \dfrac{C}{2\pi}$

47. $E = mc^2$ for m $\quad m = \dfrac{E}{c^2}$

48. $V = \pi r^2 h$ for h $\quad h = \dfrac{V}{\pi r^2}$

49. $y = mx + b$ for m $\quad m = \dfrac{y-b}{x}$

50. $P = C + MC$ for M $\quad M = \dfrac{P-C}{C}$

51. $A = \frac{1}{2}h(a+b)$ for a $\quad a = \dfrac{2A}{h} - b$

52. $A = \frac{1}{2}h(a+b)$ for b $\quad b = \dfrac{2A}{h} - a$

53. $S = P + Prt$ for r $\quad r = \dfrac{S-P}{Pt}$

54. $S = P + Prt$ for t $\quad t = \dfrac{S-P}{Pr}$

55. $Ax + By = C$ for x $\quad x = \dfrac{C-By}{A}$

56. $Ax + By = C$ for y $\quad y = \dfrac{C-Ax}{B}$

57. $a_n = a_1 + (n-1)d$ for n $\quad n = \dfrac{a_n - a_1}{d} + 1$ or $n = \dfrac{a_n - a_1 + d}{d}$

58. $a_n = a_1 + (n-1)d$ for d $\quad d = \dfrac{a_n - a_1}{n-1}$

Critical Thinking Exercises

59. The price of a dress is reduced by 40%. When the dress still does not sell, it is reduced by 40% of the reduced price. If the price of the dress after both reductions is $72, what was the original price? $200

60. In a film, the actor Charles Coburn plays an elderly "uncle" character criticized for marrying a woman when he is 3 times her age. He wittily replies, "Ah, but in 20 years time I shall only be twice her age." How old is the "uncle" and the woman? uncle: 60 years old; woman: 20 years old

61. Suppose that we agree to pay you 8¢ for every problem in this chapter that you solve correctly and fine you 5¢ for every problem done incorrectly. If at the end of 26 problems we do not owe each other any money, how many problems did you solve correctly? 10 problems

62. A thief steals a number of rare plants from a nursery. On the way out, the thief meets three security guards, one after another. To each security guard, the thief is forced to give one-half the plants that he still has, plus 2 more. Finally, the thief leaves the nursery with 1 lone palm. How many plants were originally stolen? 36 plants

Group Exercise

63. One of the best ways to learn how to *solve* a word problem in algebra is to *design* word problems of your own. Creating a word problem makes you very aware of precisely how much information is needed to solve the problem. You must also focus on the best way to present information to a reader and on how much information to give. As you write your problem, you gain skills that will help you solve problems created by others.

The group should design five different word problems that can be solved using linear equations. All of the problems should be on different topics. For example, the group should not have more than one problem on price reduction. The group should turn in both the problems and their algebraic solutions.

(If you're not sure where to begin, consider using some of the data in the Blitzer Bonus on page 129 comparing life in 1969 and 2010.)   Answers will vary.

2.4 Ratios, Rates, and Proportions

What am I supposed to learn?

After you have read this section, you should be able to:

1. Write ratios as fractions.

2. Write rates as fractions.

3. Write unit rates.

4. Solve proportions.

5. Solve problems using proportions.

The possibility of seeing a blue whale, the largest mammal ever to grace the earth, increases the excitement of gazing out over the ocean's swell of waves. Blue whales were hunted to near extinction in the last half of the nineteenth and the first half of the twentieth centuries. Using a method for estimating wildlife populations that we discuss in this section, by the mid-1960s it was determined that the world population of blue whales was less than 1000. This led the International Whaling Commission to ban the killing of blue whales to prevent their extinction. A dramatic increase in blue whale sightings indicates an ongoing increase in their population and the success of the killing ban. In this section, you will see how modeling with *ratios* and *proportions* played a significant role in preventing the extinction of blue whales. We open with the connection between fractions and ratios.

Ratios as Fractions

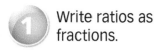

1. Write ratios as fractions.

A **ratio** is a comparison of two quantities by division. Given two numbers, a and b, the ratio of a to b can be written as $\frac{a}{b}$. Thus, a ratio is a fraction. There are three common notations for expressing ratios.

Expressing the Ratio of a to b

a to b	$\dfrac{a}{b}$	$a : b$
Separate the numbers by the word *to*.	Fraction notation	Separate the numbers by a colon.

In this section, we will use fraction notation to express ratios, and we will write the fractions in simplest form.

Writing Ratios as Fractions

1. Write the ratio in fraction notation. The first number of the ratio is the numerator of the fraction. The second number of the ratio is the denominator of the fraction.

2. Simplify the fraction, if possible.

Example 1   **Big Bites: Writing Ratios**

The bar graph in **Figure 2.6** shows the number of teeth in six selected animals.

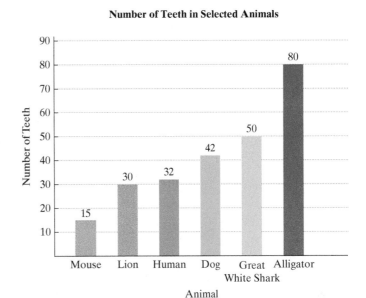

Figure 2.6

Source: Science World

Write the following ratios in fraction notation:

a. the ratio of the number of teeth in a mouse to the number of teeth in a lion

b. the ratio of the number of teeth in an alligator to the number of teeth in a great white shark.

Solution

To write these ratios as fractions, the first number of the ratio is the numerator and the second number is the denominator.

a. The ratio of the number of teeth in a mouse to the number of teeth in a lion is

Number of teeth in a mouse
Number of teeth in a lion

$$\frac{15 \text{ teeth}}{30 \text{ teeth}} = \frac{15 \cancel{\text{ teeth}}}{30 \cancel{\text{ teeth}}}$$ Divide out the common unit.

$$= \frac{1}{2}.$$ Divide the numerator and the denominator by the common factor, 15.

The ratio is $\frac{1}{2}$.

b. The ratio of the number of teeth in an alligator to the number of teeth in a great white shark is

Number of teeth in an alligator
Number of teeth in a great white shark

$$\frac{80 \text{ teeth}}{50 \text{ teeth}} = \frac{80 \cancel{\text{ teeth}}}{50 \cancel{\text{ teeth}}}$$ Divide out the common unit.

$$= \frac{8}{5}.$$ Divide the numerator and the denominator by the common factor, 10.

The ratio is $\frac{8}{5}$.

Great Question!

Can I express the ratio in Example 1(b), $\frac{8}{5}$, as $1\frac{3}{5}$?

No. If a ratio is an improper fraction, do not write the fraction as a mixed number. Unlike fraction notation, a mixed number does not show a comparison of two quantities by division.

✓ **Check Point 1** Use the bar graph in **Figure 2.6** on the previous page to write the following ratios in fraction notation:

a. the ratio of the number of teeth in a human to the number of teeth in a dog $\frac{16}{21}$

▶ **b.** the ratio of the number of teeth in a great white shark to the number of teeth in a lion. $\frac{5}{3}$

Rates

Write rates as fractions.

A **rate** is a ratio that compares two quantities that have different units. For example, suppose that a cyclist can bike 6 miles in 20 minutes. We can write this rate as a fraction:

$$\frac{6 \text{ miles}}{20 \text{ minutes}} = \frac{3 \text{ miles}}{10 \text{ minutes}}$$ Units are different and do not divide out.

Notice that rates are written in fraction notation with the units included. We include the units because they are different and therefore do not divide out.

Table 2.3 shows words that are commonly used to separate the numerator and the denominator when describing rates.

Table 2.3 Words Used to Describe Rates

Word	Example	Description
in	$\frac{3 \text{ miles}}{10 \text{ minutes}}$	3 miles in 10 minutes
for; for every	$\frac{25 \text{ horses}}{2 \text{ acres}}$	25 horses for 2 acres 25 horses for every 2 acres
on	$\frac{119 \text{ apple trees}}{3 \text{ acres}}$	119 apple trees on 3 acres
per	$\frac{387 \text{ violent crimes}}{100{,}000 \text{ people}}$	387 violent crimes per 100,000 people

Example 2 **Writing and Simplifying Rates**

Write each rate as a fraction in simplest form:

a. 55 hits for 180 times at bat

b. 260 miles on 8 gallons of gasoline

c. 114 robberies per 100,000 residents.

Solution

a. 55 hits for 180 times at bat is

$$\frac{55 \text{ hits}}{180 \text{ times at bat}} = \frac{11 \text{ hits}}{36 \text{ times at bat}}.$$

Divide the numerator and the denominator by the common factor, 5.

b. 260 miles on 8 gallons of gasoline is

$$\frac{260 \text{ miles}}{8 \text{ gallons}} = \frac{65 \text{ miles}}{2 \text{ gallons}}.$$

Divide the numerator and the denominator by the common factor, 4.

c. 114 robberies per 100,000 residents is

$$\frac{114 \text{ robberies}}{100,000 \text{ residents}} = \frac{57 \text{ robberies}}{50,000 \text{ residents}}.$$

Divide the numerator and the denominator by the common factor, 2.

 **Check Point 2**   Write each rate as a fraction in simplest form:

a. $333 for 6 tickets $\frac{\$111}{2 \text{ tickets}}$

b. 18 miles in 30 minutes $\frac{3 \text{ miles}}{5 \text{ minutes}}$

c. 6 pounds of fertilizer per 2000 square feet. $\frac{3 \text{ pounds}}{1000 \text{ square feet}}$

Unit Rates

 Write unit rates.

A **unit rate** is a rate that has a denominator of 1. A familiar example of a unit rate is "miles per hour." For example, 35 miles per hour means 35 miles per 1 hour and can be written as $\frac{35 \text{ miles}}{1 \text{ hour}}$.

> **Writing a Unit Rate**
>
> 1. Write the rate in fraction notation with the units included.
> 2. Divide the numerator and the denominator by the quantity in the denominator. This results in a unit rate with a denominator of 1.
>
> The word "per" is used to describe the division in a unit rate.

Example 3   **Writing Unit Rates**

Write each of the following as a unit rate:

a. 144 miles in 3 hours

b. 60¢ for a 4-ounce can of juice.

Solution

a. 144 miles in 3 hours is

$$\frac{144 \text{ miles}}{3 \text{ hours}}.$$

Write the rate in fraction notation with the units included.

$$= \frac{144 \text{ miles} \div 3}{3 \text{ hours} \div 3}$$

Divide the numerator and the denominator by the quantity in the denominator.

$$= \frac{48 \text{ miles}}{1 \text{ hour}}$$

$$= 48 \text{ miles per hour}$$

$$\begin{array}{r} 48 \\ 3\overline{)144} \\ \underline{12} \\ 24 \\ \underline{24} \\ 0 \end{array}$$

The unit rate is 48 miles per hour.

b. 60¢ for a 4-ounce can of juice is

$$\frac{60¢}{4 \text{ ounces}}.$$

Write the rate in fraction notation with the units included.

$$= \frac{60¢ \div 4}{4 \text{ ounces} \div 4}$$

Divide the numerator and the denominator by the quantity in the denominator.

$$= \frac{15¢}{1 \text{ ounce}}$$

$$= 15¢ \text{ per ounce}$$

```
    15
4)60
    4
   ──
   20
   20
   ──
    0
```

▷ The unit rate is 15¢ per ounce.

The unit rate in Example 3(b), 15¢ per ounce, involves a cost per unit. We call this a **unit price**. A unit price allows you to compare different brands and make a choice among various products of different sizes. When shopping at the supermarket, the best value among comparable brands is the product with the lowest unit price.

 **Check Point 3**   Write each of the following as a unit rate:

a. 80 books in 5 months   16 books per month

▷ **b.** 98¢ for a 7-ounce can of juice.   14¢ per ounce

Discover for Yourself

Which can of juice has the better unit price, the one for 60¢ in Example 3(b) or the one for 98¢ in Check Point 3(b)? Assuming that each unit price applies to the same juice, which is the better value?

Proportions

Solve proportions.

A **proportion** is a statement that two ratios or rates are equal. For example,

$$\frac{80 \text{ cellphone thefts}}{1000 \text{ people}} = \frac{4 \text{ cellphone thefts}}{50 \text{ people}}$$

is a proportion stating that 80 cellphone thefts per 1000 people is equivalent to 4 cellphone thefts per 50 people. We read this proportion as

"80 cellphone thefts **is to** 1000 people
as
4 cellphone thefts **is to** 50 people."

Let's use algebra and variables to explore proportions. We know that a proportion states that two ratios are equal. If the ratios are $\frac{a}{b}$ and $\frac{c}{d}$, then the proportion is

$$\frac{a}{b} = \frac{c}{d}.$$

We can clear this equation of fractions by multiplying both sides by bd:

$$\frac{a}{b} = \frac{c}{d}$$ This is the given proportion.

$$bd \cdot \frac{a}{b} = bd \cdot \frac{c}{d}$$ Multiply both sides by bd ($b \neq 0$ and $d \neq 0$). Then simplify.

On the left, $\frac{\cancel{b}d}{1} \cdot \frac{a}{\cancel{b}} = da = ad$. On the right, $\frac{b\cancel{d}}{1} \cdot \frac{c}{\cancel{d}} = bc$.

$$ad = bc.$$

We see that the following principle is true for any proportion:

The cross-products principle: $ad = bc$

The Cross-Products Principle for Proportions

If $\dfrac{a}{b} = \dfrac{c}{d}$,   then $ad = bc$.   ($b \neq 0$ and $d \neq 0$)

The cross products ad and bc are equal.

For example, since $\frac{2}{3} = \frac{6}{9}$, we see that $2 \cdot 9 = 3 \cdot 6$, or $18 = 18$. We can also use $\frac{2}{3} = \frac{6}{9}$ and conclude that $3 \cdot 6 = 2 \cdot 9$. When using the cross-products principle, it does not matter on which side of the equation each product is placed.

If three of the numbers in a proportion are known, the value of the missing quantity can be found by using the cross-products principle. This idea is illustrated in Example 4(a).

Example 4   Solving Proportions

Solve each proportion and check:

a. $\dfrac{63}{x} = \dfrac{7}{5}$ **b.** $\dfrac{20}{x - 10} = \dfrac{30}{x}$.

Solution

Cross products

a.
$$\frac{63}{x} = \frac{7}{5}$$ This is the given proportion.

$$63 \cdot 5 = 7x$$ Apply the cross-products principle.

$$315 = 7x$$ Simplify.

$$\frac{315}{7} = \frac{7x}{7}$$ Divide both sides by 7.

$$45 = x$$ Simplify.

The solution set is $\{45\}$.

Check

$$\frac{63}{45} \stackrel{?}{=} \frac{7}{5}$$ Substitute 45 for x in $\frac{63}{x} = \frac{7}{5}$.

$$\frac{7 \cdot \cancel{9}}{5 \cdot \cancel{9}} \stackrel{?}{=} \frac{7}{5}$$ Reduce $\frac{63}{45}$ to lowest terms.

$$\frac{7}{5} = \frac{7}{5}$$ This true statement verifies that the solution set is $\{45\}$.

b. $\dfrac{20}{x - 10} = \dfrac{30}{x}$ This is the given proportion.

$20x = 30(x - 10)$ Apply the cross-products principle.

$20x = 30x - 30 \cdot 10$ Use the distributive property.

$20x = 30x - 300$ Simplify.

$20x - 30x = 30x - 300 - 30x$ Subtract 30x from both sides.

$-10x = -300$ Simplify.

$\dfrac{-10x}{-10} = \dfrac{-300}{-10}$ Divide both sides by -10.

$x = 30$ Simplify.

The solution set is $\{30\}$.

Check

$\dfrac{20}{30 - 10} \overset{?}{=} \dfrac{30}{30}$ Substitute 30 for x in $\dfrac{20}{x - 10} = \dfrac{30}{x}$.

$\dfrac{20}{20} \overset{?}{=} \dfrac{30}{30}$ Subtract: $30 - 10 = 20$.

$1 = 1$ This true statement verifies that the solution set is $\{30\}$.

✓ **Check Point 4**   Solve each proportion and check:

a. $\dfrac{10}{x} = \dfrac{2}{3}$ $\{15\}$ **b.** $\dfrac{22}{60 - x} = \dfrac{2}{x}$. $\{5\}$

Applications of Proportions

⑤ Solve problems using proportions.

We now turn to practical application problems that can be solved by modeling with proportions. Here is a procedure for solving these problems:

> **Solving Applied Problems Using Proportions**
>
> 1. Read the problem and represent the unknown quantity by x (or any letter).
> 2. Set up a proportion that models the problem's conditions. List the given ratio on one side and the ratio with the unknown quantity on the other side. Each respective quantity should occupy the same corresponding position on each side of the proportion.
> 3. Drop units and apply the cross-products principle.
> 4. Solve for x and answer the question.

Example 5   Applying Proportions: Calculating Taxes

The property tax on a house with an assessed value of $480,000 is $5760. Determine the property tax on a house with an assessed value of $600,000, assuming the same tax rate.

Solution

Step 1   Represent the unknown by x. Let $x =$ the tax on the $600,000 house.

Great Question!

Are there other proportions that I can use in step 2 to model the problem's conditions?

Yes. Here are three other correct proportions you can use:

- $\dfrac{\$480,000\,\text{value}}{\$5760\,\text{tax}} = \dfrac{\$600,000\,\text{value}}{\$x\,\text{tax}}$

- $\dfrac{\$480,000\,\text{value}}{\$600,000\,\text{value}} = \dfrac{\$5760\,\text{tax}}{\$x\,\text{tax}}$

- $\dfrac{\$600,000\,\text{value}}{\$480,000\,\text{value}} = \dfrac{\$x\,\text{tax}}{\$5760\,\text{tax}}$

Each proportion gives the same cross product obtained in step 3.

Step 2 Set up a proportion. We will set up a proportion comparing taxes to assessed value.

$$\underbrace{\frac{\text{Tax on \$480,000 house}}{\text{Assessed value (\$480,000)}}}_{} \quad \text{equals} \quad \underbrace{\frac{\text{Tax on \$600,000 house}}{\text{Assessed value (\$600,000)}}}_{}$$

$$\text{Given ratio}\left\{\frac{\$5760}{\$480,000}\right. = \frac{\$x \,\leftarrow\, \text{Unknown}}{\$600,000 \,\leftarrow\, \text{Given quantity}}$$

Step 3 Drop the units and apply the cross-products principle. We drop the dollar signs and begin to solve for x.

$$\frac{5760}{480,000} = \frac{x}{600,000} \qquad \text{This is the proportion that models the problem's conditions.}$$

$$480,000x = (5760)(600,000) \qquad \text{Apply the cross-products principle.}$$

$$480,000x = 3,456,000,000 \qquad \text{Multiply.}$$

Step 4 Solve for x and answer the question.

$$\frac{480,000x}{480,000} = \frac{3,456,000,000}{480,000} \qquad \text{Divide both sides by 480,000.}$$

$$x = 7200 \qquad \text{Simplify.}$$

The property tax on the $600,000 house is $7200.

Check Point 5 The property tax on a house with an assessed value of $250,000 is $3500. Determine the property tax on a house with an assessed value of $420,000, assuming the same tax rate. $5880

Sampling in Nature

The method that was used to estimate the blue whale population described in the section opener is called the **capture-recapture method**. Because it is impossible to count each individual animal within a population, wildlife biologists randomly catch and tag a given number of animals. Sometime later they capture a second sample of animals and count the number of recaptured tagged animals. The total size of the wildlife population is then estimated using the following proportion:

$$\underset{\substack{\text{Initially}\\\text{unknown}\\(x)}}{\underbrace{\frac{\text{Original number of tagged animals}}{\text{Total number of animals in the population}}}} = \left.\underbrace{\frac{\text{Number of recaptured tagged animals}}{\text{Number of animals in second sample}}}_{}\right\}\begin{array}{c}\text{Known}\\\text{ratio}\end{array}$$

Although this is called the capture-recapture method, it is not necessary to recapture animals in order to observe whether or not they are tagged. This could be done from a distance, with binoculars for instance.

Example 6   Applying Proportions: Estimating Wildlife Population

Wildlife biologists catch, tag, and then release 135 deer back into a wildlife refuge. Two weeks later they observe a sample of 140 deer, 30 of which are tagged. Assuming the ratio of tagged deer in the sample holds for all deer in the refuge, approximately how many deer are in the refuge?

Solution

Step 1   Represent the unknown by x. Let $x =$ the total number of deer in the refuge.

Step 2   Set up a proportion.

$$\underset{\text{Unknown} \longrightarrow}{\underset{\substack{\text{Original number} \\ \text{of tagged deer} \\ \overline{\text{Total number}} \\ \text{of deer}}}{\vphantom{x}}} \underset{\text{equals}}{=} \left.\underset{\substack{\text{Number of tagged deer} \\ \text{in the observed sample} \\ \overline{\text{Total number of deer}} \\ \text{in the observed sample}}}{\vphantom{x}}\right\}\begin{array}{l}\text{Known} \\ \text{ratio}\end{array}$$

$$\frac{135}{x} = \frac{30}{140}$$

Steps 3 and 4   Apply the cross-products principle, solve, and answer the question.

$$\frac{135}{x} = \frac{30}{140} \qquad \text{This is the proportion that models the problem's conditions.}$$

$$(135)(140) = 30x \qquad \text{Apply the cross-products principle.}$$

$$18{,}900 = 30x \qquad \text{Multiply.}$$

$$\frac{18{,}900}{30} = \frac{30x}{30} \qquad \text{Divide both sides by 30.}$$

$$630 = x \qquad \text{Simplify.}$$

There are approximately 630 deer in the refuge.

✓ **Check Point 6**   Wildlife biologists catch, tag, and then release 120 deer back into a wildlife refuge. Two weeks later they observe a sample of 150 deer, 25 of which are tagged. Assuming the ratio of tagged deer in the sample holds for all deer in the refuge, approximately how many deer are in the refuge?   720 deer

Achieving Success

An effective way to understand something is to explain it to someone else. You can do this by using the exercises in the Concept and Vocabulary Check that ask you to respond with verbal or written explanations. Speaking about a new concept uses a different part of your brain than thinking about the concept. Explaining new ideas verbally will quickly reveal any gaps in your understanding. It will also help you to remember new concepts for longer periods of time.

Concept and Vocabulary Check

Exercises in the Concept and Vocabulary Check are intended for group and class discussions.

In Exercises 1–6, fill in each blank so that the resulting statement is true.

1. A comparison of two quantities by division is called a/an ____ratio____. If the quantities have different units, the comparison is called a/an ____rate____.

2. A rate that has a denominator of 1 is called a/an ____unit____ rate.

3. To write a rate as a rate with a denominator of 1, divide the numerator and the denominator by the quantity in the ____denominator____.

4. The word ____per____ is used to describe the division in a rate with a denominator of 1.

5. A statement that two ratios are equal is called a/an ____proportion____.

6. The cross-products principle states that if $\dfrac{a}{b} = \dfrac{c}{d}$ $(b \neq 0$ and $d \neq 0)$ then ____$ad = bc$____.

In Exercises 7–10, determine whether each statement is true or false. If the statement is false, make the necessary change(s) to produce a true statement. Changes to false statements will vary.

7. A statement that two proportions are equal is called a ratio. false

8. The ratio $\frac{2}{3}$ means the same thing as the ratio $\frac{3}{2}$. false

9. The ratio $\frac{3}{17}$ is in simplest form. true

10. An example of a unit rate is $\frac{100 \text{ miles}}{4 \text{ hours}}$. false

Respond to Exercises 11–14 using verbal or written explanations. 11–14. Answers will vary.

11. What is the difference between a ratio and a proportion?

12. Which of the following ratios is a rate?

$$\frac{260 \text{ miles}}{8 \text{ miles}} \qquad \frac{260 \text{ miles}}{8 \text{ gallons}}$$

What is the difference between these ratios?

13. Explain how to use the rate

$$\frac{90¢}{5 \text{ ounces}}$$

to write a unit rate.

14. Explain how to solve the following proportion: $\frac{x}{8} = \frac{5}{10}$.

Exercise Set 2.4

Practice Exercises

In Exercises 1–10, write each ratio in fraction notation. Express the fraction in simplest form.

1. 8 to 20 $\frac{2}{5}$

2. 15 to 25 $\frac{3}{5}$

3. 22:14 $\frac{11}{7}$

4. 26:16 $\frac{13}{8}$

5. 6 hours to 24 hours $\frac{1}{4}$

6. 8 hours to 24 hours $\frac{1}{3}$

7. 120 miles to 80 miles $\frac{3}{2}$

8. 150 miles to 60 miles $\frac{5}{2}$

9. 11 yards to 3 yards $\frac{11}{3}$

10. 14 inches to 5 inches $\frac{14}{5}$

In Exercises 11–20, write each rate as a fraction in simplest form.

11. 24 points in 36 minutes

12. 36 points in 42 minutes

13. 159 miles on 6 gallons of gasoline

14. 180 miles on 8 gallons of gasoline

15. 964 calories in 8 servings

16. 963 calories in 6 servings

17. 210 robberies per 100,000 residents

18. 215 robberies per 100,000 residents

19. 75 students for every 9 teachers

20. 32 meals for every 24 passengers

11. $\frac{2 \text{ points}}{3 \text{ minutes}}$

12. $\frac{6 \text{ points}}{7 \text{ minutes}}$

13. $\frac{53 \text{ miles}}{2 \text{ gallons}}$

14. $\frac{45 \text{ miles}}{2 \text{ gallons}}$

15. $\frac{241 \text{ calories}}{2 \text{ servings}}$

16. $\frac{321 \text{ calories}}{2 \text{ servings}}$

17. $\frac{21 \text{ robberies}}{10,000 \text{ residents}}$

18. $\frac{43 \text{ robberies}}{20,000 \text{ residents}}$

19. $\frac{25 \text{ students}}{3 \text{ teachers}}$

20. $\frac{4 \text{ meals}}{3 \text{ passengers}}$

In Exercises 21–32, write each rate as a unit rate. Express the answer as a fraction with a denominator of 1. Then express the answer in words using the word "per."

21. 126 miles in 3 hours

22. 162 miles in 3 hours

23. 154 roses for every 7 vases

24. 108 roses for every 6 vases

25. 133 calories in 19 crackers

26. 128 calories in 16 crackers

27. 50 parking spaces for 50 apartments

28. 65 parking spaces for 65 apartments

29. 3612 branches per 42 trees

30. 3432 branches per 44 trees

31. 96¢ for an 8-ounce can of juice

32. 78¢ for a 6-ounce can of juice

21. $\frac{42 \text{ miles}}{1 \text{ hour}} = 42$ miles per hour

22. $\frac{54 \text{ miles}}{1 \text{ hour}} = 54$ miles per hour

23. $\frac{22 \text{ roses}}{1 \text{ vase}} = 22$ roses per vase

24. $\frac{18 \text{ roses}}{1 \text{ vase}} = 18$ roses per vase

25. $\frac{7 \text{ calories}}{1 \text{ cracker}} = 7$ calories per cracker

26. $\frac{8 \text{ calories}}{1 \text{ cracker}} = 8$ calories per cracker

29. $\frac{86 \text{ branches}}{1 \text{ tree}} = 86$ branches per tree

30. $\frac{78 \text{ branches}}{1 \text{ tree}} = 78$ branches per tree

31. $\frac{12¢}{1 \text{ ounce}} = 12¢$ per ounce

32. $\frac{13¢}{1 \text{ ounce}} = 13¢$ per ounce

In Exercises 33–46, solve each proportion and check.

33. $\frac{24}{x} = \frac{12}{7}$ {14}

34. $\frac{56}{x} = \frac{8}{7}$ {49}

35. $\frac{x}{6} = \frac{18}{4}$ {27}

36. $\frac{x}{32} = \frac{3}{24}$ {4}

37. $\frac{-3}{8} = \frac{x}{40}$ {−15}

38. $\frac{-3}{8} = \frac{6}{x}$ {−16}

39. $\frac{x}{12} = -\frac{3}{4}$ {−9}

40. $\frac{x}{64} = -\frac{9}{16}$ {−36}

41. $\frac{x-2}{12} = \frac{8}{3}$ {34}

42. $\frac{x-4}{10} = \frac{3}{5}$ {10}

43. $\frac{x}{7} = \frac{x+14}{5}$ {−49}

44. $\frac{x}{5} = \frac{x-3}{2}$ {5}

45. $\frac{y+10}{10} = \frac{y-2}{4}$ {10}

46. $\frac{2}{y-5} = \frac{3}{y+6}$ {27}

27. $\frac{1 \text{ parking space}}{1 \text{ apartment}} = 1$ parking space per apartment

28. $\frac{1 \text{ parking space}}{1 \text{ apartment}} = 1$ parking space per apartment

Practice Plus

In Exercises 47–52, solve each proportion for x.

47. $\dfrac{x}{a} = \dfrac{b}{c}$ $\quad x = \dfrac{ab}{c}$

48. $\dfrac{a}{x} = \dfrac{b}{c}$ $\quad x = \dfrac{ac}{b}$

49. $\dfrac{a+b}{c} = \dfrac{x}{d}$ $\quad x = \dfrac{ad+bd}{c}$

50. $\dfrac{a-b}{c} = \dfrac{x}{d}$ $\quad x = \dfrac{ad-bd}{c}$

51. $\dfrac{x+a}{a} = \dfrac{b+c}{c}$ $\quad x = \dfrac{ab}{c}$

52. $\dfrac{ax-b}{b} = \dfrac{c-d}{d}$ $\quad x = \dfrac{bc}{ad}$

Application Exercises

Reefer Madness. *As the nation takes a softer stance on marijuana, the National Survey on Drug Use and Health found more Americans are using pot. The bar graph shows the number of people per 1000, by age group, who reported using marijuana regularly in 2000 and in 2012. Use the information displayed by the graph to solve Exercises 53–54. Express ratios as fractions in simplest form.*

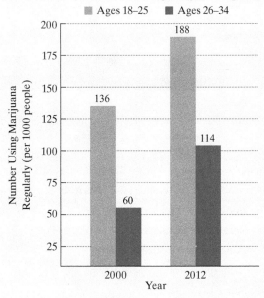

Number of Americans Using Marijuana Regularly

■ Ages 18–25   ■ Ages 26–34

Source: National Survey on Drug Use and Health

53. a. Find the ratio of the number of Americans ages 18 to 25 who regularly used marijuana in 2012 to the number in the same age group who regularly used the drug in 2000. $\frac{47}{34}$

b. Find the ratio of the number of Americans ages 26 to 34 who regularly used marijuana in 2012 to the number in the same age group who regularly used the drug in 2000.

c. Use cross products to determine if the ratios in parts (a) and (b) are proportional. not proportional

54. a. Find the ratio of the number of Americans ages 18 to 25 who regularly used marijuana in 2000 to the number ages 26 to 34 who regularly used the drug in the same year.

b. Find the ratio of the number of Americans ages 18 to 25 who regularly used marijuana in 2012 to the number ages 26 to 34 who regularly used the drug in the same year.

c. Use cross products to determine if the ratios in parts (a) and (b) are proportional. not proportional

53. b. $\frac{19}{10}$ **54. a.** $\frac{34}{15}$ **b.** $\frac{94}{57}$

55. a. $\dfrac{\$7}{1\,\text{CD}}$; \$7 per CD   **b.** $\dfrac{\$6}{1\,\text{CD}}$; \$6 per CD

55. A music store sells used CDs.

a. The store charges \$28 for 4 CDs. Write the unit rate, or the unit price, as a fraction. What is the price per CD?

b. The store charges \$42 for 7 CDs. Write the unit rate, or the unit price, as a fraction. What is the price per CD?

c. Use the unit prices in parts (a) and (b) to determine which package of CDs is the better deal. \$42 for 7 CDs

56. A nursery is having a sale on apple trees.

a. The store charges \$96 for 8 apple trees. Write the unit rate, or the unit price, as a fraction. What is the price per tree? $\dfrac{\$12}{1\,\text{tree}}$; \$12 per tree

b. The store charges \$132 for 12 apple trees. Write the unit rate, or the unit price, as a fraction. What is the price per tree? $\dfrac{\$11}{1\,\text{tree}}$; \$11 per tree

c. Use the unit prices in parts (a) and (b) to determine which offer is the better deal. \$132 for 12 apple trees

Solve Exercises 57–64 using proportions.

Drug dosage is frequently based on a patient's weight. For example, 150 milligrams of the drug Didronel (used to treat irregular bone formation) should be administered daily for every 20 pounds of a patient's weight. Use this information to solve Exercises 57–58.

57. a. What is the daily dosage for a woman who weighs 130 pounds? 975 milligrams

b. If this woman is to receive one 400-milligram tablet of Didronel every 12 hours, is she receiving enough medication? no

58. a. What is the daily dosage for a man who weighs 210 pounds? 1575 milligrams

b. If this man is to receive one 400-milligram tablet of Didronel every 6 hours, is he receiving enough medication? yes

59. The property tax on a house with an assessed value of \$520,000 is \$7280. Determine the property tax on a house with an assessed value of \$650,000, assuming the same tax rate. \$9100

60. The property tax on a house with an assessed value of \$350,000 is \$4200. Determine the property tax on a house with an assessed value of \$720,000, assuming the same tax rate. \$8640

61. An alligator's tail length is proportional to its body length. An alligator with a body length of 4 feet has a tail length of 3.6 feet. What is the tail length of an alligator whose body length is 6 feet? 5.4 feet

|←——— Body length ———→|←——— Tail length ———→|

62. An object's weight on the moon is proportional to its weight on Earth. Neil Armstrong, the first person to step on the moon on July 20, 1969, weighed 360 pounds on Earth (with all of his equipment on) and 60 pounds on the moon. What is the moon weight of a person who weighs 186 pounds on Earth? 31 pounds

63. St. Paul Island in Alaska has 12 fur seal rookeries (breeding places). In 1961, to estimate the fur seal pup population in the Gorbath rookery, 4963 fur seal pups were tagged in early August. In late August, a sample of 900 pups was observed and 218 of these were found to have been previously tagged. Estimate the total number of fur seal pups in this rookery.

64. To estimate the number of bass in a lake, wildlife biologists tagged 50 bass and released them in the lake. Later they netted 108 bass and found that 27 of them were tagged. Approximately how many bass are in the lake?   200 bass

Critical Thinking Exercises

65. The front sprocket on a bicycle has 60 teeth and the rear sprocket has 20 teeth. For mountain biking, an owner needs a 5-to-1 front-to-rear ratio. If only one of the sprockets is to be replaced, describe the two ways in which this can be done.

63. 20,489 fur seal pups   65. Either change the rear to 12 teeth or change the front to 100 teeth.

66. A baseball player's batting average is the ratio of the number of hits to the number of times at bat. A baseball player has 10 hits out of 20 times at bat. How many consecutive times at bat must a hit be made to raise the player's batting average to 0.600?   5 consecutive times at bat

Technology Exercises

Use a calculator to solve Exercises 67–68. Round your answer to two decimal places.

67. Solve: $\dfrac{7.32}{x} = \dfrac{-19.03}{28}$ $\{-10.77\}$

68. On a map, 2 inches represent 13.47 miles. How many miles does a person plan to travel if the distance on the map is 9.85 inches?   approximately 66.34 miles

---

2.5 | Modeling Using Variation

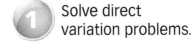

What am I supposed to learn?

After you have read this section, you should be able to:

 Solve direct variation problems.

 Solve inverse variation problems.

 Solve combined variation problems.

Have you ever wondered how telecommunication companies estimate the number of phone calls expected per day between two cities? The formula

$$C = \frac{0.02P_1P_2}{d^2}$$

shows that the daily number of phone calls, C, increases as the populations of the cities, P_1 and P_2, in thousands, increase and decreases as the distance, d, between the cities increases.

Certain formulas occur so frequently in applied situations that they are given special names. Variation formulas show how one quantity changes in relation to other quantities. Quantities can vary *directly* or *inversely*. In this section, we look at situations that can be modeled by each of these kinds of variation.

Direct Variation

① Solve direct variation problems.

When you swim underwater, the pressure in your ears depends on the depth at which you are swimming. The formula

$$p = 0.43d$$

describes the water pressure, p, in pounds per square inch, at a depth of d feet. We can use this formula to determine the pressure in your ears at various depths:

If $d = 20, p = 0.43(20) = 8.6$.   At a depth of 20 feet, water pressure is 8.6 pounds per square inch.

Doubling the depth doubles the pressure.

If $d = 40, p = 0.43(40) = 17.2$.   At a depth of 40 feet, water pressure is 17.2 pounds per square inch.

Doubling the depth doubles the pressure.

If $d = 80, p = 0.43(80) = 34.4$.   At a depth of 80 feet, water pressure is 34.4 pounds per square inch.

The formula $p = 0.43d$ illustrates that water pressure is a constant multiple of your underwater depth. If your depth is doubled, the pressure is doubled; if your depth is tripled, the pressure is tripled; and so on. Because of this, the pressure in your ears is said to **vary directly** as your underwater depth. The **equation of variation** is

$$p = 0.43d.$$

Generalizing our discussion of pressure and depth, we obtain the following statement:

Direct Variation

If a situation is described by an equation in the form

$$y = kx,$$

where k is a nonzero constant, we say that **y varies directly as x**. The number k is called the **constant of variation**.

Problems involving direct variation can be solved using the following procedure. This procedure applies to direct variation problems, as well as to the other kinds of variation problems that we will discuss.

Solving Variation Problems

1. Write an equation that models the given English statement.

2. Substitute the given pair of values into the equation in step 1 and solve for k, the constant of variation.

3. Substitute the value of k into the equation in step 1.

4. Use the equation from step 3 to answer the problem's question.

Example 1   Solving a Direct Variation Problem

The volume of blood, B, in a person's body varies directly as body weight, W. A person who weighs 160 pounds has approximately 5 quarts of blood. Estimate the number of quarts of blood in a person who weighs 200 pounds.

Solution

Step 1 Write an equation. We know that *y varies directly as x* is expressed as

$$y = kx.$$

By changing letters, we can write an equation that models the following English statement: The volume of blood, B, varies directly as body weight, W.

$$B = kW$$

Step 2 Use the given values to find k. A person who weighs 160 pounds has approximately 5 quarts of blood. Substitute 160 for W and 5 for B in the direct variation equation. Then solve for k.

$B = kW$	The volume of blood varies directly as body weight.
$5 = k \cdot 160$	Substitute 160 for W and 5 for B.
$\dfrac{5}{160} = \dfrac{k \cdot 160}{160}$	Divide both sides by 160.
$0.03125 = k$	Express $\frac{5}{160}$, or $\frac{1}{32}$, in decimal form.

Step 3   Substitute the value of k into the equation.

$$B = kW \qquad \text{Use the equation from step 1.}$$

$$B = 0.03125W \qquad \text{Replace } k, \text{ the constant of variation, with } 0.03125.$$

Step 4   Answer the problem's question. We are interested in estimating the number of quarts of blood in a person who weighs 200 pounds. Substitute 200 for W in $B = 0.03125W$ and solve for B.

$$B = 0.03125W \qquad \text{This is the equation from step 3.}$$

$$B = 0.03125(200) \qquad \text{Substitute 200 for } W.$$

$$= 6.25 \qquad \text{Multiply.}$$

A person who weighs 200 pounds has approximately 6.25 quarts of blood.

✓ **Check Point 1**   The number of gallons of water, W, used when taking a shower varies directly as the time, t, in minutes, in the shower. A shower lasting 5 minutes uses 30 gallons of water. How much water is used in a shower lasting 11 minutes?   66 gallons

A direct variation situation can involve variables to higher powers. For example, y can vary directly as x^2 ($y = kx^2$) or as x^3 ($y = kx^3$).

Direct Variation with Powers

y **varies directly as the nth power of x** if there exists some nonzero constant k such that

$$y = kx^n.$$

Example 2   Solving a Direct Variation Problem

The distance, s, that a body falls from rest varies directly as the square of the time, t, of the fall. If skydivers fall 64 feet in 2 seconds, how far will they fall in 4.5 seconds?

Solution

Step 1   Write an equation. We know that y *varies directly as the square of x* is expressed as

$$y = kx^2.$$

By changing letters, we can write an equation that models the following English statement: Distance, s, varies directly as the square of time, t, of the fall.

$$s = kt^2$$

Step 2   Use the given values to find k. Skydivers fall 64 feet in 2 seconds. Substitute 64 for s and 2 for t in the direct variation equation. Then solve for k.

$$s = kt^2 \qquad \text{Distance varies directly as the square of time.}$$

$$64 = k \cdot 2^2 \qquad \text{Skydivers fall 64 feet in 2 seconds.}$$

$$64 = 4k \qquad \text{Simplify: } 2^2 = 4.$$

$$\frac{64}{4} = \frac{4k}{4} \qquad \text{Divide both sides by 4.}$$

$$16 = k \qquad \text{Simplify.}$$

Step 3   Substitute the value of k into the equation.

$$s = kt^2 \qquad \text{Use the equation from step 1.}$$

$$s = 16t^2 \qquad \text{Replace } k, \text{ the constant of variation, with 16.}$$

Step 4 Answer the problem's question. How far will the skydivers fall in 4.5 seconds? Substitute 4.5 for t in $s = 16t^2$ and solve for s.

$$s = 16(4.5)^2 = 16(20.25) = 324$$

Thus, in 4.5 seconds, the skydivers will fall 324 feet.

✓ Check Point 2 The weight of a great white shark varies directly as the cube of its length. A great white shark caught off Catalina Island, California, was 15 feet long and weighed 2025 pounds. What was the weight of the 25-foot-long shark in the novel *Jaws*?   ▶ 9375 pounds

Inverse Variation

2 Solve inverse variation problems.

The distance from San Francisco to Los Angeles is 420 miles. The time that it takes to drive from San Francisco to Los Angeles depends on the rate at which one drives and is given by

$$\text{Time} = \frac{420}{\text{Rate}}.$$

For example, if you average 30 miles per hour, the time for the drive is

$$\text{Time} = \frac{420}{30} = 14,$$

or 14 hours. If you average 50 miles per hour, the time for the drive is

$$\text{Time} = \frac{420}{50} = 8.4,$$

or 8.4 hours. As your rate (or speed) increases, the time for the trip decreases and vice versa.

We can express the time for the 420-mile San Francisco–Los Angeles trip using t for time and r for rate:

$$t = \frac{420}{r}.$$

This equation is an example of an **inverse variation** equation. Time, t, **varies inversely** as rate, r. When two quantities vary inversely, one quantity increases as the other decreases and vice versa.

Generalizing, we obtain the following statement:

> **Inverse Variation**
>
> If a situation is described by an equation in the form
>
> $$y = \frac{k}{x},$$
>
> where k is a nonzero constant, we say that **y varies inversely as x**. The number k is called the **constant of variation**.

We use the same procedure to solve inverse variation problems as we did to solve direct variation problems. Example 3 illustrates this procedure.

Doubling the pressure halves the volume.

Example 3 Solving an Inverse Variation Problem

When you use a spray can and press the valve at the top, you decrease the pressure of the gas in the can. This decrease of pressure causes the volume of the gas in the can to increase. Because the gas needs more room than is provided in the can, it expands in spray form through the small hole near the valve. In general, if the temperature is constant, the pressure, P, of a gas in a container varies inversely as the volume, V, of the container. The pressure of a gas sample in a container whose volume is 8 cubic inches is 12 pounds per square inch. If the sample expands to a volume of 22 cubic inches, what is the new pressure of the gas?

Solution

Step 1 Write an equation. We know that *y varies inversely as x* is expressed as

$$y = \frac{k}{x}.$$

By changing letters, we can write an equation that models the following English statement: The pressure, *P*, of a gas in a container varies inversely as the volume, *V*.

$$P = \frac{k}{V}$$

Step 2 Use the given values to find *k*. The pressure of a gas sample in a container whose volume is 8 cubic inches is 12 pounds per square inch. Substitute 12 for *P* and 8 for *V* in the inverse variation equation. Then solve for *k*.

$$P = \frac{k}{V} \qquad \text{Pressure varies inversely as volume.}$$

$$12 = \frac{k}{8} \qquad \text{The pressure in an 8 cubic-inch container is 12 pounds per square inch.}$$

$$12 \cdot 8 = \frac{k}{8} \cdot 8 \qquad \text{Multiply both sides by 8.}$$

$$96 = k \qquad \text{Simplify.}$$

Step 3 Substitute the value of *k* into the equation.

$$P = \frac{k}{V} \qquad \text{Use the equation from step 1.}$$

$$P = \frac{96}{V} \qquad \text{Replace k, the constant of variation, with 96.}$$

Step 4 Answer the problem's question. We need to find the pressure when the volume expands to 22 cubic inches. Substitute 22 for *V* and solve for *P*.

$$P = \frac{96}{V} = \frac{96}{22} = 4\frac{4}{11}$$

When the volume is 22 cubic inches, the pressure of the gas is $4\frac{4}{11}$ pounds per square inch.

✓ **Check Point 3** The length of a violin string varies inversely as the frequency of its vibrations. A violin string 8 inches long vibrates at a frequency of 640 cycles per second. What is the frequency of a 10-inch string? 512 cycles per second

Combined Variation

③ **Solve combined variation problems.**

In **combined variation**, direct and inverse variation occur at the same time. For example, as the advertising budget, *A*, of a company increases, its monthly sales, *S*, also increase. Monthly sales vary directly as the advertising budget:

$$S = kA.$$

By contrast, as the price of the company's product, *P*, increases, its monthly sales, *S*, decrease. Monthly sales vary inversely as the price of the product:

$$S = \frac{k}{P}.$$

We can combine these two variation equations into one combined equation:

$$S = \frac{kA}{P}.$$

Monthly sales, *S*, vary directly as the advertising budget, *A*, and inversely as the price of the product, *P*.

The following example illustrates an application of combined variation.

Example 4   **Solving a Combined Variation Problem**

The owners of Rollerblades Plus determine that the monthly sales, S, of its skates vary directly as its advertising budget, A, and inversely as the price of the skates, P. When \$60,000 is spent on advertising and the price of the skates is \$40, the monthly sales are 12,000 pairs of rollerblades.

a. Write an equation of variation that describes this situation.

b. Determine monthly sales if the amount of the advertising budget is increased to \$70,000.

Solution

a. Write an equation.

$$S = \frac{kA}{P}$$

> Translate "sales vary directly as the advertising budget and inversely as the skates' price."

Use the given values to find k.

$$S = \frac{kA}{P}$$ Sales vary directly as the advertising budget, A, and inversely as the skates' price, P.

$$12{,}000 = \frac{k(60{,}000)}{40}$$ When \$60,000 is spent on advertising $(A = 60{,}000)$ and the price is \$40 $(P = 40)$, monthly sales are 12,000 units $(S = 12{,}000)$.

$$12{,}000 = k \cdot 1500$$ Divide 60,000 by 40.

$$\frac{12{,}000}{1500} = \frac{k \cdot 1500}{1500}$$ Divide both sides of the equation by 1500.

$$8 = k$$ Simplify.

Therefore, the equation of variation that models monthly sales is

$$S = \frac{8A}{P}.$$ Substitute 8 for k in $S = \frac{kA}{P}$.

b. The advertising budget is increased to \$70,000, so $A = 70{,}000$. The skates' price is still \$40, so $P = 40$.

$$S = \frac{8A}{P}$$ This is the equation that models monthly sales from part (a).

$$S = \frac{8(70{,}000)}{40}$$ Substitute 70,000 for A and 40 for P.

$$S = 14{,}000$$ Simplify.

With a \$70,000 advertising budget and \$40 price, the company can expect to sell 14,000 pairs of rollerblades in a month (up from 12,000).

✔ **Check Point 4**   The number of minutes needed to solve an Exercise Set of variation problems varies directly as the number of problems and inversely as the number of people working to solve the problems. It takes 4 people 32 minutes to solve 16 problems. How many minutes will it take 8 people to solve 24 problems?

24 minutes

Achieving Success

When using your professor's office hours, show up prepared. If you are having difficulty with a concept or problem, bring your work so that your instructor can determine where you are having trouble. If you miss a lecture, read the appropriate section in the textbook, borrow class notes, and attempt the assigned homework before your office visit. Because this text has an accompanying video lesson for every section, you might find it helpful to view the CD-ROM covering the material you missed. It is not realistic to expect your professor to rehash all or part of a class lecture during office hours.

Concept and Vocabulary Check

Exercises in the Concept and Vocabulary Check are intended for group and class discussions.

In Exercises 1–4, fill in each blank so that the resulting statement is true.

1. y varies directly as x can be modeled by the equation ____$y = kx$____.

2. y varies directly as the nth power of x can be modeled by the equation ____$y = kx^n$____.

3. y varies inversely as x can be modeled by the equation ____$y = \frac{k}{x}$____.

4. y varies directly as x and inversely as z can be modeled by the equation ____$y = \frac{kx}{z}$____.

In Exercises 5–8, determine whether each statement is true or false. If the statement is false, make the necessary change(s) to produce a true statement. Changes to false statements will vary.

5. A man's weight, W, varies directly as the cube of his height, H, so

$$W = k^3 H. \quad \text{false}$$

6. Radiation machines, used to treat tumors, produce an intensity of radiation, R, that varies inversely as the square of the distance from the machine, D, so

$$R = k\sqrt{D}. \quad \text{false}$$

7. Body-mass index, BMI, varies directly as one's weight, w, in pounds, and inversely as the square of one's height, h, in inches, so,

$$\text{BMI} = \frac{kh^2}{w}. \quad \text{false}$$

8. The average number of daily phone calls, C, between two cities varies directly as the product of their populations, P_1 and P_2, and inversely as the square of the distance, d, between them, so

$$C = \frac{kP_1P_2}{d^2}. \quad \text{true}$$

Respond to Exercises 9–12 using verbal or written explanations.

9. What does it mean if two quantities vary directly?

10. In your own words, explain how to solve a variation problem.

11. What does it mean if two quantities vary inversely?

12. Explain what is meant by combined variation. Give an example with your explanation. 9–12. Answers will vary.

Exercise Set 2.5

Practice Exercises

Use the four-step procedure for solving variation problems given on page 148 to solve Exercises 1–12.

1. y varies directly as x. $y = 65$ when $x = 5$. Find y when $x = 12$.   156

2. y varies directly as x. $y = 45$ when $x = 5$. Find y when $x = 13$.   117

3. y varies directly as x^2. $y = 24$ when $x = 2$. Find y when $x = 5$.   150

4. y varies directly as x^2. $y = 45$ when $x = 3$. Find y when $x = 10$.   500

5. y varies inversely as x. $y = 12$ when $x = 5$. Find y when $x = 2$.   30

6. y varies inversely as x. $y = 6$ when $x = 3$. Find y when $x = 9$.   2

7. y varies inversely as $\sqrt{x}$. $y = 10$ when $x = 4$. Find y when $x = 100$.   2

8. y varies inversely as $\sqrt{x}$. $y = 20$ when $x = 25$. Find y when $x = 4$.   50

9. y varies directly as x and inversely as z. $y = 100$ when $x = 5$ and $z = 10$. Find y when $x = 3$ and $z = 60$.   10

10. y varies directly as x and inversely as z. $y = 20$ when $x = 5$ and $z = 100$. Find y when $x = 8$ and $z = 16$.   200

11. y varies directly as x and inversely as the square of z. $y = 20$ when $x = 50$ and $z = 5$. Find y when $x = 3$ and $z = 6$. $\frac{5}{6}$

12. a varies directly as b and inversely as the square of c. $a = 7$ when $b = 9$ and $c = 6$. Find a when $b = 4$ and $c = 8$. $\frac{7}{4}$

Practice Plus

In Exercises 13–18, write an equation that expresses each relationship. Then solve the equation for y.

13. x varies directly as y. $x = ky; y = \dfrac{x}{k}$

14. x^2 varies directly as y. $x^2 = ky; y = \dfrac{x^2}{k}$

15. x varies inversely as y. $x = \dfrac{k}{y}; y = \dfrac{k}{x}$

16. x^2 varies inversely as y. $x^2 = \dfrac{k}{y}; y = \dfrac{k}{x^2}$

17. x varies directly as y and inversely as z. $x = \dfrac{ky}{z}; y = \dfrac{xz}{k}$

18. x^2 varies directly as y and inversely as z. $x^2 = \dfrac{ky}{z}; y = \dfrac{x^2 z}{k}$

Application Exercises

Use the four-step procedure for solving variation problems given on page 148 to solve Exercises 19–28.

19. The height that a ball bounces varies directly as the height from which it was dropped. A tennis ball dropped from 12 inches bounces 8.4 inches. From what height was the tennis ball dropped if it bounces 56 inches?   80 inches

20. The distance that a spring will stretch varies directly as the force applied to the spring. A force of 12 pounds is needed to stretch a spring 9 inches. What force is required to stretch the spring 15 inches?   20 pounds

21. If all men had identical body types, their weight would vary directly as the cube of their height. Shown on the right is Robert Wadlow, who reached a record height of 8 feet 11 inches (107 inches) before his death at age 22. If a man who is 5 feet 10 inches tall (70 inches) with the same body type as Mr. Wadlow weighs 170 pounds, what was Robert Wadlow's weight shortly before his death?   approximately 607 pounds

22. The number of houses that can be served by a water pipe varies directly as the square of the diameter of the pipe. A water pipe that has a 10-centimeter diameter can supply 50 houses.

    a. How many houses can be served by a water pipe that has a 30-centimeter diameter?   450 houses

    b. What size water pipe is needed for a new subdivision of 1250 houses?   50 cm

23. The figure shows that a bicyclist tips the cycle when making a turn. The angle B, formed by the vertical direction and the bicycle, is called the banking angle. The banking angle varies inversely as the cycle's turning radius. When the turning radius is 4 feet, the banking angle is 28°. What is the banking angle when the turning radius is 3.5 feet?   32°

24. The water temperature of the Pacific Ocean varies inversely as the water's depth. At a depth of 1000 meters, the water temperature is 4.4° Celsius. What is the water temperature at a depth of 5000 meters?   0.88°C

25. Radiation machines, used to treat tumors, produce an intensity of radiation that varies inversely as the square of the distance from the machine. At 3 meters, the radiation intensity is 62.5 milliroentgens per hour. What is the intensity at a distance of 2.5 meters? 90 milliroentgens per hour

26. The illumination provided by a car's headlight varies inversely as the square of the distance from the headlight. A car's headlight produces an illumination of 3.75 footcandles at a distance of 40 feet. What is the illumination when the distance is 50 feet? 2.4 footcandles

27. Body-mass index, or BMI, takes both weight and height into account when assessing whether an individual is underweight or overweight. BMI varies directly as one's weight, in pounds, and inversely as the square of one's height, in inches. In adults, normal values for the BMI are between 20 and 25, inclusive. Values below 20 indicate that an individual is underweight and values above 30 indicate that an individual is obese. A person who weighs 180 pounds and is 5 feet, or 60 inches, tall has a BMI of 35.15. What is the BMI, to the nearest tenth, for a 170-pound person who is 5 feet 10 inches tall. Is this person overweight? BMI: 24.4; not overweight

28. One's intelligence quotient, or IQ, varies directly as a person's mental age and inversely as that person's chronological age. A person with a mental age of 25 and a chronological age of 20 has an IQ of 125. What is the chronological age of a person with a mental age of 40 and an IQ of 80? 50

Heart rates and life spans of most mammals can be modeled using inverse variation. The bar graph shows the average heart rate and the average life span of five mammals. You will use the data to solve Exercises 29–30.

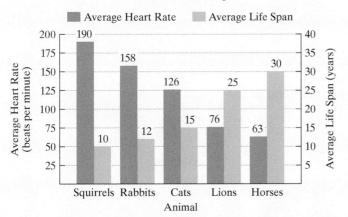

Heart Rate and Life Span

Source: The Handy Science Answer Book, Visible Ink Press, 2003.

29. **a.** A mammal's average life span, L, in years, varies inversely as its average heart rate, R, in beats per minute. Use the data shown for horses to write the equation that models this relationship. $L = \dfrac{1890}{R}$
   b. Is the inverse variation equation in part (a) an exact model or an approximate model for the data shown for lions? an approximate model
   c. Elephants have an average heart rate of 27 beats per minute. Determine their average life span. 70 years

30. **a.** A mammal's average life span, L, in years, varies inversely as its average heart rate, R, in beats per minute. Use the data shown for cats to write the equation that models this relationship. $L = \dfrac{1890}{R}$
   b. Is the inverse variation equation in part (a) an exact model or an approximate model for the data shown for squirrels? an approximate model
   c. Mice have an average heart rate of 634 beats per minute. Determine their average life span, rounded to the nearest year. 3 years

Critical Thinking Exercises

31. Sound intensity varies inversely as the square of the distance from the sound source. If you are in a movie theater and you change your seat to one that is twice as far from the speakers, how does the new sound intensity compare to that of your original seat? $\frac{1}{4}$ of what it was originally

32. Many people claim that as they get older, time seems to pass more quickly. Suppose that the perceived length of a period of time is inversely proportional to your age. How long will a year seem to be when you are three times as old as you are now?

33. In a hurricane, the wind pressure varies directly as the square of the wind velocity. If wind pressure is a measure of a hurricane's destructive capacity, what happens to this destructive power when the wind speed doubles?

34. The heat generated by a stove element varies directly as the square of the voltage and inversely as the resistance. If the voltage remains constant, what needs to be done to triple the amount of heat generated? The resistance should be divided by 3.

35. Galileo's telescope brought about revolutionary changes in astronomy. A comparable leap in our ability to observe the universe took place as a result of the Hubble Space Telescope. The space telescope can see stars and galaxies whose brightness is $\frac{1}{50}$ of the faintest objects now observable using ground-based telescopes. Use the fact that the brightness of a point source, such as a star, varies inversely as the square of its distance from an observer to show that the space telescope can see about seven times farther than a ground-based telescope.
   Distance is increased by $\sqrt{50}$, or about 7.07, for the space telescope.

Group Exercise

36. Begin by deciding on a product that interests the group because you are now in charge of advertising this product. Demand for the product varies directly as the amount spent on advertising and inversely as the price of the product. However, as more money is spent on advertising, the price of your product rises. Under what conditions would members recommend an increased expense in advertising? Once you've determined what your product is, write formulas for the given conditions and experiment with hypothetical numbers. What other factors might you take into consideration in terms of your recommendation? How do these factors affect the demand for your product? Answers will vary.

32. $\frac{1}{3}$ of a year   33. The wind pressure is 4 times more destructive.

2.6 Linear Inequalities in One Variable

What am I supposed to learn?

After you have read this section, you should be able to:

① Graph subsets of real numbers on a number line.

② Solve linear inequalities.

③ Identify linear inequalities with no solution or all real numbers as solutions.

④ Solve applied problems using linear inequalities.

Rent-a-Heap, a car rental company, charges $125 per week plus $0.20 per mile to rent one of their cars. Suppose you are limited by how much money you can spend for the week: You can spend at most $335. If we let x represent the number of miles you drive the heap in a week, we can write an inequality that models the given conditions:

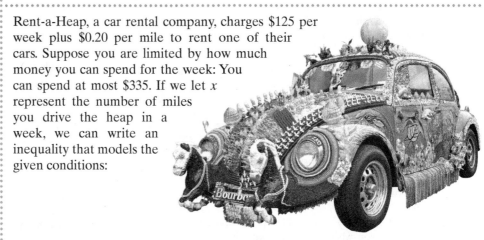

The weekly charge of $125	plus	the charge of $0.20 per mile for x miles	must be less than or equal to	$335.
125	+	0.20x	≤	335.

Notice that the highest exponent on the variable is 1. Such an inequality is called a *linear inequality in one variable*. The symbol between the two sides of an inequality can be ≤ (is less than or equal to), < (is less than), ≥ (is greater than or equal to), or > (is greater than).

In this section, we will study how to solve linear inequalities such as $125 + 0.20x \leq 335$. **Solving an inequality** is the process of finding the set of numbers that makes the inequality a true statement. These numbers are called the **solutions** of the inequality and we say that they **satisfy** the inequality. The set of all solutions is called the **solution set** of the inequality. We begin by discussing how to represent these solution sets, which are subsets of real numbers, on a number line.

A Brief Review • Representing Sets

- Sets can be designated by word descriptions, the roster method (a listing within braces, separating elements with commas), or set-builder notation:

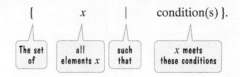

{	x	|	condition(s) }.
The set of	all elements x	such that	x meets these conditions

- Examples

Word Description	Roster Method	Set-Builder Notation
The set of weekend days	{Saturday, Sunday}	$\{x \mid x$ is a weekend day$\}$
The set of letters in the word *rear*	{r, e, a}	$\{x \mid x$ is a letter in the word *rear*$\}$

Graphing Subsets of Real Numbers on a Number Line

 Graph subsets of real numbers on a number line.

Table 2.4 shows how to represent various subsets of real numbers on a number line. Open dots (∘) indicate that a number is not included in a set. Closed dots (•) indicate that a number is included in a set.

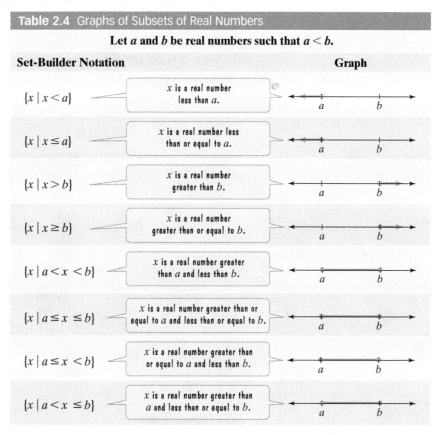

Table 2.4 Graphs of Subsets of Real Numbers

Let a and b be real numbers such that $a < b$.

Set-Builder Notation		Graph
$\{x \mid x < a\}$	x is a real number less than a.	
$\{x \mid x \le a\}$	x is a real number less than or equal to a.	
$\{x \mid x > b\}$	x is a real number greater than b.	
$\{x \mid x \ge b\}$	x is a real number greater than or equal to b.	
$\{x \mid a < x < b\}$	x is a real number greater than a and less than b.	
$\{x \mid a \le x \le b\}$	x is a real number greater than or equal to a and less than or equal to b.	
$\{x \mid a \le x < b\}$	x is a real number greater than or equal to a and less than b.	
$\{x \mid a < x \le b\}$	x is a real number greater than a and less than or equal to b.	

Example 1   Graphing Subsets of Real Numbers

Graph each set:

a. $\{x \mid x < 3\}$ **b.** $\{x \mid x \ge -1\}$ **c.** $\{x \mid -1 < x \le 3\}$.

Solution

a. $\{x \mid x < 3\}$ x is a real number less than **3**.

b. $\{x \mid x \ge -1\}$ x is a real number greater than or equal to −1.

c. $\{x \mid -1 < x \le 3\}$ x is a real number greater than −1 and less than or equal to **3**.

✓ **Check Point 1**   Graph each set:

a. $\{x \mid x < 4\}$ *   **b.** $\{x \mid x \ge -2\}$ *   **c.** $\{x \mid -4 \le x < 1\}$.   *

*See Answers to Selected Exercises.

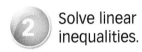

Solve linear inequalities.

Solving Linear Inequalities in One Variable

We know that a linear equation in x can be expressed as $ax + b = 0$. A **linear inequality in x** can be written in one of the following forms:

$$ax + b < 0, \quad ax + b \leq 0, \quad ax + b > 0, \quad ax + b \geq 0.$$

In each form, $a \neq 0$.

Back to our question that opened this section: How many miles can you drive your Rent-a-Heap car if you can spend at most $335? We answer the question by solving

$$0.20x + 125 \leq 335$$

for x. The solution procedure is nearly identical to that for solving

$$0.20x + 125 = 335.$$

Our goal is to get x by itself on the left side. We do this by subtracting 125 from both sides to isolate $0.20x$:

$$0.20x + 125 \leq 335 \qquad \text{This is the given inequality.}$$

$$0.20x + 125 - 125 \leq 335 - 125 \qquad \text{Subtract 125 from both sides.}$$

$$0.20x \leq 210. \qquad \text{Simplify.}$$

Finally, we isolate x from $0.20x$ by dividing both sides of the inequality by 0.20:

$$\frac{0.20x}{0.20} \leq \frac{210}{0.20} \qquad \text{Divide both sides by 0.20.}$$

$$x \leq 1050. \qquad \text{Simplify.}$$

With at most $335 per week to spend, you can travel at most 1050 miles.

We started with the inequality $0.20x + 125 \leq 335$ and obtained the inequality $x \leq 1050$ in the final step. Both of these inequalities have the same solution set, namely $\{x \mid x \leq 1050\}$. Inequalities such as these, with the same solution set, are said to be **equivalent**.

We isolated x from $0.20x$ by dividing both sides of $0.20x \leq 210$ by 0.20, a positive number. Let's see what happens if we divide both sides of an inequality by a negative number. Consider the inequality $10 < 14$. Divide 10 and 14 by -2:

$$\frac{10}{-2} = -5 \quad \text{and} \quad \frac{14}{-2} = -7.$$

Because -5 lies to the right of -7 on the number line, -5 is greater than -7:

$$-5 > -7.$$

Notice that the direction of the inequality symbol is reversed:

$$10 < 14 \qquad \text{Dividing by } -2 \text{ changes}$$
$$\uparrow\downarrow \qquad \text{the direction of the}$$
$$-5 > -7. \qquad \text{inequality symbol.}$$

In general, **when we multiply or divide both sides of an inequality by a negative number, the direction of the inequality symbol is reversed**. When we reverse the direction of the inequality symbol, we say that we change the *sense* of the inequality.

We can summarize our discussion with the following statement:

Great Question!

What are some common English phrases and sentences that I can model with linear inequalities?

English phrases such as "at least" and "at most" can be modeled by inequalities.

English Sentence	Inequality
x is at least 5.	$x \geq 5$
x is at most 5.	$x \leq 5$
x is between 5 and 7.	$5 < x < 7$
x is no more than 5.	$x \leq 5$
x is no less than 5.	$x \geq 5$

Solving Linear Inequalities

The procedure for solving linear inequalities is the same as the procedure for solving linear equations, with one important exception: When multiplying or dividing both sides of the inequality by a negative number, reverse the direction of the inequality symbol, changing the sense of the inequality.

Example 2   Solving a Linear Inequality

Solve and graph the solution set: $4x - 7 \geq 5$.

Solution

Our goal is to get x by itself on the left side. We do this by first getting $4x$ by itself, adding 7 to both sides.

$$4x - 7 \geq 5 \qquad \text{This is the given inequality.}$$
$$4x - 7 + 7 \geq 5 + 7 \qquad \text{Add 7 to both sides.}$$
$$4x \geq 12 \qquad \text{Simplify.}$$

Next, we isolate x from $4x$ by dividing both sides by 4. The inequality symbol stays the same because we are dividing by a positive number.

$$\frac{4x}{4} \geq \frac{12}{4} \qquad \text{Divide both sides by 4.}$$
$$x \geq 3 \qquad \text{Simplify.}$$

The solution set consists of all real numbers that are greater than or equal to 3, expressed in set-builder notation as $\{x \mid x \geq 3\}$. The graph of the solution set is shown as follows:

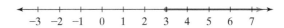

We cannot check all members of an inequality's solution set, but we can take a few values to get an indication of whether or not it is correct. In Example 2, we found that the solution set of $4x - 7 \geq 5$ is $\{x \mid x \geq 3\}$. Show that 3 and 4 satisfy the inequality, whereas 2 does not.

Check Point 2   Solve and graph the solution set: $5x - 3 \leq 17$. $\{x \mid x \leq 4\}*$

Example 3   Solving Linear Inequalities

Solve and graph the solution set:

a. $\dfrac{1}{3}x < 5$ 　　　　　　　　　　**b.** $-3x < 21$.

Solution

In each case, our goal is to isolate x. In the first inequality, this is accomplished by multiplying both sides by 3. In the second inequality, we can do this by dividing both sides by -3.

a. 　　$\dfrac{1}{3}x < 5$ 　　　　This is the given inequality.

$$3 \cdot \frac{1}{3}x < 3 \cdot 5 \qquad \text{Isolate } x \text{ by multiplying by 3 on both sides.}$$

　　　　　　　　　The symbol $<$ stays the same because we are multiplying both sides by a positive number.

　　　　$x < 15$ 　　　　Simplify.

The solution set is $\{x \mid x < 15\}$. The graph of the solution set is shown as follows:

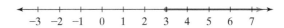

*See Answers to Selected Exercises.

b. $-3x < 21$ This is the given inequality.

$$\frac{-3x}{-3} > \frac{21}{-3}$$ Isolate x by dividing by -3 on both sides.

The symbol $<$ must be reversed because we are dividing both sides by a negative number.

$$x > -7$$ Simplify.

The solution set is $\{x \mid x > -7\}$. The graph of the solution set is shown as follows:

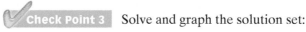

$$\begin{array}{cccccccccccc} -8 & -7 & -6 & -5 & -4 & -3 & -2 & -1 & 0 & 1 & 2 \end{array}$$

✔ **Check Point 3**   Solve and graph the solution set:

a. $\frac{1}{4}x < 2$ $\{x \mid x < 8\}$*

▶ **b.** $-6x < 18.$ $\{x \mid x > -3\}$*

Example 4   Solving a Linear Inequality

Solve and graph the solution set: $6x - 12 > 8x + 2$.

Solution

We will get x by itself on the left side. We begin by subtracting $8x$ from both sides so that the variable term appears on the left.

$$6x - 12 > 8x + 2$$ This is the given inequality.

$$6x - 8x - 12 > 8x - 8x + 2$$ Subtract 8x on both sides with the goal of isolating x on the left.

$$-2x - 12 > 2$$ Simplify.

Next, we get $-2x$ by itself, adding 12 to both sides.

$$-2x - 12 + 12 > 2 + 12$$ Add 12 to both sides.

$$-2x > 14$$ Simplify.

In order to solve $-2x > 14$, we isolate x from $-2x$ by dividing both sides by -2. The direction of the inequality symbol must be reversed because we are dividing by a negative number.

$$\frac{-2x}{-2} < \frac{14}{-2}$$ Divide both sides by -2 and change the sense of the inequality.

$$x < -7$$ Simplify.

The solution set is $\{x \mid x < -7\}$. The graph of the solution set is shown as follows:

$$\begin{array}{cccccccccccc} -9 & -8 & -7 & -6 & -5 & -4 & -3 & -2 & -1 & 0 & 1 \end{array}$$

✔ **Check Point 4**   Solve and graph the solution set: $7x - 3 > 13x + 33$.

$\{x \mid x < -6\}$*

*See Answers to Selected Exercises.

Example 5   Solving a Linear Inequality

Solve and graph the solution set:

$$2(x - 3) + 5x \le 8(x - 1).$$

Solution

Begin by simplifying the algebraic expression on each side.

$$2(x - 3) + 5x \le 8(x - 1)$$ This is the given inequality.

$$2x - 6 + 5x \le 8x - 8$$ Use the distributive property.

$$7x - 6 \le 8x - 8$$ Add like terms on the left: $2x + 5x = 7x$.

We will get x by itself on the left side. Subtract $8x$ from both sides.

$$7x - 8x - 6 \le 8x - 8x - 8$$

$$-x - 6 \le -8$$

Next, we get $-x$ by itself, adding 6 to both sides.

$$-x - 6 + 6 \le -8 + 6$$

$$-x \le -2$$

To isolate x, we must eliminate the negative sign in front of the x. Because $-x$ means $-1x$, we can do this by dividing both sides of the inequality by -1. This reverses the direction of the inequality symbol.

$$\frac{-x}{-1} \ge \frac{-2}{-1}$$ Divide both sides by -1 and change the sense of the inequality.

$$x \ge 2$$ Simplify.

The solution set is $\{x \mid x \ge 2\}$. The graph of the solution set is shown as follows:

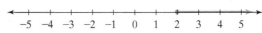

Check Point 5   Solve and graph the solution set:

$$2(x - 3) - 1 \le 3(x + 2) - 14.$$ $\{x \mid x \ge 1\}*$

Great Question!

Do I have to solve $7x - 6 \le 8x - 8$ by isolating the variable on the left?

No. You can solve

$$7x - 6 \le 8x - 8$$

by isolating x on the right side. Subtract $7x$ from both sides and add 8 to both sides:

$$7x - 6 - 7x \le 8x - 8 - 7x$$
$$-6 \le x - 8$$
$$-6 + 8 \le x - 8 + 8$$
$$2 \le x.$$

This last inequality means the same thing as

$$x \ge 2.$$

Solution sets, in this case $\{x \mid x \ge 2\}$, are expressed with the variable on the left and the constant on the right.

In our next example, the inequality has three parts:

$$-3 < 2x + 1 \le 3.$$

$2x + 1$ is greater than -3 and less than or equal to **3**.

By performing the same operation on all three parts of the inequality, our goal is to **isolate x in the middle**.

*See Answers to Selected Exercises.

Example 6   **Solving a Three-Part Inequality**

Solve and graph the solution set:

$$-3 < 2x + 1 \leq 3.$$

Solution

We would like to isolate x in the middle. We can do this by first subtracting 1 from all three parts of the inequality. Then we isolate x from $2x$ by dividing all three parts of the inequality by 2.

$-3 < 2x + 1 \leq 3$	This is the given inequality.
$-3 - 1 < 2x + 1 - 1 \leq 3 - 1$	Subtract 1 from all three parts.
$-4 < 2x \leq 2$	Simplify.
$\dfrac{-4}{2} < \dfrac{2x}{2} \leq \dfrac{2}{2}$	Divide each part by 2.
$-2 < x \leq 1$	Simplify.

The solution set consists of all real numbers greater than -2 and less than or equal to 1, represented by $\{x \mid -2 < x \leq 1\}$. The graph is shown as follows:

Check Point 6   Solve and graph the solution set on a number line: $1 \leq 2x + 3 < 11$. $\{x \mid -1 \leq x < 4\}*$

Inequalities with Unusual Solution Sets

③   Identify linear inequalities with no solution or all real numbers as solutions.

We have seen that some equations have no solution. This is also true for some inequalities. An example of such an inequality is

$$x > x + 1.$$

There is no number that is greater than itself plus 1. This inequality has no solution. Its solution set is $\varnothing$, the empty set.

By contrast, some inequalities are true for all real numbers. An example of such an inequality is

$$x < x + 1.$$

Every real number is less than itself plus 1. The solution set is expressed in set-builder notation as $\{x \mid x \text{ is a real number}\}$.

Recognizing Inequalities with No Solution or All Real Numbers as Solutions

If you attempt to solve an inequality with no solution or one that is true for every real number, you will eliminate the variable.

- An inequality with no solution results in a false statement, such as $0 > 1$. The solution set is $\varnothing$, the empty set.
- An inequality that is true for every real number results in a true statement, such as $0 < 1$. The solution set is $\{x \mid x \text{ is a real number}\}$.

---

*See Answers to Selected Exercises.

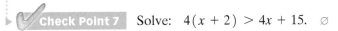 **Example 7**   Solving a Linear Inequality

Solve: $3(x + 1) > 3x + 5$.

Solution

$$3(x + 1) > 3x + 5 \qquad \text{This is the given inequality.}$$
$$3x + 3 > 3x + 5 \qquad \text{Apply the distributive property.}$$
$$3x + 3 - 3x > 3x + 5 - 3x \qquad \text{Subtract 3x from both sides.}$$
$$3 > 5 \qquad \text{Simplify.}$$

Keep reading. 3 > 5 is not the solution.

The original inequality is equivalent to the statement $3 > 5$, which is false for every value of x. The inequality has no solution. The solution set is $\varnothing$, the empty set.

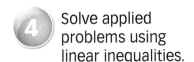 **Check Point 7**   Solve: $4(x + 2) > 4x + 15$. $\varnothing$

Example 8   Solving a Linear Inequality

Solve: $2(x + 5) \le 5x - 3x + 14$.

Solution

$$2(x + 5) \le 5x - 3x + 14 \qquad \text{This is the given inequality.}$$
$$2x + 10 \le 5x - 3x + 14 \qquad \text{Apply the distributive property.}$$
$$2x + 10 \le 2x + 14 \qquad \text{Combine like terms.}$$
$$2x + 10 - 2x \le 2x + 14 - 2x \qquad \text{Subtract 2x from both sides.}$$
$$10 \le 14 \qquad \text{Simplify.}$$

Keep reading. 10 ≤ 14 is not the solution.

The original inequality is equivalent to the statement $10 \le 14$, which is true for every value of x. The solution is the set of all real numbers, expressed in set-builder notation as $\{x \mid x \text{ is a real number}\}$.

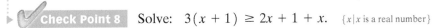 **Check Point 8**   Solve: $3(x + 1) \ge 2x + 1 + x$. $\{x \mid x \text{ is a real number}\}$

④ **Solve applied problems using linear inequalities.**

As you know, different professors may use different grading systems to determine your final course grade. Some professors require a final examination; others do not. In our next example, a final exam is required *and* it counts as two grades.

Example 9   An Application: Final Course Grade

To earn an A in a course, you must have a final average of at least 90%. On the first four examinations, you have grades of 86%, 88%, 92%, and 84%. If the final examination counts as two grades, what must you get on the final to earn an A in the course?

Solution

We will use our five-step strategy for solving algebraic word problems.

Steps 1 and 2   Represent unknown quantities in terms of x. Let

$$x = \text{your grade on the final examination.}$$

Step 3   Write an inequality in *x* that models the conditions. The average of the six grades is found by adding the grades and dividing the sum by 6.

$$\text{Average} = \frac{86 + 88 + 92 + 84 + x + x}{6}$$

Because the final counts as two grades, the *x* (your grade on the final examination) is added twice. This is also why the sum is divided by 6.

To get an A, your average must be at least 90. This means that your average must be greater than or equal to 90.

Your average		must be greater than or equal to		90.

$$\frac{86 + 88 + 92 + 84 + x + x}{6} \qquad \geq \qquad 90$$

Step 4   Solve the inequality and answer the problem's question.

$$\frac{86 + 88 + 92 + 84 + x + x}{6} \geq 90 \qquad \text{This is the inequality that models the given conditions.}$$

$$\frac{350 + 2x}{6} \geq 90 \qquad \text{Combine like terms in the numerator.}$$

$$6\left(\frac{350 + 2x}{6}\right) \geq 6(90) \qquad \text{Multiply both sides by 6, clearing the fraction.}$$

$$350 + 2x \geq 540 \qquad \text{Multiply.}$$

$$350 + 2x - 350 \geq 540 - 350 \qquad \text{Subtract 350 from both sides.}$$

$$2x \geq 190 \qquad \text{Simplify.}$$

$$\frac{2x}{2} \geq \frac{190}{2} \qquad \text{Divide both sides by 2.}$$

$$x \geq 95 \qquad \text{Simplify.}$$

You must get at least 95% on the final examination to earn an A in the course.

Step 5   Check. We can perform a partial check by computing the average with any grade that is at least 95. We will use 96. If you get 96% on the final examination, your average is

$$\frac{86 + 88 + 92 + 84 + 96 + 96}{6} = \frac{542}{6} = 90\frac{1}{3}.$$

Because $90\frac{1}{3} > 90$, you earn an A in the course.

✓ **Check Point 9**   To earn a B in a course, you must have a final average of at least 80%. On the first three examinations, you have grades of 82%, 74%, and 78%. If the final examination counts as two grades, what must you get on the final to earn a B in the course?   at least 83%

Achieving Success

Organizing and creating your own compact chapter summaries can reinforce what you know and help with the retention of this information. Imagine that your professor will permit two index cards of notes (3 by 5; front and back) on all exams. Organize and create such a two-card summary for the test on this chapter. Begin by determining what information you would find most helpful to include on the cards. Take as long as you need to create the summary. Based on how effective you find this strategy, you may decide to use the technique to help prepare for future exams.

Concept and Vocabulary Check

Exercises in the Concept and Vocabulary Check are intended for group and class discussions.

In Exercises 1–6, fill in each blank so that the resulting statement is true.

1. On a number line, an open dot indicates that a number <u>is not included</u> in a solution set, and a closed dot indicates that a number <u>is included</u> in a solution set.

2. If an inequality's solution set consists of all real numbers, x, that are less than a, the solution set is represented in set-builder notation as <u>$\{x \mid x < a\}$</u>.

3. If an inequality's solution set consists of all real numbers, x, that are greater than a and less than or equal to b, the solution set is represented in set-builder notation as <u>$\{x \mid a < x \le b\}$</u>.

4. When multiplying or dividing both sides of an inequality by a negative number, _____ of the inequality symbol. reverse the direction, or change the sense

5. If the solution of an inequality results in a false statement such as $0 > 1$, the solution set is <u>$\varnothing$</u>.

6. If the solution of an inequality in x results in a true statement such as $0 < 1$, the solution set is <u>$\{x \mid x$ is a real number$\}$</u>.

In Exercises 7–10, determine whether each statement is true or false. If the statement is false, make the necessary change(s) to produce a true statement.   Changes to false statements will vary.

7. The inequality $x - 3 > 0$ is equivalent to $x < 3$.   false

8. The statement "x is at most 5" is written $x < 5$.   false

9. The inequality $-4x < -20$ is equivalent to $x > -5$.   false

10. The statement "the sum of x and 6% of x is at least 80" is modeled by $x + 0.06x \ge 80$.   true

Respond to Exercises 11–14 using verbal or written explanations.   11–14. Answers will vary.

11. When graphing the solutions of an inequality, what is the difference between an open dot and a closed dot?

12. When solving an inequality, when is it necessary to change the direction of the inequality symbol? Give an example.

13. Describe ways in which solving a linear inequality is similar to solving a linear equation.

14. Describe ways in which solving a linear inequality is different than solving a linear equation.

Exercise Set 2.6

35. $\{x \mid x \le 3\}$* 36. $\{x \mid x \le -7\}$* 37. $\{x \mid x < 16\}$* 38. $\{x \mid x < 19\}$* 39. $\{x \mid x > -3\}$* 41. $\{x \mid x \ge -2\}$* 47. $\{x \mid x > \frac{11}{3}\}$* 48. $\{x \mid x \ge 4\}$* 49. $\{x \mid x > 2\}$*

Practice Exercises

In Exercises 1–12, graph each set of real numbers on a number line.

1. $\{x \mid x > 6\}$ *
2. $\{x \mid x > -2\}$ *
3. $\{x \mid x < -4\}$ *
4. $\{x \mid x < 0\}$ *
5. $\{x \mid x \ge -3\}$ *
6. $\{x \mid x \ge -5\}$ *
7. $\{x \mid x \le 4\}$ *
8. $\{x \mid x \le 7\}$ *
9. $\{x \mid -2 < x \le 5\}$ *
10. $\{x \mid -3 \le x < 7\}$ *
11. $\{x \mid -1 < x < 4\}$ *
12. $\{x \mid -7 \le x \le 0\}$ *

In Exercises 13–66, solve each inequality and graph the solution set on a number line.

13. $x - 3 > 2$ $\{x \mid x > 5\}$*
14. $x + 1 < 5$ $\{x \mid x < 4\}$*
15. $x + 4 \le 9$ $\{x \mid x \le 5\}$*
16. $x - 5 \ge 1$ $\{x \mid x \ge 6\}$*
17. $x - 3 < 0$ $\{x \mid x < 3\}$*
18. $x + 4 \ge 0$ $\{x \mid x \ge -4\}$*
19. $4x < 20$ $\{x \mid x < 5\}$*
20. $6x \ge 18$ $\{x \mid x \ge 3\}$*
21. $3x \ge -15$ $\{x \mid x \ge -5\}$*
22. $7x < -21$ $\{x \mid x < -3\}$*
23. $2x - 3 > 7$ $\{x \mid x > 5\}$*
24. $3x + 2 \le 14$ $\{x \mid x \le 4\}$*
25. $3x + 3 < 18$ $\{x \mid x < 5\}$*
26. $8x - 4 > 12$ $\{x \mid x > 2\}$*
27. $\frac{1}{2}x < 4$ $\{x \mid x < 8\}$*
28. $\frac{1}{2}x > 3$ $\{x \mid x > 6\}$*
29. $\frac{x}{3} > -2$ $\{x \mid x > -6\}$*
30. $\frac{x}{4} < -1$ $\{x \mid x < -4\}$*
31. $-3x < 15$ $\{x \mid x > -5\}$*
32. $-7x > 21$ $\{x \mid x < -3\}$*

33. $-3x \ge -15$ $\{x \mid x \le 5\}$*
34. $-7x \le -21$ $\{x \mid x \ge 3\}$*
35. $3x + 4 \le 2x + 7$
36. $2x + 9 \le x + 2$
37. $5x - 9 < 4x + 7$
38. $3x - 8 < 2x + 11$
39. $-2x - 3 < 3$
40. $14 - 3x > 5$ $\{x \mid x < 3\}$*
41. $3 - 7x \le 17$
42. $5 - 3x \ge 20$ $\{x \mid x \le -5\}$*
43. $-x < 4$ $\{x \mid x > -4\}$*
44. $-x > -3$ $\{x \mid x < 3\}$*
45. $5 - x \le 1$ $\{x \mid x \ge 4\}$*
46. $3 - x \ge -3$ $\{x \mid x \le 6\}$*
47. $2x - 5 > -x + 6$
48. $6x - 2 \ge 4x + 6$
49. $2x - 5 < 5x - 11$
50. $4x - 7 > 9x - 2$
51. $3(x + 1) - 5 < 2x + 1$
52. $4(x + 1) + 2 \ge 3x + 6$
53. $8x + 3 > 3(2x + 1) - x + 5$ $\{x \mid x > \frac{5}{3}\}$*
54. $7 - 2(x - 4) < 5(1 - 2x)$ $\{x \mid x < -\frac{5}{4}\}$*
55. $\frac{x}{4} - \frac{3}{2} \le \frac{x}{2} + 1$
56. $\frac{3x}{10} + 1 \ge \frac{1}{5} - \frac{x}{10}$
57. $1 - \frac{x}{2} > 4$ $\{x \mid x < -6\}$*
58. $7 - \frac{4}{5}x < \frac{3}{5}$ $\{x \mid x > 8\}$*
59. $6 < x + 3 < 8$
60. $7 < x + 5 < 11$
61. $-3 \le x - 2 < 1$
62. $-6 < x - 4 \le 1$
63. $-11 < 2x - 1 \le -5$
64. $3 \le 4x - 3 < 19$
65. $-3 \le \frac{2}{3}x - 5 < -1$
66. $-6 \le \frac{1}{2}x - 4 < -3$

50. $\{x \mid x < -1\}$*   51. $\{x \mid x < 3\}$*   52. $\{x \mid x \ge 0\}$*   55. $\{x \mid x \ge -10\}$*   56. $\{x \mid x \ge -2\}$*   59. $\{x \mid 3 < x < 5\}$*   60. $\{x \mid 2 < x < 6\}$*   61. $\{x \mid -1 \le x < 3\}$*

See Answers to Selected Exercises.   62. $\{x \mid -2 < x \le 5\}$ 63. $\{x \mid -5 < x \le -2\}$*   64. $\{x \mid \frac{3}{2} \le x < \frac{11}{2}\}$*   65. $\{x \mid 3 \le x < 6\}$*   66. $\{x \mid -4 \le x < 2\}$*

69. $\{x \mid x$ is a real number$\}$ **70.** $\{x \mid x$ is a real number$\}$ **73.** $\{x \mid x$ is a real number$\}$ **74.** $\{x \mid x$ is a real number$\}$ **75.** $\{x \mid x \leq 0\}$ **76.** $\{x \mid x \leq 0\}$

In Exercises 67–76, solve each inequality.

67. $4x - 4 < 4(x - 51)$ $\varnothing$ **68.** $3x - 5 < 3(x - 2)$ $\varnothing$

69. $x + 3 < x + 7$ **70.** $x + 4 < x + 10$

71. $7x \leq 7(x - 2)$ $\varnothing$ **72.** $3x + 1 \leq 3(x - 2)$ $\varnothing$

73. $2(x + 3) > 2x + 1$ **74.** $5(x + 4) > 5x + 10$

75. $5x - 4 \leq 4(x - 1)$ **76.** $6x - 3 \leq 3(x - 1)$

Practice Plus

In Exercises 77–80, write an inequality with x isolated on the left side that is equivalent to the given inequality.

77. $Ax + By > C$; Assume $A > 0$. $x > \dfrac{C - By}{A}$

78. $Ax + By \leq C$; Assume $A > 0$. $x \leq \dfrac{C - By}{A}$

79. $Ax + By > C$; Assume $A < 0$. $x < \dfrac{C - By}{A}$

80. $Ax + By \leq C$; Assume $A < 0$. $x \geq \dfrac{C - By}{A}$

In Exercises 81–86, use set-builder notation to describe all real numbers satisfying the given conditions.

81. A number increased by 5 is at least two times the number.

82. A number increased by 12 is at least four times the number.

83. Twice the sum of four and a number is at most 36. $\{x \mid x \leq 14\}$

84. Three times the sum of five and a number is at most 48.

85. If the quotient of three times a number and five is increased by four, the result is no more than 34. $\{x \mid x \leq 50\}$

86. If the quotient of three times a number and four is decreased by three, the result is no less than 9. $\{x \mid x \geq 16\}$

Application Exercises

The graphs show that the three components of love, namely passion, intimacy, and commitment, progress differently over time. Passion peaks early in a relationship and then declines. By contrast, intimacy and commitment build gradually. Use the graphs to solve Exercises 87–94. Assume that x represents years in a relationship.

The Course of Love Over Time

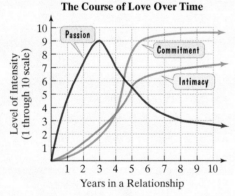

Source: R. J. Sternberg, A Triangular Theory of Love, *Psychological Review*, 93, 119–135.

87. Use set-builder notation to write an inequality that expresses for which years in a relationship intimacy is greater than commitment. $\{x \mid 0 < x < 4\}$

88. Use set-builder notation to write an inequality that expresses for which years in a relationship passion is greater than or equal to intimacy. $\{x \mid 0 \leq x \leq 5\}$

89. What is the relationship between passion and intimacy for $\{x \mid 5 \leq x < 7\}$?   intimacy $\geq$ passion or passion $\leq$ intimacy

90. What is the relationship between intimacy and commitment for $\{x \mid 4 \leq x < 7\}$?

91. What is the relationship between passion and commitment for $\{x \mid 6 < x < 8\}$?   commitment $>$ passion or passion $<$ commitment

92. What is the relationship between passion and commitment for $\{x \mid 7 < x < 9\}$?   commitment $>$ passion or passion $<$ commitment

93. What is the maximum level of intensity for passion? After how many years in a relationship does this occur?   9; after 3 years

94. After approximately how many years do levels of intensity for commitment exceed the maximum level of intensity for passion?   after approximately $5\frac{1}{2}$ years

In more U.S. marriages, spouses have different faiths. The bar graph shows the percentage of households with an interfaith marriage in 1988 and 2008. Also shown is the percentage of households in which a person of faith is married to someone with no religion.

Percentage of U.S. Households in Which Married Couples Do Not Share the Same Faith

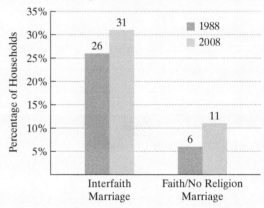

Source: General Social Survey, University of Chicago

The formula

$$I = \frac{1}{4}x + 26$$

models the percentage of U.S. households with an interfaith marriage, I, x years after 1988. The formula

$$N = \frac{1}{4}x + 6$$

models the percentage of U.S. households in which a person of faith is married to someone with no religion, N, x years after 1988. Use these models to solve Exercises 95–96.

95. a. In which years will more than 33% of U.S. households have an interfaith marriage?   years after 2016

   b. In which years will more than 14% of U.S. households have a person of faith married to someone with no religion?   years after 2020

   c. Based on your answers to parts (a) and (b), in which years will more than 33% of households have an interfaith marriage and more than 14% have a faith/no religion marriage?   years after 2020

81. $\{x \mid x \leq 5\}$ **82.** $\{x \mid x \leq 4\}$ **84.** $\{x \mid x \leq 11\}$ **90.** commitment $\geq$ intimacy or intimacy $\leq$ commitment

96. a. In which years will more than 34% of U.S. households have an interfaith marriage?   years after 2020

b. In which years will more than 15% of U.S. households have a person of faith married to someone with no religion?   years after 2024

c. Based on your answers to parts (a) and (b), in which years will more than 34% of households have an interfaith marriage and more than 15% have a faith/no religion marriage?   years after 2024

97. On two examinations, you have grades of 86 and 88. There is an optional final examination, which counts as one grade. You decide to take the final in order to get a course grade of A, meaning a final average of at least 90.

a. What must you get on the final to earn an A in the course?   at least 96

b. By taking the final, if you do poorly, you might risk the B that you have in the course based on the first two exam grades. If your final average is less than 80, you will lose your B in the course. Describe the grades on the final that will cause this to happen.   less than 66

98. On three examinations, you have grades of 88, 78, and 86. There is still a final examination, which counts as one grade.

a. In order to get an A, your average must be at least 90. If you get 100 on the final, compute your average and determine if an A in the course is possible.

b. To earn a B in the course, you must have a final average of at least 80. What must you get on the final to earn a B in the course?   at least 68

99. A car can be rented from Continental Rental for $80 per week plus 25 cents for each mile driven. How many miles can you travel if you can spend at most $400 for the week?   at most 1280 miles

100. A car can be rented from Basic Rental for $60 per week plus 50 cents for each mile driven. How many miles can you travel if you can spend at most $600 for the week?

101. An elevator at a construction site has a maximum capacity of 3000 pounds. If the elevator operator weighs 245 pounds and each cement bag weighs 95 pounds, up to how many bags of cement can be safely lifted on the elevator in one trip?   at most 29 bags

102. An elevator at a construction site has a maximum capacity of 2800 pounds. If the elevator operator weighs 265 pounds and each cement bag weighs 65 pounds, up to how many bags of cement can be safely lifted on the elevator in one trip?   at most 39 bags

98. a. 88; An A in the course is not possible.   **100.** at most 1080 miles

103. A basic cellphone plan costs $20 per month for 60 calling minutes. Additional time costs $0.40 per minute. The formula   between 80 and 110 minutes, inclusive

$$C = 20 + 0.40(x - 60)$$

gives the monthly cost for this plan, C, for x calling minutes, where $x > 60$. How many calling minutes are possible for a monthly cost of at least $28 and at most $40?

104. The formula for converting Fahrenheit temperature, F, to Celsius temperature, C, is   59°F to 95°F, inclusive

$$C = \frac{5}{9}(F - 32).$$

If Celsius temperature ranges from 15° to 35°, inclusive, what is the range for the Fahrenheit temperature?

Critical Thinking Exercises

105. A car can be rented from Basic Rental for $260 per week with no extra charge for mileage. Continental charges $80 per week plus 25 cents for each mile driven to rent the same car. How many miles should be driven in a week to make the rental cost for Basic Rental a better deal than Continental's?   more than 720 miles

106. Membership in a fitness club costs $500 yearly plus $1 per hour spent working out. A competing club charges $440 yearly plus $1.75 per hour for use of their equipment. How many hours must a person work out yearly to make membership in the first club cheaper than membership in the second club?   More than 80 hours

107. A company manufactures and sells personalized stationery. The weekly fixed cost is $3000 and it cost $3.00 to produce each package of stationery. The selling price is $5.50 per package. How many packages of stationery must be produced and sold each week for the company to generate a profit?   more than 1200 packages

108. What's wrong with this argument? Suppose x and y represent two real numbers, where $x > y$.

$2 > 1$	This is a true statement.
$2(y - x) > 1(y - x)$	Multiply both sides by $y - x$.
$2y - 2x > y - x$	Use the distributive property.
$y - 2x > -x$	Subtract y from both sides.
$y > x$	Add $2x$ to both sides.

The final inequality, $y > x$, is impossible because we were initially given $x > y$.

Since $x > y$, $y - x < 0$. Thus, when both sides were multiplied by $y - x$, the sense of the inequality should have been changed.

Chapter 2 Summary

Definitions and Concepts	**Examples**

Section 2.1 Algebraic Expressions and Formulas

An algebraic expression combines variables and numbers using addition, subtraction, multiplication, division, powers, or roots.

Evaluating an algebraic expression means finding its value for a given value of the variable or for given values of the variables. Once these values are substituted, follow the order of operations agreement:

1. Perform operations within the innermost parentheses and work outward. If the algebraic expression involves a fraction, treat the numerator and the denominator as if they were each enclosed in parentheses.
2. Evaluate all exponential expressions.
3. Perform multiplications and divisions as they occur, working from left to right.
4. Perform additions and subtractions as they occur, working from left to right.

• Evaluate $x^3 - 5(x - 2)^2$ when $x = -4$.

$$\begin{aligned} x^3 - 5(x - 2)^2 &= (-4)^3 - 5(-4 - 2)^2 \\ &= (-4)^3 - 5(-6)^2 \\ &= -64 - 5(36) \\ &= -64 - 180 = -244 \end{aligned}$$

Additional Examples to Review

Example 1, page 103; Example 2, page 103; Example 3, page 103

An equation is a statement that two expressions are equal. Formulas are equations that express relationships among two or more variables. Mathematical modeling is the process of finding formulas to describe real-world phenomena. Such formulas, together with the meaning assigned to the variables, are called mathematical models. The formulas are said to model, or describe, the relationships among the variables.

• The formula

$$h = -16t^2 + 200t + 4$$

models the height, h, in feet, of fireworks t seconds after launch. What is the height after 2 seconds?

$$\begin{aligned} h &= -16(2)^2 + 200(2) + 4 \\ &= -16(4) + 200(2) + 4 \\ &= -64 + 400 + 4 = 340 \end{aligned}$$

The height after 2 seconds is 340 feet.

Additional Example to Review

Example 4, page 104

Terms of an algebraic expression are separated by addition. Like terms have the same variables with the same exponents on the variables. To add or subtract like terms, add or subtract the coefficients and copy the common variable.

An algebraic expression is simplified when parentheses have been removed and like terms have been combined.

• Simplify: $7(3x - 4) - (10x - 5)$.

$$\begin{aligned} 7(3x - 4) - (10x - 5) &= 21x - 28 - 10x + 5 \\ &= 21x - 10x - 28 + 5 \\ &= 11x - 23 \end{aligned}$$

Additional Examples to Review

Example 5, page 106; Example 6, page 107; Example 7, page 107

Section 2.2 Linear Equations in One Variable

A linear equation in one variable can be written in the form $ax + b = 0$, where a and b are real numbers, and $a \neq 0$.

Solving a linear equation is the process of finding the set of numbers that makes the equation a true statement. These numbers are the solutions. The set of all such solutions is the solution set.

Equivalent equations have the same solution set. The addition and multiplication properties are used to generate equivalent equations.

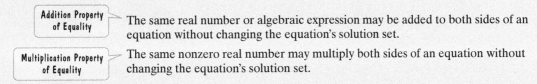

Addition Property of Equality — The same real number or algebraic expression may be added to both sides of an equation without changing the equation's solution set.

Multiplication Property of Equality — The same nonzero real number may multiply both sides of an equation without changing the equation's solution set.

Solving a Linear Equation

1. Simplify the algebraic expression on each side by removing grouping symbols and combining like terms.
2. Collect all the variable terms on one side and all the constants, or numerical terms, on the other side.
3. Isolate the variable and solve.
4. Check the proposed solution in the original equation.

- Solve: $4(x - 5) = 2x - 14$.

$$4x - 20 = 2x - 14$$
$$4x - 2x - 20 = 2x - 2x - 14$$
$$2x - 20 = -14$$
$$2x - 20 + 20 = -14 + 20$$
$$2x = 6$$
$$\frac{2x}{2} = \frac{6}{2}$$
$$x = 3 \quad \text{Checking gives } -8 = -8, \text{ so } \{3\} \text{ is the solution set.}$$

Additional Examples to Review

Example 1, page 114; Example 2, page 115; Example 3, page 116; Example 4, page 117

If an equation contains fractions, begin by multiplying both sides of the equation by the least common denominator of the fractions in the equation, thereby clearing fractions.

- Solve: $\dfrac{x}{5} + \dfrac{1}{2} = \dfrac{x}{2} - 1$.

$$10\left(\frac{x}{5} + \frac{1}{2}\right) = 10\left(\frac{x}{2} - 1\right)$$
$$10 \cdot \frac{x}{5} + 10 \cdot \frac{1}{2} = 10 \cdot \frac{x}{2} - 10 \cdot 1$$
$$2x + 5 = 5x - 10$$
$$-3x = -15$$
$$x = 5 \quad \text{Checking gives } \tfrac{3}{2} = \tfrac{3}{2}, \text{ so } \{5\} \text{ is the solution set.}$$

Additional Examples to Review

Example 5, page 118; Example 6, page 119

If a false statement (such as $-6 = 7$) is obtained in solving an equation, the equation has no solution. The solution set is $\varnothing$, the empty set.

If a true statement (such as $-6 = -6$) is obtained in solving an equation, the equation has infinitely many solutions. The solution set is the set of all real numbers, written $\{x \mid x \text{ is a real number}\}$.

- Solve: $3x + 2 = 3(x + 5)$.

$$3x + 2 = 3x + 15$$
$$3x + 2 - 3x = 3x + 15 - 3x$$
$$2 = 15 \quad \text{(false)}$$

No solution: $\varnothing$

- Solve: $2(x + 4) = x + x + 8$.

$$2x + 8 = 2x + 8$$
$$2x + 8 - 2x = 2x + 8 - 2x$$
$$8 = 8 \quad \text{(true)}$$

Solution set: $\{x \mid x \text{ is a real number}\}$

Additional Examples to Review

Example 7, page 120; Example 8, page 120

Section 2.3 Applications of Linear Equations

A Strategy for Solving Word Problems Using Linear Equations

Step 1 Read the problem carefully several times until you can state in your own words what is given and what the problem is looking for. Let x (or any variable) represent one of the unknown quantities in the problem.

Step 2 If necessary, write expressions for any other unknown quantities in the problem in terms of x.

Step 3 Write an equation in x that models the verbal conditions of the problem.

Step 4 Solve the equation and answer the problem's question.

Step 5 Check the solution *in the original wording* of the problem, not in the equation obtained from the words.

- After a 60% reduction, a graphing calculator sold for $32. What was the original price?
  Let $x =$ the original price.

Original price	minus	60% reduction	=	reduced price
x	$-$	$0.6x$	$=$	32

$$0.4x = 32$$
$$\frac{0.4x}{0.4} = \frac{32}{0.4}$$
$$x = 80$$

The original price was $80. Check this amount using the first sentence in the problem's conditions.

Additional Examples to Review

Example 1, page 126; Example 2, page 127; Example 3, page 129; Example 4, page 130

Solving a formula for a variable means rewriting the formula so that the variable is isolated on one side of the equation.

- Solve for r: $E = I(R + r)$.

$$E = IR + Ir \quad \boxed{\text{We need to isolate } r.}$$
$$E - IR = Ir$$
$$\frac{E - IR}{I} = r$$

Additional Examples to Review

Example 5, page 132; Example 6, page 132

Section 2.4 Ratios, Rates, and Proportions

A ratio is a comparison of two quantities by division.

Expressing the Ratio of *a* to *b*

a to b $\dfrac{a}{b}$ $a : b$

The ratio of 15 feet to 12 feet:

$$\frac{15 \text{ feet}}{12 \text{ feet}} = \frac{15 \cancel{\text{ feet}}}{12 \cancel{\text{ feet}}} = \frac{15}{12} = \frac{3 \cdot 5}{3 \cdot 4} = \frac{5}{4}$$

Additional Example to Review

Example 1, page 137

A rate is a ratio that compares two quantities that have different units.

6 teachers for every 100 students:

$$\frac{6 \text{ teachers}}{100 \text{ students}} = \frac{3 \text{ teachers}}{50 \text{ students}}$$

Additional Example to Review

Example 2, page 138

A unit rate is a rate that has a denominator of 1.

Writing a Unit Rate

1. Write the rate in fraction notation with the units included.
2. Divide the numerator and the denominator by the quantity in the denominator. This results in a unit rate with a denominator of 1.

The word "per" is used to describe the division in a unit rate.

Write as a unit rate: 208 miles on 8 gallons of gas.

$$\frac{208 \text{ miles}}{8 \text{ gallons}} = \frac{208 \text{ miles} \div 8}{8 \text{ gallons} \div 8} = \frac{26 \text{ miles}}{1 \text{ gallon}}$$

The unit rate is $\dfrac{26 \text{ miles}}{1 \text{ gallon}}$, or 26 miles per gallon.

Additional Example to Review

Example 3, page 139

A proportion is a statement in the form $\dfrac{a}{b} = \dfrac{c}{d}$.

The cross-products principle states that if $\dfrac{a}{b} = \dfrac{c}{d}$, then $ad = bc$.

- Solve: $\dfrac{-3}{4} = \dfrac{x}{12}$.

$$4x = -3(12)$$
$$4x = -36$$
$$x = -9 \quad \text{Checking gives } -\frac{3}{4} = -\frac{3}{4}, \text{ so } \{-9\} \text{ is the solution set.}$$

Additional Example to Review

Example 4, page 141

Solving Applied Problems Using Proportions

1. Read the problem and represent the unknown quantity by x (or any letter).
2. Set up a proportion that models the problem's conditions. List the given ratio on one side and the ratio with the unknown quantity on the other side. Each respective quantity should occupy the same corresponding position on each side of the proportion.
3. Drop units and apply the cross-products principle.
4. Solve for x and answer the question.

- 30 elk are tagged and released. Sometime later, a sample of 80 elk are observed and 10 are tagged. How many elk are there?

$$x = \text{number of elk}$$

$$\boxed{\text{Tagged}} \searrow \frac{30}{x} = \frac{10}{80}$$
$$\boxed{\text{Total}} \nearrow$$

$$10x = 30 \cdot 80$$
$$10x = 2400$$
$$x = 240$$

There are 240 elk.

Additional Examples to Review

Example 5, page 142; Example 6, page 143

Section 2.5 Modeling Using Variation

Statements of Variation

In each statement, the number k is the constant of variation.

y varies directly as x: $\quad y = kx$

y varies directly as x^n: $\quad y = kx^n$

y varies inversely as x: $\quad y = \dfrac{k}{x}$

y varies inversely as x^n: $\quad y = \dfrac{k}{x^n}$

y varies directly as x and inversely as z: $\quad y = \dfrac{kx}{z}$

Solving Variation Problems

1. Write an equation that models the given English statement.
2. Substitute the given pair of values into the equation in step 1 and solve for k, the constant of variation.
3. Substitute the value of k into the equation in step 1.
4. Use the equation from step 3 to answer the problem's question.

• The time that it takes you to drive a certain distance varies inversely as your driving rate. Averaging 40 miles per hour, it takes you 10 hours to drive the distance. How long would the trip take averaging 50 miles per hour?

1. $t = \dfrac{k}{r}$ ← Time, t, varies inversely as rate, r.

2. It takes 10 hours at 40 miles per hour.

$$10 = \dfrac{k}{40}$$
$$k = 10(40) = 400$$

3. $t = \dfrac{400}{r}$

4. How long does it take at 50 miles per hour? Substitute 50 for r.

$$t = \dfrac{400}{50} = 8$$

It takes 8 hours at 50 miles per hour.

Additional Examples to Review

Example 1, page 148; Example 2, page 149; Example 3, page 150; Example 4, page 152

Section 2.6 Linear Inequalities in One Variable

Graphs of Subsets of Real Numbers

Set-Builder Notation	Graph
$\{x \mid x < a\}$	
$\{x \mid x \le a\}$	
$\{x \mid x > b\}$	
$\{x \mid x \ge b\}$	
$\{x \mid a < x < b\}$	
$\{x \mid a \le x \le b\}$	
$\{x \mid a \le x < b\}$	
$\{x \mid a < x \le b\}$	

• $\{x \mid x \le 1\}$

• $\{x \mid x > -1\}$

• $\{x \mid -2 \le x < 3\}$

Additional Example to Review

Example 1, page 157

A linear inequality in one variable can be written in one of the following forms, where $a \neq 0$:

$$ax + b < 0, \quad ax + b \leq 0, \quad ax + b > 0, \quad ax + b \geq 0.$$

The procedure for solving linear inequalities is the same as the procedure for solving linear equations, with one important exception: When multiplying or dividing both sides of the inequality by a negative number, reverse the direction of the inequality symbol, changing the sense of the inequality.

- Solve:
$$x + 4 \geq 6x - 16.$$
$$x + 4 - 6x \geq 6x - 16 - 6x$$
$$-5x + 4 \geq -16$$
$$-5x + 4 - 4 \geq -16 - 4$$
$$-5x \geq -20$$
$$\frac{-5x}{-5} \leq \frac{-20}{-5}$$
$$x \leq 4$$

Solution set: $\{x \mid x \leq 4\}$

Additional Examples to Review

Example 2, page 159; Example 3, page 159; Example 4, page 160; Example 5, page 161

An inequality in x with three parts is solved by isolating x in the middle.

- Solve:
$$-9 < 5x + 1 \leq 16.$$
$$-9 - 1 < 5x + 1 - 1 \leq 16 - 1$$
$$-10 < 5x \leq 15$$
$$\frac{-10}{5} < \frac{5x}{5} \leq \frac{15}{5}$$
$$-2 < x \leq 3$$

Solution set: $\{x \mid -2 < x \leq 3\}$

Additional Example to Review

Example 6, page 162

If a false statement (such as $3 > 5$) is obtained in solving an inequality, the inequality has no solution. The solution set is $\varnothing$, the empty set.

If a true statement (such as $3 \leq 5$) is obtained in solving an inequality, the inequality has infinitely many solutions. The solution set is the set of all real numbers, written $\{x \mid x \text{ is a real number}\}$.

- Solve:
$$2(2x + 5) > 4x + 13.$$
$$4x + 10 > 4x + 13$$
$$4x - 4x + 10 > 4x - 4x + 13$$
$$10 > 13 \quad \text{(false)}$$

No solution: $\varnothing$

- Solve:
$$3(x - 1) + 6 \geq 7x - 4x.$$
$$3x - 3 + 6 \geq 3x$$
$$3x + 3 \geq 3x$$
$$3x - 3x + 3 \geq 3x - 3x$$
$$3 \geq 0 \quad \text{(true)}$$

Solution set: $\{x \mid x \text{ is a real number}\}$

Additional Examples to Review

Example 7, page 163; Example 8, page 163

Review Exercises

15. b. $15,180; reasonably well   **15. c.** $15,154; reasonably well

Section 2.1 Algebraic Expressions and Formulas

In Exercises 1–3, evaluate the algebraic expression for the given value of the variable.

1. $6x + 9; x = 4$ 33

2. $7x^2 + 4x - 5; x = -2$ 15

3. $6 + 2(x - 8)^3; x = 5$ -48

4. The diversity index, from 0 (no diversity) to 100, measures the chance that two randomly selected people are a different race or ethnicity. The diversity index in the United States varies widely from region to region, from as high as 81 in Hawaii to as low as 11 in Vermont. The bar graph shows the national diversity index for the United States for four years in the period from 1980 through 2010.

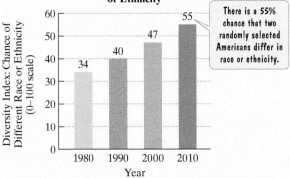

Chance That Two Randomly Selected Americans Are a Different Race or Ethnicity

There is a 55% chance that two randomly selected Americans differ in race or ethnicity.

Source: USA Today

The data in the graph can be modeled by the formula

$$D = 0.005x^2 + 0.55x + 34,$$

where D is the national diversity index in the United States x years after 1980. According to the formula, what was the U.S. diversity index in 2010? How does this compare with the index displayed by the graph?   55; It's the same.

In Exercises 5–7, simplify each algebraic expression.

5. $5(2x - 3) + 7x$ $17x - 15$

6. $3(4y - 5) - (7y - 2)$ $5y - 13$

7. $2(x^2 + 5x) + 3(4x^2 - 3x)$ $14x^2 + x$

Section 2.2 Linear Equations in One Variable

In Exercises 8–14, solve each equation.

8. $4x + 9 = 33$ $\{6\}$

9. $5x - 3 = x + 5$ $\{2\}$

10. $3(x + 4) = 5x - 12$ $\{12\}$

11. $2(x - 2) + 3(x + 5) = 2x - 2$ $\left\{-\frac{13}{3}\right\}$

12. $\frac{2x}{3} = \frac{x}{6} + 1$ $\{2\}$

13. $7x + 5 = 5(x + 3) + 2x$ $\varnothing$

14. $7x + 13 = 2(2x - 5) + 3x + 23$ $\{x | x \text{ is a real number}\}$

Section 2.3 Applications of Linear Equations

15. The line graph shows the cost of inflation. What cost $10,000 in 1982 would cost the amount shown by the graph in subsequent years.

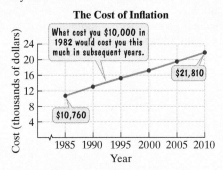

The Cost of Inflation

What cost you $10,000 in 1982 would cost you this much in subsequent years.

$21,810

$10,760

Source: U.S. Bureau of Labor Statistics

Here are two mathematical models for the data shown by the graph. In each formula, C represents the cost x years after 1985 of what cost $10,000 in 1982.

Model 1 $C = 438x + 10,800$

Model 2 $C = 0.3x^2 + 430x + 10,824$

a. Use the graph to estimate the cost in 1995, to the nearest thousand dollars, of what cost $10,000 in 1982. $15,000

b. Use model 1 to determine the cost in 1995. How well does this describe your estimate from part (a)?

c. Use model 2 to determine the cost in 1995. How well does this describe your estimate from part (a)?

d. Use model 1 to determine in which year the cost will be $28,320 for what cost $10,000 in 1982.   2025

16. Compared with peers a decade ago, young people spend 79 more minutes of free time each day listening to music, watching TV and movies, playing video games, and hanging out online. The bar graph shows daily media consumption for U.S. children and teens ages 8 to 18.

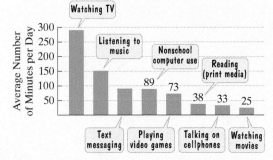

Average Number of Minutes per Day U.S. Children and Teens, Ages 8 to 18, Spend with Various Media

Watching TV

Listening to music

Nonschool computer use

Reading (print media)

89   73   38   33   25

Text messaging   Playing video games   Talking on cellphones   Watching movies

Source: Kaiser Family Foundation

Time spent per day watching TV exceeds time spent text messaging by 180 minutes. Time spent listening to music exceeds time spent text messaging by 61 minutes. Combined, these three activities consume 511 minutes (more than $8\frac{1}{2}$ hours) each day. Find the average number of minutes per day that young people ages 8 to 18 spend on each of these activities.
TV: 270 min; music: 151 min; texting: 90 min

29. $\dfrac{42 \text{ miles}}{1 \text{ hour}}$ or 42 miles per hour   **30.** $\dfrac{\$12}{1 \text{ glass}}$ or $12 per glass

17. The bar graph shows the average price of a movie ticket for selected years from 1980 through 2010. The graph indicates that in 1980, the average movie ticket price was $2.69. For the period from 1980 through 2010, the price increased by approximately $0.15 per year. If this trend continues, by which year will the average price of a movie ticket be $8.69? by 40 years after 1980, or in 2020

Average Price of a U.S. Movie Ticket

1980 *Ordinary People* Ticket Price $2.69

2010 *Alice in Wonderland* Ticket Price $7.85

Sources: Motion Picture Association of America, National Association of Theater Owners (NATO), and Bureau of Labor Statistics (BLS)

18. You are choosing between two texting plans. One plan has a monthly fee of $15 with a charge of $0.05 per text. The other plan has a monthly fee of $5 with a charge of $0.07 per text. For how many text messages will the cost for the two plans be the same? 500 text messages

19. After a 20% price reduction, a cordless phone sold for $48. What was the phone's price before the reduction? $60

20. A salesperson earns $300 per week plus 5% commission of sales. How much must be sold to earn $800 in a week? $10,000 in sales

In Exercises 21–24, solve each formula for the specified variable.

21. $Ax - By = C$ for x **22.** $A = \frac{1}{2}bh$ for h $h = \dfrac{2A}{b}$

23. $A = \dfrac{B + C}{2}$ for B **24.** $vt + gt^2 = s$ for g $g = \dfrac{s - vt}{t^2}$

21. $x = \dfrac{By + C}{A}$ **23.** $B = 2A - C$

Section 2.4 Ratios, Rates, and Proportions

In Exercises 25–26, write each ratio as a fraction in simplest form.

25. 10 to 8 $\frac{5}{4}$ **26.** 14 yards to 21 yards $\frac{2}{3}$

In Exercises 27–28, write each rate as a fraction in simplest form.

27. 16 patients for 6 nurses $\dfrac{8 \text{ patients}}{3 \text{ nurses}}$

28. 225 miles on 6 gallons of gasoline $\dfrac{75 \text{ miles}}{2 \text{ gallons}}$

In Exercises 29–30, write each rate as unit rate.

29. 210 miles in 5 hours   **30.** 8 wine glasses for $96

In Exercises 31–34, solve each proportion.

31. $\dfrac{3}{x} = \dfrac{15}{25}$ $\{5\}$ **32.** $\dfrac{-7}{5} = \dfrac{91}{x}$ $\{-65\}$

33. $\dfrac{x + 2}{3} = \dfrac{4}{5}$ $\left\{\frac{2}{5}\right\}$ **34.** $\dfrac{5}{x + 7} = \dfrac{3}{x + 3}$ $\{3\}$

35. If a school board determines that there should be 3 teachers for every 50 students, how many teachers are needed for an enrollment of 5400 students? 324 teachers

36. To determine the number of trout in a lake, a conservationist catches 112 trout, tags them, and returns them to the lake. Later, 82 trout are caught, and 32 of them are found to be tagged. How many trout are in the lake? 287 trout

Section 2.5 Modeling Using Variation

37. Many areas of Northern California depend on the snowpack of the Sierra Nevada mountain range for their water supply. The volume of water produced from melting snow varies directly as the volume of snow. Meteorologists have determined that 250 cubic centimeters of snow will melt to 28 cubic centimeters of water. How much water does 1200 cubic centimeters of melting snow produce? 134.4 cubic centimeters

38. The distance that a body falls from rest varies directly as the square of the time of the fall. If skydivers fall 144 feet in 3 seconds, how far will they fall in 10 seconds? 1600 feet

39. The pitch of a musical tone varies inversely as its wavelength. A tone has a pitch of 660 vibrations per second and a wavelength of 1.6 feet. What is the pitch of a tone that has a wavelength of 2.4 feet? 440 vibrations per second

40. The loudness of a stereo speaker, measured in decibels, varies inversely as the square of your distance from the speaker. When you are 8 feet from the speaker, the loudness is 28 decibels. What is the loudness when you are 4 feet from the speaker? 112 decibels

41. The time required to assemble computers varies directly as the number of computers assembled and inversely as the number of workers. If 30 computers can be assembled by 6 workers in 10 hours, how long would it take 5 workers to assemble 40 computers? 16 hours

Section 2.6 Linear Inequalities in One Variable

In Exercises 42–50, solve each inequality and graph the solution set on a number line. It is not necessary to provide graphs if the inequality has no solution or is true for all real numbers.

42. $2x - 5 < 3$ $\{x | x < 4\}$*   **43.** $\frac{x}{2} > -4$ $\{x | x > -8\}$*

44. $3 - 5x \le 18$ $\{x | x \ge -3\}$*   **45.** $4x + 6 < 5x$ $\{x | x > 6\}$*

46. $6x - 10 \ge 2(x + 3)$ **47.** $4x + 3(2x - 7) \le x - 3$

48. $-1 < 4x + 2 \le 6$ **49.** $2(2x + 4) > 4(x + 2) - 6$

50. $-2(x - 4) \le 3x + 1 - 5x$ ∅   **49.** $\{x | x$ is a real number$\}$

51. To pass a course, a student must have an average on three examinations of at least 60. If a student scores 42 and 74 on the first two tests, what must be earned on the third test to pass the course? at least 64

*See Answers to Selected Exercises.

46. $\{x | x \ge 4\}$*   **47.** $\{x | x \le 2\}$*   **48.** $\{x | -\frac{3}{4} < x \le 1\}$*

Chapter 2 Test

1. Evaluate $x^3 - 4(x - 1)^2$ when $x = -2$.   −44

2. Simplify: $5(3x - 2) - (x - 6)$. $14x - 4$

In Exercises 3–6, solve each equation.

3. $12x + 4 = 7x - 21$ {−5}

4. $3(2x - 4) = 9 - 3(x + 1)$ {2}

5. $3(x - 4) + x = 2(6 + 2x)$ ∅

6. $\dfrac{x}{5} - 2 = \dfrac{x}{3}$ {−15}

7. Solve for y: $By - Ax = A$. $y = \dfrac{Ax + A}{B}$

8. The bar graph shows the percentage of American adults reporting personal gun ownership for selected years from 1980 through 2010.

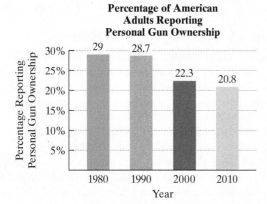

Percentage of American Adults Reporting Personal Gun Ownership

Source: General Social Survey

Here are two mathematical models for the data shown by the graph. In each formula, p represents the percentage of American adults who reported personal gun ownership x years after 1980.

| Model 1 | $p = -0.3x + 30$ |
| Model 2 | $p = -0.003x^2 - 0.22x + 30$ |

a. According to model 1, what percentage of American adults reported personal gun ownership in 2010? Does this underestimate or overestimate the percentage shown by the graph? By how much?

b. According to model 2, what percentage of American adults reported personal gun ownership in 2010? Does this underestimate or overestimate the percentage shown by the graph? By how much?

c. If trends shown by the data continue, use model 1 to determine in which year 17.7% of American adults will report personal gun ownership.   2021
   a. 21%; overestimates by 0.2   **b.** 20.7%; underestimates by 0.1

9. Write as a unit rate: 84¢ for a 7-ounce can of juice.
   $\dfrac{12¢}{1 \text{ oz}}$ or 12¢ per ounce

In Exercises 10–11, solve each proportion.

10. $\dfrac{5}{8} = \dfrac{x}{12}$ $\left\{\dfrac{15}{2}\right\}$

11. $\dfrac{x + 5}{8} = \dfrac{x + 2}{5}$ {3}

12. In this exercise, you will determine the median, or middlemost, salaries for Americans with bachelor's degrees and 10 to 20 years experience for three college majors. For those with 10 to 20 years experience, the median salary of a computer science major exceeds that of a political science major by $18 thousand and the median salary of an economics major exceeds that of a political science major by $21 thousand. Combined, three people with each of these majors and 10 to 20 years experience earn $273 thousand. Find the median salary for people with each major and 10 to 20 years experience.

13. You bought a new car for $50,750. Its value is decreasing by $5500 per year. After how many years will its value be $12,250?   after 7 years

14. You are choosing between two texting plans. Plan A charges $25 per month for unlimited texting. Plan B has a monthly fee of $13 with a charge of $0.06 per text. For how many text messages will the costs for the two plans be the same?

15. After a 60% reduction, a jacket sold for $20. What was the jacket's price before the reduction? $50

16. Park rangers catch, tag, and release 200 tule elk back into a wildlife refuge. Two weeks later they observe a sample of 150 elk, of which 5 are tagged. Assuming that the ratio of tagged elk in the sample holds for all elk in the refuge, how many elk are there in the park?   6000 tule elk

17. The Mach number is a measurement of speed named after the man who suggested it, Ernst Mach (1838–1916). The speed of an aircraft varies directly as its Mach number. Shown here are two aircraft. Use the figures for the Concorde to determine the Blackbird's speed.   2442 miles per hour

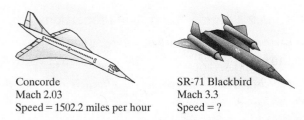

Concorde
Mach 2.03
Speed = 1502.2 miles per hour

SR-71 Blackbird
Mach 3.3
Speed = ?

18. The intensity of light received at a source varies inversely as the square of the distance from the source. A particular light has an intensity of 20 foot-candles at 15 feet. What is the light's intensity at 10 feet?   45 foot-candles

In Exercises 19–21, solve each inequality and graph the solution set on a number line.

19. $6 - 9x \geq 33$ $\{x | x \leq -3\}$*

20. $4x - 2 > 2(x + 6)$ $\{x | x > 7\}$*

21. $-3 \leq 2x + 1 < 6$ $\left\{x | -2 \leq x < \dfrac{5}{2}\right\}$*

22. A student has grades on three examinations of 76, 80, and 72. What must the student earn on a fourth examination in order to have an average of at least 80?   at least 92

12. computer science: $96,000; political science: $78,000; economics: $99,000
14. 200 text messages
*See Answers to Selected Exercises.

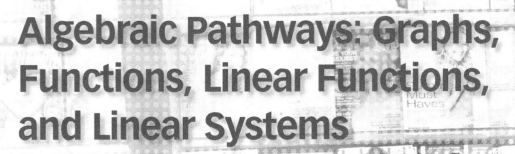

Algebraic Pathways: Graphs, Functions, Linear Functions, and Linear Systems

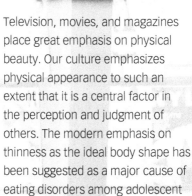

3

Television, movies, and magazines place great emphasis on physical beauty. Our culture emphasizes physical appearance to such an extent that it is a central factor in the perception and judgment of others. The modern emphasis on thinness as the ideal body shape has been suggested as a major cause of eating disorders among adolescent girls.

Cultural values of physical attractiveness change over time. During the 1950s, actress Jayne Mansfield embodied the postwar ideal: curvy, buxom, and big-hipped. Men, too, have been caught up in changes of how they "ought" to look. The 1960s' ideal was the soft and scrawny hippie. Today's ideal man is tough and muscular.

Given the importance of culture in setting standards of attractiveness, how can you establish a healthy weight range for your age and height? In this chapter, we will use systems of inequalities to explore these skin-deep issues.

Here's where you'll find these applications:

You'll find a weight for various ages using the models (mathematical, not fashion) in Example 4 of Section 3.5 and Exercises 43–46 in Exercise Set 3.5. Exercises 49–50 use graphs and a formula for body-mass index to indicate whether you are obese, overweight, borderline overweight, normal weight, or underweight.

3.1 : Graphing and Functions

What am I supposed to learn?

After you have read this section, you should be able to:

1. Plot points in the rectangular coordinate system.

2. Graph equations in the rectangular coordinate system.

3. Use function notation.

4. Graph functions.

5. Use the vertical line test.

6. Obtain information about a function from its graph.

The beginning of the seventeenth century was a time of innovative ideas and enormous intellectual progress in Europe. English theatergoers enjoyed a succession of exciting new plays by Shakespeare. William Harvey proposed the radical notion that the heart was a pump for blood rather than the center of emotion. Galileo, with his new-fangled invention called the telescope, supported the theory of Polish astronomer Copernicus that the Sun, not the Earth, was the center of the solar system. Monteverdi was writing the world's first grand operas. French mathematicians Pascal and Fermat invented a new field of mathematics called probability theory.

Into this arena of intellectual electricity stepped French aristocrat René Descartes (1596–1650). Descartes (pronounced "day cart"), propelled by the creativity surrounding him, developed a new branch of mathematics that brought together algebra and geometry in a unified way—a way that visualized numbers as points on a graph, equations as geometric figures, and geometric figures as equations. This new branch of mathematics, called *analytic geometry*, established Descartes as one of the founders of modern thought and among the most original mathematicians and philosophers of any age. We begin this section by looking at Descartes's deceptively simple idea, called the **rectangular coordinate system** or (in his honor) the **Cartesian coordinate system**.

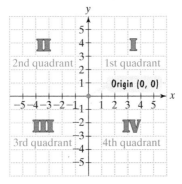

Figure 3.1 The rectangular coordinate system

Great Question!

What's the significance of the word *ordered* when describing a pair of real numbers?

The phrase *ordered pair* is used because order is important. The order in which coordinates appear makes a difference in a point's location. This is illustrated in **Figure 3.2**.

Points and Ordered Pairs

Descartes used two number lines that intersect at right angles at their zero points, as shown in **Figure 3.1**. The horizontal number line is the ***x*-axis**. The vertical number line is the ***y*-axis**. The point of intersection of these axes is their zero points, called the **origin**. Positive numbers are shown to the right and above the origin. Negative numbers are shown to the left and below the origin. The axes divide the plane into four quarters, called **quadrants**. The points located on the axes are not in any quadrant.

Each point in the rectangular coordinate system corresponds to an **ordered pair** of real numbers, (x, y). Examples of such pairs are $(-5, 3)$ and $(3, -5)$. The first number in each pair, called the ***x*-coordinate**, denotes the distance and direction from the origin along the x-axis. The second number in each pair, called the ***y*-coordinate**, denotes vertical distance and direction along a line parallel to the y-axis or along the y-axis itself.

Figure 3.2 shows how we **plot**, or locate, the points corresponding to the ordered pairs $(-5, 3)$ and $(3, -5)$. We plot $(-5, 3)$ by going 5 units from 0 to the left along the x-axis. Then we go 3 units up parallel to the y-axis. We plot $(3, -5)$ by going 3 units from 0 to the right along the x-axis and 5 units down parallel to the y-axis. The phrase "the points corresponding to the ordered pairs $(-5, 3)$ and $(3, -5)$" is often abbreviated as "the points $(-5, 3)$ and $(3, -5)$."

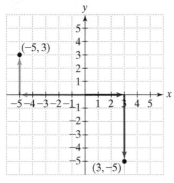

Figure 3.2 Plotting $(-5, 3)$ and $(3, -5)$

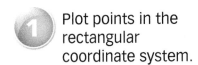

Plot points in the rectangular coordinate system.

Example 1 **Plotting Points in the Rectangular Coordinate System**

Plot the points: $A(-3, 5), B(2, -4), C(5, 0), D(-5, -3), E(0, 4)$, and $F(0, 0)$.

Solution

See **Figure 3.3**. We move from the origin and plot the points in the following way:

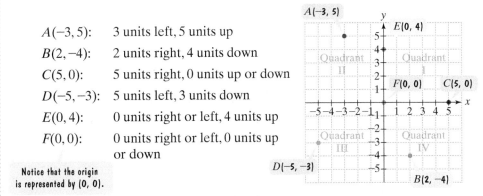

$A(-3, 5)$:   3 units left, 5 units up

$B(2, -4)$:   2 units right, 4 units down

$C(5, 0)$:   5 units right, 0 units up or down

$D(-5, -3)$:   5 units left, 3 units down

$E(0, 4)$:   0 units right or left, 4 units up

$F(0, 0)$:   0 units right or left, 0 units up or down

Notice that the origin is represented by (0, 0).

Figure 3.3 Plotting points

Check Point 1   Plot the points: $A(-2, 4), B(4, -2), C(-3, 0)$, and $D(0, -3)$. *

Graphs of Equations

Graph equations in the rectangular coordinate system.

A relationship between two quantities can sometimes be expressed as an **equation in two variables**, such as

$$y = 4 - x^2.$$

A **solution of an equation in two variables**, x and y, is an ordered pair of real numbers with the following property: When the x-coordinate is substituted for x and the y-coordinate is substituted for y in the equation, we obtain a true statement. For example, consider the equation $y = 4 - x^2$ and the ordered pair $(3, -5)$. When 3 is substituted for x and -5 is substituted for y, we obtain the statement $-5 = 4 - 3^2$, or $-5 = 4 - 9$, or $-5 = -5$. Because this statement is true, the ordered pair $(3, -5)$ is a solution of the equation $y = 4 - x^2$. We also say that $(3, -5)$ **satisfies** the equation.

We can generate as many ordered-pair solutions as desired to $y = 4 - x^2$ by substituting numbers for x and then finding the corresponding values for y. For example, suppose we let $x = 3$:

Start with x.	Compute y.	Form the ordered pair (x, y).
x	$y = 4 - x^2$	**Ordered Pair (x, y)**
3	$y = 4 - 3^2 = 4 - 9 = -5$	$(3, -5)$
Let $x = 3$.		$(3, -5)$ is a solution of $y = 4 - x^2$.

The **graph of an equation in two variables** is the set of all points whose coordinates satisfy the equation. One method for graphing such equations is the **point-plotting method**. First, we find several ordered pairs that are solutions of the equation. Next, we plot these ordered pairs as points in the rectangular coordinate system. Finally, we connect the points with a smooth curve or line. This often gives us a picture of all ordered pairs that satisfy the equation.

*See Answers to Selected Exercises.

Example 2 **Graphing an Equation Using the Point-Plotting Method**

Graph $y = 4 - x^2$. Select integers for x, starting with -3 and ending with 3.

Solution

For each value of x, we find the corresponding value for y.

Start with x.   Compute y.   Form the ordered pair (x, y).

We selected integers from -3 to 3, inclusive, to include three negative numbers, 0, and three positive numbers. We also wanted to keep the resulting computations for y relatively simple.

x	$y = 4 - x^2$	Ordered Pair (x, y)
-3	$y = 4 - (-3)^2 = 4 - 9 = -5$	$(-3, -5)$
-2	$y = 4 - (-2)^2 = 4 - 4 = 0$	$(-2, 0)$
-1	$y = 4 - (-1)^2 = 4 - 1 = 3$	$(-1, 3)$
0	$y = 4 - 0^2 = 4 - 0 = 4$	$(0, 4)$
1	$y = 4 - 1^2 = 4 - 1 = 3$	$(1, 3)$
2	$y = 4 - 2^2 = 4 - 4 = 0$	$(2, 0)$
3	$y = 4 - 3^2 = 4 - 9 = -5$	$(3, -5)$

Now we plot the seven points and join them with a smooth curve, as shown in **Figure 3.4**. The graph of $y = 4 - x^2$ is a curve where the part of the graph to the right of the y-axis is a reflection of the part to the left of it and vice versa. The arrows on the left and the right of the curve indicate that it extends indefinitely in both directions.

Figure 3.4 The graph of $y = 4 - x^2$

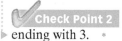 **Check Point 2** Graph $y = 4 - x$. Select integers for x, starting with -3 and ending with 3. *

Part of the beauty of the rectangular coordinate system is that it allows us to "see" formulas and visualize the solution to a problem. This idea is demonstrated in Example 3.

Example 3 **An Application Using Graphs of Equations**

The toll to a bridge costs $2.50. Commuters who use the bridge frequently have the option of purchasing a monthly discount pass for $21.00. With the discount pass, the toll is reduced to $1.00. The monthly cost, y, of using the bridge x times can be described by the following formulas:

Without the discount pass:

$$y = 2.5x$$

The monthly cost, y, is $2.50 times the number of times, x, that the bridge is used.

With the discount pass:

$$y = 21 + 1 \cdot x$$
$$y = 21 + x.$$

The monthly cost, y, is $21 for the discount pass plus $1 times the number of times, x, that the bridge is used.

a. Let $x = 0, 2, 4, 10, 12, 14,$ and 16. Make a table of values for each equation showing seven solutions for the equation.

b. Graph the equations in the same rectangular coordinate system.

c. What are the coordinates of the intersection point for the two graphs? Interpret the coordinates in practical terms.

*See Answers to Selected Exercises.

Solution

a. Tables of values showing seven solutions for each equation follow.

Without the Discount Pass

x	$y = 2.5x$	(x, y)
0	$y = 2.5(0) = 0$	$(0, 0)$
2	$y = 2.5(2) = 5$	$(2, 5)$
4	$y = 2.5(4) = 10$	$(4, 10)$
10	$y = 2.5(10) = 25$	$(10, 25)$
12	$y = 2.5(12) = 30$	$(12, 30)$
14	$y = 2.5(14) = 35$	$(14, 35)$
16	$y = 2.5(16) = 40$	$(16, 40)$

With the Discount Pass

x	$y = 21 + x$	(x, y)
0	$y = 21 + 0 = 21$	$(0, 21)$
2	$y = 21 + 2 = 23$	$(2, 23)$
4	$y = 21 + 4 = 25$	$(4, 25)$
10	$y = 21 + 10 = 31$	$(10, 31)$
12	$y = 21 + 12 = 33$	$(12, 33)$
14	$y = 21 + 14 = 35$	$(14, 35)$
16	$y = 21 + 16 = 37$	$(16, 37)$

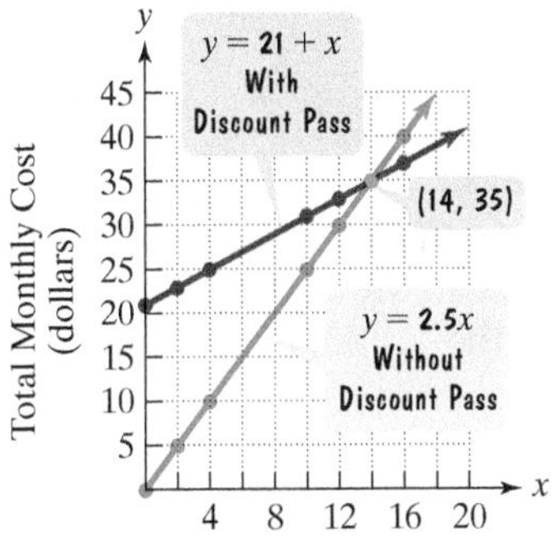

b. Now we are ready to graph the two equations. Because the x- and y-coordinates are nonnegative, it is only necessary to use the origin, the positive portions of the x- and y-axes, and the first quadrant of the rectangular coordinate system. The x-coordinates begin at 0 and end at 16. We will let each tick mark on the x-axis represent two units. However, the y-coordinates begin at 0 and get as large as 40 in the formula that describes the monthly cost without the discount pass. So that our y-axis does not get too long, we will let each tick mark on the y-axis represent five units. Using this setup and the two tables of values, we construct the graphs of $y = 2.5x$ and $y = 21 + x$, shown in **Figure 3.5**.

c. The graphs intersect at $(14, 35)$. This means that if the bridge is used 14 times in a month, the total monthly cost without the discount pass is the same as the total monthly cost with the discount pass, namely $35.

In **Figure 3.5**, look at the two graphs to the right of the intersection point $(14, 35)$. The red graph of $y = 21 + x$ lies below the blue graph of $y = 2.5x$. This means that if the bridge is used more than 14 times in a month $(x > 14)$, the (red) monthly cost, y, with the discount pass is less than the (blue) monthly cost, y, without the discount pass.

Figure 3.5 Options for a toll

Check Point 3 The toll to a bridge costs $2.00. If you use the bridge x times in a month, the monthly cost, y, is $y = 2x$. With a $10 discount pass, the toll is reduced to $1.00. The monthly cost, y, of using the bridge x times in a month with the discount pass is $y = 10 + x$.

a. Let $x = 0, 2, 4, 6, 8, 10,$ and 12. Make tables of values showing seven solutions of $y = 2x$ and seven solutions of $y = 10 + x$. *

b. Graph the equations in the same rectangular coordinate system. *

c. What are the coordinates of the intersection point for the two graphs? Interpret the coordinates in practical terms. $(10, 20)$; When the bridge is used 10 times during a month, the cost is $20 with or without the discount pass.

Functions

③ Use function notation.

Reconsider one of the equations from Example 3, $y = 2.5x$. Recall that this equation describes the monthly cost, y, of using the bridge x times, with a toll of $2.50 each time the bridge is used. The monthly cost, y, depends on the number of times the bridge is used, x. For each value of x, there is one and only one value of y. **If an equation in two variables (x and y) yields precisely one value of y for each value of x, we say that y is a function of x.**

The notation $y = f(x)$ indicates that the variable y is a function of x. The notation $f(x)$ is read "f of x."

*See Answers to Selected Exercises.

For example, the formula for the cost of the bridge

$$y = 2.5x$$

can be expressed in function notation as

$$f(x) = 2.5x.$$

We read this as "f of x is equal to $2.5x$." If, say, x equals 10 (meaning that the bridge is used 10 times), we can find the corresponding value of y (monthly cost) using the equation $f(x) = 2.5x$.

$$f(x) = 2.5x$$
$$f(10) = 2.5(10) \qquad \text{To find } f(10), \text{ read "} f \text{ of 10," replace } x \text{ with 10.}$$
$$= 25$$

Because $f(10) = 25$ (f of 10 equals 25), this means that if the bridge is used 10 times in a month, the total monthly cost is $25.

Table 3.1 compares our previous notation with the new notation of functions.

Table 3.1 Function Notation	
"y Equals" Notation	**"f(x) Equals" Notation**
$y = 2.5x$	$f(x) = 2.5x$
If $x = 10$, $y = 2.5(10) = 25.$	$f(10) = 2.5(10) = 25$ f of 10 equals 25.

In our next example, we will apply function notation to three different functions. It would be awkward to call all three functions f. We will call the first function f, the second function g, and the third function h. These are the letters most frequently used to name functions.

Example 4   Using Function Notation

Find each of the following:

a. $f(4)$ for $f(x) = 2x + 3$
b. $g(-2)$ for $g(x) = 2x^2 - 1$
c. $h(-5)$ for $h(r) = r^3 - 2r^2 + 5$.

Solution

a. $f(x) = 2x + 3$ This is the given function.
    $f(4) = 2 \cdot 4 + 3$ To find f of 4, replace x with 4.
       $= 8 + 3$ Multiply: $2 \cdot 4 = 8$.
    $f(4) = 11$ *f of 4 is 11.*   Add.

b. $g(x) = 2(x)^2 - 1$ This is the given function.
   $g(-2) = 2(-2)^2 - 1$ To find g of -2, replace x with -2.
       $= 2(4) - 1$ Evaluate the exponential expression: $(-2)^2 = 4$.
       $= 8 - 1$ Multiply: $2(4) = 8$.
   $g(-2) = 7$ *g of −2 is 7.*   Subtract.

c. $h(r) = r^3 - 2r^2 + 5$ The function's name is h and r represents the function's input.
   $h(-5) = (-5)^3 - 2(-5)^2 + 5$ To find h of -5, replace each occurrence of r with -5.
       $= -125 - 2(25) + 5$ Evaluate exponential expressions: $(-5)^3 = -125$ and $(-5)^2 = 25$.
       $= -125 - 50 + 5$ Multiply: $2(25) = 50$.
   $h(-5) = -170$ *h of −5 is −170.* $-125 - 50 = -175$ and $-175 + 5 = -170$.

Great Question!

Doesn't $f(x)$ indicate that I need to multiply f and x?

The notation $f(x)$ does *not* mean "f times x." The notation describes the "output" for the function f when the "input" is x. Think of $f(x)$ as another name for y.

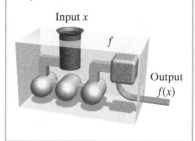

Input x

f

Output $f(x)$

Check Point 4 Find each of the following:

a. $f(6)$ for $f(x) = 4x + 5$ 29

b. $g(-5)$ for $g(x) = 3x^2 - 10$ 65

c. $h(-4)$ for $h(r) = r^2 - 7r + 2.$ 46

Example 5   An Application Involving Function Notation

Tailgaters beware: If your car is going 35 miles per hour on dry pavement, your required stopping distance is 160 feet, or the width of a football field. At 65 miles per hour, the distance required is 410 feet, or well over the length of a football field. **Figure 3.6** shows stopping distances for cars at various speeds on dry roads and on wet roads. **Figure 3.7** uses a line graph to represent stopping distances at various speeds on dry roads.

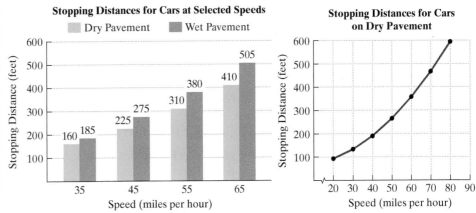

Figure 3.6
Source: National Highway Traffic Safety Administration

Figure 3.7

a. Use the line graph in **Figure 3.7** to estimate a car's required stopping distance at 60 miles per hour on dry pavement. Round to the nearest ten feet.

b. The function

$$f(x) = 0.0875x^2 - 0.4x + 66.6$$

models a car's required stopping distance, $f(x)$, in feet, on dry pavement when the car is traveling at x miles per hour. Use this function to find the required stopping distance at 60 miles per hour. Round to the nearest foot.

Solution

a. The required stopping distance at 60 miles per hour is estimated using the point shown in **Figure 3.8**. The second coordinate of this point extends slightly more than midway between 300 and 400 on the vertical axis. Thus, 360 is a reasonable estimate. We conclude that at 60 miles per hour on dry pavement, the required stopping distance is approximately 360 feet.

b. Now we use the given function to determine the required stopping

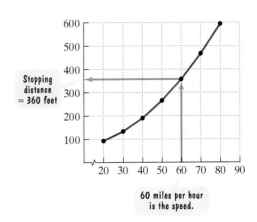

Figure 3.8

Technology

On most calculators, here is how to find

$0.0875(60)^2 - 0.4(60) + 66.6.$

Many Scientific Calculators

.0875 $\times$ 60 x^2 $-$

.4 $\times$ 60 $+$ 66.6 $=$

Many Graphing Calculators

.0875 $\times$ 60 $\wedge$ 2 $-$

.4 $\times$ 60 $+$ 66.6 $\boxed{\text{ENTER}}$

distance at 60 miles per hour. We need to find $f(60)$. The arithmetic gets somewhat "messy," so it is probably a good idea to use a calculator.

$f(x) = 0.0875x^2 - 0.4x + 66.6$ This function models stopping distance, $f(x)$, at x miles per hour.

$f(60) = 0.0875(60)^2 - 0.4(60) + 66.6$ Replace each x with 60.

$\quad\quad = 0.0875(3600) - 0.4(60) + 66.6$ Use the order of operations, first evaluating the exponential expression.

$\quad\quad = 315 - 24 + 66.6$ Perform the multiplications.

$\quad\quad = 357.6$ Subtract and add as indicated.

$\quad\quad \approx 358$ As directed, we've rounded to the nearest foot.

We see that $f(60) \approx 358$—that is, f of 60 is approximately 358. The model indicates that the required stopping distance on dry pavement at 60 miles per hour is approximately 358 feet.

✓ Check Point 5

a. Use the line graph in **Figure 3.7** on the previous page to estimate a car's required stopping distance at 40 miles per hour on dry pavement. Round to the nearest ten feet.   190 feet

b. Use the function in Example 5(b), $f(x) = 0.0875x^2 - 0.4x + 66.6$, to find the required stopping distance at 40 miles per hour. Round to the nearest foot.   191 feet

Graphing Functions

4 Graph functions.

The **graph of a function** is the graph of its ordered pairs. In our next example, we will graph two functions.

Example 6   **Graphing Functions**

Graph the functions $f(x) = 2x$ and $g(x) = 2x + 4$ in the same rectangular coordinate system. Select integers for x from -2 to 2, inclusive.

Solution

For each function, we use the suggested values for x to create a table of some of the coordinates. These tables are shown below. Then, we plot the five points in each table and connect them, as shown in **Figure 3.9**. The graph of each function is a straight line. Do you see a relationship between the two graphs? The graph of g is the graph of f shifted vertically up 4 units.

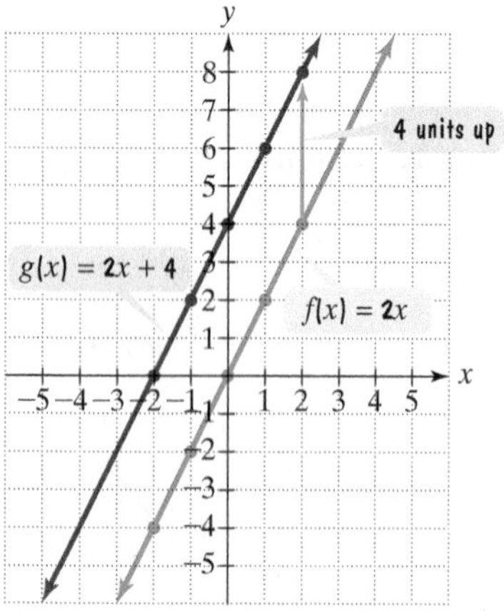

Figure 3.9

x	$f(x) = 2x$	(x, y) or $(x, f(x))$	x	$g(x) = 2x + 4$	(x, y) or $(x, g(x))$
-2	$f(-2) = 2(-2) = -4$	$(-2, -4)$	-2	$g(-2) = 2(-2) + 4 = 0$	$(-2, 0)$
-1	$f(-1) = 2(-1) = -2$	$(-1, -2)$	-1	$g(-1) = 2(-1) + 4 = 2$	$(-1, 2)$
0	$f(0) = 2 \cdot 0 = 0$	$(0, 0)$	0	$g(0) = 2 \cdot 0 + 4 = 4$	$(0, 4)$
1	$f(1) = 2 \cdot 1 = 2$	$(1, 2)$	1	$g(1) = 2 \cdot 1 + 4 = 6$	$(1, 6)$
2	$f(2) = 2 \cdot 2 = 4$	$(2, 4)$	2	$g(2) = 2 \cdot 2 + 4 = 8$	$(2, 8)$

Choose x.   Compute $f(x)$ by evaluating f at x.   Form the ordered pair.   Choose x.   Compute $g(x)$ by evaluating g at x.   Form the ordered pair.

✓ Check Point 6   Graph the functions $f(x) = 2x$ and $g(x) = 2x - 3$ in the same rectangular coordinate system. Select integers for x from -2 to 2, inclusive. How is the graph of g related to the graph of f?   *

*See Answers to Selected Exercises.

Use the vertical line test.

Technology

A graphing calculator is a powerful tool that quickly generates the graph of an equation in two variables. Here is the graph of $y = 4 - x^2$ that we drew by hand in **Figure 3.4** on page 180.

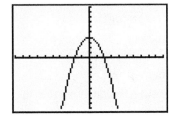

What differences do you notice between this graph and the graph we drew by hand? This graph seems a bit "jittery." Arrows do not appear on the left and right ends of the graph. Furthermore, numbers are not given along the axes. For the graph shown above, the x-axis extends from -10 to 10 and the y-axis also extends from -10 to 10. The distance represented by each consecutive tick mark is one unit. We say that the **viewing window** is $[-10, 10, 1]$ by $[-10, 10, 1]$.

To graph an equation in x and y using a graphing calculator, enter the equation, which must be solved for y, and specify the size of the viewing window. The size of the viewing window sets minimum and maximum values for both the x- and y-axes. Enter these values, as well as the values between consecutive tick marks, on the respective axes. The $[-10, 10, 1]$ by $[-10, 10, 1]$ viewing window used above is called the **standard viewing window**.

The Vertical Line Test

Not every graph in the rectangular coordinate system is the graph of a function. The definition of a function specifies that no value of x can be paired with two or more different values of y. Consequently, if a graph contains two or more different points with the same first coordinate, the graph cannot represent a function. This is illustrated in **Figure 3.10**. Observe that points sharing a common first coordinate are vertically above or below each other.

This observation is the basis of a useful test for determining whether a graph defines y as a function of x. The test is called the **vertical line test**.

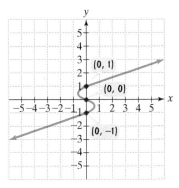

Figure 3.10 y is not a function of x because 0 is paired with three values of y, namely, 1, 0, and -1.

> **The Vertical Line Test for Functions**
>
> If any vertical line intersects a graph in more than one point, the graph does not define y as a function of x.

Example 7 Using the Vertical Line Test

Use the vertical line test to identify graphs in which y is a function of x.

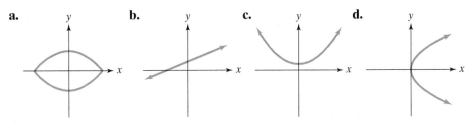

Solution

y is a function of x for the graphs in (b) and (c).

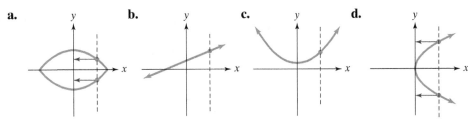

a. **y is not a function** of x.
Two values of y correspond to one x-value.

b. **y is a function** of x.

c. **y is a function** of x.

d. **y is not a function** of x.
Two values of y correspond to one x-value.

Check Point 7 Use the vertical line test to identify graphs in which y is a function of x.

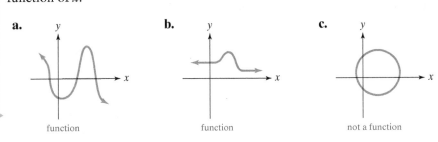

a. function

b. function

c. not a function

6 Obtain information about a function from its graph.

Obtaining Information from Graphs

Example 8 illustrates how to obtain information about a function from its graph.

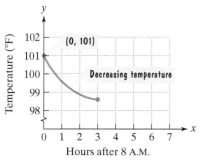

Figure 3.11 Body temperature from 8 A.M. through 3 P.M.

Example 8 Analyzing the Graph of a Function

Too late for that flu shot now! It's only 8 A.M. and you're feeling lousy. Fascinated by the way that algebra models the world (your author is projecting a bit here), you construct a graph showing your body temperature from 8 A.M. through 3 P.M. You decide to let x represent the number of hours after 8 A.M. and y represent your body temperature at time x. The graph is shown in **Figure 3.11**. The symbol $\dagger$ on the y-axis shows that there is a break in values between 0 and 98. Thus, the first tick mark on the y-axis represents a temperature of 98°F.

a. What is your temperature at 8 A.M.?

b. During which period of time is your temperature decreasing?

c. Estimate your minimum temperature during the time period shown. How many hours after 8 A.M. does this occur? At what time does this occur?

d. During which period of time is your temperature increasing?

e. Part of the graph is shown as a horizontal line segment. What does this mean about your temperature and when does this occur?

f. Explain why the graph defines y as a function of x.

Solution

a. Because x is the number of hours after 8 A.M., your temperature at 8 A.M. corresponds to $x = 0$. Locate 0 on the horizontal axis and look at the point on the graph above 0. **Figure 3.12** shows that your temperature at 8 A.M. is 101°F.

b. Your temperature is decreasing when the graph falls from left to right. This occurs between $x = 0$ and $x = 3$, also shown in **Figure 3.12**. Because x represents the number of hours after 8 A.M., your temperature is decreasing between 8 A.M. and 11 A.M.

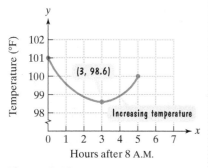

Figure 3.13

c. Your minimum temperature can be found by locating the lowest point on the graph. This point lies above 3 on the horizontal axis, shown in **Figure 3.13**. The y-coordinate of this point falls more than midway between 98 and 99, at approximately 98.6. The lowest point on the graph, $(3, 98.6)$, shows that your minimum temperature, 98.6°F, occurs 3 hours after 8 A.M., at 11 A.M.

d. Your temperature is increasing when the graph rises from left to right. This occurs between $x = 3$ and $x = 5$, shown in **Figure 3.13**. Because x represents the number of hours after 8 A.M., your temperature is increasing between 11 A.M. and 1 P.M.

e. The horizontal line segment shown in **Figure 3.14** indicates that your temperature is neither increasing nor decreasing. Your temperature remains the same, 100°F, between $x = 5$ and $x = 7$. Thus, your temperature is at a constant 100°F between 1 P.M. and 3 P.M.

f. The complete graph of your body temperature from 8 A.M. through 3 P.M. is shown in **Figure 3.14**. No vertical line can be drawn that intersects this blue graph more than once. By the vertical line test, the graph defines y as a function of x. In practical terms, this means that your body temperature is a

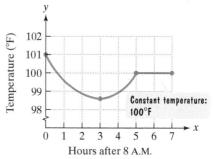

Figure 3.14

function of time. Each hour (or fraction of an hour) after 8 A.M., represented by *x*, yields precisely one body temperature, represented by *y*.

> **Check Point 8** When a physician injects a drug into a patient's muscle, the concentration of the drug in the body, measured in milligrams per 100 milliliters, depends on the time elapsed after the injection, measured in hours. **Figure 3.15** shows the graph of the drug concentration over time, where *x* represents hours after the injection and *y* represents the drug concentration at time *x*.
>
> **a.** During which period of time is the drug concentration increasing?   0 to 3 hours
>
> **b.** During which period of time is the drug concentration decreasing?   3 to 13 hours
>
> **c.** What is the drug's maximum concentration and when does this occur?
>
> **d.** What happens by the end of 13 hours?   None of the drug is left in the body.
>
> ▸ **e.** Explain why the graph defines *y* as a function of *x*.   No vertical line intersects the graph in more than one point.

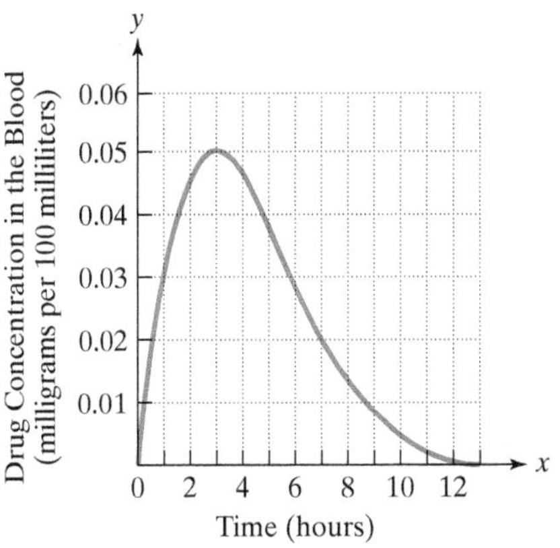

c. 0.05 mg per 100 ml; after 3 hours

Figure 3.15

Functions and Graphs for Cellphone Plans

A cellphone company offers the following plan:

- $20 per month buys 60 minutes.
- Additional time costs $0.40 per minute.

We can represent this plan mathematically by writing the total monthly cost, *C*, as a function of the number of calling minutes, *t*.

$$C(t) = \begin{cases} 20 & \text{if } 0 \le t \le 60 \\ 20 + 0.40(t - 60) & \text{if } t > 60 \end{cases}$$

The cost is $20 for up to and including 60 calling minutes.

The cost is $20 plus $0.40 per minute for additional time for more than 60 calling minutes.

$20 for first 60 minutes $0.40 per minute   times the number of calling minutes exceeding 60

A function that is defined by two (or more) equations is called a **piecewise function**. Many cellphone plans can be represented with piecewise functions. The graph of the piecewise function described above is shown in **Figure 3.16**.

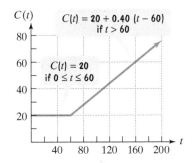

$C(t) = 20 + 0.40(t - 60)$ if $t > 60$

$C(t) = 20$ if $0 \le t \le 60$

Figure 3.16

Example 9   Using a Function for a Cellphone Plan

Use the function that describes the cellphone plan

$$C(t) = \begin{cases} 20 & \text{if } 0 \le t \le 60 \\ 20 + 0.40(t - 60) & \text{if } t > 60 \end{cases}$$

to find and interpret each of the following:

a. $C(30)$ **b.** $C(100)$.

$$C(t) = \begin{cases} 20 & \text{if } 0 \le t \le 60 \\ 20 + 0.40\,(t - 60) & \text{if } t > 60 \end{cases}$$

The piecewise function describing a cellphone plan (repeated)

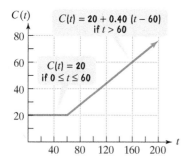

Figure 3.16 (repeated)

Solution

a. To find $C(30)$, we let $t = 30$. Because 30 lies between 0 and 60, we use the first line of the piecewise function.

$$C(t) = 20 \qquad \text{\small This is the function's equation for } 0 \le t \le 60.$$

$$C(30) = 20 \qquad \text{\small Replace } t \text{ with 30. Regardless of this function's input, the constant output is 20.}$$

This means that with 30 calling minutes, the monthly cost is $20. This can be visually represented by the point $(30, 20)$ on the first piece of the graph in **Figure 3.16**. Take a moment to identify this point on the graph.

b. To find $C(100)$, we let $t = 100$. Because 100 is greater than 60, we use the second line of the piecewise function.

$$C(t) = 20 + 0.40\,(t - 60) \qquad \text{\small This is the function's equation for } t > 60.$$

$$C(100) = 20 + 0.40\,(100 - 60) \qquad \text{\small Replace } t \text{ with 100.}$$

$$= 20 + 0.40\,(40) \qquad \text{\small Subtract within parentheses: } 100 - 60 = 40.$$

$$= 20 + 16 \qquad \text{\small Multiply: } 0.40\,(40) = 16.$$

$$= 36 \qquad \text{\small Add: } 20 + 16 = 36.$$

Thus, $C(100) = 36$. This means that with 100 calling minutes, the monthly cost is $36. This can be visually represented by the point $(100, 36)$ on the second piece of the graph in **Figure 3.16**. Take a moment to identify this point on the graph.

✓ **Check Point 9**   Use the function in Example 9 to find and interpret each of the following:

a. $C(40)$ 20; With 40 calling minutes, the monthly cost is $20; (40, 20)   **b.** $C(80)$.   28; With 80 calling minutes, the monthly cost is $28; (80, 28)

▶ Identify your solutions by points on the graph in **Figure 3.16**.

Achieving Success

We have seen that **the best way to achieve success in math is through practice**. Keeping up with your homework, preparing for tests, asking questions of your professor, reading your textbook, and attending all classes will help you learn the material and boost your confidence. Use language in a proactive way that reflects a sense of responsibility for your own successes and failures.

Reactive Language	Proactive Language
I'll try.	I'll do it.
That's just the way I am.	I can do better than that.
There's not a thing I can do.	I have options for improvement.
I have to.	I choose to.
I can't.	I can find a way.

Concept and Vocabulary Check

Exercises in the Concept and Vocabulary Check are intended for group and class discussions.

In Exercises 1–9, fill in each blank so that the resulting statement is true.

1. In the rectangular coordinate system, the horizontal number line is called the ____x-axis____.

2. In the rectangular coordinate system, the vertical number line is called the ____y-axis____.

3. In the rectangular coordinate system, the point of intersection of the horizontal axis and the vertical axis is called the ____origin____.

4. The axes of the rectangular coordinate system divide the plane into regions, called ____quadrants____. There are ____four____ of these regions.

5. The first number in an ordered pair such as $(3, 8)$ is called the ___x-coordinate___. The second number in such an ordered pair is called the ___y-coordinate___.

6. The ordered pair $(1, 3)$ is a/an ___solution___ of the equation $y = 5x - 2$ because when 1 is substituted for x and 3 is substituted for y, we obtain a true statement. We also say that $(1, 3)$ ___satisfies___ the equation.

7. If an equation in two variables (x and y) yields precisely one value of _____y_____ for each value of _____x_____, we say that y is a/an ___function___ of x.

8. If $f(x) = 3x + 5$, we can find $f(6)$ by replacing _____x_____ with _____6_____.

9. If any vertical line intersects a graph _more than once_, the graph does not define y as a/an ___function___ of x.

In Exercises 10–15, determine whether each statement is true or false. If the statement is false, make the necessary change(s) to produce a true statement.   Changes to false statements will vary.

10. If $(2, 5)$ satisfies an equation, then $(5, 2)$ also satisfies the equation.   false

11. The ordered pair $(3, 4)$ satisfies the equation

$$2y - 3x = -6.$$ false

12. No two ordered pairs of a function can have the same x-coordinate and different y-coordinates.   true

13. No two ordered pairs of a function can have the same y-coordinate and different x-coordinates.   false

14. A vertical line can intersect the graph of a function at more than one point.   false

15. A horizontal line can intersect the graph of a function at more than one point.   true

Respond to Exercises 16–21 using verbal or written explanations.

16. What is the rectangular coordinate system?

17. Explain how to plot a point in the rectangular coordinate system. Give an example with your explanation.

18. Explain why $(5, -2)$ and $(-2, 5)$ do not represent the same ordered pair.

19. Explain how to graph an equation in the rectangular coordinate system.

20. What is a function?

21. Explain how the vertical line test is used to determine whether a graph represents a function.   16–21. Answers will vary.

Exercise Set 3.1

Practice Exercises   1 –19 (odd)

In Exercises 1–20, plot the given point in a rectangular coordinate system.

1. $(1, 4)$ *
2. $(2, 5)$ *
3. $(-2, 3)$ *
4. $(-1, 4)$ *
5. $(-3, -5)$ *
6. $(-4, -2)$ *
7. $(4, -1)$ *
8. $(3, -2)$ *
9. $(-4, 0)$ *
10. $(-5, 0)$ *
11. $(0, -3)$ *
12. $(0, -4)$ *
13. $(0, 0)$ *
14. $\left(-3, -1\frac{1}{2}\right)$ *
15. $\left(-2, -3\frac{1}{2}\right)$ *
16. $(-5, -2.5)$ *
17. $(3.5, 4.5)$ *
18. $(2.5, 3.5)$ *
19. $(1.25, -3.25)$ *
20. $(2.25, -4.25)$ *

Graph each equation in Exercises 21–32. Select integers for x from -3 to 3, inclusive.

21. $y = x^2 - 2$ *
22. $y = x^2 + 2$ *
23. $y = x - 2$ *
24. $y = x + 2$ *
25. $y = 2x + 1$ *
26. $y = 2x - 4$ *
27. $y = -\frac{1}{2}x$ *
28. $y = -\frac{1}{2}x + 2$ *
29. $y = x^3$ *
30. $y = x^3 - 1$ *
31. $y = |x| + 1$ *
32. $y = |x| - 1$ *

In Exercises 33–46, evaluate each function at the given value of the variable.

33. $f(x) = x - 4$ a. $f(8)$ 4   b. $f(1)$ -3
34. $f(x) = x - 6$ a. $f(9)$ 3   b. $f(2)$ -4
35. $f(x) = 3x - 2$ a. $f(7)$ 19   b. $f(0)$ -2
36. $f(x) = 4x - 3$ a. $f(7)$ 25   b. $f(0)$ -3
37. $g(x) = x^2 + 1$ a. $g(2)$ 5   b. $g(-2)$ 5
38. $g(x) = x^2 + 4$ a. $g(3)$ 13   b. $g(-3)$ 13
39. $g(x) = -x^2 + 2$ a. $g(4)$ -14   b. $g(-3)$ -7
40. $g(x) = -x^2 + 1$ a. $g(5)$ -24   b. $g(-4)$ -15
41. $h(r) = 3r^2 + 5$ a. $h(4)$ 53   b. $h(-1)$ 8
42. $h(r) = 2r^2 - 4$ a. $h(5)$ 46   b. $h(-1)$ -2
43. $f(x) = 2x^2 + 3x - 1$ a. $f(3)$ 26   b. $f(-4)$ 19
44. $f(x) = 3x^2 + 4x - 2$ a. $f(2)$ 18   b. $f(-1)$ -3
45. $f(x) = \dfrac{x}{|x|}$ a. $f(6)$ 1   b. $f(-6)$ -1
46. $f(x) = \dfrac{|x|}{x}$ a. $f(5)$ 1   b. $f(-5)$ -1

*See Answers to Selected Exercises.

In Exercises 47–54, evaluate $f(x)$ for the given values of x. Then use the ordered pairs $(x, f(x))$ from your table to graph the function.

47. $f(x) = x^2 - 1$ *

x	$f(x) = x^2 - 1$
−2	3
−1	0
0	−1
1	0
2	3

48. $f(x) = x^2 + 1$ *

x	$f(x) = x^2 + 1$
−2	5
−1	2
0	1
1	2
2	5

49. $f(x) = x - 1$ *

x	$f(x) = x - 1$
−2	−3
−1	−2
0	−1
1	0
2	1

50. $f(x) = x + 1$ *

x	$f(x) = x + 1$
−2	−1
−1	0
0	1
1	2
2	3

51. $f(x) = (x - 2)^2$ *

x	$f(x) = (x - 2)^2$
0	4
1	1
2	0
3	1
4	4

52. $f(x) = (x + 1)^2$ *

x	$f(x) = (x + 1)^2$
−3	4
−2	1
−1	0
0	1
1	4

53. $f(x) = x^3 + 1$ *

x	$f(x) = x^3 + 1$
−3	−26
−2	−7
−1	0
0	1
1	2

54. $f(x) = (x + 1)^3$ *

x	$f(x) = (x + 1)^3$
−3	−8
−2	−1
−1	0
0	1
1	8

For Exercises 55–62, use the vertical line test to identify graphs in which y is a function of x.

55.

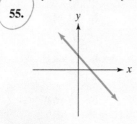

function

56.

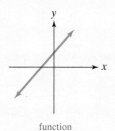

function

57.

function

58.

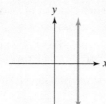

not a function

59.

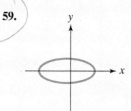

not a function

60.

not a function

61.

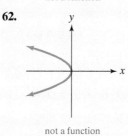

function

62.

not a function

Practice Plus

In Exercises 63–64, let $f(x) = x^2 - x + 4$ and $g(x) = 3x - 5$.

63. Find $g(1)$ and $f(g(1))$.   −2; 10

64. Find $g(-1)$ and $f(g(-1))$.   −8; 76

In Exercises 65–66, let f and g be defined by the following table:

x	$f(x)$	$g(x)$
−2	6	0
−1	3	4
0	−1	1
1	−4	−3
2	0	−6

65. Find $\sqrt{f(-1) - f(0)} - [g(2)]^2 + f(-2) \div g(2) \cdot g(-1)$.   −38

66. Find $|f(1) - f(0)| - [g(1)]^2 + g(1) \div f(-1) \cdot g(2)$.   0

In Exercises 67–70, write each English sentence as an equation in two variables. Then graph the equation.

67. The y-value is four more than twice the x-value. $y = 2x + 4$ *

68. The y-value is the difference between four and twice the x-value. $y = 4 - 2x$ *

69. The y-value is three decreased by the square of the x-value. $y = 3 - x^2$ *

70. The y-value is two more than the square of the x-value. $y = x^2 + 2$ *

Application Exercises

A football is thrown by a quarterback to a receiver. The points in the figure show the height of the football, in feet, above the ground in terms of its distance, in yards, from the quarterback. Use this information to solve Exercises 71–76.

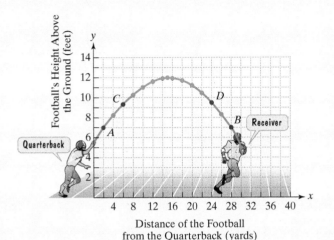

Distance of the Football
from the Quarterback (yards)

71. Find the coordinates of point *A*. Then interpret the coordinates in terms of the information given.

72. Find the coordinates of point *B*. Then interpret the coordinates in terms of the information given.

73. Estimate the coordinates of point *C*. (6, 9.25)

74. Estimate the coordinates of point *D*. (24, 9.5)

75. What is the football's maximum height? What is its distance from the quarterback when it reaches its maximum height?   12 feet; 15 yards

76. What is the football's height when it is caught by the receiver? What is the receiver's distance from the quarterback when he catches the football?   5 feet; 30 yards

The wage gap is used to compare the status of women's earnings relative to men's. The wage gap is expressed as a percent and is calculated by dividing the median, or middlemost, annual earnings for women by the median annual earnings for men. The bar graph shows the wage gap for selected years from 1980 through 2010.

Median Women's Earnings as a Percentage of Median Men's Earnings in the United States

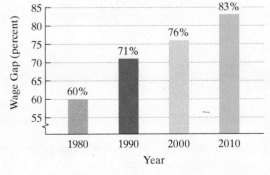

Year

Source: Bureau of Labor Statistics

71. (2, 7); The football is 7 feet above the ground when it is 2 yards from the quarterback.
72. (28, 7); The football is 7 feet above the ground when it is 28 yards from the quarterback.
*See Answers to Selected Exercises.

The function $G(x) = -0.01x^2 + x + 60$ models the wage gap, as a percent, x years after 1980. The graph of function G is shown below. Use this information to solve Exercises 77–78.

The Graph of a Function Modeling the Data

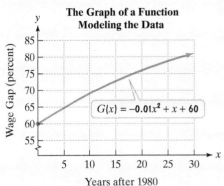

Years after 1980

77. a. 81; According to the function, women's earnings were 81% of men's earnings 30 years after 1980, or in 2010.; (30, 81)

77. a. Find and interpret $G(30)$. Identify this information as a point on the graph of the function.

  b. Does $G(30)$ overestimate or underestimate the actual data shown by the bar graph? By how much?
  underestimates by 2%

78. a. Find and interpret $G(10)$. Identify this information as a point on the graph of the function.

  b. Does $G(10)$ overestimate or underestimate the actual data shown by the bar graph? By how much?
  underestimates by 2%

Critical Thinking Exercises

A cellphone company offers the following plans. Also given are the piecewise functions that model these plans. Use this information to solve Exercises 79–80.

Plan A

• *$30 per month buys 120 minutes.*

• *Additional time costs $0.30 per minute.*

$$C(t) = \begin{cases} 30 & \text{if} \quad 0 \le t \le 120 \\ 30 + 0.30(t - 120) & \text{if} \quad t > 120 \end{cases}$$

Plan B

• *$40 per month buys 200 minutes.*

• *Additional time costs $0.30 per minute.*

$$C(t) = \begin{cases} 40 & \text{if} \quad 0 \le t \le 200 \\ 40 + 0.30(t - 200) & \text{if} \quad t > 200 \end{cases}$$

79. Simplify the algebraic expression in the second line of the piecewise function for plan A. Then use point-plotting to graph the function. $0.30t - 6$ *

80. Simplify the algebraic expression in the second line of the piecewise function for plan B. Then use point-plotting to graph the function. $0.30t - 20$ *

In Exercises 81–82, write a piecewise function that models each cellphone billing plan. Then graph the function.

81. $30 per month buys 500 minutes. Additional time costs $0.25 per minute.   *

82. $60 per month buys 2000 minutes. Additional time costs $0.10 per minute.   *

78. a. 69; According to the function, women's earnings were 69% of men's earnings 10 years after 1980, or in 1990.; (10, 69)

In Exercises 83–86, select the graph that best illustrates each story.

83. An airplane flew from Miami to San Francisco. b

a.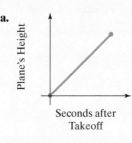

b.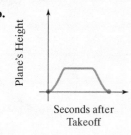

c.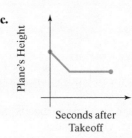

d.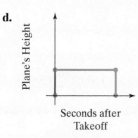

84. At noon, you begin to breathe in. a

a.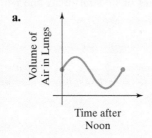

b.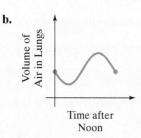

c.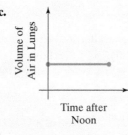

d.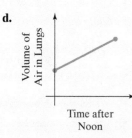

85. Measurements are taken of a person's height from birth to age 100. c

a.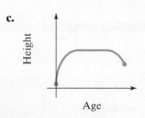

b.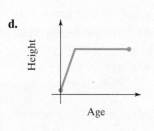

c.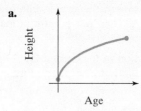

d.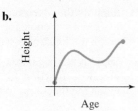

86. You begin your bike ride by riding down a hill. Then you ride up another hill. Finally, you ride along a level surface before coming to a stop. b

a.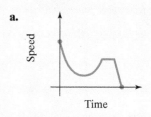

b.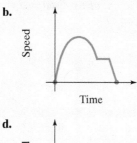

c.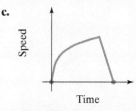

d.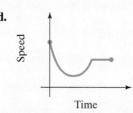

In Exercises 87–90, use the graphs of f and g to find each number.

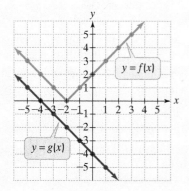

87. $f(-1) + g(-1)$ –2

88. $f(1) + g(1)$ –2

89. $f(g(-1))$ 1

90. $f(g(1))$ 3

Technology Exercise

91. Use a graphing calculator to verify the graphs that you drew by hand in Exercises 47–54.   Answers will vary.

Group Exercise

92. (For assistance with this exercise, refer to the discussion of cellphone plans beginning on page 187, as well as to Exercises 79–80.) Group members who have cellphone plans should describe the total monthly cost of the plan as follows:

$_____ per month buys _____ minutes.
Additional time costs $_____ per minute.

(For simplicity, ignore other charges.) The group should select any three plans, from "basic" to "premier." For each plan selected, write a piecewise function that describes the plan and graph the function. Graph the three functions in the same rectangular coordinate system. Now examine the graphs. For any given number of calling minutes, the best plan is the one whose graph is lowest at that point. Compare the three calling plans. Is one plan always a better deal than the other two? If not, determine the interval of calling minutes for which each plan is the best deal. (You can check out cellphone plans by visiting www.point.com.)
Answers will vary.

3.2 : Linear Functions and Their Graphs

What am I supposed to learn?

After you have read this section, you should be able to:

 Use intercepts to graph a linear equation.

2 Calculate slope.

3 Use the slope and *y*-intercept to graph a line.

4 Graph horizontal or vertical lines.

5 Interpret slope as rate of change.

6 Use slope and *y*-intercept to model data.

It's hard to believe that this gas-guzzler, with its huge fins and overstated design, was available in 1957 for approximately $1800. Sadly, its elegance quickly faded, depreciating by $300 per year, often sold for scrap just six years after its glorious emergence from the dealer's showroom.

From these casual observations, we can obtain a mathematical model and its graph. The model is

$$y = -300x + 1800.$$

| The car is depreciating by $300 per year for *x* years. | The new car is worth $1800. |

In this model, *y* is the car's value after *x* years. **Figure 3.17** shows the equation's graph. Using function notation, we can rewrite the equation as

$$f(x) = -300x + 1800.$$

A function such as this, whose graph is a straight line, is called a **linear function**. In this section, we will study linear functions and their graphs.

Graphing Using Intercepts

There is another way that we can write the equation

$$y = -300x + 1800.$$

We will collect the *x*- and *y*-terms on the left side. This is done by adding 300*x* to both sides:

$$300x + y = 1800.$$

All equations of the form $Ax + By = C$ are straight lines when graphed, as long as *A* and *B* are not both zero. Such equations are called **linear equations in two variables**. We can quickly obtain the graph for equations in this form when none of *A*, *B*, or *C* is zero by finding the points where the graph intersects the *x*-axis and the *y*-axis. The *x*-coordinate of the point where the graph intersects the *x*-axis is called the **x-intercept**. The *y*-coordinate of the point where the graph intersects the *y*-axis is called the **y-intercept**.

The graph of $300x + y = 1800$ in **Figure 3.17** intersects the *x*-axis at $(6, 0)$, so the *x*-intercept is 6. The graph intersects the *y*-axis at $(0, 1800)$, so the *y*-intercept is 1800.

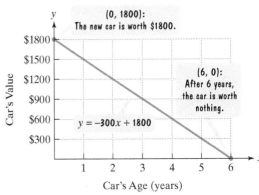

Figure 3.17

 Use intercepts to graph a linear equation.

Locating Intercepts

To locate the *x*-intercept, set $y = 0$ and solve the equation for *x*.

To locate the *y*-intercept, set $x = 0$ and solve the equation for *y*.

An equation of the form $Ax + By = C$ as described above can be graphed by finding the *x*- and *y*-intercepts, plotting the intercepts, and drawing a straight line through these points. When graphing using intercepts, it is a good idea to use a third point, a checkpoint, before drawing the line. A checkpoint can be obtained by selecting a value for *x*, other than 0 or the *x*-intercept, and finding the corresponding value for *y*. The checkpoint should lie on the same line as the *x*- and *y*-intercepts. If it does not, recheck your work and find the error.

Example 1 **Using Intercepts to Graph a Linear Equation**

Graph: $3x + 2y = 6$.

Solution

Note that $3x + 2y = 6$ is of the form $Ax + By = C$.

$$3x + 2y = 6$$

$$A = 3 \quad B = 2 \quad C = 6$$

In this case, none of A, B, or C is zero.

Find the x-intercept by letting $y = 0$ and solving for x.	**Find the y-intercept by letting $x = 0$ and solving for y.**
$3x + 2y = 6$	$3x + 2y = 6$
$3x + 2 \cdot 0 = 6$	$3 \cdot 0 + 2y = 6$
$3x = 6$	$2y = 6$
$x = 2$	$y = 3$

The x-intercept is 2, so the line passes through the point $(2, 0)$. The y-intercept is 3, so the line passes through the point $(0, 3)$.

For our checkpoint, we choose a value for x other than 0 or the x-intercept, 2. We will let $x = 1$ and find the corresponding value for y.

$3x + 2y = 6$	This is the given equation.
$3 \cdot 1 + 2y = 6$	Substitute 1 for x.
$3 + 2y = 6$	Simplify.
$2y = 3$	Subtract 3 from both sides.
$y = \dfrac{3}{2}$	Divide both sides by 2.

The checkpoint is the ordered pair $\left(1, \frac{3}{2}\right)$, or $(1, 1.5)$.

The three points in **Figure 3.18** lie along the same line. Drawing a line through the three points results in the graph of $3x + 2y = 6$. The arrowheads at the ends of the line show that the line continues indefinitely in both directions.

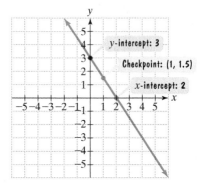

Figure 3.18 The graph of $3x + 2y = 6$

> **✓ Check Point 1** Graph: $2x + 3y = 6$. *

Slope

2 Calculate slope.

Mathematicians have developed a useful measure of the steepness of a line, called the *slope* of the line. Slope compares the vertical change (the **rise**) to the horizontal change (the **run**) when moving from one fixed point to another along the line. To calculate the slope of a line, we use a ratio that compares the change in y (the rise) to the change in x (the run).

Definition of Slope

The **slope** of the line through the distinct points (x_1, y_1) and (x_2, y_2) is

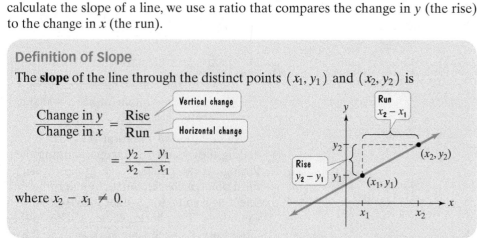

$$\frac{\text{Change in } y}{\text{Change in } x} = \frac{\text{Rise}}{\text{Run}}$$

$$= \frac{y_2 - y_1}{x_2 - x_1}$$

where $x_2 - x_1 \neq 0$.

It is common notation to let the letter m represent the slope of a line. The letter m is used because it is the first letter of the French verb *monter*, meaning "to rise," or "to ascend."

Example 2   Using the Definition of Slope

Find the slope of the line passing through each pair of points:

a. $(-3, -1)$ and $(-2, 4)$
b. $(-3, 4)$ and $(2, -2)$.

Solution

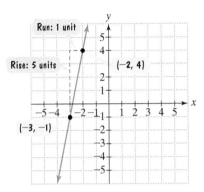

Figure 3.19 Visualizing a slope of 5

a. Let $(x_1, y_1) = (-3, -1)$ and $(x_2, y_2) = (-2, 4)$. We obtain the slope as follows:

$$m = \frac{\text{Change in } y}{\text{Change in } x} = \frac{y_2 - y_1}{x_2 - x_1} = \frac{4 - (-1)}{-2 - (-3)} = \frac{5}{1} = 5.$$

The situation is illustrated in **Figure 3.19**. The slope of the line is 5, indicating that there is a vertical change, a rise, of 5 units for each horizontal change, a run, of 1 unit. The slope is positive and the line rises from left to right.

Great Question!

When using the definition of slope, how do I know which point to call (x_1, y_1) and which point to call (x_2, y_2)?

When computing slope, it makes no difference which point you call (x_1, y_1) and which point you call (x_2, y_2). If we let $(x_1, y_1) = (-2, 4)$ and $(x_2, y_2) = (-3, -1)$, the slope is still 5:

$$m = \frac{\text{Change in } y}{\text{Change in } x} = \frac{y_2 - y_1}{x_2 - x_1} = \frac{-1 - 4}{-3 - (-2)} = \frac{-5}{-1} = 5.$$

However, you should not subtract in one order in the numerator $(y_2 - y_1)$ and then in the opposite order in the denominator $(x_1 - x_2)$. The slope is *not* -5:

$$\frac{-1 - 4}{-2 - (-3)} \quad \frac{-5}{1} = -5. \quad \text{Incorrect}$$

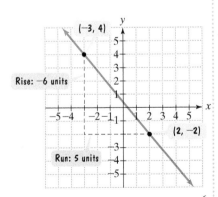

Figure 3.20 Visualizing a slope of $-\frac{6}{5}$

b. We can let $(x_1, y_1) = (-3, 4)$ and $(x_2, y_2) = (2, -2)$. The slope of the line shown in **Figure 3.20** is computed as follows:

$$m = \frac{\text{Change in } y}{\text{Change in } x} = \frac{y_2 - y_1}{x_2 - x_1} = \frac{-2 - 4}{2 - (-3)} = \frac{-6}{5} = -\frac{6}{5}.$$

The slope of the line is $-\frac{6}{5}$. For every vertical change of -6 units (6 units down), there is a corresponding horizontal change of 5 units. The slope is negative and the line falls from left to right.

Check Point 2   Find the slope of the line passing through each pair of points:

a. $(-3, 4)$ and $(-4, -2)$ $m = 6$
b. $(4, -2)$ and $(-1, 5)$. $m = -\frac{7}{5}$

Example 2 illustrates that a line with a positive slope is rising from left to right and a line with a negative slope is falling from left to right. By contrast, a horizontal line neither rises nor falls and has a slope of zero. A vertical line has no horizontal change, so $x_2 - x_1 = 0$ in the formula for slope. Because we cannot divide by zero, the slope of a vertical line is undefined. This discussion is summarized in **Table 3.2**.

Table 3.2 Possibilities for a Line's Slope

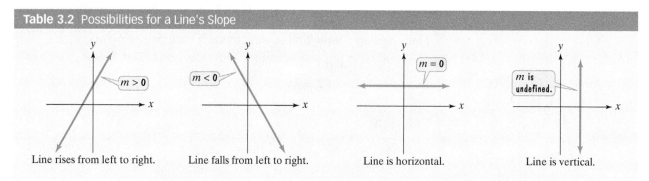

| Line rises from left to right. | Line falls from left to right. | Line is horizontal. | Line is vertical. |

The Slope-Intercept Form of the Equation of a Line

Use the slope and y-intercept to graph a line.

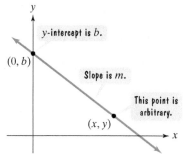

Figure 3.21 A line with slope m and y-intercept b

We can use the definition of slope to write the equation of any nonvertical line with slope m and y-intercept b. Because the y-intercept is b, the point $(0, b)$ lies on the line. Now, let (x, y) represent any other point on the line, shown in **Figure 3.21**. Keep in mind that the point (x, y) is arbitrary and is not in one fixed position. By contrast, the point $(0, b)$ is fixed.

Regardless of where the point (x, y) is located, the steepness of the line in **Figure 3.21** remains the same. Thus, the ratio for slope stays a constant m. This means that for all points along the line

$$m = \frac{\text{Change in } y}{\text{Change in } x} = \frac{y - b}{x - 0} = \frac{y - b}{x}.$$

We can clear the fraction by multiplying both sides by x, the denominator. Note that x is not zero since (x, y) is distinct from $(0, b)$, the only point on the line with first coordinate 0.

$$m = \frac{y - b}{x} \qquad \text{This is the slope of the line in Figure 3.21.}$$

$$mx = \frac{y - b}{x} \cdot x \qquad \text{Multiply both sides by } x.$$

$$mx = y - b \qquad \text{Simplify: } \frac{y-b}{x} \cdot x = y - b.$$

$$mx + b = y - b + b \qquad \text{Add } b \text{ to both sides and solve for } y.$$

$$mx + b = y \qquad \text{Simplify.}$$

Now, if we reverse the two sides, we obtain the *slope-intercept form* of the equation of a line.

Slope-Intercept Form of the Equation of a Line

The **slope-intercept form of the equation** of a nonvertical line with slope m and y-intercept b is

$$y = mx + b.$$

The slope-intercept form of a line's equation, $y = mx + b$, can be expressed in function notation by replacing y with $f(x)$:

$$f(x) = mx + b.$$

We have seen that functions in the form $f(x) = mx + b$ are called **linear functions**. Thus, in the equation of a linear function, the x-coefficient is the line's slope and the constant term is the y-intercept. Here are two examples:

$$y = 2x - 4 \qquad\qquad f(x) = \frac{1}{2}x + 2.$$

The slope is 2.　　The y-intercept is −4.　　The slope is $\frac{1}{2}$.　　The y-intercept is 2.

If a linear function's equation is in slope-intercept form, we can use the y-intercept and the slope to obtain its graph.

Great Question!

If the slope is an integer, such as 2, why should I express it as $\frac{2}{1}$ for graphing purposes?

Writing the slope, m, as a fraction allows you to identify the rise (the fraction's numerator) and the run (the fraction's denominator).

Graphing $y = mx + b$ Using the Slope and y-Intercept

1. Plot the point containing the y-intercept on the y-axis. This is the point $(0, b)$.
2. Obtain a second point using the slope, m. Write m as a fraction, and use rise over run, starting at the point containing the y-intercept, to plot this point.
3. Use a straightedge to draw a line through the two points. Draw arrowheads at the ends of the line to show that the line continues indefinitely in both directions.

Example 3　Graphing by Using the Slope and y-Intercept

Graph the linear function $y = \frac{2}{3}x + 2$ by using the slope and y-intercept.

Solution

The equation of the linear function is in the form $y = mx + b$. We can find the slope, m, by identifying the coefficient of x. We can find the y-intercept, b, by identifying the constant term.

$$y = \frac{2}{3}x + 2$$

The slope is $\frac{2}{3}$.　　The y-intercept is 2.

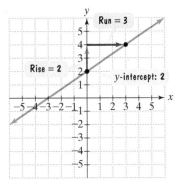

Figure 3.22 The graph of $y = \frac{2}{3}x + 2$

Now that we have identified the slope and the y-intercept, we use the three-step procedure in the box above to graph the equation.

Step 1　Plot the point containing the y-intercept on the y-axis. The y-intercept is 2. We plot $(0, 2)$, shown in **Figure 3.22**.

Step 2　Obtain a second point using the slope, m. Write m as a fraction, and use rise over run, starting at the point containing the y-intercept, to plot this point. The slope, $\frac{2}{3}$, is already written as a fraction:

$$m = \frac{2}{3} = \frac{\text{Rise}}{\text{Run}}.$$

We plot the second point on the line by starting at $(0, 2)$, the first point. Based on the slope, we move 2 units *up* (the rise) and 3 units to the *right* (the run). This puts us at a second point on the line, $(3, 4)$, shown in **Figure 3.22**.

Step 3　Use a straightedge to draw a line through the two points. The graph of $y = \frac{2}{3}x + 2$ is shown in **Figure 3.22**.

Check Point 3 Graph the linear function $y = \frac{3}{5}x + 1$ by using the slope and y-intercept. *

*See Answers to Selected Exercises.

Earlier in this section, we considered linear functions written in the form $Ax + By = C$. We used x- and y-intercepts, as well as checkpoints, to graph these functions. It is also possible to obtain the graphs by using the slope and y-intercept. To do this, begin by solving $Ax + By = C$ for y. This will put the equation in slope-intercept form. Then use the three-step procedure to graph the equation. This is illustrated in Example 4.

Example 4   Graphing by Using the Slope and y-Intercept

Graph the linear function $2x + 5y = 0$ by using the slope and y-intercept.

Solution

We put the equation in slope-intercept form by solving for y.

$$2x + 5y = 0 \qquad \text{This is the given equation.}$$
$$2x - 2x + 5y = 0 - 2x \qquad \text{Subtract } 2x \text{ from both sides.}$$
$$5y = -2x + 0 \qquad \text{Simplify.}$$
$$\frac{5y}{5} = \frac{-2x + 0}{5} \qquad \text{Divide both sides by 5.}$$
$$y = \frac{-2x}{5} + \frac{0}{5} \qquad \text{Divide each term in the numerator by 5.}$$
$$y = -\frac{2}{5}x + 0 \qquad \text{Simplify. Equivalently, } f(x) = -\tfrac{2}{5}x + 0.$$

Now that the equation is in slope-intercept form, we can use the slope and y-intercept to obtain its graph. Examine the slope-intercept form:

$$y = -\frac{2}{5}x + 0.$$

slope: $-\tfrac{2}{5}$ y-intercept: 0

Note that the slope is $-\frac{2}{5}$ and the y-intercept is 0. Use the y-intercept to plot $(0, 0)$ on the y-axis. Then locate a second point by using the slope.

$$m = -\frac{2}{5} = \frac{-2}{5} = \frac{\text{Rise}}{\text{Run}}$$

Because the rise is -2 and the run is 5, move *down* 2 units and to the *right* 5 units, starting at the point $(0, 0)$. This puts us at a second point on the line, $(5, -2)$. The ▸ graph of $2x + 5y = 0$ is the line drawn through these points, shown in **Figure 3.23**.

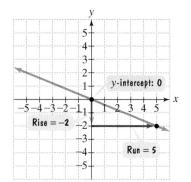

Figure 3.23 The graph of $2x + 5y = 0$, or $y = -\frac{2}{5}x$

The equation $2x + 5y = 0$ in Example 4 is of the form $Ax + By = C$ with $C = 0$. If you try graphing $2x + 5y = 0$ by using intercepts, you will find that the x-intercept is 0 and the y-intercept is 0. This means that the graph passes through the origin. A second point must be found to graph the line. In Example 4, the line's slope gave us the second point.

✓ **Check Point 4** Graph the linear function $3x + 4y = 0$ by using the slope and ▸ y-intercept. *

Equations of Horizontal and Vertical Lines

④ **Graph horizontal or vertical lines.**

If a line is horizontal, its slope is zero: $m = 0$. Thus, the equation $y = mx + b$ becomes $y = b$, where b is the y-intercept. All horizontal lines have equations of the form $y = b$.

*See Answers to Selected Exercises.

Example 5   Graphing a Horizontal Line

Graph $y = -4$ in the rectangular coordinate system.

Solution

All ordered pairs that are solutions of $y = -4$ have a value of y that is always -4. Any value can be used for x. (Think of $y = -4$ as $0x + 1y = -4$.) In the table at the right, we have selected three of the possible values for x: -2, 0, and 3. The table shows that three ordered pairs that are solutions of $y = -4$ are $(-2, -4)$, $(0, -4)$, and $(3, -4)$. Drawing a line that passes through the three points gives the horizontal line shown in **Figure 3.24**.

x	$y = -4$	(x, y)
-2	-4	$(-2, -4)$
0	-4	$(0, -4)$
3	-4	$(3, -4)$

For all choices of x, y is a constant -4.

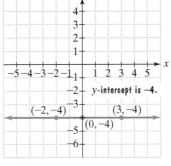

Figure 3.24 The graph of $y = -4$ or $f(x) = -4$

Check Point 5   Graph $y = 3$ in the rectangular coordinate system.   *

Next, let's see what we can discover about the graph of an equation of the form $x = a$ by looking at an example.

Example 6   Graphing a Vertical Line

Graph $x = 2$ in the rectangular coordinate system.

Solution

All ordered pairs that are solutions of $x = 2$ have a value of x that is always 2. Any value can be used for y. (Think of $x = 2$ as $1x + 0y = 2$.) In the table on the right, we have selected three of the possible values for y: -2, 0, and 3. The table shows that three ordered pairs that are solutions of $x = 2$ are $(2, -2)$, $(2, 0)$, and $(2, 3)$. Drawing a line that passes through the three points gives the vertical line shown in **Figure 3.25**.

For all choices of y,

x is always 2.

$x = 2$	y	(x, y)
2	-2	$(2, -2)$
2	0	$(2, 0)$
2	3	$(2, 3)$

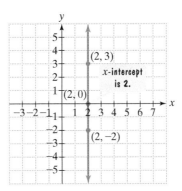

Figure 3.25 The graph of $x = 2$

Does a vertical line represent the graph of a linear function? No. Look at the graph of $x = 2$ in **Figure 3.25**. A vertical line drawn through $(2, 0)$ intersects the graph infinitely many times. This shows that infinitely many outputs are associated with the input 2. **No vertical line represents a linear function.** All other lines are graphs of functions.

Horizontal and Vertical Lines

The graph of $y = b$ or $f(x) = b$ is a horizontal line. The y-intercept is b.

The graph of $x = a$ is a vertical line. The x-intercept is a.

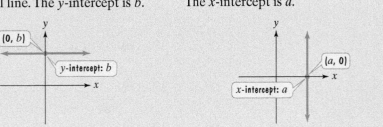

Check Point 6   Graph $x = -2$ in the rectangular coordinate system.   *

*See Answers to Selected Exercises.

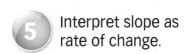

Interpret slope as rate of change.

Slope as Rate of Change

Slope is defined as the ratio of a change in y to a corresponding change in x. Our next example shows how slope can be interpreted as a **rate of change** in an applied situation.

Example 7   Slope as a Rate of Change

The line graphs in **Figure 3.26** show the percentage of American men and women ages 20 to 24 who were married from 1970 through 2010. Find the slope of the line segment representing women. Describe what the slope represents.

Figure 3.26
Source: U.S. Census Bureau

Solution

We let x represent a year and y the percentage of American women ages 20 to 24 who were married in that year. The two points shown on the line segment for women have the following coordinates:

$$(1970, 65) \quad \text{and} \quad (2010, 21).$$

In 1970, 65% of American women ages 20 to 24 were married.	In 2010, 21% of American women ages 20 to 24 were married.

Now we compute the slope.

$$m = \frac{\text{Change in } y}{\text{Change in } x} = \frac{21 - 65}{2010 - 1970}$$

The unit in the numerator is the *percentage of married women ages 20 to 24.*

$$= \frac{-44}{40} = -1.1$$

The unit in the denominator is *year.*

The slope indicates that for the period from 1970 through 2010, the percentage of married women ages 20 to 24 decreased by 1.1 per year. The rate of change is -1.1% per year.

✔ **Check Point 7**. Find the slope of the line segment representing men in **Figure 3.26**. Use your answer to complete this statement: -0.85

For the period from 1970 through 2010, the percentage of married men ages 20 to 24 decreased by ____0.85____ per year. The rate of change is ___-0.85%___ per ____year____.

Modeling Data with the Slope-Intercept Form of the Equation of a Line

Use slope and y-intercept to model data.

The slope-intercept form for equations of lines is useful for obtaining mathematical models for data that fall on or near a line. For example, the bar graph in **Figure 3.27(a)** at the top of the next page shows the percentage of the U.S. population who had graduated from high school and from college in 1960 and 2010. The data are displayed as points in a rectangular coordinate system in **Figure 3.27(b)**.

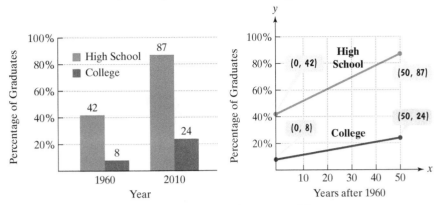

Figure 3.27(a) Figure 3.27(b)
Source: James M. Henslin, *Essentials of Sociology*, Ninth Edition, Pearson, 2011.

Example 8 illustrates how we can use the equation $y = mx + b$ to obtain a model for the data and make predictions about what might occur in the future.

Example 8   Modeling with the Slope-Intercept Form of the Equation

a. Use the two points for high school in **Figure 3.27(b)** to find a function in the form $H(x) = mx + b$ that models the percentage of high school graduates in the U.S. population, $H(x)$, x years after 1960.

b. Use the model to project the percentage of high school graduates in 2020.

Solution

a. We will use the line segment for high school using the points $(0, 42)$ and $(50, 87)$ to obtain a model. We need values for m, the slope, and b, the y-intercept.

$$y = mx + b$$

$$m = \frac{\text{Change in } y}{\text{Change in } x}$$

The point (0, 42) lies on the line segment, so the y-intercept is 42: $b = 42$.

$$= \frac{87 - 42}{50 - 0} = 0.9$$

The percentage of the U.S. population who had graduated from high school, $H(x)$, x years after 1960 can be modeled by the linear function

$$H(x) = 0.9x + 42.$$

The slope, 0.9, indicates an increase in the percentage of high school graduates of 0.9% per year from 1960 through 2010.

b. Now let's use the model to project the percentage of high school graduates in 2020. Because 2020 is 60 years after 1960, substitute 60 for x in $H(x) = 0.9x + 42$ and evaluate the function at 60.

$$H(60) = 0.9(60) + 42 = 54 + 42 = 96$$

Our model projects that 96% of the U.S. population will have graduated from high school in 2020.

Check Point 8

a. Use the two points for college in **Figure 3.27(b)** on the previous page to find a function in the form $C(x) = mx + b$ that models the percentage of college graduates in the U.S. population, $C(x)$, x years after 1960. $C(x) = 0.32x + 8$

▶ **b.** Use the model to project the percentage of college graduates in 2020.   27.2%

Achieving Success

The Secret of Math Success

What's the secret of math success? The bar graph in **Figure 3.28** shows that Japanese teachers and students are more likely than their American counterparts to believe that the key to doing well in math is working hard. Americans tend to think that either you have mathematical intelligence or you don't. Alan Bass, author of *Math Study Skills* (Pearson Education, 2008), strongly disagrees with this American perspective:

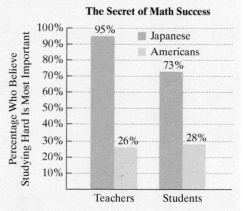

The Secret of Math Success

Figure 3.28
Source: Wade and Tavris, *Psychology*, Ninth Edition, Pearson, 2008.

"Human beings are easily intelligent enough to understand the basic principles of math. I cannot repeat this enough, but I'll try . . .

Poor performance in math is not due to a lack of intelligence! The fact is that the key to success in math is in taking an intelligent approach. Students come up to me and say, 'I'm just not good at math.' Then I ask to see their class notebooks and they show me a chaotic mess of papers jammed into a folder. In math, that's a lot like taking apart your car's engine, dumping the heap of disconnected parts back under the hood, and then going to a mechanic to ask why it won't run. Students come to me and say, 'I'm just not good at math.' Then I ask them about their study habits and they say, 'I have to do all my studying on the weekend.' In math, that's a lot like trying to do all your eating or sleeping on the weekends and wondering why you're so tired and hungry all the time. **How you approach math is much more important than how smart you are.**"

—Alan Bass

Concept and Vocabulary Check

Exercises in the Concept and Vocabulary Check are intended for group and class discussions.

In Exercises 1–8, fill in each blank so that the resulting statement is true.

1. The x-coordinate of the point where a graph crosses the x-axis is called the ___x-intercept___.

2. The point $(0, 3)$ lies along a line, so 3 is the ___y-intercept___ of that line.

3. The slope of the line through the distinct points (x_1, y_1) and (x_2, y_2) is ___$\frac{y_2 - y_1}{x_2 - x_1}$___.

4. The slope-intercept form of the equation of a line is ___$y = mx + b$___, where m represents the ___slope___ and b represents the ___y-intercept___.

5. The slope of the linear function whose equation is $f(x) = -4x + 3$ is ___−4___ and the y-intercept of its graph is ___3___.

6. In order to graph the line whose equation is $y = \frac{2}{5}x + 3$, begin by plotting the point ___$(0, 3)$___. From this point, we move ___2___ units up (the rise) and ___5___ units to the right (the run).

7. The graph of $y = b$ is a/an ___horizontal___ line.

8. The graph of $x = a$ is a/an ___vertical___ line.

In Exercises 9–12, determine whether each statement is true or false. If the statement is false, make the necessary change(s) to produce a true statement.   Changes to false statements will vary.

9. The equation $y = mx + b$ shows that no line can have a y-intercept that is numerically equal to its slope.   false

10. Every line in the rectangular coordinate system has an equation that can be expressed in slope-intercept form.   false

11. The line $3x + 2y = 5$ has slope $-\frac{3}{2}$.   true

12. The line $2y = 3x + 7$ has a y-intercept of 7.   false

Respond to Exercises 13–20 using verbal or written explanations.

13. Describe how to find the x-intercept of a linear equation.

14. Describe how to find the y-intercept of a linear equation.

15. What is the slope of a line?

16. Describe how to calculate the slope of a line passing through two points.

17. Describe how to graph a line using the slope and y-intercept. Provide an original example with your description.

18. What does it mean if the slope of a line is 0?

19. What does it mean if the slope of a line is undefined?

20. Explain why the y-values can be any number for the equation $x = 5$. How is this shown in the graph of the equation?
   13–20. Answers will vary.

Exercise Set 3.2

Practice Exercises

In Exercises 1–8, use the x- and y-intercepts to graph each linear equation.

1. $x - y = 3$ *

2. $x + y = 4$ *

3. $3x - 4y = 12$ *

4. $2x - 5y = 10$ *

5. $2x + y = 6$ *

6. $x + 3y = 6$ *

7. $5x = 3y - 15$ *

8. $3x = 2y + 6$ *

In Exercises 9–20, calculate the slope of the line passing through the given points. If the slope is undefined, so state. Then indicate whether the line rises, falls, is horizontal, or is vertical.

9. $(2, 6)$ and $(3, 5)$ -1; falls

10. $(4, 2)$ and $(3, 4)$ -2; falls

11. $(-2, 1)$ and $(2, 2)$ $\frac{1}{4}$; rises

12. $(-1, 3)$ and $(2, 4)$ $\frac{1}{3}$; rises

13. $(-2, 4)$ and $(-1, -1)$ -5; falls

14. $(6, -4)$ and $(4, -2)$ -1; falls

15. $(5, 3)$ and $(5, -2)$ undefined; vertical

16. $(3, -4)$ and $(3, 5)$ undefined; vertical

17. $(2, 0)$ and $(0, 8)$ -4; falls

18. $(3, 0)$ and $(0, -9)$ 3; rises

19. $(5, 1)$ and $(-2, 1)$ 0; horizontal

20. $(-2, 3)$ and $(1, 3)$ 0; horizontal

In Exercises 21–32, graph each linear function using the slope and y-intercept.

21. $y = 2x + 3$ *

22. $y = 2x + 1$ *

23. $y = -2x + 4$ *

24. $y = -2x + 3$ *

25. $y = \frac{1}{2}x + 3$ *

26. $y = \frac{1}{2}x + 2$ *

27. $f(x) = \frac{2}{3}x - 4$ *

28. $f(x) = \frac{3}{4}x - 5$ *

29. $y = -\frac{3}{4}x + 4$ *

30. $y = -\frac{2}{3}x + 5$ *

31. $f(x) = -\frac{5}{3}x$ *

32. $f(x) = -\frac{4}{3}x$ *

In Exercises 33–40,

  a. *Put the equation in slope-intercept form by solving for y.*
  b. *Identify the slope and the y-intercept.*
  c. *Use the slope and y-intercept to graph the line.*

33. $3x + y = 0$ *

34. $2x + y = 0$ *

35. $3y = 4x$ *

36. $4y = 5x$ *

37. $2x + y = 3$ *

38. $3x + y = 4$ *

39. $7x + 2y = 14$ *

40. $5x + 3y = 15$ *

In Exercises 41–48, graph each horizontal or vertical line.

41. $y = 4$ *

42. $y = 2$ *

43. $y = -2$ *

44. $y = -3$ *

45. $x = 2$ *

46. $x = 4$ *

47. $x + 1 = 0$ *

48. $x + 5 = 0$ *

Practice Plus

In Exercises 49–52, find the slope of the line passing through each pair of points or state that the slope is undefined. Assume that all variables represent positive real numbers. Then indicate whether the line through the points rises, falls, is horizontal, or is vertical.

49. $(0, a)$ and $(b, 0)$ $m = -\frac{a}{b}$; falls

50. $(-a, 0)$ and $(0, -b)$ $m = -\frac{b}{a}$; falls

51. (a, b) and $(a, b + c)$ undefined slope; vertical

52. $(a - b, c)$ and $(a, a + c)$ $m = \frac{a}{b}$; rises

*See Answers to Selected Exercises.

53. $m = -\dfrac{A}{B}; b = \dfrac{C}{B}$ **54.** $m = \dfrac{A}{B}; b = \dfrac{C}{B}$

In Exercises 53–54, find the slope and y-intercept of each line whose equation is given. Assume that $B \neq 0$.

53. $Ax + By = C$ **54.** $Ax = By - C$

In Exercises 55–56, find the value of y if the line through the two given points is to have the indicated slope.

55. $(3, y)$ and $(1, 4)$, $m = -3$ −2

56. $(-2, y)$ and $(4, -4)$, $m = \frac{1}{3}$ −6

Use the figure to make the lists in Exercises 57–58.

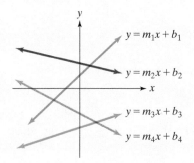

57.   List the slopes $m_1, m_2, m_3,$ and m_4 in order of decreasing size. m_1, m_3, m_2, m_4

58.   List the y-intercepts $b_1, b_2, b_3,$ and b_4 in order of decreasing size. b_2, b_1, b_4, b_3

Application Exercises

59.   Older, Calmer. As we age, daily stress and worry decrease and happiness increases, according to an analysis of 340,847 U.S. adults, ages 18–85, in the journal *Proceedings of the National Academy of Sciences.* The graphs show a portion of the research.

Percentage of Americans Reporting "a Lot" of Stress, by Age

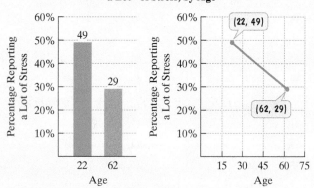

Source: National Academy of Sciences

a. Find the slope of the line passing through the two points shown by the voice balloons. Express the slope as a decimal.   −0.5

b. Use your answer from part (a) to complete the statement:

For each year of aging, the percentage of Americans reporting "a lot" of stress decreases by ____0.5____%. The rate of decrease is ____0.5____ % per __year of aging__ .

60.   Exercise is useful not only in preventing depression, but also as a treatment. The graphs at the top of the next column show the percentage of patients with depression in remission

when exercise (brisk walking) was used as a treatment. (The control group that engaged in no exercise had 11% of the patients in remission.)

Exercise and Percentage of Patients with Depression in Remission

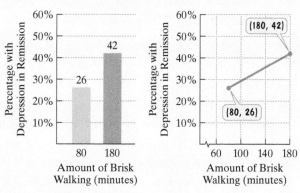

Source: Newsweek, March 26, 2007

a. Find the slope of the line passing through the two points shown by the voice balloons. Express the slope as a decimal.   0.16

b. Use your answer from part (a) to complete this statement:

For each minute of brisk walking, the percentage of patients with depression in remission increased by ____0.16____%. The rate of change is ____0.16____ % per __minute of brisk walking__ .

Grade Inflation in U.S. High Schools. *In recent decades, high school teachers have given higher and higher grades to students. The bar graph shows the percentage of grades of A and C for high school students entering college.*

High School Grades of Students Entering College

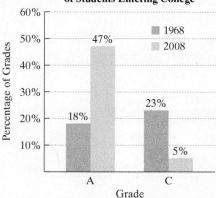

Source: UCLA Higher Education Research Institute

In Exercises 61–62, find a linear function in slope-intercept form that models the given description. Each function should model the percentage of the particular high school grade, P(x), of students entering college x years after 1968.

61.   In 1968, 18% of the grades for students entering college were A (A+, A, or A−). This has increased at an average rate of approximately 0.725% per year since then. $P(x) = 0.725x + 18$

62.   In 1968, 23% of the grades for students entering college were C (C+, C or C−). This has decreased at an average rate of approximately 0.45% per year since then. $P(x) = -0.45x + 23$

Big (Lack of) Men on Campus. *The bar graph shows the number of bachelor's degrees, in thousands, awarded to men and women in the United States for four selected years from 1980 to 2010. The trend indicated by the graphs is among the hottest topics of debate among college-admissions officers. Some private liberal arts colleges have quietly begun special efforts to recruit men—including admissions preferences for them.*

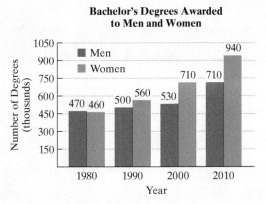

Bachelor's Degrees Awarded to Men and Women

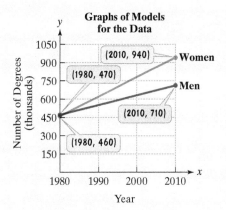

Graphs of Models for the Data

Source: Department of Education

Exercises 63–64 involve the graphs of models for the data shown in the rectangular coordinate system. **63. a.** $W(x) = 16x + 460$

63. a. Use the two points for women shown by the blue voice balloons to find a function in the form $W(x) = mx + b$ that models the number of bachelor's degrees, $W(x)$, in thousands, awarded to women x years after 1980.

    b. Use the model from part (a) to project the number of bachelor's degress that will be awarded to women in 2020.
1,100,000

64. a. Use the two points for men shown by the red voice balloons to find a function in the form $M(x) = mx + b$ that models the number of bachelor's degrees, $M(x)$, in thousands, awarded to men x years after 1980. $M(x) = 8x + 470$

    b. Use the model from part (a) to project the number of bachelor's degrees that will be awarded to men in 2020.
790,000

Critical Thinking Exercises

In Exercises 65–66, find the coefficients that must be placed in each shaded area so that the function's graph will be a line satisfying the specified conditions.

65. ▨x + ▨y = 12; x-intercept = −2; y-intercept = 4   −6; 3

66. ▨x + ▨y = 12; y-intercept = −6; slope = $\dfrac{1}{2}$.   1; −2

67. Prove that the equation of a line passing through $(a, 0)$ and $(0, b)$ $(a \neq 0, b \neq 0)$ can be written in the form $\dfrac{x}{a} + \dfrac{y}{b} = 1$. Why is this called the *intercept form* of a line? *

68. The relationship between Celsius temperature, C, and Fahrenheit temperature, F, can be described by a linear equation in the form $F = mC + b$. The graph of this equation contains the point $(0, 32)$: Water freezes at $0°C$ or at $32°F$. The line also contains the point $(100, 212)$: Water boils at $100°C$ or at $212°F$. Write the linear equation expressing Fahrenheit temperature in terms of Celsius temperature. $F = \dfrac{9}{5}C + 32$

Technology Exercises   **69–70.** Answers will vary.

69. Use a graphing utility to verify any three of your hand-drawn graphs in Exercises 21–32.

70. Use a graphing utility to verify any three of your hand-drawn graphs in Exercises 33–40. Solve the equation for y before entering it.

3.3 The Point-Slope Form of the Equation of a Line; Scatter Plots and Regression Lines

What am I supposed to learn?

After you have read this section, you should be able to:

1. Use the point-slope form to write equations of a line.

2. Write linear equations that model data and make predictions.

3. Make a scatter plot for a table of data items.

4. Interpret information given in a scatter plot.

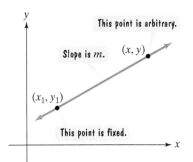

Figure 3.29 A line passing through (x_1, y_1) with slope m

Have you ever watched any movies or television shows from the 1940s or 1950s? If so, are you surprised by the number of people smoking cigarettes? At that time, there was little awareness of the relationship between tobacco use and numerous diseases. Cigarette smoking was seen as a healthy way to relax and help digest a hearty meal. Then, in 1964, a linear equation changed everything. To understand the mathematics behind this turning point in public health, we explore another form of a line's equation.

Point-Slope Form

We can use the slope of a line to obtain another useful form of the line's equation. Consider a nonvertical line that has slope m and contains the point (x_1, y_1). Now let (x, y) represent any other point on the line, shown in **Figure 3.29**. Keep in mind that the point (x, y) is arbitrary and is not in one fixed position. By contrast, the point (x_1, y_1) is fixed.

Regardless of where the point (x, y) is located on the line, the steepness of the line in **Figure 3.29** remains the same. Thus, the ratio for slope stays a constant m. This means that for all points (x, y) along the line,

$$m = \frac{\text{Change in } y}{\text{Change in } x} = \frac{y - y_1}{x - x_1}.$$

We can clear the fraction by multiplying both sides by $x - x_1$, the least common denominator, where $x - x_1 \neq 0$.

$$m = \frac{y - y_1}{x - x_1} \qquad \text{This is the slope of the line in Figure 3.29.}$$

$$m(x - x_1) = \frac{y - y_1}{x - x_1} \cdot (x - x_1) \qquad \text{Multiply both sides by } x - x_1.$$

$$m(x - x_1) = y - y_1 \qquad \text{Simplify: } \frac{y - y_1}{x - x_1} \cdot (x - x_1) = y - y_1.$$

Now, if we reverse the two sides, we obtain the **point-slope form** of the equation of a line.

Great Question!

When using $y - y_1 = m(x - x_1)$, for which variables do I substitute numbers?

When writing the point-slope form of a line's equation, you will never substitute numbers for x and y. You will substitute values for x_1, y_1, and m.

> **Point-Slope Form of the Equation of a Line**
>
> The **point-slope form of the equation** of a nonvertical line with slope m that passes through the point (x_1, y_1) is
>
> $$y - y_1 = m(x - x_1).$$

For example, the point-slope form of the equation of the line passing through $(1, 4)$ with slope 2 $(m = 2)$ is

$$y - 4 = 2(x - 1).$$

Using the Point-Slope Form to Write a Line's Equation

 Use the point-slope form to write equations of a line.

If we know the slope of a line and a point not containing the y-intercept through which the line passes, the point-slope form is the equation that we should use. Once we have obtained this equation, it is customary to solve for y and write the equation in slope-intercept form. Examples 1 and 2 illustrate these ideas.

Example 1 Writing the Point-Slope Form and the Slope-Intercept Form

Write the point-slope form and the slope-intercept form of the equation of the line with slope 4 that passes through the point $(-1, 3)$.

Solution

We begin with the point-slope form of the equation of a line with $m = 4$, $x_1 = -1$, and $y_1 = 3$.

$$y - y_1 = m(x - x_1) \quad \text{This is the point-slope form of the equation.}$$
$$y - 3 = 4[x - (-1)] \quad \text{Substitute: } (x_1, y_1) = (-1, 3) \text{ and } m = 4.$$
$$y - 3 = 4(x + 1) \quad \text{We now have the point-slope form of the equation of the given line.}$$

Now we solve this equation for y and write an equivalent equation in slope-intercept form $(y = mx + b)$.

We need to isolate y.

$$y - 3 = 4(x + 1) \quad \text{This is the point-slope form of the equation.}$$
$$y - 3 = 4x + 4 \quad \text{Use the distributive property.}$$
$$y = 4x + 7 \quad \text{Add 3 to both sides.}$$

The slope-intercept form of the line's equation is $y = 4x + 7$.

✓ **Check Point 1** Write the point-slope form and the slope-intercept form of the equation of the line with slope 6 that passes through the point $(2, -5)$.
$y + 5 = 6(x - 2); y = 6x - 17$

Example 2 Writing the Point-Slope Form and the Slope-Intercept Form

A line passes through the points $(4, -3)$ and $(-2, 6)$. (See **Figure 3.30**.) Find the equation of the line

a. in point-slope form. **b.** in slope-intercept form.

Solution

a. To use the point-slope form, we need to find the slope. The slope is the change in the y-coordinates divided by the corresponding change in the x-coordinates.

$$m = \frac{6 - (-3)}{-2 - 4} = \frac{9}{-6} = -\frac{3}{2} \quad \text{This is the definition of slope using } (4, -3) \text{ and } (-2, 6).$$

We can take either point on the line to be (x_1, y_1). Let's use $(x_1, y_1) = (4, -3)$. Now, we are ready to write the point-slope form of the equation.

$$y - y_1 = m(x - x_1) \quad \text{This is the point-slope form of the equation.}$$
$$y - (-3) = -\frac{3}{2}(x - 4) \quad \text{Substitute: } (x_1, y_1) = (4, -3) \text{ and } m = -\frac{3}{2}.$$
$$y + 3 = -\frac{3}{2}(x - 4) \quad \text{Simplify.}$$

This equation is one point-slope form of the equation of the line shown in **Figure 3.30**.

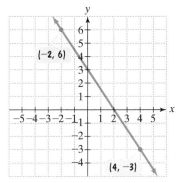

Figure 3.30

Great Question!

If I'm given two points on a line, which point do I use for (x_1, y_1) when I'm writing the point-slope equation?

You can use either point. Rework Example 2 using $(-2, 6)$ for (x_1, y_1) in part (a). Once you solve for y, you should obtain the same slope-intercept equation as the one shown in the last line of the solution to part (b).

b. Now, we solve the equation $y + 3 = -\dfrac{3}{2}(x - 4)$ for y and write an equivalent equation in slope-intercept form $(y = mx + b)$.

> We need to isolate y.

$$y + 3 = -\frac{3}{2}(x - 4)$$ This is the point-slope form of the equation.

$$y + 3 = -\frac{3}{2}x + 6$$ Use the distributive property.

$$y = -\frac{3}{2}x + 3$$ Subtract 3 from both sides.

This equation is the slope-intercept form of the equation of the line shown in **Figure 3.30** on the previous page.

✓ **Check Point 2**   A line passes through the points $(-2, -1)$ and $(-1, -6)$. Find the equation of the line

a. in point-slope form. $y + 1 = -5(x + 2)$ or $y + 6 = -5(x + 1)$

▶ **b.** in slope-intercept form. $y = -5x - 11$

The many forms of a line's equation can be a bit overwhelming. **Table 3.3** summarizes the various forms and contains the most important things you should remember about each form.

Table 3.3 Equations of Lines

Form	What You Should Know
$Ax + By = C$	• Graph equations in this form using intercepts (x-intercept: set $y = 0$; y-intercept: set $x = 0$) and a checkpoint.
$y = b$	• Graph equations in this form as horizontal lines with b as the y-intercept.
$x = a$	• Graph equations in this form as vertical lines with a as the x-intercept.
Slope-Intercept Form $y = mx + b$	• Graph equations in this form using the y-intercept, b, and the slope, m. • Start with this form when writing a linear equation if you know a line's slope and y-intercept.
Point-Slope Form $y - y_1 = m(x - x_1)$	• Start with this form when writing a linear equation if you know the slope of the line and a point on the line not containing the y-intercept or two points on the line, neither of which contains the y-intercept. Calculate the slope using $$m = \frac{\text{Change in } y}{\text{Change in } x} = \frac{y_2 - y_1}{x_2 - x_1}.$$ Although you begin with point-slope form, you usually solve for y and convert to slope-intercept form.

Applications

2 Write linear equations that model data and make predictions.

Linear equations are useful for modeling data that fall on or near a line. For example, the bar graph in **Figure 3.31(a)** at the top of the next page gives the median age of the U.S. population in the indicated year. (The median age is the age in the middle when all the ages of the U.S. population are arranged from youngest to oldest.) The data are displayed as a set of five points in a rectangular coordinate system in **Figure 3.31(b)**.

The Graying of America: Median Age of the United States Population

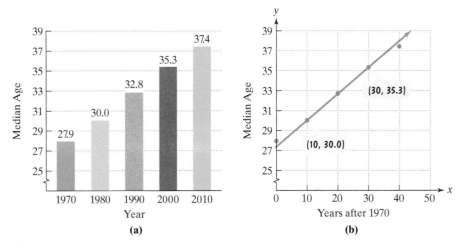

Figure 3.31
Source: U.S. Census Bureau

A set of points representing data is called a **scatter plot**. Also shown on the scatter plot in **Figure 3.31(b)** is a line that passes through or near the five points. By writing the equation of this line, we can obtain a model of the data and make predictions about the median age of the U.S. population in the future.

Example 3 ▸ Modeling the Graying of America

Write the slope-intercept form of the equation of the line shown in **Figure 3.31(b)**. Use the function to predict the median age of the U.S. population in 2020.

Solution

The line in **Figure 3.31(b)** passes through $(10, 30.0)$ and $(30, 35.3)$. We start by finding its slope.

$$m = \frac{\text{Change in } y}{\text{Change in } x} = \frac{35.3 - 30.0}{30 - 10} = \frac{5.3}{20} = 0.265$$

The slope indicates that each year the median age of the U.S. population is increasing by 0.265 year.

Now, we write the slope-intercept form of the equation of the line.

$$y - y_1 = m(x - x_1) \qquad \text{Begin with the point-slope form.}$$
$$y - 30.0 = 0.265(x - 10) \qquad \text{Either ordered pair can be } (x_1, y_1). \text{ Let}$$
$$(x_1, y_1) = (10, 30.0). \text{ From above, } m = 0.265.$$
$$y - 30.0 = 0.265x - 2.65 \qquad \text{Apply the distributive property.}$$
$$y = 0.265x + 27.35 \qquad \text{Add 30 to both sides and solve for } y.$$

A linear function that models the median age of the U.S. population, y, x years after 1970 is

$$y = 0.265x + 27.35 \quad \text{or} \quad f(x) = 0.265x + 27.35.$$

Now, let's use this equation to predict the median age in 2020. Because 2020 is 50 years after 1970, substitute 50 for x and compute y.

$$f(50) = 0.265(50) + 27.35 = 40.6$$

Our model predicts that the median age of the U.S. population in 2020 will be 40.6.

✓ **Check Point 3** Use the data points $(10, 30.0)$ and $(20, 32.8)$ from **Figure 3.31(b)** to write the slope-intercept form of an equation that models the median age of the U.S. population x years after 1970. Use this model to predict the median age in 2020.
▸ $y = 0.28x + 27.2; 41.2$

③   Make a scatter plot for a table of data items.

Scatter Plots and Correlation

Is there a relationship between education and prejudice? With increased education, does a person's level of prejudice tend to increase, decrease, or stay the same? For each person in our sample, we will record the number of years of school completed and the score on a test measuring prejudice. Higher scores on this 1-to-10 test indicate greater prejudice. Using x to represent years of education and y to represent scores on a test measuring prejudice, **Table 3.4** shows these two quantities for a random sample of ten people.

Table 3.4 Recording Two Quantities in a Sample of Ten People										
Respondent	**A**	**B**	**C**	**D**	**E**	**F**	**G**	**H**	**I**	**J**
Years of education (x)	12	5	14	13	8	10	16	11	12	4
Score on prejudice test (y)	1	7	2	3	5	4	1	2	3	10

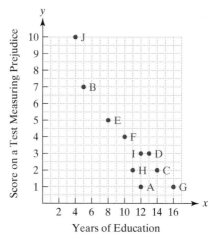

Figure 3.32 A scatter plot for education-prejudice data

We can make a scatter plot of the data in **Table 3.4** by drawing a horizontal axis to represent years of education and a vertical axis to represent scores on a test measuring prejudice. We then represent each of the ten respondents with a single point on the graph. For example, the dot for respondent A is located to represent 12 years of education on the horizontal axis and 1 on the prejudice test on the vertical axis. Plotting each of the ten pairs of data in a rectangular coordinate system results in the scatter plot shown in **Figure 3.32**.

A scatter plot like the one in **Figure 3.32** can be used to determine whether two quantities are related. If there is a clear relationship, the quantities are said to be **correlated**. The scatter plot shows a downward trend among the data points, although there are a few exceptions. People with increased education tend to have a lower score on the test measuring prejudice. **Correlation** is used to determine if there is a relationship between two variables and, if so, the strength and direction of that relationship.

Correlation and Causal Connections

Correlations can often be seen when data items are displayed on a scatter plot. Although the scatter plot in **Figure 3.32** indicates a correlation between education and prejudice, we cannot conclude that increased education causes a person's level of prejudice to decrease. There are at least three possible explanations:

1. The correlation between increased education and decreased prejudice is simply a coincidence.
2. Education usually involves classrooms with a variety of different kinds of people. Increased exposure to diversity in the classroom setting, which accompanies increased levels of education, might be an underlying cause for decreased prejudice.
3. Education, the process of acquiring knowledge, requires people to look at new ideas and see things in different ways. Thus, education causes one to be more tolerant and less prejudiced.

> This list represents three possibilities. Perhaps you can provide a better explanation about decreasing prejudice with increased education.

Establishing that one thing causes another is extremely difficult, even if there is a strong correlation between these things. For example, as the air temperature increases, there is an increase in the number of people stung by jellyfish at the beach. This does not mean that an increase in air temperature causes more people to be stung. It might mean that because it is hotter, more people go into the water. With an increased number of swimmers, more people are likely to be stung. In short, correlation is not necessarily causation.

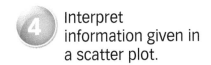

Interpret information given in a scatter plot.

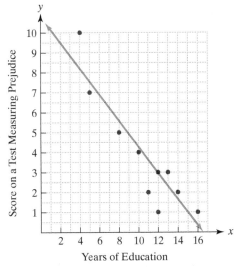

Figure 3.33 A scatter plot with a regression line

Regression Lines and Correlation Coefficients

Figure 3.33 shows the scatter plot for the education-prejudice data. Also shown is a straight line that seems to approximately "fit" the data points. Most of the data points lie either near or on this line. A line that best fits the data points in a scatter plot is called a **regression line**. The regression line is the particular line in which the spread of the data points around it is as small as possible.

A measure that is used to describe the strength and direction of a relationship between variables whose data points lie on or near a line is called the **correlation coefficient**, designated by r. **Figure 3.34** shows scatter plots and correlation coefficients. Variables are **positively correlated** if they tend to increase or decrease together, as in **Figure 3.34(a), (b),** and **(c)**. By contrast, variables are **negatively correlated** if one variable tends to decrease while the other increases, as in **Figure 3.34(e), (f),** and **(g)**. **Figure 3.34** illustrates that a correlation coefficient, r, is a number between -1 and 1, inclusive. **Figure 3.34(a)** shows a value of 1. This indicates a **perfect positive correlation** in which all points in the scatter plot lie precisely on the regression line that rises from left to right. **Figure 3.34(g)** shows a value of -1. This indicates a **perfect negative correlation** in which all points in the scatter plot lie precisely on the regression line that falls from left to right.

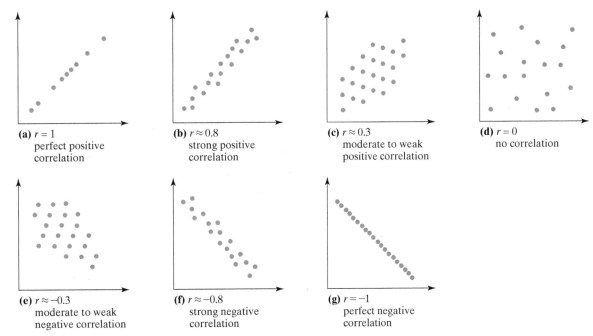

(a) $r = 1$
perfect positive
correlation

(b) $r \approx 0.8$
strong positive
correlation

(c) $r \approx 0.3$
moderate to weak
positive correlation

(d) $r = 0$
no correlation

(e) $r \approx -0.3$
moderate to weak
negative correlation

(f) $r \approx -0.8$
strong negative
correlation

(g) $r = -1$
perfect negative
correlation

Figure 3.34 Scatter plots and correlation coefficients

Take another look at **Figure 3.34**. If r is between 0 and 1, as in **(b)** and **(c)**, the two variables are positively correlated, but not perfectly. Although all the data points will not lie on the regression line, as in **(a)**, an increase in one variable tends to be accompanied by an increase in the other.

Negative correlations are also illustrated in **Figure 3.34**. If r is between 0 and -1, as in **(e)** and **(f)**, the two variables are negatively correlated, but not perfectly. Although all the data points will not lie on the regression line, as in **(g)**, an increase in one variable tends to be accompanied by a decrease in the other.

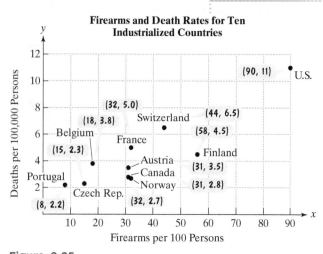

Figure 3.35
Source: International Action Network on Small Arms

Example 4   Interpreting a Correlation Coefficient

Math provides an objective way to look at social and political issues. For example, the points in the scatter plot in **Figure 3.35** show the number of firearms per 100 persons and the number of deaths per 100,000 persons for the ten industrialized countries with the highest death rates. The correlation coefficient for the data is 0.89. What does this correlation coefficient indicate about the strength and direction of the relationship between firearms per 100 persons and deaths per 100,000 persons?

Solution

The correlation coefficient $r = 0.89$ is close to 1, indicating a strong, but not perfect, positive correlation. There is a moderately strong positive relationship between the number of firearms per 100 persons and the number of deaths per 100,000 persons. As the number of firearms per 100 persons increases, the number of deaths per 100,000 persons also tends to increase.

Check Point 4   The points in the scatter plot in **Figure 3.36** show the number of hours per week spent watching TV, by annual income. The correlation coefficient for the data is -0.86. What does this correlation coefficient indicate about the strength and direction of the relationship between annual income and hours per week watching TV?

There is a moderately strong negative relationship between annual income and hours per week watching TV.

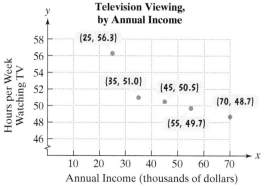

Figure 3.36
Source: Nielsen Media Research

How to Obtain the Correlation Coefficient and the Equation of the Regression Line

The easiest way to find the correlation coefficient and the equation of the regression line is to use a graphing or statistical calculator. Graphing calculators have statistical menus that enable you to enter the x and y data items for the variables. Based on this information, you can instruct the calculator to display a scatter plot, the equation of the regression line, and the correlation coefficient.

Shown below are the data involving the number of years of school, x, completed by ten randomly selected people and their scores on a test measuring prejudice, y. Recall that higher scores on the measure of prejudice (1 to 10) indicate greater levels of prejudice.

Respondent	A	B	C	D	E	F	G	H	I	J
Years of education (x)	12	5	14	13	8	10	16	11	12	4
Score on prejudice test (y)	1	7	2	3	5	4	1	2	3	10

After entering the x and y data items into a graphing calculator, we obtain the scatter plot of the data and the regression line, shown in **Figure 3.37**.

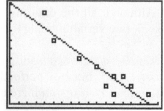

Figure 3.37

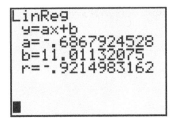

Figure 3.38

Also displayed by the calculator is the regression line's equation and the correlation coefficient, r, shown in **Figure 3.38**. The value of r, approximately -0.92, is fairly close to -1 and indicates a strong negative correlation. This means that the more education a person has, the less prejudiced that person is (based on scores on the test measuring levels of prejudice). Observe that the slope of the line is approximately -0.69. With a negative correlation coefficient, it makes sense that the slope of the regression line is negative. This line falls from left to right, indicating a negative correlation.

Blitzer Bonus

Cigarettes and Lung Cancer

This scatter plot shows a relationship between cigarette consumption among males and deaths due to lung cancer per million males. The data are from 11 countries and date back to a 1964 report by the U.S. Surgeon General. The scatter plot can be modeled by a line whose slope indicates an increasing death rate from lung cancer with increased cigarette consumption. At that time, the tobacco industry argued that in spite of this regression line, tobacco use is not the cause of cancer. Recent data do, indeed, show a causal effect between tobacco use and numerous diseases.

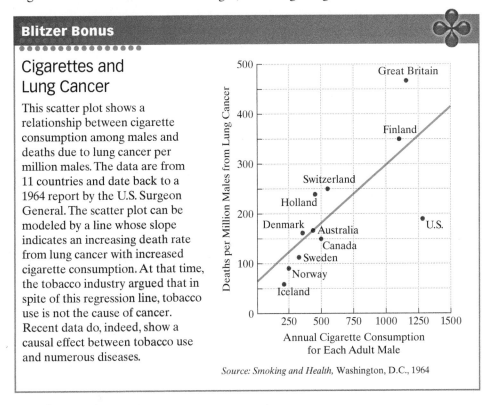

Source: Smoking and Health, Washington, D.C., 1964

Achieving Success

Because concepts in mathematics build on each other, **it is extremely important that you complete all homework assignments.** This requires more than attempting a few of the assigned exercises. When it comes to assigned homework, you need to do four things and to do these things consistently throughout any math course:

1. Attempt to work *every* assigned problem.
2. Check your answers.
3. Correct your errors.
4. Ask for help with the problems you have attempted, but do not understand.

Concept and Vocabulary Check

Exercises in the Concept and Vocabulary Check are intended for group and class discussions.

In Exercises 1–4, fill in each blank so that the resulting statement is true.

1. The point-slope form of the equation of a line is $\underline{\ y - y_1 = m(x - x_1)\ }$, where (x_1, y_1) represents $\underline{\ \text{a point on the line}\ }$ and m represents the line's $\underline{\ \ \text{slope}\ \ }$.

2. A set of points representing data is called a/an $\underline{\ \ \text{scatter plot}\ \ }$.

3. The line that best fits a set of points is called a/an $\underline{\ \ \text{regression line}\ \ }$.

4. A measure that is used to describe the strength and direction of a relationship between variables whose data points lie on or near a line is called the $\underline{\ \text{correlation coefficient}\ }$, ranging from $r = \underline{\ \ -1\ \ }$ to $r = \underline{\ \ 1\ \ }$.

In Exercises 5–8, determine whether each statement is true or false. If the statement is false, make the necessary change(s) to produce a true statement.   Changes to false statements will vary.

5. The line whose equation is $y - 3 = 7(x + 2)$ passes through $(-3, 2)$.   false

6. The point-slope form can be applied to obtain the equation of the line through the points $(2, -5)$ and $(2, 6)$.   false

7. If $r = 0$, there is no correlation between two variables.   true

8. If $r = 1$, changes in one variable cause changes in the other variable.   false

Respond to Exercises 9–14 using verbal or written explanations.

9. Describe how to write the equation of a line if its slope and a point on the line are known.   **9–14.** Answers will vary.

10. Describe how to write the equation of a line if two points on the line are known.

11. What is a scatter plot?

12. How does a scatter plot indicate that two variables are correlated?

13. What is meant by a regression line?

14. When all points in a scatter plot fall on the regression line, what is the value of the correlation coefficient? Describe what this means.

Exercise Set 3.3

Practice Exercises

Write the point-slope form of the equation of the line satisfying each of the conditions in Exercises 1–28. Then use the point-slope form of the equation to write the slope-intercept form of the equation.

1. Slope = 3, passing through $(2, 5)$ $y - 5 = 3(x - 2); y = 3x - 1$

2. Slope = 6, passing through $(3, 1)$ $y - 1 = 6(x - 3); y = 6x - 17$

3. Slope = 5, passing through $(-2, 6)$

4. Slope = 7, passing through $(-4, 9)$

5. Slope = -8, passing through $(-3, -2)$

6. Slope = -4, passing through $(-5, -2)$

7. Slope = -12, passing through $(-8, 0)$

8. Slope = -11, passing through $(0, -3)$

9. Slope = -1, passing through $\left(-\frac{1}{2}, -2\right)$

10. Slope = -1, passing through $\left(-4, -\frac{1}{4}\right)$

11. Slope = $\frac{1}{2}$, passing through the origin $y - 0 = \frac{1}{2}(x - 0); y = \frac{1}{2}x$

12. Slope = $\frac{1}{3}$, passing through the origin $y - 0 = \frac{1}{3}(x - 0); y = \frac{1}{3}x$

13. Slope = $-\frac{2}{3}$, passing through $(6, -2)$

14. Slope = $-\frac{3}{5}$, passing through $(10, -4)$

15. Passing through $(1, 2)$ and $(5, 10)$

16. Passing through $(3, 5)$ and $(8, 15)$

17. Passing through $(-3, 0)$ and $(0, 3)$

18. Passing through $(-2, 0)$ and $(0, 2)$

19. Passing through $(-3, -1)$ and $(2, 4)$

20. Passing through $(-2, -4)$ and $(1, -1)$

21. Passing through $(-4, -1)$ and $(3, 4)$

22. Passing through $(-6, 1)$ and $(2, -5)$

23. Passing through $(-3, -1)$ and $(4, -1)$

24. Passing through $(-2, -5)$ and $(6, -5)$

25. Passing through $(2, 4)$ with x-intercept $= -2$

26. Passing through $(1, -3)$ with x-intercept $= -1$

27. x-intercept $= -\frac{1}{2}$ and y-intercept $= 4$

28. x-intercept $= 4$ and y-intercept $= -2$

In Exercises 29–32, make a scatter plot for the given data. Use the scatter plot to describe whether or not the variables appear to be related.

29.

x	1	6	4	3	7	2
y	2	5	3	3	4	1

30.

x	2	1	6	3	4
y	4	5	10	8	9

31.

x	8	6	1	5	4	10	3
y	2	4	10	5	6	2	9

32.

x	4	5	2	1
y	1	3	5	4

3. $y - 6 = 5(x + 2); y = 5x + 16$ **4.** $y - 9 = 7(x + 4); y = 7x + 37$ **5.** $y + 2 = -8(x + 3); y = -8x - 26$ **6.** $y + 2 = -4(x + 5); y = -4x - 22$

7. $y - 0 = -12(x + 8); y = -12x - 96$ **8.** $y + 3 = -11(x - 0); y = -11x - 3$ **9.** $y + 2 = -1\left(x + \frac{1}{2}\right); y = -x - \frac{5}{2}$ **10.** $y + \frac{1}{4} = -1(x + 4); y = -x - \frac{17}{4}$

13. $y + 2 = -\frac{2}{3}(x - 6); y = -\frac{2}{3}x + 2$ **14.** $y + 4 = -\frac{3}{5}(x - 10); y = -\frac{3}{5}x + 2$ **15.** $y - 2 = 2(x - 1)$ or $y - 10 = 2(x - 5); y = 2x$

16. $y - 5 = 2(x - 3)$ or $y - 15 = 2(x - 8); y = 2x - 1$ **17.** $y - 0 = 1(x + 3)$ or $y - 3 = 1(x - 0); y = x + 3$ **18.** $y - 0 = 1(x + 2)$ or $y - 2 = 1(x - 0); y = x + 2$

19. $y + 1 = 1(x + 3)$ or $y - 4 = 1(x - 2); y = x + 2$ **20.** $y + 4 = 1(x + 2)$ or $y + 1 = 1(x - 1); y = x - 2$ **21.** $y + 1 = \frac{5}{7}(x + 4)$ or $y - 4 = \frac{5}{7}(x - 3); y = \frac{5}{7}x + \frac{13}{7}$

22. $y - 1 = -\frac{3}{4}(x + 6)$ or $y + 5 = -\frac{3}{4}(x - 2); y = -\frac{3}{4}x - \frac{7}{2}$ **23.** $y + 1 = 0(x + 3)$ or $y + 1 = 0(x - 4); y = -1$ **24.** $y + 5 = 0(x + 2)$ or $y + 5 = 0(x - 6); y = -5$

25. $y - 4 = 1(x - 2)$ or $y - 0 = 1(x + 2); y = x + 2$ **26.** $y + 3 = -\frac{3}{2}(x - 1)$ or $y - 0 = -\frac{3}{2}(x + 1); y = -\frac{3}{2}x - \frac{3}{2}$ **27.** $y - 0 = 8\left(x + \frac{1}{2}\right)$ or $y - 4 = 8(x - 0); y = 8x + 4$

28. $y - 0 = \frac{1}{2}(x - 4)$ or $y + 2 = \frac{1}{2}(x - 0); y = \frac{1}{2}x - 2$

*See Answers to Selected Exercises.

Use the scatter plots shown, labeled (a)–(f), to solve Exercises 33–36.

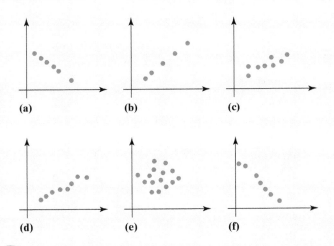

(a)   (b)   (c)

(d)   (e)   (f)

33. Which scatter plot indicates a perfect negative correlation? a

34. Which scatter plot indicates a perfect positive correlation? b

35. In which scatter plot is $r = 0.9$? d

36. In which scatter plot is $r = 0.01$? e

Practice Plus

Two nonintersecting lines that lie in the same rectangular coordinate system are parallel. If two lines are parallel, the ratio of the vertical change to the horizontal change is the same for both lines. Because two parallel lines have the same "steepness," they must have the same slope.

If two nonvertical lines are parallel, then they have the same slope.

In Exercises 37–42, use this information to write equations in point-slope form and slope-intercept form of the line satisfying the given conditions.

37. The line passes through $(-3, 2)$ and is parallel to the line whose equation is $y = 4x + 1$. $y - 2 = 4(x + 3); y = 4x + 14$

38. The line passes through $(5, -3)$ and is parallel to the line whose equation is $y = 2x + 1$. $y + 3 = 2(x - 5); y = 2x - 13$

39. The line passes through $(-1, -5)$ and is parallel to the line whose equation is $3x + y = 6$. $y + 5 = -3(x + 1); y = -3x - 8$

40. The line passes through $(-4, -7)$ and is parallel to the line whose equation is $6x + y = 8$. $y + 7 = -6(x + 4); y = -6x - 31$

41. The line has an x-intercept at -4 and is parallel to the line containing $(3, 1)$ and $(2, 6)$. $y - 0 = -5(x + 4); y = -5x - 20$

42. The line has an x-intercept at -6 and is parallel to the line containing $(4, -3)$ and $(2, 2)$. $y - 0 = -\frac{5}{2}(x + 6); y = -\frac{5}{2}x - 15$

Application Exercises

Americans are getting married later in life, or not getting married at all. In 2010, more than half of Americans ages 25 through 29 were unmarried. The bar graph shows the percentage of never-married men and women in this age group for four selected years. The data are displayed as two sets of four points each, one scatter plot for the percentage of never-married American men and one for the percentage of never-married American women. Also shown for each scatter plot is a line that passes through or near the four points. Use these lines to solve Exercises 43–44.

Percentage of United States Population Never Married, Ages 25–29

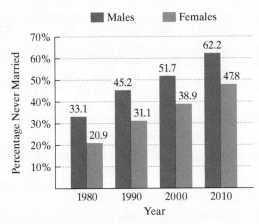

Source: U.S. Census Bureau

43. In this exercise, you will use the blue line for the women shown on the scatter plot to develop a model for the percentage of never-married American females ages 25–29.

   a. Use the two points whose coordinates are shown by the voice balloons to find the point-slope form of the equation of the line that models the percentage of never-married American females ages 25–29, y, x years after 1980. $y - 31.1 = 0.78(x - 10)$ or $y - 38.9 = 0.78(x - 20)$

   b. Write the equation from part (a) in slope-intercept form. Use function notation. $f(x) = 0.78x + 23.3$

   c. Use the linear function to predict the percentage of never-married American females, ages 25–29, in 2020.   54.5%

44. In this exercise, you will use the red line for the men shown on the scatter plot to develop a model for the percentage of never-married American males ages 25–29.

   a. Use the two points whose coordinates are shown by the voice balloons to find the point-slope form of the equation of the line that models the percentage of never-married American males ages 25–29, y, x years after 1980. $y - 45.2 = 0.65(x - 10)$ or $y - 51.7 = 0.65(x - 20)$

   b. Write the equation from part (a) in slope-intercept form. Use function notation. $f(x) = 0.65x + 38.7$

   c. Use the linear function to predict the percentage of never-married American males, ages 25–29, in 2015.   61.45%

Just as money doesn't buy happiness for individuals, the two don't necessarily go together for countries either. However, the scatter plot does show a relationship between a country's annual per capita income and the percentage of people in that country who call themselves "happy." Use the scatter plot to determine whether each of the statements in Exercises 45–52 is true or false.

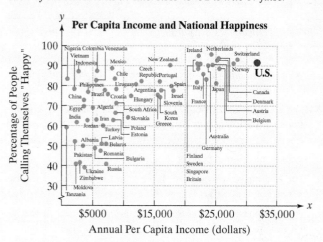

Per Capita Income and National Happiness

Source: Richard Layard, *Happiness: Lessons from a New Science,* Penguin, 2005.

45. There is no correlation between per capita income and the percentage of people who call themselves "happy." false

46. There is an almost-perfect positive correlation between per capita income and the percentage of people who call themselves "happy." false

47. There is a positive correlation between per capita income and the percentage of people who call themselves "happy." true

48. As per capita income decreases, the percentage of people who call themselves "happy" also tends to decrease. true

49. The country with the lowest per capita income has the least percentage of people who call themselves "happy." false

50. The country with the highest per capita income has the greatest percentage of people who call themselves "happy." false

51. A reasonable estimate of the correlation coefficient for the data is 0.8. false

52. A reasonable estimate of the correlation coefficient for the data is −0.3. false

Critical Thinking Exercises

For the pairs of quantities in Exercises 53–56, describe whether a scatter plot will show a positive correlation, a negative correlation, or no correlation. If there is a correlation, is it strong, moderate, or weak? Explain your answers. Explanations will vary.

53. Height and weight positive

54. Number of days absent and grade in a course negative

55. Height and grade in a course no correlation

56. Hours of video games played and grade in a course negative

57. Give an example of two variables with a strong positive correlation and explain why this is so. 57–59. Answers will vary.

58. Give an example of two variables with a strong negative correlation and explain why this is so.

59. Give an example of two variables with a strong correlation, where neither variable is the cause of the other.

60. d. $y = -0.55x + 55.62; -0.91$; Answers will vary.

*See Answers to Selected Exercises.

Technology Exercises

60. Consider the following data:

Literacy and Hunger

	Percentage Who Are	
	Literate	**Undernourished**
Country	x	y
Cuba	100	2
Egypt	71	4
Ethiopia	36	46
Grenada	96	7
Italy	98	2
Jamaica	80	9
Jordan	91	6
Pakistan	50	24
Russia	99	3
Togo	53	24
Uganda	67	19

Source: The Penguin State of the World Atlas, 2008

a. Make a scatter plot for the data. *

b. Use the scatter plot to determine the correlation coefficient that best describes the relationship between the variables. Select from the following options: $r = -0.9$
$r = 1; r = 0.9; r = 0.4; r = 0; r = -0.4; r = -0.9; r = -1.$

c. Enter the x and y data items into a graphing calculator and obtain the scatter plot of the data and the graph of the regression line. *

d. Use your calculator to obtain the regression line's equation and the correlation coefficient. Round all values to two decimal places. Is the value of r consistent with your answer from part (b)?

e. Use the equation of the regression line from part (d) to answer the following question: What percentage of people, to the nearest percent, can we anticipate are undernourished in a country where 60% of the people are literate? 23%

Group Exercise   61. Answers will vary.

61. The group should select two variables, related to people in your college, that it believes have a strong positive or negative correlation. Once these variables have been determined,

a. Collect at least 30 ordered pairs of data (x, y) from a sample of people on your campus.

b. Draw a scatter plot for the data collected.

c. Does the scatter plot indicate a positive correlation, a negative correlation, or no relationship between the variables?

d. Use a graphing calculator to determine r. Does the value of r reinforce the impression conveyed by the scatter plot?

e. Use a graphing calculator to determine the equation of the regression line.

f. Use the regression line's equation to make a prediction about a y-value given an x-value.

g. Are the results of this project consistent with the group's original belief about the correlation between the variables, or are there some surprises in the data collected?

3.4 Systems of Linear Equations in Two Variables

What am I suppose to learn?

After you have read this section, you should be able to:

① Decide whether an ordered pair is a solution of a linear system.

② Solve linear systems by graphing.

③ Solve linear systems by substitution.

④ Solve linear systems by addition.

⑤ Identify systems that do not have exactly one ordered-pair solution.

⑥ Solve problems using systems of linear equations.

A study of college students showed that students who procrastinate typically have more symptoms of physical illness. Researchers identified college students who generally were procrastinators or nonprocrastinators. The students were asked to report throughout the semester how many symptoms of physical illness they had experienced. **Figure 3.39** shows that by late in the semester, all students experienced increases in symptoms. Early in the semester, procrastinators reported fewer symptoms, but late in the semester, as work came due, they reported more symptoms than their nonprocrastinating peers.

The data in **Figure 3.39** can be analyzed using a pair of linear models in two variables. The figure shows that by week 6, both groups reported the same number of symptoms of illness, an average of approximately 3.5 symptoms per group. In this section, you will learn two algebraic methods, called *substitution* and *addition*, that will reinforce this graphic observation, verifying $(6, 3.5)$ as the point of intersection.

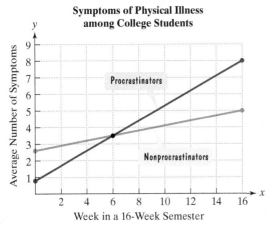

Figure 3.39
Source: Gerrig and Zimbardo, *Psychology and Life*, 18th Edition. Allyn and Bacon, 2008.

Systems of Linear Equations and Their Solutions

① **Decide whether an ordered pair is a solution of a linear system.**

We have seen that all equations in the form $Ax + By = C$, A and B not both zero, are straight lines when graphed. Two such equations are called a **system of linear equations** or a **linear system**. A **solution to a system of linear equations in two variables** is an ordered pair that satisfies both equations in the system. For example, $(3, 4)$ satisfies the system

$$\begin{cases} x + y = 7 & \text{(3 + 4 is, indeed, 7.)} \\ x - y = -1. & \text{(3 − 4 is, indeed, −1.)} \end{cases}$$

Thus, $(3, 4)$ satisfies both equations and is a solution of the system. The solution can be described by saying that $x = 3$ and $y = 4$. The solution can also be described using set notation. The solution set of the system is $\{(3, 4)\}$ — that is, the set consisting of the ordered pair $(3, 4)$.

A system of linear equations can have exactly one solution, no solution, or infinitely many solutions. We begin with systems having exactly one solution.

Example 1 Determining Whether an Ordered Pair Is a Solution of a System

Determine whether $(1, 2)$ is a solution of the system:

$$\begin{cases} 2x - 3y = -4 \\ 2x + y = 4. \end{cases}$$

Solution

Because 1 is the x-coordinate and 2 is the y-coordinate of $(1, 2)$, we replace x with 1 and y with 2.

$2x - 3y = -4$	$2x + y = 4$
$2(1) - 3(2) \overset{?}{=} -4$	$2(1) + 2 \overset{?}{=} 4$
$2 - 6 \overset{?}{=} -4$	$2 + 2 \overset{?}{=} 4$
$-4 = -4,$ true	$4 = 4,$ true

The pair $(1, 2)$ satisfies both equations: It makes each equation true. Thus, the pair is a solution of the system.

✓ **Check Point 1** Determine whether $(-4, 3)$ is a solution of the system:

$$\begin{cases} x + 2y = 2 \\ x - 2y = 6. \end{cases} \quad \text{not a solution}$$

Solving Linear Systems by Graphing

2 Solve linear systems by graphing.

The solution to a system of linear equations can be found by graphing both of the equations in the same rectangular coordinate system. For a system with one solution, **the coordinates of the point of intersection of the lines is the system's solution.**

Example 2 Solving a Linear System by Graphing

Solve by graphing:

$$\begin{cases} x + 2y = 2 \\ x - 2y = 6. \end{cases}$$

Solution

We find the solution by graphing both $x + 2y = 2$ and $x - 2y = 6$ in the same rectangular coordinate system. We will use intercepts to graph each equation.

x-intercept: Set $y = 0$.
$x + 2 \cdot 0 = 2$
$x = 2$

y-intercept: Set $x = 0$.
$0 + 2y = 2$
$2y = 2$
$y = 1$

The line passes through $(2, 0)$.   The line passes through $(0, 1)$.

We graph $x + 2y = 2$ as a blue line in **Figure 3.40**.

$x - 2y = 6$

x-intercept: Set $y = 0$.
$x - 2 \cdot 0 = 6$
$x = 6$

y-intercept: Set $x = 0$.
$0 - 2y = 6$
$-2y = 6$
$y = -3$

The line passes through $(6, 0)$.   The line passes through $(0, -3)$.

We graph $x - 2y = 6$ as a red line in **Figure 3.40**.

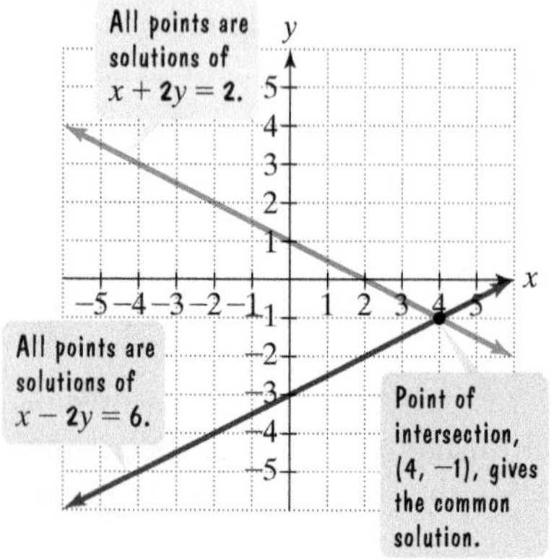

The system is graphed in **Figure 3.40**. To ensure that the graph is accurate, check the coordinates of the intersection point, $(4, -1)$, in both equations.

All points are solutions of $x + 2y = 2$.

All points are solutions of $x - 2y = 6$.

Point of intersection, $(4, -1)$, gives the common solution.

Figure 3.40 Visualizing a system's solution

We replace x with 4 and y with -1.

$$x + 2y = 2$$
$$4 + 2(-1) \stackrel{?}{=} 2$$
$$4 + (-2) \stackrel{?}{=} 2$$
$$2 = 2, \quad \text{true}$$

$$x - 2y = 6$$
$$4 - 2(-1) \stackrel{?}{=} 6$$
$$4 - (-2) \stackrel{?}{=} 6$$
$$4 + 2 \stackrel{?}{=} 6$$
$$6 = 6, \quad \text{true}$$

The pair $(4, -1)$ satisfies both equations—that is, it makes each equation true. This verifies that the system's solution set is $\{(4, -1)\}$.

✓ **Check Point 2** Solve by graphing:

$$\begin{cases} 2x + 3y = 6 \\ 2x + y = -2. \end{cases} \quad \{(-3,4)\}^*$$

Solving Linear Systems by the Substitution Method

Finding the solution to a linear system by graphing equations may not be easy to do. For example, a solution of $\left(-\frac{2}{3}, \frac{157}{29}\right)$ would be difficult to "see" as an intersection point on a graph.

Let's consider a method that does not depend on finding a system's solution visually: the substitution method. This method involves converting the system to one equation in one variable by an appropriate substitution.

> **Solving Linear Systems by Substitution**
>
> 1. Solve either of the equations for one variable in terms of the other. (If one of the equations is already in this form, you can skip this step.)
> 2. Substitute the expression found in step 1 into the *other* equation. This will result in an equation in one variable.
> 3. Solve the equation containing one variable.
> 4. Back-substitute the value found in step 3 into the equation from step 1. Simplify and find the value of the remaining variable.
> 5. Check the proposed solution in both of the system's given equations.

Example 3 **Solving a System by Substitution**

Solve by the substitution method:

$$\begin{cases} y = -x - 1 \\ 4x - 3y = 24. \end{cases}$$

Solution

Step 1 Solve either of the equations for one variable in terms of the other. This step has already been done for us. The first equation, $y = -x - 1$, is solved for y in terms of x.

Step 2 Substitute the expression from step 1 into the other equation. We substitute the expression $-x - 1$ for y into the other equation:

$$y = \boxed{-x - 1} \qquad 4x - 3\boxed{y} = 24 \qquad \text{Substitute } -x - 1 \text{ for } y.$$

*See Answers to Selected Exercises.

Great Question!

Can I use a rough sketch on scratch paper to solve a linear system by graphing?

No. When solving linear systems by graphing, neatly drawn graphs are essential for determining points of intersection.

- Use rectangular coordinate graph paper.
- Use a ruler or straightedge.
- Use a pencil with a sharp point.

3 Solve linear systems by substitution.

Great Question!

In the first step of the substitution method, how do I know which variable to isolate and in which equation?

You can choose both the variable and the equation. If possible, solve for a variable whose coefficient is 1 or -1 to avoid working with fractions.

Substituting $-x - 1$ for y in $4x - 3y = 24$ gives us an equation in one variable, namely

$$4x - 3(-x - 1) = 24.$$

The variable y has been eliminated.

Step 3   Solve the resulting equation containing one variable.

$$
\begin{aligned}
4x - 3(-x - 1) &= 24 && \text{This is the equation containing one variable.}\\
4x + 3x + 3 &= 24 && \text{Apply the distributive property.}\\
7x + 3 &= 24 && \text{Combine like terms.}\\
7x &= 21 && \text{Subtract 3 from both sides.}\\
x &= 3 && \text{Divide both sides by 7.}
\end{aligned}
$$

Step 4   Back-substitute the obtained value into the equation from step 1. We now know that the x-coordinate of the solution is 3. To find the y-coordinate, we back-substitute the x-value into the equation from step 1.

$$y = -x - 1 \qquad \text{This is the equation from step 1.}$$

Substitute 3 for x.

$$
\begin{aligned}
y &= -3 - 1 \\
y &= -4 && \text{Simplify.}
\end{aligned}
$$

With $x = 3$ and $y = -4$, the proposed solution is $(3, -4)$.

Step 5   Check. Check the proposed solution, $(3, -4)$, in both of the system's given equations. Replace x with 3 and y with -4.

$$
\begin{array}{ll}
y = -x - 1 & 4x - 3y = 24 \\
-4 \overset{?}{=} -3 - 1 & 4(3) - 3(-4) \overset{?}{=} 24 \\
-4 = -4, \quad \text{true} & 12 + 12 \overset{?}{=} 24 \\
& 24 = 24, \quad \text{true}
\end{array}
$$

The pair $(3, -4)$ satisfies both equations. The system's solution set is $\{(3, -4)\}$.

Check Point 3   Solve by the substitution method:

$$
\begin{cases}
y = 3x - 7 \\
5x - 2y = 8.
\end{cases}
\qquad \{(6, 11)\}
$$

Technology

A graphing calculator can be used to solve the system in Example 3. Graph each equation and use the intersection feature. The calculator displays the solution $(3, -4)$ as

$$x = 3, y = -4.$$

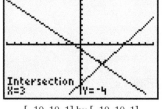

Intersection
X=3   Y=-4

$[-10, 10, 1]$ by $[-10, 10, 1]$

Example 4   Solving a System by Substitution

Solve by the substitution method:

$$
\begin{cases}
5x - 4y = 9 \\
x - 2y = -3.
\end{cases}
$$

Solution

Step 1   Solve either of the equations for one variable in terms of the other. We begin by isolating one of the variables in either of the equations. By solving for x in the second equation, which has a coefficient of 1, we can avoid fractions.

$$
\begin{aligned}
x - 2y &= -3 && \text{This is the second equation in the given system.}\\
x &= 2y - 3 && \text{Solve for x by adding $2y$ to both sides.}
\end{aligned}
$$

Step 2   Substitute the expression from step 1 into the other equation. We substitute $2y - 3$ for x in the first equation.

$$x = \boxed{2y - 3} \qquad 5\boxed{x} - 4y = 9$$

Substituting $2y - 3$ for x in $5x - 4y = 9$ gives us an equation in one variable, namely

$$5(2y - 3) - 4y = 9.$$

The variable x has been eliminated.

Step 3   Solve the resulting equation containing one variable.

$5(2y - 3) - 4y = 9$	*This is the equation containing one variable.*
$10y - 15 - 4y = 9$	*Apply the distributive property.*
$6y - 15 = 9$	*Combine like terms: $10y - 4y = 6y$.*
$6y = 24$	*Add 15 to both sides.*
$y = 4$	*Divide both sides by 6.*

Step 4   Back-substitute the obtained value into the equation from step 1. Now that we have the y-coordinate of the solution, we back-substitute 4 for y in the equation $x = 2y - 3$.

$x = 2y - 3$	*Use the equation obtained in step 1.*
$x = 2(4) - 3$	*Substitute 4 for y.*
$x = 8 - 3$	*Multiply.*
$x = 5$	*Subtract.*

With $x = 5$ and $y = 4$, the proposed solution is $(5, 4)$.

Step 5   Check. Take a moment to show that $(5, 4)$ satisfies both given equations, $5x - 4y = 9$ and $x - 2y = -3$. The solution set is $\{(5, 4)\}$.

 **Check Point 4**   Solve by the substitution method:

$$\begin{cases} 3x + 2y = -1 \\ x - y = 3. \end{cases} \quad \{(1, -2)\}$$

Great Question!

If my solution satisfies one of the equations in the system, do I have to check the solution in the other equation?

Yes. Get into the habit of checking ordered-pair solutions in both equations of the system.

Solving Linear Systems by the Addition Method

④ Solve linear systems by addition.

The substitution method is most useful if one of the given equations has an isolated variable. A third, and frequently the easiest, method for solving a linear system is the addition method. Like the substitution method, the addition method involves eliminating a variable and ultimately solving an equation containing only one variable. However, this time we eliminate a variable by adding the equations.

For example, consider the following system of linear equations:

$$\begin{cases} 3x - 4y = 11 \\ -3x + 2y = -7. \end{cases}$$

When we add these two equations, the x-terms are eliminated. This occurs because the coefficients of the x-terms, 3 and -3, are opposites (additive inverses) of each other:

$$\begin{cases} 3x - 4y = 11 \\ \underline{-3x + 2y = -7} \end{cases}$$

Add: $\qquad -2y = 4$ The sum is an equation in one variable.

$\qquad\qquad y = -2$ *Divide both sides by -2 and solve for y.*

Now we can back-substitute -2 for y into one of the original equations to find x. It does not matter which equation we use: We will obtain the same value for x in either case. If we use either equation, we can show that $x = 1$ and the solution $(1, -2)$ satisfies both equations in the system.

Great Question!

Isn't the addition method also called the elimination method?

Although the addition method is also known as the elimination method, variables are eliminated when using both the substitution and addition methods. The name *addition method* specifically tells us that the elimination of a variable is accomplished by adding equations.

When we use the addition method, we want to obtain two equations whose sum is an equation containing only one variable. The key step is to **obtain, for one of the variables, coefficients that differ only in sign.** To do this, we may need to multiply one or both equations by some nonzero number so that the coefficients of one of the variables, x or y, become opposites. Then when the two equations are added, this variable is eliminated.

Solving Linear Systems by Addition

1. If necessary, rewrite both equations in the form $Ax + By = C$.
2. If necessary, multiply either equation or both equations by appropriate nonzero numbers so that the sum of the x-coefficients or the sum of the y-coefficients is 0.
3. Add the equations in step 2. The sum is an equation in one variable.
4. Solve the equation in one variable.
5. Back-substitute the value obtained in step 4 into either of the given equations and solve for the other variable.
6. Check the solution in both of the original equations.

Example 5   Solving a System by the Addition Method

Solve by the addition method:

$$\begin{cases} 3x + 2y = 48 \\ 9x - 8y = -24. \end{cases}$$

Solution

Step 1 Rewrite both equations in the form $Ax + By = C$. Both equations are already in this form. Variable terms appear on the left and constants appear on the right.

Step 2 If necessary, multiply either equation or both equations by appropriate numbers so that the sum of the x-coefficients or the sum of the y-coefficients is 0. We can eliminate x or y. Let's eliminate x. Consider the terms in x in each equation, that is, $3x$ and $9x$. To eliminate x, we can multiply each term of the first equation by -3 and then add the equations.

$$\begin{cases} 3x + 2y = 48 \quad \xrightarrow{\text{Multiply by } -3.} \\ 9x - 8y = -24 \quad \xrightarrow{\text{No change}} \end{cases} \begin{cases} -9x - 6y = -144 \\ \underline{9x - 8y = -24} \\ {-14y} = -168 \end{cases}$$

Step 3 Add the equations.   Add:

Step 4 Solve the equation in one variable. We solve $-14y = -168$ by dividing both sides by -14.

$$\frac{-14y}{-14} = \frac{-168}{-14} \qquad \text{Divide both sides by } -14.$$
$$y = 12 \qquad \text{Simplify.}$$

Step 5 Back-substitute and find the value for the other variable. We can back-substitute 12 for y into either one of the given equations. We'll use the first one.

$$3x + 2y = 48 \qquad \text{This is the first equation in the given system.}$$
$$3x + 2(12) = 48 \qquad \text{Substitute 12 for } y.$$
$$3x + 24 = 48 \qquad \text{Multiply.}$$
$$3x = 24 \qquad \text{Subtract 24 from both sides.}$$
$$x = 8 \qquad \text{Divide both sides by 3.}$$

We found that $y = 12$ and $x = 8$. The proposed solution is $(8, 12)$.

Step 6 Check. Take a few minutes to show that $(8, 12)$ satisfies both of the original equations in the system. The solution set is $\{(8, 12)\}$.

✔ **Check Point 5** Solve by the addition method:

$$\begin{cases} 4x + 5y = 3 \\ 2x - 3y = 7. \quad \{(2, -1)\} \end{cases}$$

Example 6 Solving a System by the Addition Method

Solve by the addition method:

$$\begin{cases} 7x = 5 - 2y \\ 3y = 16 - 2x. \end{cases}$$

Solution

Step 1 Rewrite both equations in the form *Ax + By = C*. We first arrange the system so that variable terms appear on the left and constants appear on the right. We obtain

$$\begin{cases} 7x + 2y = 5 \qquad \text{Add } 2y \text{ to both sides of the first equation.} \\ 2x + 3y = 16. \quad \text{Add } 2x \text{ to both sides of the second equation.} \end{cases}$$

Step 2 If necessary, multiply either equation or both equations by appropriate numbers so that the sum of the *x*-coefficients or the sum of the *y*-coefficients is 0. We can eliminate *x* or *y*. Let's eliminate *y* by multiplying the first equation by 3 and the second equation by −2.

$$\begin{cases} 7x + 2y = 5 \quad \xrightarrow{\text{Multiply by 3.}} \\ 2x + 3y = 16 \quad \xrightarrow{\text{Multiply by } -2.} \end{cases} \quad \begin{cases} 21x + 6y = 15 \\ -4x - 6y = -32 \end{cases}$$

Step 3 Add the equations.

$$\text{Add:} \quad \frac{\begin{aligned} 21x + 6y &= 15 \\ -4x - 6y &= -32 \end{aligned}}{\begin{aligned} 17x + 0y &= -17 \\ 17x &= -17 \end{aligned}}$$

Step 4 Solve the equation in one variable. We solve $17x = -17$ by dividing both sides by 17.

$$\frac{17x}{17} = \frac{-17}{17} \qquad \text{Divide both sides by 17.}$$

$$x = -1 \qquad \text{Simplify.}$$

Step 5 Back-substitute and find the value for the other variable. We can back-substitute −1 for *x* into either one of the given equations. We'll use the second one.

$$3y = 16 - 2x \qquad \text{This is the second equation in the given system.}$$

$$3y = 16 - 2(-1) \qquad \text{Substitute } -1 \text{ for x.}$$

$$3y = 16 + 2 \qquad \text{Multiply.}$$

$$3y = 18 \qquad \text{Add.}$$

$$y = 6 \qquad \text{Divide both sides by 3.}$$

With $x = -1$ and $y = 6$, the proposed solution is $(-1, 6)$.

Step 6 Check. Take a moment to show that $(-1, 6)$ satisfies both given equations. The solution is $(-1, 6)$ and the solution set is $\{(-1, 6)\}$.

✔ **Check Point 6** Solve by the addition method:

$$\begin{cases} 3x = 2 - 4y \\ 5y = -1 - 2x. \quad \{(2, -1)\} \end{cases}$$

Linear Systems Having No Solution or Infinitely Many Solutions

⑤ Identify systems that do not have exactly one ordered-pair solution.

We have seen that a system of linear equations in two variables represents a pair of lines. The lines either intersect at one point, are parallel, or are identical. Thus, there are three possibilities for the number of solutions to a system of two linear equations.

The Number of Solutions to a System of Two Linear Equations

The number of solutions to a system of two linear equations in two variables is given by one of the following. (See **Figure 3.41**.)

Number of Solutions	What This Means Graphically
Exactly one ordered-pair solution	The two lines intersect at one point.
No solution	The two lines are parallel.
Infinitely many solutions	The two lines are identical.

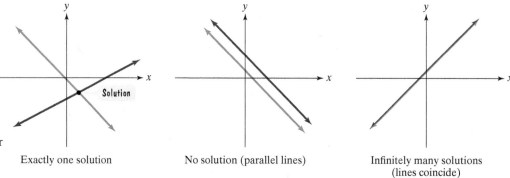

Exactly one solution   No solution (parallel lines)   Infinitely many solutions (lines coincide)

Figure 3.41 Possible graphs for a system of two linear equations in two variables

Example 7   A System with No Solution

Solve the system:

$$\begin{cases} 4x + 6y = 12 \\ 6x + 9y = 12. \end{cases}$$

Solution

Because no variable is isolated, we will use the addition method. To obtain coefficients of x that differ only in sign, we multiply the first equation by 3 and the second equation by -2.

$$\begin{cases} 4x + 6y = 12 \quad \text{Multiply by 3.} \\ 6x + 9y = 12 \quad \text{Multiply by } -2. \end{cases} \longrightarrow \begin{cases} 12x + 18y = 36 \\ -12x - 18y = -24 \end{cases}$$

Add: $ 0 = 12$

There are no values of x and y for which $0 = 12$. No values of x and y satisfy $0x + 0y = 12$.

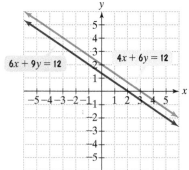

Figure 3.42 The graph of a system with no solution

The false statement $0 = 12$ indicates that the system has no solution. The solution set is the empty set, $\varnothing$.

The lines corresponding to the two equations in Example 7 are shown in **Figure 3.42**. The lines are parallel and have no point of intersection.

✓ **Check Point 7**   Solve the system:

$$\begin{cases} x + 2y = 4 \\ 3x + 6y = 13. \end{cases} \quad \varnothing$$

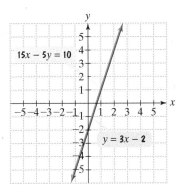

Figure 3.43 The graph of a system with infinitely many solutions

Great Question!

The system in Example 8 has infinitely many solutions. Does that mean that any ordered pair of numbers is a solution?

No. Although the system in Example 8 has infinitely many solutions, this does not mean that any ordered pair of numbers you can form will be a solution. The ordered pair (x, y) must satisfy one of the system's equations, $y = 3x - 2$ or $15x - 5y = 10$, and there are infinitely many such ordered pairs. Because the graphs are coinciding lines, the ordered pairs that are solutions of one of the equations are also solutions of the other equation.

Example 8 A System with Infinitely Many Solutions

Solve the system:

$$\begin{cases} y = 3x - 2 \\ 15x - 5y = 10. \end{cases}$$

Solution

Because the variable y is isolated in $y = 3x - 2$, the first equation, we can use the substitution method. We substitute the expression for y into the second equation.

$$y = \boxed{3x - 2} \qquad 15x - 5\boxed{y} = 10 \qquad \text{Substitute } 3x - 2 \text{ for } y.$$

The substitution results in an equation in one variable.

$$15x - 5(3x - 2) = 10 \qquad \text{Apply the distributive property.}$$
$$15x - 15x + 10 = 10$$

This statement is true for all values of x and y.

$$10 = 10 \qquad \text{Simplify.}$$

In our final step, both variables have been eliminated and the resulting statement, $10 = 10$, is true. This true statement indicates that the system has infinitely many solutions. The solution set consists of all points (x, y) lying on either of the coinciding lines, $y = 3x - 2$ or $15x - 5y = 10$, as shown in **Figure 3.43**.

We express the solution set for the system in one of two equivalent ways:

$$\{(x, y) \mid y = 3x - 2\} \qquad \text{or} \qquad \{(x, y) \mid 15x - 5y = 10\}.$$

The set of all ordered pairs (x, y) such that $y = 3x - 2$ The set of all ordered pairs (x, y) such that $15x - 5y = 10$

Check Point 8 Solve the system:

$$\begin{cases} y = 4x - 4 \quad \{(x, y) \mid y = 4x - 4\} \text{ or} \\ 8x - 2y = 8. \quad \{(x, y) \mid 8x - 2y = 8\} \end{cases}$$

Linear Systems Having No Solution or Infinitely Many Solutions

If both variables are eliminated when solving a system of linear equations by substitution or addition, one of the following is true:

1. There is no solution if the resulting statement is false.

2. There are infinitely many solutions if the resulting statement is true.

Modeling with Systems of Equations: Making Money (and Losing It)

What does every entrepreneur, from a kid selling lemonade to Bill Gates, want to do? Generate profit, of course. The profit made is the money taken in, or the revenue, minus the money spent, or the cost.

Solve problems using systems of linear equations.

Revenue and Cost Functions

A company produces and sells x units of a product. Its **revenue** is the money generated by selling x units of the product. Its **cost** is the cost of producing x units of the product.

Revenue Function

$$R(x) = (\text{price per unit sold})x$$

Cost Function

$$C(x) = \text{fixed cost} + (\text{cost per unit produced})x$$

The point of intersection of the graphs of the revenue and cost functions is called the **break-even point**. The x-coordinate of the point reveals the number of units that a company must produce and sell so that money coming in, the revenue, is equal to money going out, the cost. The y-coordinate of the break-even point gives the amount of money coming in and going out. Example 9 illustrates the use of the substitution method in determining a company's break-even point.

Example 9 Finding a Break-Even Point

Technology is now promising to bring light, fast, and beautiful wheelchairs to millions of disabled people. A company is planning to manufacture these radically different wheelchairs. Fixed cost will be $500,000 and it will cost $400 to produce each wheelchair. Each wheelchair will be sold for $600.

a. Write the cost function, C, of producing x wheelchairs.

b. Write the revenue function, R, from the sale of x wheelchairs.

c. Determine the break-even point. Describe what it means.

Solution

a. The cost function is the sum of the fixed cost and variable cost.

$$\underbrace{\text{Fixed cost of}}_{\$500,000} \quad \text{plus} \quad \underbrace{\text{Variable cost: }\$400 \text{ for}}_{\text{each chair produced}}$$

$$C(x) = 500,000 + 400x$$

b. The revenue function is the money generated from the sale of x wheelchairs. We are given that each wheelchair will be sold for $600.

$$\underbrace{\text{Revenue per chair, }\$600, \text{ times}}_{} \quad \underbrace{\text{the number of chairs sold}}_{}$$

$$R(x) = 600x$$

c. The break-even point occurs where the graphs of C and R intersect. Thus, we find this point by solving the system

$$\begin{cases} C(x) = 500,000 + 400x \\ R(x) = 600x \end{cases} \quad \text{or} \quad \begin{cases} y = 500,000 + 400x \\ y = 600x. \end{cases}$$

Using substitution, we can substitute $600x$ for y in the first equation:

$600x = 500,000 + 400x$ Substitute $600x$ for y in $y = 500,000 + 400x$.

$200x = 500,000$ Subtract $400x$ from both sides.

$x = 2500$ Divide both sides by 200.

Back-substituting 2500 for x in either of the system's equations (or functions), $C(x) = 500,000 + 400x$ or $R(x) = 600x$, we obtain

$$R(2500) = 600(2500) = 1,500,000.$$

We used $R(x) = 600x$.

The break-even point is $(2500, 1,500,000)$. This means that the company will break even if it produces and sells 2500 wheelchairs. At this level, the money coming in is equal to the money going out: $1,500,000.

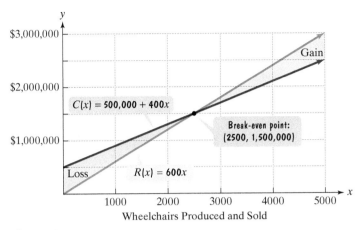

Figure 3.44

Figure 3.44 shows the graphs of the revenue and cost functions for the wheelchair business. Similar graphs and models apply no matter how small or large a business venture may be.

The intersection point confirms that the company breaks even by producing and selling 2500 wheelchairs. Can you see what happens for $x < 2500$? The red cost graph lies above the blue revenue graph. The cost is greater than the revenue and the business is losing money. Thus, if they sell fewer than 2500 wheelchairs, the result is a *loss*. By contrast, look at what happens for $x > 2500$. The blue revenue graph lies above the red cost graph. The revenue is greater than the cost and the business is making money. Thus, if they sell more than 2500 wheelchairs, the result is a *gain*.

✓ **Check Point 9** A company that manufactures running shoes has a fixed cost of $300,000. Additionally, it costs $30 to produce each pair of shoes. They are sold at $80 per pair. **a.** $C(x) = 300,000 + 30x$ **b.** $R(x) = 80x$

a. Write the cost function, C, of producing x pairs of running shoes.

b. Write the revenue function, R, from the sale of x pairs of running shoes.

▸ **c.** Determine the break-even point. Describe what it means. $(6000, 480,000)$; The company will break even if it produces and sells 6000 pairs of shoes. At this level, the money coming in is equal to the money going out: $480,000.

The profit generated by a business is the money taken in (its revenue) minus the money spent (its cost). Thus, once a business has modeled its revenue and cost with a system of equations, it can determine its *profit function*, $P(x)$.

> **The Profit Function**
>
> The profit, $P(x)$, generated after producing and selling x units of a product is given by the **profit function**
>
> $$P(x) = R(x) - C(x),$$
>
> where R and C are the revenue and cost functions, respectively.

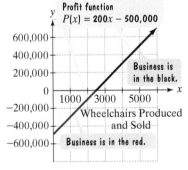

Figure 3.45

The profit function for the wheelchair business in Example 9 is

$$\begin{aligned} P(x) &= R(x) - C(x) \\ &= 600x - (500,000 + 400x) \\ &= 200x - 500,000. \end{aligned}$$

The graph of this profit function is shown in **Figure 3.45**. The red portion lies below the x-axis and shows a loss when fewer than 2500 wheelchairs are sold. The business is "in the red." The black portion lies above the x-axis and shows a gain when more than 2500 wheelchairs are sold. The wheelchair business is "in the black."

Achieving Success

Form a study group with other students in your class. Working in small groups often serves as an excellent way to learn and reinforce new material. Set up helpful procedures and guidelines for the group. "Talk" math by discussing and explaining the concepts and exercises to one another.

Concept and Vocabulary Check

Exercises in the Concept and Vocabulary Check
are intended for group and class discussions.

In Exercises 1–9, fill in each blank so that the resulting statement is true.

1. A solution to a system of linear equations in two variables is an ordered pair that <u>satisfies both equations in the system</u>

2. When solving a system of linear equations by graphing, the system's solution is determined by locating <u>the intersection point</u>

3. When solving
$$\begin{cases} 3x - 2y = 5 \\ y = 3x - 3 \end{cases}$$
by the substitution method, we obtain $x = \frac{1}{3}$, so the solution set is <u>$\{(\frac{1}{3}, -2)\}$</u>.

4. When solving
$$\begin{cases} 2x + 10y = 9 \\ 8x + 5y = 7 \end{cases}$$
by the addition method, we can eliminate y by multiplying the second equation by <u>-2</u> and then adding the equations.

5. When solving
$$\begin{cases} 4x - 3y = 15 \\ 3x - 2y = 10 \end{cases}$$
by the addition method, we can eliminate y by multiplying the first equation by 2 and the second equation by <u>-3</u> and then adding the equations.

6. When solving
$$\begin{cases} 12x - 21y = 24 \\ 4x - 7y = 7 \end{cases}$$
by the addition method, we obtain $0 = 3$, so the solution set is <u>$\varnothing$</u>. If you attempt to solve such a system by graphing, you will obtain two lines that are <u>parallel</u>.

7. When solving
$$\begin{cases} x = 3y + 2 \\ 5x - 15y = 10 \end{cases}$$
by the substitution method, we obtain $10 = 10$, so the solution set is <u>$\{(x,y)|x = 3y + 2\}$ or $\{(x,y)|5x - 15y = 10\}$</u>. If you attempt to solve such a system by graphing, you will obtain two lines that <u>are identical or coincide</u>

8. A company's <u>revenue</u> function is the money generated by selling x units of its product. The difference between this function and the company's cost function is called its <u>profit</u> function.

9. A company has a graph that shows the money it generates by selling x units of its product. It also has a graph that shows its cost of producing x units of its product. The point of intersection of these graphs is called the company's <u>break-even point</u>.

In Exercises 10–13, determine whether each statement is true or false. If the statement is false, make the necessary change(s) to produce a true statement.   Changes to false statements will vary.

10. Unlike the graphing method, where solutions cannot be seen, the substitution method provides a way to visualize solutions as intersection points.   false

11. To solve the system
$$\begin{cases} 2x - y = 5 \\ 3x + 4y = 7 \end{cases}$$
by substitution, replace y in the second equation with $5 - 2x$.   false

12. If the two equations in a linear system are $5x - 3y = 7$ and $4x + 9y = 11$, multiplying the first equation by 4, the second by 5, and then adding equations will eliminate x.   false

13. If x can be eliminated by the addition method, y cannot be eliminated by using the original equations of the system.   false

Respond to Exercises 14–23 using verbal or written explanations.

14. What is a system of linear equations? Provide an example with your description.   14–23. Answers will vary.

15. What is the solution to a system of linear equations?

16. Explain how to solve a system of equations using graphing.

17. Explain how to solve a system of equations using the substitution method. Use $y = 3 - 3x$ and $3x + 4y = 6$ to illustrate your explanation.

18. Explain how to solve a system of equations using the addition method. Use $3x + 5y = -2$ and $2x + 3y = 0$ to illustrate your explanation.

19. What is the disadvantage to solving a system of equations using the graphing method?

20. When is it easier to use the addition method rather than the substitution method to solve a system of equations?

21. When using the addition or substitution method, how can you tell whether a system of linear equations has infinitely many solutions? What is the relationship between the graphs of the two equations?

22. When using the addition or substitution method, how can you tell whether a system of linear equations has no solution? What is the relationship between the graphs of the two equations?

23. Describe the break-even point for a business.

Exercise Set 3.4

Practice Exercises

In Exercises 1–4, determine whether the given ordered pair is a solution of the system.

1. $(2, 3)$ solution
$$\begin{cases} x + 3y = 11 \\ x - 5y = -13 \end{cases}$$

2. $(-3, 5)$ solution
$$\begin{cases} 9x + 7y = 8 \\ 8x - 9y = -69 \end{cases}$$

3. $(2, 5)$ not a solution
$$\begin{cases} 2x + 3y = 17 \\ x + 4y = 16 \end{cases}$$

4. $(8, 5)$ not a solution
$$\begin{cases} 5x - 4y = 20 \\ 3y = 2x + 1 \end{cases}$$

In Exercises 5–12, solve each system by graphing. Check the coordinates of the intersection point in both equations.

5. $\begin{cases} x + y = 6 \\ x - y = 2 \end{cases}$ $\{(4, 2)\}*$

6. $\begin{cases} x + y = 2 \\ x - y = 4 \end{cases}$ $\{(3, -1)\}*$

7. $\begin{cases} 2x - 3y = 6 \\ 4x + 3y = 12 \end{cases}$ $\{(3, 0)\}*$

8. $\begin{cases} 4x + y = 4 \\ 3x - y = 3 \end{cases}$ $\{(1, 0)\}*$

9. $\begin{cases} y = x + 5 \\ y = -x + 3 \end{cases}$ $\{(-1, 4)\}*$

10. $\begin{cases} y = x + 1 \\ y = 3x - 1 \end{cases}$ $\{(1, 2)\}*$

11. $\begin{cases} y = -x - 1 \\ 4x - 3y = 24 \end{cases}$ $\{(3, -4)\}*$

12. $\begin{cases} y = 3x - 4 \\ 2x + y = 1 \end{cases}$ $\{(1, -1)\}*$

In Exercises 13–24, solve each system by the substitution method. Be sure to check all proposed solutions.

13. $\begin{cases} x + y = 4 \\ y = 3x \end{cases}$ $\{(1, 3)\}$

14. $\begin{cases} x + y = 6 \\ y = 2x \end{cases}$ $\{(2, 4)\}$

15. $\begin{cases} x + 3y = 8 \\ y = 2x - 9 \end{cases}$ $\{(5, 1)\}$

16. $\begin{cases} 2x - 3y = -13 \\ y = 2x + 7 \end{cases}$ $\{(-2, 3)\}$

17. $\begin{cases} x + 3y = 5 \\ 4x + 5y = 13 \end{cases}$ $\{(2, 1)\}$

18. $\begin{cases} 3x + y = -18 \\ 5x - 2y = -8 \end{cases}$ $\{(-4, -6)\}$

19. $\begin{cases} 2x - y = -5 \\ x + 5y = 14 \end{cases}$ $\{(-1, 3)\}$

20. $\begin{cases} 2x + 3y = 11 \\ x - 4y = 0 \end{cases}$ $\{(4, 1)\}$

21. $\begin{cases} 2x - y = 3 \\ 5x - 2y = 10 \end{cases}$ $\{(4, 5)\}$

22. $\begin{cases} -x + 3y = 10 \\ 2x + 8y = -6 \end{cases}$ $\{(-7, 1)\}$

23. $\begin{cases} x + 8y = 6 \\ 2x + 4y = -3 \end{cases}$ $\{(-4, \frac{5}{4})\}$

24. $\begin{cases} -4x + y = -11 \\ 2x - 3y = 5 \end{cases}$ $\{(\frac{14}{5}, \frac{1}{5})\}$

In Exercises 25–36, solve each system by the addition method. Be sure to check all proposed solutions.

25. $\begin{cases} x + y = 1 \\ x - y = 3 \end{cases}$ $\{(2, -1)\}$

26. $\begin{cases} x + y = 6 \\ x - y = -2 \end{cases}$ $\{(2, 4)\}$

27. $\begin{cases} 2x + 3y = 6 \\ 2x - 3y = 6 \end{cases}$ $\{(3, 0)\}$

28. $\begin{cases} 3x + 2y = 14 \\ 3x - 2y = 10 \end{cases}$ $\{(4, 1)\}$

29. $\begin{cases} x + 2y = 2 \\ -4x + 3y = 25 \end{cases}$ $\{(-4, 3)\}$

30. $\begin{cases} 2x - 7y = 2 \\ 3x + y = -20 \end{cases}$ $\{(-6, -2)\}$

31. $\begin{cases} 4x + 3y = 15 \\ 2x - 5y = 1 \end{cases}$ $\{(3, 1)\}$

32. $\begin{cases} 3x - 7y = 13 \\ 6x + 5y = 7 \end{cases}$ $\{(2, -1)\}$

33. $\begin{cases} 3x - 4y = 11 \\ 2x + 3y = -4 \end{cases}$ $\{(1, -2)\}$

34. $\begin{cases} 2x + 3y = -16 \\ 5x - 10y = 30 \end{cases}$ $\{(-2, -4)\}$

35. $\begin{cases} 2x = 3y - 4 \\ -6x + 12y = 6 \end{cases}$ $\{(-5, -2)\}$

36. $\begin{cases} 5x = 4y - 8 \\ 3x + 7y = 14 \end{cases}$ $\{(0, 2)\}$

In Exercises 37–44, solve by the method of your choice. Identify systems with no solution and systems with infinitely many solutions, using set notation to express their solution sets.

37. $\begin{cases} x = 9 - 2y \\ x + 2y = 13 \end{cases}$ $\varnothing$

38. $\begin{cases} 6x + 2y = 7 \\ y = 2 - 3x \end{cases}$ $\varnothing$

39. $\begin{cases} y = 3x - 5 \\ 21x - 35 = 7y \end{cases}$

40. $\begin{cases} 9x - 3y = 12 \\ y = 3x - 4 \end{cases}$

41. $\begin{cases} 3x - 2y = -5 \\ 4x + y = 8 \end{cases}$ $\{(1, 4)\}$

42. $\begin{cases} 2x + 5y = -4 \\ 3x - y = 11 \end{cases}$ $\{(3, -2)\}$

43. $\begin{cases} x + 3y = 2 \\ 3x + 9y = 6 \end{cases}$ $\{(x, y) | x + 3y = 2\}$ or $\{(x, y) | 3x + 9y = 6\}$

44. $\begin{cases} 4x - 2y = 2 \\ 2x - y = 1 \end{cases}$ $\{(x, y) | 4x - 2y = 2\}$ or $\{(x, y) | 2x - y = 1\}$

39. $\{(x, y) | y = 3x - 5\}$ or $\{(x, y) | 21x - 35 = 7y\}$ **40.** $\{(x, y) | y = 3x - 4\}$ or $\{(x, y) | 9x - 3y = 12\}$

61. a. $y = 0.45x + 0.8$ **b.** $y = 0.15x + 2.6$ **c.** week 6; 3.5 symptoms; by the intersection point $(6, 3.5)$

Practice Plus

Use the graphs of the linear functions to solve Exercises 45–46.

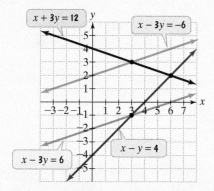

45. Write the linear system whose solution set is $\{(6, 2)\}$. Express each equation in the system in slope-intercept form.   *

46. Write the linear system whose solution set is $\emptyset$. Express each equation in the system in slope-intercept form.   *

In Exercises 47–48, solve each system for x and y, expressing either value in terms of a or b, if necessary. Assume that $a \neq 0$ and $b \neq 0$.

47. $\begin{cases} 5ax + 4y = 17 \\ ax + 7y = 22 \end{cases}$ $\{(\frac{1}{a}, 3)\}$ **48.** $\begin{cases} 4ax + by = 3 \\ 6ax + 5by = 8 \end{cases}$ $\{(\frac{1}{2a}, \frac{1}{b})\}$

49. For the linear function $f(x) = mx + b$, $f(-2) = 11$ and $f(3) = -9$. Find m and b. $m = -4, b = 3$

50. For the linear function $f(x) = mx + b$, $f(-3) = 23$ and $f(2) = -7$. Find m and b. $m = -6, b = 5$

Application Exercises

The figure shows the graphs of the cost and revenue functions for a company that manufactures and sells small radios. Use the information in the figure to solve Exercises 51–56.

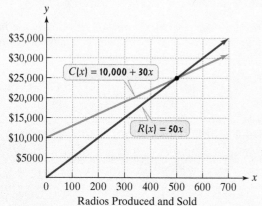

Radios Produced and Sold

53. -6000; When the company produces and sells 200 radios, the loss is \$6000.

51. How many radios must be produced and sold for the company to break even?   500 radios

52. More than how many radios must be produced and sold for the company to have a profit?   more than 500 radios

53. Use the formulas shown in the voice balloons to find $R(200) - C(200)$. Describe what this means for the company.

54. Use the formulas shown in the voice balloons to find $R(300) - C(300)$. Describe what this means for the company. -4000; When the company produces and sells 300 radios, the loss is \$4000.

55. a. Use the formulas shown in the voice balloons to write the company's profit function, P, from producing and selling x radios. $P(x) = 20x - 10{,}000$

b. Find the company's profit if 10,000 radios are produced and sold.   \$190,000

56. a. Use the formulas shown in the voice balloons to write the company's profit function, P, from producing and selling x radios. $P(x) = 20x - 10{,}000$

b. Find the company's profit if 20,000 radios are produced and sold.   \$390,000

Exercises 57–60 describe a number of business ventures. For each exercise,

    a. *Write the cost function, C.*
    b. *Write the revenue function, R.*
    c. *Determine the break-even point. Describe what it means.*

57. A company that manufactures small canoes has a fixed cost of \$18,000. It costs \$20 to produce each canoe. The selling price is \$80 per canoe. (In solving this exercise, let x represent the number of canoes produced and sold.)   *

58. A company that manufactures bicycles has a fixed cost of \$100,000. It costs \$100 to produce each bicycle. The selling price is \$300 per bike. (In solving this exercise, let x represent the number of bicycles produced and sold.)   *

59. You invest in a new play. The cost includes an overhead of \$30,000, plus production costs of \$2500 per performance. A sold-out performance brings in \$3125. (In solving this exercise, let x represent the number of sold-out performances.)   *

60. You invested \$30,000 and started a business writing greeting cards. Supplies cost 2¢ per card and you are selling each card for 50¢. (In solving this exercise, let x represent the number of cards produced and sold.)   *

61. We opened this section with a study showing that late in the semester, procrastinating students reported more symptoms of physical illness than their nonprocrastinating peers.

a. At the beginning of the semester, procrastinators reported an average of 0.8 symptoms, increasing at a rate of 0.45 symptoms per week. Write a function that models the average number of symptoms, y, after x weeks.

b. At the beginning of the semester, nonprocrastinators reported an average of 2.6 symptoms, increasing at a rate of 0.15 symptoms per week. Write a function that models the average number of symptoms, y, after x weeks.

c. By which week in the semester did both groups report the same number of symptoms of physical illness? For that week, how many symptoms were reported by each group? How is this shown in **Figure 3.39** on page 217?

62. a. In 1960, 5% of U.S. adults lived alone, increasing at a rate of 0.2% per year. Write a function that models the percentage of U.S. adults living alone, y, x years after 1960.

b. In 1960, 47% of U.S. adults were married, living with kids, decreasing at a rate of 0.4% per year. Write a function that models the percentage of married U.S. adults living with kids, y, x years after 1960. $y = -0.4x + 47$

c. Use the models from parts (a) and (b) to project the year in which the percentage of adults living alone will be the same as the percentage of married adults living with kids. What percentage of U.S. adults will belong to each group during that year?   2030; 19%

62. a. $y = 0.2x + 5$

Use a system of linear equations to solve Exercises 63–68.

The graph shows the four candy bars with the highest fat content, representing grams of fat and calories in each bar. Exercises 63–66 are based on the graph.

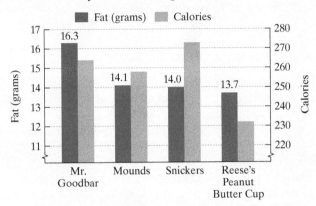

Candy Bars with the Highest Fat Content

Source: Krantz and Sveum, *The World's Worsts*, Harper Collins, 2005.

63. Mr. Goodbar: 264 cal; Mounds: 258 cal

63. One Mr. Goodbar and two Mounds bars contain 780 calories. Two Mr. Goodbars and one Mounds bar contain 786 calories. Find the caloric content of each candy bar.

64. One Snickers bar and two Reese's Peanut Butter Cups contain 737 calories. Two Snickers bars and one Reese's Peanut Butter Cup contain 778 calories. Find the caloric content of each candy bar.   Snickers: 273 cal; Reese's: 232 cal

65. A collection of Halloween candy contains a total of five Mr. Goodbars and Mounds bars. Chew on this: The grams of fat in these candy bars exceed the daily maximum desirable fat intake of 70 grams by 7.1 grams. How many bars of each kind of candy are contained in the Halloween collection?   3 Mr. Goodbars and 2 Mounds

66. A collection of Halloween candy contains a total of 12 Snickers bars and Reese's Peanut Butter Cups. Chew on this: The grams of fat in these candy bars exceed twice the daily maximum desirable fat intake of 70 grams by 26.5 grams. How many bars of each kind of candy are contained in the Halloween collection?   7 Snickers and 5 Reese's

67. A hotel has 200 rooms. Those with kitchen facilities rent for $100 per night and those without kitchen facilities rent for $80 per night. On a night when the hotel was completely occupied, revenues were $17,000. How many of each type of room does the hotel have?   50 rooms with kitchens and 150 without

68. A new restaurant is to contain two-seat tables and four-seat tables. Fire codes limit the restaurant's maximum occupancy to 56 customers. If the owners have hired enough servers to handle 17 tables of customers, how many of each kind of table should they purchase?   two-seat table: 6; four-seat table: 11

Critical Thinking Exercises

69. Write a system of equations having $\{(-2, 7)\}$ as a solution set. (More than one system is possible.)   Answers will vary.

70. One apartment is directly above a second apartment. The resident living downstairs calls his neighbor living above him and states, "If one of you is willing to come downstairs, we'll have the same number of people in both apartments." The upstairs resident responds, "We're all too tired to move. Why don't one of you come up here? Then we will have twice as many people up here as you've got down there." How many people are in each apartment?   downstairs: 5 people; upstairs: 7 people

71. A set of identical twins can only be distinguished by the characteristic that one always tells the truth and the other always lies. One twin tells you of a lucky number pair: "When I multiply my first lucky number by 3 and my second lucky number by 6, the addition of the resulting numbers produces a sum of 12. When I add my first lucky number and twice my second lucky number, the sum is 5." Which twin is talking?   the twin who always lies

---

3.5 : Linear Inequalities in Two Variables

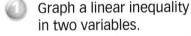
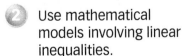
What am I supposed to learn?

After you have read this section, you should be able to:

1. Graph a linear inequality in two variables.

2. Use mathematical models involving linear inequalities.

3. Graph a system of linear inequalities.

We opened the chapter noting that the modern emphasis on thinness as the ideal body shape has been suggested as a major cause of eating disorders. In this section, as well as in the Exercise Set, we use systems of linear inequalities in two variables that will enable you to identify a healthy weight range for various heights and ages.

Linear Inequalities in Two Variables and Their Solutions

We have seen that equations in the form $Ax + By = C$, where A and B are not both zero, are straight lines when graphed. If we change the symbol = to >, <, ≥, or ≤, we obtain a **linear inequality in two variables**. Some examples of linear inequalities in two variables are $x + y > 2$, $3x - 5y \leq 15$, and $2x - y < 4$.

A **solution of an inequality in two variables**, x and y, is an ordered pair of real numbers with the following property: When the x-coordinate is substituted for x and the y-coordinate is substituted for y in the inequality, we obtain a true statement. For example, $(3, 2)$ is a solution of the inequality $x + y > 1$. When 3 is substituted for x and 2 is substituted for y, we obtain the true statement $3 + 2 > 1$, or $5 > 1$. Because there are infinitely many pairs of numbers that have a sum greater than 1, the inequality $x + y > 1$ has infinitely many solutions. Each ordered-pair solution is said to **satisfy** the inequality. Thus, $(3, 2)$ satisfies the inequality $x + y > 1$.

The Graph of a Linear Inequality in Two Variables

Graph a linear inequality in two variables.

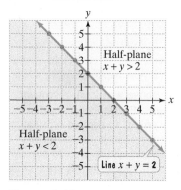

Figure 3.46

We know that the graph of an equation in two variables is the set of all points whose coordinates satisfy the equation. Similarly, the **graph of an inequality in two variables** is the set of all points whose coordinates satisfy the inequality.

Let's use **Figure 3.46** to get an idea of what the graph of a linear inequality in two variables looks like. Part of the figure shows the graph of the linear equation $x + y = 2$. The line divides the points in the rectangular coordinate system into three sets. First, there is the set of points along the line satisfying $x + y = 2$. Next, there is the set of points in the green region above the line. Points in the green region satisfy the linear inequality $x + y > 2$. Finally, there is the set of points in the purple region below the line. Points in the purple region satisfy the linear inequality $x + y < 2$.

A **half-plane** is the set of all the points on one side of a line. In **Figure 3.46**, the green region is a half-plane. The purple region is also a half-plane. A half-plane is the graph of a linear inequality that involves $>$ or $<$. The graph of an inequality that involves $\geq$ or $\leq$ is a half-plane and a line. A solid line is used to show that a line is part of a graph. A dashed line is used to show that a line is not part of a graph.

> **Graphing a Linear Inequality in Two Variables**
>
> 1. Replace the inequality symbol with an equal sign and graph the corresponding linear equation. Draw a solid line if the original inequality contains a $\leq$ or $\geq$ symbol. Draw a dashed line if the original inequality contains a $<$ or $>$ symbol.
> 2. Choose a test point from one of the half-planes. (Do not choose a point on the line.) Substitute the coordinates of the test point into the inequality.
> 3. If a true statement results, shade the half-plane containing this test point. If a false statement results, shade the half-plane not containing this test point.

Example 1   Graphing a Linear Inequality in Two Variables

Graph: $3x - 5y \geq 15$.

Solution

Step 1   Replace the inequality symbol by = and graph the linear equation. We need to graph $3x - 5y = 15$. We can use intercepts to graph this line.

We set $y = 0$ to find the x-intercept.	We set $x = 0$ to find the y-intercept.
$3x - 5y = 15$	$3x - 5y = 15$
$3x - 5 \cdot 0 = 15$	$3 \cdot 0 - 5y = 15$
$3x = 15$	$-5y = 15$
$x = 5$	$y = -3$

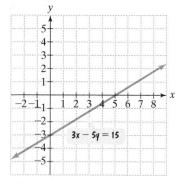

Figure 3.47 Preparing to graph $3x - 5y \geq 15$

The x-intercept is 5, so the line passes through $(5, 0)$. The y-intercept is -3, so the line passes through $(0, -3)$. Using the intercepts, the line is shown in **Figure 3.47** as a solid line. This is because the inequality $3x - 5y \geq 15$ contains a $\geq$ symbol, in which equality is included.

Step 2   Choose a test point from one of the half-planes and not from the line. Substitute its coordinates into the inequality. The line $3x - 5y = 15$ divides the plane into three parts—the line itself and two half-planes. The points in one half-plane satisfy $3x - 5y > 15$. The points in the other half-plane satisfy $3x - 5y < 15$. We need to find which half-plane belongs to the solution of $3x - 5y \geq 15$. To do so, we test a point from either half-plane. The origin, $(0, 0)$, is the easiest point to test.

$$3x - 5y \geq 15 \qquad \text{This is the given inequality.}$$

$$3 \cdot 0 - 5 \cdot 0 \overset{?}{\geq} 15 \qquad \text{Test } (0,0) \text{ by substituting } 0 \text{ for } x \text{ and } 0 \text{ for } y.$$

$$0 - 0 \overset{?}{\geq} 15 \qquad \text{Multiply.}$$

$$0 \geq 15 \qquad \text{This statement is false.}$$

Step 3   If a false statement results, shade the half-plane not containing the test point. Because 0 is not greater than or equal to 15, the test point, $(0, 0)$, is not part of the solution set. Thus, the half-plane below the solid line $3x - 5y = 15$ is part of the solution set. The solution set is the line and the half-plane that does not contain the point $(0, 0)$, indicated by shading this half-plane. The graph is shown using green shading and a blue line in **Figure 3.48**.

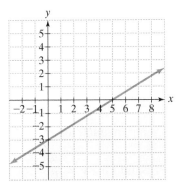

Figure 3.48 The graph of $3x - 5y \geq 15$

> **Check Point 1**   Graph: $2x - 4y \geq 8$.   *

When graphing a linear inequality, test a point that lies in one of the half-planes and *not on the line dividing the half-planes*. The test point $(0, 0)$ is convenient because it is easy to calculate when 0 is substituted for each variable. However, if $(0, 0)$ lies on the dividing line and not in a half-plane, a different test point must be selected.

Example 2   Graphing a Linear Inequality in Two Variables

Graph: $y > -\dfrac{2}{3}x$.

Solution

Step 1   Replace the inequality symbol by = and graph the linear equation. Because we are interested in graphing $y > -\frac{2}{3}x$, we begin by graphing $y = -\frac{2}{3}x$. We can use the slope and the y-intercept to graph this linear function.

$$y = -\frac{2}{3}x + 0$$

$$\text{Slope} = \frac{-2}{3} = \frac{\text{rise}}{\text{run}} \qquad y\text{-intercept} = 0$$

The y-intercept is 0, so the line passes through $(0, 0)$. Using the y-intercept and the slope, the line is shown in **Figure 3.49** as a dashed line. This is because the inequality $y > -\frac{2}{3}x$ contains a $>$ symbol, in which equality is not included.

Step 2   Choose a test point from one of the half-planes and not from the line. Substitute its coordinates into the inequality. We cannot use $(0, 0)$ as a test point because it lies on the line and not in a half-plane. Let's use $(1, 1)$, which lies in the half-plane above the line.

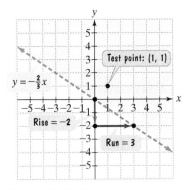

Figure 3.49 The graph of $y > -\frac{2}{3}x$

$$y > -\frac{2}{3}x \qquad \text{This is the given inequality.}$$

$$1 \overset{?}{>} -\frac{2}{3} \cdot 1 \qquad \text{Test } (1, 1) \text{ by substituting 1 for } x \text{ and 1 for } y.$$

$$1 > -\frac{2}{3} \qquad \text{This statement is true.}$$

Step 3   If a true statement results, shade the half-plane containing the test point. Because 1 is greater than $-\frac{2}{3}$, the test point, $(1, 1)$, is part of the solution set. All the points on the same side of the line $y = -\frac{2}{3}x$ as the point $(1, 1)$ are members of the solution set. The solution set is the half-plane that contains the point $(1, 1)$, indicated by shading this half-plane. The graph is shown using green shading and a dashed blue line in **Figure 3.49**.

*See Answers to Selected Exercises.

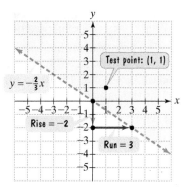

Figure 3.49 (repeated)

Great Question!

When is it important to use test points to graph linear inequalities?

Continue using test points to graph inequalities in the form $Ax + By > C$ or $Ax + By < C$. The graph of $Ax + By > C$ can lie above or below the line given by $Ax + By = C$, depending on the values of A and B. The same comment applies to the graph of $Ax + By < C$.

 **Check Point 2**   Graph: $y > -\dfrac{3}{4}x$. *

Graphing Linear Inequalities without Using Test Points

You can graph inequalities in the form $y > mx + b$ or $y < mx + b$ without using test points. The inequality symbol indicates which half-plane to shade.

- If $y > mx + b$, shade the half-plane above the line $y = mx + b$.
- If $y < mx + b$, shade the half-plane below the line $y = mx + b$.

Observe how this is illustrated in **Figure 3.49**. The graph of $y > -\frac{2}{3}x$ is the half-plane above the line $y = -\frac{2}{3}x$.

It is also not necessary to use test points when graphing inequalities involving half-planes on one side of a vertical or a horizontal line.

For the Vertical Line $x = a$:

- If $x > a$, shade the half-plane to the right of $x = a$.
- If $x < a$, shade the half-plane to the left of $x = a$.

For the Horizontal Line $y = b$:

- If $y > b$, shade the half-plane above $y = b$.
- If $y < b$, shade the half-plane below $y = b$.

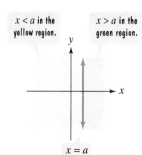

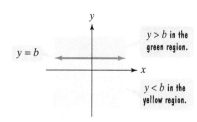

Example 3   Graphing Inequalities without Using Test Points

Graph each inequality in a rectangular coordinate system:

a. $y \le -3$

b. $x > 2$.

Solution

a. $y \le -3$

Graph $y = -3$, a horizontal line with y-intercept -3. The line is solid because equality is included in $y \le -3$. Because of the less than part of $\le$, shade the half-plane below the horizontal line.

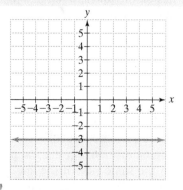

b. $x > 2$

Graph $x = 2$, a vertical line with x-intercept 2. The line is dashed because equality is not included in $x > 2$. Because of $>$, the greater than symbol, shade the half-plane to the right of the vertical line.

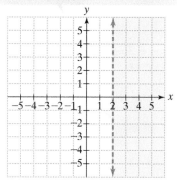

Check Point 3   Graph each inequality in a rectangular coordinate system:

a. $y > 1$ *   **b.** $x \le -2$. *

*See Answers to Selected Exercises.

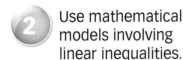

Use mathematical models involving linear inequalities.

Modeling with Systems of Linear Inequalities

Just as two or more linear equations make up a system of linear equations, two or more linear inequalities make up a **system of linear inequalities**. A **solution of a system of linear inequalities** in two variables is an ordered pair that satisfies each inequality in the system.

Example 4   Does Your Weight Fit You?

The latest guidelines, which apply to both men and women, give healthy weight ranges, rather than specific weights, for your height. **Figure 3.50** shows the healthy weight region for various heights for people between the ages of 19 and 34, inclusive.

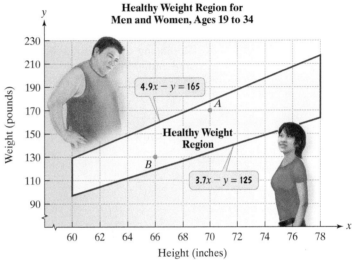

Figure 3.50
Source: U.S. Department of Health and Human Services

If x represents height, in inches, and y represents weight, in pounds, the healthy weight region in **Figure 3.50** can be modeled by the following system of linear inequalities:

$$\begin{cases} 4.9x - y \geq 165 \\ 3.7x - y \leq 125. \end{cases}$$

Show that point A in **Figure 3.50** is a solution of the system of inequalities that describes healthy weight.

Solution

Point A has coordinates $(70, 170)$. This means that if a person is 70 inches tall, or 5 feet 10 inches, and weighs 170 pounds, then that person's weight is within the healthy weight region. We can show that $(70, 170)$ satisfies the system of inequalities by substituting 70 for x and 170 for y in each inequality in the system.

$$4.9x - y \geq 165 \qquad\qquad 3.7x - y \leq 125$$
$$4.9(70) - 170 \overset{?}{\geq} 165 \qquad\qquad 3.7(70) - 170 \overset{?}{\leq} 125$$
$$343 - 170 \overset{?}{\geq} 165 \qquad\qquad 259 - 170 \overset{?}{\leq} 125$$
$$173 \geq 165, \quad \text{true} \qquad\qquad 89 \leq 125, \quad \text{true}$$

The coordinates $(70, 170)$ make each inequality true. Thus, $(70, 170)$ satisfies the system for the healthy weight region and is a solution of the system.

✓ **Check Point 4**   Show that point B in **Figure 3.50** is a solution of the system of inequalities that describes healthy weight.

Point $B = (66, 130)$; $4.9(66) - 130 \geq 165$, or $193.4 \geq 165$, is true; $3.7(66) - 130 \leq 125$, or $114.2 \leq 125$, is true.

3 Graph a system of linear inequalities.

Graphing Systems of Linear Inequalities

The **solution set of a system of linear inequalities in two variables** is the set of all ordered pairs that satisfy each inequality in the system. Thus, to graph a system of inequalities in two variables, begin by graphing each individual inequality in the same rectangular coordinate system. Then find the region, if there is one, that is common to every graph in the system. This region of intersection gives a picture of the system's solution set.

Example 5 Graphing a System of Linear Inequalities

Graph the solution set of the system:

$$\begin{cases} x - y < 1 \\ 2x + 3y \geq 12. \end{cases}$$

Solution

Replacing each inequality symbol in $x - y < 1$ and $2x + 3y \geq 12$ with an equal sign indicates that we need to graph $x - y = 1$ and $2x + 3y = 12$. We can use intercepts to graph these lines.

$x - y = 1$

x-intercept: $x - 0 = 1$ *(Set y = 0 in each equation.)*
$x = 1$
The line passes through $(1, 0)$.

y-intercept: $0 - y = 1$ *(Set x = 0 in each equation.)*
$-y = 1$
$y = -1$
The line passes through $(0, -1)$.

$2x + 3y = 12$

x-intercept: $2x + 3 \cdot 0 = 12$
$2x = 12$
$x = 6$
The line passes through $(6, 0)$.

y-intercept: $2 \cdot 0 + 3y = 12$
$3y = 12$
$y = 4$
The line passes through $(0, 4)$.

Now we are ready to graph the solution set of the system of linear inequalities.

Graph $x - y < 1$. The blue line, $x - y = 1$, is dashed: Equality is not included in $x - y < 1$. Because $(0, 0)$ makes the inequality true $(0 - 0 < 1$, or $0 < 1$, is true), shade the half-plane containing $(0, 0)$ in yellow.

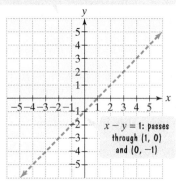

The graph of $x - y < 1$

Add the graph of $2x + 3y \geq 12$. The red line, $2x + 3y = 12$, is solid: Equality is included in $2x + 3y \geq 12$. Because $(0, 0)$ makes the inequality false $(2 \cdot 0 + 3 \cdot 0 \geq 12$, or $0 \geq 12$, is false), shade the half-plane not containing $(0, 0)$ using green vertical shading.

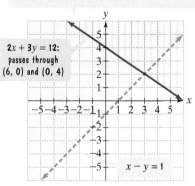

Adding the graph of $2x + 3y \geq 12$

The solution set of the system is graphed as the intersection (the overlap) of the two half-planes. This is the region in which the yellow shading and the green vertical shading overlap.

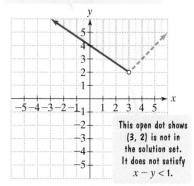

The graph of $x - y < 1$ and $2x + 3y \geq 12$

Check Point 5 Graph the solution set of the system:

$$\begin{cases} x + 2y > 4 \\ 2x - 3y \leq -6. \end{cases} \quad *$$

*See Answers to Selected Exercises.

Example 6 **Graphing a System of Linear Inequalities**

Graph the solution set of the system:

$$\begin{cases} x \le 4 \\ y > -2. \end{cases}$$

Solution

Graph $x \le 4$. The blue vertical line, $x = 4$, is solid. Graph $x < 4$, the half-plane to the left of the blue line, using yellow shading.

Add the graph of $y > -2$. The red horizontal line, $y = -2$, is dashed. Graph $y > -2$, the half-plane above the dashed red line, using green vertical shading.

The solution set of the system is graphed as the intersection (the overlap) of the two half-planes. This is the region in which the yellow shading and the green vertical shading overlap.

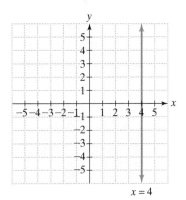

The graph of $x \le 4$

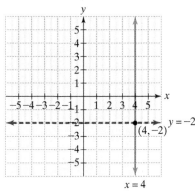

Adding the graph of $y > -2$

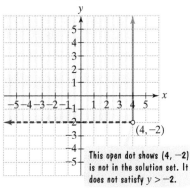

This open dot shows $(4, -2)$ is not in the solution set. It does not satisfy $y > -2$.

The graph of $x \le 4$ and $y > -2$

Check Point 6 Graph the solution set of the system:

$$\begin{cases} x < 3 \\ y \ge -1. \end{cases} \ *$$

Achieving Success

Assuming that you have done very well preparing for an exam, **there are certain things you can do that will make you a better test taker**.

- Get a good sleep the night before the exam.
- Have a good breakfast that balances protein, carbohydrates, and fruit.
- Just before the exam, briefly review the relevant material in the chapter summary.
- Bring everything you need to the exam, including two pencils, an eraser, scratch paper (if permitted), a calculator (if you're allowed to use one), water, and a watch.
- Survey the entire exam quickly to get an idea of its length.
- Read the directions to each problem carefully. Make sure that you have answered the specific question asked.
- Work the easy problems first. Then return to the hard problems you are not sure of. Doing the easy problems first will build your confidence. If you get bogged down on any one problem, you may not be able to complete the exam and receive credit for the questions you can easily answer.
- Attempt every problem. There may be partial credit even if you do not obtain the correct answer.
- Work carefully. Show your step-by-step solutions neatly. Check your work and answers.
- Watch the time. Pace yourself and be aware of when half the time is up. Determine how much of the exam you have completed. This will indicate if you're moving at a good pace or need to speed up. Prepare to spend more time on problems worth more points.
- Never turn in a test early. Use every available minute you are given for the test. If you have extra time, double check your arithmetic and look over your solutions.

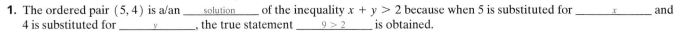

Concept and Vocabulary Check

Exercises in the Concept and Vocabulary Check are intended for group and class discussions.

In Exercises 1–4, fill in each blank so that the resulting statement is true.

1. The ordered pair $(5, 4)$ is a/an ____solution____ of the inequality $x + y > 2$ because when 5 is substituted for ____x____ and 4 is substituted for ____y____, the true statement ____$9 > 2$____ is obtained.

2. The set of all points that satisfy an inequality is called the ____graph____ of the inequality.

3. The set of all points on one side of a line is called a/an ____half-plane____.

4. Two or more linear inequalities make up a/an ____system____ of linear inequalities.

In Exercises 5–8, determine whether each statement is true or false. If the statement is false, make the necessary change(s) to produce a true statement. Changes to false statements will vary.

5. The ordered pair $(0, -3)$ satisfies $y > 2x - 3$.   false

6. The graph of $x < y + 1$ is the half-plane below the line $x = y + 1$.   false

7. In graphing $y \geq 4x$, a dashed line is used.   false

8. The graph of $x < 4$ is the half-plane to the left of the vertical line described by $x = 4$.   true

Respond to Exercises 9–14 using verbal or written explanations.

9. What is a half-plane?   **9–14.** Answers will vary.

10. What does a dashed line mean in the graph of an inequality?

11. What does a solid line mean in the graph of an inequality?

12. Explain how to graph $2x - 3y < 6$.

13. Compare the graphs of $3x - 2y > 6$ and $3x - 2y \leq 6$. Discuss similarities and differences between the graphs.

14. Describe how to solve a system of linear inequalities.

Exercise Set 3.5

Practice Exercises

In Exercises 1–22, graph each linear inequality.

1. $x + y \geq 2$ *

2. $x - y \leq 1$ *

3. $3x - y \geq 6$ *

4. $3x + y \leq 3$ *

5. $2x + 3y > 12$ *

6. $2x - 5y < 10$ *

7. $5x + 3y \leq -15$ *

8. $3x + 4y \leq -12$ *

9. $2y - 3x > 6$ *

10. $2y - x > 4$ *

11. $y > \dfrac{1}{3}x$ *

12. $y > \dfrac{1}{4}x$ *

13. $y \leq 3x + 2$ *

14. $y \leq 2x - 1$ *

15. $y < -\dfrac{1}{4}x$ *

16. $y < -\dfrac{1}{3}x$ *

17. $x \leq 2$ *

18. $x \leq -4$ *

19. $y > -4$ *

20. $y > -2$ *

21. $y \geq 0$ *

22. $x \geq 0$ *

In Exercises 23–38, graph the solution set of each system of inequalities.

23. $\begin{cases} 3x + 6y \leq 6 \\ 2x + y \leq 8 \end{cases}$ *

24. $\begin{cases} x - y \geq 4 \\ x + y \leq 6 \end{cases}$ *

25. $\begin{cases} 2x + y < 3 \\ x - y > 2 \end{cases}$ *

26. $\begin{cases} x + y < 4 \\ 4x - 2y < 6 \end{cases}$ *

27. $\begin{cases} 2x + y < 4 \\ x - y > 4 \end{cases}$ *

28. $\begin{cases} 2x - y < 3 \\ x + y < 6 \end{cases}$ *

29. $\begin{cases} x \geq 2 \\ y \leq 3 \end{cases}$ *

30. $\begin{cases} x \geq 4 \\ y \leq 2 \end{cases}$ *

31. $\begin{cases} x \leq 5 \\ y > -3 \end{cases}$ *

32. $\begin{cases} x \leq 3 \\ y > -1 \end{cases}$ *

33. $\begin{cases} x - y \leq 1 \\ x \geq 2 \end{cases}$ *

34. $\begin{cases} 4x - 5y \geq -20 \\ x \geq -3 \end{cases}$ *

35. $\begin{cases} y > 2x - 3 \\ y < -x + 6 \end{cases}$ *

36. $\begin{cases} y < -2x + 4 \\ y < x - 4 \end{cases}$ *

37. $\begin{cases} x + 2y \leq 4 \\ y \geq x - 3 \end{cases}$ *

38. $\begin{cases} x + y \leq 4 \\ y \geq 2x - 4 \end{cases}$ *

Practice Plus

In Exercises 39–40, write each sentence as an inequality in two variables. Then graph the inequality.

39. The y-variable is at least 4 more than the product of -2 and the x-variable. $y \geq -2x + 4$ *

40. The y-variable is at least 2 more than the product of -3 and the x-variable. $y \geq -3x + 2$ *

In Exercises 41–42, write the given sentences as a system of inequalities in two variables. Then graph the system.

41. The sum of the x-variable and the y-variable is at most 4. The y-variable added to the product of 3 and the x-variable does not exceed 6. $x + y \leq 4, 3x + y \leq 6$ *

42. The sum of the x-variable and the y-variable is at most 3. The y-variable added to the product of 4 and the x-variable does not exceed 6. $x + y \leq 3, 4x + y \leq 6$ *

*See Answers to Selected Exercises.

Application Exercises

The figure shows the healthy weight region for various heights for people ages 35 and older.

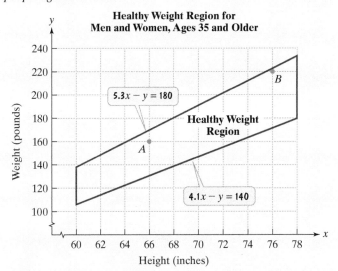

Healthy Weight Region for Men and Women, Ages 35 and Older

Source: U.S. Department of Health and Human Services

If x represents height, in inches, and y represents weight, in pounds, the healthy weight region can be modeled by the following system of linear inequalities:

$$\begin{cases} 5.3x - y \geq 180 \\ 4.1x - y \leq 140. \end{cases}$$

Use this information to solve Exercises 43–46.

47. a. $50x + 150y > 2000$

43. Show that point *A* is a solution of the system of inequalities that describes healthy weight for this age group.   *

44. Show that point *B* is a solution of the system of inequalities that describes healthy weight for this age group.   *

45. Is a person in this age group who is 6 feet tall weighing 205 pounds within the healthy weight region?   no

46. Is a person in this age group who is 5 feet 8 inches tall weighing 135 pounds within the healthy weight region?   no

47. Many elevators have a capacity of 2000 pounds.
   a. If a child averages 50 pounds and an adult 150 pounds, write an inequality that describes when *x* children and *y* adults will cause the elevator to be overloaded.
   b. Graph the inequality. Because *x* and *y* must be positive, limit the graph to quadrant I only.   *
   c. Select an ordered pair satisfying the inequality. What are its coordinates and what do they represent in this situation?   Answers will vary.

48. A patient is not allowed to have more than 330 milligrams of cholesterol per day from a diet of eggs and meat. Each egg provides 165 milligrams of cholesterol. Each ounce of meat provides 110 milligrams.   **a.** $165x + 110y \leq 330$
   a. Write an inequality that describes the patient's dietary restrictions for *x* eggs and *y* ounces of meat.
   b. Graph the inequality. Because *x* and *y* must be positive, limit the graph to quadrant I only.   *
   c. Select an ordered pair satisfying the inequality. What are its coordinates and what do they represent in this situation?   Answers will vary.

The graph of an inequality in two variables is a region in the rectangular coordinate system. Regions in coordinate systems have numerous applications. For example, the regions in the following two graphs indicate whether a person is obese, overweight, borderline overweight, normal weight, or underweight.

Each horizontal axis shows a person's age. Each vertical axis shows that person's body-mass index (BMI), computed using the following formula:

$$BMI = \frac{703W}{H^2}.$$

The variable W represents weight, in pounds. The variable H represents height, in inches. Use this information and the graphs on the right to solve Exercises 49–50.

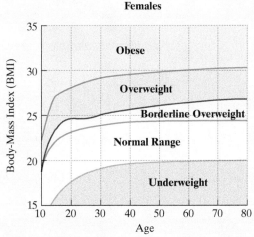

Females

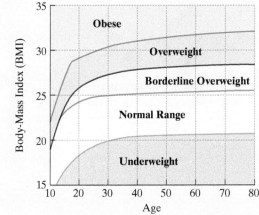

Males

Source: Centers for Disease Control and Prevention

49. A man is 20 years old, 72 inches (6 feet) tall, and weighs 200 pounds.
   a. Compute the man's BMI. Round to the nearest tenth.
   27.1
   b. Use the man's age and his BMI to locate this information as a point in the coordinate system for males. Is this person obese, overweight, borderline overweight, normal weight, or underweight?   overweight

50. A woman is 25 years old, 66 inches (5 feet, 6 inches) tall, and weighs 105 pounds.
   a. Compute the woman's BMI. Round to the nearest tenth.
   16.9
   b. Use the woman's age and her BMI to locate this information as a point in the coordinate system for females. Is this person obese, overweight, borderline overweight, normal weight, or underweight?   underweight

*See Answers to Selected Exercises.

Critical Thinking Exercises

In Exercises 51–52, write a system of inequalities for each graph.

51.

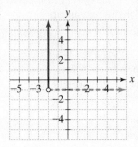

52. *

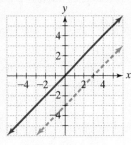

Without graphing, in Exercises 53–56, determine if each system has no solution or infinitely many solutions.

53. $\begin{cases} 3x + y < 9 \\ 3x + y > 9 \end{cases}$ no solution

54. $\begin{cases} 6x - y \le 24 \\ 6x - y > 24 \end{cases}$ no solution

55. $\begin{cases} 3x + y \le 9 \\ 3x + y \ge 9 \end{cases}$ infinitely many solutions

56. $\begin{cases} 6x - y \le 24 \\ 6x - y \ge 24 \end{cases}$ infinitely many solutions

---

Chapter 3 Summary

| **Definitions and Concepts** | **Examples** |

Section 3.1 Graphing and Functions

The rectangular coordinate system is formed using two number lines that intersect at right angles at their zero points. The horizontal line is the *x*-axis and the vertical line is the *y*-axis. Their point of intersection, $(0, 0)$, is the origin. Each point in the system corresponds to an ordered pair of real numbers, (x, y).

• Plot: $(2, 3)$, $(-5, 4)$, $(-4\ -3)$, and $(5, -2)$.

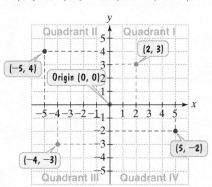

Additional Example to Review

Example 1, page 179

The graph of an equation in two variables is the set of all points whose coordinates satisfy the equation.

• Graph: $y = x^2 - 1$.

x	$y = x^2 - 1$	**Ordered Pair (x, y)**
-2	$(-2)^2 - 1 = 3$	$(-2, 3)$
-1	$(-1)^2 - 1 = 0$	$(-1, 0)$
0	$0^2 - 1 = -1$	$(0, -1)$
1	$1^2 - 1 = 0$	$(1, 0)$
2	$2^2 - 1 = 3$	$(2, 3)$

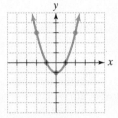

Additional Examples to Review

Example 2, page 180;
Example 3, page 180

If an equation in x and y yields one value of y for each value of x, then y is a function of x, indicated by writing $f(x)$ for y.

- $f(x) = 5x - 2;\quad f(-4) = 5(-4) - 2 = -20 - 2 = -22$
- $g(x) = 2x^2 - 5x - 1;\quad g(-3) = 2(-3)^2 - 5(-3) - 1$
$$= 2(9) - 5(-3) - 1$$
$$= 18 - (-15) - 1$$
$$= 18 + 15 - 1 = 32$$
- $h(x) = 6;\quad h(-3) = 6, h(0) = 6, h(100) = 6$

Additional Examples to Review
Example 4, page 182; Example 5, page 183

The graph of a function is the graph of its ordered pairs.

- Graph: $f(x) = |x| + 2$.
  Select integers for x from -2 to 2, inclusive.

x	$f(x) = |x| + 2$	(x, y)
-2	$f(-2) = |-2| + 2 = 2 + 2 = 4$	$(-2, 4)$
-1	$f(-1) = |-1| + 2 = 1 + 2 = 3$	$(-1, 3)$
0	$f(0) = |0| + 2 = 0 + 2 = 2$	$(0, 2)$
1	$f(2) = |1| + 2 = 1 + 2 = 3$	$(1, 3)$
2	$f(2) = |2| + 2 = 2 + 2 = 4$	$(2, 4)$

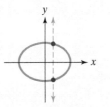

The graph of $f(x) = |x| + 2$

Additional Example to Review
Example 6, page 184

The Vertical Line Test: If any vertical line intersects a graph in more than one point, the graph does not define y as a function of x.

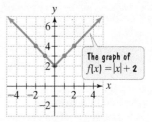

Not the graph of a function   The graph of a function

Additional Example to Review
Example 7, page 185

Section 3.2 Linear Functions and Their Graphs

A function whose graph is a straight line is a linear function.

If a graph intersects the x-axis at $(a, 0)$, then a is an x-intercept.

If a graph intersects the y-axis at $(0, b)$, then b is a y-intercept.

The graph of $Ax + By = C$, a linear equation in two variables, is a straight line as long as A and B are not both zero. If none of A, B, and C is zero, the line can be graphed using intercepts and a checkpoint. To locate the x-intercept, set $y = 0$ and solve for x. To locate the y-intercept, set $x = 0$ and solve for y.

Graph using intercepts: $4x + 3y = 12$.

x-intercept: $\quad 4x = 12$ [Line passes through (3, 0).]
(Set $y = 0$.) $\quad\quad x = 3$

y-intercept: $\quad 3y = 12$ [Line passes through (0, 4).]
(Set $x = 0$.) $\quad\quad y = 4$

Checkpoint: Let $x = 2$.

$$4 \cdot 2 + 3y = 12$$
$$8 + 3y = 12$$
$$3y = 4$$
$$y = \frac{4}{3}$$

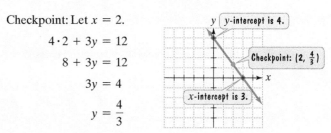

Checkpoint is $\left(2, \frac{4}{3}\right)$, or $\left(2, 1\frac{1}{3}\right)$.

Additional Example to Review

Example 1, page 194

The slope of the line through (x_1, y_1) and (x_2, y_2) is

$$m = \frac{\text{Rise}}{\text{Run}} = \frac{y_2 - y_1}{x_2 - x_1}.$$

If the slope is positive, the line rises from left to right. If the slope is negative, the line falls from left to right. The slope of a horizontal line is 0. The slope of a vertical line is undefined.

• For points $(-7, 2)$ and $(3, -4)$, the slope of the line through the points is:

$$m = \frac{\text{Change in } y}{\text{Change in } x} = \frac{-4 - 2}{3 - (-7)} = \frac{-6}{10} = -\frac{3}{5}.$$

The slope is negative, so the line falls.

• For points $(2, -5)$ and $(2, 16)$, the slope of the line through the points is:

$$m = \frac{\text{Change in } y}{\text{Change in } x} = \frac{16 - (-5)}{2 - 2} = \frac{21}{0}.$$

  [undefined]

The slope is undefined, so the line is vertical.

Additional Examples to Review

Example 2, page 195; Example 7, page 200

The equation $y = mx + b$ is the slope-intercept form of the equation of a line, in which m is the slope and b is the y-intercept. Using function notation, the equation is $f(x) = mx + b$.

• Graph: $f(x) = -\dfrac{3}{4}x + 1$.

  [Slope is $-\frac{3}{4}$.]   [y-intercept is 1.]

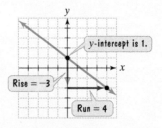

Additional Examples to Review

Example 3, page 197; Example 4, page 198; Example 8, page 201

Horizontal and Vertical Lines

1. The graph of $y = b$ is a horizontal line that intersects the y-axis at $(0, b)$.
2. The graph of $x = a$ is a vertical line that intersects the x-axis at $(a, 0)$.

- Graph: $y = 3$.

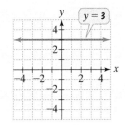

- Graph: $x = -2$.

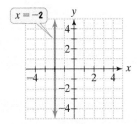

Additional Examples to Review

Example 5, page 199; Example 6, page 199

Section 3.3 The Point-Slope Form of the Equation of a Line; Scatter Plots and Regression Lines

The point-slope form of the equation of a nonvertical line with slope m that passes through the point (x_1, y_1) is

$$y - y_1 = m(x - x_1).$$

- Slope $= -4$, passing through $(-1, 5)$

  $m = -4$ $x_1 = -1$ $y_1 = 5$

  The point-slope form of the line's equation is

  $$y - 5 = -4[x - (-1)].$$

  Simplify:

  $$y - 5 = -4(x + 1).$$

Additional Example to Review

Example 1, page 207

To write the point-slope form of the line passing through two points, begin by using the points to compute the slope, m. Use either given point as (x_1, y_1) and write the point-slope equation:

$$y - y_1 = m(x - x_1).$$

Solving this equation for y gives the slope-intercept form of the line's equation.

- Write equations in point-slope form and in slope-intercept form of the line passing through $(4, 1)$ and $(3, -2)$.

  $$m = \frac{-2 - 1}{3 - 4} = \frac{-3}{-1} = 3$$

  Using $(4, 1)$ as (x_1, y_1), the point-slope form of the equation is

  $$y - 1 = 3(x - 4).$$

  Solve for y to obtain the slope-intercept form.

  $$y - 1 = 3x - 12$$
  $$y = 3x - 11$$

Additional Examples to Review

Example 2, page 207; Example 3, page 209

A plot of data points is called a scatter plot. If the points lie approximately along a line, the line that best fits the data is called a regression line.

A correlation coefficient, r, measures the strength and direction of a possible relationship between variables. If $r = 1$, there is a perfect positive correlation, and if $r = -1$, there is a perfect negative correlation. If $r = 0$, there is no relationship between the variables.

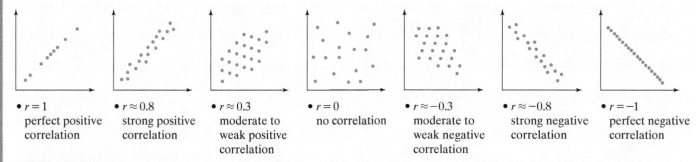

- $r = 1$
  perfect positive correlation

- $r \approx 0.8$
  strong positive correlation

- $r \approx 0.3$
  moderate to weak positive correlation

- $r = 0$
  no correlation

- $r \approx -0.3$
  moderate to weak negative correlation

- $r \approx -0.8$
  strong negative correlation

- $r = -1$
  perfect negative correlation

Additional Example to Review

Example 4, page 212

Section 3.4 Systems of Linear Equations in Two Variables

Two equations in the form $Ax + By = C$ are called a system of linear equations. A solution of the system is an ordered pair that satisfies both equations in the system.

- Determine whether $(3, -1)$ is a solution of

$$\begin{cases} 2x + 5y = 1 \\ 4x + y = 11. \end{cases}$$

Replace x with 3 and y with -1 in both equations.

$$2x + 5y = 1 \qquad\qquad 4x + y = 11$$
$$2 \cdot 3 + 5(-1) \overset{?}{=} 1 \qquad\qquad 4 \cdot 3 + (-1) \overset{?}{=} 11$$
$$6 + (-5) \overset{?}{=} 1 \qquad\qquad 12 + (-1) \overset{?}{=} 11$$
$$1 = 1, \;\text{true} \qquad\qquad 11 = 11, \;\text{true}$$

Thus, $(3, -1)$ is a solution of the system.

Additional Example to Review

Example 1, page 218

Linear systems with one solution can be solved by graphing. The coordinates of the point of intersection of the lines are the system's solution.

- Solve by graphing:

$$\begin{cases} 2x + y = 4 \\ x + y = 2. \end{cases}$$

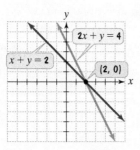

The solution set is $\{(2, 0)\}$.

Additional Example to Review

Example 2, page 218

Solving Linear Systems by the Substitution Method

1. Solve either of the equations for one variable in terms of the other. (If one of the equations is already in this form, you can skip this step.)
2. Substitute the expression found in step 1 into the *other* equation. This will result in an equation in one variable.
3. Solve the equation containing one variable.
4. Back-substitute the value found in step 3 into the equation from step 1. Simplify and find the value of the remaining variable.
5. Check the proposed solution in both of the system's given equations.

• Solve by the substitution method:

$$\begin{cases} y = 2x + 3 \\ 7x - 5y = -18. \end{cases}$$

Substitute $2x + 3$ for y in the second equation.

$$7x - 5(2x + 3) = -18$$
$$7x - 10x - 15 = -18$$
$$-3x - 15 = -18$$
$$-3x = -3$$
$$x = 1$$

Find y. Substitute 1 for x in $y = 2x + 3$.

$$y = 2 \cdot 1 + 3 = 2 + 3 = 5$$

The solution, $(1, 5)$, checks. The solution set is $\{(1, 5)\}$.

Additional Examples to Review

Example 3, page 219; Example 4, page 220

Solving Linear Systems by the Addition Method

1. If necessary, rewrite both equations in the form $Ax + By = C$.
2. If necessary, multiply either equation or both equations by appropriate nonzero numbers so that the sum of the x-coefficients or the sum of the y-coefficients is 0.
3. Add the equations in step 2. The sum is an equation in one variable.
4. Solve the equation in one variable.
5. Back-substitute the value obtained in step 4 into either of the given equations and solve for the other variable.
6. Check the solution in both of the original equations.

• Solve by the addition method:

$$\begin{cases} 2x + y = 10 \xrightarrow{\text{Multiply by } -4.} \\ 3x + 4y = 25 \xrightarrow{\text{No change}} \end{cases} \begin{cases} -8x - 4y = -40 \\ \underline{3x + 4y = 25} \\ \text{Add:} \quad -5x = -15 \\ x = 3 \end{cases}$$

Find y. Back-substitute 3 for x. Use the first equation, $2x + y = 10$.

$$2(3) + y = 10$$
$$6 + y = 10$$
$$y = 4$$

The solution, $(3, 4)$, checks. The solution set is $\{(3, 4)\}$.

Additional Examples to Review

Example 5, page 222; Example 6, page 223

When solving by substitution or addition, if the variable is eliminated and a false statement results, the linear system has no solution. If the variable is eliminated and a true statement results, the system has infinitely many solutions.

• Solve:

$$\begin{cases} x + 2y = 4 \xrightarrow{\text{Multiply by } -3.} \\ 3x + 6y = 13. \xrightarrow{\text{No change}} \end{cases} \begin{cases} -3x - 6y = -12 \\ \underline{3x + 6y = 13} \\ \text{Add:} \quad 0 = 1 \text{ false} \end{cases}$$

No solution: $\varnothing$

• Solve:

$$\begin{cases} y = 4x - 4 \\ 4x - y = 4. \end{cases}$$

Substitute $4x - 4$ for y in the second equation.

$$4x - (4x - 4) = 4$$
$$4x - 4x + 4 = 4$$
$$4 = 4, \quad \text{true}$$

Solution set: $\{(x, y) \mid y = 4x - 4\}$ or $\{(x, y) \mid 4x - y = 4\}$

Additional Examples to Review

Example 7, page 224; Example 8, page 225

Functions of Business

A company produces and sells x units of a product.

Revenue Function:

$R(x) = (\text{price per unit sold})x$

Cost Function:

$C(x) = \text{fixed cost} + (\text{cost per unit produced})x$

Profit Function:

$P(x) = R(x) - C(x)$

The point of intersection of the graphs of R and C is the break-even point. The x-coordinate of the point reveals the number of units that a company must produce and sell so that the money coming in, the revenue, is equal to the money going out, the cost. The y-coordinate gives the amount of money coming in and going out.

- A company that manufactures lamps has a fixed cost of $80,000 and it costs $20 to produce each lamp. Lamps are sold for $70.

  a. Write the cost function.

  $$C(x) = 80,000 + 20x$$

  | Fixed cost | Variable cost: $20 per lamp |

  b. Write the revenue function.

  $$R(x) = 70x$$

  | Revenue per lamp, $70, times number of lamps sold |

  c. Find the break-even point.
  Solve

  $$\begin{cases} y = 80,000 + 20x \\ y = 70x \end{cases}$$

  by substitution. Solving

  $$70x = 80,000 + 20x$$

  yields $x = 1600$. Back-substituting yields $y = 112,000$. The break-even point is $(1600, 112,000)$: The company breaks even if it sells 1600 lamps. At this level, money coming in equals money going out: $112,000.

Additional Example to Review

Example 9, page 226

Section 3.5 Linear Inequalities in Two Variables

A linear inequality in two variables can be written in the form $Ax + By > C$, $Ax + By \geq C$, $Ax + By < C$, or $Ax + By \leq C$.

Graphing a Linear Inequality in Two Variables

  1. Replace the inequality symbol with an equal sign and graph the corresponding linear equation. Draw a solid line if the original inequality contains a $\leq$ or $\geq$ symbol. Draw a dashed line if the original inequality contains a $<$ or $>$ symbol.

  2. Choose a test point from one of the half-planes. (Do not choose a point on the line.) Substitute the coordinates of the test point into the inequality.

  3. If a true statement results, shade the half-plane containing this test point. If a false statement results, shade the half-plane not containing this test point.

- Graph: $x - 2y \leq 4$.

  1. Graph $x - 2y = 4$. Use a solid line because the inequality symbol is $\leq$.

  2. Test $(0,0)$.

  $$x - 2y \leq 4$$
  $$0 - 2 \cdot 0 \overset{?}{\leq} 4$$
  $$0 \leq 4, \quad \text{true}$$

  3. The inequality is true. Shade the half-plane containing $(0,0)$.

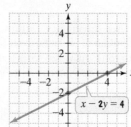

Additional Examples to Review

Example 1, page 232; Example 2, page 233

Some inequalities can be graphed without using test points, including $y > mx + b$ (the half-plane above $y = mx + b$), $y < mx + b$, $x > a$ (the half-plane to the right of $x = a$), $x < a$, $y > b$ (the half-plane above $y = b$), and $y < b$.

• Graph: $y < 2$.

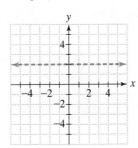

• Graph: $x \geq -1$.

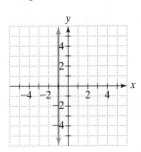

Additional Example to Review

Example 3, page 234

Two or more linear inequalities make up a system of linear inequalities. A solution is an ordered pair satisfying all inequalities in the system.

Graphing Systems of Linear Inequalities

1. Graph each inequality in the system in the same rectangular coordinate system.
2. Find the intersection of the individual graphs.

• Graph the solution set of the system:

$$\begin{cases} y \leq -2x \\ x - y \geq 3. \end{cases}$$

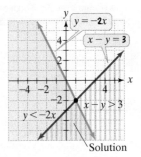

Additional Examples to Review

Example 4, page 235; Example 5, page 236; Example 6, page 237

Review Exercises

Section 3.1 Graphing and Functions

In Exercises 1–4, plot the given point in a rectangular coordinate system.

1. $(2, 5)$ *

2. $(-4, 3)$ *

3. $(-5, -3)$ *

4. $(2, -5)$ *

Graph each equation in Exercises 5–7. Let $x = -3, -2, -1, 0, 1, 2,$ and 3.

5. $y = 2x - 2$ *

6. $y = |x| + 2$ *

7. $y = x$ *

8. If $f(x) = 4x + 11$, find $f(-2)$.   3

9. If $f(x) = -7x + 5$, find $f(-3)$.   26

10. If $f(x) = 3x^2 - 5x + 2$, find $f(4)$.   30

11. If $f(x) = -3x^2 + 6x + 8$, find $f(-4)$.   −64

In Exercises 12–13, evaluate $f(x)$ for the given values of x. Then use the ordered pairs $(x, f(x))$ from your table to graph the function.

12. $f(x) = \frac{1}{2}|x|$ *

x	$f(x) = \frac{1}{2}|x|$
-6	3
-4	2
-2	1
0	0
2	1
4	2
6	3

13. $f(x) = x^2 - 2$ *

x	$f(x) = x^2 - 2$
-2	2
-1	-1
0	-2
1	-1
2	2

*See Answers to Selected Exercises.

In Exercises 14–15, use the vertical line test to identify graphs in which y is a function of x.

14.

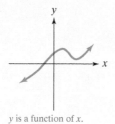

y is a function of x.

15.

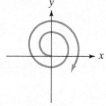

y is not a function of x.

16. Whether on the slopes or at the shore, people are exposed to harmful amounts of the sun's skin-damaging ultraviolet (UV) rays. The quadratic function

$$D(x) = 0.8x^2 - 17x + 109$$

models the average time in which skin damage begins for burn-prone people, $D(x)$, in minutes, where x is the UV index, or measure of the sun's UV intensity. The graph of D is shown for a UV index from 1 (low) to 11 (high).

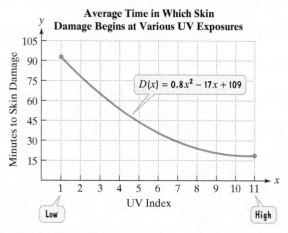

Average Time in Which Skin Damage Begins at Various UV Exposures

$D(x) = 0.8x^2 - 17x + 109$

Source: National Oceanic and Atmospheric Administration

a. Find and interpret $D(1)$. How is this shown on the graph of D?

b. Find and interpret $D(10)$. How is this shown on the graph of D?

Section 3.2 Linear Functions and Their Graphs

In Exercises 17–19, use the x- and y-intercepts to graph each linear equation.

17. $2x + y = 4$ *

18. $2x - 3y = 6$ *

19. $5x - 3y = 15$ *

In Exercises 20–23, calculate the slope of the line passing through the given points. If the slope is undefined, so state. Then indicate whether the line rises, falls, is horizontal, or is vertical.

20. $(3, 2)$ and $(5, 1)$ $-\frac{1}{2}$; falls

21. $(-1, 2)$ and $(-3, -4)$ 3; rises

22. $(-3, 4)$ and $(6, 4)$ 0; horizontal

23. $(5, 3)$ and $(5, -3)$ undefined; vertical

In Exercises 24–27, graph each linear function using the slope and y-intercept.

24. $y = 2x - 4$ *

25. $y = -\frac{2}{3}x + 5$ *

26. $f(x) = \frac{3}{4}x - 2$ *

27. $y = \frac{1}{2}x + 0$ *

*In Exercises 28–30, **a.** Write the equation in slope-intercept form; **b.** Identify the slope and the y-intercept; **c.** Use the slope and y-intercept to graph the line.*

28. $2x + y = 0$ *   **29.** $3y = 5x$ *

30. $3x + 2y = 4$ *

In Exercises 31–33, graph each horizontal or vertical line.

31. $x = 3$ *

32. $y = -4$ *

33. $x + 2 = 0$ *

34. The scatter plot indicates a relationship between the percentage of adult females in a country who are literate and the mortality of children under five. Also shown is a line that passes through or near the points.

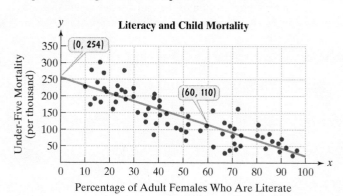

Literacy and Child Mortality

$(0, 254)$

$(60, 110)$

Source: United Nations

a. According to the graph, what is the y-intercept of the line? Describe what this represents in this situation. *

b. Use the coordinates of the two points shown to compute the slope of the line. Describe what this means in terms of the rate of change. *

c. Use the y-intercept from part (a) and the slope from part (b) to write a linear function that models child mortality, $f(x)$, per thousand, for children under five in a country where $x\%$ of adult women are literate. $f(x) = -2.4x + 254$

d. Use the function from part (c) to predict the mortality rate of children under five in a country where 50% of adult females are literate.   134 per thousand

Section 3.3 The Point-Slope Form of the Equation of a Line; Scatter Plots and Regression Lines

In Exercises 35–36, use the given conditions to write an equation for each line in point-slope form and in slope-intercept form.

35. Passing through $(-3, 2)$ with slope -6 35. $y - 2 = -6(x + 3); y = -6x - 16$

36. Passing through $(1, 6)$ and $(-1, 2)$
$y - 6 = 2(x - 1)$ or $y - 2 = 2(x + 1); y = 2x + 4$

16. a. $D(1) = 92.8$; Skin damage begins for burn-prone people after 92.8 minutes when the sun's UV index is 1.; by the point $(1, 92.8)$

16. b. $D(10) = 19$; Skin damage begins for burn-prone people after 19 minutes when the sun's UV index is 10.; by the point $(10, 19)$

*See Answers to Selected Exercises.

37. a. $y - 2.3 = 0.116(x - 15)$ or $y - 11 = 0.116(x - 90)$ **c.** ≈ 5 deaths per 100,000 persons

d. 4.3 deaths per 100,000 persons; underestimates by 0.7 death per 100,000 persons; The line passes below the point for France.

37. Shown, again, is the scatter plot that indicates a relationship between the number of firearms per 100 persons and the number of deaths per 100,000 persons for industrialized countries with the highest death rates. Also shown is a line that passes through or near the points.

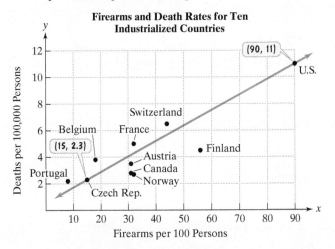

Firearms and Death Rates for Ten Industrialized Countries

Source: International Action Network on Small Arms

a. Use the two points whose coordinates are shown by the voice balloons to find an equation in point-slope form for the line that models deaths per 100,000 persons, y, as a function of firearms per 100 persons, x.

b. Write the equation in part (a) in slope-intercept form. Use function notation. $f(x) = 0.116x + 0.56$

c. France has 32 firearms per 100 persons. Use the appropriate point in the scatter plot to estimate that country's deaths per 100,000 persons.

d. Use the function from part (b) to find the number of deaths per 100,000 persons for France. Round to one decimal place. Does the function underestimate or overestimate the deaths per 100,000 persons that you estimated in part (c)? How is this shown by the line in the scatter plot?

In Exercises 38–39, make a scatter plot for the given data. Use the scatter plot to describe whether or not the variables appear to be related.

38.

x	1	3	4	6	8	9
y	1	2	3	3	5	5

39.

Country	Canada	U.S.	Mexico	Brazil	Costa Rica	Denmark	China	Egypt	Pakistan	Bangladesh	Australia	Japan	Russia
Life expectancy in years, x	81	78	76	72	77	78	73	72	64	63	82	82	66
Infant deaths per 1000 births, y	5.1	6.3	19.0	23.3	9.0	4.4	21.2	28.4	66.9	57.5	4.8	2.8	10.8

Source: U.S. Bureau of the Census International Database

Shown, again, is the scatter plot that indicates a relationship between the percentage of adult females in a country who are literate and the mortality of children under five. Also shown is the regression line. Use this information to determine whether each of the statements in Exercises 40–46 is true or false.

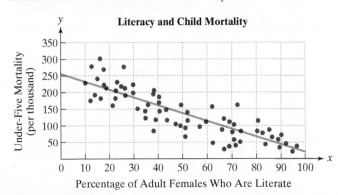

Literacy and Child Mortality

Source: United Nations

40. There is a perfect negative correlation between the percentage of adult females who are literate and under-five mortality.   false

41. As the percentage of adult females who are literate increases, under-five mortality tends to decrease.   true

42. The country with the least percentage of adult females who are literate has the greatest under-five mortality.   false

43. No two countries have the same percentage of adult females who are literate but different under-five mortalities.   false

44. There are more than 20 countries in this sample.   true

*See Answers to Selected Exercises.

45. There is no correlation between the percentage of adult females who are literate and under-five mortality.   false

46. The country with the greatest percentage of adult females who are literate has an under-five mortality rate that is less than 50 children per thousand.   true

47. Which one of the following scatter plots indicates a correlation coefficient of approximately -0.9?   c

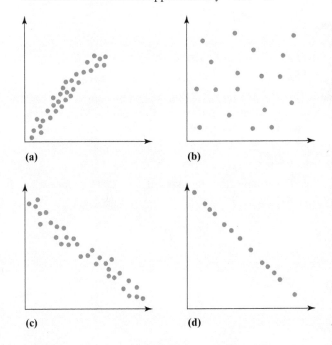

(a)   (b)

(c)   (d)

Section 3.4 Systems of Linear Equations in Two Variables

In Exercises 48–50, solve each system by graphing. Check the coordinates of the intersection point in both equations.

48. $\begin{cases} x + y = 5 \\ 3x - y = 3 \end{cases}$ $\{(2,3)\}$

49. $\begin{cases} 2x - y = -1 \\ x + y = -5 \end{cases}$ $\{(-2,-3)\}$

50. $\begin{cases} y = -x + 5 \\ 2x - y = 4 \end{cases}$ $\{(3,2)\}$

In Exercises 51–53, solve each system by the substitution method.

51. $\begin{cases} 2x + 3y = 2 \\ x = 3y + 10 \end{cases}$ $\{(4,-2)\}$

52. $\begin{cases} y = 4x + 1 \\ 3x + 2y = 13 \end{cases}$ $\{(1,5)\}$

53. $\begin{cases} x + 4y = 14 \\ 2x - y = 1 \end{cases}$ $\{(2,3)\}$

In Exercises 54–56, solve each system by the addition method.

54. $\begin{cases} x + 2y = -3 \\ x - y = -12 \end{cases}$ $\{(-9,3)\}$

55. $\begin{cases} 2x - y = 2 \\ x + 2y = 11 \end{cases}$ $\{(3,4)\}$

56. $\begin{cases} 5x + 3y = 1 \\ 3x + 4y = -6 \end{cases}$ $\{(2,-3)\}$

In Exercises 57–59, solve by the method of your choice. Identify systems with no solution and systems with infinitely many solutions, using set notation to express their solution sets.

57. $\begin{cases} y = -x + 4 \\ 3x + 3y = -6 \end{cases}$ $\varnothing$

58. $\begin{cases} 3x + y = 8 \\ 2x - 5y = 11 \end{cases}$ $\{(3,-1)\}$

59. $\begin{cases} 3x - 2y = 6 \\ 6x - 4y = 12 \end{cases}$ $\{(x,y)\,|\,3x - 2y = 6\}$ or $\{(x,y)\,|\,6x - 4y = 12\}$

60. A company is planning to manufacture computer desks. The fixed cost will be $60,000 and it will cost $200 to produce each desk. Each desk will be sold for $450.

   a. Write the cost function, C, of producing x desks.

   b. Write the revenue function, R, from the sale of x desks.

   c. Determine the break-even point. Describe what it means. $(240, 108,000)$*

60. a. $C(x) = 60,000 + 200x$ **60. b.** $R(x) = 450x$ *See Answers to Selected Exercises.

61. a. approximately $(2016, 325)$; 2016; 325 million

61. The graph shows the number of guns in private hands in the United States and the country's population, both expressed in millions, from 1995 through 2020, with projections from 2012 onward.

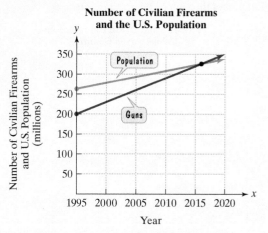

Number of Civilian Firearms and the U.S. Population

Source: *Mother Jones*, November/December 2012

   a. Use the graphs to estimate the point of intersection. In what year will there be a gun for every man, woman, and child in the United States? What will be the population and the number of firearms in that year?

   b. In 1995, there were an estimated 200 million firearms in private hands. This has increased at an average rate of 6 million firearms per year. Write a function that models the number of civilian firearms in the United States, y, in millions, x years after 1995. $y = 6x + 200$

   c. The function $y - 3x = 263$ models the U.S. population, y, in millions, x years after 1995. Use this model and the model you obtained in part (b) to determine the year in which there will be a gun for every U.S. citizen. According to the models, what will be the population and the number of firearms in that year? 2016; 326 million

   d. How well do the models in parts (b) and (c) describe the point of intersection of the graphs that you estimated in part (a)? quite well

Section 3.5 Linear Inequalities in Two Variables

In Exercises 62–68, graph each linear inequality.

62. $x - 3y \le 6$ * **63.** $2x + 3y \ge 12$ * **64.** $2x - 7y > 14$ *

65. $y > \frac{3}{5}x$ * **66.** $y \le -\frac{1}{2}x + 2$ * **67.** $x \le 2$ *

68. $y > -3$ *

In Exercises 69–74, graph the solution set of each system of linear inequalities.

69. $\begin{cases} 3x - y \le 6 \\ x + y \ge 2 \end{cases}$ *

70. $\begin{cases} x + y < 4 \\ x - y < 4 \end{cases}$ *

71. $\begin{cases} x \le 3 \\ y > -2 \end{cases}$ *

72. $\begin{cases} 4x + 6y \le 24 \\ y > 2 \end{cases}$ *

73. $\begin{cases} x + y \le 6 \\ y \ge 2x - 3 \end{cases}$ *

74. $\begin{cases} y < -x + 4 \\ y > x - 4 \end{cases}$ *

5. a. Yes; no vertical line intersects the graph in more than one point.   **b.** 0; The eagle was on the ground after 15 seconds.   **d.** between 3 and 12 seconds

Chapter 3 Test

1. Graph $y = |x| - 2$. Let $x = -3, -2, -1, 0, 1, 2,$ and 3.   *

2. If $f(x) = 3x^2 - 7x - 5$, find $f(-2)$.   21

In Exercises 3–4, use the vertical line test to identify graphs in which y is a function of x.

3.

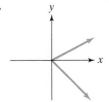

y is not a function of *x*.

4.

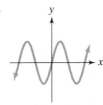

y is a function of *x*.

5. The graph shows the height, in meters, of an eagle in terms of its time, in seconds, in flight.

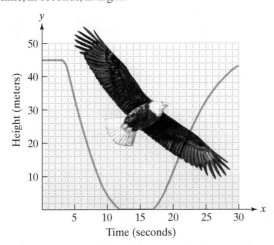

a. Is the eagle's height a function of time? Use the graph to explain why or why not.

b. Find $f(15)$. Describe what this means in practical terms.

c. What is a reasonable estimate of the eagle's maximum height?   45 meters

d. During which period of time was the eagle descending?

6. Use the *x*- and *y*-intercepts to graph $4x - 2y = -8$.   *

7. Find the slope of the line passing through $(-3, 4)$ and $(-5, -2)$. $m = 3$

In Exercises 8–9, graph each linear function using the slope and y-intercept.

8. $y = \frac{2}{3}x - 1$ *

9. $f(x) = -2x + 3$ *

10. On Super Bowl Sunday, viewers of the big game expect to be entertained by commercials that offer an upbeat mix of punch lines, animal tricks, humor, and special effects. The price tag for a 30-second ad slot also follows tradition, up to an average of $3 million in 2009 from $2.7 million in 2008. The scatter plot at the top of the next column shows the cost, in millions of dollars, of 30 seconds of ad time in the Super Bowl from 2003 through 2009. Also shown is a line that passes through or near the seven data points.

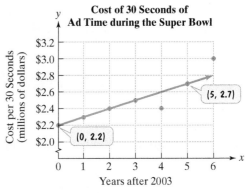

Source: TNS Media Intelligence

a. According to the graph, what is the *y*-intercept of the line? Describe what this represents in this situation.

b. Use the coordinates of the two points shown to compute the slope. What does this represent in terms of the rate of change in the cost of Super Bowl commercials?

c. Use the *y*-intercept shown and the slope from part (b) to write a linear function that models the cost of 30 seconds of ad time during the Super Bowl, $C(x)$, in millions of dollars, *x* years after 2003. $C(x) = 0.1x + 2.2$

d. Use the function from part (c) to project the cost of 30 seconds of ad time in the Super Bowl in 2018. $3.7 million

11. Write an equation in point-slope form and in slope-intercept form of the line passing through $(2, 1)$ and $(-1, -8)$.

12. The bar graph shows world population, in billions, for seven selected years from 1950 through 2010. Also shown is a scatter plot with a line passing through two of the data points.

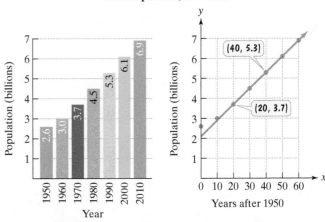

Source: U.S. Census Bureau, International Database

a. Use the two points whose coordinates are shown by the voice balloons to write the slope-intercept form of an equation that models world population, *y*, in billions, *x* years after 1950. $y = 0.08x + 2.1$

b. Use the model from part (a) to project world population in 2025.   8.1 billion

10. a. 2.2; In 2003, the cost per 30 seconds was $2.2 million.   **11.** $y - 1 = 3(x - 2)$ or $y + 8 = 3(x + 1)$; $y = 3x - 5$

   b. 0.1; The cost of a 30-second commercial increased approximately $0.1 million each year. The rate of change is $0.1 million per year.

*See Answers to Selected Exercises.

13. Make a scatter plot for the given data. Use the scatter plot to describe whether or not the variables appear to be related. *

x	1	4	3	5	2
y	5	2	2	1	4

The scatter plot in the figure shows the relationship between the number of hours that students studied for a quiz and their quiz scores. The highest quiz score is 10 and the lowest is 1. Use the scatter plot to determine whether each of the statements in Exercises 14–16 is true or false.

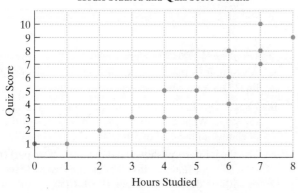

Hours Studied and Quiz Score Results

14. There is a strong positive correlation between hours studied and quiz scores.   true

15. The student who studied for one hour had a better quiz score than the student who did not study.   false

16. The student who studied the greatest amount of time had the best quiz score.   false

17. Solve by graphing:

$$\begin{cases} x + y = 6 \\ 4x - y = 4. \end{cases} \quad \{(2,4)\}$$

18. Solve by substitution:

$$\begin{cases} x = y + 4 \\ 3x + 7y = -18. \end{cases} \quad \{(1,-3)\}$$

19. Solve by addition:

$$\begin{cases} 5x + 4y = 10 \\ 3x + 5y = -7. \end{cases} \quad \{(6,-5)\}$$

20. A company is planning to produce and sell a new line of computers. The fixed cost will be $360,000 and it will cost $850 to produce each computer. Each computer will be sold for $1150.

   a. Write the cost function, C, of producing x computers.

   b. Write the revenue function, R, from the sale of x computers. $R(x) = 1150x$

   c. Determine the break-even point. Describe what it means.

Graph each linear inequality in Exercises 21–23.

21. $3x - 2y < 6$ *   **22.** $y \le \frac{1}{2}x - 1$ *

23. $y > -1$ *

24. Graph the system of linear inequalities:

$$\begin{cases} 2x - y \le 4 \\ 2x - y > -1. \end{cases} \quad *$$

20. a. $C(x) = 360,000 + 850x$ **c.** $(1200, 1,380,000)$; The company will break even if it produces and sells 1200 computers. At this level, both revenue and cost are $1,380,000.

*See Answers to Selected Exercises.

Algebraic Pathways: Polynomials, Quadratic Equations, and Quadratic Functions

4

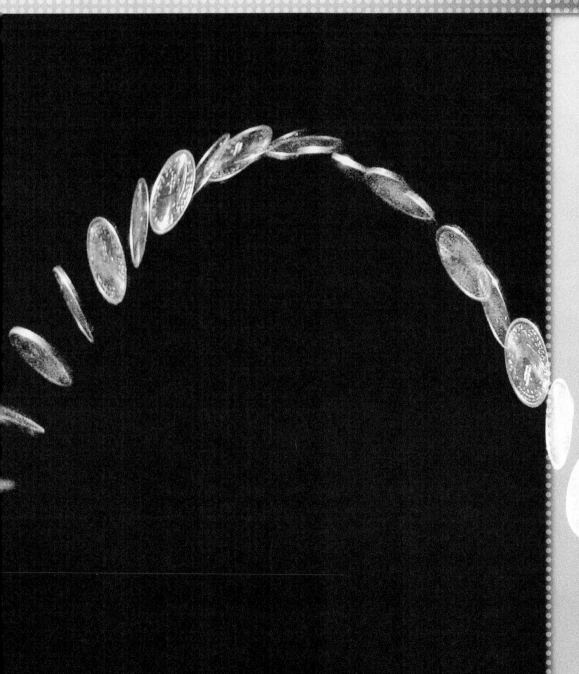

We are surrounded by evidence that the world is profoundly mathematical. After turning a somersault, a diver's path can be modeled by a function in the form $f(x) = ax^2 + bx + c$, called a *quadratic function*, as can the path of a football tossed from a quarterback to a receiver, or the path of a flipped coin. Even if you throw an object directly upward, although its path is straight and vertical, its changing height over time is described by a quadratic function. And tailgaters take note: whether you're driving a car or a truck on dry or wet roads, an array of quadratic functions that model your required stopping distances at various speeds are available to help you become a safer driver.

Here's where you'll find these applications:

The quadratic functions surrounding our long history of objects that are thrown, kicked, or hit appear throughout the chapter, including Example 4 of Section 4.3 and Example 7 of Section 4.5. Tailgaters on motorcycles should pay close attention to Exercises 78–79 in the Chapter Review Exercises.

4.1 Operations with Polynomials; Polynomial Functions

What am I supposed to learn?

After you have read this section, you should be able to:

① Understand the vocabulary of polynomials.

② Add polynomials.

③ Subtract polynomials.

④ Multiply polynomials.

⑤ Use FOIL in polynomial multiplication.

⑥ Multiply the sum and difference of two terms.

⑦ Square binomials.

⑧ Use polynomial functions.

We're born. We die. **Figure 4.1** quantifies these statements by showing the numbers of births and deaths in the United States for ten selected years.

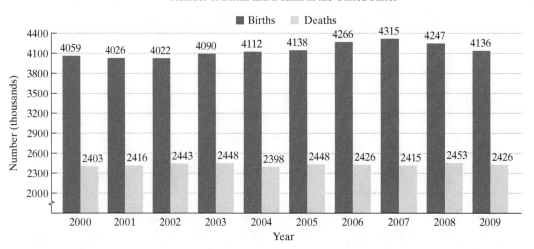

Number of Births and Deaths in the United States

Figure 4.1
Source: U.S. Department of Health and Human Services

Here are two functions that model the data in **Figure 4.1**:

$$B(x) = -2.6x^2 + 49x + 3994 \qquad D(x) = -0.6x^2 + 7x + 2412.$$

| Number of births, $B(x)$, in thousands, x years after 2000 | Number of deaths, $D(x)$, in thousands, x years after 2000 |

The algebraic expressions that appear on the right sides of the models are examples of *polynomials*. A **polynomial** is a single term or the sum of two or more terms containing variables with whole-number exponents. The polynomials above each contain three terms. In a **polynomial function**, the expression that defines the function is a polynomial. Polynomial functions are used in such diverse areas as science, business, medicine, psychology, and sociology. In this section, we consider basic ideas about polynomials and their operations.

1 Understand the vocabulary of polynomials.

How We Describe Polynomials

Consider the polynomial
$$7x^3 - 9x^2 + 13x - 6.$$
We can express this polynomial as
$$7x^3 + (-9x^2) + 13x + (-6).$$
The polynomial contains four terms. It is customary to write the terms in the order of descending powers of the variable. This is the **standard form** of a polynomial.

Some polynomials contain only one variable. Each term of such a polynomial in x is of the form ax^n. If $a \neq 0$, the **degree** of the term ax^n is n. For example, the degree of the term $7x^3$ is 3.

Great Question!

Why doesn't the constant 0 have a degree?

We can express 0 in many ways, including $0x$, $0x^2$, and $0x^3$. It is impossible to assign a unique exponent to the variable. This is why 0 has no defined degree.

> ### The Degree of ax^n
>
> If $a \neq 0$, the degree of the term ax^n is n. The degree of a nonzero constant term is 0. The constant 0 has no defined degree.

Here is an example of a polynomial and the degree of each of its four terms:
$$6x^4 - 3x^3 + 2x - 5.$$

degree 4   degree 3   degree 1   degree of nonzero constant: 0

Notice that the exponent on x for the term $2x$, meaning $2x^1$, is understood to be 1. For this reason, the degree of $2x$ is 1. You can think of -5 as $-5x^0$; thus, its degree is 0.

A polynomial is simplified when it contains no grouping symbols and no like terms. A simplified polynomial that has exactly one term is called a **monomial**. A **binomial** is a simplified polynomial that has two terms. A **trinomial** is a simplified polynomial with three terms. Simplified polynomials with four or more terms have no special names.

The **degree of a polynomial** is the greatest of the degrees of all the terms of the polynomial. For example, $4x^2 + 3x$ is a binomial of degree 2 because the degree of the first term is 2, and the degree of the other term is less than 2. Also, $7x^5 - 2x^2 + 4$ is a trinomial of degree 5 because the degree of the first term is 5, and the degrees of the other terms are less than 5. In a simplified polynomial containing only one variable, the term of the greatest degree is called the **leading term**. Its coefficient is called the **leading coefficient**.

Up to now, we have used x to represent the variable in a polynomial. However, any letter can be used. For example,

- $7x^5 - 3x^3 + 8$ is a polynomial (in x) of degree 5. Because there are three terms, the polynomial is a trinomial. The leading coefficient is 7.

- $6y^3 + 4y^2 - y + 3$ is a polynomial (in y) of degree 3. Because there are four terms, the polynomial has no special name. The leading coefficient is 6.

- $z^7 + \sqrt{2}$ is a polynomial (in z) of degree 7. Because there are two terms, the polynomial is a binomial. The leading coefficient is 1.

Adding Polynomials

2 Add polynomials.

Polynomials are added by combining like terms. For example, we can combine the monomials $-9x^3$ and $13x^3$ using addition as follows:

$$-9x^3 + 13x^3 = (-9 + 13)x^3 = 4x^3.$$

These like terms both contain x to the third power.

Add coefficients and keep the same variable factor, x^3.

Example 1 Adding Polynomials

Add: $(-6x^3 + 5x^2 + 4) + (2x^3 + 7x^2 - 10)$.

Solution

$(-6x^3 + 5x^2 + 4) + (2x^3 + 7x^2 - 10)$

$= -6x^3 + 5x^2 + 4 + 2x^3 + 7x^2 - 10$ *Remove the parentheses. Like terms are shown in the same color.*

$= -6x^3 + 2x^3 + 5x^2 + 7x^2 + 4 - 10$ *Rearrange the terms so that like terms are adjacent.*

$= \quad -4x^3 \qquad + 12x^2 \qquad - 6$ *Combine like terms.*

$= -4x^3 + 12x^2 - 6$ *This is the same sum as above, written more concisely.*

Polynomials can be added by arranging like terms in columns and combining like terms, column by column. Here's the solution to Example 1 using columns and a vertical format:

$$
\begin{array}{r}
-6x^3 + 5x^2 + 4 \\
2x^3 + 7x^2 - 10 \\
\hline
-4x^3 + 12x^2 - 6.
\end{array}
$$

 **Check Point 1** Add: $(-7x^3 + 4x^2 + 3) + (4x^3 + 6x^2 - 13)$. $-3x^3 + 10x^2 - 10$

Subtracting Polynomials

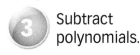 Subtract polynomials.

We subtract real numbers by adding the opposite, or additive inverse, of the number being subtracted. For example,

$$8 - 3 = 8 + (-3) = 5.$$

Similarly, we subtract one polynomial from another by adding the opposite of the polynomial being subtracted.

> **Subtracting Polynomials**
>
> To subtract two polynomials, change the sign of every term of the second polynomial. Add this result to the first polynomial.

Example 2 Subtracting Polynomials

Subtract: $(7x^3 - 8x^2 + 9x - 6) - (2x^3 - 6x^2 - 3x + 9)$.

Solution

$(7x^3 - 8x^2 + 9x - 6) - (2x^3 - 6x^2 - 3x + 9)$

Change the sign of each coefficient.

$= (7x^3 - 8x^2 + 9x - 6) + (-2x^3 + 6x^2 + 3x - 9)$ *Rewrite subtraction as addition of the additive inverse.*

$= (7x^3 - 2x^3) + (-8x^2 + 6x^2) + (9x + 3x) + (-6 - 9)$ *Group like terms.*

$= \quad 5x^3 \quad + \quad (-2x^2) \quad + \quad 12x \quad + \quad (-15)$ *Combine like terms.*

$= 5x^3 - 2x^2 + 12x - 15$ *Simplify.*

You can also subtract polynomials using a vertical format. Here's the solution to Example 2 using a vertical format. Notice that you still distribute the negative sign, thereby obtaining the opposite.

$$7x^3 - 8x^2 + 9x - 6$$
$$-(2x^3 - 6x^2 - 3x + 9)$$

Change the sign of every term.

$$\begin{aligned} 7x^3 - 8x^2 + 9x - 6 \\ -2x^3 + 6x^2 + 3x - 9 \\ \hline 5x^3 - 2x^2 + 12x - 15 \end{aligned}$$

 **Check Point 2**   Subtract: $(14x^3 - 5x^2 + x - 9) - (4x^3 - 3x^2 - 7x + 1)$.

$10x^3 - 2x^2 + 8x - 10$

Multiplying Polynomials

4 Multiply polynomials.

The product of two monomials is obtained by using properties of exponents. For example,

$$(-8x^6)(5x^3) = -8 \cdot 5x^{6+3} = -40x^9.$$

Multiply coefficients and add exponents.

A Brief Review • Differences between Adding and Multiplying Monomials

- Don't confuse adding and multiplying monomials.

  Addition:

  $$5x^4 + 6x^4 = 11x^4$$

  Multiplication:

  $$(5x^4)(6x^4) = (5 \cdot 6)(x^4 \cdot x^4)$$
  $$= 30x^{4+4}$$
  $$= 30x^8$$

- Only like terms can be added or subtracted, but unlike terms may be multiplied.

  Addition:

  $5x^4 + 3x^2$ cannot be simplified.

  Multiplication:

  $$(5x^4)(3x^2) = (5 \cdot 3)(x^4 \cdot x^2)$$
  $$= 15x^{4+2}$$
  $$= 15x^6$$

Multiplying a Monomial and a Polynomial That Is Not a Monomial

We use the distributive property to multiply a monomial and a polynomial that is not a monomial. For example,

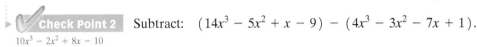

$$3x^2(2x^3 + 5x) = 3x^2 \cdot 2x^3 + 3x^2 \cdot 5x = 3 \cdot 2x^{2+3} + 3 \cdot 5x^{2+1} = 6x^5 + 15x^3.$$

Monomial   Binomial   *Multiply coefficients and add exponents.*

> **Multiplying a Monomial and a Polynomial That Is Not a Monomial**
>
> To multiply a monomial and a polynomial, use the distributive property to multiply each term of the polynomial by the monomial.

Example 3   **Multiplying a Monomial and a Trinomial**

Multiply: $4x^3(6x^5 - 2x^2 + 3)$.

Solution

$$4x^3(6x^5 - 2x^2 + 3) = 4x^3 \cdot 6x^5 - 4x^3 \cdot 2x^2 + 4x^3 \cdot 3$$

Use the distributive property.

$$= 24x^8 - 8x^5 + 12x^3$$

Multiply coefficients and add exponents.

Check Point 3   Multiply: $6x^4(2x^5 - 3x^2 + 4)$. $12x^9 - 18x^6 + 24x^4$

Multiplying Polynomials When Neither Is a Monomial

How do we multiply two polynomials if neither is a monomial? For example, consider

$$(2x + 3)(x^2 + 4x + 5).$$

Binomial   Trinomial

One way to perform $(2x + 3)(x^2 + 4x + 5)$ is to distribute $2x$ throughout the trinomial

$$2x(x^2 + 4x + 5)$$

and 3 throughout the trinomial

$$3(x^2 + 4x + 5).$$

Then combine the like terms that result.

> **Multiplying Polynomials When Neither Is a Monomial**
>
> Multiply each term of one polynomial by each term of the other polynomial. Then combine like terms.

Example 4   **Multiplying a Binomial and a Trinomial**

Multiply: $(2x + 3)(x^2 + 4x + 5)$.

Solution

$$(2x + 3)(x^2 + 4x + 5)$$

$$= 2x(x^2 + 4x + 5) + 3(x^2 + 4x + 5)$$

Multiply the trinomial by each term of the binomial.

$$= 2x \cdot x^2 + 2x \cdot 4x + 2x \cdot 5 + 3x^2 + 3 \cdot 4x + 3 \cdot 5$$

Use the distributive property.

$$= 2x^3 + 8x^2 + 10x + 3x^2 + 12x + 15$$

Multiply monomials: Multiply coefficients and add exponents.

$$= 2x^3 + 11x^2 + 22x + 15$$

Combine like terms: $8x^2 + 3x^2 = 11x^2$ and $10x + 12x = 22x$.

Another method for performing the multiplication is to use a vertical format similar to that used for multiplying whole numbers.

$$
\begin{array}{r}
x^2 + 4x + 5 \\
2x + 3 \\
\hline
3x^2 + 12x + 15 \\
2x^3 + 8x^2 + 10x \\
\hline
2x^3 + 11x^2 + 22x + 15
\end{array}
$$

$3(x^2 + 4x + 5)$

$2x(x^2 + 4x + 5)$ *Write like terms in the same column.*

Combine like terms.

 **Check Point 4**   Multiply: $(5x - 2)(3x^2 - 5x + 4)$. $15x^3 - 31x^2 + 30x - 8$

The Product of Two Binomials: FOIL

⑤ Use FOIL in polynomial multiplication.

Frequently, we need to find the product of two binomials. One way to perform this multiplication is to distribute each term in the first binomial through the second binomial. For example, we can find the product of the binomials $3x + 2$ and $4x + 5$ as follows:

$$(3x + 2)(4x + 5) = 3x(4x + 5) + 2(4x + 5)$$
$$= 3x(4x) + 3x(5) + 2(4x) + 2(5)$$
$$= 12x^2 + 15x + 8x + 10.$$

Distribute $3x$ over $4x + 5$.   Distribute 2 over $4x + 5$.

We'll combine these like terms later. For now, our interest is in how to obtain *each* of these four terms.

We can also find the product of $3x + 2$ and $4x + 5$ using a method called FOIL, which is based on our work shown above. Any two binomials can be quickly multiplied by using the FOIL method, in which **F** represents the product of the **first** terms in each binomial, **O** represents the product of the **outside** terms, **I** represents the product of the **inside** terms, and **L** represents the product of the **last**, or second, terms in each binomial. For example, we can use the FOIL method to find the product of the binomials $3x + 2$ and $4x + 5$ as follows:

first   last

F   O   I   L

$$(3x + 2)(4x + 5) = 12x^2 + 15x + 8x + 10$$

inside

outside

| Product of First terms | Product of Outside terms | Product of Inside terms | Product of Last terms |

$$= 12x^2 + 23x + 10$$ *Combine like terms.*

In general, here's how to use the FOIL method to find the product of $ax + b$ and $cx + d$:

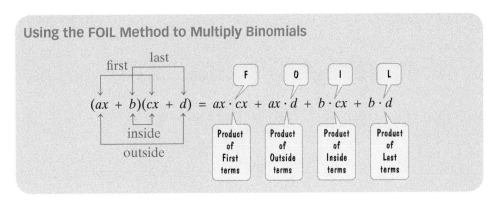

Using the FOIL Method to Multiply Binomials

first   last

F   O   I   L

$$(ax + b)(cx + d) = ax \cdot cx + ax \cdot d + b \cdot cx + b \cdot d$$

inside

outside

| Product of First terms | Product of Outside terms | Product of Inside terms | Product of Last terms |

Example 5 **Using the FOIL Method**

Multiply:

a. $(3x + 4)(5x - 3)$ **b.** $(5x^3 - 6)(4x^3 - x)$.

Solution

a.

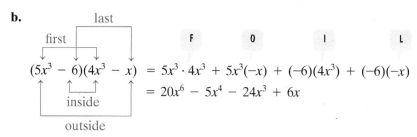

$$(3x + 4)(5x - 3) = 3x \cdot 5x + 3x(-3) + 4 \cdot 5x + 4(-3)$$
$$= 15x^2 - 9x + 20x - 12$$
$$= 15x^2 + 11x - 12$$

b.

$$(5x^3 - 6)(4x^3 - x) = 5x^3 \cdot 4x^3 + 5x^3(-x) + (-6)(4x^3) + (-6)(-x)$$
$$= 20x^6 - 5x^4 - 24x^3 + 6x$$

✓ **Check Point 5** Multiply:

a. $(7x + 5)(4x - 3)$ $28x^2 - x - 15$

▶ **b.** $(4x^3 - 5)(x^3 - 3x)$. $4x^6 - 12x^4 - 5x^3 + 15x$

Multiplying the Sum and Difference of Two Terms

6 Multiply the sum and difference of two terms.

We can use the FOIL method to multiply $A + B$ and $A - B$ as follows:

$$\overset{\text{F \quad O \quad I \quad L}}{(A + B)(A - B) = A^2 - AB + AB - B^2 = A^2 - B^2.}$$

Notice that the outside and inside products have a sum of 0 and the terms cancel. The FOIL multiplication provides us with a quick rule for multiplying the sum and difference of two terms.

The Product of the Sum and Difference of Two Terms

$$(A + B)(A - B) = A^2 - B^2$$

The product of the sum and the difference of the same two terms | is | the square of the first term minus the square of the second term.

Example 6 **Finding the Product of the Sum and Difference of Two Terms**

Multiply:

a. $(4y + 3)(4y - 3)$ **b.** $(5a^4 + 6)(5a^4 - 6)$.

Solution

Use the formula for the product of the sum and difference of two terms.

$$(A + B)(A - B) \quad = \quad A^2 \quad - \quad B^2$$

	First term squared	−	Second term squared	=	Product
a. $(4y + 3)(4y - 3)$ =	$(4y)^2$	−	3^2	=	$16y^2 - 9$
b. $(5a^4 + 6)(5a^4 - 6)$ =	$(5a^4)^2$	−	6^2	=	$25a^8 - 36$

 **Check Point 6** Multiply:

▶ **a.** $(7x + 8)(7x - 8)$ $\quad{}_{49x^2 - 64}$ **b.** $(2y^3 - 5)(2y^3 + 5)$. $\quad{}_{4y^6 - 25}$

The Square of a Binomial

7 Square binomials.

Let us find $(A + B)^2$, the square of a binomial sum. To do so, we begin with the FOIL method and look for a general rule.

$$\overset{\quad\quad\quad\quad\quad\quad\quad\quad\quad\quad F\quad\quad O\quad\quad I\quad\quad L}{(A + B)^2 = (A + B)(A + B) = A \cdot A + A \cdot B + A \cdot B + B \cdot B}$$
$$= A^2 + 2AB + B^2$$

This result implies the following rule:

> **The Square of a Binomial Sum**
>
> $$(A + B)^2 \quad = \quad A^2 \quad + \quad 2AB \quad + \quad B^2$$
>
> The square of a binomial sum / is / first term squared / plus / 2 times the product of the terms / plus / last term squared.

Great Question!

When finding $(x + 3)^2$, why can't I just write $x^2 + 3^2$, or $x^2 + 9$?

Caution! The square of a sum is *not* the sum of the squares.

$$(A + B)^2 \neq A^2 + B^2$$

The middle term 2AB is missing.

$$(x + 3)^2 \neq x^2 + 9$$

Incorrect!

Show that $(x + 3)^2$ and $x^2 + 9$ are not equal by substituting 5 for x in each expression and simplifying.

Example 7   **Finding the Square of a Binomial Sum**

Multiply:

a. $(x + 3)^2$

b. $(3x + 7)^2$.

Solution

Use the formula for the square of a binomial sum.

$$(A + B)^2 = A^2 + 2AB + B^2$$

	(First Term)^2	+	2 · Product of the Terms	+	(Last Term)^2	= Product
a. $(x + 3)^2 =$	x^2	+	$2 \cdot x \cdot 3$	+	3^2	$= x^2 + 6x + 9$
b. $(3x + 7)^2 =$	$(3x)^2$	+	$2(3x)(7)$	+	7^2	$= 9x^2 + 42x + 49$

 **Check Point 7** Multiply:

▶ **a.** $(x + 10)^2$ $\quad{}_{x^2 + 20x + 100}$ **b.** $(5x + 4)^2$. $\quad{}_{25x^2 + 40x + 16}$

A similar pattern occurs for $(A - B)^2$, the square of a binomial difference. Using the FOIL method on $(A - B)^2$, we obtain the following rule:

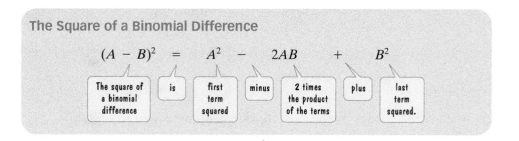

The Square of a Binomial Difference

$$(A - B)^2 = A^2 - 2AB + B^2$$

The square of a binomial difference | is | first term squared | minus | 2 times the product of the terms | plus | last term squared.

Example 8   Finding the Square of a Binomial Difference

Multiply:

a. $(x - 4)^2$ **b.** $(5y - 6)^2$.

Solution

Use the formula for the square of a binomial difference.

$$(A - B)^2 = A^2 - 2AB + B^2$$

	(First Term)^2	−	2 · Product of the Terms	+	(Last Term)^2	= Product
a. $(x - 4)^2 =$	x^2	−	$2 \cdot x \cdot 4$	+	4^2	$= x^2 - 8x + 16$
b. $(5y - 6)^2 =$	$(5y)^2$	−	$2(5y)(6)$	+	6^2	$= 25y^2 - 60y + 36$

✓ **Check Point 8**   Multiply:

a. $(x - 9)^2$ $x^2 - 18x + 81$ **b.** $(7x - 3)^2$. $49x^2 - 42x + 9$

Polynomial Functions

8) Use polynomial functions.

We have seen that a polynomial function is defined by an expression that is a polynomial. In our final example, we apply operations with polynomials to polynomial functions.

Example 9   Modeling with Polynomial Functions

We opened the section with functions that model the number of births and deaths in the United States from 2000 through 2009:

$$B(x) = -2.6x^2 + 49x + 3994 \qquad D(x) = -0.6x^2 + 7x + 2412.$$

Number of births, $B(x)$, in thousands, x years after 2000   Number of deaths, $D(x)$, in thousands, x years after 2000

a. Write a function that models the change in U.S. population, $C(x)$, for each year from 2000 through 2009.

b. Use the function from part (a) to find the change in U.S. population in 2008.

c. Does the result in part (b) overestimate or underestimate the actual population change in 2008 obtained from the data in **Figure 4.1** on page 254? By how much?

Solution

a. The change in population, $C(x)$, is the number of births minus the number of deaths: $C(x) = B(x) - D(x)$.

$C(x)$

$= B(x) - D(x)$

$= (-2.6x^2 + 49x + 3994) - (-0.6x^2 + 7x + 2412)$ Substitute the given functions.

$= -2.6x^2 + 49x + 3994 + 0.6x^2 - 7x - 2412$ Remove parentheses and change the sign of each term in the second set of parentheses.

$= (-2.6x^2 + 0.6x^2) + (49x - 7x) + (3994 - 2412)$ Group like terms.

$= -2x^2 + 42x + 1582$ Combine like terms.

The function

$$C(x) = -2x^2 + 42x + 1582$$

models the change in U.S. population, $C(x)$, in thousands, x years after 2000.

b. Because 2008 is 8 years after 2000, we substitute 8 for x in the difference function $(B - D)(x)$.

$(B - D)(x) = -2x^2 + 42x + 1582$ Use the difference function $B - D$.

$(B - D)(8) = -2(8)^2 + 42(8) + 1582$ Substitute 8 for x.

$= -2(64) + 42(8) + 1582$ Evaluate the exponential expression: $8^2 = 64$.

$= -128 + 336 + 1582$ Perform the multiplications.

$= 1790$ Add from left to right.

We see that $(B - D)(8) = 1790$. The model indicates that there was a population increase of 1790 thousand, or approximately 1,790,000 people, in 2008.

c. The data for 2008 in **Figure 4.1** on page 254 show 4247 thousand births and 2453 thousand deaths.

$$\text{population change} = \text{births} - \text{deaths}$$

$$= 4247 - 2453 = 1794$$

The actual population increase was 1794 thousand, or 1,794,000. Our model gave us an increase of 1790 thousand. Thus, the model underestimates the actual increase by $1794 - 1790$, or 4 thousand people.

Check Point 9   Use the birth and death models from Example 9.

a. Write a function that models the total number of births and deaths in the United States for the years from 2000 through 2009. $(B + D)(x) = -3.2x^2 + 56x + 6406$

b. Use the function from part (a) to find the total number of births and deaths in the United States in 2003.   6545.2 thousand

c. Does the result in part (b) overestimate or underestimate the actual number of total births and deaths in 2003 obtained from the data in **Figure 4.1** on page 254? By how much?   overestimates by 7.2 thousand

Blitzer Bonus
●●●●●●●●●●●●●●●●

Labrador Retrievers and Polynomial Multiplication

The color of a Labrador retriever is determined by its pair of genes. A single gene is inherited at random from each parent. The black-fur gene, B, is dominant. The yellow-fur gene, Y, is recessive. This means that labs with at least one black-fur gene (BB or BY) have black coats. Only labs with two yellow-fur genes (YY) have yellow coats.

Axl, your author's yellow lab, pictured below, inherited his genetic makeup from two black BY parents.

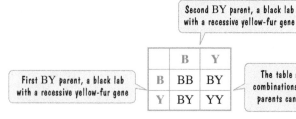

Old Dog... New Chicks

Because YY is one of four possible outcomes, the probability that a yellow lab like Axl will be the offspring of these black parents is $\frac{1}{4}$.

The probabilities suggested by the table can be modeled by the expression $\left(\frac{1}{2}B + \frac{1}{2}Y\right)^2$.

$$\left(\frac{1}{2}B + \frac{1}{2}Y\right)^2 = \left(\frac{1}{2}B\right)^2 + 2\left(\frac{1}{2}B\right)\left(\frac{1}{2}Y\right) + \left(\frac{1}{2}Y\right)^2$$

$$= \frac{1}{4}BB + \frac{1}{2}BY + \frac{1}{4}YY$$

The probability of a black lab with two dominant black genes is $\frac{1}{4}$.

The probability of a black lab with a recessive yellow gene is $\frac{1}{2}$.

The probability of a yellow lab with two recessive yellow genes is $\frac{1}{4}$.

Achieving Success

Avoid coursus interruptus.

Now that you've started pathways to college mathematics, don't interrupt the sequence until you have completed all your required math classes. You'll have better results if you take your math courses without a break. If you start, stop, and start again, it's easy to forget what you've learned and lose your momentum.

Concept and Vocabulary Check

Exercises in the Concept and Vocabulary Check are intended for group and class discussions.

In Exercises 1–4, fill in each blank so that the resulting statement is true.

1. Exponents on the variables of a polynomial must be _____whole_____ numbers.

2. A polynomial with one term is called a/an ___monomial___, a polynomial with two terms is called a/an ___binomial___, and a polynomial with three terms is called a/an ___trinomial___.

3. The degree of ax^n, $a \neq 0$, is _____n_____.

4. The degree of a polynomial is _the greatest degree_ of all its terms

In Exercises 5–8, determine whether each statement is true or false. If the statement is false, make the necessary change(s) to produce a true statement. Changes to false statements will vary.

5. $4x^3 + 7x^2 - 5x + 2x^{-1}$ is a polynomial containing four terms.   false

6. If two polynomials of degree 2 are added, the sum must be a polynomial of degree 2.   false

7. $(x^2 - 7x) - (x^2 - 4x) = -11x$ for all values of x.   false

8. All terms of a polynomial are monomials.   true

Respond to Exercises 9–20 using verbal or written explanations.

9. What is a polynomial?   **9–20.** Answers will vary.

10. Explain how to determine the degree of each term of a polynomial.

11. Explain how to determine the degree of a polynomial.

12. What is a polynomial function?

13. Explain how to add polynomials.

14. Explain how to subtract polynomials.

15. Explain how to multiply a monomial and a polynomial that is not a monomial. Give an example.

16. Explain how to multiply a binomial and a trinomial.

17. What is the FOIL method and when is it used? Give an example of the method.

18. Explain how to find the product of the sum and difference of two terms. Give an example with your explanation.

19. Explain how to square a binomial sum. Give an example.

20. Explain how to square a binomial difference. Give an example.

Exercise Set 4.1

Practice Exercises

In Exercises 1–4, is the algebraic expression a polynomial? If it is, write the polynomial in standard form.

1. $2x + 3x^2 - 5$ yes; $3x^2 + 2x - 5$

2. $2x + 3x^{-1} - 5$ no

3. $2x^2 + 3^{-1}x - 5$ yes; $2x^2 + \frac{1}{3}x - 5$

4. $x^2 - x^3 + x^4 - 5$ yes; $x^4 - x^3 + x^2 - 5$

In Exercises 5–8, find the degree of the polynomial.

5. $3x^2 - 5x + 4$ 2

6. $-4x^3 + 7x^2 - 11$ 3

7. $x^2 - 4x^3 + 9x - 12x^4 + 63$ 4

8. $x^2 - 8x^3 + 15x^4 + 91$ 4

In Exercises 9–14, perform the indicated operations. Write the resulting polynomial in standard form and indicate its degree.

9. $(-6x^3 + 5x^2 - 8x + 9) + (17x^3 + 2x^2 - 4x - 13)$

10. $(-7x^3 + 6x^2 - 11x + 13) + (19x^3 - 11x^2 + 7x - 17)$

11. $(17x^3 - 5x^2 + 4x - 3) - (5x^3 - 9x^2 - 8x + 11)$

12. $(18x^4 - 2x^3 - 7x + 8) - (9x^4 - 6x^3 - 5x + 7)$

13. $(5x^2 - 7x - 8) + (2x^2 - 3x + 7) - (x^2 - 4x - 3)$

14. $(8x^2 + 7x - 5) - (3x^2 - 4x) - (-6x^3 - 5x^2 + 3)$

In Exercises 15–56, find each product.

15. $4x^2(3x + 2)$ $12x^3 + 8x^2$

16. $5x^2(6x + 7)$ $30x^3 + 35x^2$

17. $2y(y^2 - 5y)$ $2y^3 - 10y^2$

18. $3y(y^2 - 4y)$ $3y^3 - 12y^2$

19. $5x^3(2x^5 - 4x^2 + 9)$

20. $6x^3(3x^5 - 5x^2 + 7)$

21. $(x + 1)(x^2 - x + 1)$ $x^3 + 1$

22. $(x + 5)(x^2 - 5x + 25)$ $x^3 + 125$

23. $(2x - 3)(x^2 - 3x + 5)$ $2x^3 - 9x^2 + 19x - 15$

24. $(2x - 1)(x^2 - 4x + 3)$ $2x^3 - 9x^2 + 10x - 3$

25. $(x + 7)(x + 3)$

26. $(x + 8)(x + 5)$

27. $(x - 5)(x + 3)$

28. $(x - 1)(x + 2)$ $x^2 + x - 2$

29. $(3x + 5)(2x + 1)$

30. $(7x + 4)(3x + 1)$

31. $(2x - 3)(5x + 3)$

32. $(2x - 5)(7x + 2)$

33. $(5x^2 - 4)(3x^2 - 7)$

34. $(7x^2 - 2)(3x^2 - 5)$

35. $(8x^3 + 3)(x^2 - 5)$

36. $(7x^3 + 5)(x^2 - 2)$

37. $(x + 3)(x - 3)$ $x^2 - 9$

38. $(x + 5)(x - 5)$ $x^2 - 25$

39. $(3x + 2)(3x - 2)$

40. $(2x + 5)(2x - 5)$ $4x^2 - 25$

41. $(5 - 7x)(5 + 7x)$

42. $(4 - 3x)(4 + 3x)$ $16 - 9x^2$

43. $(4x^2 + 5x)(4x^2 - 5x)$

44. $(3x^2 + 4x)(3x^2 - 4x)$

45. $(1 - y^5)(1 + y^5)$ $1 - y^{10}$

46. $(2 - y^5)(2 + y^5)$ $4 - y^{10}$

47. $(x + 2)^2$ $x^2 + 4x + 4$

48. $(x + 5)^2$ $x^2 + 10x + 25$

49. $(2x + 3)^2$ $4x^2 + 12x + 9$

50. $(3x + 2)^2$ $9x^2 + 12x + 4$

51. $(x - 3)^2$ $x^2 - 6x + 9$

52. $(x - 4)^2$ $x^2 - 8x + 16$

53. $(4x^2 - 1)^2$ $16x^4 - 8x^2 + 1$

54. $(5x^2 - 3)^2$ $25x^4 - 30x^2 + 9$

55. $(7 - 2x)^2$ $4x^2 - 28x + 49$

56. $(9 - 5x)^2$ $25x^2 - 90x + 81$

Practice Plus

In Exercises 57–64, perform the indicated operation or operations.

57. $(3x + 4)^2 - (3x - 4)^2$ $48x$

58. $(5x + 2)^2 - (5x - 2)^2$ $40x$

59. $(5x - 7)(3x - 2) - (4x - 5)(6x - 1)$ $-9x^2 + 3x + 9$

60. $(3x + 5)(2x - 9) - (7x - 2)(x - 1)$ $-x^2 - 8x - 47$

61. $(2x + 5)(2x - 5)(4x^2 + 25)$ $16x^4 - 625$

62. $(3x + 4)(3x - 4)(9x^2 + 16)$ $81x^4 - 256$

63. $\dfrac{(2x - 7)^5}{(2x - 7)^3}$ $4x^2 - 28x + 49$ **64.** $\dfrac{(5x - 3)^6}{(5x - 3)^4}$ $25x^2 - 30x + 9$

9. $11x^3 + 7x^2 - 12x - 4; 3$ **10.** $12x^3 - 5x^2 - 4x - 4; 3$ **11.** $12x^3 + 4x^2 + 12x - 14; 3$ **12.** $9x^4 + 4x^3 - 2x + 1; 4$ **13.** $6x^2 - 6x + 2; 2$
14. $6x^3 + 10x^2 + 11x - 8; 3$ **19.** $10x^8 - 20x^5 + 45x^3$ **20.** $18x^8 - 30x^5 + 42x^3$ **25.** $x^2 + 10x + 21$ **26.** $x^2 + 13x + 40$ **27.** $x^2 - 2x - 15$
29. $6x^2 + 13x + 5$ **30.** $21x^2 + 19x + 4$ **31.** $10x^2 - 9x - 9$ **32.** $14x^2 - 31x - 10$ **33.** $15x^4 - 47x^2 + 28$ **34.** $21x^4 - 41x^2 + 10$ **35.** $8x^5 - 40x^3 + 3x^2 - 15$
36. $7x^5 - 14x^3 + 5x^2 - 10$ **39.** $9x^2 - 4$ **41.** $25 - 49x^2$ **43.** $16x^4 - 25x^2$ **44.** $9x^4 - 16x^2$

Application Exercises

The bar graph shows the population of the United States, in millions, for six selected years.

Population of the United States

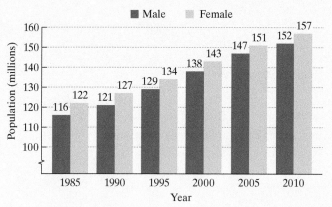

Source: U.S. Census Bureau

Here are two functions that model the data:

$$M(x) = 0.003x^2 + 1.45x + 115$$

> Male U.S. population,
> $M(x)$, in millions,
> x years after 1985

$$F(x) = 0.006x^2 + 1.30x + 121.$$

> Female U.S. population,
> $F(x)$, in millions,
> x years after 1985

Use the functions to solve Exercises 65–66.

65. a. Write a function d that models the difference between the female U.S. population and the male U.S. population for the years shown in the bar graph. $d(x) = 0.003x^2 - 0.15x + 6$

b. Use the function from part (a) to find how many more women than men there were in the U.S. population in 2005.   4.2 million

c. Does the result in part (b) overestimate or underestimate the actual difference between the female and male population in 2005 shown by the bar graph? By how much?   overestimates by 0.2 million

66. a. Write a function p that models the total U.S. population for the years shown in the bar graph.

b. Use the function from part (a) to find the total U.S. population in 1995.   264.4 million

c. Does the result in part (b) overestimate or underestimate the actual total U.S. population in 1995 shown by the bar graph? By how much?   overestimates by 1.4 million

The area, A, of a rectangle with length l and width w is given by the formula A = lw. In Exercises 67–72, use this formula to write a polynomial in standard form that models, or represents, the area of each shaded region.

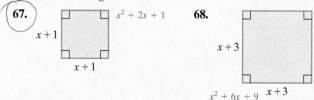

67. $x^2 + 2x + 1$ **68.**

66. a. $p(x) = 0.009x^2 + 2.75x + 236$

69. $4x^2 - 9$ **70.** $16x^2 - 9$

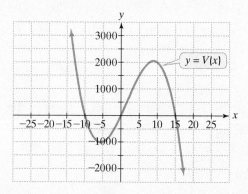

71.

72.

73. A popular model of carry-on luggage has a length that is 10 inches greater than its depth. Airline regulations require that the sum of the length, width, and depth cannot exceed 40 inches. These conditions, with the assumption that this sum *is* 40 inches, can be modeled by a function that gives the volume of the luggage, V, in cubic inches, in terms of its depth, x, in inches.

> Volume = depth · length · width: 40 − (depth + length)

$$V(x) = x \cdot (x + 10) \cdot [40 - (x + x + 10)]$$
$$V(x) = x(x + 10)(30 - 2x)$$

a. Perform the multiplications in the formula for $V(x)$ and express the formula in standard form.

b. Use the formula from part (a) to find $V(10)$. Describe what this means in practical terms.

c. The graph of the function modeling the volume of carry-on luggage is shown below. Identify your answer from part (b) as a point on the graph.   (10, 2000)

$y = V(x)$

d. Use the graph to describe realistic values for x for the volume function, where x represents the depth of the carry-on luggage. Use set-builder notation to express these realistic values. $\{x \mid 0 < x < 15\}$, although answers may vary.

73. a. $V(x) = -2x^3 + 10x^2 + 300x$ **b.** $V(10) = 2000$: Carry-on luggage with a depth of 10 inches has a volume of 2000 cubic inches.

74. Before working this exercise, be sure that you have read the Blitzer Bonus on page 264. The table shows the four combinations of color genes that a YY yellow lab and a BY black lab can pass to their offspring.

	B	**Y**
Y	BY	YY
Y	BY	YY

a. How many combinations result in a yellow lab with two recessive yellow genes? What is the probability of a yellow lab? $2; \frac{1}{2}$

b. How many combinations result in a black lab with a recessive yellow gene? What is the probability of a black lab? $2; \frac{1}{2}$

c. Find the product of Y and $\frac{1}{2}B + \frac{1}{2}Y$. How does this product model the probabilities that you determined in parts (a) and (b)?

74. c. $\frac{1}{2}BY + \frac{1}{2}Y^2$ or $\frac{1}{2}BY + \frac{1}{2}YY$; The first term describes the probability in part (b) and the second term describes the probability in part (a).

Critical Thinking Exercises

In Exercises 75–77, perform the indicated operations.

75. $[(7x^2 + 5) + 4x][(7x^2 + 5) - 4x]$ $49x^4 + 54x^2 + 25$

76. $[(3x^2 + x) + 1]^2$ $9x^4 + 6x^3 + 7x^2 + 2x + 1$

77. $(x^n + 2)(x^n - 2) - (x^n - 3)^2$ $6x^n - 13$

78. Express the area of the plane figure shown as a polynomial in standard form. $x^2 + 2x$

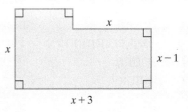

4.2 Factoring Polynomials

After you have read this section, you should be able to:

1️⃣ Factor out the greatest common factor of a polynomial.

2️⃣ Factor by grouping.

3️⃣ Factor a trinomial whose leading coefficient is 1.

4️⃣ Factor a trinomial whose leading coefficient is not 1.

5️⃣ Factor the difference of squares.

6️⃣ Factor perfect square trinomials.

7️⃣ Use a general strategy for factoring polynomials.

A two-year-old boy is asked, "Do you have a brother?" He answers, "Yes." "What is your brother's name?" "Tom." Asked if Tom has a brother, the two-year-old replies, "No." The child can go in the direction from self to brother, but he cannot reverse this direction and move from brother back to self.

As our intellects develop, we learn to reverse the direction of our thinking. Reversibility of thought is found throughout algebra. For example, we can multiply polynomials and show that

$$5x(2x + 3) = 10x^2 + 15x.$$

We can also reverse this process and express the resulting polynomial as

$$10x^2 + 15x = 5x(2x + 3).$$

Factoring a polynomial containing the sum of monomials means finding an equivalent expression that is a product.

Factoring $10x^2 + 15x$

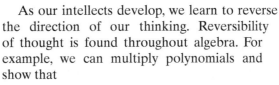

$$10x^2 + 15x = 5x(2x + 3)$$

The factors of $10x^2 + 15x$
are $5x$ and $2x + 3$.

In this section, we will be **factoring over the set of integers**, meaning that the coefficients in the factors are integers. Polynomials that cannot be factored using integer coefficients are called **irreducible over the integers**, or **prime**.

The goal in factoring a polynomial is to use one or more factoring techniques until each of the polynomial's factors, except possibly for a monomial factor, is prime or irreducible. In this situation, the polynomial is said to be **factored completely**.

We will now discuss basic techniques for factoring polynomials.

Factoring Out the Greatest Common Factor

① Factor out the greatest common factor of a polynomial.

In any factoring problem, the first step is to look for the *greatest common factor*. The **greatest common factor**, abbreviated GCF, is an expression with the greatest coefficient and of the highest degree that divides each term of the polynomial. Can you see that $7x$ is the greatest common factor of $21x^2 + 28x$? 7 is the greatest integer that divides both 21 and 28. Furthermore, x is the greatest power of x that divides x^2 and x.

The variable part of the greatest common factor always contains the *smallest* power of a variable that appears in all terms of the polynomial. For example, consider the polynomial

$$21x^2 + 28x.$$

> x^1, or x, is the variable raised to the smallest exponent.

We see that x is the variable part of the greatest common factor, $7x$.

When factoring a monomial from a polynomial, determine the greatest common factor of all terms in the polynomial. Sometimes there may not be a GCF other than 1. When a GCF other than 1 exists, we use the following procedure:

> ### Factoring a Monomial from a Polynomial
>
> **1.** Determine the greatest common factor of all terms in the polynomial.
> **2.** Express each term as the product of the GCF and its other factor.
> **3.** Use the distributive property to factor out the GCF.

Example 1 Factoring Out the Greatest Common Factor

Factor: $21x^2 + 28x$.

Solution

The GCF of the two terms of the polynomial is $7x$.

$$21x^2 + 28x$$
$$= 7x(3x) + 7x(4) \qquad \text{Express each term as the product of the GCF and its other factor.}$$
$$= 7x(3x + 4) \qquad \text{Factor out the GCF.}$$

We can check this factorization by multiplying $7x$ and $3x + 4$, obtaining the original polynomial as the answer.

 **Check Point 1** Factor: $20x^2 + 30x$. $10x(2x + 3)$

Example 2   **Factoring Out the Greatest Common Factor**

Factor:

a. $18x^3 + 27x^2$ **b.** $x^2(x + 3) + 5(x + 3)$.

Solution

a. First, determine the greatest common factor.

9 is the greatest integer that divides 18 and 27.

$$18x^3 + 27x^2$$

x^2 is the greatest expression that divides x^3 and x^2.

The GCF of the two terms of the polynomial is $9x^2$.

$$18x^3 + 27x^2$$
$$= 9x^2(2x) + 9x^2(3) \quad \text{Express each term as the product of the GCF and its other factor.}$$
$$= 9x^2(2x + 3) \quad \text{Factor out the GCF.}$$

b. In this situation, the greatest common factor is the common binomial factor $(x + 3)$. We factor out this common factor as follows:

$$x^2(x + 3) + 5(x + 3) = (x + 3)(x^2 + 5). \quad \text{Factor out the common binomial factor.}$$

Check Point 2   Factor:

a. $10x^3 - 4x^2$ $2x^2(5x - 2)$ **b.** $2x(x - 7) + 3(x - 7)$. $(x - 7)(2x + 3)$

Factoring by Grouping

 **Factor by grouping.**

The terms of some polynomials have only a greatest common factor of 1. However, by a suitable grouping of the terms, it still may be possible to factor by factoring out common factors. This process, called **factoring by grouping**, is illustrated in Example 3.

Example 3   **Factoring by Grouping**

Factor: $x^3 + 4x^2 + 3x + 12$.

Solution

There is no factor other than 1 common to all terms. However, we can group terms that have a common factor:

$$\boxed{x^3 + 4x^2} \;+\; \boxed{3x + 12}.$$

Common factor is x^2.   Common factor is 3.

We now factor the given polynomial as follows:

$$x^3 + 4x^2 + 3x + 12$$
$$= (x^3 + 4x^2) + (3x + 12) \quad \text{Group terms with common factors.}$$
$$= x^2(x + 4) + 3(x + 4) \quad \text{Factor out the greatest common factor from the grouped terms. The remaining two terms have } x + 4 \text{ as a common binomial factor.}$$
$$= (x + 4)(x^2 + 3). \quad \text{Factor out the GCF, } x + 4.$$

Great Question!

Do I have to group the terms as $(x^3 + 4x^2) + (3x + 12)$ when factoring $x^3 + 4x^2 + 3x + 12$ by grouping?

No. In Example 3, you can group the terms as follows:

$$(x^3 + 3x) + (4x^2 + 12).$$

Factor out the greatest common factor from each group and complete the factoring process. Describe what happens. What can you conclude?

Thus, $x^3 + 4x^2 + 3x + 12 = (x + 4)(x^2 + 3)$. Check the factorization by multiplying the right side of the equation using the FOIL method. Because the factorization is correct, you should obtain the original polynomial.

> ✓ **Check Point 3**   Factor: $x^3 + 5x^2 + 2x + 10$. $(x + 5)(x^2 + 2)$

Factoring by Grouping

1. Group terms that have a common monomial factor. There will usually be two groups. Sometimes the terms must be rearranged.
2. Factor out the common monomial factor from each group.
3. Factor out the remaining common binomial factor (if one exists).

---

③ Factor a trinomial whose leading coefficient is 1.

Factoring a Trinomial Whose Leading Coefficient Is 1

In Section 4.1, we used the FOIL method to multiply two binomials. The product was often a trinomial. The following are some examples:

Factored Form		F O I L		Trinomial Form
$(x + 3)(x + 4)$	$=$	$x^2 + 4x + 3x + 12$	$=$	$x^2 + 7x + 12$
$(x - 3)(x - 4)$	$=$	$x^2 - 4x - 3x + 12$	$=$	$x^2 - 7x + 12$
$(x + 3)(x - 5)$	$=$	$x^2 - 5x + 3x - 15$	$=$	$x^2 - 2x - 15$

Observe that each trinomial is of the form $x^2 + bx + c$, where the coefficient of the squared term is 1. Our goal is to start with the trinomial form and, assuming that it is factorable, return to the factored form.

The first FOIL multiplication shown in our list indicates that

$$(x + 3)(x + 4) = x^2 + 7x + 12.$$

Let's reverse the sides of this equation:

$$x^2 + 7x + 12 = (x + 3)(x + 4).$$

We can make several important observations about the factors on the right side.

$x^2 + 7x + 12 = (x + 3)(x + 4)$ $x^2 + 7x + 12 = (x + 3)(x + 4)$ $x^2 + 7x + 12 = (x + 3)(x + 4)$

I: $3x$

O: $4x$

The first term of each factor is x. The product of the First terms is $x \cdot x = x^2$.

3 and 4 are factors of 12. The product of the Last terms is $3 \cdot 4 = 12$.

The sum of the Outside and Inside products is $4x + 3x = 7x$.

These observations provide us with a procedure for factoring $x^2 + bx + c$.

A Strategy for Factoring $x^2 + bx + c$

1. Enter x as the first term of each factor.

$$(x \quad)(x \quad) = x^2 + bx + c$$

2. List all pairs of factors of the constant c.

3. Try various combinations of these factors as the second term in each set of parentheses. Select the combination in which the sum of the Outside and Inside products is equal to bx.

$$(x + \square)(x + \square) = x^2 + bx + c$$

I

O

Sum of O + I

4. Check your work by multiplying the factors using the FOIL method. You should obtain the original trinomial.

If none of the possible combinations yield an Outside product and an Inside product whose sum is equal to bx, the trinomial cannot be factored using integers and is called **prime** over the set of integers.

Example 4   Factoring a Trinomial Whose Leading Coefficient Is 1

Factor: $x^2 + 6x + 8$.

Solution

Step 1   Enter x as the first term of each factor.

$$x^2 + 6x + 8 = (x \quad)(x \quad)$$

To find the second term of each factor, we must find two integers whose product is 8 and whose sum is 6.

Step 2   List all pairs of factors of the constant, 8.

Factors of 8	8, 1	4, 2	−8, −1	−4, −2

Step 3   Try various combinations of these factors. The correct factorization of $x^2 + 6x + 8$ is the one in which the sum of the Outside and Inside products is equal to $6x$. Here is a list of the possible factorizations:

Possible Factorizations of $x^2 + 6x + 8$	Sum of Outside and Inside Products (Should Equal 6x)
$(x + 8)(x + 1)$	$x + 8x = 9x$
$(x + 4)(x + 2)$	$2x + 4x = 6x$
$(x - 8)(x - 1)$	$-x - 8x = -9x$
$(x - 4)(x - 2)$	$-2x - 4x = -6x$

This is the required middle term.

Thus, $x^2 + 6x + 8 = (x + 4)(x + 2)$.

Great Question!

Is there a way to eliminate some of the combinations of factors for $x^2 + bx + c$ when c is positive?

Yes. To factor $x^2 + bx + c$ when c is positive, find two numbers with the same sign as the middle term.

$$x^2 + 6x + 8 = (x + 2)(x + 4)$$

Same signs

$$x^2 - 5x + 6 = (x - 3)(x - 2)$$

Same signs

Using this observation, it is not necessary to list the last two factorizations in step 3 in the solution of Example 4 shown on the right.

Step 4   Check this result, $x^2 + 6x + 8 = (x + 4)(x + 2)$, by multiplying the right side using the FOIL method. You should obtain the original trinomial. Because of the commutative property, the factorization can also be expressed as

$$x^2 + 6x + 8 = (x + 2)(x + 4).$$

 **Check Point 4**   Factor: $x^2 + 5x + 6$. $(x+2)(x+3)$

Example 5   Factoring a Trinomial Whose Leading Coefficient Is 1

Factor: $x^2 + 2x - 35$.

Solution

Step 1   Enter x as the first term of each factor.

$$x^2 + 2x - 35 = (x \quad)(x \quad)$$

To find the second term of each factor, we must find two integers whose product is -35 and whose sum is 2.

Step 2   List all pairs of factors of the constant, -35.

Factors of -35	35, -1	-35, 1	-7, 5	7, -5

Step 3   Try various combinations of these factors. The correct factorization of $x^2 + 2x - 35$ is the one in which the sum of the Outside and Inside products is equal to $2x$. Here is a list of the possible factorizations:

Possible Factorizations of $x^2 + 2x - 35$	Sum of Outside and Inside Products (Should Equal $2x$)
$(x - 1)(x + 35)$	$35x - x = 34x$
$(x + 1)(x - 35)$	$-35x + x = -34x$
$(x - 7)(x + 5)$	$5x - 7x = -2x$
$(x + 7)(x - 5)$	$-5x + 7x = 2x$

This is the required middle term.

Thus, $x^2 + 2x - 35 = (x + 7)(x - 5)$ or $(x - 5)(x + 7)$.

Step 4   Verify the factorization using the FOIL method.

F   O   I   L

$$(x + 7)(x - 5) = x^2 - 5x + 7x - 35 = x^2 + 2x - 35$$

Because the product of the factors is the original trinomial, the factorization is correct.

 **Check Point 5**   Factor: $x^2 + 3x - 10$. $(x+5)(x-2)$

Factoring a Trinomial Whose Leading Coefficient Is Not 1

How do we factor a trinomial such as $3x^2 - 20x + 28$? Notice that the leading coefficient is 3. We must find two binomials whose product is $3x^2 - 20x + 28$. The product of the First terms must be $3x^2$:

$$(3x \quad)(x \quad).$$

Great Question!

Is there a way to eliminate some of the combinations of factors for $x^2 + bx + c$ when c is negative?

Yes. To factor $x^2 + bx + c$ when c is negative, find two numbers with opposite signs whose sum is the coefficient of the middle term.

$x^2 + 2x - 35 = (x + 7)(x - 5)$

Negative   Opposite signs

 4 Factor a trinomial whose leading coefficient is not 1.

From this point on, the factoring strategy is exactly the same as the one we use to factor a trinomial whose leading coefficient is 1.

> **A Strategy for Factoring $ax^2 + bx + c$**
>
> Assume, for the moment, that the greatest common factor is 1.
>
> **1.** Find two **First** terms whose product is ax^2:
>
> $$(\Box x + \quad)(\Box x + \quad) = ax^2 + bx + c.$$
>
> **2.** Find two **Last** terms whose product is c:
>
> $$(\Box x + \Box)(\Box x + \Box) = ax^2 + bx + c.$$
>
> **3.** By trial and error, perform steps 1 and 2 until the sum of the **O**utside product and **I**nside product is bx:
>
> $$(\Box x + \Box)(\Box x + \Box) = ax^2 + bx + c.$$
>
> $$\text{I}$$
> $$\text{O}$$
> $$\text{Sum of O + I}$$
>
> If no such combination exists, the polynomial is prime.

Great Question!

Should I feel discouraged if it takes me a while to get the correct factorization?

The *error* part of the factoring strategy plays an important role in the process. If you do not get the correct factorization the first time, this is not a bad thing. This error is often helpful in leading you to the correct factorization.

Example 6   Factoring a Trinomial Whose Leading Coefficient Is Not 1

Factor: $3x^2 - 20x + 28$.

Solution

Step 1   Find two First terms whose product is $3x^2$.

$$3x^2 - 20x + 28 = (3x \quad)(x \quad)$$

Step 2   Find two Last terms whose product is 28. The number 28 has pairs of factors that are either both positive or both negative. Because the middle term, $-20x$, is negative, both factors must be negative. The negative factorizations of 28 are $-1(-28)$, $-2(-14)$, and $-4(-7)$.

Step 3   Try various combinations of these factors. The correct factorization of $3x^2 - 20x + 28$ is the one in which the sum of the Outside and Inside products is equal to $-20x$. Here is a list of the possible factorizations:

Great Question!

When factoring trinomials, must I list every possible factorization before getting the correct one?

With practice, you will find that it is not necessary to list every possible factorization of the trinomial. As you practice factoring, you will be able to narrow down the list of possible factorizations to just a few. When it comes to factoring, practice makes perfect.

Possible Factorizations of $3x^2 - 20x + 28$	Sum of Outside and Inside Products (Should Equal $-20x$)
$(3x - 1)(x - 28)$	$-84x - x = -85x$
$(3x - 28)(x - 1)$	$-3x - 28x = -31x$
$(3x - 2)(x - 14)$	$-42x - 2x = -44x$
$(3x - 14)(x - 2)$	$-6x - 14x = -20x$
$(3x - 4)(x - 7)$	$-21x - 4x = -25x$
$(3x - 7)(x - 4)$	$-12x - 7x = -19x$

This is the required middle term.

Thus,

$$3x^2 - 20x + 28 = (3x - 14)(x - 2) \text{ or } (x - 2)(3x - 14).$$

Step 4 Verify the factorization, $3x^2 - 20x + 28 = (3x - 14)(x - 2)$, using the FOIL method.

$$\overset{\text{F}\qquad\quad\text{O}\qquad\qquad\quad\text{I}\qquad\qquad\text{L}}{}$$

$$(3x - 14)(x - 2) = 3x \cdot x + 3x(-2) + (-14) \cdot x + (-14)(-2)$$
$$= 3x^2 - 6x - 14x + 28$$
$$= 3x^2 - 20x + 28$$

Because this is the trinomial we started with, the factorization is correct.

✔ Check Point 6 Factor: $5x^2 - 14x + 8$. $(5x - 4)(x - 2)$

Example 7 Factoring a Trinomial Whose Leading Coefficient Is Not 1

Factor: $8y^2 - 10y - 3$.

Solution

Step 1 Find two First terms whose product is $8y^2$.

$$8y^2 - 10y - 3 \overset{?}{=} (8y\quad)(y\quad)$$
$$8y^2 - 10y - 3 \overset{?}{=} (4y\quad)(2y\quad)$$

Step 2 Find two Last terms whose product is -3. The possible factorizations are $1(-3)$ and $-1(3)$.

Step 3 Try various combinations of these factors. The correct factorization of $8y^2 - 10y - 3$ is the one in which the sum of the Outside and Inside products is equal to $-10y$. Here is a list of the possible factorizations:

Possible Factorizations of $8y^2 - 10y - 3$	Sum of Outside and Inside Products (Should Equal $-10y$)
$(8y + 1)(y - 3)$	$-24y + y = -23y$
$(8y - 3)(y + 1)$	$8y - 3y = 5y$
$(8y - 1)(y + 3)$	$24y - y = 23y$
$(8y + 3)(y - 1)$	$-8y + 3y = -5y$
$(4y + 1)(2y - 3)$	$-12y + 2y = -10y$
$(4y - 3)(2y + 1)$	$4y - 6y = -2y$
$(4y - 1)(2y + 3)$	$12y - 2y = 10y$
$(4y + 3)(2y - 1)$	$-4y + 6y = 2y$

These four factorizations are $(8y\quad)(y\quad)$ with $1(-3)$ and $-1(3)$ as factorizations of -3.

These four factorizations are $(4y\quad)(2y\quad)$ with $1(-3)$ and $-1(3)$ as factorizations of -3.

This is the required middle term.

Thus,

$$8y^2 - 10y - 3 = (4y + 1)(2y - 3) \text{ or } (2y - 3)(4y + 1).$$

Show that either of these factorizations is correct by multiplying the factors using the FOIL method. You should obtain the original trinomial.

✔ Check Point 7 Factor: $6y^2 + 19y - 7$. $(3y - 1)(2y + 7)$

Not every trinomial can be factored. For example, consider

$$6x^2 + 14x + 7 = (6x + \square)(x + \square)$$
$$6x^2 + 14x + 7 = (3x + \square)(2x + \square).$$

The possible factors for the last term are 1 and 7. However, regardless of how these factors are placed in the boxes shown, the sum of the Outside and Inside products is not equal to $14x$. Thus, the trinomial $6x^2 + 14x + 7$ cannot be factored and is prime.

Factoring the Difference of Two Squares

⑤ Factor the difference of squares.

A method for factoring the difference of two squares is obtained by reversing the product formula for the sum and difference of two terms.

The Difference of Two Squares

If A and B are real numbers, variables, or algebraic expressions, then

$$A^2 - B^2 = (A + B)(A - B).$$

In words: The difference of the squares of two terms factors as the product of the sum and the difference of those terms.

Example 8   Factoring the Difference of Two Squares

Factor:

a. $x^2 - 4$

b. $81x^2 - 49.$

Solution

We must express each term as the square of some monomial. Then we use the formula for factoring $A^2 - B^2$.

a. $x^2 - 4 = x^2 - 2^2 = (x + 2)(x - 2)$

b. $81x^2 - 49 = (9x)^2 - 7^2 = (9x + 7)(9x - 7)$

✓ **Check Point 8**   Factor:

a. $x^2 - 81$ $(x + 9)(x - 9)$

b. $36x^2 - 25.$ $(6x + 5)(6x - 5)$

We have seen that a polynomial is factored completely when it is written as the product of prime polynomials. To be sure that you have factored completely, check to see whether any factors with more than one term in the factored polynomial can be factored further. If so, continue factoring.

Example 9   A Repeated Factorization

Factor completely: $x^4 - 81.$

Solution

$x^4 - 81 = (x^2)^2 - 9^2$ *Express as the difference of two squares.*

$= (x^2 + 9)(x^2 - 9)$ *The factors are the sum and the difference of the expressions being squared.*

$= (x^2 + 9)(x^2 - 3^2)$ *The factor $x^2 - 9$ is the difference of two squares and can be factored.*

$= (x^2 + 9)(x + 3)(x - 3)$ *The factors of $x^2 - 9$ are the sum and the difference of the expressions being squared.*

Great Question!

Why isn't factoring $x^4 - 81$ as

$$(x^2 + 9)(x^2 - 9)$$

a complete factorization?

The second factor, $x^2 - 9$, is itself a difference of two squares and can be factored.

✓ **Check Point 9**   Factor completely: $81x^4 - 16.$ $(9x^2 + 4)(3x + 2)(3x - 2)$

6 Factor perfect square trinomials.

Factoring Perfect Square Trinomials

Our next factoring technique is obtained by reversing the product formulas for squaring binomials. The trinomials that are factored using this technique are called **perfect square trinomials**.

Factoring Perfect Square Trinomials

Let A and B be real numbers, variables, or algebraic expressions.

1. $A^2 + 2AB + B^2 = (A + B)^2$ **2.** $A^2 - 2AB + B^2 = (A - B)^2$

         Same sign   Same sign

The two items in the box show that perfect square trinomials, $A^2 + 2AB + B^2$ and $A^2 - 2AB + B^2$, come in two forms: one in which the coefficient of the middle term is positive and one in which the coefficient of the middle term is negative. Here's how to recognize a perfect square trinomial:

1. The first and last terms are squares of monomials or integers.
2. The middle term is twice the product of the expressions being squared in the first and last terms.

Example 10 **Factoring Perfect Square Trinomials**

Factor:

a. $x^2 + 6x + 9$ **b.** $25x^2 - 60x + 36.$

Solution

a. $x^2 + 6x + 9 = x^2 + 2 \cdot x \cdot 3 + 3^2 = (x + 3)^2$ *The middle term has a positive sign.*

                $A^2 + 2AB + B^2 = (A + B)^2$

b. We suspect that $25x^2 - 60x + 36$ is a perfect square trinomial because $25x^2 = (5x)^2$ and $36 = 6^2$. The middle term can be expressed as twice the product of $5x$ and 6.

$$25x^2 - 60x + 36 = (5x)^2 - 2 \cdot 5x \cdot 6 + 6^2 = (5x - 6)^2$$

                      $A^2 - 2AB + B^2 = (A - B)^2$

 **Check Point 10** Factor:

a. $x^2 + 14x + 49$ $(x + 7)^2$ **b.** $16x^2 - 56x + 49.$ $(4x - 7)^2$

A Strategy for Factoring Polynomials

7 Use a general strategy for factoring polynomials.

It is important to practice factoring a wide variety of polynomials so that you can quickly select the appropriate technique. A polynomial is factored completely when all its polynomial factors, except possibly for monomial factors, are prime. Because of the commutative property, the order of the factors does not matter.

> **A Strategy for Factoring a Polynomial**
> 1. If the terms have a common factor, factor out the GCF.
> 2. Determine the number of terms in the polynomial and try factoring as follows:
>   **a.** If there are two terms, can the binomial be factored as the difference of two squares?
> $$A^2 - B^2 = (A + B)(A - B)$$
>   **b.** If there are three terms, is the trinomial a perfect square trinomial? If so, factor by using one of the following formulas:
> $$A^2 + 2AB + B^2 = (A + B)^2$$
> $$A^2 - 2AB + B^2 = (A - B)^2.$$
>   If the trinomial is not a perfect square trinomial, try factoring by trial and error.
>   **c.** If there are four or more terms, try factoring by grouping.
> 3. Check to see if any factors with more than one term in the factored polynomial can be factored further. If so, factor completely.
> 4. Check by multiplying.

Example 11   Factoring a Polynomial

Factor: $3x^2 - 6x - 45$.

Solution

Step 1   If the terms have a common factor, factor out the GCF. Because 3 is common to all terms, we factor it out.
$$3x^2 - 6x - 45 = 3(x^2 - 2x - 15) \quad \text{Factor out the GCF.}$$

Step 2   Determine the number of terms and factor accordingly. The factor $x^2 - 2x - 15$ has three terms, but it is not a perfect square trinomial. We factor it using trial and error.
$$3x^2 - 6x - 45 = 3(x^2 - 2x - 15) = 3(x - 5)(x + 3)$$

Step 3   Check to see if factors can be factored further. In this case, they cannot, so we have factored completely.

Step 4   Check by multiplying.

$$3(x - 5)(x + 3) = 3(x^2 - 2x - 15) = 3x^2 - 6x - 45$$

FOIL

This is the original polynomial, so the factorization is correct.

 **Check Point 11**   Factor: $4x^2 - 16x - 48$. $4(x - 6)(x + 2)$

Example 12   Factoring a Polynomial

Factor: $7x^5 - 7x$.

Solution

Step 1   If the terms have a common factor, factor out the GCF. Because $7x$ is common to both terms, we factor it out.
$$7x^5 - 7x = 7x(x^4 - 1) \quad \text{Factor out the GCF.}$$

Step 2   Determine the number of terms and factor accordingly. Up to this point, we have $7x^5 - 7x = 7x(x^4 - 1)$. However, the factor $x^4 - 1$ has two terms. This binomial can be expressed as $(x^2)^2 - 1^2$, so it can be factored as the difference of two squares.

$$7x^5 - 7x = 7x(x^4 - 1) = 7x(x^2 + 1)(x^2 - 1)$$ Use $A^2 - B^2 = (A + B)(A - B)$ on
$$x^4 - 1: A = x^2 \text{ and } B = 1.$$

Step 3   Check to see if factors can be factored further. We note that $(x^2 - 1)$ is also the difference of two squares, $x^2 - 1^2$, so we continue factoring.

$$7x^5 - 7x = 7x(x^2 + 1)(x + 1)(x - 1)$$ Factor $x^2 - 1$ as the difference of two squares.

Step 4   Check by multiplying.

$$7x(x^2 + 1)(x + 1)(x - 1) = 7x(x^2 + 1)(x^2 - 1) = 7x(x^4 - 1) = 7x^5 - 7x$$

We obtain the original polynomial, so the factorization is correct.

▶ **Check Point 12**   Factor: $4x^5 - 64x$. $4x(x^2 + 4)(x + 2)(x - 2)$

Example 13   Factoring a Polynomial

Factor: $x^3 - 5x^2 - 4x + 20$.

Solution

Step 1   If the terms have a common factor, factor out the GCF. Other than 1, there is no common factor.

Step 2   Determine the number of terms and factor accordingly. There are four terms. We try factoring by grouping.

> Notice that it is necessary to factor out a negative number from the second grouping to obtain a common binomial factor, $x - 5$, for the two groupings.

$$x^3 - 5x^2 - 4x + 20$$
$$= (x^3 - 5x^2) + (-4x + 20)$$ Group terms with common factors.
$$= x^2(x - 5) - 4(x - 5)$$ Factor from each group.
$$= (x - 5)(x^2 - 4)$$ Factor out the common binomial factor, $x - 5$.

Step 3   Check to see if factors can be factored further. We note that $(x^2 - 4)$ is the difference of two squares, $x^2 - 2^2$, so we continue factoring.

$$x^3 - 5x^2 - 4x + 20 = (x - 5)(x + 2)(x - 2)$$ Factor $x^2 - 4$ as the difference of two squares.

We have factored completely because no factor with more than one term can be factored further.

Step 4   Check by multiplying.

$$\overset{\text{F} \quad \text{O} \quad \text{I} \quad \text{L}}{}$$
$$(x - 5)(x + 2)(x - 2) = (x - 5)(x^2 - 4) = x^3 - 4x - 5x^2 + 20$$
$$= x^3 - 5x^2 - 4x + 20$$

We obtain the original polynomial, so the factorization is correct.

▶ **Check Point 13**   Factor: $x^3 - 4x^2 - 9x + 36$. $(x - 4)(x + 3)(x - 3)$

Example 14 Factoring a Polynomial

Factor: $2x^3 - 24x^2 + 72x$.

Solution

Step 1 If the terms have a common factor, factor out the GCF. Because $2x$ is common to all terms, we factor it out.

$$2x^3 - 24x^2 + 72x = 2x(x^2 - 12x + 36) \quad \text{Factor out the GCF.}$$

Step 2 Determine the number of terms and factor accordingly. The factor $x^2 - 12x + 36$ has three terms. Is it a perfect square trinomial? Yes. The first term, x^2, is the square of a monomial. The last term, 36 or 6^2, is the square of an integer. The middle term involves twice the product of x and 6. We factor using $A^2 - 2AB + B^2 = (A - B)^2$.

$$2x^3 - 24x^2 + 72x = 2x(x^2 - 12x + 36)$$

$$= 2x(\underbrace{x^2 - 2 \cdot x \cdot 6 + 6^2}_{A^2 - 2 \ A \ B + B^2}) \quad \text{The second factor is a perfect square trinomial.}$$

$$= 2x(x - 6)^2 \quad A^2 - 2AB + B^2 = (A - B)^2$$

Step 3 Check to see if factors can be factored further. In this problem, they cannot, so we have factored completely.

Step 4 Check by multiplying. Let's verify that $2x^3 - 24x^2 + 72x = 2x(x - 6)^2$.

$$2x(x - 6)^2 = 2x(x^2 - 12x + 36) = 2x^3 - 24x^2 + 72x$$

We obtain the original polynomial, so the factorization is correct.

▶ **Check Point 14** Factor: $3x^3 - 30x^2 + 75x$. $3x(x - 5)^2$

Blitzer Bonus

Candy and Quotients of Polynomials

A few candy factoids:

- Despite the fact that eating lots of candy has negative effects on health and teeth, the average American spends \$84 per year on candy, consuming 23.9 pounds annually.

- 55% of candy sales are "impulse buys."

- The word *candy* originated with the Sanskrit word *khanda*, meaning "pieces of crystallized sugar."

- A function consisting of the quotient of two polynomials models what happens in your mouth after eating "pieces of crystallized sugar":

$$f(x) = \frac{6.5x^2 - 20.4x + 234}{x^2 + 36}.$$

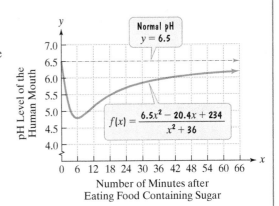

Figure 4.2

The function models the pH level, $f(x)$, of the human mouth after a person eats food containing sugar. The graph of this function is shown in **Figure 4.2**. The figure indicates that

- The normal pH level of the human mouth is 6.5.

- After eating food containing sugar, the pH level is the lowest, approximately 4.8, after 6 minutes.

- After quickly dropping below normal, the pH level of the mouth slowly begins to approach the normal level of 6.5 during the first hour.

Achieving Success

As you continue your study of mathematics, you will see how functions involving quotients of polynomials provide insights into phenomena as diverse as the aftereffects of candy on your mouth, the cost of environmental cleanup, and even our ongoing processes of learning and forgetting. Factoring is an essential skill for working with such functions. **Success in algebra cannot be achieved without a complete understanding of factoring.** Now is the time to practice factoring a wide variety of polynomials. As you approach this Exercise Set, be sure you can apply each of the factoring techniques discussed in the section. This can be achieved by working the assigned problems from Exercises 1–52. Then it's time to work problems that require you to select the appropriate technique or techniques from Exercises 53–110. The more deeply you force your brain to think about factoring by working many exercises, the better will be your chances of achieving success in future algebra courses.

Concept and Vocabulary Check

Exercises in the Concept and Vocabulary Check are intended for group and class discussions.

Here is a list of the factoring techniques that we have discussed.

a. Factoring out the GCF

b. Factoring by grouping

c. Factoring trinomials by trial and error

d. Factoring the difference of two squares
$A^2 - B^2 = (A + B)(A - B)$

e. Factoring perfect square trinomials
$A^2 + 2AB + B^2 = (A + B)^2$
$A^2 - 2AB + B^2 = (A - B)^2$

In Exercises 1–7, fill in each blank by writing the letter of the technique (a through e) for factoring the polynomial.

1. $16x^2 - 25$ ___d___

2. $x^2 + 7x + 10$ ___c___

3. $10x^2 + 50x$ ___a___

4. $x^2 + 6x + 9$ ___e, although__ c is also correct

5. $x^3 + 5x^2 + 2x + 10$ ___b___

6. $2x^2 + 13x + 15$ ___c___

7. $9x^2 - 6x + 1$ ___e, although__ c is also correct

In Exercises 8–13, determine whether each statement is true or false. If the statement is false, make the necessary change(s) to produce a true statement. Changes to false statements will vary.

8. One factor of $x^2 + x + 20$ is $x + 5$.   false

9. A trinomial can never have two identical factors.   false

10. One factor of $y^2 + 5y - 24$ is $y - 3$.   true

11. $x^2 + 4 = (x + 2)(x + 2)$ false

12. Because $x^2 - 25 = (x + 5)(x - 5)$, then $x^2 + 25 = (x - 5)(x + 5)$.   false

13. All perfect square trinomials are squares of binomials.   true

Respond to Exercises 14–19 using verbal or written explanations.

14. Using an example, explain how to factor out the greatest common factor of a polynomial.   **14–19.** Answers will vary.

15. Suppose that a polynomial contains four terms. Explain how to use factoring by grouping to factor the polynomial.

16. Explain how to factor $3x^2 + 10x + 8$.

17. Explain how to factor the difference of two squares. Provide an example with your explanation.

18. What is a perfect square trinomial and how is it factored?

19. What does it mean to factor completely?

5. $9x^2(x^2 - 2x + 3)$ **6.** $6x^2(x^2 - 3x + 2)$ **7.** $(x + 5)(x + 3)$ **8.** $(2x + 1)(x + 4)$ **9.** $(x - 3)(x^2 + 12)$
10. $(2x + 5)(x^2 + 17)$ **11.** $(x - 2)(x^2 + 5)$ **12.** $(x - 3)(x^2 + 4)$ **13.** $(x - 1)(x^2 + 2)$ **4.2** Factoring Polynomials   **281**
14. $(x + 6)(x^2 - 2)$ **15.** $(3x - 2)(x^2 - 2)$ **16.** $(x - 1)(x^2 - 5)$ **53.** $5x(x + 2)(x - 2)$ **54.** $4x(x + 5)(x - 5)$ **57.** $5(x - 3)(x + 2)$
58. $5(x - 5)(x + 2)$ **59.** $2(x^2 + 9)(x + 3)(x - 3)$ **60.** $7(x^2 + 1)(x + 1)(x - 1)$ **61.** $(x + 2)(x + 3)(x - 3)$ **62.** $(x + 3)(x + 5)(x - 5)$

Exercise Set 4.2

Practice Exercises

In Exercises 1–10, factor out the greatest common factor.

1. $18x + 27$ $\ (9(2x + 3)$

2. $16x - 24$ $\ 8(2x - 3)$

3. $3x^2 + 6x$ $\ 3x(x + 2)$

4. $4x^2 - 8x$ $\ 4x(x - 2)$

5. $9x^4 - 18x^3 + 27x^2$

6. $6x^4 - 18x^3 + 12x^2$

7. $x(x + 5) + 3(x + 5)$

8. $x(2x + 1) + 4(2x + 1)$

9. $x^2(x - 3) + 12(x - 3)$

10. $x^2(2x + 5) + 17(2x + 5)$

In Exercises 11–16, factor by grouping. If necessary, factor out a negative number from the second grouping to obtain a common binomial factor.

11. $x^3 - 2x^2 + 5x - 10$

12. $x^3 - 3x^2 + 4x - 12$

13. $x^3 - x^2 + 2x - 2$

14. $x^3 + 6x^2 - 2x - 12$

15. $3x^3 - 2x^2 - 6x + 4$

16. $x^3 - x^2 - 5x + 5$

In Exercises 17–34, factor each trinomial, or state that the trinomial is prime.

17. $x^2 + 5x + 6$ $\ (x + 2)(x + 3)$

18. $x^2 + 8x + 15$ $\ (x + 3)(x + 5)$

19. $x^2 - 2x - 15$ $\ (x - 5)(x + 3)$

20. $x^2 - 4x - 5$ $\ (x - 5)(x + 1)$

21. $x^2 - 8x + 15$ $\ (x - 5)(x - 3)$

22. $x^2 - 14x + 45$ $\ (x - 5)(x - 9)$

23. $3x^2 - x - 2$ $\ (3x + 2)(x - 1)$

24. $2x^2 + 5x - 3$ $\ (2x - 1)(x + 3)$

25. $3x^2 - 25x - 28$ $\ (3x - 28)(x + 1)$

26. $3x^2 - 2x - 5$ $\ (3x - 5)(x + 1)$

27. $6x^2 - 11x + 4$ $\ (2x - 1)(3x - 4)$

28. $6x^2 - 17x + 12$ $\ (2x - 3)(3x - 4)$

29. $4x^2 + 16x + 15$ $\ (2x + 3)(2x + 5)$

30. $8x^2 + 33x + 4$ $\ (8x + 1)(x + 4)$

31. $9x^2 - 9x + 2$ $\ (3x - 2)(3x - 1)$

32. $9x^2 + 5x - 4$ $\ (9x - 4)(x + 1)$

33. $20x^2 + 27x - 8$ $\ (5x + 8)(4x - 1)$

34. $15x^2 - 19x + 6$ $\ (5x - 3)(3x - 2)$

In Exercises 35–44, factor the difference of two squares.

35. $x^2 - 100$ $\ (x + 10)(x - 10)$

36. $x^2 - 144$ $\ (x + 12)(x - 12)$

37. $36x^2 - 49$ $\ (6x + 7)(6x - 7)$

38. $64x^2 - 81$ $\ (8x + 9)(8x - 9)$

39. $9 - 25y^2$ $\ (3 + 5y)(3 - 5y)$

40. $36 - 49y^2$ $\ (6 + 7y)(6 - 7y)$

41. $x^4 - 16$ $\ (x^2 + 4)(x + 2)(x - 2)$

42. $x^4 - 1$ $\ (x^2 + 1)(x + 1)(x - 1)$

43. $16x^4 - 81$ $\ (4x^2 + 9)(2x + 3)(2x - 3)$

44. $81x^4 - 1$ $\ (9x^2 + 1)(3x + 1)(3x - 1)$

In Exercises 45–52, factor each perfect square trinomial.

45. $x^2 + 2x + 1$ $\ (x + 1)^2$ **46.** $x^2 + 4x + 4$ $\ (x + 2)^2$

47. $x^2 - 14x + 49$ $\ (x - 7)^2$ **48.** $x^2 - 10x + 25$ $\ (x - 5)^2$

49. $4x^2 + 4x + 1$ $\ (2x + 1)^2$ **50.** $25x^2 + 10x + 1$ $\ (5x + 1)^2$

51. $9x^2 - 6x + 1$ $\ (3x - 1)^2$ **52.** $64x^2 - 16x + 1$ $\ (8x - 1)^2$

In Exercises 53–110, factor completely, or state that the polynomial is prime.

53. $5x^3 - 20x$

54. $4x^3 - 100x$

55. $7x^3 + 7x$ $\ 7x(x^2 + 1)$

56. $6x^3 + 24x$ $\ 6x(x^2 + 4)$

57. $5x^2 - 5x - 30$

58. $5x^2 - 15x - 50$

59. $2x^4 - 162$

60. $7x^4 - 7$

61. $x^3 + 2x^2 - 9x - 18$

62. $x^3 + 3x^2 - 25x - 75$

63. $3x^3 - 24x^2 + 48x$

64. $5x^3 - 20x^2 + 20x$

65. $6x^2 + 8x$

66. $21x^2 - 35x$

67. $2y^2 - 2y - 112$

68. $6x^2 - 6x - 12$

69. $7y^4 + 14y^3 + 7y^2$

70. $2y^4 + 28y^3 + 98y^2$

71. $y^2 + 8y - 16$ $\ $ prime

72. $y^2 - 18y - 81$ $\ $ prime

73. $16y^2 - 4y - 2$

74. $32y^2 + 4y - 6$

75. $r^2 - 25r$ $\ r(r - 25)$

76. $3r^2 - 27r$ $\ 3r(r - 9)$

77. $4w^2 + 8w - 5$

78. $35w^2 - 2w - 1$

79. $x^3 - 4x$ $\ x(x + 2)(x - 2)$ **80.** $9x^3 - 9x$ $\ 9x(x + 1)(x - 1)$

81. $x^2 + 64$ $\ $ prime

82. $y^2 + 36$ $\ $ prime

83. $9y^2 + 13y + 4$

84. $20y^2 + 12y + 1$

85. $y^3 + 2y^2 - 4y - 8$

86. $y^3 + 2y^2 - y - 2$

87. $16y^2 + 24y + 9$

88. $25y^2 + 20y + 4$

89. $4y^3 - 28y^2 + 40y$

90. $7y^3 - 21y^2 + 14y$

91. $y^5 - 81y$

92. $y^5 - 16y$

93. $20a^4 - 45a^2$

94. $48a^4 - 3a^2$

95. $9x^4 + 18x^3 + 6x^2$

96. $10x^4 + 20x^3 + 15x^2$

97. $12y^2 - 11y + 2$ $\ (4y - 1)(3y - 2)$

98. $21x^2 - 25x - 4$ $\ (7x + 1)(3x - 4)$

99. $9y^2 - 64$ $\ (3y + 8)(3y - 8)$

100. $100y^2 - 49$ $\ (10y + 7)(10y - 7)$

101. $9y^2 + 64$ $\ $ prime

102. $100y^2 + 49$ $\ $ prime

103. $2y^3 + 3y^2 - 50y - 75$ $\ (2y + 3)(y + 5)(y - 5)$

104. $12y^3 + 16y^2 - 3y - 4$ $\ (3y + 4)(2y + 1)(2y - 1)$

105. $2r^3 + 30r^2 - 68r$ $\ 2r(r + 17)(r - 2)$

63. $3x(x - 4)^2$ **64.** $5x(x - 2)^2$ **65.** $2x(3x + 4)$ **66.** $7x(3x - 5)$ **67.** $2(y - 8)(y + 7)$ **68.** $6(x - 2)(x + 1)$ **69.** $7y^2(y + 1)^2$ **70.** $2y^2(y + 7)^2$
73. $2(4y + 1)(2y - 1)$ **74.** $2(8y - 3)(2y + 1)$ **77.** $(2w + 5)(2w - 1)$ **78.** $(7w + 1)(5w - 1)$ **83.** $(9y + 4)(y + 1)$ **84.** $(2y + 1)(10y + 1)$
85. $(y + 2)(y + 2)(y - 2)$ **86.** $(y + 2)(y + 1)(y - 1)$ **87.** $(4y + 3)^2$ **88.** $(5y + 2)^2$ **89.** $4y(y - 5)(y - 2)$ **90.** $7y(y - 2)(y - 1)$
91. $y(y^2 + 9)(y + 3)(y - 3)$ **92.** $y(y^2 + 4)(y + 2)(y - 2)$ **93.** $5a^2(2a + 3)(2a - 3)$ **94.** $3a^2(4a + 1)(4a - 1)$ **95.** $3x^2(3x^2 + 6x + 2)$ **96.** $5x^2(2x^2 + 4x + 3)$

111. $(x + 1)(5x - 6)(2x + 1)$

112. $(x - 1)(6x - 5)(2x + 1)$ **113.** $(x^2 + 6)(6x^2 - 1)$ **114.** $(x^2 + 5)(7x^2 - 1)$ **115.** $(x - 7 + 2a)(x - 7 - 2a)$ **116.** $(x - 6 + 3a)(x - 6 - 3a)$

106. $3r^3 - 27r^2 - 210r$ $3r(r + 5)(r - 14)$

107. $8x^5 - 2x^3$ $2x^3(2x + 1)(2x - 1)$

108. $y^9 - y^5$ $y^5(y^2 + 1)(y + 1)(y - 1)$

109. $3x^2 + 243$ $3(x^2 + 81)$

110. $27x^2 + 75$ $3(9x^2 + 25)$

Practice Plus

In Exercises 111–118, factor completely.

111. $10x^2(x + 1) - 7x(x + 1) - 6(x + 1)$

112. $12x^2(x - 1) - 4x(x - 1) - 5(x - 1)$

113. $6x^4 + 35x^2 - 6$ **114.** $7x^4 + 34x^2 - 5$

115. $(x - 7)^2 - 4a^2$ **116.** $(x - 6)^2 - 9a^2$

117. $x^2 + 8x + 16 - 25a^2$ **118.** $x^2 + 14x + 49 - 16a^2$
$(x + 4 + 5a)(x + 4 - 5a)$ $(x + 7 + 4a)(x + 7 - 4a)$

Application Exercises

119. Your computer store is having an incredible sale. The price on one model is reduced by 40%. Then the sale price is reduced by another 40%. If x is the computer's original price, the sale price can be modeled by

$$(x - 0.4x) - 0.4(x - 0.4x).$$

   a. Factor out $(x - 0.4x)$ from each term. Then simplify the resulting expression.

   b. Use the simplified expression from part (a) to answer these questions. With a 40% reduction followed by a 40% reduction, is the computer selling at 20% of its original price? If not, at what percentage of the original price is it selling?   no; 36%

120. Your local electronics store is having an end-of-the-year sale. The price on a plasma television had been reduced by 30%. Now the sale price is reduced by another 30%. If x is the television's original price, the sale price can be modeled by

$$(x - 0.3x) - 0.3(x - 0.3x).$$

   a. Factor out $(x - 0.3x)$ from each term. Then simplify the resulting expression.

   b. Use the simplified expression from part (a) to answer these questions. With a 30% reduction followed by a 30% reduction, is the television selling at 40% of its original price? If not, at what percentage of the original price is it selling?   no; 49%

In Exercises 121–124,

   a. *Write an expression for the area of the shaded region.*

   b. *Write the expression in factored form.*

121.

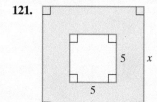

   a. $x^2 - 25$ **b.** $(x + 5)(x - 5)$

122.

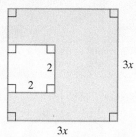

   a. $9x^2 - 4$ **b.** $(3x + 2)(3x - 2)$

123.

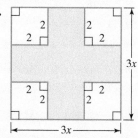

   a. $9x^2 - 16$ **b.** $(3x + 4)(3x - 4)$

124.

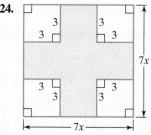

   a. $49x^2 - 36$ **b.** $(7x + 6)(7x - 6)$

Critical Thinking Exercises

125. Where is the error in this "proof" that $2 = 0$?

$a = b$	Suppose that a and b are any equal real numbers.
$a^2 = b^2$	Square both sides of the equation.
$a^2 - b^2 = 0$	Subtract b^2 from both sides.
$2(a^2 - b^2) = 2 \cdot 0$	Multiply both sides by 2.
$2(a^2 - b^2) = 0$	On the right side, $2 \cdot 0 = 0$.
$2(a + b)(a - b) = 0$	Factor $a^2 - b^2$.
$2(a + b) = 0$	Divide both sides by $a - b$.
$2 = 0$	Divide both sides by $a + b$.

In Exercises 126–129, factor each polynomial.

126. $x^2 - y^2 + 3x + 3y$ $(x + y)(x - y + 3)$

127. $x^{2n} - 25y^{2n}$ $(x^n + 5y^n)(x^n - 5y^n)$

128. $4x^{2n} + 12x^n + 9$ $(2x^n + 3)^2$

129. $(x + 3)^2 - 2(x + 3) + 1$ $[(x + 3) - 1]^2$ or $(x + 2)^2$

In Exercises 130–131, find all positive integers b so that the trinomial can be factored.

130. $x^2 + bx + 15$ 8 and 16

131. $x^2 + 4x + b$ 3 and 4

---

119. a. $(x - 0.4x)(1 - 0.4) = (0.6x)(0.6) = 0.36x$ **120. a.** $(x - 0.3x)(1 - 0.3) = (0.7x)(0.7) = 0.49x$ **125.** $a - b = 0$ and division by 0 is not permitted.

4.3 Solving Quadratic Equations by Factoring

What am I supposed to learn?

After you have read this section, you should be able to:

1. Solve quadratic equations by factoring.

2. Solve higher-degree polynomial equations by factoring.

3. Solve problems using quadratic equations.

Motion and change are the very essence of life. Moving air brushes against our faces; rain falls on our heads; birds fly past us; plants spring from the earth, grow, and then die; and rocks thrown upward reach a maximum height before falling to the ground. In this section, you will use quadratic functions and factoring strategies to model and visualize motion. Analyzing the where and when of moving objects involves equations in which the highest exponent on the variable is 2, called *quadratic equations*.

The Standard Form of a Quadratic Equation

We begin by defining a quadratic equation.

> **Definition of a Quadratic Equation**
>
> A **quadratic equation** in x is an equation that can be written in the **standard form**
>
> $$ax^2 + bx + c = 0,$$
>
> where a, b, and c are real numbers, with $a \neq 0$. A quadratic equation in x is also called a **second-degree polynomial equation** in x.

Here is an example of a quadratic equation in standard form:

$$x^2 - 12x + 27 = 0.$$

$$a = 1 \quad b = -12 \quad c = 27$$

Solving Quadratic Equations by Factoring

1. **Solve quadratic equations by factoring.**

We can factor the left side of the quadratic equation $x^2 - 12x + 27 = 0$. We obtain $(x - 3)(x - 9) = 0$. If a quadratic equation has zero on one side and a factored expression on the other side, it can be solved using the **zero-product principle**.

> **The Zero-Product Principle**
>
> If the product of two algebraic expressions is zero, then at least one of the factors is equal to zero.
>
> If $AB = 0$, then $A = 0$ or $B = 0$.

For example, consider the equation $(x - 3)(x - 9) = 0$. According to the zero-product principle, this product can be zero only if at least one of the factors is zero. We set each individual factor equal to zero and solve the resulting equations for x.

$$(x - 3)(x - 9) = 0$$
$$x - 3 = 0 \quad \text{or} \quad x - 9 = 0$$
$$x = 3 \qquad\qquad x = 9$$

The solutions of the original quadratic equation, $x^2 - 12x + 27 = 0$, are 3 and 9. The solution set is $\{3, 9\}$.

Solving a Quadratic Equation by Factoring

1. If necessary, rewrite the equation in the standard form $ax^2 + bx + c = 0$, moving all terms to one side, thereby obtaining zero on the other side.
2. Factor completely.
3. Apply the zero-product principle, setting each factor containing a variable equal to zero.
4. Solve the equations in step 3.
5. Check the solutions in the original equation.

Example 1 | **Solving a Quadratic Equation by Factoring**

Solve: $2x^2 - 5x = 12$.

Solution

Step 1 Move all terms to one side and obtain zero on the other side. Subtract 12 from both sides and write the equation in standard form.

$$2x^2 - 5x - 12 = 12 - 12$$
$$2x^2 - 5x - 12 = 0$$

Step 2 Factor.

$$(2x + 3)(x - 4) = 0$$

Steps 3 and 4 Set each factor equal to zero and solve the resulting equations.

$$2x + 3 = 0 \quad \text{or} \quad x - 4 = 0$$
$$2x = -3 \qquad\qquad x = 4$$
$$x = -\frac{3}{2}$$

Step 5 Check the solutions in the original equation.

Check $-\dfrac{3}{2}$:	**Check 4:**
$2x^2 - 5x = 12$	$2x^2 - 5x = 12$
$2\left(-\dfrac{3}{2}\right)^2 - 5\left(-\dfrac{3}{2}\right) \stackrel{?}{=} 12$	$2(4)^2 - 5(4) \stackrel{?}{=} 12$
$2\left(\dfrac{9}{4}\right) - 5\left(-\dfrac{3}{2}\right) \stackrel{?}{=} 12$	$2(16) - 5(4) \stackrel{?}{=} 12$
$\dfrac{9}{2} + \dfrac{15}{2} \stackrel{?}{=} 12$	$32 - 20 \stackrel{?}{=} 12$
$\dfrac{24}{2} \stackrel{?}{=} 12$	$12 = 12, \quad \text{true}$
$12 = 12, \quad \text{true}$	

The solutions are $-\frac{3}{2}$ and 4, and the solution set is $\left\{-\frac{3}{2}, 4\right\}$.

▶ ✔ **Check Point 1** Solve: $2x^2 - 9x = 5$. $\left\{-\frac{1}{2}, 5\right\}$

Great Question!

What's the difference between factoring a polynomial and solving a quadratic equation by factoring?

Here's an example that illustrates the difference:

Factoring a Polynomial	**Solving a Quadratic Equation**
Factor: $2x^2 - 5x - 12$.	Solve: $2x^2 - 5x - 12 = 0$.
This is not an equation. There is no equal sign.	This is an equation. There is an equal sign.
Solution: $(2x + 3)(x - 4)$	Solution: $(2x + 3)(x - 4) = 0$
Stop! Avoid the common error of setting each factor equal to zero.	$2x + 3 = 0$ or $x - 4 = 0$
	$x = -\dfrac{3}{2}$ $\qquad x = 4$
	The solution set is $\left\{-\dfrac{3}{2}, 4\right\}$.

A polynomial function of the form $y = ax^2 + bx + c$ or $f(x) = ax^2 + bx + c$, $a \neq 0$, is called a **quadratic function**. An example of a quadratic function is $y = 2x^2 - 5x - 12$ or $f(x) = 2x^2 - 5x - 12$.

There is an important relationship between a quadratic equation in standard form, such as

$$2x^2 - 5x - 12 = 0$$

and a quadratic function, such as

$$y = 2x^2 - 5x - 12.$$

The solutions of $ax^2 + bx + c = 0$ correspond to the x-intercepts of the graph of the quadratic function $y = ax^2 + bx + c$. For example, you can visualize the solutions of $2x^2 - 5x - 12 = 0$ by looking at the x-intercepts of the graph of the quadratic function $y = 2x^2 - 5x - 12$. The graph, shaped like a bowl, is shown in **Figure 4.3**. The solutions of the equation $2x^2 - 5x - 12 = 0$, $-\frac{3}{2}$ and 4, appear as the graph's x-intercepts.

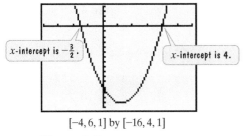

x-intercept is $-\frac{3}{2}$. x-intercept is 4.

$[-4, 6, 1]$ by $[-16, 4, 1]$

Figure 4.3

Example 2 Solving Quadratic Equations by Factoring

Solve:

a. $5x^2 = 20x$ **b.** $x^2 + 4 = 8x - 12$ **c.** $(x - 7)(x + 5) = -20$.

Solution

a. $\qquad 5x^2 = 20x$ This is the given equation.

$\qquad 5x^2 - 20x = 0$ Subtract 20x from both sides and write the equation in standard form.

$\qquad 5x(x - 4) = 0$ Factor.

$\qquad 5x = 0$ or $x - 4 = 0$ Set each factor equal to 0.

$\qquad x = 0 \qquad\qquad x = 4$ Solve the resulting equations.

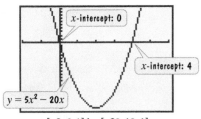

[-2, 6, 1] by [-20, 10, 1]

Figure 4.4 The solution set of $5x^2 = 20x$, or $5x^2 - 20x = 0$, is $\{0, 4\}$.

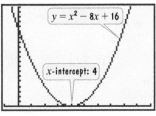

[-1, 10, 1] by [0, 20, 1]

Figure 4.5 The solution set of $x^2 + 4 = 8x - 12$, or $x^2 - 8x + 16 = 0$, is $\{4\}$.

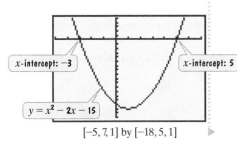

[-5, 7, 1] by [-18, 5, 1]

Figure 4.6 The solution set of $(x - 7)(x + 5) = -20$, or $x^2 - 2x - 15 = 0$, is $\{-3, 5\}$.

Check by substituting 0 and 4 into the given equation. The graph of $y = 5x^2 - 20x$, obtained with a graphing utility, is shown in **Figure 4.4**. The x-intercepts are 0 and 4. This verifies that the solutions are 0 and 4, and the solution set is $\{0, 4\}$.

b.

$x^2 + 4 = 8x - 12$	This is the given equation.
$x^2 - 8x + 16 = 0$	Write the equation in standard form by subtracting 8x and adding 12 on both sides.
$(x - 4)(x - 4) = 0$	Factor.
$x - 4 = 0$ or $x - 4 = 0$	Set each factor equal to 0.
$x = 4$ $\qquad x = 4$	Solve the resulting equations.

Notice that there is only one solution (or, if you prefer, a repeated solution). The trinomial $x^2 - 8x + 16$ is a perfect square trinomial that could have been factored as $(x - 4)^2$. The graph of $y = x^2 - 8x + 16$, obtained with a graphing utility, is shown in **Figure 4.5**. The graph has only one x-intercept at 4. This verifies that the equation's solution is 4 and the solution set is $\{4\}$.

c. Be careful! Although the left side of $(x - 7)(x + 5) = -20$ is factored, we cannot use the zero-product principle. Why not? The right side of the equation is not 0. So we begin by multiplying the factors on the left side of the equation. Then we add 20 to both sides to obtain 0 on the right side.

$(x - 7)(x + 5) = -20$	This is the given equation.
$x^2 - 2x - 35 = -20$	Use the FOIL method to multiply on the left side.
$x^2 - 2x - 15 = 0$	Add 20 to both sides.
$(x + 3)(x - 5) = 0$	Factor.
$x + 3 = 0$ or $x - 5 = 0$	Set each factor equal to 0.
$x = -3$ $\qquad x = 5$	Solve the resulting equations.

Check by substituting -3 and 5 into the given equation. The graph of $y = x^2 - 2x - 15$, obtained with a graphing utility, is shown in **Figure 4.6**. The x-intercepts are -3 and 5. This verifies that the solutions are -3 and 5, and the solution set is $\{-3, 5\}$.

Great Question!

When you solved $5x^2 = 20x$ in part (a), why didn't you first divide both sides of the equation by x? And in part (c), how come you didn't immediately use the zero-product principle on $(x - 7)(x + 5) = -20$ and set $x - 7$ and $x + 5$ equal to 0?

Avoid the following errors:

$5x^2 = 20x$

$\dfrac{5x^2}{x} = \dfrac{20x}{x}$

$5x = 20$

$x = 4$

Never divide both sides of an equation by x. Division by zero is undefined and x may be zero. Indeed, the solutions for this equation (Example 2a) are 0 and 4. Dividing both sides by x does not permit us to find both solutions.

$(x - 7)(x + 5) = -20$

$x - 7 = -20$ or $x + 5 = -20$

$x = -13$ or $x = -25$

The zero-product principle cannot be used because the right side of the equation is not equal to 0.

✓ **Check Point 2**   Solve:

a. $3x^2 = 2x$ $\left\{0, \frac{2}{3}\right\}$ **b.** $x^2 + 7 = 10x - 18$ $\{5\}$ **c.** $(x - 2)(x + 3) = 6.$ $\{-4, 3\}$

2 Solve higher-degree polynomial equations by factoring.

Polynomial Equations

A **polynomial equation** is the result of setting two polynomials equal to each other. The equation is in **standard form** if one side is 0 and the polynomial on the other side is in standard form, that is, in descending powers of the variable. The **degree of a polynomial equation** in standard form is the same as the highest degree of the terms in the polynomial. Here are examples of three polynomial equations:

$$2x^2 + 7x = 4 \qquad\qquad x^3 + x^2 = 4x + 4$$
$$3x - 21 = 0 \qquad 2x^2 + 7x - 4 = 0 \qquad x^3 + x^2 - 4x - 4 = 0.$$

This equation is of degree 1 because 1 is the highest degree.

This equation is of degree 2 because 2 is the highest degree.

This equation is of degree 3 because 3 is the highest degree.

Notice that a polynomial equation of degree 1 is a linear equation. A polynomial equation of degree 2 is a quadratic equation.

Some polynomial equations of degree 3 or higher can be solved by moving all terms to one side, thereby obtaining 0 on the other side. Once the equation is in standard form, factor and then set each factor equal to 0.

Example 3   Solving a Polynomial Equation by Factoring

Solve by factoring: $x^3 + x^2 = 4x + 4$.

Solution

Step 1   Move all terms to one side and obtain zero on the other side. Subtract $4x$ and subtract 4 from both sides.

$$x^3 + x^2 - 4x - 4 = 4x + 4 - 4x - 4$$
$$x^3 + x^2 - 4x - 4 = 0$$

Step 2   Factor. Use factoring by grouping. Group terms that have a common factor.

$$\boxed{x^3 + x^2} + \boxed{-4x - 4} = 0$$

Common factor is x^2.   Common factor is -4.

$$x^2(x + 1) - 4(x + 1) = 0 \qquad \text{Factor } x^2 \text{ from the first two terms and } -4 \text{ from the last two terms.}$$

$$(x + 1)(x^2 - 4) = 0 \qquad \text{Factor out the common binomial, } x + 1, \text{ from each term.}$$

$$(x + 1)(x + 2)(x - 2) = 0 \qquad \text{Factor completely by factoring } x^2 - 4 \text{ as the difference of two squares.}$$

Steps 3 and 4   Set each factor equal to zero and solve the resulting equation.

$$x + 1 = 0 \quad \text{or} \quad x + 2 = 0 \quad \text{or} \quad x - 2 = 0$$
$$x = -1 \qquad\qquad x = -2 \qquad\qquad x = 2$$

Step 5   Check the solutions in the original equation. Check the three solutions, $-1, -2,$ and 2, by substituting them into the original equation. Can you verify that the solutions are $-1, -2,$ and 2, and the solution set is $\{-2, -1, 2\}$?

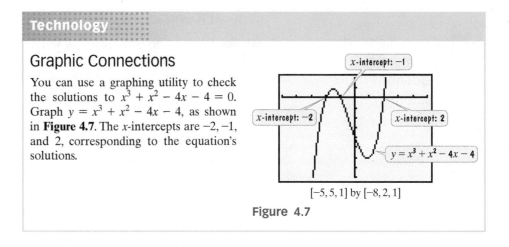

Technology

Graphic Connections

You can use a graphing utility to check the solutions to $x^3 + x^2 - 4x - 4 = 0$. Graph $y = x^3 + x^2 - 4x - 4$, as shown in **Figure 4.7**. The x-intercepts are $-2, -1$, and 2, corresponding to the equation's solutions.

x-intercept: −1

x-intercept: −2

x-intercept: 2

$y = x^3 + x^2 - 4x - 4$

$[-5, 5, 1]$ by $[-8, 2, 1]$

Figure 4.7

✓ **Check Point 3**   Solve by factoring: $2x^3 + 3x^2 = 8x + 12$. $\left\{-2, -\frac{3}{2}, 2\right\}$

Applications of Quadratic Equations

③ Solve problems using quadratic equations.

Solving quadratic equations by factoring can be used to answer questions about variables contained in mathematical models.

Example 4   **Modeling Motion**

You throw a ball straight up from a rooftop 384 feet high with an initial speed of 32 feet per second. The function

$$s(t) = -16t^2 + 32t + 384$$

describes the ball's height above the ground, $s(t)$, in feet, t seconds after you throw it. The ball misses the rooftop on its way down and eventually strikes the ground. How long will it take for the ball to hit the ground?

Solution

The ball hits the ground when $s(t)$, its height above the ground, is 0 feet. Thus, we substitute 0 for $s(t)$ in the given function and solve for t.

$s(t) = -16t^2 + 32t + 384$	This is the function that models the ball's height.
$0 = -16t^2 + 32t + 384$	Substitute 0 for $s(t)$.
$0 = -16(t^2 - 2t - 24)$	Factor out −16.
$0 = -16(t - 6)(t + 4)$	Factor $t^2 - 2t - 24$, the trinomial.

Do not set the constant, −16, equal to zero: $-16 \neq 0$.

$t - 6 = 0$	or $t + 4 = 0$	Set each variable factor equal to 0.
$t = 6$	$t = -4$	Solve for t.

Because we begin describing the ball's height at $t = 0$, we discard the solution $t = -4$. The ball hits the ground after 6 seconds.

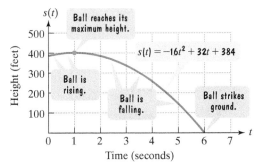

Figure 4.8

Figure 4.8 shows the graph of the quadratic function $s(t) = -16t^2 + 32t + 384$. The horizontal axis is labeled t, for the ball's time in motion. The vertical axis is labeled $s(t)$, for the ball's height above the ground at time t. Because time and height are both positive, the function is graphed in quadrant I only.

The graph visually shows what we discovered algebraically: The ball hits the ground after 6 seconds. The graph also reveals that the ball reaches its maximum height, 400 feet, after 1 second. Then the ball begins to fall.

Check Point 4 Use the function $s(t) = -16t^2 + 32t + 384$ to determine when the ball's height is 336 feet. Identify your meaningful solution as a point on the graph in **Figure 4.8**. *after 3 seconds; (3, 336)*

A Brief Review • Area and Volume

* The area of a two-dimensional figure is the number of square units, such as square inches (in.^{2}) or square feet (ft^2), that it takes to fill the interior of the figure.
  Example:

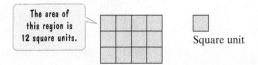

* The area, A, of a rectangle with length l and width w is given by the formula $A = lw$.
  Example:

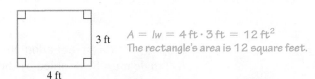

* The volume of a three-dimensional figure is the number of cubic units, such as cubic inches (in.^{3}) or cubic feet (ft^3), that it takes to fill the interior of the figure.
  Example:

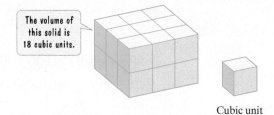

* The volume, V, of a rectangular solid with length l, width w, and height h is given by the formula $V = lwh$.
  Example:

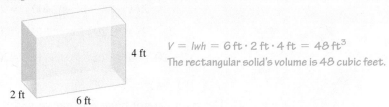

A complete discussion of area and volume can be found in Chapters 5 and 6.

In our next example, we use our five-step strategy from Section 2.3 for solving algebraic word problems.

Example 5 **Solving a Problem Involving Landscape Design**

A rectangular garden measures 80 feet by 60 feet. A large path of uniform width is to be added along both shorter sides and one longer side of the garden. The landscape designer doing the work wants to double the garden's area with the addition of this path. How wide should the path be?

Solution

Step 1   Let x represent one of the unknown quantities. We will let

$$x = \text{the width of the path.}$$

The situation is illustrated in **Figure 4.9**. The figure shows the original 80-by-60 foot rectangular garden and the path of width x added along both shorter sides and one longer side.

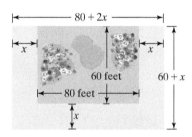

Figure 4.9 The garden's area is to be doubled by adding the path.

Step 2   Represent other unknown quantities in terms of x. Because the path is added along both shorter sides and one longer side, **Figure 4.9** shows that

$$80 + 2x = \text{the length of the new, expanded rectangle}$$

$$60 + x = \text{the width of the new, expanded rectangle.}$$

Step 3   Write an equation in x that models the conditions. The area of the rectangle must be doubled by the addition of the path.

The area, or length times width, of the new, expanded rectangle	must be	twice that of	the area of the garden.
$(80 + 2x)(60 + x)$	$=$	$2 \cdot 80$	$\cdot \ 60$

Step 4   Solve the equation and answer the question.

$(80 + 2x)(60 + x) = 2 \cdot 80 \cdot 60$	This is the equation that models the problem's conditions.
$4800 + 200x + 2x^2 = 9600$	Multiply. Use FOIL on the left side.
$2x^2 + 200x - 4800 = 0$	Subtract 9600 from both sides and write the equation in standard form.
$2(x^2 + 100x - 2400) = 0$	Factor out 2, the GCF.
$2(x - 20)(x + 120) = 0$	Factor the trinomial.
$x - 20 = 0 \quad \text{or} \quad x + 120 = 0$	Set each variable factor equal to 0.
$x = 20 \quad \text{or} \quad x = -120$	Solve for x.

The path cannot have a negative width. Because -120 is geometrically impossible, we use $x = 20$. The width of the path should be 20 feet.

Step 5   Check the proposed solution in the original wording of the problem. Has the landscape architect doubled the garden's area with the 20-foot-wide path? The area of the garden is 80 feet times 60 feet, or 4800 square feet. Because $80 + 2x$ and $60 + x$ represent the length and width of the expanded rectangle,

$$80 + 2x = 80 + 2 \cdot 20 = 120 \text{ feet is the expanded rectangle's length.}$$
$$60 + x = 60 + 20 \quad = 80 \text{ feet is the expanded rectangle's width.}$$

The area of the expanded rectangle is 120 feet times 80 feet, or 9600 square feet. This is double the area of the garden, 4800 square feet, as specified by the problem's conditions.

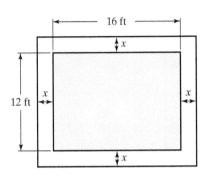

✓ **Check Point 5**   A rectangular garden measures 16 feet by 12 feet. A path of uniform width is to be added so as to surround the entire garden. The landscape artist doing the work wants the garden and path to cover an area of 320 square feet. ▶ How wide should the path be?   2 feet

Achieving Success

In college, it is recommended that students study and do homework for at least two hours for each hour of class time. For college students, finding the necessary time to study is not always easy. In order to make education a top priority, efficient time management is essential. You can continue to improve your time management by identifying your biggest time-wasters. Do you really need to spend hours on the phone, update your Facebook page throughout the night, or watch that sitcom rerun? My biggest time-wasters:_____

Concept and Vocabulary Check

Exercises in the Concept and Vocabulary Check are intended for group and class discussions.

In Exercises 1–4, fill in each blank so that the resulting statement is true.

1. A quadratic equation in x can be written in the standard form ___$ax^2 + bx + c = 0$___, where $a \neq 0$.

2. The zero-product principle states that if $AB = 0$, then ___$A = 0$ or $B = 0$___.

3. The solutions of a quadratic equation correspond to the ___x-intercepts___ of the graph of a quadratic function.

4. The polynomial equation $x^3 + x^2 - 4x - 4 = 0$ is written in standard form because one side is 0 and the polynomial on the other side is in ___descending___ powers of the variable.

In Exercises 5–8, determine whether each statement is true or false. If the statement is false, make the necessary change(s) to produce a true statement.   Changes to false statements will vary.

5. If $(x + 3)(x - 4) = 2$, then $x + 3 = 0$ or $x - 4 = 0$.   false

6. The solutions of the equation $4(x - 5)(x + 3) = 0$ are 4, 5, and –3.   false

7. Quadratic equations solved by factoring always have two different solutions.   false

8. Both 0 and $-\pi$ are solutions of the equation $x(x + \pi) = 0$.   true

Respond to Exercises 9–16 using verbal or written explanations.

9. What is a quadratic equation?   **9–16.** Answers will vary.

10. What is the zero-product principle?

11. Explain how to solve $x^2 - x = 6$.

12. Describe the relationship between the solutions of a quadratic equation and the graph of the corresponding quadratic function.

13. What is a polynomial equation? When is it in standard form?

14. What is the degree of a polynomial equation? What are polynomial equations of degree 1 and degree 2, respectively, called?

15. Explain how to solve $x^3 + x^2 = x + 1$.

16. A toy rocket is launched vertically upward. Using a quadratic equation, we find that the rocket will reach a height of 220 feet at 2.5 seconds and again at 5.5 seconds. How can this be?

27. $\{-3, -2\}$ **28.** $\{-6, 3\}$ **29.** $\{4\}$ **30.** $\{-5, 4\}$ **37.** $\{-5, -4, 5\}$ **38.** $\{-1, 1, 2\}$ **39.** $\{-5, 1, 5\}$ **40.** $\{-4, -2, 4\}$

Exercise Set 4.3

Practice Exercises

In Exercises 1–36, use factoring to solve each quadratic equation. Check by substitution or by using a graphing utility and identifying x-intercepts.

1. $x^2 + x - 12 = 0$ $\{-4, 3\}$ **2.** $x^2 - 2x - 15 = 0$ $\{-3, 5\}$

3. $x^2 + 6x = 7$ $\{-7, 1\}$ **4.** $x^2 - 4x = 45$ $\{-5, 9\}$

5. $3x^2 + 10x - 8 = 0$ $\left\{-4, \frac{2}{3}\right\}$ **6.** $2x^2 - 5x - 3 = 0$ $\left\{-\frac{1}{2}, 3\right\}$

7. $5x^2 = 8x - 3$ $\left\{\frac{3}{5}, 1\right\}$ **8.** $7x^2 = 30x - 8$ $\left\{\frac{2}{7}, 4\right\}$

9. $3x^2 = 2 - 5x$ $\left\{-2, \frac{1}{3}\right\}$ **10.** $5x^2 = 2 + 3x$ $\left\{-\frac{2}{5}, 1\right\}$

11. $x^2 = 8x$ $\{0, 8\}$ **12.** $x^2 = 4x$ $\{0, 4\}$

13. $3x^2 = 5x$ $\left\{0, \frac{5}{3}\right\}$ **14.** $2x^2 = 5x$ $\left\{0, \frac{5}{2}\right\}$

15. $x^2 + 4x + 4 = 0$ $\{-2\}$ **16.** $x^2 + 6x + 9 = 0$ $\{-3\}$

17. $x^2 = 14x - 49$ $\{7\}$ **18.** $x^2 = 12x - 36$ $\{6\}$

19. $9x^2 = 30x - 25$ $\left\{\frac{5}{3}\right\}$ **20.** $4x^2 = 12x - 9$ $\left\{\frac{3}{2}\right\}$

21. $x^2 - 25 = 0$ $\{-5, 5\}$ **22.** $x^2 - 49 = 0$ $\{-7, 7\}$

23. $9x^2 = 100$ $\left\{-\frac{10}{3}, \frac{10}{3}\right\}$ **24.** $4x^2 = 25$ $\left\{-\frac{5}{2}, \frac{5}{2}\right\}$

25. $x(x - 3) = 18$ $\{-3, 6\}$ **26.** $x(x - 4) = 21$ $\{-3, 7\}$

27. $(x - 3)(x + 8) = -30$ **28.** $(x - 1)(x + 4) = 14$

29. $x(x + 8) = 16(x - 1)$ **30.** $x(x + 9) = 4(2x + 5)$

31. $(x + 1)^2 - 5(x + 2) = 3x + 7$ $\{-2, 8\}$

32. $(x + 1)^2 = 2(x + 5)$ $\{-3, 3\}$

33. $x(8x + 1) = 3x^2 - 2x + 2$ $\left\{-1, \frac{2}{5}\right\}$

34. $2x(x + 3) = -5x - 15$ $\left\{-3, -\frac{5}{2}\right\}$

35. $\dfrac{x^2}{18} + \dfrac{x}{2} + 1 = 0$ $\{-6, -3\}$

36. $\dfrac{x^2}{4} - \dfrac{5x}{2} + 6 = 0$ $\{4, 6\}$

In Exercises 37–46, use factoring to solve each polynomial equation. Check by substitution or by using a graphing utility and identifying x-intercepts.

37. $x^3 + 4x^2 - 25x - 100 = 0$ **38.** $x^3 - 2x^2 - x + 2 = 0$

39. $x^3 - x^2 = 25x - 25$ **40.** $x^3 + 2x^2 = 16x + 32$

41. $3x^4 - 48x^2 = 0$ $\{-4, 0, 4\}$ **42.** $5x^4 - 20x^2 = 0$ $\{-2, 0, 2\}$

43. $x^4 - 4x^3 + 4x^2 = 0$ $\{0, 2\}$ **44.** $x^4 - 6x^3 + 9x^2 = 0$ $\{0, 3\}$

45. $2x^3 + 16x^2 + 30x = 0$ **46.** $3x^3 - 9x^2 - 30x = 0$
$\{-5, -3, 0\}$ $\{-2, 0, 5\}$

In Exercises 47–50, determine the x-intercepts of the graph of each quadratic function. Then match the function with its graph shown in the next column, labeled (a)–(d). Each graph is shown in the next column in a $[-10, 10, 1]$ by $[-10, 10, 1]$ viewing rectangle.

47. $y = x^2 - 6x + 8$ 2 and 4; d

48. $y = x^2 - 2x - 8$ −2 and 4; b

49. $y = x^2 + 6x + 8$ −4 and −2; c

50. $y = x^2 + 2x - 8$ −4 and 2; a

a.

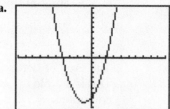

b.

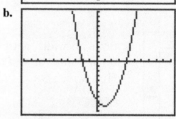

c.

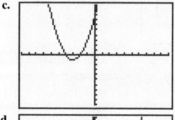

d.

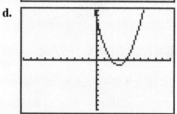

Practice Plus

In Exercises 51–54, solve each polynomial equation.

51. $x(x + 1)^3 - 42(x + 1)^2 = 0$ $\{-7, -1, 6\}$

52. $x(x - 2)^3 - 35(x - 2)^2 = 0$ $\{-5, 2, 7\}$

53. $-4x[x(3x - 2) - 8](25x^2 - 40x + 16) = 0$ $\left\{-\frac{4}{3}, 0, \frac{4}{5}, 2\right\}$

54. $-7x[x(2x - 5) - 12](9x^2 + 30x + 25) = 0$ $\left\{-\frac{5}{3}, -\frac{3}{2}, 0, 4\right\}$

In Exercises 55–58, find all values of c satisfying the given conditions.

55. $f(x) = x^2 - 4x - 27$ and $f(c) = 5$. −4 and 8

56. $f(x) = 5x^2 - 11x + 6$ and $f(c) = 4$. $\frac{1}{5}$ and 2

57. $f(x) = 2x^3 + x^2 - 8x + 2$ and $f(c) = 6$. $-2, -\frac{1}{2},$ and 2

58. $f(x) = x^3 + 4x^2 - x + 6$ and $f(c) = 10$. −4, −1, and 1

In Exercises 59–62, find all numbers satisfying the given conditions.

59. The product of the number decreased by 1 and increased by 4 is 24. −7 and 4

60. The product of the number decreased by 6 and increased by 2 is 20. −4 and 8

61. If 5 is subtracted from 3 times the number, the result is the square of 1 less than the number. 2 and 3

62. If the square of the number is subtracted from 61, the result is the square of 1 more than the number. −6 and 5

Application Exercises

A gymnast dismounts the uneven parallel bars at a height of 8 feet with an initial upward velocity of 8 feet per second. The function

$$s(t) = -16t^2 + 8t + 8$$

describes the height of the gymnast's feet above the ground, $s(t)$, in feet, t seconds after dismounting. The graph of the function is shown, with unlabeled tick marks along the horizontal axis. Use the function to solve Exercises 63–64.

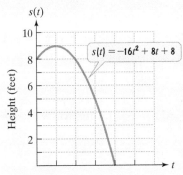

63. How long will it take the gymnast to reach the ground? Use this information to provide a number on each tick mark along the horizontal axis in the figure shown.   1 second; 0.25, 0.5, 0.75, 1, 1.25

64. When will the gymnast be 8 feet above the ground? Identify the solution(s) as one or more points on the graph.   after 0 seconds, the moment of dismount, and after $\frac{1}{2}$ second; $(0,8)$ and $(0.5,8)$

In a round-robin chess tournament, each player is paired with every other player once. The function

$$f(x) = \frac{x^2 - x}{2}$$

models the number of chess games, $f(x)$, that must be played in a round-robin tournament with x chess players. Use this function to solve Exercises 65–66.

65. In a round-robin chess tournament, 21 games were played. How many players were entered in the tournament?   7

66. In a round-robin chess tournament, 36 games were played. How many players were entered in the tournament?   9

The graph of the quadratic function in Exercises 65–66 is shown. Use the graph to solve Exercises 67–68.

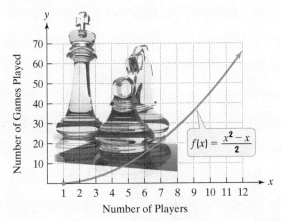

67. Identify your solution to Exercise 65 as a point on the graph.

68. Identify your solution to Exercise 66 as a point on the graph.

67. $(7, 21)$ **68.** $(9, 36)$

69. The length of a rectangular sign is 3 feet longer than the width. If the sign's area is 54 square feet, find its length and width.   length: 9 ft; width: 6 ft

70. A rectangular parking lot has a length that is 3 yards greater than the width. The area of the parking lot is 180 square yards. Find the length and the width.   length: 15 yd; width: 12 yd

71. Each side of a square is lengthened by 3 inches. The area of this new, larger square is 64 square inches. Find the length of a side of the original square.   5 in.

72. Each side of a square is lengthened by 2 inches. The area of this new, larger square is 36 square inches. Find the length of a side of the original square.   4 in.

73. A pool measuring 10 meters by 20 meters is surrounded by a path of uniform width, as shown in the figure. If the area of the pool and the path combined is 600 square meters, what is the width of the path?   5 m

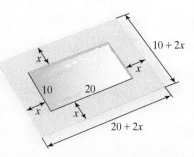

74. A vacant rectangular lot is being turned into a community vegetable garden measuring 15 meters by 12 meters. A path of uniform width is to surround the garden. If the area of the lot is 378 square meters, find the width of the path surrounding the garden.   3 m

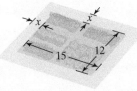

75. As part of a landscaping project, you put in a flower bed measuring 10 feet by 12 feet. You plan to surround the bed with a uniform border of low-growing plants.

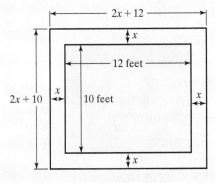

a. Write a polynomial that describes the area of the uniform border that surrounds your flower bed. *Hint:* The area of the border is the area of the large rectangle shown in the figure minus the area of the flower bed. $4x^2 + 44x$

b. The low-growing plants surrounding the flower bed require 1 square foot each when mature. If you have 168 of these plants, how wide a strip around the flower bed should you prepare for the border?   3 ft

76. As part of a landscaping project, you put in a flower bed measuring 20 feet by 30 feet. To finish off the project, you are putting in a uniform border of pine bark around the outside of the rectangular garden. You have enough pine bark to cover 336 square feet. How wide should the border be?   3 ft

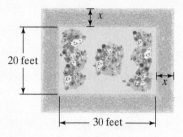

77. A machine produces open boxes using square sheets of metal. The figure illustrates that the machine cuts equal-sized squares measuring 2 inches on a side from the corners and then shapes the metal into an open box by turning up the sides. If each box must have a volume of 200 cubic inches, find the length and width of the open box.   Length and width are both 10 inches.

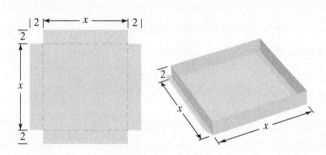

78. A machine produces open boxes using square sheets of metal. The machine cuts equal-sized squares measuring 3 inches on a side from the corners and then shapes the metal into an open box by turning up the sides. If each box must have a volume of 75 cubic inches, find the length and width of the open box.   Length and width are both 5 inches.

Critical Thinking Exercises

In Exercises 79–80, solve each equation.

79. $3^{x^2-9x+20} = 1$ {4,5}

80. $(x^2 - 5x + 5)^3 = 1$ {1,4}

81. Write a quadratic equation in standard form whose solutions are −3 and 5. $x^2 - 2x - 15 = 0$

Technology Exercises

In Exercises 82–85, use the x-intercepts of the graph of y in a $[-10,10,1]$ by $[-13,10,1]$ viewing rectangle to solve the quadratic equation. Check by substitution.

82. Use the graph of $y = x^2 + 3x - 4$ to solve
$$x^2 + 3x - 4 = 0.   \{-4,1\}$$

83. Use the graph of $y = x^2 + x - 6$ to solve
$$x^2 + x - 6 = 0.   \{-3,2\}$$

84. Use the graph of $y = (x - 2)(x + 3) - 6$ to solve
$$(x - 2)(x + 3) - 6 = 0.   \{-4,3\}$$

85. Use the graph of $y = x^2 - 2x + 1$ to solve
$$x^2 - 2x + 1 = 0.   \{1\}$$

4.4 Solving Quadratic Equations by the Square Root Property and the Quadratic Formula

What am I supposed to learn?

After you have read this section, you should be able to:

① Solve quadratic equations using the square root property.

② Solve quadratic equations using the quadratic formula.

③ Determine the most efficient method to use when solving a quadratic equation.

④ Use the quadratic formula to solve problems.

I'm very well acquainted, too, with matters mathematical,
I understand equations, both simple and quadratical.
About binomial theorem I'm teeming with a lot of news,
With many cheerful facts about the square of the hypotenuse.

—Gilbert and Sullivan, *The Pirates of Penzance*

Equations quadratical? Cheerful news about the square of the hypotenuse? You've come to the right place. In this section, we will present a formula that will enable you to solve a quadratic equation, $ax^2 + bx + c = 0$, even if the trinomial $ax^2 + bx + c$ cannot be factored. (Yes, it's *quadratic* and not *quadratical*, despite the latter's rhyme with mathematical.) In Chapter 6 (Section 6.2), we look at an application of quadratic equations, introducing (cheerfully, of course) the Pythagorean Theorem and the square of the hypotenuse.

1 Solve quadratic equations using the square root property.

The Square Root Property

Let's begin with a relatively simple quadratic equation:

$$x^2 = 9.$$

The value of x must be a number whose square is 9. There are two such numbers:

$$x = \sqrt{9} = 3 \quad \text{or} \quad x = -\sqrt{9} = -3.$$

Thus, the solutions of $x^2 = 9$ are 3 and -3. This is an example of the **square root property**.

The Square Root Property

If u is an algebraic expression and d is a positive real number, then $u^2 = d$ is equivalent to $u = \sqrt{d}$ or $u = -\sqrt{d}$:

$$\text{If } u^2 = d, \quad \text{then } u = \sqrt{d} \quad \text{or} \quad u = -\sqrt{d}.$$

Equivalently,

$$\text{If } u^2 = d, \quad \text{then } u = \pm\sqrt{d}.$$

Notice that $u = \pm\sqrt{d}$ is a shorthand notation to indicate that $u = \sqrt{d}$ or $u = -\sqrt{d}$. Although we usually read $u = \pm\sqrt{d}$ as "u equals plus or minus the square root of d," we actually mean that u is the positive square root of d or the negative square root of d.

Example 1   Solving a Quadratic Equation by the Square Root Property

Solve: $3x^2 = 18$.

Solution

To apply the square root property, we need a squared expression by itself on one side of the equation.

$$3x^2 = 18$$

We want x^2 by itself.

We can get x^2 by itself if we divide both sides by 3.

$$3x^2 = 18 \qquad \text{This is the original equation.}$$

$$\frac{3x^2}{3} = \frac{18}{3} \qquad \text{Divide both sides by 3.}$$

$$x^2 = 6 \qquad \text{Simplify.}$$

$$x = \sqrt{6} \quad \text{or} \quad x = -\sqrt{6} \qquad \text{Apply the square root property.}$$

Now let's check these proposed solutions in the original equation. Because the equation has an x^2-term and no x-term, we can check both values, $\pm\sqrt{6}$, at once.

Check $\sqrt{6}$ and $-\sqrt{6}$:

$$3x^2 = 18 \qquad \text{This is the original equation.}$$

$$3(\pm\sqrt{6})^2 \stackrel{?}{=} 18 \qquad \text{Substitute the proposed solutions.}$$

$$3 \cdot 6 \stackrel{?}{=} 18 \qquad (\pm\sqrt{6})^2 = 6$$

$$18 = 18, \qquad \text{true}$$

The solutions are $-\sqrt{6}$ and $\sqrt{6}$. The solution set is $\{-\sqrt{6}, \sqrt{6}\}$ or $\{\pm\sqrt{6}\}$.

✓ **Check Point 1** Solve: $4x^2 = 28$. $\{\pm\sqrt{7}\}$

In this section, we will express irrational solutions in simplified radical form, rationalizing denominators when possible.

Example 2 Solving a Quadratic Equation by the Square Root Property

Solve: $2x^2 - 7 = 0$.

Solution

To solve by the square root property, we isolate the squared expression on one side of the equation.

$$2x^2 - 7 = 0$$

We want x^2 by itself.

$$2x^2 - 7 = 0 \qquad \text{This is the original equation.}$$
$$2x^2 = 7 \qquad \text{Add 7 to both sides.}$$
$$x^2 = \frac{7}{2} \qquad \text{Divide both sides by 2.}$$
$$x = \sqrt{\frac{7}{2}} \quad \text{or} \quad x = -\sqrt{\frac{7}{2}} \qquad \text{Apply the square root property.}$$

Because the proposed solutions are opposites, we can rationalize both denominators at once:

$$\pm\sqrt{\frac{7}{2}} = \pm\frac{\sqrt{7}}{\sqrt{2}} \cdot \frac{\sqrt{2}}{\sqrt{2}} = \pm\frac{\sqrt{14}}{2}.$$

Substitute these values into the original equation and verify that the solutions are $-\frac{\sqrt{14}}{2}$ and $\frac{\sqrt{14}}{2}$. The solution set is $\left\{-\frac{\sqrt{14}}{2}, \frac{\sqrt{14}}{2}\right\}$ or $\left\{\pm\frac{\sqrt{14}}{2}\right\}$.

✓ **Check Point 2** Solve: $3x^2 - 11 = 0$. $\left\{\pm\frac{\sqrt{33}}{3}\right\}$

Can we solve an equation such as $(x - 1)^2 = 5$ using the square root property? Yes. The equation is in the form $u^2 = d$, where u^2, the squared expression, is by itself on the left side and d is positive.

$$(x - 1)^2 \qquad = \qquad 5$$

This is u^2 in $u^2 = d$ with $u = x - 1$.

This is d in $u^2 = d$ with $d = 5$.

Example 3 Solving a Quadratic Equation by the Square Root Property

Solve by the square root property: $(x - 1)^2 = 5$.

Solution

$$(x - 1)^2 = 5 \qquad \text{This is the original equation.}$$

$$x - 1 = \sqrt{5} \quad \text{or} \quad x - 1 = -\sqrt{5} \qquad \text{Apply the square root property.}$$

$$x = 1 + \sqrt{5} \qquad x = 1 - \sqrt{5} \qquad \text{Add 1 to both sides in each equation.}$$

Check $1 + \sqrt{5}$:

$$(x - 1)^2 = 5$$
$$(1 + \sqrt{5} - 1)^2 \overset{?}{=} 5$$
$$(\sqrt{5})^2 \overset{?}{=} 5$$
$$5 = 5, \quad \text{true}$$

Check $1 - \sqrt{5}$:

$$(x - 1)^2 = 5$$
$$(1 - \sqrt{5} - 1)^2 \overset{?}{=} 5$$
$$(-\sqrt{5})^2 \overset{?}{=} 5$$
$$5 = 5, \quad \text{true}$$

The solutions are $1 \pm \sqrt{5}$, and the solution set is $\{1 + \sqrt{5}, 1 - \sqrt{5}\}$ or $\{1 \pm \sqrt{5}\}$.

▶ ✔ **Check Point 3** Solve: $(x - 3)^2 = 10$. $\{3 \pm \sqrt{10}\}$

The Quadratic Formula

2 Solve quadratic equations using the quadratic formula.

The solutions of a quadratic equation cannot always be found by factoring. Some trinomials are difficult to factor, and others cannot be factored (that is, they are prime). Furthermore, not every quadratic equation is of the form $u^2 = d$, where solutions can be obtained by the square root property. However, there is a formula that can be used to solve all quadratic equations, whether or not they contain factorable trinomials or are of the form $u^2 = d$. The formula is called the *quadratic formula*.

Great Question!

Is it ok if I write
$$x = -b \pm \frac{\sqrt{b^2 - 4ac}}{2a}?$$

No. The entire numerator of the quadratic formula must be divided by $2a$. Always write the fraction bar all the way across the numerator.

$$x = \frac{-b \pm \sqrt{b^2 - 4ac}}{2a}$$

The Quadratic Formula

The solutions of a quadratic equation in the form $ax^2 + bx + c = 0$, with $a \neq 0$, are given by the **quadratic formula**

$$x = \frac{-b \pm \sqrt{b^2 - 4ac}}{2a}.$$

x equals negative b plus or minus the square root of $b^2 - 4ac$, all divided by $2a$.

To use the quadratic formula, be sure that the quadratic equation is expressed with all terms on one side and zero on the other side. It may be necessary to begin by rewriting the equation in this form. Then determine the numerical values for a (the coefficient of the x^2-term), b (the coefficient of the x-term), and c (the constant term). Substitute the values of a, b, and c into the quadratic formula and evaluate the expression. The $\pm$ sign indicates that there are two solutions of the equation.

Example 4 Solving a Quadratic Equation Using the Quadratic Formula

Solve using the quadratic formula: $2x^2 + 9x - 5 = 0$.

Solution

The given equation, $2x^2 + 9x - 5 = 0$, is in the desired form, with all terms on one side and zero on the other side. Begin by identifying the values for a, b, and c.

$$2x^2 + 9x - 5 = 0$$

$$a = 2 \quad b = 9 \quad c = -5$$

Substituting these values into the quadratic formula and simplifying gives the equation's solutions.

$$x = \frac{-b \pm \sqrt{b^2 - 4ac}}{2a}$$ Use the quadratic formula.

$$x = \frac{-9 \pm \sqrt{9^2 - 4(2)(-5)}}{2(2)}$$ Substitute the values for a, b, and c: $a = 2$, $b = 9$, and $c = -5$.

$$= \frac{-9 \pm \sqrt{81 + 40}}{4}$$ $9^2 - 4(2)(-5) = 81 - (-40) = 81 + 40$

$$= \frac{-9 \pm \sqrt{121}}{4}$$ Add under the radical sign.

$$= \frac{-9 \pm 11}{4}$$ $\sqrt{121} = 11$

Now we will evaluate this expression in two different ways to obtain the two solutions. On the left, we will *add* 11 to -9. On the right, we will *subtract* 11 from -9.

$$x = \frac{-9 + 11}{4} \quad \text{or} \quad x = \frac{-9 - 11}{4}$$

$$= \frac{2}{4} = \frac{1}{2} \qquad\qquad = \frac{-20}{4} = -5$$

Check the solutions, -5 and $\frac{1}{2}$, by substituting them into the original equation. The solution set is $\left\{-5, \frac{1}{2}\right\}$.

 **Check Point 4** Solve using the quadratic formula: $8x^2 + 2x - 1 = 0$. $\left\{-\frac{1}{2}, \frac{1}{4}\right\}$

The quadratic equation in Example 4 has rational solutions, namely -5 and $\frac{1}{2}$. The equation can also be solved by factoring. Take a few minutes to do this now and convince yourself that you will arrive at the same two solutions.

Any quadratic equation that has rational solutions can be solved by factoring or using the quadratic formula. However, quadratic equations with irrational solutions cannot be solved by factoring. These equations can be readily solved using the quadratic formula.

Great Question!

What's the bottom line on whether I can use factoring to solve a quadratic equation?

Compute $b^2 - 4ac$, which appears under the radical sign in the quadratic formula. If $b^2 - 4ac$ is a perfect square, such as 4, 25, or 121, then the equation can be solved by factoring.

Example 5 ### Solving a Quadratic Equation Using the Quadratic Formula

Solve using the quadratic formula: $2x^2 = 4x + 1$.

Solution

The quadratic equation must have zero on one side to identify the values for a, b, and c. To move all terms to one side and obtain zero on the right, we subtract $4x + 1$ from both sides. Then we can identify the values for a, b, and c.

Technology

Using a Calculator to Approximate $\dfrac{4 + \sqrt{24}}{4}$:

Many Scientific Calculators

$\boxed{(}\ \boxed{4}\ \boxed{+}\ \boxed{24}\ \boxed{\sqrt{}}\ \boxed{)}\ \boxed{\div}\ \boxed{4}\ \boxed{=}$

Many Graphing Calculators

$\boxed{(}\ \boxed{4}\ \boxed{+}\ \boxed{\sqrt{}}\ \boxed{24}\ \boxed{)}\ \boxed{\div}\ \boxed{4}\ \boxed{\text{ENTER}}$

(See the voice balloon in the Technology box on page 301.)

Correct to the nearest tenth,

$$\frac{4 + \sqrt{24}}{4} \approx 2.2.$$

$$2x^2 = 4x + 1 \qquad \text{This is the given equation.}$$
$$2x^2 - 4x - 1 = 0 \qquad \text{Subtract } 4x + 1 \text{ from both sides.}$$

$$a = 2 \qquad b = -4 \qquad c = -1$$

Substituting $a = 2$, $b = -4$, and $c = -1$ into the quadratic formula and simplifying gives the equation's solutions.

$$x = \frac{-b \pm \sqrt{b^2 - 4ac}}{2a} \qquad \text{Use the quadratic formula.}$$

$$x = \frac{-(-4) \pm \sqrt{(-4)^2 - 4(2)(-1)}}{2(2)} \qquad \begin{array}{l}\text{Substitute the values for } a, b, \text{ and } c\text{: } a = 2,\\ b = -4, \text{ and } c = -1.\end{array}$$

$$= \frac{4 \pm \sqrt{16 - (-8)}}{4} \qquad \begin{array}{l}-(-4) = 4, (-4)^2 = (-4)(-4) = 16,\\ \text{and } 4(2)(-1) = -8.\end{array}$$

$$= \frac{4 \pm \sqrt{24}}{4} \qquad 16 - (-8) = 16 + 8 = 24$$

The solutions are $\dfrac{4 + \sqrt{24}}{4}$ and $\dfrac{4 - \sqrt{24}}{4}$. These solutions are irrational numbers.

You can use a calculator to obtain a decimal approximation for each solution. However, in situations such as this that do not involve applications, it is best to leave the irrational solutions in radical form as exact answers. In some cases, we can simplify this radical form. Using methods for simplifying square roots discussed in Section 1.4, we can simplify $\sqrt{24}$:

$$\sqrt{24} = \sqrt{4 \cdot 6} = \sqrt{4}\sqrt{6} = 2\sqrt{6}.$$

Now we can use this result to simplify the two solutions. First, use the distributive property to factor out 2 from both terms in the numerator. Then, divide the numerator and the denominator by 2.

$$x = \frac{4 \pm \sqrt{24}}{4} = \frac{4 \pm 2\sqrt{6}}{4} = \frac{\overset{1}{2}(2 \pm \sqrt{6})}{\underset{2}{4}} = \frac{2 \pm \sqrt{6}}{2}$$

In simplified radical form, the equation's solution set is

$$\left\{ \frac{2 + \sqrt{6}}{2}, \frac{2 - \sqrt{6}}{2} \right\}.$$

Great Question!

The simplification of the irrational solutions in Example 5 was kind of tricky. Any suggestions to guide the process?

Many students use the quadratic formula correctly until the last step, where they make an error in simplifying the solutions. Be sure to factor the numerator before dividing the numerator and the denominator by the greatest common factor.

$$\frac{4 \pm 2\sqrt{6}}{4} = \frac{2(2 \pm \sqrt{6})}{4} = \frac{\overset{1}{2}(2 \pm \sqrt{6})}{\underset{2}{4}} = \frac{2 \pm \sqrt{6}}{2}$$

You cannot divide just one term in the numerator and the denominator by their greatest common factor.

Incorrect!

$$\frac{\overset{1}{4} \pm 2\sqrt{6}}{\underset{1}{4}} = 1 \pm 2\sqrt{6} \qquad \frac{4 \pm \overset{1}{2}\sqrt{6}}{\underset{2}{4}} = \frac{4 \pm \sqrt{6}}{2}$$

Examples 4 and 5 illustrate that the solutions of quadratic equations can be rational or irrational numbers. In Example 4, the expression under the square root was 121, a perfect square ($\sqrt{121} = 11$), and we obtained rational solutions. In Example 5, this expression was 24, which is not a perfect square (although we simplified $\sqrt{24}$ to $2\sqrt{6}$), and we obtained irrational solutions. If the expression under the square root simplifies to a negative number, then the quadratic equation has **no real solution**. The solution set consists of *imaginary numbers*, discussed in the Blitzer Bonus on page 61.

 **Check Point 5** Solve using the quadratic formula: $2x^2 = 6x - 1$.

$$\left\{\frac{3 + \sqrt{7}}{2}, \frac{3 - \sqrt{7}}{2}\right\}$$

Determining Which Method to Use

 Determine the most efficient method to use when solving a quadratic equation.

All quadratic equations can be solved by the quadratic formula. However, if an equation is in the form $u^2 = d$, such as $x^2 = 5$ or $(2x + 3)^2 = 8$, it is faster to use the square root property, taking the square root of both sides. If the equation is not in the form $u^2 = d$, write the quadratic equation in standard form ($ax^2 + bx + c = 0$). Try to solve the equation by factoring. If $ax^2 + bx + c$ cannot be factored, then solve the quadratic equation by the quadratic formula.

Table 4.1 summarizes our observations about which technique to use when solving a quadratic equation.

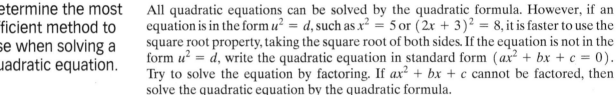

Table 4.1 Determining the Most Efficient Technique to Use When Solving a Quadratic Equation

Description and Form of the Quadratic Equation	Most Efficient Solution Method	Example
$ax^2 + bx + c = 0$ and $ax^2 + bx + c$ can be factored easily.	Factor and use the zero-product principle.	$3x^2 + 5x - 2 = 0$ $(3x - 1)(x + 2) = 0$ $3x - 1 = 0$ or $x + 2 = 0$ $x = \dfrac{1}{3}$ $\qquad$ $x = -2$
$ax^2 + c = 0$ The quadratic equation has no x-term. ($b = 0$)	Solve for x^2 and apply the square root property.	$4x^2 - 7 = 0$ $4x^2 = 7$ $x^2 = \dfrac{7}{4}$ $x = \pm\dfrac{\sqrt{7}}{2}$
$u^2 = d$; u is a first-degree polynomial.	Use the square root property.	$(x + 4)^2 = 5$ $x + 4 = \pm\sqrt{5}$ $x = -4 \pm \sqrt{5}$
$ax^2 + bx + c = 0$ and $ax^2 + bx + c$ cannot be factored or the factoring is too difficult.	Use the quadratic formula: $x = \dfrac{-b \pm \sqrt{b^2 - 4ac}}{2a}.$	$x^2 - 2x - 6 = 0$ $a = 1 \quad b = -2 \quad c = -6$ $x = \dfrac{-(-2) \pm \sqrt{(-2)^2 - 4(1)(-6)}}{2(1)}$ $= \dfrac{2 \pm \sqrt{4 - 4(1)(-6)}}{2(1)}$ $= \dfrac{2 \pm \sqrt{28}}{2} = \dfrac{2 \pm \sqrt{4}\sqrt{7}}{2}$ $= \dfrac{2 \pm 2\sqrt{7}}{2} = \dfrac{2(1 \pm \sqrt{7})}{2}$ $= 1 \pm \sqrt{7}$

Use the quadratic
formula to solve
problems.

Applications

Quadratic equations can be solved to answer questions about variables contained in
quadratic functions.

Example 6   Blood Pressure and Age

The graphs in **Figure 4.10** illustrate that a person's normal systolic blood pressure,
measured in millimeters of mercury (mm Hg), depends on his or her age. The
function

$$P(A) = 0.006A^2 - 0.02A + 120$$

models a man's normal systolic pressure, $P(A)$, at age A.

a. Find the age, to the nearest year, of a man whose normal systolic blood pressure
is 125 mm Hg.

b. Use the graphs in **Figure 4.10** to describe the differences between the normal
systolic blood pressures of men and women as they age.

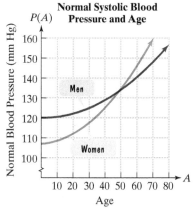

Figure 4.10

Solution

a. We are interested in the age of a man with a normal systolic blood pressure
of 125 millimeters of mercury. Thus, we substitute 125 for $P(A)$ in the given
function for men. Then we solve for A, the man's age.

$$P(A) = 0.006A^2 - 0.02A + 120 \quad \text{This is the given function for men.}$$

$$125 = 0.006A^2 - 0.02A + 120 \quad \text{Substitute 125 for } P(A).$$

$$0 = 0.006A^2 - 0.02A - 5 \quad \text{Subtract 125 from both sides and write the quadratic equation in standard form.}$$

$$a = 0.006 \quad b = -0.02 \quad c = -5$$

Because the trinomial on the right side of the equation appears difficult to factor
and is, in fact, prime, we solve using the quadratic formula.

Notice that the
variable is A,
rather than the
usual x.

$$A = \frac{-b \pm \sqrt{b^2 - 4ac}}{2a} \quad \text{Use the quadratic formula.}$$

$$= \frac{-(-0.02) \pm \sqrt{(-0.02)^2 - 4(0.006)(-5)}}{2(0.006)} \quad \text{Substitute the values for } a, b, \text{ and } c: a = 0.006, b = -0.02, \text{ and } c = -5.$$

$$= \frac{0.02 \pm \sqrt{0.1204}}{0.012} \quad \text{Use a calculator to simplify the expression under the square root.}$$

$$\approx \frac{0.02 \pm 0.347}{0.012} \quad \text{Use a calculator: } \sqrt{0.1204} \approx 0.347.$$

$$A \approx \frac{0.02 + 0.347}{0.012} \quad \text{or} \quad A \approx \frac{0.02 - 0.347}{0.012}$$

$$A \approx 31 \qquad\qquad A \approx -27 \quad \text{Use a calculator and round to the nearest integer.}$$

Reject this solution.
Age cannot be negative.

Technology

On most calculators, here is how
to approximate

$$\frac{0.02 + \sqrt{0.1204}}{0.012}.$$

Many Scientific Calculators

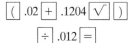

Many Graphing Calculators

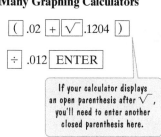

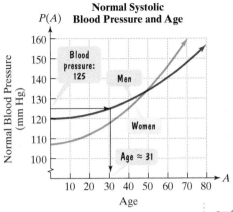

Normal Systolic Blood Pressure and Age

Figure 4.11

The positive solution, $A \approx 31$, indicates that 31 is the approximate age of a man whose normal systolic blood pressure is 125 mm Hg. This is illustrated by the black lines with the arrows on the red graph representing men in **Figure 4.11**.

b. Take a second look at the graphs in **Figure 4.11**. Before approximately age 50, the blue graph representing women's normal systolic blood pressure lies below the red graph representing men's normal systolic blood pressure. Thus, up to age 50, women's normal systolic blood pressure is lower than men's, although it is increasing at a faster rate. After age 50, women's normal systolic blood pressure is higher than men's.

✓ **Check Point 6** The function $P(A) = 0.01A^2 + 0.05A + 107$ models a woman's normal systolic blood pressure, $P(A)$, at age A. Use this function to find the age, to the nearest year, of a woman whose normal systolic blood pressure is 115 mm Hg. Use the blue graph in **Figure 4.11** to verify your solution.

approximately 26 years old; The point (26, 115) lies approximately on the blue graph.

Blitzer Bonus

Art, Nature, and Quadratic Equations

A **golden rectangle** can be a rectangle of any size, but its long side must be Φ times as long as its short side, where $\Phi \approx 1.6$. Artists often use golden rectangles in their work because they are considered to be more visually pleasing than other rectangles.

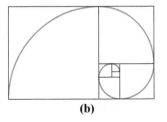

If a golden rectangle is divided into a square and a rectangle, as in **Figure 4.12(a)**, the smaller rectangle is a golden rectangle. If the smaller golden rectangle is divided again, the same is true of the yet smaller rectangle, and so on. The process of repeatedly dividing each golden rectangle in this manner is illustrated in **Figure 4.12(b)**. We've also created a spiral by connecting the opposite corners of all the squares with a smooth curve. This spiral matches the spiral shape of the chambered nautilus shell shown in **Figure 4.12(c)**. The shell spirals out at an ever-increasing rate that is governed by this geometry.

In *The Bathers at Asnières*, by the French impressionist Georges Seurat (1859–1891), the artist positions parts of the painting as though they were inside golden rectangles.

Georges Seurat, "The Bathers at Asnières," 1883–1884. Oil on canvas, $79\frac{1}{2} \times 118\frac{1}{2}$ in. Erich Lessing/Art Resource, NY.

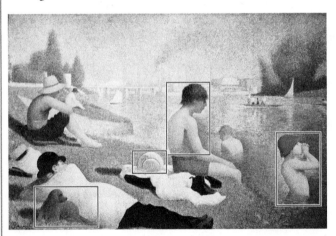

Golden Rectangle A

| Square | Golden Rectangle B |

(a)

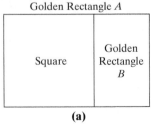

(b)

(c)

Figure 4.12

In the Exercise Set that follows, you will use the golden rectangles in **Figure 4.12(a)** to obtain an exact value for Φ, the ratio of the long side to the short side in a golden rectangle of any size. Your model will involve a quadratic equation that can be solved by the quadratic formula. (See Exercise 87.)

Achieving Success

Two Ways to Stay Sharp

- **Concentrate on one task at a time.** Do not multitask. Doing several things at once can cause confusion and can take longer to complete the tasks than tackling them sequentially.
- **Get enough sleep.** Fatigue impedes the ability to learn and do complex tasks.

Concept and Vocabulary Check

Exercises in the Concept and Vocabulary Check are intended for group and class discussions.

In Exercises 1–4, fill in each blank so that the resulting statement is true.

1. The square root property states that if $u^2 = d$, then ___$u = \pm \sqrt{d}$___.

2. The solutions of $ax^2 + bx + c = 0, a \neq 0$, are given by ___$x = \dfrac{-b \pm \sqrt{b^2 - 4ac}}{2a}$___, called the ___quadratic formula___.

3. The most efficient method for solving $(x + 1)^2 = 7$ is ___the square root property___.

4. The most efficient method for solving $x^2 - 2x - 10 = 0$ is ___the quadratic formula___.

In Exercises 5–8, determine whether each statement is true or false. If the statement is false, make the necessary change(s) to produce a true statement.   Changes to false statements will vary.

5. The solutions of $3x^2 - 5 = 0$ are $-\dfrac{\sqrt{5}}{3}$ and $\dfrac{\sqrt{5}}{3}$.   false

6. In using the quadratic formula to solve the quadratic equation $5x^2 = 2x - 7$, we have $a = 5, b = 2$, and $c = -7$.   false

7. The quadratic formula can be expressed as

$$x = -b \pm \dfrac{\sqrt{b^2 - 4ac}}{2a}.$$ false

8. The solutions $\dfrac{4 \pm \sqrt{3}}{2}$ can be simplified to $2 \pm \sqrt{3}$.   false

Respond to Exercises 9–12 using verbal or written explanations.

9. What is the square root property?   9–12. Answers will vary.

10. Explain how to solve $(x - 1)^2 = 16$ using the square root property.

11. What is the quadratic formula and why is it useful?

12. Explain how to solve $x^2 + 6x + 8 = 0$ using the quadratic formula.

Exercise Set 4.4

Practice Exercises

In Exercises 1–30, solve each quadratic equation by the square root property. If possible, simplify radicals or rationalize denominators.

1. $x^2 = 16$ $\{\pm 4\}$

2. $x^2 = 100$ $\{\pm 10\}$

3. $y^2 = 81$ $\{\pm 9\}$

4. $y^2 = 144$ $\{\pm 12\}$

5. $x^2 = 7$ $\{\pm \sqrt{7}\}$

6. $x^2 = 13$ $\{\pm \sqrt{13}\}$

7. $x^2 = 50$ $\{\pm 5\sqrt{2}\}$

8. $x^2 = 27$ $\{\pm 3\sqrt{3}\}$

9. $5x^2 = 20$ $\{\pm 2\}$

10. $3x^2 = 75$ $\{\pm 5\}$

11. $4y^2 = 49$ $\left\{\pm \dfrac{7}{2}\right\}$

12. $16y^2 = 25$ $\left\{\pm \dfrac{5}{4}\right\}$

13. $2x^2 + 1 = 51$ $\{\pm 5\}$

14. $3x^2 - 1 = 47$ $\{\pm 4\}$

15. $3x^2 - 2 = 0$ $\left\{\pm \dfrac{\sqrt{6}}{3}\right\}$

16. $3x^2 - 5 = 0$ $\left\{\pm \dfrac{\sqrt{15}}{3}\right\}$

17. $5z^2 - 7 = 0$ $\left\{\pm \dfrac{\sqrt{35}}{5}\right\}$

18. $5z^2 - 2 = 0$ $\left\{\pm \dfrac{\sqrt{10}}{5}\right\}$

19. $(x - 3)^2 = 16$ $\{-1, 7\}$

20. $(x - 2)^2 = 25$ $\{-3, 7\}$

21. $(x + 5)^2 = 121$ $\{-16, 6\}$

22. $(x + 6)^2 = 144$ $\{-18, 6\}$

23. $(3x + 2)^2 = 9$ $\left\{-\dfrac{5}{3}, \dfrac{1}{3}\right\}$

24. $(2x + 1)^2 = 49$ $\{-4, 3\}$

25. $(x - 5)^2 = 3$ $\{5 \pm \sqrt{3}\}$

26. $(x - 3)^2 = 15$ $\{3 \pm \sqrt{15}\}$

27. $(y + 8)^2 = 11$ $\{-8 \pm \sqrt{11}\}$

28. $(y + 7)^2 = 5$ $\{-7 \pm \sqrt{5}\}$

29. $(z - 4)^2 = 18$ $\{4 \pm 3\sqrt{2}\}$

30. $(z - 6)^2 = 12$ $\{6 \pm 2\sqrt{3}\}$

31. $\{-5, -3\}$ **32.** $\{-6, -2\}$

304 **Chapter 4** Algebraic Pathways: Polynomials, Quadratic Equations, and Quadratic Functions

34. $\left\{\dfrac{-5 + \sqrt{17}}{2}, \dfrac{-5 - \sqrt{17}}{2}\right\}$

35. $\{-2 + \sqrt{10}, -2 - \sqrt{10}\}$ **36.** $\{-1 + \sqrt{5}, -1 - \sqrt{5}\}$ **37.** $\{-2 + \sqrt{11}, -2 - \sqrt{11}\}$ **38.** $\{-2 + \sqrt{3}, -2 - \sqrt{3}\}$ **41.** $\left\{-\dfrac{2}{3}, \dfrac{3}{2}\right\}$ **43.** $\{1 + \sqrt{11}, 1 - \sqrt{11}\}$

Solve the equations in Exercises 31–50 using the quadratic formula.

31. $x^2 + 8x + 15 = 0$ **32.** $x^2 + 8x + 12 = 0$

33. $x^2 + 5x + 3 = 0$ **34.** $x^2 + 5x + 2 = 0$

35. $x^2 + 4x = 6$ **36.** $x^2 + 2x = 4$

37. $x^2 + 4x - 7 = 0$ **38.** $x^2 + 4x + 1 = 0$

39. $x^2 - 3x = 18$ $\{-3, 6\}$ **40.** $x^2 - 3x = 10$ $\{-2, 5\}$

41. $6x^2 - 5x - 6 = 0$ **42.** $9x^2 - 12x - 5 = 0$ $\left\{-\dfrac{1}{3}, \dfrac{5}{3}\right\}$

43. $x^2 - 2x - 10 = 0$ **44.** $x^2 + 6x - 10 = 0$

45. $x^2 - x = 14$ **46.** $x^2 - 5x = 10$

47. $6x^2 + 6x + 1 = 0$ **48.** $3x^2 = 5x - 1$

49. $4x^2 = 12x - 9$ $\left\{\dfrac{3}{2}\right\}$ **50.** $9x^2 + 6x + 1 = 0$ $\left\{-\dfrac{1}{3}\right\}$

In Exercises 51–70, solve each equation by the method of your choice. Simplify irrational solutions, if possible.

51. $2x^2 - x = 1$ $\left\{-\dfrac{1}{2}, 1\right\}$ **52.** $3x^2 - 4x = 4$ $\left\{-\dfrac{2}{3}, 2\right\}$

53. $5x^2 + 2 = 11x$ $\left\{\dfrac{1}{5}, 2\right\}$ **54.** $5x^2 = 6 - 13x$ $\left\{-3, \dfrac{2}{5}\right\}$

55. $3x^2 = 60$ $\{\pm 2\sqrt{5}\}$ **56.** $2x^2 = 250$ $\{\pm 5\sqrt{5}\}$

57. $x^2 - 2x = 1$ $\{1 \pm \sqrt{2}\}$ **58.** $2x^2 + 3x = 1$ $\left\{\dfrac{-3 \pm \sqrt{17}}{4}\right\}$

59. $(2x + 3)(x + 4) = 1$ **60.** $(2x - 5)(x + 1) = 2$

61. $(3x - 4)^2 = 16$ $\left\{0, \dfrac{8}{3}\right\}$ **62.** $(2x + 7)^2 = 25$ $\{-6, -1\}$

63. $3x^2 - 12x + 12 = 0$ $\{2\}$ **64.** $9 - 6x + x^2 = 0$ $\{3\}$

65. $4x^2 - 16 = 0$ $\{-2, 2\}$ **66.** $3x^2 - 27 = 0$ $\{-3, 3\}$

67. $x^2 + 9x = 0$ $\{-9, 0\}$ **68.** $x^2 - 6x = 0$ $\{0, 6\}$

69. $(3x - 2)^2 = 10$ $\left\{\dfrac{2 \pm \sqrt{10}}{3}\right\}$ **70.** $(4x - 1)^2 = 15$ $\left\{\dfrac{1 \pm \sqrt{15}}{4}\right\}$

Practice Plus

59. $\left\{\dfrac{-11 \pm \sqrt{33}}{4}\right\}$ **60.** $\left\{\dfrac{3 \pm \sqrt{65}}{4}\right\}$

In Exercises 71–78, solve each equation by the method of your choice.

71. $\dfrac{3x^2}{4} - \dfrac{5x}{2} - 2 = 0$ $\left\{-\dfrac{2}{3}, 4\right\}$ **72.** $\dfrac{x^2}{3} - \dfrac{x}{2} - \dfrac{3}{2} = 0$ $\left\{-\dfrac{3}{2}, 3\right\}$

73. $(x - 1)(3x + 2) = -7(x - 1)$ $\{-3, 1\}$

74. $x(x + 1) = 4 - (x + 2)(x + 2)$ $\left\{-\dfrac{5}{2}, 0\right\}$

75. $(2x - 6)(x + 2) = 5(x - 1) - 12$ $\left\{1, \dfrac{5}{2}\right\}$

76. $7x(x - 2) = 3 - 2(x + 4)$ $\left\{\dfrac{5}{7}, 1\right\}$

77. $2x^2 - 9x - 3 = 9 - 9x$ $\{\pm \sqrt{6}\}$

78. $3x^2 - 6x - 3 = 12 - 6x$ $\{\pm \sqrt{5}\}$ **33.** $\left\{\dfrac{-5 + \sqrt{13}}{2}, \dfrac{-5 - \sqrt{13}}{2}\right\}$

79. When the sum of 6 and twice a positive number is subtracted from the square of the number, 0 results. Find the number.

80. When the sum of 1 and twice a negative number is subtracted from twice the square of the number, 0 results. Find the number.

79. $1 + \sqrt{7}$ **80.** $\left\{\dfrac{1 - \sqrt{3}}{2}\right\}$

Application Exercises

The function $s(t) = 16t^2$ models the distance, $s(t)$, in feet, that an object falls in t seconds. Use this function and the square root property to solve Exercises 81–82. Express answers in simplified radical form. Then use your calculator to find a decimal approximation to the nearest tenth of a second.

81. A sky diver jumps from an airplane and falls for 4800 feet before opening a parachute. For how many seconds was the diver in a free fall? $10\sqrt{3}$ seconds ≈ 17.3 seconds

82. A sky diver jumps from an airplane and falls for 3200 feet before opening a parachute. For how many seconds was the diver in a free fall? $10\sqrt{2}$ seconds ≈ 14.1 seconds

A substantial percentage of the United States population is foreign-born. The bar graph shows the percentage of foreign-born Americans for selected years from 1920 through 2010.

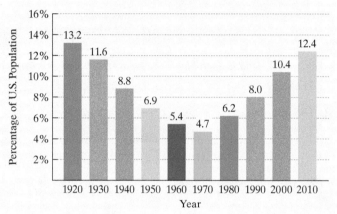

Percentage of the United States Population That Was Foreign-Born, 1920–2010

Source: U.S. Census Bureau

The percentage, p, of the United States population that was foreign-born x years after 1920 can be modeled by the formula

$$p = 0.004x^2 - 0.36x + 14.$$

Use the formula to solve Exercises 83–84.

83. a. According to the model, what percentage of the U.S. population was foreign-born in 2000? Does the model underestimate or overestimate the actual number displayed by the bar graph? By how much?

   b. If trends shown by the model continue, in which year will 18% of the U.S. population be foreign-born?   2020

84. a. According to the model, what percentage of the U.S. population was foreign-born in 1990? Does the model underestimate or overestimate the actual number displayed by the bar graph? By how much?

   b. If trends shown by the model continue, in which year will 23% of the U.S. population be foreign-born? Round to the nearest year.   2030

85. The length of a rectangle is 3 meters longer than the width. If the area is 36 square meters, find the rectangle's dimensions. Round to the nearest tenth of a meter.

86. The length of a rectangle is 2 meters longer than the width. If the area is 10 square meters, find the rectangle's dimensions. Round to the nearest tenth of a meter.

87. If you have not yet done so, read the Blitzer Bonus on page 302. In this exercise, you will use the golden rectangles shown to obtain an exact value for Φ, the ratio of the long side to the short side in a golden rectangle of any size.

Golden Rectangle *A*

44. $\{-3 + \sqrt{19}, -3 - \sqrt{19}\}$ **45.** $\left\{\dfrac{1 + \sqrt{57}}{2}, \dfrac{1 - \sqrt{57}}{2}\right\}$ **46.** $\left\{\dfrac{5 + \sqrt{65}}{2}, \dfrac{5 - \sqrt{65}}{2}\right\}$ **47.** $\left\{\dfrac{-3 + \sqrt{3}}{6}, \dfrac{-3 - \sqrt{3}}{6}\right\}$ **48.** $\left\{\dfrac{5 + \sqrt{13}}{6}, \dfrac{5 - \sqrt{13}}{6}\right\}$

83. a. 10.8; overestimates by 0.4   **84. a.** 8.4; overestimates by 0.4   **85.** width: 4.7 meters; length: 7.7 meters   **86.** width: 2.3 meters; length: 4.3 meters

a. The golden ratio in rectangle *A,* or the ratio of the long side to the short side, can be modeled by $\frac{\Phi}{1}$. Write a fractional expression that models the golden ratio in rectangle *B.* $\frac{1}{\Phi-1}$

b. Set the expression for the golden ratio in rectangle *A* equal to the expression for the golden ratio in rectangle *B.* Solve the resulting proportion using the quadratic formula. Express Φ as an exact value in simplified radical form. $\Phi = \frac{1+\sqrt{5}}{2}$

c. Use your solution from part (b) to complete this statement: The ratio of the long side to the short side in a golden rectangle of any size is _____ to 1. $\frac{1+\sqrt{5}}{2}$

Critical Thinking Exercises

88. Solve: $x^4 - 8x^2 + 15 = 0$. $\{\pm\sqrt{3}, \pm\sqrt{5}\}$

89. Solve: $x^2 + 2\sqrt{3}x - 9 = 0$. $\{-3\sqrt{3}, \sqrt{3}\}$

90. Solve for *t*: $s = -16t^2 + v_0t$. $t = \frac{v_0 \pm \sqrt{v_0^2 - 64s}}{32}$

91. The radicand of the quadratic formula, $b^2 - 4ac$, can be used to determine whether $ax^2 + bx + c = 0$ has solutions that are rational, irrational, or not real numbers. Explain how this works. Is it possible to determine the kinds of answers that one will obtain to a quadratic equation without actually solving the equation? Explain. Answers will vary.

92. A rectangular vegetable garden is 5 feet wide and 9 feet long. The garden is to be surrounded by a tile border of uniform width. If there are 40 square feet of tile for the border, how wide, to the nearest tenth of a foot, should it be? about 1.2 feet

4.5 Quadratic Functions and Their Graphs

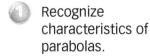
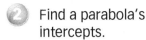

What am I supposed to learn?

After you have read this section, you should be able to:

1 Recognize characteristics of parabolas.

2 Find a parabola's intercepts.

3 Find a parabola's vertex.

4 Graph quadratic functions.

5 Solve problems involving a quadratic function's minimum or maximum value.

Many sports involve objects that are thrown, kicked, or hit, and then proceed with no additional force of their own. Such objects are called **projectiles**. Paths of projectiles, as well as their heights over time, can be modeled by quadratic functions. We have seen that a **quadratic function** is any function of the form

$$f(x) = ax^2 + bx + c,$$

where *a*, *b*, and *c* are real numbers, with $a \neq 0$. A quadratic function is a polynomial function whose greatest exponent is 2 when written in standard form. In this section, you will learn to use graphs of quadratic functions to gain a visual understanding of the algebra that describes football, baseball, basketball, the shot put, and other projectile sports.

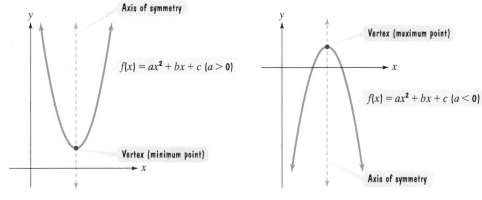

Graphs of Quadratic Functions

The graph of any quadratic function is called a **parabola**. Parabolas are shaped like bowls or inverted bowls, as shown in **Figure 4.13**. If the coefficient of x^2 (the value of *a* in $ax^2 + bx + c$) is positive, the parabola opens upward. If the coefficient of x^2 is negative, the parabola opens downward. The **vertex** (or turning point) of the parabola is the lowest point on the graph when it opens upward and the highest point on the graph when it opens downward.

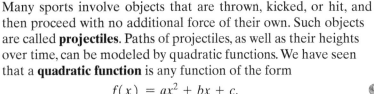
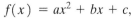

$a > 0$: Parabola opens upward. $a < 0$: Parabola opens downward.

Figure 4.13 Characteristics of graphs of quadratic functions

Look at the unusual image of the word *mirror* shown to the left. The artist, Scott Kim, has created the image so that the two halves of the whole are mirror images of each other. A parabola shares this kind of symmetry, in which a vertical line through the vertex divides the figure in half. Parabolas are symmetric with respect to this line, called the **axis of symmetry**. If a parabola is folded along its axis of symmetry, the two halves match exactly.

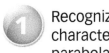 **1** Recognize characteristics of parabolas.

Example 1 **Using Point Plotting to Graph a Parabola**

Consider the quadratic function $f(x) = x^2 + 4x + 3$.

a. Is the graph a parabola that opens upward or downward?

b. Use point plotting to graph the parabola. Select integers from −5 to 1, inclusive, for x.

Solution

a. To determine whether a parabola opens upward or downward, we begin by identifying a, the coefficient of x^2. The following voice balloons show the values for $a, b,$ and c in $f(x) = x^2 + 4x + 3$. Notice that we wrote x^2 as $1x^2$.

$$f(x) = 1x^2 + 4x + 3$$

| a, the coefficient of x^2, is 1. | b, the coefficient of x, is 4. | c, the constant term, is 3. |

When a is greater than 0, a parabola opens upward. When a is less than 0, a parabola opens downward. Because $a = 1$, which is greater than 0, the parabola opens upward.

b. To use point plotting to graph the parabola, we first make a table of x- and y-coordinates.

x	$f(x) = x^2 + 4x + 3$	(x, y)
−5	$f(-5) = (-5)^2 + 4(-5) + 3 = 8$	$(-5, 8)$
−4	$f(-4) = (-4)^2 + 4(-4) + 3 = 3$	$(-4, 3)$
−3	$f(-3) = (-3)^2 + 4(-3) + 3 = 0$	$(-3, 0)$
−2	$f(-2) = (-2)^2 + 4(-2) + 3 = -1$	$(-2, -1)$
−1	$f(-1) = (-1)^2 + 4(-1) + 3 = 0$	$(-1, 0)$
0	$f(0) = 0^2 + 4(0) + 3 = 3$	$(0, 3)$
1	$f(1) = 1^2 + 4(1) + 3 = 8$	$(1, 8)$

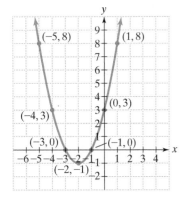

Figure 4.14 The graph of $f(x) = x^2 + 4x + 3$

Then we plot the points and connect them with a smooth curve. The graph of $f(x) = x^2 + 4x + 3$ is shown in **Figure 4.14**.

✓ **Check Point 1** Consider the quadratic function $f(x) = x^2 - 6x + 8$.

a. Is the graph a parabola that opens upward or downward? upward

b. Use point plotting to graph the parabola. Select integers from 0 to 6, inclusive, for x. *

*See Answers to Selected Exercises.

Several points are important when graphing a quadratic function. These points, labeled in **Figure 4.15**, are the x-intercepts (although not every parabola has two x-intercepts), the y-intercept, and the vertex. Let's see how we can locate these points.

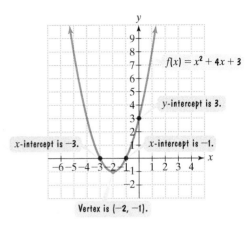

Figure 4.15 Useful points in graphing a parabola

Finding a Parabola's x-Intercepts

 Find a parabola's intercepts.

At each point where a parabola crosses the x-axis, the value of y, or $f(x)$, equals 0. Thus, the x-intercepts can be found by replacing $f(x)$ with 0 in $f(x) = ax^2 + bx + c$. Use factoring or the quadratic formula to solve the resulting quadratic equation for x.

Example 2 Finding a Parabola's x-Intercepts

Find the x-intercepts for the parabola whose equation is $f(x) = x^2 + 4x + 3$.

Solution

Replace $f(x)$ with 0 in $f(x) = x^2 + 4x + 3$. We obtain $0 = x^2 + 4x + 3$, or $x^2 + 4x + 3 = 0$. We can solve this quadratic equation by factoring.

$$x^2 + 4x + 3 = 0$$
$$(x + 3)(x + 1) = 0$$
$$x + 3 = 0 \quad \text{or} \quad x + 1 = 0$$
$$x = -3 \qquad x = -1$$

Thus, the x-intercepts are -3 and -1. The parabola passes through $(-3, 0)$ and $(-1, 0)$, as shown in **Figure 4.15**.

Check Point 2 Find the x-intercepts for the parabola whose equation is $f(x) = x^2 - 6x + 8$. 2 and 4

Finding a Parabola's y-Intercept

At the point where a parabola crosses the y-axis, the value of x equals 0. Thus, the y-intercept can be found by replacing x with 0 in $f(x) = ax^2 + bx + c$. Simple arithmetic will produce a value for $f(0)$, which is the y-intercept.

Example 3 Finding a Parabola's y-Intercept

Find the y-intercept for the parabola whose equation is $f(x) = x^2 + 4x + 3$.

Solution

Replace x with 0 in $f(x) = x^2 + 4x + 3$.

$$f(0) = 0^2 + 4 \cdot 0 + 3 = 0 + 0 + 3 = 3$$

The y-intercept is 3. The parabola passes through $(0, 3)$, as shown in **Figure 4.15**.

✓ **Check Point 3** Find the y-intercept for the parabola whose equation is
▶ $f(x) = x^2 - 6x + 8.$ 8

Finding a Parabola's Vertex

3 Find a parabola's vertex.

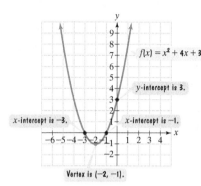

Vertex is $(-2, -1)$.

Figure 4.15 (repeated)
Useful points in graphing a parabola

Keep in mind that a parabola's vertex is its turning point. The x-coordinate of the vertex for the parabola in **Figure 4.15**, -2, is midway between the x-intercepts, -3 and -1. If a parabola has two x-intercepts, they are found by solving $ax^2 + bx + c = 0$. The solutions of this equation,

$$x = \frac{-b - \sqrt{b^2 - 4ac}}{2a} \quad \text{and} \quad x = \frac{-b + \sqrt{b^2 - 4ac}}{2a},$$

are the x-intercepts. The value of x midway between these intercepts is $x = -\dfrac{b}{2a}$.
This equation can be used to find the x-coordinate of the vertex even when no x-intercepts exist.

The Vertex of a Parabola

For a parabola whose equation is $f(x) = ax^2 + bx + c,$

1. The x-coordinate of the vertex is $-\dfrac{b}{2a}.$
2. The y-coordinate of the vertex is found by substituting the x-coordinate into the parabola's equation and evaluating the function at this value of x.

Example 4 **Finding a Parabola's Vertex**

Find the vertex for the parabola whose equation is $f(x) = x^2 + 4x + 3.$

Solution

We know that the x-coordinate of the vertex is $x = -\dfrac{b}{2a}.$ Let's identify the numbers $a, b,$ and c in the given equation, which is in the form $f(x) = ax^2 + bx + c.$

$$f(x) = x^2 + 4x + 3$$

$a = 1 \qquad b = 4 \qquad c = 3$

Substitute the values of a and b into the equation for the x-coordinate:

$$x = -\frac{b}{2a} = -\frac{4}{2 \cdot 1} = -\frac{4}{2} = -2$$

The x-coordinate of the vertex is -2. We substitute -2 for x into the equation of the function, $f(x) = x^2 + 4x + 3,$ to find the y-coordinate:

$$f(-2) = (-2)^2 + 4(-2) + 3 = 4 + (-8) + 3 = -1.$$

The vertex is $(-2, -1),$ as shown in **Figure 4.15**.

✓ **Check Point 4** Find the vertex for the parabola whose equation is
▶ $f(x) = x^2 - 6x + 8.$ $(3, -1)$

4 Graph quadratic functions.

A Strategy for Graphing Quadratic Functions

Here is a procedure to sketch the graph of the quadratic function $f(x) = ax^2 + bx + c$:

Graphing Quadratic Functions

The graph of $f(x) = ax^2 + bx + c$, called a parabola, can be graphed using the following steps:

1. Determine whether the parabola opens upward or downward. If $a > 0$, it opens upward. If $a < 0$, it opens downward.
2. Determine the vertex of the parabola. The x-coordinate is $-\dfrac{b}{2a}$. The y-coordinate is found by substituting the x-coordinate into the parabola's equation and evaluating the function at this value of x.
3. Find any x-intercepts by replacing $f(x)$ with 0. Solve the resulting quadratic equation for x. The real solutions of $ax^2 + bx + c = 0$ are the x-intercepts.
4. Find the y-intercept by replacing x with 0 and computing $f(0)$. Because $f(0) = a \cdot 0^2 + b \cdot 0 + c$ simplifies to c, the y-intercept is c, the constant term, and the parabola passes through $(0, c)$.
5. Plot the intercepts and the vertex.
6. Connect these points with a smooth curve.

Example 5   Graphing a Quadratic Function

Graph the quadratic function: $f(x) = x^2 - 2x - 3$.

Solution

We can graph this function by following the steps in the box.

Step 1   Determine how the parabola opens. Note that a, the coefficient of x^2, is 1. Thus, $a > 0$; this positive value tells us that the parabola opens upward.

Step 2   Find the vertex. We know that the x-coordinate of the vertex is $-\dfrac{b}{2a}$. Let's identify the numbers a, b, and c in the given equation, which is in the form $f(x) = ax^2 + bx + c$.

$$f(x) = x^2 - 2x - 3$$

$$a = 1 \qquad b = -2 \qquad c = -3$$

Now we substitute the values of a and b, $a = 1$ and $b = -2$, into the expression for the x-coordinate:

$$x\text{-coordinate of vertex} = -\frac{b}{2a} = -\frac{-2}{2(1)} = -\left(\frac{-2}{2}\right) = -(-1) = 1.$$

The x-coordinate of the vertex is 1. We can substitute 1 for x in the equation of the function, $f(x) = x^2 - 2x - 3$, to find the y-coordinate:

$$y\text{-coordinate of vertex} = f(1) = 1^2 - 2 \cdot 1 - 3 = 1 - 2 - 3 = -4.$$

The vertex is $(1, -4)$.

Step 3   Find the x-intercepts. Replace $f(x)$ with 0 in $f(x) = x^2 - 2x - 3$. We obtain $0 = x^2 - 2x - 3$ or $x^2 - 2x - 3 = 0$. We can solve this quadratic equation by factoring.

$$x^2 - 2x - 3 = 0$$
$$(x - 3)(x + 1) = 0$$
$$x - 3 = 0 \quad \text{or} \quad x + 1 = 0$$
$$x = 3 \qquad\qquad x = -1$$

The x-intercepts are 3 and -1. The parabola passes through $(3, 0)$ and $(-1, 0)$.

Step 4   Find the y-intercept. Replace x with 0 in $f(x) = x^2 - 2x - 3$:

$$f(0) = 0^2 - 2 \cdot 0 - 3 = 0 - 0 - 3 = -3.$$

The y-intercept is -3, which is the constant term in the function's equation. The parabola passes through $(0, -3)$.

Steps 5 and 6   Plot the intercepts and the vertex. Connect these points with a smooth curve. The intercepts and the vertex are shown as the four labeled points in **Figure 4.16**. Also shown is the graph of the quadratic function, obtained by connecting the points with a smooth curve.

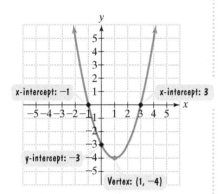

Figure 4.16 The graph of $f(x) = x^2 - 2x - 3$

Check Point 5   Graph the quadratic function: $f(x) = x^2 + 6x + 5.$ *

Example 6   Graphing a Quadratic Function

Graph the quadratic function: $f(x) = -x^2 + 4x - 1.$

Solution

Step 1   Determine how the parabola opens. Note that a, the coefficient of x^2, is -1. Thus, $a < 0$; this negative value tells us that the parabola opens downward.

Step 2   Find the vertex. The x-coordinate of the vertex is $-\dfrac{b}{2a}$.

$$f(x) = -x^2 + 4x - 1$$

$$x\text{-coordinate of vertex} = -\frac{b}{2a} = -\frac{4}{2(-1)} = -\left(\frac{4}{-2}\right) = -(-2) = 2$$

Substitute 2 for x in $f(x) = -x^2 + 4x - 1$ to find the y-coordinate:

$$y\text{-coordinate of vertex} = f(2) = -2^2 + 4 \cdot 2 - 1 = -4 + 8 - 1 = 3.$$

The vertex is $(2, 3)$.

Step 3   Find the x-intercepts. Replace $f(x)$ with 0 in $f(x) = -x^2 + 4x - 1$. We obtain $0 = -x^2 + 4x - 1$ or $-x^2 + 4x - 1 = 0$. This quadratic equation cannot be solved by factoring. We will use the quadratic formula to solve it with $a = -1$, $b = 4$, and $c = -1$.

$$x = \frac{-b \pm \sqrt{b^2 - 4ac}}{2a} = \frac{-4 \pm \sqrt{4^2 - 4(-1)(-1)}}{2(-1)} = \frac{-4 \pm \sqrt{16 - 4}}{-2}$$

To locate the x-intercepts, we need decimal approximations. Thus, there is no need to simplify the radical form of the solutions.

$$x = \frac{-4 + \sqrt{12}}{-2} \approx 0.3 \quad \text{or} \quad x = \frac{-4 - \sqrt{12}}{-2} \approx 3.7$$

*See Answers to Selected Exercises.

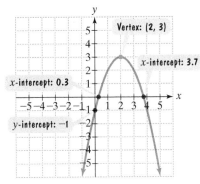

Figure 4.17 The graph of
$f(x) = -x^2 + 4x - 1$

The x-intercepts are approximately 0.3 and 3.7. The parabola passes through $(0.3, 0)$ and $(3.7, 0)$.

Step 4 Find the y-intercept. Replace x with 0 in $f(x) = -x^2 + 4x - 1$:

$$f(0) = -0^2 + 4 \cdot 0 - 1 = -1.$$

The y-intercept is -1. The parabola passes through $(0, -1)$.

Steps 5 and 6 Plot the intercepts and the vertex. Connect these points with a smooth curve. The intercepts and the vertex are shown as the four labeled points in **Figure 4.17**. Also shown is the graph of the quadratic function, obtained by connecting the points with a smooth curve.

▶ ✓ **Check Point 6** Graph the quadratic function: $f(x) = -x^2 - 2x + 5.$ *

Applications of Quadratic Functions

⑤ Solve problems involving a quadratic function's minimum or maximum value.

Many applied problems involve finding the maximum or minimum value of a quadratic function, as well as where this value occurs.

Example 7 The Parabolic Path of a Punted Football

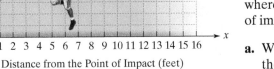

Figure 4.18

Figure 4.18 shows that when a football is kicked, the nearest defensive player is 6 feet from the point of impact with the kicker's foot. The height of the punted football, $f(x)$, in feet, can be modeled by

$$f(x) = -0.01x^2 + 1.18x + 2,$$

where x is the ball's horizontal distance, in feet, from the point of impact with the kicker's foot.

a. What is the maximum height of the punt and how far from the point of impact does this occur?

b. How far must the nearest defensive player, who is 6 feet from the kicker's point of impact, reach to block the punt?

c. If the ball is not blocked by the defensive player, how far down the field will it go before hitting the ground?

d. Graph the function that models the football's parabolic path.

Solution

a. We begin by identifying the numbers a, b, and c in the function's equation.

$$f(x) = -0.01x^2 + 1.18x + 2$$

$$a = -0.01 \quad b = 1.18 \quad c = 2$$

Because $a < 0$, the function has a maximum that occurs at $x = -\dfrac{b}{2a}$.

$$x = -\frac{b}{2a} = -\frac{1.18}{2(-0.01)} = -(-59) = 59$$

This means that the maximum height of the punt occurs 59 feet from the kicker's point of impact. The maximum height of the punt is

$$f(59) = -0.01(59)^2 + 1.18(59) + 2 = 36.81,$$

or 36.81 feet.

*See Answers to Selected Exercises.

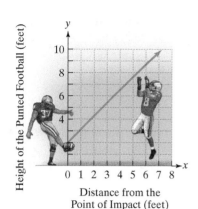

Figure 4.18 (repeated)

$$-0.01x^2 + 1.18x + 2 = 0$$

$a = -0.01$ $b = 1.18$ $c = 2$

The equation for determining the ball's maximum horizontal distance

b. **Figure 4.18** shows that the defensive player is 6 feet from the kicker's point of impact. To block the punt, he must touch the football along its parabolic path. This means that we must find the height of the ball 6 feet from the kicker. Replace x with 6 in the given function, $f(x) = -0.01x^2 + 1.18x + 2$.

$$f(6) = -0.01(6)^2 + 1.18(6) + 2 = -0.36 + 7.08 + 2 = 8.72$$

The defensive player must reach 8.72 feet above the ground to block the punt.

c. Assuming that the ball is not blocked by the defensive player, we are interested in how far down the field it will go before hitting the ground. We are looking for the ball's horizontal distance, x, when its height above the ground, $f(x)$, is 0 feet. To find this x-intercept, replace $f(x)$ with 0 in $f(x) = -0.01x^2 + 1.18x + 2$. We obtain $0 = -0.01x^2 + 1.18x + 2$, or $-0.01x^2 + 1.18x + 2 = 0$. This quadratic equation cannot be solved by factoring. We will use the quadratic formula to solve it.

Use a calculator to evaluate the radicand.

$$x = \frac{-b \pm \sqrt{b^2 - 4ac}}{2a} = \frac{-1.18 \pm \sqrt{(1.18)^2 - 4(-0.01)(2)}}{2(-0.01)} = \frac{-1.18 \pm \sqrt{1.4724}}{-0.02}$$

$$x = \frac{-1.18 + \sqrt{1.4724}}{-0.02} \quad \text{or} \quad x = \frac{-1.18 - \sqrt{1.4724}}{-0.02}$$

$$x \approx -1.7 \qquad\qquad x \approx 119.7 \quad \text{Use a calculator and round to the nearest tenth.}$$

Reject this value. We are interested in the football's height corresponding to horizontal distances from its point of impact onward, or $x \geq 0$.

If the football is not blocked by the defensive player, it will go approximately 119.7 feet down the field before hitting the ground.

d. In terms of graphing the model for the football's parabolic path, $f(x) = -0.01x^2 + 1.18x + 2$, we have already determined the vertex and the approximate x-intercept.

vertex: $(59, 36.81)$

The ball's maximum height, 36.81 feet, occurs at a horizontal distance of 59 feet.

x-intercept: 119.7

The ball's maximum horizontal distance is approximately 119.7 feet.

Figure 4.18 indicates that the y-intercept is 2, meaning that the ball is kicked from a height of 2 feet. Let's verify this value by replacing x with 0 in $f(x) = -0.01x^2 + 1.18x + 2$.

$$f(0) = -0.01 \cdot 0^2 + 1.18 \cdot 0 + 2 = 0 + 0 + 2 = 2$$

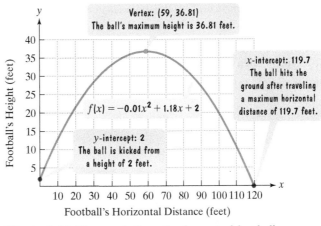

Figure 4.19 The parabolic path of a punted football

Using the vertex, $(59, 36.81)$, the x-intercept, 119.7, and the y-intercept, 2, the graph of the equation that models the football's parabolic path is shown in **Figure 4.19**. The graph is shown only for $x \geq 0$, indicating horizontal distances that begin at the football's impact with the kicker's foot and end with the ball hitting the ground.

✓ Check Point 7   An archer's arrow follows a parabolic path. The height of the arrow, $f(x)$, in feet, can be modeled by

$$f(x) = -0.005x^2 + 2x + 5,$$

where x is the arrow's horizontal distance from the archer, in feet.

a. What is the maximum height of the arrow and how far from its release does this occur?   205 feet; 200 feet

b. Find the horizontal distance the arrow travels before it hits the ground. Round to the nearest foot.   402 feet

▸ **c.** Graph the function that models the arrow's parabolic path.   *

Achieving Success

Address your stress. Stress levels can help or hinder performance. The parabola in **Figure 4.20** serves as a model that shows people under both low stress and high stress perform worse than their moderate-stress counterparts.

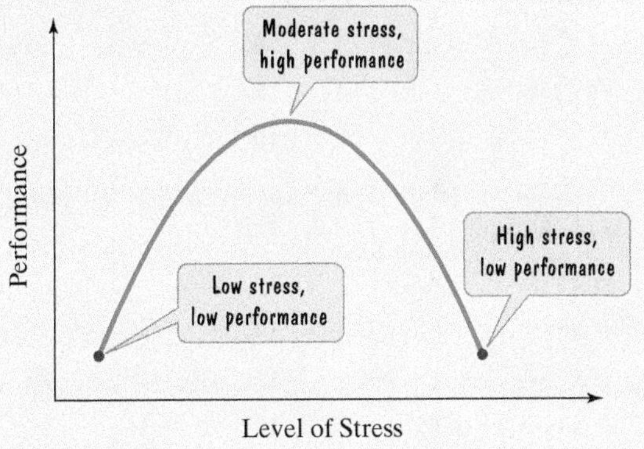

Figure 4.20
Source: Herbert Benson, *Your Maximum Mind,* Random House, 1987.

Concept and Vocabulary Check

Exercises in the Concept and Vocabulary Check are intended for group and class discussions.

In Exercises 1–4, fill in each blank so that the resulting statement is true.

1. The graph of $f(x) = ax^2 + bx + c$ opens upward if _____ $a > 0$ _____ and opens downward if _____ $a < 0$ _____ .

2. The x-intercepts for the graph of $f(x) = ax^2 + bx + c$ can be found by determining the real solutions of the equation _____ $ax^2 + bx + c = 0$ _____ .

3. The y-intercept for the graph of $f(x) = ax^2 + bx + c$ can be determined by replacing x with _____ 0 _____ and computing _____ $f(0)$ _____ .

4. The x-coordinate of the vertex of the parabola whose equation is $f(x) = ax^2 + bx + c$ is _____ $-\dfrac{b}{2a}$ _____ . The y-coordinate of the vertex is found by substituting _____ $-\dfrac{b}{2a}$ or the x-coordinate _____ into the parabola's equation and evaluating the function at this value of x.

*See Answers to Selected Exercises.

In Exercises 5–8, determine whether each statement is true or false. If the statement is false, make the necessary change(s) to produce a true statement. Changes to false statements will vary.

5. The x-coordinate of the vertex of the parabola whose equation is $f(x) = ax^2 + bx + c$ is $\dfrac{b}{2a}$. false

6. If a parabola has only one x-intercept, then the x-intercept is also the vertex. true

7. There is no relationship between the graph of $f(x) = ax^2 + bx + c$ and the number of real solutions of the equation $ax^2 + bx + c = 0$. false

8. If $f(x) = 4x^2 - 40x + 4$, then the vertex is the highest point on the graph. false

Respond to Exercises 9–13 using verbal or written explanations.

9. What is a parabola? Describe its shape. 9–13. Answers will vary.

10. Explain how to decide whether a parabola opens upward or downward.

11. If a parabola has two x-intercepts, explain how to find them.

12. Explain how to find a parabola's y-intercept.

13. Describe how to find a parabola's vertex.

Exercise Set 4.5

Practice Exercises

In Exercises 1–4, determine if the parabola whose equation is given opens upward or downward.

1. $f(x) = x^2 - 4x + 3$ **2.** $f(x) = x^2 - 6x + 5$

3. $f(x) = -2x^2 + x + 6$ **4.** $f(x) = -2x^2 - 4x + 6$

In Exercises 5–10, find the x-intercepts for the parabola whose equation is given. If the x-intercepts are irrational numbers, round your answers to the nearest tenth.

5. $f(x) = x^2 - 4x + 3$ **6.** $f(x) = x^2 - 6x + 5$

7. $f(x) = -x^2 + 8x - 12$ **8.** $f(x) = -x^2 - 2x + 3$

9. $f(x) = x^2 + 2x - 4$ **10.** $f(x) = x^2 + 8x + 14$

In Exercises 11–18, find the y-intercept for the parabola whose equation is given.

11. $f(x) = x^2 - 4x + 3$ 3 **12.** $f(x) = x^2 - 6x + 5$ 5

13. $f(x) = -x^2 + 8x - 12$ **14.** $f(x) = -x^2 - 2x + 3$ 3

15. $f(x) = x^2 + 2x - 4$ -4 **16.** $f(x) = x^2 + 8x + 14$ 14

17. $f(x) = x^2 + 6x$ 0 **18.** $f(x) = x^2 + 8x$ 0

In Exercises 19–24, find the vertex for the parabola whose equation is given.

19. $f(x) = x^2 - 4x + 3$ **20.** $f(x) = x^2 - 6x + 5$

21. $f(x) = 2x^2 + 4x - 6$ **22.** $f(x) = -2x^2 - 4x - 2$

23. $f(x) = x^2 + 6x$ **24.** $f(x) = x^2 + 8x$

In Exercises 25–36, graph each quadratic function.

25. $f(x) = x^2 + 8x + 7$ * **26.** $f(x) = x^2 + 10x + 9$ *

27. $f(x) = x^2 - 2x - 8$ * **28.** $f(x) = x^2 + 4x - 5$ *

29. $f(x) = -x^2 + 4x - 3$ * **30.** $f(x) = -x^2 + 2x + 3$ *

31. $f(x) = x^2 - 1$ * **32.** $f(x) = x^2 - 4$ *

33. $f(x) = x^2 + 2x + 1$ * **34.** $f(x) = x^2 - 2x + 1$ *

35. $f(x) = -2x^2 + 4x + 5$ * **36.** $f(x) = -3x^2 + 6x - 2$ *

Practice Plus

In Exercises 37–43, find the vertex for the parabola whose equation is given by first writing the equation in the form $f(x) = ax^2 + bx + c$.

37. $f(x) = (x - 3)^2 + 2$ **38.** $f(x) = (x - 4)^2 + 3$

39. $f(x) = (x + 5)^2 - 4$ **40.** $f(x) = (x + 6)^2 - 5$

41. $f(x) = 2(x - 1)^2 - 3$ **42.** $f(x) = 2(x - 1)^2 - 4$

43. $f(x) = -3(x + 2)^2 + 5$ $(-2, 5)$ **42.** $(1, -4)$

44. Generalize your work in Exercises 37–43 and complete the following statement: For a parabola whose equation is $f(x) = a(x - h)^2 + k$, the vertex is the point ___(h, k)___.

Application Exercises

An athlete whose event is the shot put releases the shot with the same initial velocity, but at different angles. The figure shows the parabolic paths for shots released at angles of 35° and 65°. Exercises 45–46 are based on the functions that model the parabolic paths.

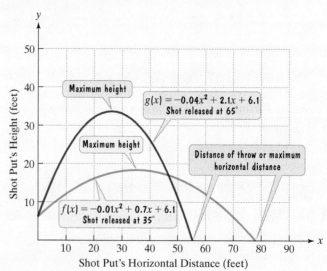

Shot Put's Height (feet)

Maximum height

$g(x) = -0.04x^2 + 2.1x + 6.1$
Shot released at 65°

Maximum height

Distance of throw or maximum horizontal distance

$f(x) = -0.01x^2 + 0.7x + 6.1$
Shot released at 35°

Shot Put's Horizontal Distance (feet)

1. upward 2. upward 3. downward 4. downward 5. 1 and 3 6. 1 and 5 7. 2 and 6 8. -3 and 1 9. -3.2 and 1.2 10. -5.4 and -2.6 13. -12
19. $(2, -1)$ 20. $(3, -4)$ 21. $(-1, -8)$ 22. $(-1, 0)$ 23. $(-3, -9)$ 24. $(-4, -16)$ 37. $(3, 2)$ 38. $(4, 3)$ 39. $(-5, -4)$ 40. $(-6, -5)$ 41. $(1, -3)$
*See Answers to Selected Exercises.

45. Refer to the blue graph at the bottom of the previous page. When the shot whose path is shown by the blue graph is released at an angle of 35°, its height, $f(x)$, in feet, can be modeled by

$$f(x) = -0.01x^2 + 0.7x + 6.1,$$

where x is the shot's horizontal distance, in feet, from its point of release. Use this model to solve parts (a) through (c) and verify your answers using the blue graph.

   a. What is the maximum height of the shot and how far from its point of release does this occur?   18.35 ft; 35 ft

   b. What is the shot's maximum horizontal distance, to the nearest tenth of a foot, or the distance of the throw?   77.8 ft

   c. From what height was the shot released?   6.1 ft

46. Refer to the red graph at the bottom of the previous page. When the shot whose path is shown by the red graph is released at an angle of 65°, its height, $g(x)$, in feet, can be modeled by

$$g(x) = -0.04x^2 + 2.1x + 6.1,$$

where x is the shot's horizontal distance, in feet, from its point of release. Use this model to solve parts (a) through (c) and verify your answers using the red graph.

   a. What is the maximum height, to the nearest tenth of a foot, of the shot and how far from its point of release does this occur?   approximately 33.7 ft; 26.25 ft

   b. What is the shot's maximum horizontal distance, to the nearest tenth of a foot, or the distance of the throw?   55.3 ft

   c. From what height was the shot released?   6.1 ft

47. A ball is thrown upward and outward from a height of 6 feet. The height of the ball, $f(x)$, in feet, can be modeled by

$$f(x) = -0.8x^2 + 2.4x + 6,$$

where x is the ball's horizontal distance, in feet, from where it was thrown.

   a. What is the maximum height of the ball and how far from where it was thrown does this occur?   7.8 ft; 1.5 ft

   b. How far does the ball travel horizontally before hitting the ground? Round to the nearest tenth of a foot.   4.6 ft

   c. Graph the function that models the ball's parabolic path.   *

48. A ball is thrown upward and outward from a height of 6 feet. The height of the ball, $f(x)$, in feet, can be modeled by

$$f(x) = -0.8x^2 + 3.2x + 6,$$

where x is the ball's horizontal distance, in feet, from where it was thrown.

   a. What is the maximum height of the ball and how far from where it was thrown does this occur?   9.2 ft; 2 ft

   b. How far does the ball travel horizontally before hitting the ground? Round to the nearest tenth of a foot.   5.4 ft

   c. Graph the function that models the ball's parabolic path.   *

49. You have 120 feet of fencing to enclose a rectangular plot that borders on a river. If you do not fence the side along the river, find the length and width of the plot that will maximize the area. What is the largest area that can be enclosed?

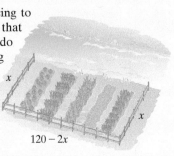

$120 - 2x$

50. You have 100 yards of fencing to enclose a rectangular area. Find the dimensions of the rectangle that maximize the enclosed area. What is the maximum area?   25 yd by 25 yd; 625 sq yd

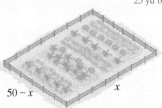

$50 - x$ x

Critical Thinking Exercises

51. Among all pairs of numbers whose sum is 16, find a pair whose product is as large as possible. What is the maximum product?   8 and 8; 64

52. Among all pairs of numbers whose sum is 20, find a pair whose product is as large as possible. What is the maximum product?   10 and 10; 100

53. Graph $x + y = 2$ and $y = x^2 - 4x + 4$ in the same rectangular coordinate system. What are the coordinates of the points of intersection? Show that each ordered pair is a solution of the system

$$\begin{cases} x + y = 2 \\ y = x^2 - 4x + 4. \end{cases}$$ (1, 1) and (2, 0) *

54. You have 80 yards of fencing to enclose a rectangular region. Find the dimensions of the rectangle that maximize the enclosed area. What is the maximum area?   20 yd by 20 yd; 400 sq yd

55. A parabola has x-intercepts at 3 and 7, a y-intercept at -21, and $(5, 4)$ for its vertex. Write the parabola's equation. $y = -x^2 + 10x - 21$

Technology Exercise

56. Use a graphing utility to verify any five of your hand drawn graphs in Exercises 25–36.

Group Exercise

57. Each group member should consult an almanac, newspaper, magazine, or the Internet to find data that initially increase and then decrease, or vice versa, and therefore can be modeled by a quadratic function. Group members should select the two sets of data that are most interesting and relevant. For each data set selected,   Answers will vary.

   a. Use the quadratic regression feature of a graphing utility to find the quadratic function that best fits the data.

   b. Use the equation of the quadratic function to make a prediction from the data. What circumstances might affect the accuracy of your prediction?

   c. Use the equation of the quadratic function to write and solve a problem involving maximizing or minimizing the function.

49. length: 60 ft; width: 30 ft; 1800 sq ft

Chapter 4 Summary

Definitions and Concepts	Example

Section 4.1 Operations with Polynomials; Polynomial Functions

A polynomial is a single term or the sum of two or more terms containing variables with whole number exponents. A monomial is a polynomial with exactly one term; a binomial has exactly two terms; a trinomial has exactly three terms. The degree of a polynomial is the greatest of the powers of all the terms. The standard form of a polynomial is written in descending powers of the variable. In a polynomial function, the expression that defines the function is a polynomial.

To add polynomials, add like terms.

- $(6x^3 + 5x^2 - 7x) + (-9x^3 + x^2 + 6x)$
  $= (6x^3 - 9x^3) + (5x^2 + x^2) + (-7x + 6x)$
  $= -3x^3 + 6x^2 - x$

Additional Example to Review

Example 1, page 256

To subtract two polynomials, change the sign of every term of the second polynomial. Add this result to the first polynomial.

- $(5y^3 - 9y^2 - 4) - (3y^3 - 12y^2 - 5)$
  $= (5y^3 - 9y^2 - 4) + (-3y^3 + 12y^2 + 5)$
  $= (5y^3 - 3y^3) + (-9y^2 + 12y^2) + (-4 + 5)$
  $= 2y^3 + 3y^2 + 1$

Additional Example to Review

Example 2, page 256

To multiply a monomial and a polynomial that is not a monomial, use the distributive property to multiply each term of the polynomial by the monomial.

- $2x^4(3x^2 - 6x + 5)$
  $= 2x^4 \cdot 3x^2 - 2x^4 \cdot 6x + 2x^4 \cdot 5$
  $= 6x^6 - 12x^5 + 10x^4$

Additional Example to Review

Example 3, page 258

To multiply polynomials when neither is a monomial, multiply each term of one polynomial by each term of the other polynomial. Then combine like terms.

- $(2x + 3)(5x^2 - 4x + 2)$

  $= 2x(5x^2 - 4x + 2) + 3(5x^2 - 4x + 2)$
  $= 10x^3 - 8x^2 + 4x + 15x^2 - 12x + 6$
  $= 10x^3 + 7x^2 - 8x + 6$

Additional Example to Review

Example 4, page 258

FOIL

The FOIL method may be used when multiplying two binomials: First terms multiplied. Outside terms multiplied. Inside terms multiplied. Last terms multiplied.

$$\begin{array}{cccc} \boxed{\text{F}} & \boxed{\text{O}} & \boxed{\text{I}} & \boxed{\text{L}} \end{array}$$

- $(3x + 7)(2x - 5) = 3x \cdot 2x + 3x(-5) + 7 \cdot 2x + 7(-5)$
$$= 6x^2 - 15x + 14x - 35$$
$$= 6x^2 - x - 35$$

Additional Example to Review

Example 5, page 260

The Product of the Sum and Difference of Two Terms

$$(A + B)(A - B) = A^2 - B^2$$

- $(4x + 7)(4x - 7) = (4x)^2 - 7^2$
$$= 16x^2 - 49$$

Additional Example to Review

Example 6, page 260

The Square of a Binomial Sum

$$(A + B)^2 = A^2 + 2AB + B^2$$

- $(x^2 + 6)^2 = (x^2)^2 + 2 \cdot x^2 \cdot 6 + 6^2$
$$= x^4 + 12x^2 + 36$$

Additional Example to Review

Example 7, page 261

The Square of a Binomial Difference

$$(A - B)^2 = A^2 - 2AB + B^2$$

- $(9x - 3)^2 = (9x)^2 - 2 \cdot 9x \cdot 3 + 3^2$
$$= 81x^2 - 54x + 9$$

Additional Example to Review

Example 8, page 262

Section 4.2 Factoring Polynomials

Factoring a polynomial consisting of the sum of monomials means finding an equivalent expression that is a product. Polynomials that cannot be factored using integer coefficients are called prime polynomials over the integers. The greatest common factor, GCF, is an expression that divides every term of the polynomial. The GCF is the product of the largest common numerical factor and the variable of lowest degree common to every term of the polynomial. To factor a monomial from a polynomial, express each term as the product of the GCF and its other factor. Then use the distributive property to factor out the GCF.

- $20x^3 + 30x^2 = 10x^2(2x) + 10x^2(3) = 10x^2(2x + 3)$
- $x^2(x + 4) + 7(x + 4) = (x + 4)(x^2 + 7)$

Additional Examples to Review

Example 1, page 268; Example 2, page 269

To factor by grouping, factor out the GCF from each group. Then factor out the remaining common factor.

- $x^3 + 2x^2 + 7x + 14 = (x^3 + 2x^2) + (7x + 14)$
$$= x^2(x + 2) + 7(x + 2)$$
$$= (x + 2)(x^2 + 7)$$

Additional Example to Review

Example 3, page 269

To factor a trinomial of the form $x^2 + bx + c$, find two numbers whose product is c and whose sum is b. The factorization is

$$(x + \text{one number})(x + \text{other number}).$$

To factor $ax^2 + bx + c$, try various combinations of factors of ax^2 and c until a middle term of bx is obtained for the sum of the outside and inside products.

The Difference of Two Squares

$$A^2 - B^2 = (A + B)(A - B)$$

Perfect Square Trinomials

$$A^2 + 2AB + B^2 = (A + B)^2$$
$$A^2 - 2AB + B^2 = (A - B)^2$$

A Factoring Strategy

1. Factor out the GCF.
2. **a.** If two terms, try
$$A^2 - B^2 = (A + B)(A - B).$$
   b. If three terms, try
$$A^2 + 2AB + B^2 = (A + B)^2$$
$$A^2 - 2AB + B^2 = (A - B)^2.$$
   If not a perfect square trinomial, try factoring by trial and error.
   c. If four terms, try factoring by grouping.
3. See if any factors can be factored further.
4. Check by multiplying.

• Factor: $x^2 + 9x + 20$.

Find two numbers whose product is 20 and whose sum is 9. The numbers are 4 and 5.

$$x^2 + 9x + 20 = (x + 4)(x + 5)$$

Additional Examples to Review

Example 4, page 271; Example 5, page 272

• Factor: $2x^2 + 7x - 15$.

Factors of $2x^2$: $2x, x$

Factors of -15: 1 and -15, -1 and 15, 3 and -5, -3 and 5

$$(2x - 3)(x + 5)$$

Sum of outside and inside products should equal $7x$.

$$10x - 3x = 7x$$

Thus, $2x^2 + 7x - 15 = (2x - 3)(x + 5)$.

Additional Examples to Review

Example 6, page 273; Example 7, page 274

• $16x^2 - 25 = (4x)^2 - 5^2 = (4x + 5)(4x - 5)$

Additional Examples to Review

Example 8, page 275; Example 9, page 275

• $x^2 + 20x + 100 = x^2 + 2 \cdot x \cdot 10 + 10^2 = (x + 10)^2$
• $9x^2 - 30x + 25 = (3x)^2 - 2 \cdot 3x \cdot 5 + 5^2 = (3x - 5)^2$

Additional Example to Review

Example 10, page 276

• Factor: $3x^4 + 12x^3 - 3x^2 - 12x$.

The GCF is $3x$.

$$3x^4 + 12x^3 - 3x^2 - 12x$$
$$= 3x(x^3 + 4x^2 - x - 4)$$

Four terms: Try grouping.

Factor out a negative number from the second grouping to obtain a common binomial factor, $x + 4$, for the two groupings.

$$= 3x[x^2(x + 4) - 1(x + 4)]$$
$$= 3x(x + 4)(x^2 - 1)$$

This can be factored further.

$$= 3x(x + 4)(x + 1)(x - 1)$$

Additional Examples to Review

Example 11, page 277; Example 12, page 277; Example 13, page 278; Example 14, page 279

Section 4.3 Solving Quadratic Equations by Factoring

A quadratic equation in x can be written in the standard form

$$ax^2 + bx + c = 0, \quad a \neq 0.$$

A polynomial equation is the result of setting two polynomials equal to each other. The equation is in standard form if one side is 0 and the polynomial on the other side is in standard form, that is, in descending powers of the variable. In standard form, its degree is the highest degree of any term in the equation. A polynomial equation of degree 1 is a linear equation and of degree 2 a quadratic equation. Some polynomial equations can be solved by writing the equation in standard form, factoring, and then using the zero-product principle: If a product is 0, then at least one of the factors is equal to 0.

- Solve: $5x^2 + 7x = 6$.

$$5x^2 + 7x - 6 = 0$$
$$(5x - 3)(x + 2) = 0$$
$$5x - 3 = 0 \quad \text{or} \quad x + 2 = 0$$
$$5x = 3 \qquad\qquad x = -2$$
$$x = \frac{3}{5}$$

The solutions are -2 and $\frac{3}{5}$, and the solution set is $\left\{-2, \frac{3}{5}\right\}$. (The solutions are the x-intercepts of the graph of the quadratic function $f(x) = 5x^2 + 7x - 6$.)

- Solve: $x^3 - 9x^2 = x - 9$.

$$x^3 - 9x^2 - x + 9 = 0$$
$$x^2(x - 9) - 1(x - 9) = 0$$
$$(x - 9)(x^2 - 1) = 0$$
$$(x - 9)(x + 1)(x - 1) = 0$$
$$x - 9 = 0 \quad \text{or} \quad x + 1 = 0 \quad \text{or} \quad x - 1 = 0$$
$$x = 9 \qquad\qquad x = -1 \qquad\qquad x = 1$$

The solution set is $\{-1, 1, 9\}$.

Additional Examples to Review

Example 1, page 284; Example 2, page 285; Example 3, page 287; Example 4, page 288; Example 5, page 290

Section 4.4 Solving Quadratic Equations by the Square Root Property and the Quadratic Formula

The Square Root Property

If u is an algebraic expression and d is a positive real number, then

$$\text{If } u^2 = d, \quad \text{then } u = \sqrt{d} \quad \text{or} \quad u = -\sqrt{d}.$$

Equivalently,

$$\text{If } u^2 = d, \quad \text{then } u = \pm\sqrt{d}.$$

- Solve: $3x^2 = 7$.

$$x^2 = \frac{7}{3}$$
$$x = \pm\sqrt{\frac{7}{3}} = \pm\frac{\sqrt{7}}{\sqrt{3}} = \pm\frac{\sqrt{7}}{\sqrt{3}} \cdot \frac{\sqrt{3}}{\sqrt{3}}$$
$$= \pm\frac{\sqrt{21}}{3}$$

Solution set: $\left\{\pm\dfrac{\sqrt{21}}{3}\right\}$

- Solve: $(x - 6)^2 = 50$.

$$x - 6 = \pm\sqrt{50}$$
$$x = 6 \pm \sqrt{50}$$
$$= 6 \pm \sqrt{25 \cdot 2}$$
$$= 6 \pm 5\sqrt{2}$$

Solution set: $\{6 \pm 5\sqrt{2}\}$

Additional Examples to Review

Example 1, page 295; Example 2, page 296; Example 3, page 296

The solutions of a quadratic equation in standard form

$$ax^2 + bx + c = 0, \quad a \neq 0,$$

are given by the quadratic formula

$$x = \frac{-b \pm \sqrt{b^2 - 4ac}}{2a}.$$

- Solve using the quadratic formula:

$$2x^2 = 6x - 3.$$

First, write the equation in standard form by subtracting $6x$ and adding 3 on both sides.

$$2x^2 - 6x + 3 = 0$$

$a = 2$ $b = -6$ $c = 3$

$$x = \frac{-b \pm \sqrt{b^2 - 4ac}}{2a} = \frac{-(-6) \pm \sqrt{(-6)^2 - 4 \cdot 2 \cdot 3}}{2 \cdot 2}$$

$$= \frac{6 \pm \sqrt{36 - 24}}{4}$$

$$= \frac{6 \pm \sqrt{12}}{4}$$

$$= \frac{6 \pm \sqrt{4 \cdot 3}}{4}$$

$$= \frac{6 \pm 2\sqrt{3}}{4}$$

$$= \frac{2(3 \pm \sqrt{3})}{2 \cdot 2}$$

$$= \frac{3 \pm \sqrt{3}}{2}$$

Solution set: $\left\{ \dfrac{3 \pm \sqrt{3}}{2} \right\}$

Additional Examples to Review

Example 4, page 297; Example 5, page 298; Table 4.1, page 300; Example 6, page 301

Section 4.5 Quadratic Functions and Their Graphs

The graph of the quadratic function $f(x) = ax^2 + bx + c$, called a parabola, can be graphed using the following steps:

1. If $a > 0$, the parabola opens upward. If $a < 0$, it opens downward.
2. Find the vertex, the lowest point if the parabola opens upward and the highest point if it opens downward. The x-coordinate of the vertex is $-\dfrac{b}{2a}$. Substitute this value into the quadratic function's equation and evaluate to find the y-coordinate.
3. Find any x-intercepts by letting $f(x) = 0$ and solving the resulting equation.
4. Find the y-intercept by letting $x = 0$ and computing $f(0)$.
5. Plot the intercepts and the vertex.
6. Connect these points with a smooth curve.

- Graph: $f(x) = x^2 - 2x - 8$.

$a = 1$ $b = -2$ $c = -8$

- $a > 0$, so the parabola opens upward.

- Vertex: x-coordinate $= -\dfrac{b}{2a} = -\dfrac{(-2)}{2 \cdot 1} = 1$

  y-coordinate $= f(1) = 1^2 - 2 \cdot 1 - 8 = -9$

  Vertex is $(1, -9)$.

- x-intercepts: Let $f(x) = 0$.

$$x^2 - 2x - 8 = 0$$
$$(x - 4)(x + 2) = 0$$
$$x - 4 = 0 \quad \text{or} \quad x + 2 = 0$$
$$x = 4 \qquad\qquad x = -2$$

The parabola passes through $(4, 0)$ and $(-2, 0)$.

- y-intercept: Let $x = 0$.

$$f(0) = 0^2 - 2 \cdot 0 - 8 = 0 - 0 - 8 = -8$$

The parabola passes through $(0, -8)$.

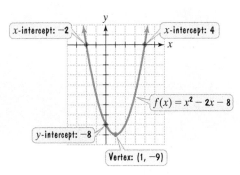

Additional Examples to Review

Example 5, page 309; Example 6, page 310; Example 7, page 311

Review Exercises

Section 4.1 Operations with Polynomials; Polynomial Functions

In Exercises 1–2, perform the indicated operations. Write the resulting polynomial in standard form and indicate its degree.

1. $(-6x^3 + 7x^2 - 9x + 3) + (14x^3 + 3x^2 - 11x - 7)$

2. $(13x^4 - 8x^3 + 2x^2) - (5x^4 - 3x^3 + 2x^2 - 6)$

In Exercises 3–8, find each product.

3. $8x^2(5x^3 - 2x - 1)$

4. $(3x - 2)(4x^2 + 3x - 5)$

5. $(3x - 5)(2x + 1)$

6. $(4x + 5)(4x - 5)$

7. $(2x + 5)^2$ $4x^2 + 20x + 25$

8. $(3x - 4)^2$ $9x^2 - 24x + 16$

9. The bar graph shows the amount, in billions of dollars, that the United States government spent on human resources and total budget outlays for six selected years. (Human resources include education, health, Medicare, Social Security, and veterans benefits and services.)

Federal Budget Expenditures on Human Resources

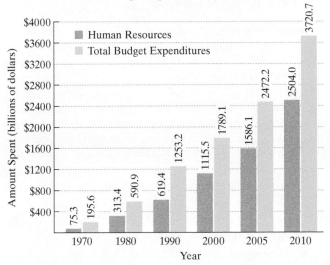

Source: Office of Management and Budget

The data can be modeled by the following functions:

Total budget expenditures → $B(x) = 2.1x^2 - 3.5x + 296$

Amount spent on human resources → $H(x) = 1.8x^2 - 15.9x + 160.$

Each function gives the amount, $B(x)$ and $H(x)$, in billions of dollars, x years after 1970.

a. Write a function d that gives the difference between total budget expenditures, in billions of dollars, and the amount spent on human resources, in billions of dollars, x years after 1970. $d(x) = 0.3x^2 + 12.4x + 136$

b. Use the function from part (a) to find the difference between total budget expenditures and the amount spent on human resources in 1980. $290 billion

c. Does the function value in part (b) overestimate or underestimate the difference between total budget expenditures and the amount spent on human resources shown by the graph in 1980? By how much?

                          overestimates by $12.5 billion

Section 4.2 Factoring Polynomials

In Exercises 10–11, factor out the greatest common factor.

10. $12x^3 + 16x^2 - 400x$ $4x(3x^2 + 4x - 100)$

11. $x^2(x - 5) + 13(x - 5)$ $(x - 5)(x^2 + 13)$

In Exercises 12–13, factor by grouping.

12. $x^3 + 3x^2 + 2x + 6$ $(x^2 + 2)(x + 3)$

13. $x^3 + 5x + x^2 + 5$ $(x^2 + 5)(x + 1)$

In Exercises 14–25, factor each trinomial completely, or state that the trinomial is prime.

14. $x^2 - 3x + 2$ $(x - 2)(x - 1)$

15. $x^2 - x - 20$ $(x - 5)(x + 4)$

16. $x^2 + 19x + 48$ $(x + 3)(x + 16)$

17. $x^2 - 6x + 8$ $(x - 4)(x - 2)$

1. $8x^3 + 10x^2 - 20x - 4; 3$ 2. $8x^4 - 5x^3 + 6; 4$ 3. $40x^5 - 16x^3 - 8x^2$ 4. $12x^3 + x^2 - 21x + 10$ 5. $6x^2 - 7x - 5$ 6. $16x^2 - 25$

26. $(2x + 1)(2x - 1)$ **27.** $(9 + 10y)(9 - 10y)$ **28.** $(z^2 + 4)(z + 2)(z - 2)$ **35.** $(5y + 2)(2y + 1)$ **39.** $(x - 3)^2(x + 3)$ **41.** $x(2x + 5)(x + 7)$

18. $x^2 + 5x - 9$ prime

19. $3x^2 + 17x + 10$ $(x + 5)(3x + 2)$

20. $5y^2 - 17y + 6$ $(y - 3)(5y - 2)$

21. $4x^2 + 4x - 15$ $(2x + 5)(2x - 3)$

22. $5y^2 + 11y + 4$ prime

23. $8x^2 + 8x - 6$ $2(2x + 3)(2x - 1)$

24. $2x^3 + 7x^2 - 72x$ $x(2x - 9)(x + 8)$

25. $12y^3 + 28y^2 + 8y$ $4y(3y + 1)(y + 2)$

In Exercises 26–28, factor the difference of two squares.

26. $4x^2 - 1$ **27.** $81 - 100y^2$ **28.** $z^4 - 16$

In Exercises 29–33, factor any perfect square trinomials, or state that the polynomial is prime.

29. $x^2 + 22x + 121$ $(x + 11)^2$

30. $x^2 - 16x + 64$ $(x - 8)^2$ **31.** $9y^2 + 48y + 64$ $(3y + 8)^2$

32. $16x^2 - 40x + 25$ $(4x - 5)^2$ **33.** $25x^2 + 15x + 9$ prime

In Exercises 34–54, factor completely, or state that the polynomial is prime.

34. $x^3 - 8x^2 + 7x$ $x(x - 7)(x - 1)$

35. $10y^2 + 9y + 2$ **36.** $128 - 2y^2$ $2(8 + y)(8 - y)$

37. $9x^2 + 6x + 1$ $(3x + 1)^2$ **38.** $20x^7 - 36x^3$ $4x^3(5x^4 - 9)$

39. $x^3 - 3x^2 - 9x + 27$ **40.** $y^2 + 16$ prime

41. $2x^3 + 19x^2 + 35x$ **42.** $3x^3 - 30x^2 + 75x$

43. $4y^4 - 36y^2$ **44.** $5x^2 + 20x - 105$

45. $9x^2 + 8x - 3$ prime **46.** $10x^5 - 44x^4 + 16x^3$

47. $100y^2 - 49$ **48.** $9x^5 - 18x^4$ $9x^4(x - 2)$

49. $x^4 - 1$ **50.** $6x^2 + 11x - 10$

51. $3x^4 - 12x^2$ **52.** $x^2 - x - 90$

53. $32y^3 + 32y^2 + 6y$ **54.** $2y^2 - 16y + 32$ $2(y - 4)^2$

Section 4.3 Solving Quadratic Equations by Factoring

In Exercises 55–59, use factoring to solve each polynomial equation.

55. $x^2 + 6x + 5 = 0$ $\{-5, -1\}$

56. $3x^2 = 22x - 7$ $\left\{\frac{1}{3}, 7\right\}$ **57.** $(x + 3)(x - 2) = 50$

58. $3x^2 = 12x$ $\{0, 4\}$ **59.** $x^3 + 5x^2 = 9x + 45$

60. The length of a rectangular sign is 3 feet longer than the width. If the sign has space for 54 square feet of advertising, find its length and its width. length: 9 ft; width 6 ft

61. A painting measuring 10 inches by 16 inches is surrounded by a frame of uniform width. If the combined area of the painting and frame is 280 square inches, determine the width of the frame. 2 in.

Section 4.4 Solving Quadratic Equations by the Square Root Property and the Quadratic Formula

In Exercises 62–69, solve each quadratic equation by the square root property. If possible, simplify radicals or rationalize denominators.

62. $x^2 = 64$ $\{-8, 8\}$ **63.** $x^2 = 17$ $\{\pm\sqrt{17}\}$

64. $2x^2 = 150$ $\{\pm 5\sqrt{3}\}$ **65.** $(x - 3)^2 = 9$ $\{0, 6\}$

66. $(y + 4)^2 = 5$ $\{-4 \pm \sqrt{5}\}$ **67.** $3y^2 - 5 = 0$ $\left\{\pm\frac{\sqrt{15}}{3}\right\}$

68. $(2x - 7)^2 = 25$ $\{1, 6\}$ **69.** $(x + 5)^2 = 12$

In Exercises 70–72, solve each equation using the quadratic formula. Simplify irrational solutions, if possible.

70. $2x^2 + 5x - 3 = 0$ $\left\{-3, \frac{1}{2}\right\}$ **71.** $x^2 = 2x + 4$ $\{1 \pm \sqrt{5}\}$

72. $3x^2 + 5 = 9x$ $\left\{\frac{9 \pm \sqrt{21}}{6}\right\}$

In Exercises 73–77, solve each equation by the method of your choice. Simplify irrational solutions, if possible.

73. $2x^2 - 11x + 5 = 0$ $\left\{\frac{1}{2}, 5\right\}$ **74.** $(3x + 5)(x - 3) = 5$

75. $3x^2 - 7x + 1 = 0$ **76.** $x^2 - 9 = 0$ $\{-3, 3\}$

77. $(2x - 3)^2 = 5$ $\left\{\frac{3 \pm \sqrt{5}}{2}\right\}$ **74.** $\left\{-2, \frac{10}{3}\right\}$ **75.** $\left\{\frac{7 \pm \sqrt{37}}{6}\right\}$

78. The graph shows stopping distances for motorcycles at various speeds on dry roads and on wet roads.

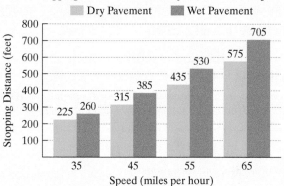

Stopping Distances for Motorcycles at Selected Speeds

Source: National Highway Traffic Safety Administration

The functions

$$f(x) = 0.125x^2 - 0.8x + 99$$

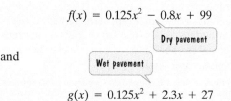

and

$$g(x) = 0.125x^2 + 2.3x + 27$$

model a motorcycle's stopping distance, $f(x)$ or $g(x)$, in feet, traveling at x miles per hour. Function f models stopping distance on dry pavement and function g models stopping distance on wet pavement.

a. Use function g to find the stopping distance on wet pavement for a motorcycle traveling at 35 miles per hour. Round to the nearest foot. Does your rounded answer overestimate or underestimate the stopping distance shown by the graph? By how many feet? 261 ft; overestimates by 1 ft

b. Use function f to determine a motorcycle's speed requiring a stopping distance on dry pavement of 267 feet. 40 miles per hours

79. The graphs of the functions in Exercise 78 are shown at the top of the next page for $\{x \mid x \geq 30\}$.

a. How is your answer to Exercise 78(a) shown on the graph of g? by the point $(35, 261)$

42. $3x(x - 5)^2$ **43.** $4y^2(y + 3)(y - 3)$ **44.** $5(x + 7)(x - 3)$ **46.** $2x^3(5x - 2)(x - 4)$ **47.** $(10y + 7)(10y - 7)$ **49.** $(x^2 + 1)(x + 1)(x - 1)$
50. $(3x - 2)(2x + 5)$ **51.** $3x^2(x + 2)(x - 2)$ **52.** $(x - 10)(x + 9)$ **53.** $2y(4y + 3)(4y + 1)$ **57.** $\{-8, 7\}$ **59.** $\{-5, -3, 3\}$ **69.** $\{-5 \pm 2\sqrt{3}\}$

b. How is your answer to Exercise 78(b) shown on the graph of f?   by the point $(40, 267)$

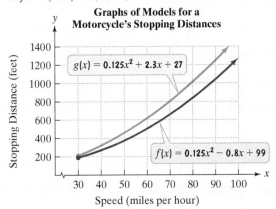

Graphs of Models for a Motorcycle's Stopping Distances

$g(x) = 0.125x^2 + 2.3x + 27$

$f(x) = 0.125x^2 - 0.8x + 99$

80. A baseball is hit by a batter. The function

$$s(t) = -16t^2 + 140t + 3$$

models the ball's height above the ground, $s(t)$, in feet, t seconds after it is hit. How long will it take for the ball to strike the ground? Round to the nearest tenth of a second.   about 8.8 seconds

Section 4.5 Quadratic Functions and Their Graphs

In Exercises 81–84, graph each quadratic function.

81. $f(x) = x^2 - 6x - 7$ *   **82.** $f(x) = -x^2 - 2x + 3$ *

83. $f(x) = -3x^2 + 6x + 1$ *

84. $f(x) = x^2 - 4x$ *

85. Fireworks are launched into the air. The function

$$f(t) = -16t^2 + 200t + 4$$

models the fireworks' height, $f(t)$, in feet, t seconds after they are launched. When should the fireworks explode so that they go off at the greatest height? What is that height?   6.25 seconds; 629 feet

86. A quarterback tosses a football to a receiver 40 yards downfield. The function

$$f(x) = -0.025x^2 + x + 6$$

models the football's height, $f(x)$, in feet, when it is x yards from the quarterback.

a. How many yards from the quarterback does the football reach its greatest height? What is that height?   20 yards; 16 feet

b. If a defender is 38 yards from the quarterback, how far must he reach to deflect or catch the ball?   7.9 feet

c. If the football is neither deflected by the defender nor caught by the receiver, how many yards will it go, to the nearest tenth of a yard, before hitting the ground?   45.3 yards

d. Graph the quadratic function that models the football's parabolic path.   *

Chapter 4 Test

In Exercises 1–8, perform the indicated operations.

1. $(7x^3 + 3x^2 - 5x - 11) + (6x^3 - 2x^2 + 4x - 13)$

2. $(9x^3 - 6x^2 - 11x - 4) - (4x^3 - 8x^2 - 13x + 5)$

3. $6x^2(8x^3 - 5x - 2)$ $48x^5 - 30x^3 - 12x^2$

4. $(3x + 2)(x^2 - 4x - 3)$ $3x^3 - 10x^2 - 17x - 6$

5. $(3y + 7)(2y - 9)$ $6y^2 - 13y - 63$

6. $(7x + 5)(7x - 5)$ $49x^2 - 25$

7. $(x^2 + 3)^2$ $x^4 + 6x^2 + 9$

8. $(5x - 3)^2$ $25x^2 - 30x + 9$

In Exercises 9–17, factor completely, or state that the polynomial is prime.

9. $x^2 - 9x + 18$ $(x - 3)(x - 6)$

10. $x^3 + 2x^2 + 3x + 6$ $(x^2 + 3)(x + 2)$

11. $15y^4 - 35y^3 + 10y^2$ $5y^2(3y - 1)(y - 2)$

12. $25x^2 - 9$ $(5x + 3)(5x - 3)$

13. $x^2 + 4$ prime

14. $36x^2 - 84x + 49$ $(6x - 7)^2$

15. $7x^2 - 50x + 7$ $(7x - 1)(x - 7)$

16. $4y^2 - 36$ $4(y + 3)(y - 3)$

17. $4x^2 + 12x + 9$ $(2x + 3)^2$

In Exercises 18–23, solve each equation by the method of your choice. Simplify irrational solutions, if possible.

18. $3x^2 + 5x + 1 = 0$ $\left\{\frac{-5 \pm \sqrt{13}}{6}\right\}$

19. $(3x - 5)(x + 2) = -6$ $\left\{-\frac{4}{3}, 1\right\}$

20. $(2x + 1)^2 = 36$ $\left\{-\frac{7}{2}, \frac{5}{2}\right\}$

21. $2x^2 = 6x - 1$ $\left\{\frac{3 \pm \sqrt{7}}{2}\right\}$

22. $2x^2 + 9x = 5$ $\left\{-5, \frac{1}{2}\right\}$

23. $x^3 - 4x^2 - x + 4 = 0$ $\{-1, 1, 4\}$

In Exercises 24–25, graph each quadratic function.

24. $f(x) = x^2 + 2x - 8$ *

25. $f(x) = -2x^2 + 16x - 24$ *

A baseball player hits a pop fly into the air. The function
$$s(t) = -16t^2 + 64t + 5$$
models the ball's height above the ground, $s(t)$, in feet, t seconds after it is hit. Use the function to solve Exercises 26–27.

26. When does the baseball reach its maximum height? What is that height?   after 2 seconds; 69 feet

27. After how many seconds does the baseball hit the ground? Round to the nearest tenth of a second.   after about 4.1 seconds

28. The length of a rectangular sign is 3 feet longer than the width. If the sign has space for 40 square feet of advertising, find its length and its width.   length: 8 ft; width: 5 ft

1. $13x^3 + x^2 - x - 24$ **2.** $5x^3 + 2x^2 + 2x - 9$

See Answers to Selected Exercises.

Geometric Pathways: Measurement

You are feeling crowded in. Perhaps it would be a good time to move to a state with more elbow room. But which state? You can look up the population of each state, but that does not take into account the amount of land the population occupies. How is this land measured and how can you use this measure to select a place where wildlife outnumber humans?

In this chapter, we explore ways of measuring things in our English system, as well as in the metric system. Knowing how units of measure are used to describe your world can help you understand issues ranging from population density to the size of viruses to recognizing the potential dangers of alcohol.

Here's where you'll find these applications:

- Measuring viruses is discussed in the Blitzer Bonus on page 331.
- Finding a state with lots of room to spread out is explored in Example 2 of Section 5.2.
- Measuring alcohol consumption is addressed in the Blitzer Bonus on page 349.

5.1 Measuring Length; The Metric System

What am I supposed to learn?

After you have read this section, you should be able to:

1 Use dimensional analysis to change units of measurement.

2 Understand and use metric prefixes.

3 Convert units within the metric system.

4 Use dimensional analysis to change to and from the metric system.

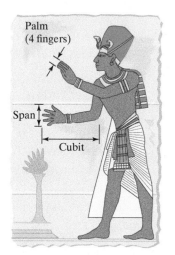

Palm (4 fingers)

Span

Cubit

Linear units of measure were originally based on parts of the body. The Egyptians used the palm (equal to four fingers), the span (a handspan), and the cubit (length of forearm).

1 Use dimensional analysis to change units of measurement.

Have you seen any of the *Jurassic Park* films? The popularity of these movies reflects our fascination with dinosaurs and their incredible size. From end to end, the largest dinosaur from the Jurassic period, which lasted from 208 to 146 million years ago, was about 88 feet. To **measure** an object such as a dinosaur is to assign a number to its size. The number representing its measure from end to end is called its **length**. Measurements are used to describe properties of length, area, volume, weight, and temperature. Over the centuries, people have developed systems of measurement that are now accepted in most of the world.

Length

Every measurement consists of two parts: a number and a unit of measure. For example, if the length of a dinosaur is 88 feet, the number is 88 and the unit of measure is the foot. Many different units are commonly used in measuring length. The foot is from a system of measurement called the **English system**, which is generally used in the United States. In this system of measurement, length is expressed in such units as inches, feet, yards, and miles.

The result obtained from measuring length is called a **linear measurement** and is stated in **linear units**.

Linear Units of Measure: The English System
12 inches (in.) = 1 foot (ft)
3 feet = 1 yard (yd)
36 inches = 1 yard
5280 feet = 1 mile (mi)

Because many of us are familiar with the measures in the box, we find it simple to change from one measure to another, say from feet to inches. We know that there are 12 inches in a foot. To convert from 5 feet to a measure in inches, we multiply by 12. Thus, 5 feet = 5 × 12 inches = 60 inches.

Another procedure used to convert from one unit of measurement to another is called **dimensional analysis**. Dimensional analysis uses *unit fractions*. A **unit fraction** has two properties: The numerator and denominator contain different units and the value of the unit fraction is 1. Here are some examples of unit fractions:

$$\frac{12 \text{ in.}}{1 \text{ ft}}; \quad \frac{1 \text{ ft}}{12 \text{ in.}}; \quad \frac{3 \text{ ft}}{1 \text{ yd}}; \quad \frac{1 \text{ yd}}{3 \text{ ft}}; \quad \frac{5280 \text{ ft}}{1 \text{ mi}}; \quad \frac{1 \text{ mi}}{5280 \text{ ft}}.$$

In each unit fraction, the numerator and denominator are equal measures, making the value of the fraction 1.

Let's see how to convert 5 feet to inches using dimensional analysis.

$$5 \text{ ft} = ? \text{ in.}$$

We need to eliminate feet and introduce inches. The unit we need to introduce, inches, must appear in the numerator of the fraction. The unit we need to eliminate, feet, must appear in the denominator. Therefore, we choose the unit fraction with inches in the numerator and feet in the denominator. The units divide out as follows:

$$5 \text{ ft} = \frac{5 \text{ ft}}{1} \cdot \frac{12 \text{ in.}}{1 \text{ ft}} = 5 \cdot 12 \text{ in.} = 60 \text{ in.}$$

unit fraction

Dimensional Analysis

To convert a measurement to a different unit, multiply by a unit fraction (or by unit fractions). The given unit of measurement should appear in the denominator of the unit fraction so that this unit cancels upon multiplication. The unit of measurement that needs to be introduced should appear in the numerator of the fraction; this unit will be retained upon multiplication.

Example 1   Using Dimensional Analysis to Change Units of Measurement

Convert:

a. 40 inches to feet   **b.** 13,200 feet to miles   **c.** 9 inches to yards.

Solution

a. Because we want to convert 40 inches to feet, feet should appear in the numerator and inches in the denominator. We use the unit fraction

$$\frac{1 \text{ ft}}{12 \text{ in.}}$$

and proceed as follows:

This period ends the sentence and is not part of the abbreviated unit.

$$40 \text{ in.} = \frac{40 \text{ in.}}{1} \cdot \frac{1 \text{ ft}}{12 \text{ in.}} = \frac{40}{12} \text{ ft} = 3\tfrac{1}{3} \text{ ft or } 3.\overline{3} \text{ ft.}$$

b. To convert 13,200 feet to miles, miles should appear in the numerator and feet in the denominator. We use the unit fraction

$$\frac{1 \text{ mi}}{5280 \text{ ft}}$$

and proceed as follows:

$$13,200 \text{ ft} = \frac{13,200 \text{ ft}}{1} \cdot \frac{1 \text{ mi}}{5280 \text{ ft}} = \frac{13,200}{5280} \text{ mi} = 2\tfrac{1}{2} \text{ mi or } 2.5 \text{ mi.}$$

c. To convert 9 inches to yards, yards should appear in the numerator and inches in the denominator. We use the unit fraction

$$\frac{1 \text{ yd}}{36 \text{ in.}}$$

and proceed as follows:

$$9 \text{ in.} = \frac{9 \text{ in.}}{1} \cdot \frac{1 \text{ yd}}{36 \text{ in.}} = \frac{9}{36} \text{ yd} = \tfrac{1}{4} \text{ yd or } 0.25 \text{ yd.}$$

In each part of Example 1, we converted from a smaller unit to a larger unit. Did you notice that this results in a smaller number in the converted unit of measure? **Converting to a larger unit always produces a smaller number. Converting to a smaller unit always produces a larger number.**

Great Question!

When should I use a period for abbreviated units of measurement?

The abbreviations for units of measurement are written without a period, such as ft for feet. The exception is the abbreviation for inches (in.), which might get confused with the word "in" if it did not have a period.

✓ **Check Point 1**   Convert:

a. 78 inches to feet   6.5 ft   **b.** 17,160 feet to miles   3.25 mi

c. 3 inches to yards. $\frac{1}{12}$ yd

Length and the Metric System

② Understand and use metric prefixes.

Although the English system of measurement is most commonly used in the United States, most industrialized countries use the metric system of measurement. One of the advantages of the metric system is that units are based on powers of ten, making it much easier than the English system to change from one unit of measure to another.

The basic unit for linear measure in the metric system is the meter (m). A meter is slightly longer than a yard, approximately 39 inches. Prefixes are used to denote a multiple or part of a meter. **Table 5.1** summarizes the more commonly used metric prefixes and their meanings.

Table 5.1 Commonly Used Metric Prefixes

Prefix	Symbol	Meaning
kilo	k	1000 × base unit
hecto	h	100 × base unit
deka	da	10 × base unit
deci	d	$\frac{1}{10}$ of base unit
centi	c	$\frac{1}{100}$ of base unit
milli	m	$\frac{1}{1000}$ of base unit

Great Question!

Does the "i" in the prefixes ending in "i" (deci, centi, milli) have any significance?

Prefixes ending in "i" are all fractional parts of one unit.

The prefixes kilo, centi, and milli are used more frequently than hecto, deka, and deci. **Table 5.2** applies all six prefixes to the meter. The first part of the symbol indicates the prefix and the second part (m) indicates meter.

Table 5.2 Commonly Used Units of Linear Measure in the Metric System

Symbol	Unit	Meaning
km	kilometer	1000 meters
hm	hectometer	100 meters
dam	dekameter	10 meters
m	meter	1 meter
dm	decimeter	0.1 meter
cm	centimeter	0.01 meter
mm	millimeter	0.001 meter

Kilometer is pronounced kil'-oh-met-er with the accent on the FIRST syllable. If pronounced correctly, kilometers should sound something like "kill all meters."

In the metric system, the kilometer is used to measure distances comparable to those measured in miles in the English system. One kilometer is approximately 0.6 mile, and one mile is approximately 1.6 kilometers.

kilodollar

hectodollar

dekadollar

dollar

decidollar   centidollar

Like our system of money, the metric system is based on powers of ten.

Metric units of centimeters and millimeters are used to measure what the English system measures in inches. **Figure 5.1** shows that a centimeter is less than half an inch; there are 2.54 centimeters in an inch. The smaller markings on the bottom scale are millimeters. A millimeter is approximately the thickness of a dime. The length of a bee or a fly may be measured in millimeters.

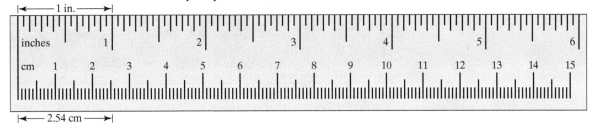

Figure 5.1

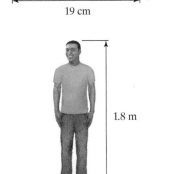

Those of us born in the United States have a good sense of what a length in the English system tells us about an object. An 88-foot dinosaur is huge, about 15 times the height of a 6-foot man. But what sense can we make of knowing that a whale is 25 meters long? The following lengths and the given approximations can help give you a feel for metric units of linear measure.

(1 meter ≈ 39 inches   1 kilometer ≈ 0.6 mile)

Item	Approximate Length
Width of lead in pencil	2 mm or 0.08 in.
Width of an adult's thumb	2 cm or 0.8 in.
Height of adult male	1.8 m or 6 ft
Typical room height	2.5 m or 8.3 ft
Length of medium-size car	5 m or 16.7 ft
Height of Empire State Building	381 m or 1270 ft
Average depth of ocean	4 km or 2.5 mi
Length of Manhattan Island	18 km or 11.25 mi
Distance from New York City to San Francisco	4800 km or 3000 mi
Radius of Earth	6378 km or 3986 mi
Distance from Earth to the moon	384,401 km or 240,251 mi

③ **Convert units within the metric system.**

Although dimensional analysis can be used to convert from one unit to another within the metric system, there is an easier, faster way to accomplish this conversion. The procedure is based on the observation that successively smaller units involve division by 10 and successively larger units involve multiplication by 10.

Changing Units within the Metric System

Use the chart to find equivalent measures of length:

Multiply by 10 for each step to the right.

×10

km hm dam m dm cm mm
÷10

Divide by 10 for each step to the left.

1. To change from a larger unit to a smaller unit (moving to the right in the diagram), multiply by 10 for each step to the right. Thus, move the decimal point in the given quantity one place to the right for each smaller unit until the desired unit is reached.

2. To change from a smaller unit to a larger unit (moving to the left in the diagram), divide by 10 for each step to the left. Thus, move the decimal point in the given quantity one place to the left for each larger unit until the desired unit is reached.

Great Question!

Is there a way to help me remember the metric units for length from largest to smallest?

The following sentence should help:

King	Henry	died	Monday	drinking	chocolate	milk.
km	hm	dam	m	dm	cm	mm

Blitzer Bonus

The First Meter

The French first defined the meter in 1791, calculating its length as a romantic one ten-millionth of a line running from the Equator through Paris to the North Pole. Today's meter, officially accepted in 1983, is equal to the length of the path traveled by light in a vacuum during the time interval of 1/299,794,458 of a second.

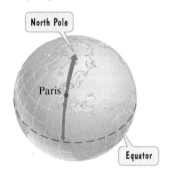

1 meter = 1/10,000,000 of the distance from the North Pole to the Equator on the meridian through Paris

Example 2 Changing Units within the Metric System

a. Convert 504.7 meters to kilometers.

b. Convert 27 meters to centimeters.

c. Convert 704 mm to hm.

d. Convert 9.71 dam to dm.

Solution

a. To convert from meters to kilometers, we start at meters and move three steps to the left to obtain kilometers:

$$\text{km hm dam m dm cm mm.}$$

Hence, we move the decimal point three places to the left:

$$504.7 \text{ m} = 0.5047 \text{ km.}$$

Thus, 504.7 meters converts to 0.5047 kilometer. Changing from a smaller unit of measurement (meter) to a larger unit of measurement (kilometer) results in an answer with a smaller number of units.

b. To convert from meters to centimeters, we start at meters and move two steps to the right to obtain centimeters:

$$\text{km hm dam m dm cm mm.}$$

Hence, we move the decimal point two places to the right:

$$27 \text{ m} = 2700 \text{ cm.}$$

Thus, 27 meters converts to 2700 centimeters. Changing from a larger unit of measurement (meter) to a smaller unit of measurement (centimeter) results in an answer with a larger number of units.

c. To convert from mm (millimeters) to hm (hectometers), we start at mm and move five steps to the left to obtain hm:

$$\text{km hm dam m dm cm mm.}$$

Hence, we move the decimal point five places to the left:

$$704 \text{ mm} = 0.00704 \text{ hm.}$$

d. To convert from dam (dekameters) to dm (decimeters), we start at dam and move two places to the right to obtain dm:

$$\text{km hm dam m dm cm mm.}$$

Hence, we move the decimal point two places to the right:

$$9.71 \text{ dam} = 971 \text{ dm.}$$

In Example 2(b), we showed that 27 meters converts to 2700 centimeters. This is the average length of the California blue whale, the longest of the great whales. Blue whales can have lengths that exceed 30 meters, making them over 100 feet long.

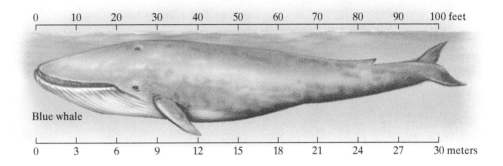

Blue whale

 **Check Point 2**

a. Convert 8000 meters to kilometers.   8 km

b. Convert 53 meters to millimeters.   53,000 mm

c. Convert 604 cm to hm.   0.0604 hm

d. Convert 6.72 dam to cm.   6720 cm

Blitzer Bonus

Viruses and Metric Prefixes

Viruses are measured in attometers. An attometer is one quintillionth of a meter, or 10^{-18} meter, symbolized am. If a virus measures 1 am, you can place 10^{15} of them across a penciled 1 millimeter dot. If you were to enlarge each of these viruses to the size of the dot, they would stretch far into space, almost reaching Saturn.

Here is a list of all twenty metric prefixes. When applied to the meter, they range from the yottameter (10^{24} meters) to the yoctometer (10^{-24} meters).

Larger Than the Basic Unit

Prefix	Symbol	Power of Ten	English Name
yotta-	Y	+24	septillion
zetta-	Z	+21	sextillion
exa-	E	+18	quintillion
peta-	P	+15	quadrillion
tera-	T	+12	trillion
giga-	G	+9	billion
mega-	M	+6	million
kilo-	k	+3	thousand
hecto-	h	+2	hundred
deca-	da	+1	ten

Smaller Than the Basic Unit

Prefix	Symbol	Power of Ten	English Name
deci-	d	−1	tenth
centi-	c	−2	hundredth
milli-	m	−3	thousandth
micro-	μ	−6	millionth
nano-	n	−9	billionth
pico-	p	−12	trillionth
femto-	f	−15	quadrillionth
atto-	a	−18	quintillionth
zepto-	z	−21	sextillionth
yocto-	y	−24	septillionth

4 Use dimensional analysis to change to and from the metric system.

Although dimensional analysis is not necessary when changing units within the metric system, it is a useful tool when converting to and from the metric system. Some conversions are given in **Table 5.3**.

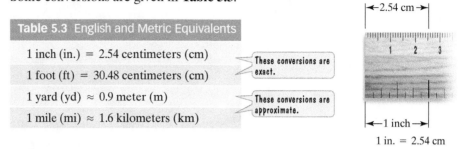

Table 5.3 English and Metric Equivalents	
1 inch (in.) = 2.54 centimeters (cm)	These conversions are exact.
1 foot (ft) = 30.48 centimeters (cm)	
1 yard (yd) ≈ 0.9 meter (m)	These conversions are approximate.
1 mile (mi) ≈ 1.6 kilometers (km)	

1 in. = 2.54 cm

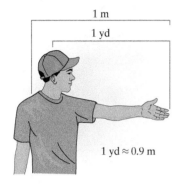

1 yd ≈ 0.9 m

Example 3 Using Dimensional Analysis to Change to and from the Metric System

a. Convert 8 inches to centimeters.

b. Convert 125 miles to kilometers.

c. Convert 26,800 millimeters to inches.

Solution

a. To convert 8 inches to centimeters, we use a unit fraction with centimeters in the numerator and inches in the denominator:

$$\frac{2.54 \text{ cm}}{1 \text{ in.}}.$$ Table 5.3 shows that 1 in. = 2.54 cm.

We proceed as follows.

$$8 \text{ in.} = \frac{8 \text{ in.}}{1} \cdot \frac{2.54 \text{ cm}}{1 \text{ in.}} = 8(2.54) \text{ cm} = 20.32 \text{ cm}$$

b. To convert 125 miles to kilometers, we use a unit fraction with kilometers in the numerator and miles in the denominator:

$$\frac{1.6 \text{ km}}{1 \text{ mi}}.$$ Table 5.3 shows that 1 mi ≈ 1.6 km.

Thus,

$$125 \text{ mi} \approx \frac{125 \text{ mi}}{1} \cdot \frac{1.6 \text{ km}}{1 \text{ mi}} = 125(1.6) \text{ km} = 200 \text{ km}.$$

1 mi ≈ 1.6 km

c. To convert 26,800 millimeters to inches, we observe that **Table 5.3** has only a conversion factor between inches and centimeters. We begin by changing millimeters to centimeters:

$$26,800 \text{ mm} = 2680.0 \text{ cm}.$$

Now we need to convert 2680 centimeters to inches. We use a unit fraction with inches in the numerator and centimeters in the denominator:

$$\frac{1 \text{ in.}}{2.54 \text{ cm}}.$$

Thus,

$$26,800 \text{ mm} = 2680 \text{ cm} = \frac{2680 \text{ cm}}{1} \cdot \frac{1 \text{ in.}}{2.54 \text{ cm}} = \frac{2680}{2.54} \text{in.} \approx 1055 \text{ in.}$$

The measure 1055 inches is equivalent to about 88 feet, the length of the largest dinosaur from the Jurassic period. The diplodocus, a plant eater, was 26.8 meters, approximately 88 feet, long.

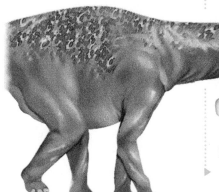

 Check Point 3

a. Convert 8 feet to centimeters. 243.84 cm
b. Convert 20 meters to yards. ≈22.22 yd
▶ c. Convert 30 meters to inches. ≈1181.1 in.

So far, we have used dimensional analysis to change units of length. Dimensional analysis may also be used to convert other kinds of measures, such as speed.

Blitzer Bonus
●●●●●●●●●●●●●●●●

The Mars Climate Orbiter

If you think that using dimensional analysis to change to and from the metric system is no big deal, consider this: The Mars Climate Orbiter, launched on December 11, 1998, was lost when it crashed into Mars about a year later due to failure to properly convert English units into metric units. The price tag: $125 million!

Example 4 Using Dimensional Analysis

a. The speed limit on many highways in the United States is 55 miles per hour (mi/hr). How many kilometers per hour (km/hr) is this?
b. If a high-speed train in Japan is capable of traveling at 200 kilometers per hour, how many miles per hour is this?

Solution

a. To change miles per hour to kilometers per hour, we need to concentrate on changing miles to kilometers, so we need a unit fraction with kilometers in the numerator and miles in the denominator:

$$\frac{1.6 \text{ km}}{1 \text{ mi}}.$$ Table 5.3 shows that 1 mi ≈ 1.6 km.

Thus,

$$\frac{55 \text{ mi}}{\text{hr}} \approx \frac{55 \text{ mi}}{\text{hr}} \cdot \frac{1.6 \text{ km}}{1 \text{ mi}} = 55(1.6)\frac{\text{km}}{\text{hr}} = 88 \text{ km/hr}.$$

This shows that 55 miles per hour is approximately 88 kilometers per hour.

b. To change 200 kilometers per hour to miles per hour, we must convert kilometers to miles. We need a unit fraction with miles in the numerator and kilometers in the denominator:

$$\frac{1 \text{ mi}}{1.6 \text{ km}}.$$ Table 5.3 shows that 1 mi ≈ 1.6 km.

Thus,

$$\frac{200 \text{ km}}{\text{hr}} \approx \frac{200 \text{ km}}{\text{hr}} \cdot \frac{1 \text{ mi}}{1.6 \text{ km}} = \frac{200 \text{ mi}}{1.6 \text{ hr}} = 125 \text{ mi/hr}.$$

A train capable of traveling at 200 kilometers per hour can therefore travel at about 125 miles per hour.

Check Point 4 A road in Europe has a speed limit of 60 kilometers per hour.
▶ Approximately how many miles per hour is this? ≈37.5 mi/hr

Achieving Success

Be obsessive about neatness.

In all your math work, including class notes, homework, and tests, be ultra neat. A 5 that looks like a 6, a plus that looks like a minus, a point that is only approximately graphed in rectangular coordinates, or a dm that looks like a cm can create confusion like you wouldn't believe.

Concept and Vocabulary Check

Exercises in the Concept and Vocabulary Check are intended for group and class discussions.

In Exercises 1–4, fill in each blank so that the resulting statement is true.

1. The result obtained from measuring length is called a/an ____linear____ measurement and is stated in ____linear____ units.

2. In the English system, ____12____ in. = 1 ft, ____3____ ft = 1 yd, ____36____ in. = 1 yd, and ____5280____ ft = 1 mi.

3. Fractions such as $\dfrac{12 \text{ in.}}{1 \text{ ft}}$ and $\dfrac{1 \text{ yd}}{3 \text{ ft}}$ are called ____unit____ fractions. The value of such fractions is ____1____.

4. In the metric system, 1 km = ____1000____ m, 1 hm = ____100____ m, 1 dam = ____10____ m, 1 dm = ____0.1____ m, 1 cm = ____0.01____ m, and 1 mm = ____0.001____ m.

In Exercises 5–8, determine whether each statement is true or false. If the statement is false, make the necessary change(s) to produce a true statement.　Changes to false statements will vary.

5. Dimensional analysis uses powers of ten to convert from one unit of measurement to another.　false

6. One of the advantages of the English system is that units are based on powers of ten.　false

7. There are 2.54 inches in a centimeter.　false

8. The height of an adult male is approximately 6 meters.　false

Respond to Exercises 9–12 using verbal or written explanations.　9–12. Answers will vary.

9. Describe the two parts of a measurement.

10. Describe how to use dimensional analysis to convert 20 inches to feet.

11. Describe advantages of the metric system over the English system.

12. Explain how to change units within the metric system.

Exercise Set 5.1

Practice Exercises

In Exercises 1–16, use dimensional analysis to convert the quantity to the indicated unit. If necessary, round the answer to two decimal places.

1. 30 in. to ft　2.5 ft

2. 100 in. to ft　$8\frac{1}{3}$ ft ≈ 8.33 ft

3. 30 ft to in.　360 in.

4. 100 ft to in.　1200 in.

5. 6 in. to yd　$\frac{1}{6}$ yd ≈ 0.17 yd

6. 21 in. to yd　$\frac{7}{12}$ yd ≈ 0.58 yd

7. 6 yd to in.　216 in.

8. 21 yd to in.　756 in.

9. 6 yd to ft　18 ft

10. 12 yd to ft　36 ft

11. 6 ft to yd　2 yd

12. 12 ft to yd　4 yd

13. 23,760 ft to mi　4.5 mi

14. 19,800 ft to mi　3.75 mi

15. 0.75 mi to ft　3960 ft

16. 0.25 mi to ft　1320 ft

In Exercises 17–26, use the diagram in the box on page 329 to convert the given measurement to the unit indicated.

17. 5 m to cm   500 cm

18. 8 dam to m   80 m

19. 16.3 hm to m   1630 m

20. 0.37 hm to m   37 m

21. 317.8 cm to hm   0.03178 hm

22. 8.64 hm to cm   86,400 cm

23. 0.023 mm to m   0.000023 m

24. 0.00037 km to cm   37 cm

25. 2196 mm to dm   21.96 dm

26. 71 dm to km   0.0071 km

In Exercises 27–44, use the following English and metric equivalents, along with dimensional analysis, to convert the given measurement to the unit indicated.

English and Metric Equivalents

$$1 \text{ in.} = 2.54 \text{ cm}$$
$$1 \text{ ft} = 30.48 \text{ cm}$$
$$1 \text{ yd} \approx 0.9 \text{ m}$$
$$1 \text{ mi} \approx 1.6 \text{ km}$$

27. 14 in. to cm   35.56 cm

28. 26 in. to cm   66.04 cm

29. 14 cm to in.   ≈5.51 in.

30. 26 cm to in.   ≈10.24 in.

31. 265 mi to km   ≈424 km

32. 776 mi to km   ≈1241.6 km

33. 265 km to mi   ≈165.625 mi

34. 776 km to mi   ≈485 mi

35. 12 m to yd   ≈13.33 yd

36. 20 m to yd   ≈22.22 yd

37. 14 dm to in.   ≈55.12 in.

38. 1.2 dam to in.   ≈472.44 in.

39. 160 in. to dam   0.4064 dam

40. 180 in. to hm   0.04572 hm

41. 5 ft to m   1.524 m

42. 8 ft to m   2.4384 m

43. 5 m to ft   ≈16.40 ft

44. 8 m to ft   ≈26.25 ft

Use 1 mi ≈ 1.6 km to solve Exercises 45–48.

45. Express 96 kilometers per hour in miles per hour.   ≈60 mi/hr

46. Express 104 kilometers per hour in miles per hour.   ≈65 mi/hr

47. Express 45 miles per hour in kilometers per hour.   ≈72 km/hr

48. Express 50 miles per hour in kilometers per hour.   ≈80 km/hr

Practice Plus

In Exercises 49–52, use the unit fractions

$$\frac{36 \text{ in.}}{1 \text{ yd}} \quad \text{and} \quad \frac{2.54 \text{ cm}}{1 \text{ in.}}.$$

49. Convert 5 yd to cm.   457.2 cm

50. Convert 8 yd to cm.   731.52 cm

51. Convert 762 cm to yd. $8\frac{1}{3}$ yd

52. Convert 1016 cm to yd. $11\frac{1}{9}$ yd

In Exercises 53–54, use the unit fractions

$$\frac{5280 \text{ ft}}{1 \text{ mi}}, \quad \frac{12 \text{ in.}}{1 \text{ ft}}, \quad \text{and} \quad \frac{2.54 \text{ cm}}{1 \text{ in.}}.$$

53. Convert 30 mi to km.   48.28032 km

54. Convert 50 mi to km.   80.4672 km

55. Use unit fractions to express 120 miles per hour in feet per second.   176 ft/sec

56. Use unit fractions to express 100 miles per hour in feet per second. $146\frac{2}{3}$ ft/sec

Application Exercises

In Exercises 57–66, selecting from millimeter, meter, and kilometer, determine the best unit of measure to express the given length.

57. A person's height   meter

58. The length of a football field   meter

59. The length of a bee   millimeter

60. The distance from New York City to Washington, D.C.   kilometer

61. The distance around a one-acre lot   meter

62. The length of a car   meter

63. The width of a book   millimeter

64. The altitude of an airplane   kilometer

65. The diameter of a screw   millimeter

66. The width of a human foot   millimeter

In Exercises 67–74, select the best estimate for the measure of the given item.

67. The length of a pen   b
  a. 30 cm   **b.** 19 cm   **c.** 19 mm

68. The length of this page   c
  a. 2.5 mm   **b.** 25 mm   **c.** 250 mm

69. The height of a skyscraper   a
  a. 325 m   **b.** 32.5 km   **c.** 325 km   **d.** 3250 km

70. The length of a pair of pants   b
  a. 700 cm   **b.** 70 cm   **c.** 7 cm

71. The height of a room   c
  a. 4 mm   **b.** 4 cm   **c.** 4 m   **d.** 4 dm

72. The length of a rowboat   c
  a. 4 cm   **b.** 4 dm   **c.** 4 m   **d.** 4 dam

73. The width of an electric cord   a
  a. 4 mm   **b.** 4 cm   **c.** 4 dm   **d.** 4 m

74. The dimensions of a piece of paper   b
  a. 22 mm by 28 mm   **b.** 22 cm by 28 cm
  c. 22 dm by 28 dm   **d.** 22 m by 28 m

75. A baseball diamond measures 27 meters along each side. If a batter scored two home runs in a game, how many kilometers did the batter run?   0.216 km

76. If you jog six times around a track that is 700 meters long, how many kilometers have you covered?   4.2 km

77. The distance from the Earth to the sun is about 93 million miles. What is this distance in kilometers?   ≈148.8 million km

78. The distance from New York City to Los Angeles is 4690 kilometers. What is the distance in miles?   ≈2931.25 mi

Exercises 79–80 give the approximate length of some of the world's longest rivers. In each exercise, determine which is the longer river and by how many kilometers.

79. Nile: 4130 miles; Amazon: 6400 kilometers   Nile; ≈208 km

80. Yangtze: 3940 miles; Mississippi: 6275 kilometers
Yangtze; ≈29 km

Exercises 81–82 give the approximate height of some of the world's tallest mountains. In each exercise, determine which is the taller mountain and by how many meters. Round to the nearest meter.

81. K2: 8611 meters; Everest: 29,035 feet   Everest; ≈239 m

82. Lhotse: 8516 meters; Kangchenjunga: 28,170 feet
Kangchenjunga; ≈70 m

Exercises 83–84 give the average rainfall of some of the world's wettest places. In each exercise, determine which location has the greater average rainfall and by how many inches. Round to the nearest inch.

83. Debundscha (Cameroon): 10,280 millimeters; Waialeale (Hawaii): 451 inches   Waialeale; ≈46 in.

84. Mawsynram (India): 11,870 millimeters; Cherrapunji (India): 498 inches   Cherrapunji; ≈31 in.

(Source for Exercises 79–84: Russell Ash, *The Top 10 of Everything 2009*)

Critical Thinking Exercises

85. You jog 500 meters in a given period of time. The next day, you jog 500 yards over the same time period. On which day was your speed faster? Explain your answer.
the first day; Answers will vary.

86. What kind of difficulties might arise if the United States immediately eliminated all units of measure in the English system and replaced the system by the metric system?

87. The United States is the only Westernized country that does not use the metric system as its primary system of measurement. What reasons might be given for continuing to use the English system?   86–87. Answers will vary.

In Exercises 88–92, convert to an appropriate metric unit so that the numerical expression in the given measure does not contain any zeros.

88. 6000 cm   6 dam

89. 900 m   9 hm

90. 7000 dm   7 hm

91. 11,000 mm   11 m or 1.1 dam

92. 0.0002 km   2 dm

5.2 Measuring Area and Volume

What am I supposed to learn?

After you have read this section, you should be able to:

1. Use square units to measure area.

2. Use dimensional analysis to change units for area.

3. Use cubic units to measure volume.

4. Use English and metric units to measure capacity.

1 Use square units to measure area.

Are you feeling a bit crowded in? Although there are more people on the East Coast of the United States than there are bears, there are places in the Northwest where bears outnumber humans. The most densely populated state is New Jersey, averaging 1195.5 people per square mile. The least densely populated state is Alaska, averaging 1.2 persons per square mile. The U.S. average is 87.4 persons per square mile.

A square mile is one way of measuring the **area** of a state. A state's area is the region within its boundaries. Its **population density** is its population divided by its area. In this section, we discuss methods for measuring both area and volume.

Measuring Area

In order to measure a region that is enclosed by boundaries, we begin by selecting a *square unit*. A **square unit** is a square, each of whose sides is one unit in length, illustrated in **Figure 5.2**. The region in **Figure 5.2** is said to have an area of **one square unit**. The side of the square can be 1 inch, 1 centimeter, 1 meter, 1 foot, or one of any linear unit of measure. The corresponding

Figure 5.2
One square unit

units of area are the square inch (in.²), the square centimeter (cm²), the square meter (m²), the square foot (ft²), and so on. **Figure 5.3** illustrates 1 square inch and 1 square centimeter, drawn to actual size.

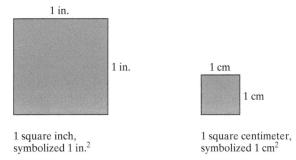

1 square inch, symbolized 1 in.²

1 square centimeter, symbolized 1 cm²

Figure 5.3 Common units of measurement for area, drawn to actual size

Square Unit

Figure 5.4

Example 1 · Measuring Area

What is the area of the region shown in **Figure 5.4**?

Solution

We can determine the area of the region by counting the number of square units contained within the region. There are 12 such units. Therefore, the area of the region is 12 square units.

✓ **Check Point 1** What is the area of the region represented by the first two rows in **Figure 5.4**? 8 square units

Although there are 12 inches in one foot and 3 feet in one yard, these numerical relationships are not the same for square units.

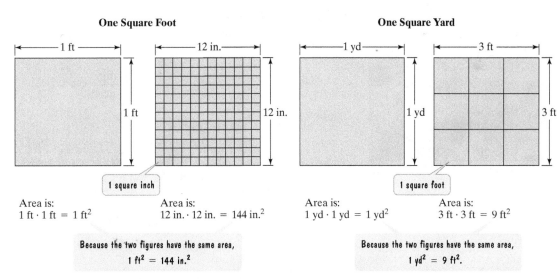

One Square Foot

Area is:
1 ft · 1 ft = 1 ft²

Area is:
12 in. · 12 in. = 144 in.²

One Square Yard

Area is:
1 yd · 1 yd = 1 yd²

Area is:
3 ft · 3 ft = 9 ft²

Because the two figures have the same area,
1 ft² = 144 in.²

Because the two figures have the same area,
1 yd² = 9 ft².

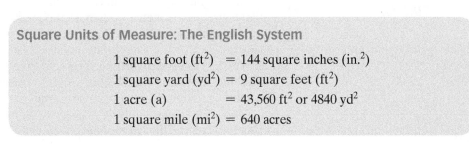

Square Units of Measure: The English System

1 square foot (ft²) = 144 square inches (in.²)

1 square yard (yd²) = 9 square feet (ft²)

1 acre (a)   = 43,560 ft² or 4840 yd²

1 square mile (mi²) = 640 acres

Great Question!

Which square units of measure are frequently used in everyday situations?

A small plot of land is usually measured in square feet, rather than a small fraction of an acre. Curiously, square yards are rarely used in measuring land. However, square yards are commonly used for carpet and flooring measures.

Example 2   Using Square Units to Compute Population Density

After Alaska, Wyoming is the least densely populated state. The population of Wyoming is 568,158 and its area is 97,814 square miles. What is Wyoming's population density?

Solution

We compute the population density by dividing Wyoming's population by its area.

$$\text{population density} = \frac{\text{population}}{\text{area}} = \frac{568,\!158 \text{ people}}{97,\!814 \text{ square miles}}$$

Using a calculator and rounding to the nearest tenth, we obtain a population density of 5.8 people per square mile. This means that there is an average of only 5.8 people for each square mile of area.

Check Point 2   The population of California is 37,691,912 and its area is 158,633 square miles. What is California's population density? Round to the nearest tenth.   237.6 people per square mile

2   Use dimensional analysis to change units for area.

Example 3   Using Dimensional Analysis on Units of Area

Your author wrote *Pathways to College Mathematics* in Point Reyes National Seashore, 40 miles north of San Francisco. The national park consists of 75,000 acres with miles of pristine surf-washed beaches, forested ridges, and bays bordered by white cliffs. How large is the national park in square miles?

Solution

We use the fact that 1 square mile = 640 acres to set up our unit fraction:

$$\frac{1 \text{ mi}^2}{640 \text{ acres}}.$$

Thus,

$$75,\!000 \text{ acres} = \frac{75,\!000 \text{ acres}}{1} \cdot \frac{1 \text{ mi}^2}{640 \text{ acres}} = \frac{75,\!000}{640} \text{ mi}^2 \approx 117 \text{ mi}^2.$$

The area of Point Reyes National Seashore is approximately 117 square miles.

Check Point 3   The National Park Service administers approximately 84,000,000 acres of national parks. How large is this in square miles?   131,250 mi^2

In Section 5.1, we saw that in most other countries, the system of measurement that is used is the metric system. In the metric system, the square centimeter is used instead of the square inch. The square meter replaces the square foot and the square yard.

The English system uses the acre and the square mile to measure large land areas. The metric system uses the hectare (symbolized ha and pronounced "hectair"). A hectare is about the area of two football fields placed side by side, approximately 2.5 acres. One square mile of land consists of approximately 260 hectares. Just as the hectare replaces the acre, the square kilometer is used instead of the square mile. One square kilometer is approximately 0.38 square mile.

Some basic approximate conversions for units of area are given in **Table 5.4**.

Table 5.4 English and Metric Equivalents for Area
1 square inch (in.²) $\approx$ 6.5 square centimeters (cm²)
1 square foot (ft²) $\approx$ 0.09 square meter (m²)
1 square yard (yd²) $\approx$ 0.8 square meter (m²)
1 square mile (mi²) $\approx$ 2.6 square kilometers (km²)
1 acre $\approx$ 0.4 hectare (ha)

Example 4   Using Dimensional Analysis on Units of Area

A property in Italy is advertised at $545,000 for 6.8 hectares.

a. Find the area of the property in acres.

b. What is the price per acre?

Solution

a. Using **Table 5.4**, we see that 1 acre $\approx$ 0.4 hectare. To convert 6.8 hectares to acres, we use a unit fraction with acres in the numerator and hectares in the denominator.

$$6.8 \text{ ha} \approx \frac{6.8 \text{ ha}}{1} \cdot \frac{1 \text{ acre}}{0.4 \text{ ha}} = \frac{6.8}{0.4} \text{ acres} = 17 \text{ acres}$$

The area of the property is approximately 17 acres.

b. The price per acre is the total price, $545,000, divided by the number of acres, 17.

$$\text{price per acre} = \frac{\$545,000}{17 \text{ acres}} \approx \$32,059/\text{acre}$$

The price is approximately $32,059 per acre.

✓ **Check Point 4**   A property in Northern California is on the market at $415,000 for 1.8 acres.

a. Find the area of the property in hectares. $\approx 0.72 \text{ ha}$

▶ **b.** What is the price per hectare? $\approx \$576,389 \text{ per hectare}$

Measuring Volume

③ Use cubic units to measure volume.

A shoe box and a basketball are examples of three-dimensional figures. **Volume** refers to the amount of space occupied by such figures. In order to measure this space, we begin by selecting a *cubic unit*. Two such cubic units are shown in **Figure 5.5**.

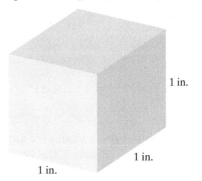

1 in.
1 in.
1 in.

1 cubic inch, symbolized 1 in.³

1 cm
1 cm
1 cm

1 cubic centimeter, symbolized 1 cm³

Figure 5.5 Common units of measurement for volume

The edges of a cube all have the same length. Other cubic units used to measure volume include 1 cubic foot (1 ft³) and 1 cubic meter (1 m³). One way to measure the volume of a solid is to calculate the number of cubic units contained in its interior.

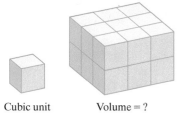

Cubic unit   Volume = ?

Figure 5.6

Example 5   Measuring Volume

What is the volume of the solid shown in **Figure 5.6**?

Solution

We can determine the volume of the solid by counting the number of cubic units contained within the region. Because we have drawn a solid three-dimensional figure on a flat two-dimensional page, some of the small cubic units in the back right are hidden. The figures below show how the cubic units are used to fill the inside of the solid.

Do these figures help you to see that there are 18 cubic units inside the solid? The volume of the solid is 18 cubic units.

Check Point 5   What is the volume of the region represented by the bottom row of blocks in **Figure 5.6**?   9 cubic units

We have seen that there are 3 feet in a yard, but 9 square feet in a square yard. Neither of these relationships holds for cubic units. **Figure 5.7** illustrates that there are 27 cubic feet in a cubic yard. Furthermore, there are 1728 cubic inches in a cubic foot.

Great Question!

I'm having difficulty seeing the detail in Figure 5.7. Can you help me out?

Cubing numbers is helpful:

$$3 \text{ ft} = 1 \text{ yd}$$
$$(3 \text{ ft})^3 = (1 \text{ yd})^3$$
$$3 \cdot 3 \cdot 3 \text{ ft}^3 = 1 \cdot 1 \cdot 1 \text{ yd}^3.$$

Conclusion: $27 \text{ ft}^3 = 1 \text{ yd}^3$

$$12 \text{ in.} = 1 \text{ ft}$$
$$(12 \text{ in.})^3 = (1 \text{ ft})^3$$
$$12 \cdot 12 \cdot 12 \text{ in.}^3 = 1 \cdot 1 \cdot 1 \text{ ft}^3$$

Conclusion: $1728 \text{ in.}^3 = 1 \text{ ft}^3$

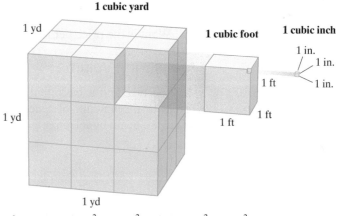

1 cubic yard

1 yd   1 yd   1 yd

1 cubic foot   **1 cubic inch**

1 in.   1 in.

1 ft   1 in.

1 ft   1 ft

Figure 5.7 $27 \text{ ft}^3 = 1 \text{ yd}^3$ and $1728 \text{ in.}^3 = 1 \text{ ft}^3$

The measure of volume also includes the amount of fluid that a three-dimensional object can hold. This is often called the object's **capacity**. For example, we often refer to the capacity, in gallons, of a gas tank. A cubic yard has a capacity of about 200 gallons and a cubic foot has a capacity of about 7.48 gallons.

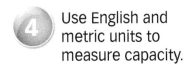

Use English and metric units to measure capacity.

Table 5.5 contains information about standard units of capacity in the English system.

Table 5.5 English Units for Capacity	
2 pints (pt)	= 1 quart (qt)
4 quarts	= 1 gallon (gal)
1 gallon	= 128 ounces (oz)
1 cup (c)	≈ 8 ounces

Volume in Cubic Units	Capacity
1 cubic yard	about 200 gallons
1 cubic foot	about 7.48 gallons
231 cubic inches	about 1 gallon

Example 6   Volume and Capacity in the English System

A swimming pool has a volume of 22,500 cubic feet. How many gallons of water does the pool hold?

Solution

We use the fact that 1 cubic foot has a capacity of about 7.48 gallons to set up our unit fraction:

$$\frac{7.48 \text{ gal}}{1 \text{ ft}^3}.$$

We use this unit fraction to find the capacity of the 22,500 cubic feet.

$$22,500 \text{ ft}^3 \approx \frac{22,500 \text{ ft}^3}{1} \cdot \frac{7.48 \text{ gal}}{1 \text{ ft}^3} = 22,500(7.48) \text{ gal} = 168,300 \text{ gal}$$

The pool holds approximately 168,300 gallons of water.

✓ **Check Point 6**   A pool has a volume of 10,000 cubic feet. How many gallons of water does the pool hold?   ≈ 74,800 gal

As we have come to expect, things are simpler when the metric system is used to measure capacity. The basic unit is the **liter**, symbolized by L. A liter is slightly larger than a quart.

$$1 \text{ liter} \approx 1.0567 \text{ quarts}$$

1 liter   1 quart

The standard metric prefixes are used to denote a multiple or part of a liter. **Table 5.6** applies these prefixes to the liter.

Table 5.6 Units of Capacity in the Metric System		
Symbol	**Unit**	**Meaning**
kL	kiloliter	1000 liters
hL	hectoliter	100 liters
daL	dekaliter	10 liters
L	liter	1 liter ≈ 1.06 quarts
dL	deciliter	0.1 liter
cL	centiliter	0.01 liter
mL	milliliter	0.001 liter

The following list should help give you a feel for capacity in the metric system.

Item	Capacity
Average cup of coffee	250 mL
12-ounce can of soda	355 mL
One quart of fruit juice	0.95 L
One gallon of milk	3.78 L
Average gas tank capacity of a car (about 18.5 gallons)	70 L

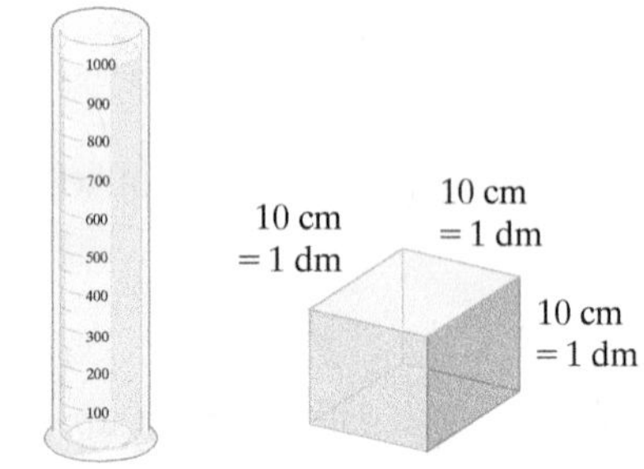

Figure 5.8 shows a 1-liter container filled with water. The water in the liter container will fit exactly into the box shown to its right. The volume of this box is 1000 cubic centimeters, or equivalently, 1 cubic decimeter. Thus,

$$1000 \text{ cm}^3 = 1 \text{ dm}^3 = 1 \text{ L}.$$

1 L = 1000 mL $1000 \text{ cm}^3 = 1 \text{ dm}^3$

Figure 5.8

Table 5.7 expands on the relationship between volume and capacity in the metric system.

Table 5.7 Volume and Capacity in the Metric System		
Volume in Cubic Units		**Capacity**
1 cm^3	=	1 mL
$1 \text{ dm}^3 = 1000 \text{ cm}^3$	=	1 L
1 m^3	=	1 kL

A milliliter is the capacity of a cube measuring 1 centimeter on each side.

A liter is the capacity of a cube measuring 10 centimeters on each side.

Blitzer Bonus
• • • • • • • • • • • • • • • • •

Measuring Dosages of Medicine

Table 5.7 indicates that

$$1 \text{ cm}^3 = 1 \text{ mL}.$$

Dosages of medicine are measured using cubic centimeters, or milliliters as they are also called. (In the United States, cc, rather than cm³, denotes cubic centimeters.) When taking medication by mouth, such as cough syrup, there is an easy conversion between the English and metric systems:

$$1 \text{ teaspoon} = 5 \text{ cc}.$$

Example 7   Volume and Capacity in the Metric System

An aquarium has a volume of 36,000 cubic centimeters. How many liters of water does the aquarium hold?

Solution

We use the fact that 1000 cubic centimeters corresponds to a capacity of 1 liter to set up our unit fraction:

$$\frac{1 \text{ L}}{1000 \text{ cm}^3}.$$

We use this unit fraction to find the capacity of the 36,000 cubic centimeters.

$$36{,}000 \text{ cm}^3 = \frac{36{,}000 \text{ cm}^3}{1} \cdot \frac{1 \text{ L}}{1000 \text{ cm}^3} = \frac{36{,}000}{1000}\text{L} = 36 \text{ L}$$

The aquarium holds 36 liters of water.

Check Point 7   A fish pond has a volume of 220,000 cubic centimeters. How many liters of water does the pond hold?   220 L

Achieving Success

Eight things your math professor never wants to hear

Interaction with your professor depends on goodwill from both sides. Other than misbehaving in class (messaging on your cellphone, chatting or nodding off during lecture), here are eight common ways to get on the wrong side of your professor. In some cases, you may not even be aware that you have said something to irritate the prof. Avoid these faux pas:

- "I missed your last lecture. Did you do anything important?"
- "I missed the test because my alarm clock didn't go off. When can I take a makeup?"
- "I lost the course policy sheet. Can you email it to me? And I think I might also need the syllabus and list of assignments."
- "I'd rather go to the math lab than attend your lectures. You go too fast and it's hard to follow what you're saying."
- "My friend and I showed exactly the same work on problem 7. How come you took off 3 points on my test, but only 2 points on hers?"
- "Can we go over my test?"
- "I'm leaving for a skiing vacation. Can I take the final early?"
- "I'd do anything for an A."

Concept and Vocabulary Check

Exercises in the Concept and Vocabulary Check are intended for group and class discussions.

In Exercises 1–6, fill in each blank so that the resulting statement is true.

1. Area is measured in ___square___ units and volume is measured in ___cubic___ units.

2. In the English system, $1 \text{ ft}^2 = $ ___144___ in.^2 and $1 \text{ yd}^2 = $ ___9___ ft^2.

3. Because $1 \text{ mi}^2 = 640$ acres, the unit fraction needed to convert from acres to square miles is $\frac{1 \text{ mi}^2}{640 \text{ acres}}$ and the unit fraction needed to convert from square miles to acres is _____ $\cdot \frac{640 \text{ acres}}{1 \text{ mi}^2}$.

4. In the English system, ___2___ pt = 1 qt and ___4___ qt = 1 gal.

5. The amount of fluid that a three-dimensional object can hold is called the object's ___capacity___, measured in the metric system using a basic unit called the ___liter___.

6. A state's population density is its population divided by its ___area___.

In Exercises 7–9, determine whether each statement is true or false. If the statement is false, make the necessary change(s) to produce a true statement.   Changes to false statements will vary.

7. Because there are 3 feet in one yard, there are also 3 square feet in one square yard.   false

8. The English system uses in.^2 to measure large land areas.   false

9. One quart is approximately 1.06 liters.   false

Respond to Exercises 10–15 using verbal or written explanations.   10–15. Answers will vary.

10. Describe how area is measured. Explain why linear units cannot be used.

11. New Mexico has a population density of 17 people per square mile. Describe what this means. Use an almanac or the Internet to compare this population density to that of the state in which you are now living.

12. Describe the difference between the following problems: How much fencing is needed to enclose a garden? How much fertilizer is needed for the garden?

13. Describe how volume is measured. Explain why linear or square units cannot be used.

14. For a swimming pool, what is the difference between the following units of measure: cubic feet and gallons? For each unit, write a sentence about the pool that makes the use of the unit appropriate.

15. If there are 10 decimeters in a meter, explain why there are not 10 cubic decimeters in a cubic meter.

Exercise Set 5.2

Practice Exercises

In Exercises 1–4, use the given figure to find its area in square units.

1.

16 square units

2.
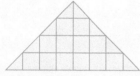

15 square units

3.

8 square units

4.

16 square units

*In Exercises 5–12, use **Table 5.4** on page 339, along with dimensional analysis, to convert the given area measurement to the square unit indicated. Where necessary, round answers to two decimal places.*

5. 14 cm^2 to in.^{2} ≈2.15 in.^{2} **6.** 20 m^2 to ft^2 ≈222.22 ft^2

7. 30 m^2 to yd^2 ≈37.5 yd^2 **8.** 14 mi^2 to km^2 ≈36.4 km^2

9. 10.2 ha to acres ≈25.5 acres **10.** 20.6 ha to acres ≈51.5 acres

11. 14 in.^2 to cm^2 ≈91 cm^2 **12.** 20 in.^2 to cm^2 ≈130 cm^2

In Exercises 13–14, use the given figure to find its volume in cubic units.

13.

24 cubic units

14.
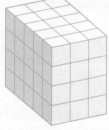

60 cubic units

*In Exercises 15–22, use **Table 5.5** on page 341, along with dimensional analysis, to convert the given measurement of volume or capacity to the unit indicated. Where necessary, round answers to two decimal places.*

15. $10,000 \text{ ft}^3$ to gal   ≈74,800 gal

16. $25,000 \text{ ft}^3$ to gal   ≈187,000 gal

17. 8 yd^3 to gal   ≈1600 gal

18. 35 yd^3 to gal   ≈7000 gal

19. 2079 in.^3 to gal   ≈9 gal

20. 6237 in.^3 to gal   ≈27 gal

21. 2700 gal to yd^3 ≈13.5 yd^3

22. 1496 gal to ft^3 ≈200 ft^3

*In Exercises 23–32, use **Table 5.7** on page 342, along with dimensional analysis, to convert the given measurement of volume or capacity to the unit indicated.*

23. $45,000 \text{ cm}^3$ to L 45 L   **24.** $75,000 \text{ cm}^3$ to L 75 L

25. 17 cm^3 to mL 17 mL   **26.** 19 cm^3 to mL 19 mL

27. 1.5 L to cm^3 1500 cm^3 **28.** 4.5 L to cm^3 4500 cm^3

29. 150 mL to cm^3 150 cm^3 **30.** 250 mL to cm^3 250 cm^3

31. 12 kL to dm^3 12,000 dm^3 **32.** 16 kL to dm^3 16,000 dm^3

Practice Plus

The bar graph shows the resident population and the land area of the United States for selected years from 1800 through 2010. Use the information shown by the graph to solve Exercises 33–36.

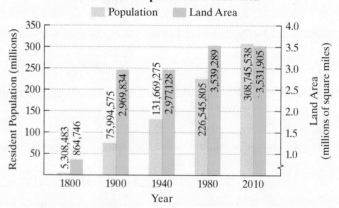

U.S. Resident Population and Land Area

Source: U.S. Bureau of the Census

33. a. Find the population density of the United States, to the nearest tenth, in 1900 and in 2010.   *

   b. Find the percent increase in population density, to the nearest tenth of a percent, from 1900 to 2010.   241.4%

34. a. Find the population density of the United States, to the nearest tenth, in 1800 and in 2010.   *

   b. Find the percent increase in population density, to the nearest tenth of a percent, from 1800 to 2010.   1332.8%

*See Answers to Selected Exercises.

(In Exercises 35–36, refer to the bar graph on the previous page.)

35. Find the population density of the United States, to the nearest tenth, expressed in people per square kilometer, in 1940.   17.0 people per square kilometer

36. Find the population density of the United States, to the nearest tenth, expressed in people per square kilometer, in 1980.   24.6 people per square kilometer

Application Exercises

In Exercises 37–38, find the population density, to the nearest tenth, for each state. Which state has the greater population density? How many more people per square mile inhabit the state with the greater density than inhabit the state with the lesser density?

37. Illinois population: 12,869,257   area: 57,914 mi^2

Ohio population: 11,544,951   area: 44,826 mi^2 *

38. New York population: 19,465,197   area: 54,555 mi^2

Rhode Island population: 1,051,302   area: 1545 mi^2 *

In Exercises 39–40, use the fact that 1 square mile = 640 acres to find the area of each national park to the nearest square mile.

39. Everglades National Park (Florida): 1,509,154 acres   ≈2358 mi^2

40. Yosemite National Park (California): 761,268 acres   ≈1189 mi^2

41. A property that measures 8 hectares is for sale.
   a. How large is the property in acres?   ≈20 acres
   b. If the property is selling for $250,000, what is the price per acre?   ≈$12,500 per acre

42. A property that measures 100 hectares is for sale.
   a. How large is the property in acres?   ≈250 acres
   b. If the property is selling for $350,000, what is the price per acre?   ≈$1400 per acre

In Exercises 43–46, selecting from square centimeters, square meters, or square kilometers, determine the best unit of measure to express the area of the object described.

43. The top of a desk   square centimeters

44. A dollar bill   square centimeters

45. A national park   square kilometers

46. The wall of a room   square meters

In Exercises 47–50, select the best estimate for the measure of the area of the object described.

47. The area of the floor of a room   b
   a. 25 cm^2 **b.** 25 m^2 **c.** 25 km^2

48. The area of a television screen   b
   a. 2050 mm^2 **b.** 2050 cm^2 **c.** 2050 m^2

49. The area of the face of a small coin   b
   a. 6 mm^2 **b.** 6 cm^2 **c.** 6 dm^2

50. The area of a parcel of land in a large metropolitan area on which a house can be built   b
   a. 900 cm^2 **b.** 900 m^2 **c.** 900 ha

51. A swimming pool has a volume of 45,000 cubic feet. How many gallons of water does the pool hold?   ≈336,600 gal

52. A swimming pool has a volume of 66,000 cubic feet. How many gallons of water does the pool hold?   ≈493,680 gal

53. A container of grapefruit juice has a volume of 4000 cubic centimeters. How many liters of juice does the container hold?   4 L

54. An aquarium has a volume of 17,500 cubic centimeters. How many liters of water does the aquarium hold?   17.5 L

Exercises 55–56 give the approximate area of some of the world's largest island countries. In each exercise, determine which country has the greater area and by how many square kilometers.

55. Philippines: 300,000 km^2; Japan: 145,900 mi^2 Japan; ≈79,000 km^2

56. Iceland: 103,000 km^2; Cuba: 42,800 mi^2 Cuba; ≈8000 km^2

Exercises 57–58 give the approximate area of some of the world's largest islands. In each exercise, determine which island has the greater area and by how many square miles. Round to the nearest square mile.

57. Baffin Island, Canada: 194,574 mi^2;
Sumatra, Indonesia: 443,070 km^2 Baffin Island; ≈24,162 mi^2

58. Honshu, Japan: 87,805 mi^2;
Victoria Island, Canada: 217,300 km^2 Honshu; ≈4228 mi^2

(Source for Exercises 55–58: Russell Ash, The Top 10 of Everything 2009)

Critical Thinking Exercises

59. Singapore has the highest population density of any country: 46,690 people per 1000 hectares. How many people are there per square mile?   ≈11,952.64 people per square mile

60. Nebraska has a population density of 23.8 people per square mile and a population of 1,842,641. What is the area of Nebraska? Round to the nearest square mile.   ≈77,422 mi^2

61. A high population density is a condition common to extremely poor and extremely rich locales. Explain why this is so.   Answers will vary.

62. Although Alaska is the least densely populated state, over 90% of its land is protected federal property that is off-limits to settlement. A resident of Anchorage, Alaska, might feel hemmed in. In terms of "elbow room," what other important factor must be considered when calculating a state's population density?   how the population is distributed

63. Does an adult's body contain approximately 6.5 liters, kiloliters, or milliliters of blood? Explain your answer.

64. Is the volume of a coin approximately 1 cubic centimeter, 1 cubic millimeter, or 1 cubic decimeter? Explain your answer.   1 cubic centimeter; Answers will vary.

63. 6.5 liters; Answers will vary.

Group Exercise

65. If you could select any place in the world, where would you like to live? Look up the population and land area of your ideal place, and compute its population density. Group members should share places to live and population densities. What trend, if any, does the group observe?   Answers will vary.

*See Answers to Selected Exercises.

5.3 Measuring Weight and Temperature

What am I supposed to learn?

After you have read this section, you should be able to:

1. Apply metric prefixes to units of weight.

2. Convert units of weight within the metric system.

3. Use relationships between volume and weight within the metric system.

4. Use dimensional analysis to change units of weight to and from the metric system.

5. Understand temperature scales.

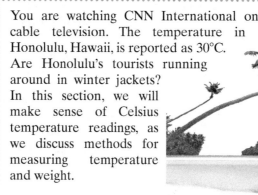

You are watching CNN International on cable television. The temperature in Honolulu, Hawaii, is reported as 30°C. Are Honolulu's tourists running around in winter jackets? In this section, we will make sense of Celsius temperature readings, as we discuss methods for measuring temperature and weight.

Measuring Weight

You step on the scale at the doctor's office to check your weight, discovering that you are 150 pounds. Compare this to your weight on the moon: 25 pounds. Why the difference? **Weight** is the measure of the gravitational pull on an object. The gravitational pull on the moon is only about one-sixth the gravitational pull on Earth. Although your weight varies depending on the force of gravity, your mass is exactly the same in all locations. **Mass** is a measure of the quantity of matter in an object, determined by its molecular structure. On Earth, as your weight increases, so does your mass. In this section, measurements are assumed to involve everyday situations on the surface of Earth. Thus, we will treat weight and mass as equivalent, and refer strictly to weight.

> **Units of Weight: The English System**
>
> 16 ounces (oz) = 1 pound (lb)
>
> 2000 pounds (lb) = 1 ton (T)

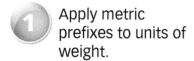

1. Apply metric prefixes to units of weight.

The basic metric unit of weight is the **gram** (g), used for very small objects such as a coin, a candy bar, or a teaspoon of salt. A nickel has a weight of about 5 grams.

As with meters, prefixes are used to denote a multiple or part of a gram. **Table 5.8** applies the common metric prefixes to the gram. The first part of the symbol indicates the prefix and the second part (g) indicates gram.

Weight of pineapple is 1 kg, or 1000 g.

Symbol	Unit	Meaning
kg	kilogram	1000 grams
hg	hectogram	100 grams
dag	dekagram	10 grams
g	gram	1 gram
dg	decigram	0.1 gram
cg	centigram	0.01 gram
mg	milligram	0.001 gram

Table 5.8 Commonly Used Units of Weight in the Metric System

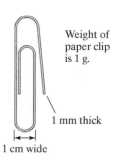

Weight of paper clip is 1 g.

1 mm thick

1 cm wide

In the metric system, the kilogram is the comparable unit to the pound in the English system. **A weight of 1 kilogram is approximately 2.2 pounds.** Thus, an average man has a weight of about 75 kilograms. Objects that we measure in pounds are measured in kilograms in most countries.

A milligram, equivalent to 0.001 gram, is an extremely small unit of weight and is used extensively in the pharmaceutical industry. If you look at the label on a bottle of tablets, you will see that the amounts of different substances in each tablet are expressed in milligrams.

The weight of a very heavy object is expressed in terms of the metric tonne (t), which is equivalent to 1000 kilograms, or about 2200 pounds. This is 10 percent more than the English ton (T) of 2000 pounds.

We change units of weight within the metric system exactly the same way that we changed units of length.

2 Convert units of weight within the metric system.

Changing Units of Weight within the Metric System
Use the following diagram to find equivalent measures of weight:

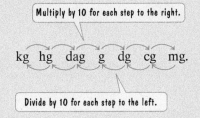

Multiply by 10 for each step to the right.

kg hg dag g dg cg mg.

Divide by 10 for each step to the left.

Example 1 Changing Units within the Metric System

a. Convert 8.7 dg to mg.
b. Convert 950 mg to g.

Solution

a. To convert from dg (decigrams) to mg (milligrams), we start at dg and move two steps to the right:

kg hg dag g dg cg mg.

Hence, we move the decimal point two places to the right:

$$8.7 \text{ dg} = 870 \text{ mg}.$$

b. To convert from mg (milligrams) to g (grams), we start at mg and move three steps to the left:

kg hg dag g dg cg mg.

Hence, we move the decimal point three places to the left:

$$950 \text{ mg} = 0.950 \text{ g}.$$

 **Check Point 1**

a. Convert 4.2 dg to mg.   420 mg
▶ **b.** Convert 620 cg to g.   6.2 g

3 Use relationships between volume and weight within the metric system.

We have seen a convenient relationship in the metric system between volume and capacity:

$$1000 \text{ cm}^3 = 1 \text{ dm}^3 = 1 \text{ L}.$$

This relationship can be extended to include weight based on the following:

One kilogram of water has a volume of 1 liter.

Thus,

$$1000 \text{ cm}^3 = 1 \text{ dm}^3 = 1 \text{ L} = 1 \text{ kg}.$$

Table 5.9 shows the relationships between volume and weight of water in the metric system.

Table 5.9 Volume and Weight of Water in the Metric System					
Volume		**Capacity**		**Weight**	
1 cm^3	=	1 mL	=	1 g	
$1 \text{ dm}^3 = 1000 \text{ cm}^3$	=	1 L	=	1 kg	
1 m^3	=	1 kL	=	1000 kg = 1 t	

Example 2   Volume and Weight in the Metric System

An aquarium holds 0.25 m^3 of water. How much does the water weigh?

Solution

We use the fact that 1 m^3 of water $= 1000 \text{ kg}$ of water to set up our unit fraction:

$$\frac{1000 \text{ kg}}{1 \text{ m}^3}.$$

Thus,

$$0.25 \text{ m}^3 = \frac{0.25 \text{ m}^3}{1} \cdot \frac{1000 \text{ kg}}{1 \text{ m}^3} = 250 \text{ kg}.$$

The water weighs 250 kilograms.

 Check Point 2   An aquarium holds 0.145 m^3 of water. How much does the water weigh?   145 kg

4 Use dimensional analysis to change units of weight to and from the metric system.

A problem like Example 2 involves more awkward computation in the English system. For example, if you know the aquarium's volume in cubic feet, you must also know that 1 cubic foot of water weighs about 62.5 pounds to determine the water's weight.

Dimensional analysis is a useful tool when converting units of weight between the English and metric systems. Some basic approximate conversions are given in **Table 5.10**.

Table 5.10 Weight: English and Metric Equivalents
1 ounce (oz) $\approx$ 28 grams (g)
1 pound (lb) $\approx$ 0.45 kilogram (kg)
1 ton (T) $\approx$ 0.9 tonne (t)

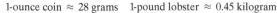
1-ounce coin ≈ 28 grams   1-pound lobster ≈ 0.45 kilogram

1-ton bison ≈ 0.9 tonne

Example 3   Using Dimensional Analysis

a. Convert 160 pounds to kilograms.   **b.** Convert 300 grams to ounces.

Blitzer Bonus
● ● ● ● ● ● ● ● ● ● ● ● ● ● ●

The SI System

The metric system is officially called the Système International d'Unités, abbreviated the SI system. Liberia, Myanmar, and the United States are the only countries in the world not using the SI system.

In 1975, the United States prepared to adopt the SI system with the Metric Conversion Act. However, due to concerns in grassroots America, Congress made the conversion voluntary, effectively chasing it off. But it's not as bleak as you think; from metric tools to milligram measures of medication, we have been edging, inching, even centimetering toward the metric system.

Solution

a. To convert 160 pounds to kilograms, we use a unit fraction with kilograms in the numerator and pounds in the denominator:

$$\frac{0.45 \text{ kg}}{1 \text{ lb}} \cdot \qquad \text{Table 5.10 on the previous page shows that } 1 \text{ lb} \approx 0.45 \text{ kg.}$$

Thus,

$$160 \text{ lb} \approx \frac{160 \text{ lb}}{1} \cdot \frac{0.45 \text{ kg}}{1 \text{ lb}} = 160(0.45) \text{ kg} = 72 \text{ kg.}$$

b. To convert 300 grams to ounces, we use a unit fraction with ounces in the numerator and grams in the denominator:

$$\frac{1 \text{ oz}}{28 \text{ g}} \cdot \qquad \text{Table 5.10 on the previous page shows that } 1 \text{ oz} \approx 28 \text{ g.}$$

Thus,

$$300 \text{ g} \approx \frac{300 \text{ g}}{1} \cdot \frac{1 \text{ oz}}{28 \text{ g}} = \frac{300}{28} \text{ oz} \approx 10.7 \text{ oz.}$$

✓ **Check Point 3**

a. A man weighs 186 pounds. Convert his weight to kilograms. ≈ 83.7 kg

b. For each kilogram of weight, 1.2 milligrams of a drug is to be given. What dosage should a 186-pound man be given? ≈ 100.44 mg

---

Blitzer Bonus ✿
● ● ● ● ● ● ● ● ● ● ● ● ● ● ●

Using the Metric System to Measure Blood Alcohol Concentration

In Chapter 2, we presented a formula for determining blood alcohol concentration. Blood alcohol concentration (BAC) is measured in grams of alcohol per 100 milliliters of blood. To put this measurement into perspective, a 175-pound man has approximately 5 liters (5000 milliliters) of blood and a 12-ounce can of 6%-alcohol-by-volume beer contains about 15 grams of alcohol. Based on the time it takes for alcohol to be absorbed into the bloodstream, as well as its elimination at a rate of 10–15 grams per hour, **Figure 5.9** shows the effect on BAC of a number of drinks on individuals within weight classes. It is illegal to drive with a BAC at 0.08 g/100 mL, or 0.08 gram of alcohol per 100 milliliters of blood, or greater.

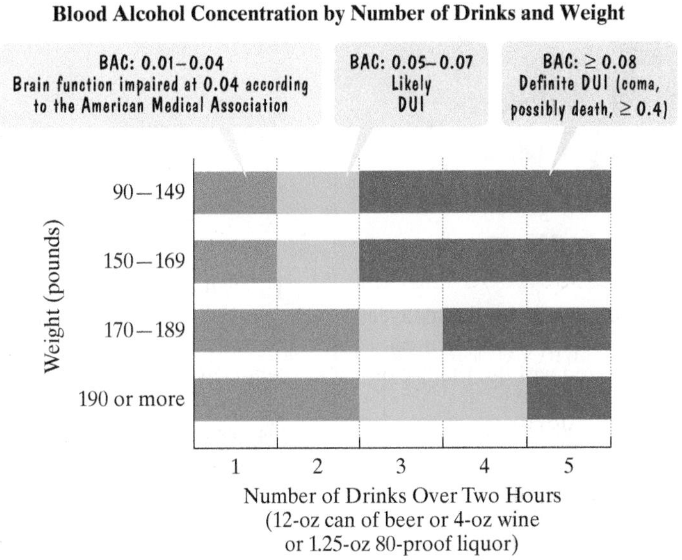

Blood Alcohol Concentration by Number of Drinks and Weight

Figure 5.9 *Source: Patrick McSharry, Everyday Numbers, Random House, 2002.*

Measuring Temperature

⑤ Understand temperature scales.

You'll be leaving the cold of winter for a vacation to Hawaii. CNN International reports a temperature in Hawaii of 30°C. Should you pack a winter coat?

The idea of changing from Celsius readings—or is it Centigrade?—to familiar Fahrenheit readings can be disorienting. Reporting a temperature of 30°C doesn't have the same impact as the Fahrenheit equivalent of 86 degrees (don't pack the coat). Why these annoying temperature scales?

The **Fahrenheit temperature scale**, the one we are accustomed to, was established in 1714 by the German physicist Gabriel Daniel Fahrenheit. He took a mixture of salt and ice, then thought to be the coldest possible temperature, and called it 0 degrees. He called the temperature of the human body 96 degrees, dividing the space between 0 and 96 into 96 parts. Fahrenheit was wrong about body temperature. It was later found to be 98.6 degrees. On his scale, water froze (without salt) at 32 degrees and boiled at 212 degrees. The symbol ° was used to replace the word *degree*.

Twenty years later, the Swedish scientist Anders Celsius introduced another temperature scale. He set the freezing point of water at 0° and its boiling point at 100°, dividing the space into 100 parts. Degrees were called centigrade until 1948, when the name was officially changed to honor its inventor. However, *centigrade* is still commonly used in the United States.

Figure 5.10 shows a thermometer that measures temperatures in both degrees Celsius (°C is the scale on the left) and degrees Fahrenheit (°F is the scale on the right). The thermometer should help orient you if you need to know what a temperature in °C means. For example, if it is 40°C, find the horizontal line representing this temperature on the left. Now read across to the °F scale on the right. The reading is above 100°, indicating heat wave conditions.

The following formulas can be used to convert from one temperature scale to the other:

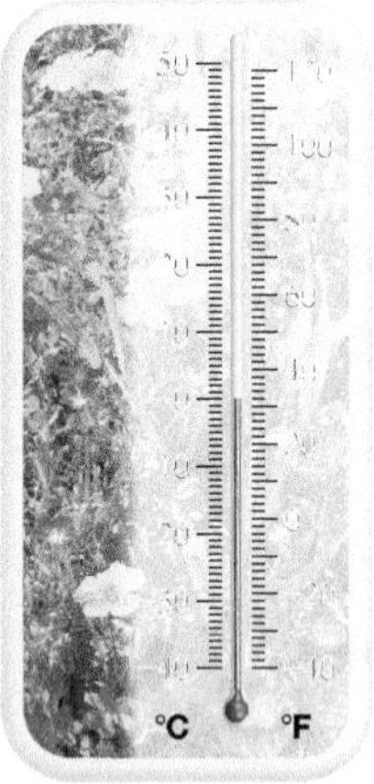

Figure 5.10 The Celsius scale is on the left and the Fahrenheit scale is on the right.

> **From Celsius to Fahrenheit**
> $$F = \frac{9}{5}C + 32$$

> **From Fahrenheit to Celsius**
> $$C = \frac{5}{9}(F - 32)$$

Great Question!

Because $\frac{9}{5} = 1.8$, can I use 1.8 instead of $\frac{9}{5}$ in the formula $F = \frac{9}{5}C + 32$?

Yes. The formula used to convert from Celsius to Fahrenheit can be expressed without the use of fractions:

$$F = 1.8C + 32.$$

Some students find this form of the formula easier to memorize.

Example 4   Converting from Celsius to Fahrenheit

The bills from your European vacation have you feeling a bit feverish, so you decide to take your temperature. The thermometer reads 37°C. Should you panic?

Solution

Use the formula

$$F = \frac{9}{5}C + 32$$

to convert 37°C from °C to °F. Substitute 37 for C in the formula and find the value of F.

$$F = \frac{9}{5}(37) + 32 = 66.6 + 32 = 98.6$$

No need to panic! Your temperature is 98.6° F, which is perfectly normal.

▶ ✓ **Check Point 4**   Convert 50°C from °C to °F.   122°F

Example 5 Converting from Fahrenheit to Celsius

The temperature on a warm spring day is 77°F. Find the equivalent temperature on the Celsius scale.

Solution

Use the formula

$$C = \frac{5}{9}(F - 32)$$

to convert 77°F from °F to °C. Substitute 77 for F in the formula and find the value of C.

$$C = \frac{5}{9}(77 - 32) = \frac{5}{9}(45) = \frac{5}{\underset{1}{9}}\left(\frac{\overset{5}{45}}{1}\right) = 25$$

Thus, 77°F is equivalent to 25°C.

 **Check Point 5** Convert 59°F from °F to °C. 15°C

Because temperature is a measure of heat, scientists do not find negative temperatures meaningful in their work. In 1948, the British physicist Lord Kelvin introduced a third temperature scale. He put 0 degrees at absolute zero, the coldest possible temperature, at which there is no heat and molecules stop moving. **Figure 5.11** illustrates the three temperature scales.

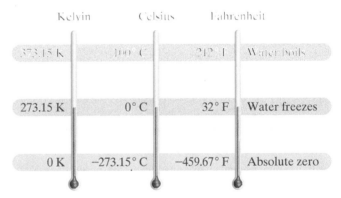

Figure 5.11 The three temperature scales

Lake Baikal, Siberia, is one of the coldest places on Earth, reaching −76°F (−60°C) in winter. The lowest temperature possible is absolute zero. Scientists have cooled atoms to a few millionths of a degree above absolute zero.

Figure 5.11 shows that water freezes at 273.15 K (read "K" or "Kelvins," not "degrees Kelvin") and boils at 373.15 K. The Kelvin scale is the same as the Celsius scale, except in its starting (zero) point. This makes it easy to go back and forth from Celsius to Kelvin.

From Celsius to Kelvin
$$K = C + 273.15$$

From Kelvin to Celsius
$$C = K - 273.15$$

Blitzer Bonus

Running a 5 K Race?

A 5 K race means a race at 5 Kelvins. This is a race so cold that no one would be able to move because all the participants would be frozen solid! The proper symbol for a race five kilometers long is a 5 km race.

Kelvin's scale was embraced by the scientific community. Today, it is the final authority, as scientists define Celsius and Fahrenheit in terms of Kelvins.

Achieving Success

Analyze the errors you make on quizzes and tests.

For each error, write out the correct solution along with a description of the concept needed to solve the problem correctly. Do your mistakes indicate gaps in understanding concepts or do you at times believe that you are just not a good test taker? Are you repeatedly making the same kinds of mistakes on tests? Keeping track of errors should increase your understanding of the material, resulting in improved test scores.

Concept and Vocabulary Check

Exercises in the Concept and Vocabulary Check are intended for group and class discussions.

In Exercises 1–6, fill in each blank so that the resulting statement is true.

1. In the English system, ____16____ oz = 1 lb and ____2000____ lb = 1 T.

2. The basic metric unit of weight is the ____gram____, used for very small objects such as a coin, a candy bar, or a teaspoon of salt.

3. A weight of 1 kilogram is approximately ____2.2____ pounds.

4. One kilogram of water has a volume of ____1____ liter(s).

5. On the Fahrenheit temperature scale, water freezes at ____32____ degrees and boils at ____212____ degrees.

6. On the Celsius temperature scale, water freezes at ____0____ degrees and boils at ____100____ degrees.

In Exercises 7–10, determine whether each statement is true or false. If the statement is false, make the necessary change(s) to produce a true statement.   Changes to false statements will vary.

7. 1 gram = 1000 kilograms   false

8. 1 gram ≈ 28 ounces   false

9. The formula $F = \frac{9}{5}C + 32$ is used to convert from Fahrenheit to Celsius.   false

10. The formula $C = \frac{1}{9}(F - 32)$ is used to convert from Fahrenheit to Celsius.   false

Respond to Exercises 11–16 using verbal or written explanations.   11–16. Answers will vary.

11. Describe the difference between weight and mass.

12. Explain how to use dimensional analysis to convert 200 pounds to kilograms.

13. Why do you think that countries using the metric system prefer the Celsius scale over the Fahrenheit scale?

14. Describe in words how to convert from Celsius to Fahrenheit.

15. Describe in words how to convert from Fahrenheit to Celsius.

16. If you decide to travel outside the United States, which one of the two temperature conversion formulas should you take? Explain your answer.

Exercise Set 5.3

Practice Exercises

In Exercises 1–10, convert the given weight to the unit indicated.

1. 7.4 dg to mg   740 mg
2. 6.9 dg to mg   690 mg
3. 870 mg to g   0.87 g
4. 640 mg to g   0.64 g
5. 8 g to cg   800 cg
6. 7 g to cg   700 cg
7. 18.6 kg to g   18,600 g
8. 0.37 kg to g   370 g
9. 0.018 mg to g   0.000018 g
10. 0.029 mg to g   0.000029 g

*In Exercises 11–18, use **Table 5.9** on page 348 to convert the given measurement of water to the unit indicated.*

11. 0.05 m³ to kg   50 kg
12. 0.02 m³ to kg   20 kg
13. 4.2 kg to cm³   4200 cm³
14. 5.8 kg to cm³   5800 cm³
15. 1100 m³ to t   1100 t
16. 1500 t to m³   1500 m³
17. 0.04 kL to g   40,000 g
18. 0.03 kL to g   30,000 g

In Exercises 19–30, use the following equivalents, along with dimensional analysis, to convert the given weight to the unit indicated. When necessary, round answers to two decimal places.

$$16 \text{ oz} = 1 \text{ lb}$$
$$2000 \text{ lb} = 1 \text{ T}$$
$$1 \text{ oz} \approx 28 \text{ g}$$
$$1 \text{ lb} \approx 0.45 \text{ kg}$$
$$1 \text{ T} \approx 0.9 \text{ t}$$

19. 36 oz to lb   2.25 lb   **20.** 26 oz to lb   1.625 lb

21. 36 oz to g   ≈1008 g   **22.** 26 oz to g   ≈728 g

23. 540 lb to kg   ≈243 kg   **24.** 220 lb to kg   ≈99 kg

25. 80 lb to g   ≈36,000 g   **26.** 150 lb to g   ≈67,500 g

27. 540 kg to lb   ≈1200 lb   **28.** 220 kg to lb   ≈488.89 lb

29. 200 t to T   ≈222.22 T   **30.** 100 t to T   ≈111.11 T

In Exercises 31–38, convert the given Celsius temperature to its equivalent temperature on the Fahrenheit scale. Where appropriate, round to the nearest tenth of a degree.

31. 10°C   50°F   **32.** 20°C   68°F

33. 35°C   95°F   **34.** 45°C   113°F

35. 57°C   134.6°F   **36.** 98°C   208.4°F

37. −5°C   23°F   **38.** −10°C   14°F

In Exercises 39–50, convert the given Fahrenheit temperature to its equivalent temperature on the Celsius scale. Where appropriate, round to the nearest tenth of a degree.

39. 68°F   20°C   **40.** 86°F   30°C

41. 41°F   5°C   **42.** 50°F   10°C

43. 72°F   22.2°C   **44.** 90°F   32.2°C

45. 23°F   −5°C   **46.** 14°F   −10°C

47. 350°F   176.7°C   **48.** 475°F   246.1°C

49. −22°F   −30°C   **50.** −31°F   −35°C

Practice Plus

51. The nine points shown below represent Celsius temperatures and their equivalent Fahrenheit temperatures. Also shown is a line that passes through the points.

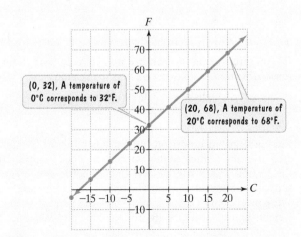

a. Use the coordinates of the two points identified by the voice balloons to compute the line's slope. Express the answer as a fraction reduced to lowest terms. What does this mean about the change in Fahrenheit temperature for each degree change in Celsius temperature?   *

b. Use the slope-intercept form of the equation of a line, $y = mx + b$, the slope from part (a), and the y-intercept shown by the graph to derive the formula used to convert from Celsius to Fahrenheit. $F = \dfrac{9}{5}C + 32$

52. Solve the formula used to convert from Celsius to Fahrenheit for C and derive the other temperature conversion formula. $C = \dfrac{5}{9}(F - 32)$

Application Exercises

In Exercises 53–59, selecting from milligram, gram, kilogram, and tonne, determine the best unit of measure to express the given item's weight.

53. A bee   milligram

54. This book   kilogram

55. A tablespoon of salt   gram

56. A Boeing 747   tonne

57. A stacked washer-dryer   kilogram

58. A pen   gram

59. An adult male   kilogram

In Exercises 60–66, select the best estimate for the weight of the given item.

60. A newborn infant's weight   d
   a. 3000 kg   **b.** 300 kg   **c.** 30 kg   **d.** 3 kg

61. The weight of a nickel   b
   a. 5 kg   **b.** 5 g   **c.** 5 mg

62. A person's weight   a
   a. 60 kg   **b.** 60 g   **c.** 60 dag

63. The weight of a box of cereal   a
   a. 0.5 kg   **b.** 0.5 g   **c.** 0.5 t

64. The weight of a glass of water   b
   a. 400 dg   **b.** 400 g   **c.** 400 dag   **d.** 400 hg

65. The weight of a regular-size car   c
   a. 1500 dag   **b.** 1500 hg   **c.** 1500 kg   **d.** 15,000 kg

66. The weight of a bicycle   b
   a. 140 kg   **b.** 140 hg   **c.** 140 dag   **d.** 140 g

67. Six items purchased at a grocery store weigh 14 kilograms. One of the items is detergent weighing 720 grams. What is the total weight, in kilograms, of the other five items?   13.28 kg

68. If a nickel weighs 5 grams, how many nickels are there in 4 kilograms of nickels?   800 nickels

69. If the cost to mail a letter is 44 cents for mail weighing up to one ounce and 24 cents for each additional ounce or fraction of an ounce, find the cost of mailing a letter that weighs 85 grams. $1.16

70. Using the information given below the pictured finback whale, estimate the weight, in tons and kilograms, of the killer whale.
Answers will vary. Sample answer: 12.5 T; 11,340 kg

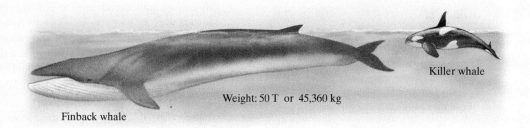

Killer whale

Weight: 50 T or 45,360 kg

Finback whale

71. Which is more economical: purchasing the economy size of a detergent at 3 kilograms for $3.15 or purchasing the regular size at 720 grams for 60¢?
Purchasing the regular size is more economical.

Exercises 72–73 ask you to determine drug dosage by a patient's weight. Use the fact that 1 lb ≈ 0.45 kg.

72. The prescribed dosage of a drug is 10 mg/kg daily, meaning that 10 milligrams of the drug should be administered daily for each kilogram of a patient's weight. How many 400-milligram tablets should be given each day to a patient who weighs 175 pounds?   2

73. The prescribed dosage of a drug is 15 mg/kg daily, meaning that 15 milligrams of the drug should be administered daily for each kilogram of a patient's weight. How many 200-milligram tablets should be given each day to a patient who weighs 120 pounds?   4

The label on a bottle of Emetrol ("for food or drink indiscretions") reads

> *Each 5 mL teaspoonful contains glucose, 1.87 g; levulose, 1.87 g; and phosphoric acid, 21.5 mg.*

Use this information to solve Exercises 74–75.

74. a. Find the amount of glucose in the recommended dosage of two teaspoons.   3.74 g

   b. If the bottle contains 4 ounces, find the quantity of glucose in the bottle. (1 oz ≈ 30 mL)   ≈44.88 g

75. a. Find the amount of phosphoric acid in the recommended dosage of two teaspoons.   43 mg

   b. If the bottle contains 4 ounces, find the quantity of phosphoric acid in the bottle. (1 oz ≈ 30 mL)   ≈516 mg

In Exercises 76–79, select the best estimate of the Celsius temperature of

76. A very hot day.   c
   a. 85°C   **b.** 65°C   **c.** 35°C   **d.** 20°C

77. A warm winter day in Washington, D.C.   a
   a. 10°C   **b.** 30°C   **c.** 50°C   **d.** 70°C

78. A setting for a home thermostat.   a
   a. 20°C   **b.** 40°C   **c.** 60°C   **d.** 80°C

79. The oven temperature for cooking a roast.   c
   a. 80°C   **b.** 100°C   **c.** 175°C   **d.** 350°C

Exercises 80–81 give the average temperature of some of the world's hottest places. In each exercise, determine which location has the hotter average temperature and by how many degrees Fahrenheit. Round to the nearest tenth of a degree.

80. Assab, Eritrea: 86.8°F; Dalol, Ethiopia: 34.6°C   Dalol; ≈7.5°F

81. Berbera, Somalia: 86.2°F; Néma, Mauritania: 30.3°C
Néma; ≈0.3°F

Exercises 82–83 give the average temperature of some of the world's coldest places. In each exercise, determine which location has the colder average temperature and by how many degrees Celsius. Round to the nearest tenth of a degree.

82. Plateau, Antarctica: −56.7°C;
   Amundsen-Scott, Antarctica: −56.2°F   Plateau; 7.7°C

83. Eismitte, Greenland: −29.2°C;
   Resolute, Canada: −11.6°F   Eismitte; ≈5°C

(Source for Exercises 80–83: Russell Ash, *The Top 10 of Everything 2009*)

Critical Thinking Exercises

In Exercises 84–91, determine whether each statement is true or false. If the statement is false, make the necessary change(s) to produce a true statement.   85. false

84. A 4-pound object weighs more than a 2000-gram object.   false

85. A 100-milligram object weighs more than a 2-ounce object.

86. A 50-gram object weighs more than a 2-ounce object.   false

87. A 10-pound object weighs more than a 4-kilogram object.   true

88. Flour selling at 3¢ per gram is a better buy than flour selling at 55¢ per pound.   false

89. The measures

   32,600 g,   32.1 kg,   4 lb,   36 oz

   are arranged in order, from greatest to least weight.   true

90. If you are taking aspirin to relieve cold symptoms, a reasonable dose is 2 kilograms four times a day.   false

91. A large dog weighs about 350 kilograms.   false

Group Exercise

92. Present a group report on the current status of the metric system in the United States. At present, does it appear that the United States will convert to the metric system? Who supports the conversion and who opposes it? Summarize each side's position. Give examples of how our current system of weights and measures is an economic liability. What are the current obstacles to metric conversion?   Answers will vary.

Chapter 5 Summary

| **Definitions and Concepts** | **Examples** |

Section 5.1 Measuring Length; The Metric System

The result obtained from measuring length is called a linear measurement, stated in linear units.

Linear Units: The English System

   $12 \text{ in.} = 1 \text{ ft}, 3 \text{ ft} = 1 \text{ yd}, 36 \text{ in.} = 1 \text{ yd}, 5280 \text{ ft} = 1 \text{ mi}$

Dimensional Analysis

Multiply the given measurement by a unit fraction with the unit of measurement that needs to be introduced in the numerator and the unit of measurement that needs to be eliminated in the denominator.

- Convert 50 inches to feet.

$$50 \text{ in.} = \frac{50 \text{ in.}}{1} \cdot \frac{1 \text{ ft}}{12 \text{ in.}} = \frac{50}{12} \text{ ft} = 4\frac{1}{6} \text{ ft or } 4.1\overline{6} \text{ ft}$$

- Convert 4 inches to yards.

$$4 \text{ in.} = \frac{4 \text{ in.}}{1} \cdot \frac{1 \text{ yd}}{36 \text{ in.}} = \frac{4}{36} \text{ yd} = \frac{1}{9} \text{ yd or } 0.\overline{1} \text{ yd}$$

Additional Example to Review

Example 1, page 327

Linear Units: The Metric System

The basic unit is the meter (m), approximately 39 inches.

   $1 \text{ km} = 1000 \text{ m}, 1 \text{ hm} = 100 \text{ m}, 1 \text{ dam} = 10 \text{ m},$
   $1 \text{ dm} = 0.1 \text{ m}, 1 \text{ cm} = 0.01 \text{ m}, 1 \text{ mm} = 0.001 \text{ m}$

Changing Linear Units within the Metric System

$\times 10$

km hm dam m dm cm mm

$\div 10$

- $905 \text{ mm} = 0.00905 \text{ hm}$
- $86 \text{ m} = 8600 \text{ cm}$
- $57 \text{ m} = 0.057 \text{ km}$

Additional Concept and Example to Review

Approximate Metric Lengths, table on page 329;
Example 2, page 330

English and Metric Equivalents for Length

   $1 \text{ inch (in.)} = 2.54 \text{ centimeters (cm)}$
   $1 \text{ foot (ft)} = 30.48 \text{ centimeters (cm)}$
   $1 \text{ yard (yd)} \approx 0.9 \text{ meter (m)}$
   $1 \text{ mile (mi)} \approx 1.6 \text{ kilometers (km)}$

- Convert 381 centimeters to inches.

$$381 \text{ cm} = \frac{381 \text{ cm}}{1} \cdot \frac{1 \text{ in.}}{2.54 \text{ cm}} = \frac{381}{2.54} \text{ in.} = 150 \text{ in.}$$

- Convert 50 miles per hour to kilometers per hour.

$$\frac{50 \text{ mi}}{\text{hr}} \approx \frac{50 \text{ mi}}{\text{hr}} \cdot \frac{1.6 \text{ km}}{1 \text{ mi}} = 50(1.6)\frac{\text{km}}{\text{hr}} = 80 \text{ km/hr}$$

Additional Examples to Review

Example 3, page 332; Example 4, page 333

Section 5.2 Measuring Area and Volume

The area measure of a plane region is the number of square units contained in the given region.

Square Units of Measure: The English System

   $1 \text{ square foot (ft}^2) = 144 \text{ square inches (in.}^2)$
   $1 \text{ square yard (yd}^2) = 9 \text{ square feet (ft}^2)$
   $1 \text{ acre (a)} = 43{,}560 \text{ ft}^2 \text{ or } 4840 \text{ yd}^2$
   $1 \text{ square mile (mi}^2) = 640 \text{ acres}$

$$\text{Population density} = \frac{\text{population}}{\text{area}}$$

- Jamaica's population: 2,825,928
  Jamaica's area: 4244 mi^2
  Jamaica's population density

$$= \frac{\text{population}}{\text{area}} = \frac{2{,}825{,}928}{4244} \approx 665.9$$

In Jamaica, there is an average of approximately 665.9 people for each square mile of land.

Additional Examples to Review

Example 1, page 337; Example 2, page 338;
Example 3, page 338

English and Metric Equivalents for Area

1 square inch (in.^{2})	$\approx$ 6.5 square centimeters (cm^2)
1 square foot (ft^2)	$\approx$ 0.09 square meter (m^2)
1 square yard (yd^2)	$\approx$ 0.8 square meter (m^2)
1 square mile (mi^2)	$\approx$ 2.6 square kilometers (km^2)
1 acre	$\approx$ 0.4 hectare (ha)

- Convert 26 hectares to acres.

$$26 \text{ ha} \approx \frac{26 \text{ ha}}{1} \cdot \frac{1 \text{ acre}}{0.4 \text{ ha}} = \frac{26}{0.4} \text{ acres} = 65 \text{ acres}$$

Additional Example to Review

Example 4, page 339

The volume measure of a three-dimensional figure is the number of cubic units contained in its interior.

Capacity refers to the amount of fluid that a three-dimensional object can hold. English units for capacity include pints, quarts, and gallons: 2 pt = 1 qt; 4 qt = 1 gal; one cubic yard has a capacity of about 200 gallons.

English Units for Capacity

2 pints (pt) = 1 quart (qt)
4 quarts = 1 gallon (gal)
1 gallon = 128 ounces (oz)
1 cup (c) = 8 ounces

Volume in Cubic Units	Capacity
1 cubic yard	about 200 gallons
1 cubic foot	about 7.48 gallons
231 cubic inches	about 1 gallon

- A pool has a volume of 20,000 cubic feet. How many gallons of water does the pool hold?

$$20,000 \text{ ft}^3 \approx \frac{20,000 \text{ ft}^3}{1} \cdot \frac{7.48 \text{ gal}}{1 \text{ ft}^3} = 20,000(7.48) \text{ gal}$$
$$= 149,600 \text{ gal}$$

Additional Examples to Review

Example 5, page 340; Example 6, page 341

The basic unit for capacity in the metric system is the liter (L). One liter is about 1.06 quarts. Prefixes for the liter are the same as throughout the metric system.

Units of Capacity in the Metric System

Symbol	Unit	Meaning
kL	kiloliter	1000 liters
hL	hectoliter	100 liters
daL	dekaliter	10 liters
L	liter	1 liter $\approx$ 1.06 quarts
dL	deciliter	0.1 liter
cL	centiliter	0.01 liter
mL	milliliter	0.001 liter

Item	Capacity
Average cup of coffee	250 mL
12-ounce can of soda	355 mL
One quart of fruit juice	0.95 L
One gallon of milk	3.78 L
Average gas tank capacity of a car (about 18.5 gallons)	70 L

Volume and Capacity in the Metric System

Volume in Cubic Units		Capacity
1 cm^3	=	1 mL
1 dm^3 = 1000 cm^3	=	1 L
1 m^3	=	1 kL

- A pond has a volume of 300,000 cubic centimeters. How many liters of water does the pond hold?

$$300{,}000 \text{ cm}^3 = \frac{300{,}000 \text{ cm}^3}{1} \cdot \frac{1 \text{ L}}{1000 \text{ cm}^3} = \frac{300{,}000}{1000} \text{ L} = 300 \text{ L}$$

Additional Example to Review

Example 7, page 342

Section 5.3 Measuring Weight and Temperature

Weight: The English System

$$16 \text{ oz} = 1 \text{ lb}, \quad 2000 \text{ lb} = 1 \text{ T}$$

Units of Weight: The Metric System

The basic unit is the gram (g).

$$1000 \text{ grams } (1 \text{ kg}) \approx 2.2 \text{ lb},$$

$$1 \text{ kg} = 1000 \text{ g}, 1 \text{ hg} = 100 \text{ g}, 1 \text{ dag} = 10 \text{ g},$$

$$1 \text{ dg} = 0.1 \text{ g}, 1 \text{ cg} = 0.01 \text{ g}, 1 \text{ mg} = 0.001 \text{ g}$$

Changing Units of Weight within the Metric System

×10

kg   hg   dag   g   dg   cg   mg

÷10

One kilogram of water has a volume of 1 liter.

Volume and Weight of Water in the Metric System

Volume		Capacity		Weight
1 cm³	=	1 mL	=	1 g
1 dm³ = 1000 cm³	=	1 L	=	1 kg
1 m³	=	1 kL	=	1000 kg = 1 t

English and Metric Equivalents for Weight

1 ounce (oz) ≈ 28 grams (g)	
1 pound (lb) ≈ 0.45 kilogram (kg)	
1 ton (T) ≈ 0.9 tonne (t)	

- 1285 g = 1.285 kg
- 4.3 dg = 430 mg

Additional Example to Review

Example 1, page 347

- An aquarium holds 0.35 m³ of water. How much does the water weigh?

$$0.35 \text{ m}^3 = \frac{0.35 \text{ m}^3}{1} \cdot \frac{1000 \text{ kg}}{1 \text{ m}^3}$$

$$= 0.35(1000) \text{ kg} = 350 \text{ kg}$$

Additional Example to Review

Example 2, page 348

- Convert 300 pounds to kilograms.

$$300 \text{ lb} \approx \frac{300 \text{ lb}}{1} \cdot \frac{0.45 \text{ kg}}{1 \text{ lb}} = 300(0.45) \text{ kg} = 135 \text{ kg}$$

- Convert 196 grams to ounces.

$$196 \text{ g} \approx \frac{196 \text{ g}}{1} \cdot \frac{1 \text{ oz}}{28 \text{ g}} = \frac{196}{28} \text{ oz} = 7 \text{ oz}$$

Additional Example to Review

Example 3, page 348

Converting between Temperature Scales

Celsius to Fahrenheit: $F = \frac{9}{5}C + 32$

Fahrenheit to Celsius: $C = \frac{5}{9}(F - 32)$

- Convert 20°C from °C to °F.

$$F = \frac{9}{5}C + 32 = \frac{9}{5}(20) + 32 = 36 + 32 = 68: \quad 20°C = 68°F$$

- Convert −49°F from °F to °C.

$$C = \frac{5}{9}(F - 32) = \frac{5}{9}(-49 - 32) = \frac{5}{9}(-81) = \frac{5}{\underset{1}{9}}\left(\frac{\overset{9}{-81}}{1}\right)$$

$$= -45: \quad -49°F = -45°C$$

Additional Examples to Review

Example 4, page 350; Example 5, page 351

Review Exercises

Section 5.1 Measuring Length; The Metric System

In Exercises 1–4, use dimensional analysis to convert the quantity to the indicated unit.

1. 69 in. to ft 5.75 ft

2. 9 in. to yd 0.25 yd

3. 21 ft to yd 7 yd

4. 13,200 ft to mi 2.5 mi

In Exercises 5–10, convert the given linear measurement to the metric unit indicated.

5. 22.8 m to cm 2280 cm

6. 7 dam to m 70 m

7. 19.2 hm to m 1920 m

8. 144 cm to hm 0.0144 hm

9. 0.5 mm to m 0.0005 m

10. 18 cm to mm 180 mm

In Exercises 11–16, use the given English and metric equivalents, along with dimensional analysis, to convert the given measurement to the unit indicated. Where necessary, round answers to two decimal places.

11. 23 in. to cm 58.42 cm

12. 19 cm to in. ≈7.48 in.

13. 330 mi to km ≈528 km

14. 600 km to mi ≈375 mi

15. 14 m to yd ≈15.56 yd

16. 12 m to ft ≈39.37 ft

| 1 in. = 2.54 cm |
| 1 ft = 30.48 cm |
| 1 yd ≈ 0.9 m |
| 1 mi ≈ 1.6 km |

17. Express 45 kilometers per hour in miles per hour. ≈28.13 mi/hr

18. Express 60 miles per hour in kilometers per hour. ≈96 km/hr

19. Arrange from smallest to largest: 0.024 km, 2400 m, 24,000 cm. 0.024 km, 24,000 cm, 2400 m

20. If you jog six times around a track that is 800 meters long, how many kilometers have you covered? 4.8 km

Section 5.2 Measuring Area and Volume

21. Use the given figure to find its area in square units. 24 square units

22. Singapore, with an area of 268 square miles and a population of 4,425,700, is one of the world's most densely populated countries. Find Singapore's population density, to the nearest tenth. Describe what this means. 16,513.8*

23. Acadia National Park on the coast of Maine consists of 47,453 acres. How large is the national park in square miles? Round to the nearest square mile. (1 mi² = 640 a) 74 mi²

24. Given 1 acre ≈ 0.4 hectare, use dimensional analysis to find the size of a property in acres measured at 7.2 hectares. ≈18 acres

25. Using 1 ft² ≈ 0.09 m², convert 30 m² to ft². ≈333.33 ft²

26. Using 1 mi² ≈ 2.6 km², convert 12 mi² to km². ≈31.2 km²

27. Which one of the following is a reasonable measure for the area of a flower garden in a person's yard? a
  a. 100 m²   **b.** 0.4 ha   **c.** 0.01 km²

28. Use the given figure to find its volume in cubic units. 24 cubic units

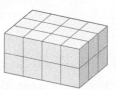

29. A swimming pool has a volume of 33,600 cubic feet. Given that 1 cubic foot has a capacity of about 7.48 gallons, how many gallons of water does the pool hold? ≈251,328 gal

30. An aquarium has a volume of 76,000 cubic centimeters. How many liters of water does the aquarium hold? 76 L

31. The capacity of a one-quart container of juice is approximately c
  a. 0.1 kL   **b.** 0.5 L   **c.** 1 L   **d.** 1 mL

32. There are 3 feet in a yard. Explain why there are not 3 square feet in a square yard. If helpful, illustrate your explanation with a diagram. Answers will vary.

33. Explain why the area of Texas could not be measured in cubic miles. Answers will vary.

Section 5.3 Measuring Weight and Temperature

In Exercises 34–37, convert the given weight to the unit indicated.

34. 12.4 dg to mg   1240 mg   **35.** 12 g to cg   1200 cg

36. 0.012 mg to g   0.000012 g   **37.** 450 mg to kg   0.00045 kg

*In Exercises 38–39, use **Table 5.9** on page 348 to convert the given measurement of water to the unit or units indicated.*   **40.** ≈ 94.5 kg

38. 50 kg to cm³   50,000 cm³

39. 4 kL to dm³ to g   4000 dm³, 4,000,000 g

40. Using 1 lb ≈ 0.45 kg, convert 210 pounds to kilograms.

41. Using 1 oz ≈ 28 g, convert 392 grams to ounces.   ≈ 14 oz

42. If you are interested in your weight in the metric system, would it be best to report it in milligrams, grams, or kilograms? Explain why the unit you selected would be most appropriate. Explain why each of the other two units is not the best choice for reporting your weight.   kilograms; Answers will vary.

43. Given 16 oz = 1 lb, use dimensional analysis to convert 36 ounces to pounds.   2.25 lb

In Exercises 44–45, select the best estimate for the weight of the given item.

44. A dollar bill: a
   a. 1 g   **b.** 10 g   **c.** 1 kg   **d.** 4 kg

45. A hamburger: c
   a. 3 kg   **b.** 1 kg   **c.** 200 g   **d.** 5 g

In Exercises 46–50, convert the given Celsius temperature to its equivalent temperature on the Fahrenheit scale.

46. 15°C   59°F   **47.** 100°C   212°F

48. 5°C   41°F   **49.** 0°C   32°F

50. −25°C   −13°F

In Exercises 51–56, convert the given Fahrenheit temperature to its equivalent temperature on the Celsius scale.

51. 59°F   15°C   **52.** 41°F   5°C   **53.** 212°F   100°C

54. 98.6°F   37°C   **55.** 0°F   ≈ −17.8°C   **56.** 14°F   −10°C

57. Is a decrease of 15° Celsius more or less than a decrease of 15° Fahrenheit? Explain your answer.   more; Answers will vary.

Chapter 5 Test

1. Change 807 mm to hm.   0.00807 hm

2. Given 1 inch = 2.54 centimeters, use dimensional analysis to change 635 centimeters to inches.   250 in.

3. If you jog eight times around a track that is 600 meters long, how many kilometers have you covered?   4.8 km

In Exercises 4–6, write the most reasonable metric unit for length in each blank. Select from mm, cm, m, and km.

4. A human thumb is 20 ____mm____ wide.

5. The height of the table is 45 ____cm____.

6. The towns are 60 ____km____ apart.

7. If 1 mile ≈ 1.6 kilometers, express 80 miles per hour in kilometers per hour.   ≈ 128 km/hr

8. How many times greater is a square yard than a square foot?   9 times

9. Australia has a population of 20,090,400 and an area of 2,967,908 square miles. Find Australia's population density to the nearest tenth. Describe what this means.

10. Given 1 acre ≈ 0.4 hectare, use dimensional analysis to find the area of a property measured at 18 hectares.   ≈ 45 acres

11. The area of a dollar bill is approximately   b
   a. 10 cm²   **b.** 100 cm²   **c.** 1000 cm²   **d.** 1 m²

12. There are 10 decimeters in a meter. Explain why there are not 10 cubic decimeters in a cubic meter. How many times greater is a cubic meter than a cubic decimeter?

13. A swimming pool has a volume of 10,000 cubic feet. Given that 1 cubic foot has a capacity of about 7.48 gallons, how many gallons of water does the pool hold?   ≈ 74,800 gal

14. The capacity of a pail used to wash floors is approximately   b
   a. 3 L   **b.** 12 L   **c.** 80 L   **d.** 2 kL

15. Change 137 g to kg.   0.137 kg

16. Using 1 lb ≈ 0.45 kg, convert 90 pounds to kilograms.   ≈ 40.5 kg

In Exercises 17–18, write the most reasonable metric unit for weight in each blank. Select from mg, g, and kg.

17. My suitcase weighs 20 ____kg____.

18. I took a 350 ____mg____ aspirin.

19. Convert 30°C to Fahrenheit.   86°F

20. Convert 176°F to Celsius.   80°C

21. Comfortable room temperature is approximately   d
   a. 70°C   **b.** 50°C   **c.** 30°C   **d.** 20°C

9. 6.8 people per square mile; In Australia, there is an average of 6.8 people for each square mile.   **12.** Answers will vary.; 1000 times

Geometric Pathways

Geometry is the study of the space you live in and the shapes that surround you. You're even made of it! The human lung consists of nearly 300 spherical air sacs, geometrically designed to provide the greatest surface area within the limited volume of our bodies. Viewed in this way, geometry becomes an intimate experience.

For thousands of years, people have studied geometry in some way to obtain a better understanding of the world in which they live. A study of the shape of your world will provide you with many practical applications and perhaps help to increase your appreciation of its beauty.

Here's where you'll find these applications:

- Applications of geometry are presented throughout the chapter.
- A relationship between geometry and the visual arts is developed in Section 6.3 (Tessellations: pages 386–387).

6.1   Points, Lines, Planes, and Angles

What am I supposed to learn?

After you have read this section, you should be able to:

1 Understand points, lines, and planes as the basis of geometry.

2 Solve problems involving angle measures.

3 Solve problems involving angles formed by parallel lines and transversals.

The San Francisco Museum of Modern Art was constructed in 1995 to illustrate how art and architecture can enrich one another. The exterior involves geometric shapes, symmetry, and unusual facades. Although there are no windows, natural light streams in through a truncated cylindrical skylight that crowns the building. The architect worked with a scale model of the museum at the site and observed how light hit it during different times of the day. These observations were used to cut the cylindrical skylight at an angle that maximizes sunlight entering the interior.

Angles play a critical role in creating modern architecture. They are also fundamental in the study of geometry. The word "geometry" means "earth measure." Because it involves the mathematics of shapes, geometry connects mathematics to art and architecture. It also has many practical applications. You can use geometry at home when you buy carpet, build a fence, tile a floor, or determine whether a piece of furniture will fit through your doorway. In this chapter, we look at the shapes that surround us and their applications.

Points, Lines, and Planes

Points, lines, and planes make up the basis of all geometry. Stars in the night sky look like points of light. Long stretches of overhead power lines that appear to extend endlessly look like lines. The top of a flat table resembles part of a plane. However, stars, power lines, and tabletops only approximate points, lines, and planes. Points, lines, and planes do not exist in the physical world. Representations of these forms are shown in **Figure 6.1**. A **point**, represented as a small dot, has no length, width, or thickness. No object in the real world has zero size. A **line**, connecting two points along the shortest possible path, has no thickness and extends infinitely in both directions. However, no familiar everyday object is infinite in length. A **plane** is a flat surface with no thickness and no boundaries. This page resembles a plane, although it does not extend indefinitely and it does have thickness.

A line may be named using any two of its points. In **Figure 6.2(a)**, line AB can be symbolized $\overleftrightarrow{AB}$ or $\overleftrightarrow{BA}$. Any point on the line divides the line into three parts—the point and two **half-lines**. **Figure 6.2(b)** illustrates half-line AB, symbolized $\overset{\circ}{AB}$. The open circle above the A in the symbol and in the diagram indicates that point A is not included in the half-line. A **ray** is a half-line with its endpoint included. **Figure 6.2(c)** illustrates ray AB, symbolized $\overrightarrow{AB}$. The closed dot above the A in the diagram shows that point A is included in the ray. A portion of a line joining two points and including the endpoints is called a **line segment**. **Figure 6.2(d)** illustrates line segment AB, symbolized $\overline{AB}$ or $\overline{BA}$.

1 Understand points, lines, and planes as the basis of geometry.

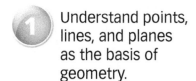

Point A

Line AB

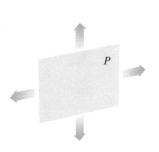

Plane P

Figure 6.1 Representing a point, a line, and a plane

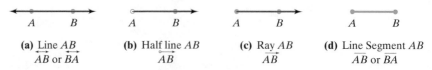

(a) Line AB
$\overleftrightarrow{AB}$ or $\overleftrightarrow{BA}$

(b) Half line AB
$\overset{\circ}{AB}$

(c) Ray AB
$\overrightarrow{AB}$

(d) Line Segment AB
$\overline{AB}$ or $\overline{BA}$

Figure 6.2 Lines, half-lines, rays, and line segments

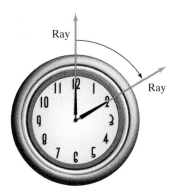

Figure 6.3 Clock with hands forming an angle

Angles

An **angle**, symbolized ∡, is formed by the union of two rays that have a common endpoint. One ray is called the **initial side** and the other the **terminal side**.

A rotating ray is often a useful way to think about angles. The ray in **Figure 6.3** rotates from 12 to 2. The ray pointing to 12 is the initial side and the ray pointing to 2 is the terminal side. The common endpoint of an angle's initial side and terminal side is the **vertex** of the angle.

Figure 6.4 shows an angle. The common endpoint of the two rays, B, is the vertex. The two rays that form the angle, $\overrightarrow{BA}$ and $\overrightarrow{BC}$, are the sides. Four ways of naming the angle are shown to the right of **Figure 6.4**.

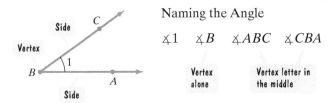

Figure 6.4 An angle: two rays with a common endpoint

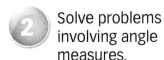

Solve problems involving angle measures.

Measuring Angles Using Degrees

Angles are measured by determining the amount of rotation from the initial side to the terminal side. One way to measure angles is in **degrees**, symbolized by a small, raised circle °. Think of the hour hand of a clock. From 12 noon to 12 midnight, the hour hand moves around in a complete circle. By definition, the ray has rotated through 360 degrees, or 360°, shown in **Figure 6.5**. Using 360° as the amount of rotation of a ray back onto itself, **a degree, 1°, is $\frac{1}{360}$ of a complete rotation**.

Figure 6.5 A complete 360° rotation

Example 1 **Using Degree Measure**

The hand of a clock moves from 12 to 2 o'clock, shown in **Figure 6.3**. Through how many degrees does it move?

Solution

We know that one complete rotation is 360°. Moving from 12 to 2 o'clock is $\frac{2}{12}$, or $\frac{1}{6}$, of a complete revolution. Thus, the hour hand moves

$$\frac{1}{6} \times 360° = \frac{360°}{6} = 60°$$

in going from 12 to 2 o'clock.

Check Point 1 The hand of a clock moves from 12 to 1 o'clock. Through how many degrees does it move? 30°

Figure 6.6 shows angles classified by their degree measurement. An **acute angle** measures less than 90° [see **Figure 6.6(a)**]. A **right angle**, one-quarter of a complete rotation, measures 90° [**Figure 6.6(b)**]. Examine the right angle—do you see a small square at the vertex? This symbol is used to indicate a right angle. An **obtuse angle** measures more than 90°, but less than 180° [**Figure 6.6(c)**]. Finally, a **straight angle**, one-half a complete rotation, measures 180° [**Figure 6.6(d)**]. The two rays in a straight angle form a straight line.

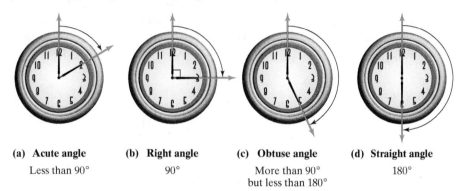

Figure 6.6 Classifying angles by their degree measurements

(a) **Acute angle**
Less than 90°

(b) **Right angle**
90°

(c) **Obtuse angle**
More than 90°
but less than 180°

(d) **Straight angle**
180°

Figure 6.7 illustrates a **protractor**, used for finding the degree measure of an angle. As shown in the figure, we measure an angle by placing the center point of the protractor on the vertex of the angle and the straight side of the protractor along one side of the angle. The measure of ∡ABC is then read as 50°. Observe that the measure is not 130° because the angle is obviously less than 90°. We indicate the angle's measure by writing $m\angle ABC = 50°$, read "the measure of angle ABC is 50°."

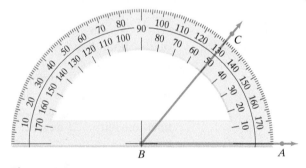

Figure 6.7 Using a protractor to measure an angle:
$m\angle ABC = 50°$

Two angles whose measures have a sum of 90° are called **complementary angles**. For example, angles measuring 70° and 20° are complementary angles because 70° + 20° = 90°. For angles such as those measuring 70° and 20°, each angle is the **complement** of the other: The 70° angle is the complement of the 20° angle and the 20° angle is the complement of the 70° angle. **The measure of the complement can be found by subtracting the angle's measure from 90°.** For example, we can find the complement of a 25° angle by subtracting 25° from 90°: 90° − 25° = 65°. Thus, an angle measuring 65° is the complement of one measuring 25°.

Example 2   Angle Measures and Complements

Use **Figure 6.8(a)** to find $m\angle DBC$.

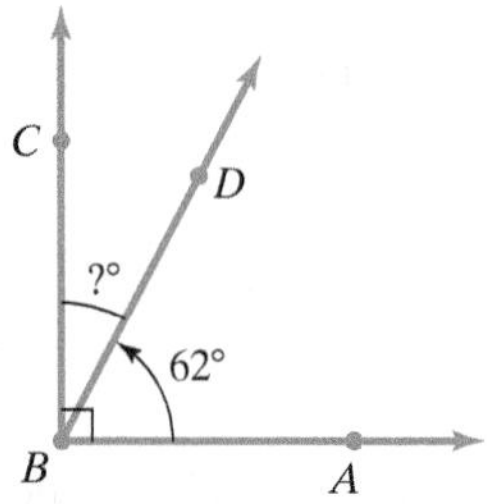

Solution

The measure of ∡DBC is not yet known. It is shown as ?° in **Figure 6.8(a)**. The acute angles ∡ABD, which measures 62°, and ∡DBC form a right angle, indicated by the square at the vertex. This means that the measures of the acute angles add up to 90°.

Figure 6.8(a)

Thus, $\angle DBC$ is the complement of the angle measuring 62°. The measure of $\angle DBC$ is found by subtracting 62° from 90°:

$$m\angle DBC = 90° - 62° = 28°.$$

The measure of $\angle DBC$ can also be found using an algebraic approach.

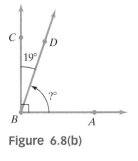

Figure 6.8(b)

$m\angle ABD + m\angle DBC = 90°$ The sum of the measures of complementary angles is 90°.

$62° + m\angle DBC = 90°$ We are given $m\angle ABD = 62°$.

$m\angle DBC = 90° - 62° = 28°$ Subtract 62° from both sides of the equation.

 **Check Point 2** In **Figure 6.8(b)**, suppose that $m\angle DBC = 19°$. Find $m\angle DBA$.
71°

Two angles whose measures have a sum of 180° are called **supplementary angles**. For example, angles measuring 110° and 70° are supplementary angles because $110° + 70° = 180°$. For angles such as those measuring 110° and 70°, each angle is the **supplement** of the other: The 110° angle is the supplement of the 70° angle, and the 70° angle is the supplement of the 110° angle. **The measure of the supplement can be found by subtracting the angle's measure from 180°.** For example, we can find the supplement of a 25° angle by subtracting 25° from 180°: $180° - 25° = 155°$. Thus, an angle measuring 155° is the supplement of one measuring 25°.

Example 3   Angle Measures and Supplements

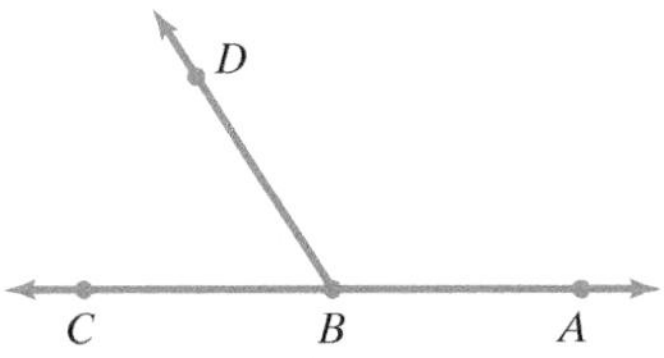

Figure 6.9

Figure 6.9 shows that $\angle ABD$ and $\angle DBC$ are supplementary angles. If $m\angle ABD$ is 66° greater than $m\angle DBC$, find the measure of each angle.

Solution

Let $m\angle DBC = x$. Because $m\angle ABD$ is 66° greater than $m\angle DBC$, then $m\angle ABD = x + 66°$. We are given that these angles are supplementary.

$m\angle DBC + m\angle ABD = 180°$ The sum of the measures of supplementary angles is 180°.

$x + (x + 66°) = 180°$ Substitute the variable expressions for the measures.

$2x + 66° = 180°$ Combine like terms: $x + x = 2x$.

$2x = 114°$ Subtract 66° from both sides.

$x = 57°$ Divide both sides by 2.

Thus, $m\angle DBC = 57°$ and $m\angle ABD = 57° + 66° = 123°$.

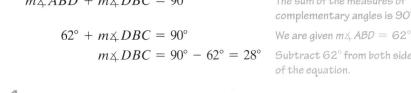 **Check Point 3** In **Figure 6.9**, if $m\angle ABD$ is 88° greater than $m\angle DBC$, find the measure of each angle. 46°; 134°

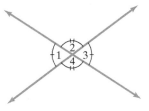

Figure 6.10

Figure 6.10 illustrates a highway sign that warns of a railroad crossing. When two lines intersect, the opposite angles formed are called **vertical angles**.

In **Figure 6.11**, there are two pairs of vertical angles. Angles 1 and 3 are vertical angles. Angles 2 and 4 are also vertical angles.

Figure 6.11

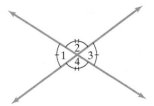

Figure 6.11 (repeated)

We can use **Figure 6.11** to show that vertical angles have the same measure. Let's concentrate on angles 1 and 3, each denoted by one tick mark. Can you see that each of these angles is supplementary to angle 2?

$$m\angle 1 + m\angle 2 = 180°$$ The sum of the measures of
$$m\angle 2 + m\angle 3 = 180°$$ supplementary angles is 180°.
$$m\angle 1 + m\angle 2 = m\angle 2 + m\angle 3$$ Substitute $m\angle 2 + m\angle 3$ for 180° in the first equation.
$$m\angle 1 = m\angle 3$$ Subtract $m\angle 2$ from both sides.

Using a similar approach, we can show that $m\angle 2 = m\angle 4$, each denoted by two tick marks in **Figure 6.11**.

> Vertical angles have the same measure.

Example 4   Using Vertical Angles

Figure 6.12 shows that the angle on the left measures 68°. Find the measures of the other three angles.

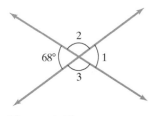

Figure 6.12

Solution

Angle 1 and the angle measuring 68° are vertical angles. Because vertical angles have the same measure,

$$m\angle 1 = 68°.$$

Angle 2 and the angle measuring 68° form a straight angle and are supplementary. Because their measures add up to 180°,

$$m\angle 2 = 180° - 68° = 112°.$$

Angle 2 and angle 3 are also vertical angles, so they have the same measure. Because the measure of angle 2 is 112°,

$$m\angle 3 = 112°.$$

✓ **Check Point 4**   In **Figure 6.12**, suppose that the angle on the left measures 57°. ▶ Find the measures of the other three angles. $m\angle 1 = 57°, m\angle 2 = 123°$, and $m\angle 3 = 123°$

Parallel Lines

Parallel lines are lines that lie in the same plane and have no points in common. If two different lines in the same plane are not parallel, they have a single point in common and are called **intersecting lines**. If the lines intersect at an angle of 90°, they are called **perpendicular lines**.

If we intersect a pair of parallel lines with a third line, called a **transversal**, eight angles are formed, as shown in **Figure 6.13**. Certain pairs of these angles have special names, as well as special properties. These names and properties are summarized in **Table 6.1**.

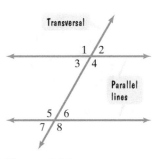

Figure 6.13

Table 6.1 Names of Angle Pairs Formed by a Transversal Intersecting Parallel Lines

Name	Description	Sketch	Angle Pairs Described	Property
Alternate interior angles	Interior angles that do not have a common vertex on alternate sides of the transversal		$\angle 3$ and $\angle 6$ $\angle 4$ and $\angle 5$	Alternate interior angles have the same measure. $m\angle 3 = m\angle 6$ $m\angle 4 = m\angle 5$
Alternate exterior angles	Exterior angles that do not have a common vertex on alternate sides of the transversal		$\angle 1$ and $\angle 8$ $\angle 2$ and $\angle 7$	Alternate exterior angles have the same measure. $m\angle 1 = m\angle 8$ $m\angle 2 = m\angle 7$
Corresponding angles	One interior and one exterior angle on the same side of the transversal		$\angle 1$ and $\angle 5$ $\angle 2$ and $\angle 6$ $\angle 3$ and $\angle 7$ $\angle 4$ and $\angle 8$	Corresponding angles have the same measure. $m\angle 1 = m\angle 5$ $m\angle 2 = m\angle 6$ $m\angle 3 = m\angle 7$ $m\angle 4 = m\angle 8$

③ Solve problems involving angles formed by parallel lines and transversals.

When two parallel lines are intersected by a transversal, the following relationships are true:

> **Parallel Lines and Angle Pairs**
>
> If parallel lines are intersected by a transversal,
>
> - alternate interior angles have the same measure,
> - alternate exterior angles have the same measure, and
> - corresponding angles have the same measure.
>
> Conversely, if two lines are intersected by a third line and a pair of alternate interior angles or a pair of alternate exterior angles or a pair of corresponding angles have the same measure, then the two lines are parallel.

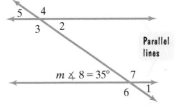

Parallel lines

$m\angle 8 = 35°$

Figure 6.14

Example 5   **Finding Angle Measures when Parallel Lines Are Intersected by a Transversal**

In **Figure 6.14**, two parallel lines are intersected by a transversal. One of the angles ($\angle 8$) has a measure of 35°. Find the measure of each of the other seven angles.

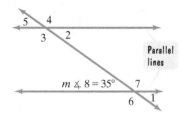

Figure 6.14 (repeated)

Solution

Look carefully at **Figure 6.14** and fill in the angle measures as you read each line in this solution.

$m\angle 1 = 35°$ $\angle 8$ and $\angle 1$ are vertical angles and vertical angles have the same measure.

$m\angle 6 = 180° - 35° = 145°$ $\angle 8$ and $\angle 6$ are supplementary.

$m\angle 7 = 145°$ $\angle 6$ and $\angle 7$ are vertical angles, so they have the same measure.

$m\angle 2 = 35°$ $\angle 8$ and $\angle 2$ are alternate interior angles, so they have the same measure.

$m\angle 3 = 145°$ $\angle 7$ and $\angle 3$ are alternate interior angles. Thus, they have the same measure.

$m\angle 5 = 35°$ $\angle 8$ and $\angle 5$ are corresponding angles. Thus, they have the same measure.

$m\angle 4 = 180° - 35° = 145°$ $\angle 4$ and $\angle 5$ are supplementary.

✔ Check Point 5   In **Figure 6.14**, suppose that $m\angle 8 = 29°$. Find the measure of each of the other seven angles.

$m\angle 1 = 29°, m\angle 2 = 29°, m\angle 3 = 151°, m\angle 4 = 151°, m\angle 5 = 29°, m\angle 6 = 151°,$ and $m\angle 7 = 151°$

Great Question!

Is there more than one way to solve Example 5?

Yes. For example, once you know that $m\angle 2 = 35°$, $m\angle 5$ is also $35°$ because $\angle 2$ and $\angle 5$ are vertical angles.

Achieving Success

Write down all the steps.

In this textbook, examples are written that provide step-by-step solutions. No steps are omitted and each step is explained to the right of the mathematics. Some professors are careful to write down every step of a problem as they are doing it in class; others aren't so fastidious. In either case, write down what the professor puts up and when you get home, fill in whatever steps have been omitted (if any). In your math work, including homework and tests, show clear step-by-step solutions. Detailed solutions help organize your thoughts and enhance understanding. Doing too many steps mentally often results in preventable mistakes.

Concept and Vocabulary Check

Exercises in the Concept and Vocabulary Check are intended for group and class discussions.

In Exercises 1–6, fill in each blank so that the resulting statement is true.

1. $\overleftrightarrow{AB}$ symbolizes _____line_____ AB, $\overrightarrow{AB}$ symbolizes _____half-line_____ AB, $\overrightarrow{AB}$ symbolizes _____ray_____ AB, and $\overline{AB}$ symbolizes _____line segment_____ AB.

2. A/an _____acute_____ angle measures less than 90°, a/an _____right_____ angle measures 90°, a/an _____obtuse_____ angle measures more than 90° and less than 180°, and a/an _____straight_____ angle measures 180°.

3. Two angles whose measures have a sum of 90° are called _____complementary_____ angles. Two angles whose measures have a sum of 180° are called _____supplementary_____ angles.

4. When two lines intersect, the opposite angles are called _____vertical_____ angles.

5. Lines that lie in the same plane and have no points in common are called _____parallel_____ lines. If these lines are intersected by a third line, called a _____transversal_____, eight angles are formed.

6. Lines that intersect at an angle of 90° are called _____perpendicular_____ lines.

In Exercises 7–12, determine whether each statement is true or false. If the statement is false, make the necessary change(s) to produce a true statement.   Changes to false statements will vary.

7. A ray extends infinitely in both directions.   false

8. An angle is formed by the union of two lines that have a common endpoint.   false

9. A degree, 1°, is $\frac{1}{90}$ of a complete rotation.   false

10. A ruler is used for finding the degree measure of an angle.   false

11. The measure of an angle's complement is found by subtracting the angle's measure from 90°.   true

12. Vertical angles have the same measure.   true

Respond to Exercises 13–20 using verbal or written explanations. **13–20.** Answers will vary.

13. Describe the differences among lines, half-lines, rays, and line segments.

14. What is an angle and what determines its size?

15. Describe each type of angle: acute, right, obtuse, and straight.

16. What are complementary angles? Describe how to find the measure of an angle's complement.

17. What are supplementary angles? Describe how to find the measure of an angle's supplement.

18. Describe the difference between perpendicular and parallel lines.

19. If two parallel lines are intersected by a transversal, describe the location of the alternate interior angles, the alternate exterior angles, and the corresponding angles.

20. If a transversal is perpendicular to one of two parallel lines, must it be perpendicular to the other parallel line as well? Explain your answer.

Exercise Set 6.1

Practice Exercises

1. The hour hand of a clock moves from 12 to 5 o'clock. Through how many degrees does it move?   150°

2. The hour hand of a clock moves from 12 to 4 o'clock. Through how many degrees does it move?   120°

3. The hour hand of a clock moves from 1 to 4 o'clock. Through how many degrees does it move?   90°

4. The hour hand of a clock moves from 1 to 7 o'clock. Through how many degrees does it move?   180°

In Exercises 5–10, use the protractor to find the measure of each angle. Then classify the angle as acute, right, straight, or obtuse.

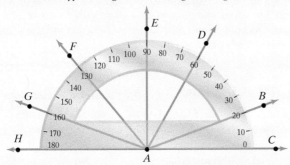

5. $\angle CAB$ 20°; acute

6. $\angle CAF$ 130°; obtuse

7. $\angle HAB$ 160°; obtuse

8. $\angle HAF$ 50°; acute

9. $\angle CAH$ 180°; straight

10. $\angle HAE$ 90°; right

In Exercises 11–14, find the measure of the angle in which a question mark with a degree symbol appears.

11.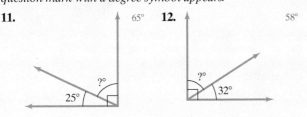

12.

13.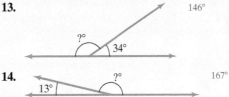

14.

In Exercises 15–20, find the measure of the complement and the supplement of each angle.

15. 48°   42°; 132°

16. 52°   38°; 128°

17. 89°   1°; 91°

18. 1°   89°; 179°

19. 37.4°   52.6°; 142.6°

20. $15\frac{1}{3}°$ $74\frac{2}{3}°; 164\frac{2}{3}°$

In Exercises 21–24, use an algebraic equation to find the measures of the two angles described. Begin by letting x represent the degree measure of the angle's complement or its supplement.

21. The measure of the angle is 12° greater than its complement.   *

22. The measure of the angle is 56° greater than its complement.   *

23. The measure of the angle is three times greater than its supplement. $x + 3x = 180°; 45°, 135°$

24. The measure of the angle is 81° more than twice that of its supplement. $x + 2x + 81° = 180°; 33°, 147°$

In Exercises 25–28, find the measures of angles 1, 2, and 3.

25.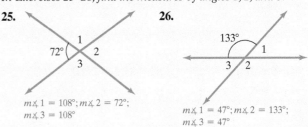

$m\angle 1 = 108°; m\angle 2 = 72°;$
$m\angle 3 = 108°$

26.

$m\angle 1 = 47°; m\angle 2 = 133°;$
$m\angle 3 = 47°$

27.

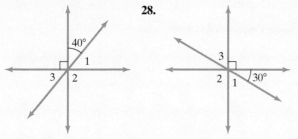

$m\angle 1 = 50°; m\angle 2 = 90°; m\angle 3 = 50°$ $m\angle 1 = 60°; m\angle 2 = 90°; m\angle 3 = 60°$

The figures for Exercises 29–30 show two parallel lines intersected by a transversal. One of the angle measures is given. Find the measure of each of the other seven angles.

29.   * **30.**

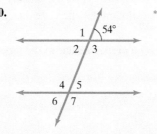

The figures for Exercises 31–34 show two parallel lines intersected by more than one transversal. Two of the angle measures are given. Find the measures of angles 1, 2, and 3.

31.   **32.**

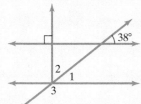

$m\angle 1 = 38°; m\angle 2 = 52°; m\angle 3 = 142°$ $m\angle 1 = 90°; m\angle 2 = 50°; m\angle 3 = 130°$

33.   **34.**

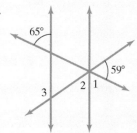

$m\angle 1 = 65°; m\angle 2 = 56°; m\angle 3 = 124°$ $m\angle 1 = 63°; m\angle 2 = 57°; m\angle 3 = 120°$

Use the following figure to determine whether each statement in Exercises 35–38 is true or false.

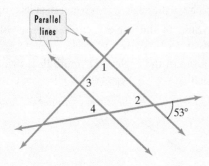

35. $m\angle 2 = 37°$ false   **36.** $m\angle 1 = m\angle 2$ false

37. $m\angle 4 = 53°$ true   **38.** $m\angle 3 = m\angle 4$ false

See Answers to Selected Exercises.

Use the following figure to determine whether each statement in Exercises 39–42 is true or false.

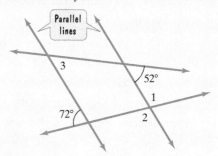

39. $m\angle 1 = 38°$ false   **40.** $m\angle 1 = 108°$ true

41. $m\angle 2 = 52°$ false   **42.** $m\angle 3 = 72°$ false

Practice Plus

In Exercises 43–46, use an algebraic equation to find the measure of each angle that is represented in terms of x.

43.   **44.**

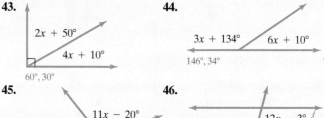

60°, 30°   146°, 34°

45.   **46.**

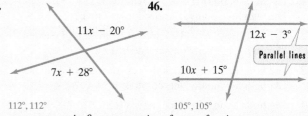

112°, 112°   105°, 105°

Because geometric figures consist of sets of points, we can apply set operations to obtain the union, ∪, or the intersection, ∩, of such figures. (See Appendix D.) The union of two geometric figures is the set of points that belongs to either of the figures or to both figures. The intersection of two geometric figures is the set of points common to both figures. In Exercises 47–54, use the line shown to find each set of points.

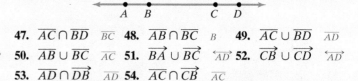

47. $\overline{AC} \cap \overline{BD}$ $\overline{BC}$ **48.** $\overline{AB} \cap \overline{BC}$ B **49.** $\overline{AC} \cup \overline{BD}$ $\overline{AD}$

50. $\overline{AB} \cup \overline{BC}$ $\overline{AC}$ **51.** $\overrightarrow{BA} \cup \overrightarrow{BC}$ $\overleftrightarrow{AD}$ **52.** $\overrightarrow{CB} \cup \overrightarrow{CD}$ $\overleftrightarrow{AD}$

53. $\overrightarrow{AD} \cap \overrightarrow{DB}$ $\overline{AD}$ **54.** $\overrightarrow{AC} \cap \overrightarrow{CB}$ $\overline{AC}$

Application Exercises

55. The picture shows the top of an umbrella in which all the angles formed by the spokes have the same measure. Find the measure of each angle. 45°

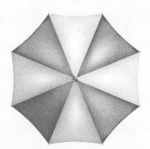

56. In the musical *Company*, composer Stephen Sondheim describes the marriage between two of the play's characters as "parallel lines who meet." What is the composer saying about this relationship? Answers will vary.

57. The picture shows a window with parallel framing in which snow has collected in the corners. What property of parallel lines is illustrated by where the snow has collected? *

In Exercises 58–59, consider the following uppercase letters from the English alphabet:

A E F H N T X Z.

58. Which letters contain parallel line segments? E, F, H, N, and Z

59. Which letters contain perpendicular line segments? E, F, H, and T

*Angles play an important role in custom bikes that are properly fitted to the biking needs of cyclists. One of the angles to help find the perfect fit is called the **hip angle**. The figure indicates that the hip angle is created when you're sitting on the bike, gripping the handlebars, and your leg is fully extended. Your hip is the vertex, with one ray extending to your shoulder and the other ray extending to the front-bottom of your foot.*

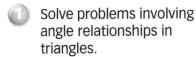

*See Answers to Selected Exercises.

The table indicates hip angles for various biking needs. Use this information to pedal through Exercises 60–63.

Hip Angle	Used For
$85° \le$ hip angle $\le 89°$	short-distance aggressive racing
$91° \le$ hip angle $\le 115°$	long-distance riding
$116° \le$ hip angle $\le 130°$	mountain biking

60. Which type or types of biking require an acute hip angle?

61. Which type or types of biking require an obtuse hip angle?

62. A racer who had an 89° hip angle decides to switch to long-distance riding. What is the maximum difference in hip angle for the two types of biking? 30°

63. A racer who had an 89° hip angle decides to switch to mountain biking. What is the minimum difference in hip angle for the two types of biking? 27°

(Source for Exercises 60–63: *Scholastic Math*, January 11, 2010)

60. short-distance aggressive racing **61.** long-distance riding and mountain biking

Critical Thinking Exercises

64. Use the figure to select a pair of complementary angles. d
   a. ∡1 and ∡4
   b. ∡3 and ∡6
   c. ∡2 and ∡5
   d. ∡1 and ∡5

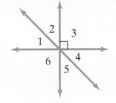

65. If $m\angle AGB = m\angle BGC$, and $m\angle CGD = m\angle DGE$, find $m\angle BGD$. 90°

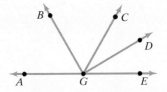

6.2 Triangles

What am I supposed to learn?

After you have read this section, you should be able to:

1. Solve problems involving angle relationships in triangles.
2. Solve problems involving similar triangles.
3. Solve problems using the Pythagorean Theorem.

Deductive reasoning is the process of proving a specific conclusion from one or more general statements. A conclusion that is proved to be true through deductive reasoning is called a **theorem**. The Greek mathematician Euclid, who lived more than 2000 years ago, used deductive reasoning. In his 13-volume book, the *Elements*, Euclid proved over 465 theorems about geometric figures. Euclid's work established deductive reasoning as a fundamental tool of mathematics. Here's looking at Euclid!

The Walter Pyramid, California State University, Long Beach

A **triangle** is a geometric figure that has three sides, all of which lie on a flat surface or plane. If you start at any point along the triangle and trace along the entire figure exactly once, you will end at the same point at which you started. Because the beginning point and ending point are the same, the triangle is called a **closed** geometric figure. Euclid used parallel lines to prove one of the most important properties of triangles: The sum of the measures of the three angles of any triangle is 180°. Here is how he did it. He began with the following general statement:

Euclid's Assumption about Parallel Lines

Given a line and a point not on the line, one and only one line can be drawn through the given point parallel to the given line.

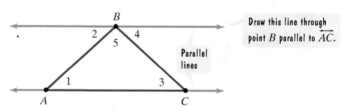

In **Figure 6.15**, triangle ABC represents any triangle. Using the general assumption given above, we draw a line through point B parallel to line AC.

Figure 6.15

Because the lines are parallel, alternate interior angles have the same measure.

$$m\angle 1 = m\angle 2 \quad \text{and} \quad m\angle 3 = m\angle 4$$

Also observe that angles 2, 5, and 4 form a straight angle.

$$m\angle 2 + m\angle 5 + m\angle 4 = 180°$$

Because $m\angle 1 = m\angle 2$, replace $m\angle 2$ with $m\angle 1$.   Because $m\angle 3 = m\angle 4$, replace $m\angle 4$ with $m\angle 3$.

$$m\angle 1 + m\angle 5 + m\angle 3 = 180°$$

Because $\angle 1$, $\angle 5$, and $\angle 3$ are the three angles of the triangle, this last equation shows that the measures of the triangle's three angles have a sum of 180°.

The Angles of a Triangle

The sum of the measures of the three angles of any triangle is 180°.

Solve problems involving angle relationships in triangles.

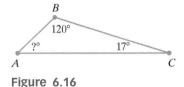

Figure 6.16

Example 1 Using Angle Relationships in Triangles

Find the measure of angle A for triangle ABC in **Figure 6.16**.

Solution

Because $m\angle A + m\angle B + m\angle C = 180°$, we obtain

$m\angle A + 120° + 17° = 180°$.   *The sum of the measures of a triangle's three angles is 180°.*

$m\angle A + 137° = 180°$ *Simplify: $120° + 17° = 137°$.*

$m\angle A = 180° - 137°$ *Find the measure of A by subtracting 137° from both sides of the equation.*

$m\angle A = 43°$ *Simplify.*

 **Check Point 1**   In **Figure 6.16**, suppose that $m\angle B = 116°$ and $m\angle C = 15°$. Find $m\angle A$.   49°

Example 2 Using Angle Relationships in Triangles

Find the measures of angles 1 through 5 in **Figure 6.17**.

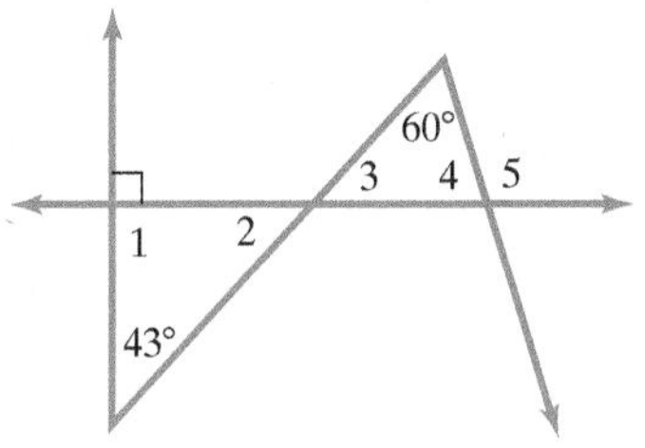

Solution

Because $\angle 1$ is supplementary to the right angle, $m\angle 1 = 90°$.

$m\angle 2$ can be found using the fact that the sum of the measures of the angles of a triangle is 180°.

$m\angle 1 + m\angle 2 + 43° = 180°$ *The sum of the measures of a triangle's three angles is 180°.*

$90° + m\angle 2 + 43° = 180°$ *We previously found that $m\angle 1 = 90°$.*

$m\angle 2 + 133° = 180°$ *Simplify: $90° + 43° = 133°$.*

$m\angle 2 = 180° - 133°$ *Subtract 133° from both sides.*

$m\angle 2 = 47°$ *Simplify.*

$m\angle 3$ can be found using the fact that vertical angles have equal measures: $m\angle 3 = m\angle 2$. Thus, $m\angle 3 = 47°$.

$m\angle 4$ can be found using the fact that the sum of the measures of the angles of a triangle is 180°. Refer to the triangle at the top of **Figure 6.17**.

$m\angle 3 + m\angle 4 + 60° = 180°$ *The sum of the measures of a triangle's three angles is 180°.*

$47° + m\angle 4 + 60° = 180°$ *We previously found that $m\angle 3 = 47°$.*

$m\angle 4 + 107° = 180°$ *Simplify: $47° + 60° = 107°$.*

$m\angle 4 = 180° - 107°$ *Subtract 107° from both sides.*

$m\angle 4 = 73°$ *Simplify.*

Finally, we can find $m\angle 5$ by observing that angles 4 and 5 form a straight angle.

$m\angle 4 + m\angle 5 = 180°$ *A straight angle measures 180°.*

$73° + m\angle 5 = 180°$ *We previously found that $m\angle 4 = 73°$.*

$m\angle 5 = 180° - 73°$ *Subtract 73° from both sides.*

$m\angle 5 = 107°$ *Simplify.*

Figure 6.17

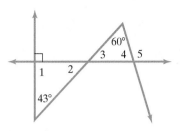

Figure 6.17 (repeated)

✓ **Check Point 2**   In **Figure 6.17**, suppose that the angle shown to measure 43° measures, instead, 36°. Further suppose that the angle shown to measure 60° measures, instead, 58°. Under these new conditions, find the measures of angles 1 through 5 in the figure. $m\angle 1 = 90°; m\angle 2 = 54°; m\angle 3 = 54°; m\angle 4 = 68°; m\angle 5 = 112°$

Triangles can be described using characteristics of their angles or their sides.

Great Question!

Does an isosceles triangle have exactly two equal sides or at least two equal sides?

We checked a variety of math textbooks and found two slightly different definitions for an isosceles triangle. One definition states that an isosceles triangle has *exactly* two sides of equal length. A second definition asserts that an isosceles triangle has *at least* two sides of equal length. (This makes an equilateral triangle, with three sides of the same length, an isosceles triangle.) In this book, we assume that an isosceles triangle has exactly two sides, but not three sides, of the same length.

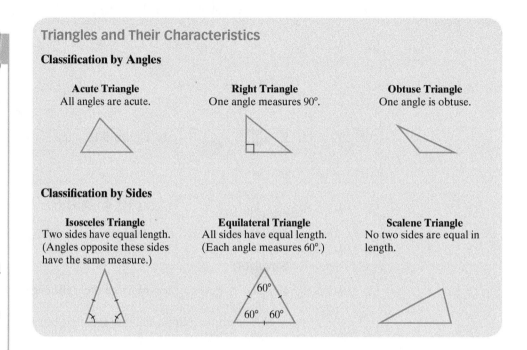

Triangles and Their Characteristics

Classification by Angles

Acute Triangle
All angles are acute.

Right Triangle
One angle measures 90°.

Obtuse Triangle
One angle is obtuse.

Classification by Sides

Isosceles Triangle
Two sides have equal length. (Angles opposite these sides have the same measure.)

Equilateral Triangle
All sides have equal length. (Each angle measures 60°.)

Scalene Triangle
No two sides are equal in length.

② **Solve problems involving similar triangles.**

Pedestrian crossing

Similar Triangles

Shown in the margin is an international road sign. This sign is shaped just like the actual sign, although its size is smaller. Figures that have the same shape, but not the same size, are used in **scale drawings**. A scale drawing always pictures the exact shape of the object that the drawing represents. Architects, engineers, landscape gardeners, and interior decorators use scale drawings in planning their work.

Figures that have the same shape, but not necessarily the same size, are called **similar figures**. In **Figure 6.18**, triangles ABC and DEF are similar. Angles A and D measure the same number of degrees and are called **corresponding angles**. Angles C and F are corresponding angles, as are angles B and E. Angles with the same number of tick marks in **Figure 6.18** are the corresponding angles.

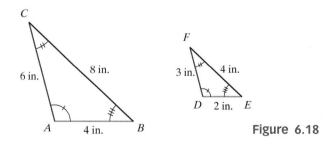

Figure 6.18

The sides opposite the corresponding angles are called **corresponding sides**. Thus, $\overline{CB}$ and $\overline{FE}$ are corresponding sides. $\overline{AB}$ and $\overline{DE}$ are also corresponding sides, as are $\overline{AC}$ and $\overline{DF}$. Corresponding angles measure the same number of degrees, but corresponding sides may or may not be the same length. For the triangles in **Figure 6.18**, each side in the smaller triangle is half the length of the corresponding side in the larger triangle.

The triangles in **Figure 6.18** illustrate what it means to be **similar triangles. Corresponding angles have the same measure and the ratios of the lengths of the corresponding sides are equal.**

$$\frac{\text{length of } \overline{AC}}{\text{length of } \overline{DF}} = \frac{6 \text{ in.}}{3 \text{ in.}} = \frac{2}{1}; \quad \frac{\text{length of } \overline{CB}}{\text{length of } \overline{FE}} = \frac{8 \text{ in.}}{4 \text{ in.}} = \frac{2}{1}; \quad \frac{\text{length of } \overline{AB}}{\text{length of } \overline{DE}} = \frac{4 \text{ in.}}{2 \text{ in.}} = \frac{2}{1}$$

In similar triangles, the lengths of the corresponding sides are proportional. Thus, in **Figure 6.18**,

> AC represents the length of $\overline{AC}$, DF the length of $\overline{DF}$, and so on.

$$\frac{AC}{DF} = \frac{CB}{FE} = \frac{AB}{DE}.$$

How can we quickly determine if two triangles are similar? **If the measures of two angles of one triangle are equal to those of two angles of a second triangle, then the two triangles are similar.** If the triangles are similar, then their corresponding sides are proportional.

Example 3   Using Similar Triangles

Explain why the triangles in **Figure 6.19** are similar. Then find the missing length, x.

Solution

Figure 6.20 shows that two angles of the small triangle are equal in measure to two angles of the large triangle. One angle pair is given to have the same measure. Another angle pair consists of vertical angles with the same measure. Thus, the triangles are similar and their corresponding sides are proportional.

$$\frac{5}{8} = \frac{7}{x}$$

We solve $\frac{5}{8} = \frac{7}{x}$ for x by applying the cross-products principle for proportions that we discussed in Section 2.4: If $\frac{a}{b} = \frac{c}{d}$, then $ad = bc$.

$5x = 8 \cdot 7$ Apply the cross-products principle.

$5x = 56$ Multiply: $8 \cdot 7 = 56$.

$\dfrac{5x}{5} = \dfrac{56}{5}$ Divide both sides by 5.

$x = 11.2$ Simplify.

The missing length, x, is 11.2 inches.

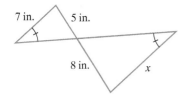

Figure 6.19

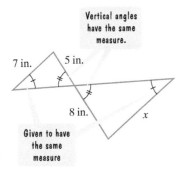

Figure 6.20

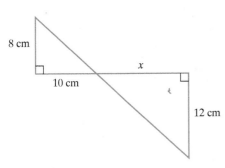

Figure 6.21

✓ **Check Point 3** Explain why the triangles in **Figure 6.21** are similar. Then find the missing length, x. Two angles of the small triangle are equal in measure to two angles of the large triangle. One angle pair is given to have the same measure (right angles). Another angle pair consists of vertical angles with the same measure.; 15 cm

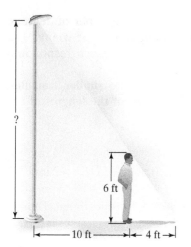

Figure 6.22

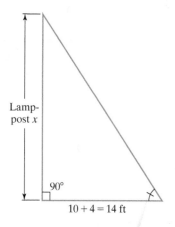

Lamp-
post x

90°

10 + 4 = 14 ft

Angle shared by
both triangles

Man
6 ft

90°

← 4 ft →

Figure 6.23

Example 4   **Problem Solving Using Similar Triangles**

A man who is 6 feet tall is standing 10 feet from the base of a lamppost (see **Figure 6.22**). The man's shadow has a length of 4 feet. How tall is the lamppost?

Solution

The drawing in **Figure 6.23** makes the similarity of the triangles easier to see. The large triangle with the lamppost on the left and the small triangle with the man on the left both contain 90° angles. They also share an angle. Thus, two angles of the large triangle are equal in measure to two angles of the small triangle. This means that the triangles are similar and their corresponding sides are proportional. We begin by letting x represent the height of the lamppost, in feet. Because corresponding sides of the two similar triangles are proportional,

Opposite the angle
with one tick
in the large △

Opposite the unmarked
angle in the large △

$$\frac{x}{6} = \frac{14}{4}.$$

Opposite the angle
with one tick
in the small △

Opposite the unmarked
angle in the small △

We solve for x by applying the cross-products principle.

$4x = 6 \cdot 14$ Apply the cross-products principle.

$4x = 84$ Multiply: $6 \cdot 14 = 84$.

$\dfrac{4x}{4} = \dfrac{84}{4}$ Divide both sides by 4.

$x = 21$ Simplify.

The lamppost is 21 feet tall.

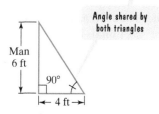 **Check Point 4**   Find the height of the lookout tower shown in **Figure 6.24** using the figure that lines up the top of the tower with the top of a stick that is 2 yards long and 3.5 yards from the line to the top of the tower.   32 yd

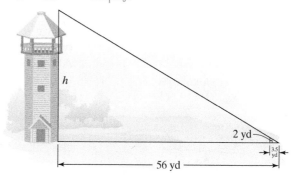

h

2 yd

3.5 yd

56 yd

Figure 6.24

The Pythagorean Theorem

3   Solve problems using the Pythagorean Theorem.

The ancient Greek philosopher and mathematician Pythagoras (approximately 582–500 B.C.) founded a school whose motto was "All is number." Pythagoras is best remembered for his work with the **right triangle**, a triangle with one angle measuring 90°. The side opposite the 90° angle is called the **hypotenuse**. The other sides are called **legs**. Pythagoras found that if he constructed squares on each of the legs, as well as a larger square on the hypotenuse, the sum of the areas of the smaller squares is equal to the area of the larger square. This is illustrated in **Figure 6.25**.

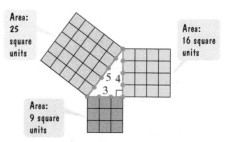

Area:
25
square
units

Area:
16 square
units

5 4
3

Area:
9 square
units

Figure 6.25 The area of the large square equals the sum of the areas of the smaller squares.

This relationship is usually stated in terms of the lengths of the three sides of a right triangle and is called the **Pythagorean Theorem**.

> **The Pythagorean Theorem**
>
> The sum of the squares of the lengths of the legs of a right triangle equals the square of the length of the hypotenuse.
>
> If the legs have lengths a and b and the hypotenuse has length c, then
>
> $$a^2 + b^2 = c^2.$$

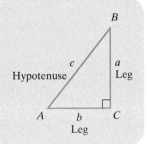

Blitzer Bonus

The Universality of Mathematics

Is mathematics discovered or invented? The "Pythagorean" Theorem, credited to Pythagoras, was known in China, from land surveying, and in Egypt, from pyramid building, centuries before Pythagoras was born. The same mathematics is often discovered/invented by independent researchers separated by time, place, and culture.

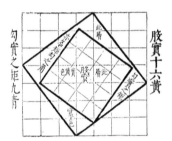

This diagram of the Pythagorean Theorem is from a Chinese manuscript dated as early as 1200 B.C.

Example 5   Using the Pythagorean Theorem

Find the length of the hypotenuse c in the right triangle shown in **Figure 6.26**.

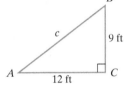

Figure 6.26

Solution

Let $a = 9$ and $b = 12$. Substituting these values into $c^2 = a^2 + b^2$ enables us to solve for c.

$c^2 = a^2 + b^2$ Use the symbolic statement of the Pythagorean Theorem.

$c^2 = 9^2 + 12^2$ Let $a = 9$ and $b = 12$.

$c^2 = 81 + 144$ $9^2 = 9 \cdot 9 = 81$ and $12^2 = 12 \cdot 12 = 144$.

$c^2 = 225$ Add.

$c = \sqrt{225} = 15$ Solve for c by taking the positive square root of 225.

The length of the hypotenuse is 15 feet.

✓ **Check Point 5**   Find the length of the hypotenuse in a right triangle whose legs have lengths 7 feet and 24 feet.   25 ft

Example 6   Using the Pythagorean Theorem

a. A wheelchair ramp with a length of 122 inches has a horizontal distance of 120 inches. How high is the ramp at the top?

b. Construction laws are very specific when it comes to access ramps for the disabled. Every vertical rise of 1 inch requires a horizontal run of 12 inches. Does this ramp satisfy the requirement?

Solution

a. The problem's conditions state that the wheelchair ramp has a length of 122 inches and a horizontal distance of 120 inches. **Figure 6.27** shows the right triangle that is formed by the ramp, the wall, and the ground. We can find x, the ramp's height, using the Pythagorean Theorem.

(leg)²	plus	(leg)²	equals	(hypotenuse)²
x^2	$+$	120^2	$=$	122^2

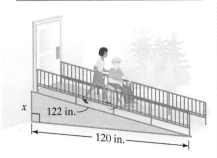

Figure 6.27

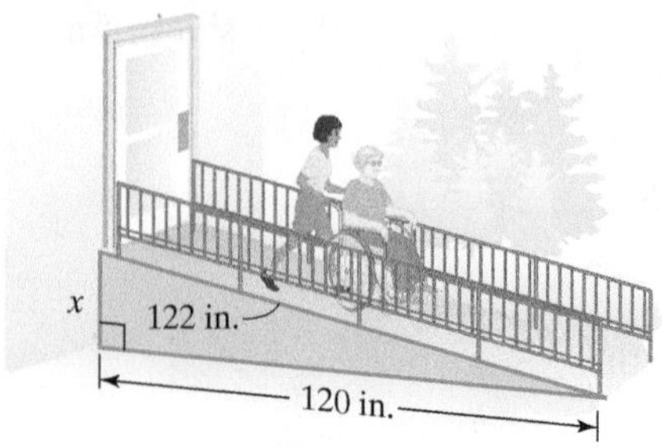

$$x^2 + 120^2 = 122^2$$

This is the equation resulting from the Pythagorean Theorem.

$$x^2 + 14,400 = 14,884$$

Square 120 and 122.

$$x^2 = 484$$

Isolate x^2 by subtracting 14,400 from both sides.

$$x = \sqrt{484} = 22$$

Solve for x by taking the positive square root of 484.

Figure 6.27 (repeated)

The ramp is 22 inches high at the top.

b. Every vertical rise of 1 inch requires a horizontal run of 12 inches. Because the ramp has a height of 22 inches, it requires a horizontal distance of 22(12) inches, or 264 inches. The horizontal distance is only 120 inches, so this ramp does not satisfy construction laws for access ramps for the disabled.

✓ **Check Point 6**   A radio tower is supported by two wires that are each 130 yards long and attached to the ground 50 yards from the base of the tower. How far from the ground are the wires attached to the tower?   120 yd

Concept and Vocabulary Check

Exercises in the Concept and Vocabulary Check are intended for group and class discussions.

In Exercises 1–8, fill in each blank so that the resulting statement is true.

1. The sum of the measures of the three angles of any triangle is _____180°_____.

2. A triangle in which each angle measures less than 90° is called a/an _____acute_____ triangle.

3. A triangle in which one angle measures more than 90° is called a/an _____obtuse_____ triangle.

4. A triangle with exactly two sides of the same length is called a/an _____isosceles_____ triangle.

5. A triangle whose sides are all the same length is called a/an _____equilateral_____ triangle.

6. A triangle that has no sides of the same length is called a/an _____scalene_____ triangle.

7. Triangles that have the same shape, but not necessarily the same size, are called _____similar_____ triangles. For such triangles, corresponding angles have _the same measure_ and the lengths of the corresponding sides are ___proportional___.

8. The Pythagorean Theorem states that in any _____right_____ triangle, the sum of the squares of the lengths of the _____legs_____ equals _the square of the length of the hypotenuse_

In Exercises 9–13, determine whether each statement is true or false. If the statement is false, make the necessary change(s) to produce a true statement.   Changes to false statements will vary.

9. Euclid's assumption about parallel lines states that given a line and a point not on the line, one and only one line can be drawn through the given point parallel to the given line.   true

10. A triangle cannot have a right angle and an obtuse angle.   true

11. Each angle of an isosceles triangle measures 60°.   false

12. If the measures of two angles of one triangle are equal to those of two angles of a second triangle, then the two triangles have the same size and shape.   false

13. In any triangle, the sum of the squares of the two shorter sides equals the square of the length of the longest side.   false

Respond to Exercises 14–21 using verbal or written explanations.   **14–21.** Answers will vary.

14. If the measures of two angles of a triangle are known, explain how to find the measure of the third angle.

15. Can a triangle contain two right angles? Explain your answer.

16. What general assumption did Euclid make about a point and a line in order to prove that the sum of the measures of the angles of a triangle is 180°?

17. What are similar triangles?

18. If the ratio of the corresponding sides of two similar triangles is 1 to $1\left(\frac{1}{1}\right)$, what must be true about the triangles?

19. What are corresponding angles in similar triangles?

20. Describe how to identify the corresponding sides in similar triangles.

21. In your own words, state the Pythagorean Theorem.

Exercise Set 6.2

Practice Exercises

In Exercises 1–4, find the measure of angle A for the triangle shown.

1.

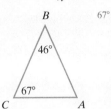

67°

2.

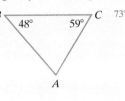

73°

3.

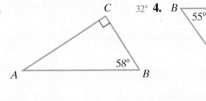

32°

4.

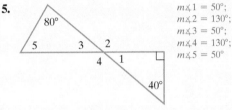

35°

In Exercises 5–6, find the measures of angles 1 through 5 in the figure shown.

5.

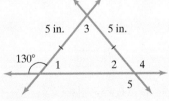

$m\angle 1 = 50°;$
$m\angle 2 = 130°;$
$m\angle 3 = 50°;$
$m\angle 4 = 130°;$
$m\angle 5 = 50°$

6.

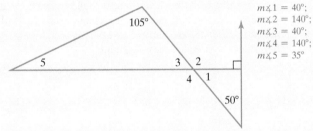

$m\angle 1 = 40°;$
$m\angle 2 = 140°;$
$m\angle 3 = 40°;$
$m\angle 4 = 140°;$
$m\angle 5 = 35°$

We have seen that isosceles triangles have two sides of equal length. The angles opposite these sides have the same measure. In Exercises 7–8, use this information to help find the measure of each numbered angle.

7.

$m\angle 1 = 50°;$
$m\angle 2 = 50°;$
$m\angle 3 = 80°;$
$m\angle 4 = 130°;$
$m\angle 5 = 130°$

8.

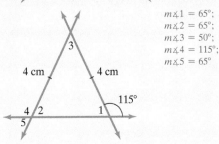

$m\angle 1 = 65°;$
$m\angle 2 = 65°;$
$m\angle 3 = 50°;$
$m\angle 4 = 115°;$
$m\angle 5 = 65°$

In Exercises 9–10, lines l and m are parallel. Find the measure of each numbered angle.

9.

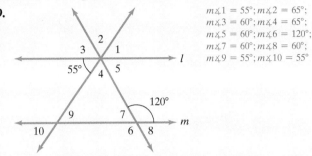

$m\angle 1 = 55°; m\angle 2 = 65°;$
$m\angle 3 = 60°; m\angle 4 = 65°;$
$m\angle 5 = 60°; m\angle 6 = 120°;$
$m\angle 7 = 60°; m\angle 8 = 60°;$
$m\angle 9 = 55°; m\angle 10 = 55°$

10.

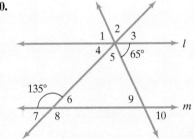

$m\angle 1 = 65°; m\angle 2 = 70°;$
$m\angle 3 = 45°; m\angle 4 = 45°;$
$m\angle 5 = 70°; m\angle 6 = 45°;$
$m\angle 7 = 45°; m\angle 8 = 135°;$
$m\angle 9 = 65°; m\angle 10 = 65°$

In Exercises 11–16, explain why the triangles are similar. Then find the missing length, x.

11.

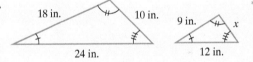

12.

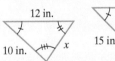

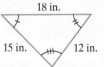

13.

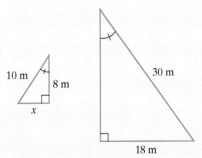

14.

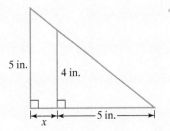

15.

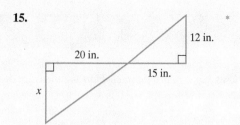

16.

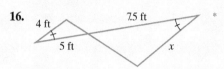

In Exercises 17–19, △ABC and △ADE are similar. Find the length of the indicated side.

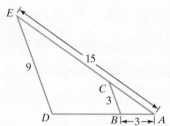

17. $\overline{CA}$ 5   **18.** $\overline{DB}$ 6   **19.** $\overline{DA}$ 9

20. In the diagram for Exercises 17–19, suppose that you are not told that △ABC and △ADE are similar. Instead, you are given that $\overleftrightarrow{ED}$ and $\overleftrightarrow{CB}$ are parallel. Under these conditions, explain why the triangles must be similar.   *

In Exercises 21–26, use the Pythagorean Theorem to find the missing length in each right triangle. Use your calculator to find square roots, rounding, if necessary, to the nearest tenth.

21.

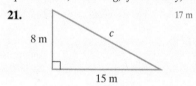

22.

23.

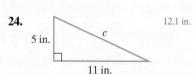

24.

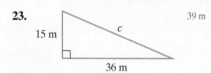

25.

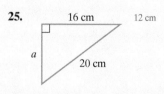

26.

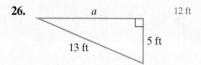

Practice Plus

Two triangles are **congruent** *if they have the same shape and the same size. In congruent triangles, the measures of corresponding angles are equal and the corresponding sides have the same length. The following triangles are congruent:*

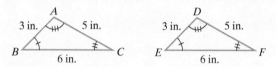

Any one of the following may be used to determine if two triangles are congruent.

Determining Congruent Triangles

1. **Side-Side-Side (SSS)**

   If the lengths of three sides of one triangle equal the lengths of the corresponding sides of a second triangle, then the two triangles are congruent.

   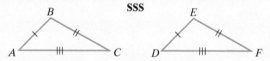

2. **Side-Angle-Side (SAS)**

   If the lengths of two sides of one triangle equal the lengths of the corresponding sides of a second triangle, and the measures of the angles between each pair of sides are equal, then the two triangles are congruent.

   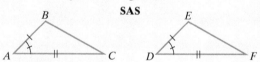

3. **Angle-Side-Angle (ASA)**

   If the measures of two angles of one triangle equal the measures of two angles of a second triangle, and the lengths of the sides between each pair of angles are equal, then the two triangles are congruent.

   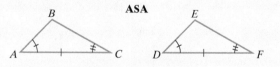

In Exercises 27–36, determine whether △I and △II are congruent. If the triangles are congruent, state the reason why, selecting from SSS, SAS, or ASA. (More than one reason may be possible.)

27.   congruent; SAS

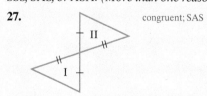

*See Answers to Selected Exercises.

28. 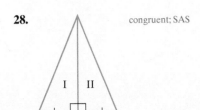 congruent; SAS

29. 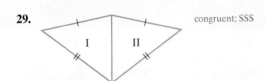 congruent; SSS

30. 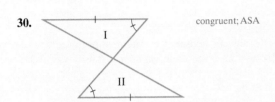 congruent; ASA

31. 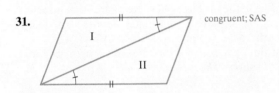 congruent; SAS

32. 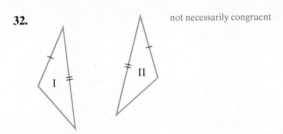 not necessarily congruent

33. 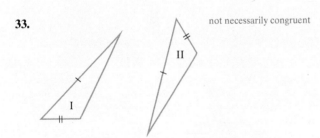 not necessarily congruent

34. 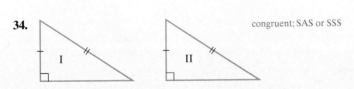 congruent; SAS or SSS

35. 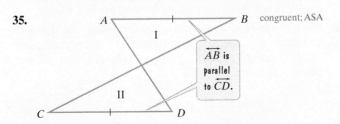 congruent; ASA

$\overrightarrow{AB}$ is parallel to $\overrightarrow{CD}$.

36. 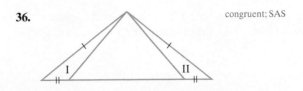 congruent; SAS

Application Exercises

Use similar triangles to solve Exercises 37–38.

37. A person who is 5 feet tall is standing 80 feet from the base of a tree and the tree casts an 86-foot shadow. The person's shadow is 6 feet in length. What is the tree's height? Round to the nearest tenth of a foot.   71.7 ft

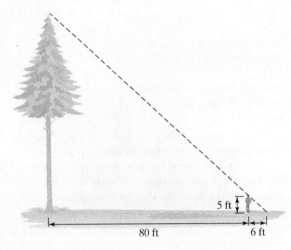

38. A tree casts a shadow 12 feet long. At the same time, a vertical rod 8 feet high casts a shadow that is 6 feet long. How tall is the tree?   16 ft

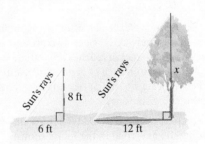

Use the Pythagorean Theorem to solve Exercises 39–46. Use your calculator to find square roots, rounding, if necessary, to the nearest tenth.

39. A baseball diamond is actually a square with 90-foot sides. What is the distance from home plate to second base?

$90\sqrt{2}$ ft $\approx$ 127.3 ft

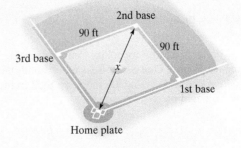

40. The base of a 20-foot ladder is 15 feet from the house. How far up the house does the ladder reach? $5\sqrt{7}$ ft ≈ 13.2 ft

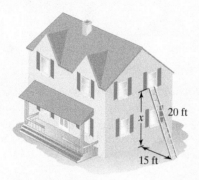

41. A flagpole has a height of 16 yards. It will be supported by three cables, each of which is attached to the flagpole at a point 4 yards below the top of the pole and attached to the ground at a point that is 9 yards from the base of the pole. Find the total number of feet of cable that will be required.   45 yd

42. A flagpole has a height of 10 yards. It will be supported by three cables, each of which is attached to the flagpole at a point 4 yards below the top of the pole and attached to the ground at a point that is 8 yards from the base of the pole. Find the total number of feet of cable that will be required.   30 yd

43. A rectangular garden bed measures 5 feet by 12 feet. A water faucet is located at one corner of the garden bed. A hose will be connected to the water faucet. The hose must be long enough to reach the opposite corner of the garden bed when stretched straight. Find the required length of hose.   13 ft

44. A rocket ascends vertically after being launched from a location that is midway between two ground-based tracking stations. When the rocket reaches an altitude of 4 kilometers, it is 5 kilometers from each of the tracking stations. Assuming that this is a locale where the terrain is flat, how far apart are the two tracking stations?   6 km

45. If construction costs are $150,000 per *kilometer*, find the cost of building the new road in the figure shown. $750,000

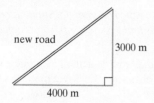

46. Picky, Picky, Picky This problem appeared on a high school exit exam:

Alex is building a ramp for a bike competition. He has two rectangular boards. One board is six meters and the other is five meters long. If the ramp has to form a right triangle, what should its height be?

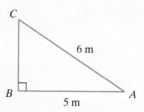

Students were asked to select the correct answer from the following options:

3 meters; 4 meters; 3.3 meters; 7.8 meters.

a. Among the available choices, which option best expresses the ramp's height? How many feet, to the nearest tenth of a foot, is this? Does a bike competition that requires riders to jump off these heights seem realistic? (ouch!)   3.3 m; 10.8 ft; no

b. Express the ramp's height to the nearest hundredth of a meter. By how many centimeters does this differ from the "correct" answer on the test? How many inches, to the nearest half inch, is this? Is it likely that a carpenter with a tape measure would make this error?   3.32 m; 2 cm; 1 in.; no

c. According to the problem, Alex has boards that measure 5 meters and 6 meters. A 6-meter board? How many feet, to the nearest tenth of a foot, is this? When was the last time you found a board of this length at Home Depot? (*Source: The New York Times,* April 24, 2005)   19.7 ft; Answers will vary.

Critical Thinking Exercises

47. Find the measure of angle R.   70°

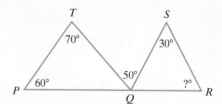

48. What is the length of $\overline{AB}$ in the accompanying figure?   21 ft

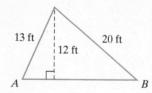

49. One angle of a triangle is twice as large as another. The measure of the third angle is 20° more than that of the smallest angle. Find the measure of each angle.   40°, 80°, 60°

50. One angle of a triangle is three times as large as another. The measure of the third angle is 30° greater than that of the smallest angle. Find the measure of each angle.   30°, 90°, 60°

51. Use a quadratic equation to solve this problem. The width of a rectangular carpet is 7 meters shorter than the length, and the diagonal is 1 meter longer than the length. What are the carpet's dimensions?   5 m by 12 m

6.3 : Polygons, Perimeter, and Tessellations

What am I supposed to learn?

After you have read this section, you should be able to:

1. Name certain polygons according to the number of sides.

2. Recognize the characteristics of certain quadrilaterals.

3. Solve problems involving a polygon's perimeter.

4. Find the sum of the measures of a polygon's angles.

5. Understand tessellations and their angle requirements.

You have just purchased a beautiful plot of land in the country, shown in **Figure 6.28**. In order to have more privacy, you decide to put fencing along each of its four sides. The cost of this project depends on the distance around the four outside edges of the plot, called its **perimeter**, as well as the cost for each foot of fencing.

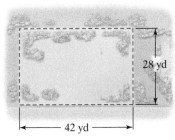

Figure 6.28

Your plot of land is a geometric figure: It has four straight sides that are line segments. The plot is on level ground, so that the four sides all lie on a flat surface, or plane. The plot is an example of a *polygon*. Any closed shape in the plane formed by three or more line segments that intersect only at their endpoints is a **polygon**.

A polygon is named according to the number of sides it has. We know that a three-sided polygon is called a **triangle**. A four-sided polygon is called a **quadrilateral**.

A polygon whose sides are all the same length and whose angles all have the same measure is called a **regular polygon**. **Table 6.2** provides the names of six polygons. Also shown are illustrations of regular polygons.

1 Name certain polygons according to the number of sides.

Table 6.2 Illustrations of Regular Polygons			
Name	**Picture**	**Name**	**Picture**
Triangle 3 sides		Hexagon 6 sides	
Quadrilateral 4 sides		Heptagon 7 sides	
Pentagon 5 sides		Octagon 8 sides	

2 Recognize the characteristics of certain quadrilaterals.

Quadrilaterals

The plot of land in **Figure 6.28** is a four-sided polygon, or a quadrilateral. However, when you first looked at the figure, perhaps you thought of the plot as a rectangular field. A **rectangle** is a special kind of quadrilateral in which both pairs of opposite sides are parallel, have the same measure, and whose angles are right angles. **Table 6.3** on the next page presents some special quadrilaterals and their characteristics.

Blitzer Bonus
●●●●●●●●●●●●●●●●

Bees and Regular Hexagons

Bees use honeycombs to store honey and house larvae. They construct honey storage cells from wax. Each cell has the shape of a regular hexagon. The cells fit together perfectly, preventing dirt or predators from entering. Squares or equilateral triangles would fit equally well, but regular hexagons provide the largest storage space for the amount of wax used.

Table 6.3 Types of Quadrilaterals

Name	Characteristics	Representation
Parallelogram	Quadrilateral in which both pairs of opposite sides are parallel and have the same measure. Opposite angles have the same measure.	
Rhombus	Parallelogram with all sides having equal length.	
Rectangle	Parallelogram with four right angles. Because a rectangle is a parallelogram, opposite sides are parallel and have the same measure.	
Square	A rectangle with all sides having equal length. Each angle measures 90°, and the square is a regular quadrilateral.	
Trapezoid	A quadrilateral with exactly one pair of parallel sides.	

3 Solve problems involving a polygon's perimeter.

Perimeter

The **perimeter**, P, of a polygon is the sum of the lengths of its sides. Perimeter is measured in linear units, such as inches, feet, yards, meters, or kilometers.

Example 1 involves the perimeter of a rectangle. Because perimeter is the sum of the lengths of the sides, the perimeter of the rectangle shown in **Figure 6.29** is $l + w + l + w$. This can be expressed as

$$P = 2l + 2w.$$

Figure 6.29
A rectangle with length l and width w

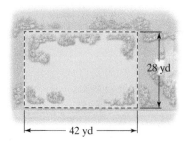

Figure 6.28 (repeated)

Example 1 An Application of Perimeter

The rectangular field we discussed at the beginning of this section (see **Figure 6.28**) has a length of 42 yards and a width of 28 yards. If fencing costs $5.25 per foot, find the cost to enclose the field with fencing.

Solution

We begin by finding the perimeter of the rectangle in yards. Using 3 ft = 1 yd and dimensional analysis, we express the perimeter in feet. Finally, we multiply the perimeter, in feet, by $5.25 because the fencing costs $5.25 per foot.

The length, l, is 42 yards and the width, w, is 28 yards. The perimeter of the rectangle is determined using the formula $P = 2l + 2w$.

$$P = 2l + 2w = 2 \cdot 42 \text{ yd} + 2 \cdot 28 \text{ yd} = 84 \text{ yd} + 56 \text{ yd} = 140 \text{ yd}$$

Because 3 ft = 1 yd, we use the unit fraction $\frac{3 \text{ ft}}{1 \text{ yd}}$ to convert from yards to feet.

$$140 \text{ yd} = \frac{140 \text{ yd}}{1} \cdot \frac{3 \text{ ft}}{1 \text{ yd}} = 140 \cdot 3 \text{ ft} = 420 \text{ ft}$$

The perimeter of the rectangle is 420 feet. Now we are ready to find the cost of the fencing. We multiply 420 feet by $5.25, the cost per foot.

$$\text{Cost} = \frac{420 \text{ feet}}{1} \cdot \frac{\$5.25}{\text{foot}} = 420(\$5.25) = \$2205$$

The cost to enclose the field with fencing is $2205.

 Check Point 1 A rectangular field has a length of 50 yards and a width of 30 yards. If fencing costs $6.50 per foot, find the cost to enclose the field with fencing. $3120

The Sum of the Measures of a Polygon's Angles

4 Find the sum of the measures of a polygon's angles.

We know that the sum of the measures of the three angles of any triangle is 180°. We can use this relationship to find the sum of the measures of the angles of any polygon. Start by drawing line segments from a single point where two sides meet so that nonoverlapping triangles are formed. This is done in **Figure 6.30**.

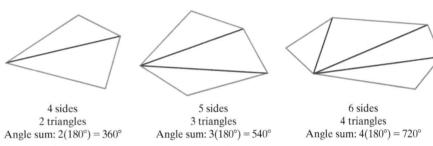

4 sides
2 triangles
Angle sum: 2(180°) = 360°

5 sides
3 triangles
Angle sum: 3(180°) = 540°

6 sides
4 triangles
Angle sum: 4(180°) = 720°

Figure 6.30

In each case, the number of triangles is two less than the number of sides of the polygon. Thus, for an n-sided polygon, there are $n - 2$ triangles. Because each triangle has an angle-measure sum of 180°, the sum of the measures for the angles in the $n - 2$ triangles is $(n - 2)180°$. Thus, the sum of the measures of the angles of an n-sided polygon is $(n - 2)180°$.

> **The Angles of a Polygon**
> The sum of the measures of the angles of a polygon with n sides is
> $$(n - 2)180°.$$

Example 2 Using the Formula for the Angles of a Polygon

a. Find the sum of the measures of the angles of an octagon.

b. **Figure 6.31** shows a regular octagon. Find the measure of angle A.

c. Find the measure of exterior angle B.

Solution

a. An octagon has eight sides. Using the formula $(n - 2)180°$ with $n = 8$, we can find the sum of the measures of its eight angles.
The sum of the measures of an octagon's angles is
$$(n - 2)180° = (8 - 2)180°$$
$$= 6 \cdot 180°$$
$$= 1080°.$$

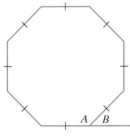

Figure 6.31 A regular octagon

Figure 6.31 (repeated)
A regular octagon

b. Examine the regular octagon in **Figure 6.31**. Note that all eight sides have the same length. Likewise, all eight angles have the same degree measure. Angle A is one of these eight angles. We find its measure by taking the sum of the measures of all eight angles, 1080°, and dividing by 8.

$$m \angle A = \frac{1080°}{8} = 135°$$

c. Because $\angle B$ is the supplement of $\angle A$,

$$m \angle B = 180° - 135° = 45°.$$

✓ **Check Point 2**

a. Find the sum of the measures of the angles of a 12-sided polygon. 1800°

▶ **b.** Find the measure of an angle of a regular 12-sided polygon. 150°

Tessellations

⑤ Understand tessellations and their angle requirements.

A relationship between geometry and the visual arts is found in an art form called *tessellations*. A **tessellation**, or **tiling**, is a pattern consisting of the repeated use of the same geometric figures to completely cover a plane, leaving no gaps and no overlaps. **Figure 6.32** shows eight tessellations, each consisting of the repeated use of two or more regular polygons.

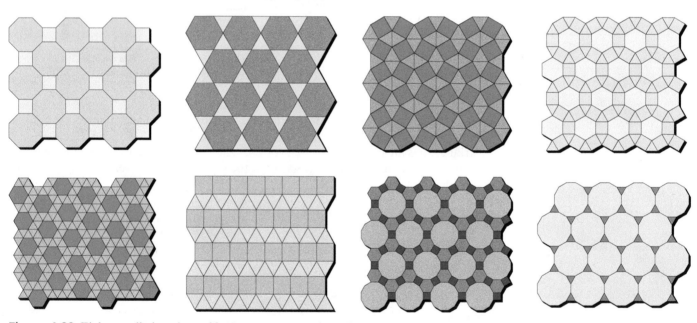

Figure 6.32 Eight tessellations formed by two or more regular polygons

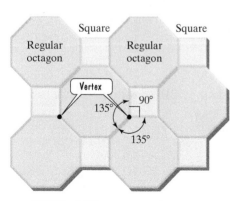

Figure 6.33

In each tessellation in **Figure 6.32**, the same types of regular polygons surround every vertex (intersection point). Furthermore, **the sum of the measures of the angles that come together at each vertex is 360°, a requirement for the formation of a tessellation**. This is illustrated in the enlarged version of the tessellation in **Figure 6.33**. If you select any vertex and count the sides of the polygons that touch it, you'll see that vertices are surrounded by two regular octagons and a square. Can you see why the sum of the angle measures at every vertex is 360°?

$$135° + 135° + 90° = 360°$$

In Example 2, we found that each angle of a regular octagon measures 135°.	Each angle of a square measures 90°.

The most restrictive condition in creating tessellations is that just *one type of regular polygon* may be used. With this restriction, there are only three possible tessellations, made from equilateral triangles or squares or regular hexagons, as shown in **Figure 6.34**.

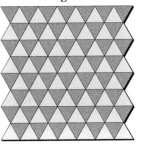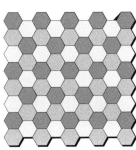

Figure 6.34 The three tessellations formed using one regular polygon

Each tessellation is possible because the angle sum at every vertex is 360°:

Six equilateral triangles at each vertex

$$60° + 60° + 60° + 60° + 60° + 60° = 360°$$

Four squares at each vertex

$$90° + 90° + 90° + 90° = 360°$$

Three regular hexagons at each vertex

$$120° + 120° + 120° = 360°.$$

Each angle measures $\dfrac{(n-2)180°}{n} = \dfrac{(6-2)180°}{6} = 120°.$

Example 3   Angle Requirements of Tessellations

Explain why a tessellation cannot be created using only regular pentagons.

Solution

Let's begin by applying $(n-2)180°$ to find the measure of each angle of a regular pentagon. Each angle measures

$$\frac{(5-2)180°}{5} = \frac{3(180°)}{5} = 108°.$$

A requirement for the formation of a tessellation is that the sum of the measures of the angles that come together at each vertex is 360°. With each angle of a regular pentagon measuring 108°, **Figure 6.35** shows that three regular pentagons fill in $3 \cdot 108° = 324°$ and leave a $360° - 324°$, or a 36°, gap. Because the 360° angle requirement cannot be met, no tessellation by regular pentagons is possible.

✓ **Check Point 3**   Explain why a tessellation cannot be created using only regular octagons.   Each angle of a regular octagon measures 135°. 360° is not a multiple of 135°.

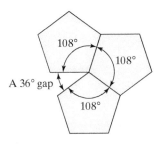

108°

108°

A 36° gap

108°

Figure 6.35

"The regular division of the plane into congruent figures evoking an association in the observer with a familiar natural object is one of these hobbies or problems . . . I have embarked on this geometric problem again and again over the years, trying to throw light on different aspects each time. I cannot imagine what my life would be like if this problem had never occurred to me; one might say that I am head over heels in love with it, and I still don't know why."
—M. C. Escher

Tessellations that are not restricted to the repeated use of regular polygons are endless in number. They are prominent in Islamic art, Italian mosaics, quilts, and ceramics. The Dutch artist M. C. Escher (1898–1972) created a dazzling array of prints, drawings, and paintings using tessellations composed of stylized interlocking animals. Escher's art reflects the mathematics that underlies all things, while creating surreal manipulations of space and perspective that make gentle fun of consensus reality.

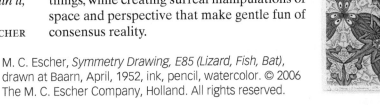

M. C. Escher, *Symmetry Drawing, E85 (Lizard, Fish, Bat)*, drawn at Baarn, April, 1952, ink, pencil, watercolor. © 2006 The M. C. Escher Company, Holland. All rights reserved.

Achieving Success

Watching lectures on YouTube should not be a substitute for going to class.

For many students, it's much easier to get to class at designated times than to watch lectures at home, where no time ever seems to be the right time. If working from the web or MyMathLab, avoid the temptation of listening to multiple lectures in a single sitting. This cramming of information is never a good way to learn. It's risky business to think you can rely on the web as a substitute for the valuable insights and interactions that occur during classroom lecture.

Concept and Vocabulary Check

Exercises in the Concept and Vocabulary Check are intended for group and class discussions.

In Exercises 1–11, fill in each blank so that the resulting statement is true.

1. The distance around the sides of a polygon is called its ___perimeter___.

2. A four-sided polygon is a/an ___quadrilateral___, a five-sided polygon is a/an ___pentagon___, a six-sided polygon is a/an ___hexagon___, a seven-sided polygon is a/an ___heptagon___, and an eight-sided polygon is a/an ___octagon___.

3. A polygon whose sides are all the same length and whose angles all have the same measure is called a/an ___regular___ polygon.

4. Opposite sides of a parallelogram are ___equal in measure___ and ___parallel___.

5. A parallelogram with all sides of equal length without any right angles is called a/an ___rhombus___.

6. A parallelogram with four right angles without all sides of equal length is called a/an ___rectangle___.

7. A parallelogram with four right angles and all sides of equal length is called a/an ___square___.

8. A four-sided figure with exactly one pair of parallel sides is called a/an ___trapezoid___.

9. The perimeter, P, of a rectangle with length l and width w is given by the formula ___$P = 2l + 2w$___.

10. The sum of the measures of the angles of a polygon with n sides is ___$(n-2)180°$___.

11. A pattern consisting of the repeated use of the same geometric figures to completely cover a plane, leaving no gaps and no overlaps, is called a/an ___tessellation___.

In Exercises 12–18, determine whether each statement is true or false. If the statement is false, make the necessary change(s) to produce a true statement. Changes to false statements will vary.

12. Every parallelogram is a rhombus. false

13. Every rhombus is a parallelogram. true

14. All squares are rectangles. true

15. Some rectangles are not squares. true

16. No triangles are polygons. false

17. Every rhombus is a regular polygon. false

18. A requirement for the formation of a tessellation is that the sum of the measures of the angles that come together at each vertex is 360°. true

Respond to Exercises 19–24 using verbal or written explanations. 19–24. Answers will vary.

19. What is a polygon?

20. Explain why rectangles and rhombuses are also parallelograms.

21. Explain why every square is a rectangle, a rhombus, a parallelogram, a quadrilateral, and a polygon.

22. Explain why a square is a regular polygon, but a rhombus is not.

23. Using words only, describe how to find the perimeter of a rectangle.

24. Describe how to find the measure of an angle of a regular pentagon.

Exercise Set 6.3

Practice Exercises

In Exercises 1–4, use the number of sides to name the polygon.

1. quadrilateral **2.**  octagon

3. 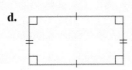 pentagon **4.**  heptagon

Use these quadrilaterals to solve Exercises 5–10.

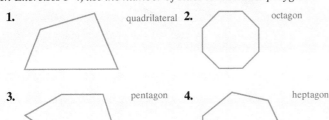

a. **b.**

c. **d.**

e.

5. Which of these quadrilaterals have opposite sides that are parallel? Name these quadrilaterals. *

6. Which of these quadrilaterals have sides of equal length that meet at a vertex? Name these quadrilaterals. *

7. Which of these quadrilaterals have right angles? Name these quadrilaterals. a: square; d: rectangle

8. Which of these quadrilaterals do not necessarily have four sides of equal length? Name these quadrilaterals. *

9. Which of these quadrilaterals is not a parallelogram? Name this quadrilateral. c: trapezoid

10. Which of these quadrilaterals is/are a regular polygon? Name this/these quadrilateral(s). a: square

In Exercises 11–20, find the perimeter of the figure named and shown. Express the perimeter using the same unit of measure that appears on the given side or sides.

11. Rectangle

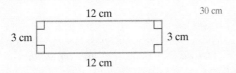

12 cm · 30 cm · 3 cm · 3 cm · 12 cm

12. Parallelogram

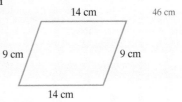

14 cm · 46 cm · 9 cm · 9 cm · 14 cm

13. Rectangle 28 yd

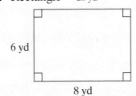

6 yd · 8 yd

14. Rectangle

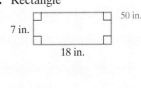

50 in. · 7 in. · 18 in.

15. Square

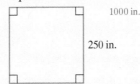

1000 in. · 250 in.

16. Square

3.5 m · 14 m

17. Triangle

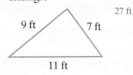

27 ft · 9 ft · 7 ft · 11 ft

18. Triangle

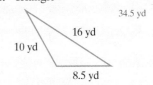

34.5 yd · 16 yd · 10 yd · 8.5 yd

19. Equilateral triangle

18 yd · 6 yd

20. Regular hexagon

4 mm · 24 mm

In Exercises 21–24, find the perimeter of the figure shown. Express the perimeter using the same unit of measure that appears on the given side or sides.

21.

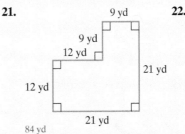

9 yd · 9 yd · 12 yd · 21 yd · 12 yd · 21 yd · 84 yd

22.

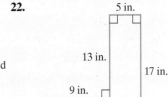

5 in. · 13 in. · 17 in. · 9 in. · 4 in. · 14 in. · 62 in.

23.

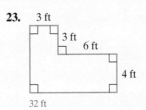

3 ft · 3 ft · 6 ft · 4 ft · 32 ft

24.

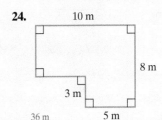

10 m · 8 m · 3 m · 5 m · 36 m

*See Answers to Selected Exercises.

25. Find the sum of the measures of the angles of a five-sided polygon. 540°

26. Find the sum of the measures of the angles of a six-sided polygon. 720°

27. Find the sum of the measures of the angles of a quadrilateral. 360°

28. Find the sum of the measures of the angles of a heptagon. 900°

In Exercises 29–30, each figure shows a regular polygon. Find the measures of angle A and angle B.

29. 108°; 72° **30.** 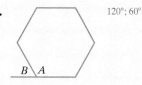 120°; 60°

In Exercises 31–32, **a.** *Find the sum of the measures of the angles for the figure given;* **b.** *Find the measures of angle A and angle B.*

31. 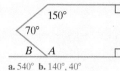 **32.**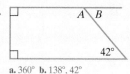

a. 540° **b.** 140°, 40°   **a.** 360° **b.** 138°, 42°

In Exercises 33–36, tessellations formed by two or more regular polygons are shown.

a. *Name the types of regular polygons that surround each vertex.*

b. *Determine the number of angles that come together at each vertex, as well as the measures of these angles.*

c. *Use the angle measures from part (b) to explain why the tessellation is possible.*

33.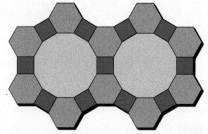

a. squares, hexagons, dodecagons
b. 3 angles;
90°, 120°, 150°
c. The sum of the measures is 360°.

34.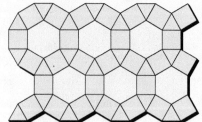

a. triangles, squares, hexagons
b. 4 angles;
60°, 90°, 90°, 120°
c. The sum of the measures is 360°.

35.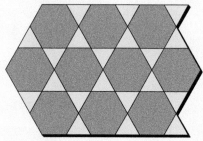

a. triangles, hexagons
b. 4 angles;
60°, 60°, 120°, 120°
c. The sum of the measures is 360°.

36.

a. triangles, dodecagons
b. 3 angles;
60°, 150°, 150°
c. The sum of the measures is 360°.

37. Can a tessellation be created using only regular nine-sided polygons? Explain your answer.

38. Can a tessellation be created using only regular ten-sided polygons? Explain your answer.

37. no; Each angle of the polygon measures 140°. 360° is not a multiple of 140°.
38. no; Each angle of the polygon measures 144°. 360° is not a multiple of 144°.

Practice Plus

In Exercises 39–42, use an algebraic equation to determine each rectangle's dimensions.

39. A rectangular field is four times as long as it is wide. If the perimeter of the field is 500 yards, what are the field's dimensions? 50 yd by 200 yd

40. A rectangular field is five times as long as it is wide. If the perimeter of the field is 288 yards, what are the field's dimensions? 24 yd by 120 yd

41. An American football field is a rectangle with a perimeter of 1040 feet. The length is 200 feet more than the width. Find the width and length of the rectangular field. 160 ft by 360 ft

42. A basketball court is a rectangle with a perimeter of 86 meters. The length is 13 meters more than the width. Find the width and length of the basketball court. 15 m by 28 m

In Exercises 43–44, use algebraic equations to find the measure of each angle that is represented in terms of x.

43. 115°, 115°, 120°, 120°

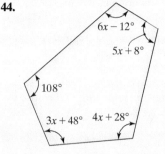

44. 108°, 108°, 108°, 108°

In the figure shown at the top of the next page, the artist has cunningly distorted the "regular" polygons to create a fraudulent tessellation with discrepancies that are too subtle for the eye to notice. In Exercises 45–46, you will use mathematics, not your eyes, to observe the irregularities.

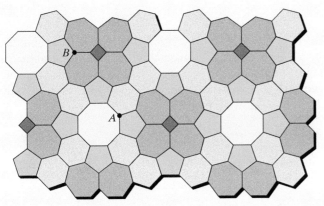

45. Find the sum of the angle measures at vertex A. Then explain why the tessellation is a fake. *

46. Find the sum of the angle measures at vertex B. Then explain why the tessellation is a fake. *

Application Exercises

47. A school playground is in the shape of a rectangle 400 feet long and 200 feet wide. If fencing costs $14 per yard, what will it cost to place fencing around the playground? $5600

48. A rectangular field is 70 feet long and 30 feet wide. If fencing costs $8 per yard, how much will it cost to enclose the field? $533.33

49. One side of a square flower bed is 8 feet long. How many plants are needed if they are to be spaced 8 inches apart around the outside of the bed? 48

50. What will it cost to place baseboard around the region shown if the baseboard costs $0.25 per foot? No baseboard is needed for the 2-foot doorway. $9.50

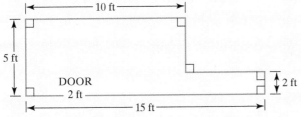

*See Answers to Selected Exercises.

Critical Thinking Exercises

In Exercises 51–52, write an algebraic expression that represents the perimeter of the figure shown.

51.

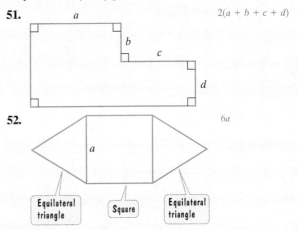

$2(a + b + c + d)$

52.

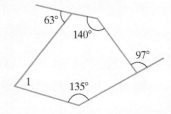

6a

Equilateral triangle   Square   Equilateral triangle

53. Find $m\angle 1$ in the figure shown. 65°

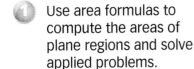

Group Exercise

54. Group members should consult sites on the Internet devoted to tessellations, or tilings, and present a report that expands upon the information in this section. Include a discussion of cultures that have used tessellations on fabrics, wall coverings, baskets, rugs, and pottery, with examples. Include the Alhambra, a fourteenth-century palace in Granada, Spain, in the presentation, as well as works by the artist M. C. Escher. Discuss the various symmetries (translations, rotations, reflections) associated with tessellations. Demonstrate how to create unique tessellations, including Escher-type patterns. Other than creating beautiful works of art, are there any practical applications of tessellations? Answers will vary.

6.4 : Area and Circumference

What am I supposed to learn?

After you have read this section, you should be able to:

1. Use area formulas to compute the areas of plane regions and solve applied problems.

2. Use formulas for a circle's circumference and area.

The size of a house is described in square feet. But how do you know from the real estate ad whether a 1200-square-foot home with the backyard pool is large enough to warrant a visit? Faced with hundreds of ads, you need some way to sort out the best bets. What does 1200 square feet mean and how is this area determined? In this section, we discuss how to compute the areas of plane regions.

1 Use area formulas to compute the areas of plane regions and solve applied problems.

Square unit of measure

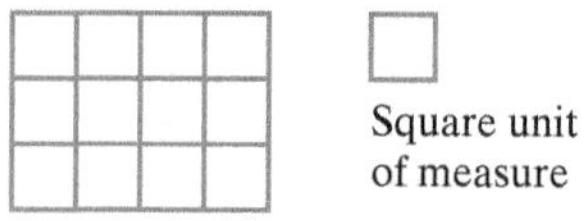

Figure 6.36 The area of the region on the left is 12 square units.

Formulas for Area

In Section 5.2, we saw that the area of a two-dimensional figure is the number of square units, such as square inches or square miles, it takes to fill the interior of the figure. For example, **Figure 6.36** shows that there are 12 square units contained within the rectangular region. The area of the region is 12 square units. Notice that the area can be determined in the following manner:

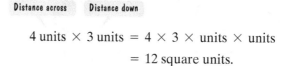

$$4 \text{ units} \times 3 \text{ units} = 4 \times 3 \times \text{units} \times \text{units}$$
$$= 12 \text{ square units.}$$

The area of a rectangular region, usually referred to as the area of a rectangle, is the product of the distance across (length) and the distance down (width).

Area of a Rectangle and a Square

The area, A, of a rectangle with length l and width w is given by the formula

$$A = lw.$$

The area, A, of a square with one side measuring s linear units is given by the formula

$$A = s^2.$$

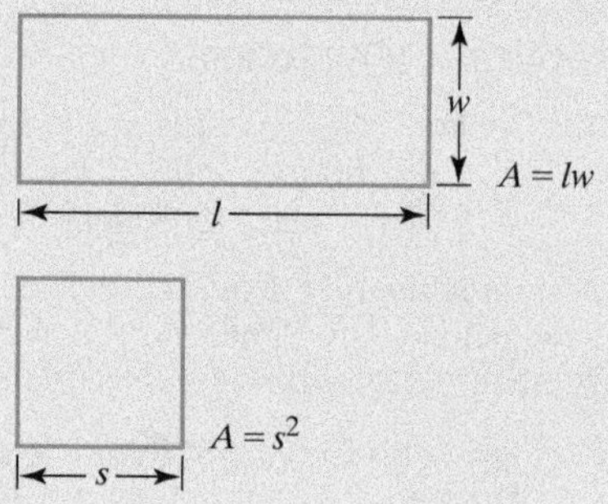

$A = lw$

$A = s^2$

Example 1   **Solving an Area Problem**

You decide to cover the path shown in **Figure 6.37** with bricks.

a. Find the area of the path.

b. If the path requires four bricks for every square foot, how many bricks are needed for the project?

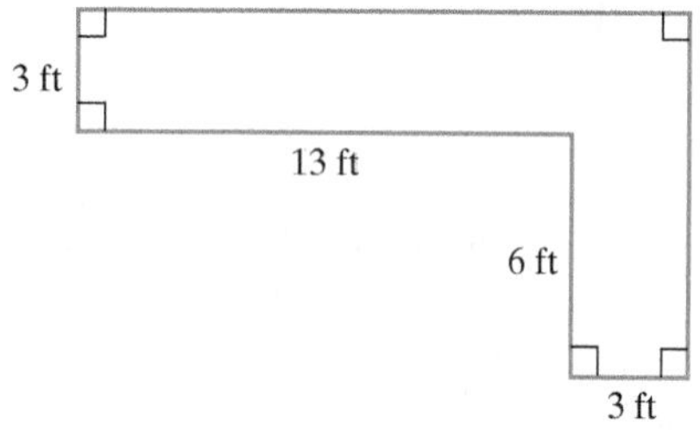

3 ft

13 ft

6 ft

3 ft

Figure 6.37

Solution

a. Because we have a formula for the area of a rectangle, we begin by drawing a dashed line that divides the path into two rectangles. One way of doing this is shown below. We then use the length and width of each rectangle to find its area. The computations for area are shown in the green and blue voice balloons.

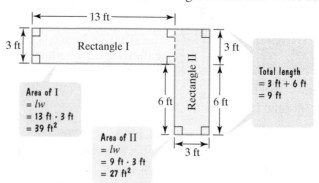

The area of the path is found by adding the areas of the two rectangles.

$$\text{Area of path} = 39 \text{ ft}^2 + 27 \text{ ft}^2 = 66 \text{ ft}^2$$

Blitzer Bonus

Appraising a House

A house is measured by an appraiser hired by a bank to help establish its value. The appraiser works from the outside, measuring off a rectangle. Then the appraiser adds the living spaces that lie outside the rectangle and subtracts the empty areas inside the rectangle. The final figure, in square feet, includes all the finished floor space in the house. Not included are the garage, outside porches, decks, or an unfinished basement.

A 1000-square-foot house is considered small, one with 2000 square feet average, and one with more than 2500 square feet pleasantly large. If a 1200-square-foot house has three bedrooms, the individual rooms might seem snug and cozy. With only one bedroom, the space may feel palatial!

Average Size of New U.S. Single-Family Homes

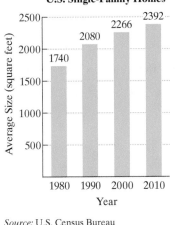

Source: U.S. Census Bureau

b. The path requires 4 bricks per square foot. The number of bricks needed for the project is the number of square feet in the path, its area, times 4.

$$\text{Number of bricks needed} = 66 \text{ ft}^2 \cdot \frac{4 \text{ bricks}}{\text{ft}^2} = 66 \cdot 4 \text{ bricks} = 264 \text{ bricks}$$

Thus, 264 bricks are needed for the project.

Check Point 1 Find the area of the path described in Example 1, rendered below as a green region, by first measuring off a large rectangle as shown. The area of the path is the area of the large rectangle (the blue and green regions combined) minus the area of the blue rectangle. Do you get the same answer as we did in Example 1(a)?

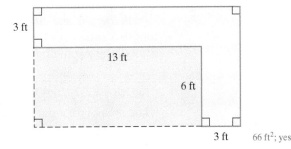

66 ft²; yes

In Section 5.2, we saw that although there are 3 linear feet in a linear yard, there are 9 square feet in a square yard. If a problem requires measurement of area in square yards and the linear measures are given in feet, to avoid errors, first convert feet to yards. Then apply the area formula. This idea is illustrated in Example 2.

Example 2 Solving an Area Problem

What will it cost to carpet a rectangular floor measuring 12 feet by 15 feet if the carpet costs $18.50 per square yard?

Solution

We begin by converting the linear measures from feet to yards.

$$12 \text{ ft} = \frac{12 \text{ ft}}{1} \cdot \frac{1 \text{ yd}}{3 \text{ ft}} = \frac{12}{3} \text{ yd} = 4 \text{ yd}$$

$$15 \text{ ft} = \frac{15 \text{ ft}}{1} \cdot \frac{1 \text{ yd}}{3 \text{ ft}} = \frac{15}{3} \text{ yd} = 5 \text{ yd}$$

Next, we find the area of the rectangular floor in square yards.

$$A = lw = 5 \text{ yd} \cdot 4 \text{ yd} = 20 \text{ yd}^2$$

Finally, we find the cost of the carpet by multiplying the cost per square yard, $18.50, by the number of square yards in the floor, 20.

$$\text{Cost of carpet} = \frac{\$18.50}{\text{yd}^2} \cdot \frac{20 \text{ yd}^2}{1} = \$18.50(20) = \$370$$

It will cost $370 to carpet the floor.

Check Point 2 What will it cost to carpet a rectangular floor measuring 18 feet by 21 feet if the carpet costs $16 per square yard? $672

We can use the formula for the area of a rectangle to develop formulas for areas of other polygons. We begin with a parallelogram, a quadrilateral with opposite sides equal and parallel. The **height** of a parallelogram is the perpendicular distance between two of the parallel sides. Height is denoted by h in **Figure 6.38**. The **base**, denoted by b, is the length of either of these parallel sides.

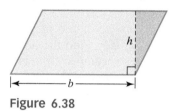

Figure 6.38   Figure 6.39

In **Figure 6.39**, the red triangular region has been cut off from the right of the parallelogram and attached to the left. The resulting figure is a rectangle with length b and width h. Because bh is the area of the rectangle, it also represents the area of the parallelogram.

Great Question!

Is the height of a parallelogram the length of one of its sides?

No. The height of a parallelogram is the perpendicular distance between two of the parallel sides. It is *not* the length of a side.

Area of a Parallelogram

The area, A, of a parallelogram with height h and base b is given by the formula

$$A = bh.$$

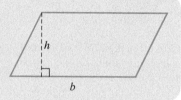

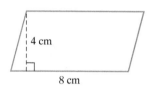

4 cm

8 cm

Figure 6.40

Example 3   Using the Formula for a Parallelogram's Area

Find the area of the parallelogram in **Figure 6.40**.

Solution

As shown in the figure, the base is 8 centimeters and the height is 4 centimeters. Thus, $b = 8$ and $h = 4$.

$$A = bh$$
$$A = 8 \text{ cm} \cdot 4 \text{ cm} = 32 \text{ cm}^2$$

The area is 32 cm².

✓ **Check Point 3**   Find the area of a parallelogram with a base of 10 inches and a height of 6 inches.   60 in.²

Figure 6.41 demonstrates how we can use the formula for the area of a parallelogram to obtain a formula for the area of a triangle. The area of the parallelogram in **Figure 6.41(a)** is given by $A = bh$. The diagonal shown in the parallelogram divides it into two triangles with the same size and shape. This means that the area of each triangle is one-half that of the parallelogram. Thus, the area of the triangle in **Figure 6.41(b)** is given by $A = \frac{1}{2}bh$.

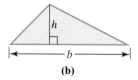

(a)

(b)

Figure 6.41

Area of a Triangle

The area, A, of a triangle with height h and base b is given by the formula

$$A = \frac{1}{2}bh.$$

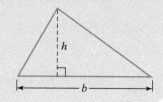

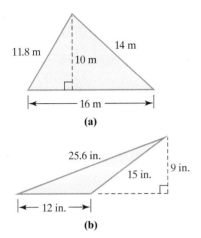

Figure 6.42

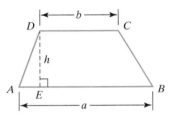

Figure 6.43

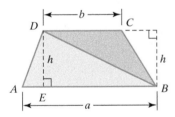

Figure 6.44

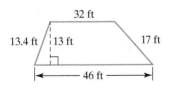

Figure 6.45

Example 4 **Using the Formula for a Triangle's Area**

Find the area of each triangle in **Figure 6.42**.

Solution

a. In **Figure 6.42(a)**, the base is 16 meters and the height is 10 meters, so $b = 16$ and $h = 10$. We do not need the 11.8 meters or the 14 meters to find the area. The area of the triangle is

$$A = \frac{1}{2}bh = \frac{1}{2} \cdot 16 \text{ m} \cdot 10 \text{ m} = 80 \text{ m}^2.$$

The area is 80 square meters.

b. In **Figure 6.42(b)**, the base is 12 inches. The base line needs to be extended to draw the height. However, we still use 12 inches for b in the area formula. The height, h, is given to be 9 inches. The area of the triangle is

$$A = \frac{1}{2}bh = \frac{1}{2} \cdot 12 \text{ in.} \cdot 9 \text{ in.} = 54 \text{ in.}^2.$$

The area of the triangle is 54 square inches.

Check Point 4   A sailboat has a triangular sail with a base of 12 feet and a height of 5 feet. Find the area of the sail.   30 ft²

The formula for the area of a triangle can be used to obtain a formula for the area of a trapezoid. Consider the trapezoid shown in **Figure 6.43**. The lengths of the two parallel sides, called the **bases**, are represented by a (the lower base) and b (the upper base). The trapezoid's height, denoted by h, is the perpendicular distance between the two parallel sides.

In **Figure 6.44**, we have drawn line segment BD, dividing the trapezoid into two triangles, shown in yellow and red. The area of the trapezoid is the sum of the areas of these triangles.

Area of trapezoid	=	Area of yellow △	plus	Area of red △
A	=	$\frac{1}{2}ah$	+	$\frac{1}{2}bh$
	=	$\frac{1}{2}h(a+b)$		

Area of a Trapezoid

The area, A, of a trapezoid with parallel bases a and b and height h is given by the formula

$$A = \frac{1}{2}h(a + b).$$

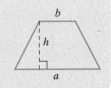

Example 5 **Finding the Area of a Trapezoid**

Find the area of the trapezoid in **Figure 6.45**.

Solution

The height, h, is 13 feet. The lower base, a, is 46 feet, and the upper base, b, is 32 feet. We do not use the 17-foot and 13.4-foot sides in finding the trapezoid's area.

$$A = \frac{1}{2}h(a + b) = \frac{1}{2} \cdot 13 \text{ ft} \cdot (46 \text{ ft} + 32 \text{ ft})$$

$$= \frac{1}{2} \cdot 13 \text{ ft} \cdot 78 \text{ ft} = 507 \text{ ft}^2$$

The area of the trapezoid is 507 square feet.

Check Point 5 Find the area of a trapezoid with bases of length 20 feet and 10 feet and height 7 feet. 105 ft²

2 Use formulas for a circle's circumference and area.

A **circle** is a set of points in the plane equally distant from a given point, its **center**. **Figure 6.46** shows two circles. The **radius** (plural: radii), r, is a line segment from the center to any point on the circle. For a given circle, all radii have the same length. The **diameter**, d, is a line segment through the center whose endpoints both lie on the circle. For a given circle, all diameters have the same length. In any circle, the **length of the diameter is twice the length of the radius**.

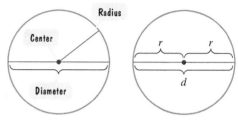

Figure 6.46

The words *radius* and *diameter* refer to both the line segments in **Figure 6.46**, as well as to their linear measures. The distance around a circle (its perimeter) is called its **circumference**, C. For all circles, if you divide the circumference by the diameter, or by twice the radius, you will get the same number. This ratio is the irrational number π and is approximately equal to 3.14:

$$\frac{C}{d} = \pi \quad \text{or} \quad \frac{C}{2r} = \pi.$$

Thus,

$$C = \pi d \quad \text{or} \quad C = 2\pi r.$$

Finding the Distance Around a Circle

The circumference, C, of a circle with diameter d and radius r is

$$C = \pi d \quad \text{or} \quad C = 2\pi r.$$

When computing a circle's circumference by hand, round π to 3.14. When using a calculator, use the $\boxed{\pi}$ key, which gives the value of π rounded to approximately 11 decimal places. In either case, calculations involving π give approximate answers. These answers can vary slightly depending on how π is rounded. The symbol $\approx$ (is approximately equal to) will be written in these calculations.

Example 6 Finding a Circle's Circumference

Find the circumference of the circle in **Figure 6.47**.

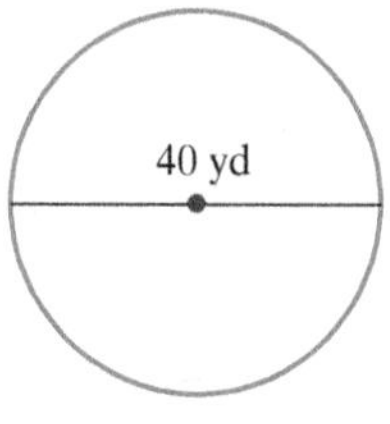

Solution

The diameter is 40 yards, so we use the formula for circumference with d in it.

$$C = \pi d = \pi(40 \text{ yd}) = 40\pi \text{ yd} \approx 125.7 \text{ yd}$$

The distance around the circle is approximately 125.7 yards.

40 yd

Figure 6.47

Check Point 6 Find the circumference of a circle whose diameter measures 10 inches. Express the answer in terms of π and then round to the nearest tenth of an inch. 10π in.; 31.4 in.

Figure 6.48

Example 7 — Using the Circumference Formula

How much trim, to the nearest tenth of a foot, is needed to go around the window shown in **Figure 6.48**?

Solution

The trim covers the 6-foot bottom of the window, the two 8-foot sides, and the half-circle (called a semicircle) on top. The length needed is

6 ft + 8 ft + 8 ft + circumference of the semicircle.

The circumference of the semicircle is half the circumference of a circle whose diameter is 6 feet.

Circumference of semicircle

$$\text{Circumference of semicircle} = \frac{1}{2}\pi d$$

$$= \frac{1}{2}\pi(6 \text{ ft}) = 3\pi \text{ ft} \approx 9.4 \text{ ft}$$

Rounding the circumference to the nearest tenth (9.4 feet), the length of trim that is needed is approximately

6 ft + 8 ft + 8 ft + 9.4 ft,

or 31.4 feet.

Check Point 7 In **Figure 6.48**, suppose that the dimensions are 10 feet and 12 feet for the window's bottom and side, respectively. How much trim, to the nearest tenth of a foot, is needed to go around the window? 49.7 ft

The irrational number π is also used to find the area of a circle in square units. This is because the ratio of a circle's area to the square of its radius is π:

$$\frac{A}{r^2} = \pi.$$

Multiplying both sides of this equation by r^2 gives a formula for determining a circle's area.

> **Finding the Area of a Circle**
> The area, A, of a circle with radius r is
> $$A = \pi r^2.$$

Example 8 — Problem Solving Using the Formula for a Circle's Area

Which one of the following is the better buy: a large pizza with a 16-inch diameter for $15.00 or a medium pizza with an 8-inch diameter for $7.50?

Solution

The better buy is the pizza with the lower price per square inch. The radius of the large pizza is $\frac{1}{2} \cdot 16$ inches, or 8 inches, and the radius of the medium pizza is $\frac{1}{2} \cdot 8$ inches, or 4 inches. The area of the surface of each circular pizza is determined using the formula for the area of a circle.

Large pizza: $A = \pi r^2 = \pi(8 \text{ in.})^2 = 64\pi \text{ in.}^2 \approx 201 \text{ in.}^2$

Medium pizza: $A = \pi r^2 = \pi(4 \text{ in.})^2 = 16\pi \text{ in.}^2 \approx 50 \text{ in.}^2$

Technology

You can use your calculator to obtain the price per square inch for each pizza in Example 8. The price per square inch for the large pizza, $\frac{15}{64\pi}$, is approximated by one of the following sequences of keystrokes:

Many Scientific Calculators

$15 \div (\ (\ 64 \times \pi\ )\ ) =$

Many Graphing Calculators

$15 \div (\ (\ 64\ \pi\ )\ )$ ENTER

The area of the large pizza is 64π square inches, or approximately 201 square inches. The area of the small pizza is 16π square inches, or approximately 50 square inches. For each pizza, the price per square inch is found by dividing the price by the area:

$$\text{Price per square inch for large pizza} = \frac{\$15.00}{64\pi \text{ in.}^2} \approx \frac{\$15.00}{201 \text{ in.}^2} \approx \frac{\$0.07}{\text{in.}^2}$$

$$\text{Price per square inch for medium pizza} = \frac{\$7.50}{16\pi \text{ in.}^2} \approx \frac{\$7.50}{50 \text{ in.}^2} = \frac{\$0.15}{\text{in.}^2}$$

The large pizza costs approximately \$0.07 per square inch and the medium pizza costs approximately \$0.15 per square inch. Thus, the large pizza is the better buy.

In Example 8, did you at first think that the price per square inch would be the same for the large and the medium pizzas? After all, the radius of the large pizza is twice that of the medium pizza, and the cost of the large is twice that of the medium. However, the large pizza's area, 64π square inches, is *four times the area* of the medium pizza, 16π square inches. Doubling the radius of a circle increases its area by a factor of 2^2, or 4. In general, if the radius of a circle is increased by k times its original linear measure, the area is multiplied by k^2. The same principle is true for any two-dimensional figure: If the shape of the figure is kept the same while linear dimensions are increased k times, the area of the larger figure is k^2 times greater than the area of the original figure.

Check Point 8 Which one of the following is the better buy: a large pizza with an 18-inch diameter for \$20.00 or a medium pizza with a 14-inch diameter for \$14.00? large pizza

Concept and Vocabulary Check

Exercises in the Concept and Vocabulary Check are intended for group and class discussions.

In Exercises 1–8, fill in each blank so that the resulting statement is true.

1. The area, A, of a rectangle with length l and width w is given by the formula ___$A = lw$___.
2. The area, A, of a square with one side measuring s linear units is given by the formula ___$A = s^2$___.
3. The area, A, of a parallelogram with height h and base b is given by the formula ___$A = bh$___.
4. The area, A, of a triangle with height h and base b is given by the formula ___$A = \frac{1}{2}bh$___.
5. The area, A, of a trapezoid with parallel bases a and b and height h is given by the formula ___$A = \frac{1}{2}h(a + b)$___.
6. The circumference, C, of a circle with diameter d is given by the formula ___$C = \pi d$___.
7. The circumference, C, of a circle with radius r is given by the formula ___$C = 2\pi r$___.
8. The area, A, of a circle with radius r is given by the formula ___$A = \pi r^2$___.

In Exercises 9–13, determine whether each statement is true or false. If the statement is false, make the necessary change(s) to produce a true statement. Changes to false statements will vary.

9. The area, A, of a rectangle with length l and width w is given by the formula $A = 2l + 2w$. false
10. The height of a parallelogram is the perpendicular distance between two of the parallel sides. true
11. The area of either triangle formed by drawing a diagonal in a parallelogram is one-half that of the parallelogram. true
12. In any circle, the length of the radius is twice the length of the diameter. false
13. The ratio of a circle's circumference to its diameter is the irrational number π. true

Respond to Exercises 14–18 using verbal or written explanations. 14–18. Answers will vary.

14. Using the formula for the area of a rectangle, explain how the formula for the area of a parallelogram ($A = bh$) is obtained.
15. Using the formula for the area of a parallelogram ($A = bh$), explain how the formula for the area of a triangle ($A = \frac{1}{2}bh$) is obtained.
16. Using the formula for the area of a triangle, explain how the formula for the area of a trapezoid is obtained.
17. Explain why a circle is not a polygon.
18. Describe the difference between the formulas needed to solve the following problems: How much fencing is needed to enclose a circular garden? How much fertilizer is needed for a circular garden?

15. C: 8π cm, ≈ 25.1 cm; A: 16π cm^2, ≈ 50.3 cm^2 **16.** C: 18π m, ≈ 56.5 m; A: 81π m^2, ≈ 254.5 m^2 **17.** C: 12π yd, ≈ 37.7 yd; A: 36π yd^2, ≈ 113.1 yd^2
18. C: 40π ft, ≈ 125.7 ft; A: 400π ft^2, ≈ 1256.6 ft^2

Exercise Set 6.4

Practice Exercises

In Exercises 1–14, use the formulas developed in this section to find the area of each figure.

1.
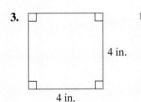
3 m
18 m^2 6 m

2.
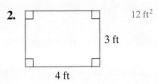
12 ft^2
3 ft
4 ft

3.
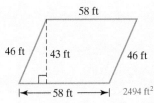
16 in.^{2}
4 in.
4 in.

4.
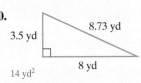
9 cm^2
3 cm
3 cm

5.
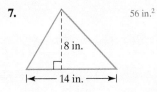
50 cm
44 cm   42 cm   44 cm
|← 50 cm →| 2100 cm^2

6.
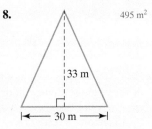
58 ft
46 ft   43 ft   46 ft
|← 58 ft →| 2494 ft^2

7.
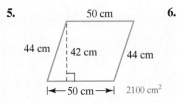
56 in.^{2}
8 in.
|← 14 in. →|

8.
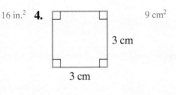
495 m^2
33 m
|← 30 m →|

9.
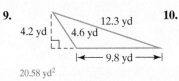
12.3 yd
4.2 yd   4.6 yd
|← 9.8 yd →|
20.58 yd^2

10.
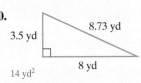
8.73 yd
3.5 yd
8 yd
14 yd^2

11.
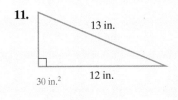
13 in.
12 in.
30 in.^{2}

12.
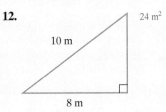
24 m^2
10 m
8 m

13.
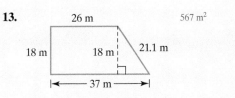
26 m
18 m   18 m   21.1 m
|← 37 m →|
567 m^2

14.
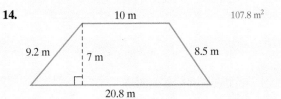
10 m
9.2 m   7 m   8.5 m
20.8 m
107.8 m^2

In Exercises 15–18, find the circumference and area of each circle. Express answers in terms of π and then round to the nearest tenth.

15.
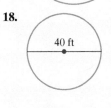
4 cm

16.

9 m

17.
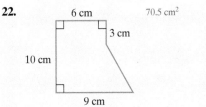
12 yd

18.

40 ft

Find the area of each figure in Exercises 19–24. Where necessary, express answers in terms of π and then round to the nearest tenth.

19.   72 m^2
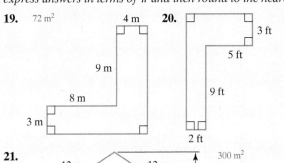
4 m
9 m
8 m
3 m

20.   39 ft^2
3 ft
5 ft
9 ft
2 ft

21.
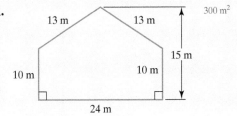
300 m^2
13 m   13 m
15 m
10 m   10 m
24 m

22.
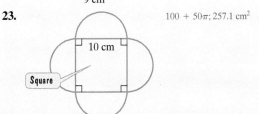
6 cm
3 cm
10 cm
9 cm
70.5 cm^2

23.
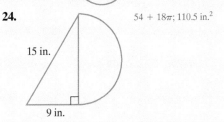
$100 + 50\pi$; 257.1 cm^2
10 cm
Square

24.
15 in.
9 in.
$54 + 18\pi$; 110.5 in.^{2}

In Exercises 25–28, find a formula for the total area, A, of each figure in terms of the variable(s) shown. Where necessary, use π in the formula.

25.

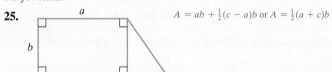

$A = ab + \frac{1}{2}(c - a)b$ or $A = \frac{1}{2}(a + c)b$

26.

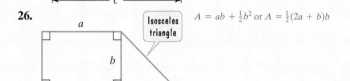

$A = ab + \frac{1}{2}b^2$ or $A = \frac{1}{2}(2a + b)b$

27.

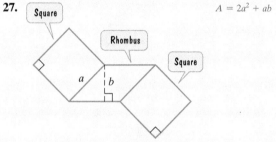

$A = 2a^2 + ab$

28.

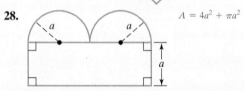

$A = 4a^2 + \pi a^2$

Practice Plus

In Exercises 29–30, find the area of each shaded region.

29.

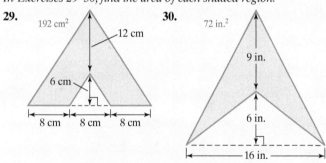

192 cm²
12 cm
6 cm
8 cm 8 cm 8 cm

30.

72 in.²
9 in.
6 in.
16 in.

In Exercises 31–34, find the area of each shaded region in terms of π.

31.

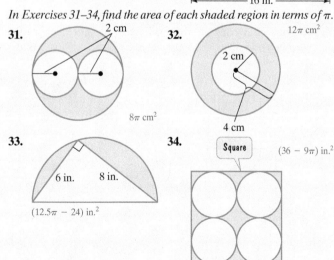

2 cm
8π cm²

32.

12π cm²
2 cm
4 cm

33.

6 in. 8 in.
(12.5π − 24) in.²

34.

Square
(36 − 9π) in.²
6 in.

In Exercises 35–36, find the perimeter and the area of each figure. Where necessary, express answers in terms of π and round to the nearest tenth.

35.

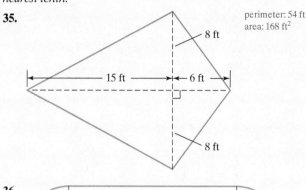

perimeter: 54 ft
area: 168 ft²
8 ft
15 ft   6 ft
8 ft

36.

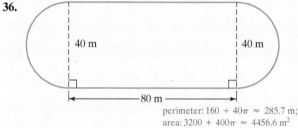

40 m   40 m
80 m

perimeter: 160 + 40π ≈ 285.7 m;
area: 3200 + 400π ≈ 4456.6 m²

Application Exercises

37. What will it cost to carpet a rectangular floor measuring 9 feet by 21 feet if the carpet costs $26.50 per square yard? $556.50

38. A plastering contractor charges $18 per square yard. What is the cost of plastering 60 feet of wall in a house with a 9-foot ceiling? $1080

39. A rectangular kitchen floor measures 12 feet by 15 feet. A stove on the floor has a rectangular base measuring 3 feet by 4 feet, and a refrigerator covers a rectangular area of the floor measuring 4 feet by 5 feet. How many square feet of tile will be needed to cover the kitchen floor not counting the area used by the stove and the refrigerator?   148 ft²

40. A rectangular room measures 12 feet by 15 feet. The entire room is to be covered with rectangular tiles that measure 3 inches by 2 inches. If the tiles are sold at ten for 30¢, what will it cost to tile the room? $129.60

41. The lot in the figure shown, except for the house, shed, and driveway, is lawn. One bag of lawn fertilizer costs $25.00 and covers 4000 square feet.

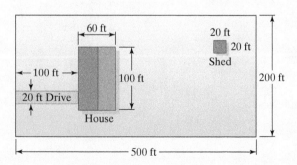

60 ft
20 ft
20 ft
Shed
100 ft
100 ft
200 ft
20 ft Drive
House
500 ft

a. Determine the minimum number of bags of fertilizer needed for the lawn.   23 bags

b. Find the total cost of the fertilizer. $575

42. Taxpayers with an office in their home may deduct a percentage of their home-related expenses. This percentage is based on the ratio of the office's area to the area of the home. A taxpayer with an office in a 2200-square-foot home maintains a 20 foot by 16 foot office. If the yearly electric bills for the home come to $4800, how much of this is deductible?   ≈ $698.18

43. You are planning to paint the house whose dimensions are shown in the figure.

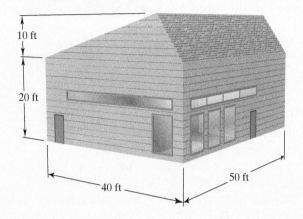

a. How many square feet will you need to paint? (There are four windows, each 8 ft by 5 ft; two windows, each 30 ft by 2 ft; and two doors, each 80 in. by 36 in., that do not require paint.)   3680 ft²

b. The paint that you have chosen is available in gallon cans only. Each can covers 500 square feet. If you want to use two coats of paint, how many cans will you need for the project?   15 cans

c. If the paint you have chosen sells for $26.95 per gallon, what will it cost to paint the house? $404.25

The diagram shows the floor plan for a one-story home. Use the given measurements to solve Exercises 44–46. (A calculator will be helpful in performing the necessary computations.)

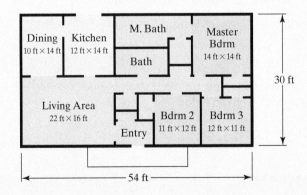

44. If construction costs $95 per square foot, find the cost of building the home. $153,900

45. If carpet costs $17.95 per square yard and is available in whole square yards only, find the cost of carpeting the three bedroom floors. $933.40

46. If ceramic tile costs $26.95 per square yard and is available in whole square yards only, find the cost of installing ceramic tile on the kitchen and dining room floors. $943.25

In Exercises 47–48, express the required calculation in terms of π and then round to the nearest tenth.

47. How much fencing is required to enclose a circular garden whose radius is 20 meters? 40π m; 125.7 m

48. A circular rug is 6 feet in diameter. How many feet of fringe is required to edge this rug? 6π ft; 18.8 ft

49. How many plants spaced every 6 inches are needed to surround a circular garden with a 30-foot radius?   377 plants

50. A stained glass window is to be placed in a house. The window consists of a rectangle, 6 feet high by 3 feet wide, with a semicircle at the top. Approximately how many feet of stripping, to the nearest tenth of a foot, will be needed to frame the window?   19.7 ft

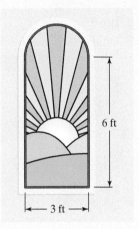

51. Which one of the following is a better buy: a large pizza with a 14-inch diameter for $12.00 or a medium pizza with a 7-inch diameter for $5.00?   large pizza

52. Which one of the following is a better buy: a large pizza with a 16-inch diameter for $12.00 or two small pizzas, each with a 10-inch diameter, for $12.00?   large pizza

Critical Thinking Exercises

53. You need to enclose a rectangular region with 200 feet of fencing. Experiment with different lengths and widths to determine the maximum area you can enclose. Which quadrilateral encloses the most area?   the square with sides of 50 ft; 2500 ft²

54. Suppose you know the cost for building a rectangular deck measuring 8 feet by 10 feet. If you decide to increase the dimensions to 12 feet by 15 feet, by how much will the cost increase?   by a factor of $\frac{9}{4}$

55. A rectangular swimming pool measures 14 feet by 30 feet. The pool is surrounded on all four sides by a path that is 3 feet wide. If the cost to resurface the path is $2 per square foot, what is the total cost of resurfacing the path? $600

56. A proposed oil pipeline will cross 16.8 miles of national forest. The width of the land needed for the pipeline is 200 feet. If the U.S. Forest Service charges the oil company $32 per acre, calculate the total cost. (1 mile = 5280 feet and 1 acre = 43,560 square feet.) $13,032.73

6.5 : Volume

What am I supposed to learn?

After you have read this section, you should be able to:

1. Use volume formulas to compute the volumes of three-dimensional figures and solve applied problems.

2. Compute the surface area of a three-dimensional figure.

You are considering going to Judge Judy's Web site and filling out a case submission form to appear on her TV show. The case involves your contractor, who promised to install a water tank that holds 500 gallons of water. Upon delivery, you noticed that the capacity was not printed anywhere, so you decided to do some measuring. The tank is shaped like a giant tuna can, with a circular top and bottom. You measured the radius of each circle to be 3 feet and you measured the tank's height to be 2 feet 4 inches. You know that 500 gallons is the capacity of a solid figure with a volume of about 67 cubic feet. Now you need some sort of method to compute the volume of the

water tank. In this section, we discuss how to compute the volumes of various solid, three-dimensional figures. Using a formula you will learn in the section, you can determine whether the evidence indicates you can win a case against the contractor if you appear on *Judge Judy*. Or do you risk joining a cast of bozos who entertain television viewers by being loudly castigated by the judge? (Before a possible ear-piercing "Baloney, sir, you're a geometric idiot!", we suggest working Exercise 45 in Exercise Set 6.5.)

Formulas for Volume

In Section 5.2, we saw that **volume** refers to the amount of space occupied by a solid object, determined by the number of cubic units it takes to fill the interior of that object. For example, **Figure 6.49** shows that there are 18 cubic units contained within the box. The volume of the box, called a **rectangular solid**, is 18 cubic units. The box has a length of 3 units, a width of 3 units, and a height of 2 units. The volume, 18 cubic units, may be determined by finding the product of the length, the width, and the height:

$$\text{Volume} = 3 \text{ units} \cdot 3 \text{ units} \cdot 2 \text{ units} = 18 \text{ units}^3.$$

In general, the volume, V, of a rectangular solid is the product of its length, l, its width, w, and its height, h:

$$V = lwh.$$

If the length, width, and height are the same, the rectangular solid is called a **cube**. Formulas for these boxlike shapes are given below.

1. Use volume formulas to compute the volumes of three-dimensional figures and solve applied problems.

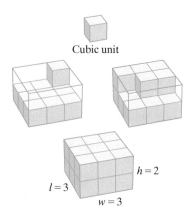

Cubic unit

$h = 2$
$l = 3$
$w = 3$

Figure 6.49
Volume = 18 cubic units

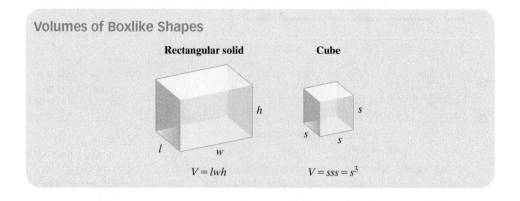

Volumes of Boxlike Shapes

Rectangular solid

h
l
w

$V = lwh$

Cube

s
s
s

$V = sss = s^3$

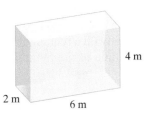

4 m

2 m

6 m

Figure 6.50

Example 1   **Finding the Volume of a Rectangular Solid**

Find the volume of the rectangular solid in **Figure 6.50**.

Solution

As shown in the figure, the length is 6 meters, the width is 2 meters, and the height is 4 meters. Thus, $l = 6$, $w = 2$, and $h = 4$.

$$V = lwh = 6 \text{ m} \cdot 2 \text{ m} \cdot 4 \text{ m} = 48 \text{ m}^3$$

The volume of the rectangular solid is 48 cubic meters.

✓ **Check Point 1**   Find the volume of a rectangular solid with length 5 feet, width 3 feet, and height 7 feet. 105 ft^3

In Section 5.2, we saw that although there are 3 feet in a yard, there are 27 cubic feet in a cubic yard. If a problem requires measurement of volume in cubic yards and the linear measures are given in feet, to avoid errors, first convert feet to yards. Then apply the volume formula. This idea is illustrated in Example 2.

Example 2   **Solving a Volume Problem**

You are about to begin work on a swimming pool in your yard. The first step is to dig a hole that is 90 feet long, 60 feet wide, and 6 feet deep. You will use a truck that can carry 10 cubic yards of dirt and charges $35 per load. How much will it cost you to have all the dirt hauled away?

Solution

We begin by converting feet to yards:

$$90 \text{ ft} = \frac{90 \text{ ft}}{1} \cdot \frac{1 \text{ yd}}{3 \text{ ft}} = \frac{90}{3} \text{ yd} = 30 \text{ yd}.$$

Similarly, 60 ft = 20 yd and 6 ft = 2 yd. Next, we find the volume of dirt that needs to be dug out and hauled off.

$$V = lwh = 30 \text{ yd} \cdot 20 \text{ yd} \cdot 2 \text{ yd} = 1200 \text{ yd}^3$$

Now, we find the number of loads necessary for the truck to haul off all the dirt. Because the truck carries 10 cubic yards, divide the number of cubic yards of dirt by 10.

$$\text{Number of truckloads} = \frac{1200 \text{ yd}^3}{\dfrac{10 \text{ yd}^3}{\text{load}}} = \frac{1200 \text{ yd}^3}{1} \cdot \frac{\text{load}}{10 \text{ yd}^3} = \frac{1200}{10} \text{ loads} = 120 \text{ loads}$$

Because the truck charges $35 per load, the cost to have all the dirt hauled away is the number of loads, 120, times the cost per load, $35.

$$\text{Cost to haul all dirt away} = \frac{120 \text{ loads}}{1} \cdot \frac{\$35}{\text{load}} = 120(\$35) = \$4200$$

The dirt-hauling phase of the pool project will cost you $4200.

✓ **Check Point 2**   Find the volume, in cubic yards, of a cube whose edges each measure 6 feet. 8 yd^3

Figure 6.51 The volume of a pyramid is $\frac{1}{3}$ the volume of a rectangular solid having the same base and the same height.

A rectangular solid is an example of a **polyhedron**, a solid figure bounded by polygons. A rectangular solid is bounded by six rectangles, called faces. By contrast, a **pyramid** is a polyhedron whose base is a polygon and whose sides are triangles. **Figure 6.51** shows a pyramid with a rectangular base drawn inside a rectangular solid. The contents of three pyramids with rectangular bases exactly fill a rectangular solid of the same base and height. Thus, the formula for the volume of the pyramid is $\frac{1}{3}$ that of the rectangular solid.

Volume of a Pyramid

The volume, V, of a pyramid is given by the formula

$$V = \frac{1}{3}Bh,$$

where B is the area of the base and h is the height (the perpendicular distance from the top to the base).

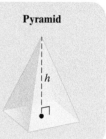

Pyramid

The Transamerica Tower's 3678 windows take cleaners one month to wash. Its foundation is sunk 15.5 m (52 ft) into the ground and is designed to move with earth tremors.

Example 3 Using the Formula for a Pyramid's Volume

Capped with a pointed spire on top of its 48 stories, the Transamerica Tower in San Francisco is a pyramid with a square base. The pyramid is 256 meters (853 feet) tall. Each side of the square base has a length of 52 meters. Although San Franciscans disliked it when it opened in 1972, they have since accepted it as part of the skyline. Find the volume of the building.

Solution

First find the area of the base, represented as B in the volume formula. Because each side of the square base is 52 meters, the area of the base is

$$B = 52 \text{ m} \cdot 52 \text{ m} = 2704 \text{ m}^2.$$

The area of the base is 2704 square meters. Because the pyramid is 256 meters tall, its height, h, is 256 meters. Now we apply the formula for the volume of a pyramid:

$$V = \frac{1}{3}Bh = \frac{1}{3} \cdot \frac{2704 \text{ m}^2}{1} \cdot \frac{256 \text{ m}}{1} = \frac{2704 \cdot 256}{3}\text{m}^3 \approx 230{,}741 \text{ m}^3.$$

▶ The volume of the building is approximately 230,741 cubic meters.

The San Francisco pyramid is relatively small compared to the Great Pyramid outside Cairo, Egypt. Built about 2550 B.C. by a labor force of 100,000, the Great Pyramid is approximately 11 times the volume of San Francisco's pyramid.

✓ **Check Point 3** A pyramid is 4 feet tall. Each side of the square base has a length of 6 feet. Find the pyramid's volume. 48 ft³

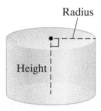

Radius

Height

Figure 6.52

Not every three-dimensional figure is a polyhedron. Take, for example, the right circular cylinder shown in **Figure 6.52**. Its shape should remind you of a soup can or a stack of coins. The right circular cylinder is so named because the top and bottom are circles, and the side forms a right angle with the top and bottom. The formula for the volume of a right circular cylinder is given at the top of the next page.

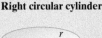

Volume of a Right Circular Cylinder

The volume, V, of a right circular cylinder is given by the formula

$$V = \pi r^2 h,$$

where r is the radius of the circle at either end and h is the height.

Right circular cylinder

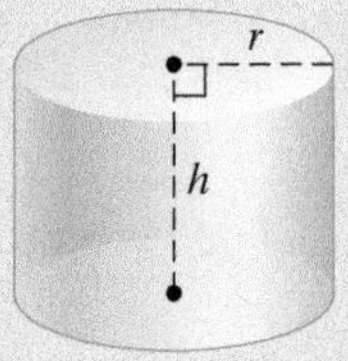

Example 4   Finding the Volume of a Cylinder

Find the volume of the cylinder in **Figure 6.53**.

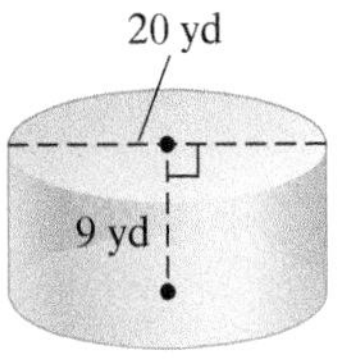

Solution

In order to find the cylinder's volume, we need both its radius and its height. Because the diameter is 20 yards, the radius is half this length, or 10 yards. The height of the cylinder is given to be 9 yards. Thus, $r = 10$ and $h = 9$. Now we apply the formula for the volume of a cylinder.

$$V = \pi r^2 h = \pi (10 \text{ yd})^2 \cdot 9 \text{ yd} = 900\pi \text{ yd}^3 \approx 2827 \text{ yd}^3$$

The volume of the cylinder is approximately 2827 cubic yards.

Check Point 4 Find the volume, to the nearest cubic inch, of a cylinder with a diameter of 8 inches and a height of 6 inches.   302 in.³

Figure 6.54 shows a **right circular cone** inside a cylinder, sharing the same circular base as the cylinder. The height of the cone, the perpendicular distance from the top to the circular base, is the same as that of the cylinder. Three such cones can occupy the same amount of space as the cylinder. Therefore, the formula for the volume of the cone is $\frac{1}{3}$ the volume of the cylinder.

Volume of a Cone

The volume, V, of a right circular cone that has height h and radius r is given by the formula

$$V = \frac{1}{3}\pi r^2 h.$$

Cone

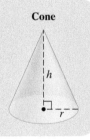

Example 5   Finding the Volume of a Cone

Find the volume of the cone in **Figure 6.55**.

Solution

The radius of the cone is 7 meters and the height is 10 meters. Thus, $r = 7$ and $h = 10$. Now we apply the formula for the volume of a cone.

$$V = \frac{1}{3}\pi r^2 h = \frac{1}{3}\pi (7 \text{ m})^2 \cdot 10 \text{ m} = \frac{490\pi}{3} \text{ m}^3 \approx 513 \text{ m}^3$$

The volume of the cone is approximately 513 cubic meters.

Check Point 5 Find the volume, to the nearest cubic inch, of a cone with a radius of 4 inches and a height of 6 inches.   101 in.³

20 yd

9 yd

Figure 6.53

Figure 6.54

10 m

7 m

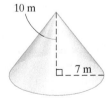

Figure 6.55

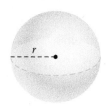

Figure 6.56

Figure 6.56 shows a *sphere*. Its shape may remind you of a basketball. The Earth is not a perfect sphere, but it's close. A **sphere** is the set of points in space equally distant from a given point, its **center**. Any line segment from the center to a point on the sphere is a **radius** of the sphere. The word *radius* is also used to refer to the length of this line segment. A sphere's volume can be found by using π and its radius.

Volume of a Sphere

Sphere

The volume, V, of a sphere of radius r is given by the formula

$$V = \frac{4}{3}\pi r^3.$$

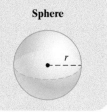

Example 6 **Applying Volume Formulas**

An ice cream cone is 5 inches deep and has a radius of 1 inch. A spherical scoop of ice cream also has a radius of 1 inch. (See **Figure 6.57**.) If the ice cream melts into the cone, will it overflow?

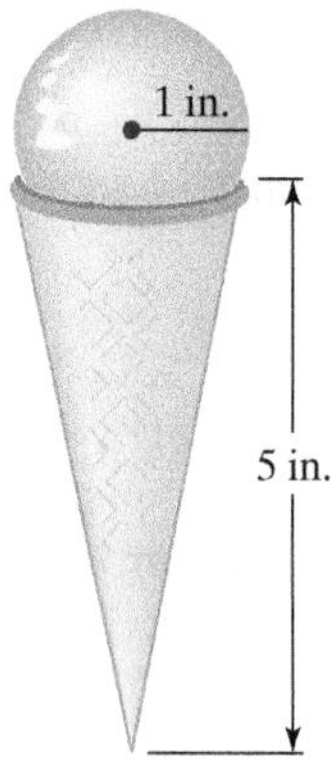

Solution

The ice cream will overflow if the volume of the ice cream, a sphere, is greater than the volume of the cone. Find the volume of each.

$$V_{\text{cone}} = \frac{1}{3}\pi r^2 h = \frac{1}{3}\pi(1 \text{ in.})^2 \cdot 5 \text{ in.} = \frac{5\pi}{3}\text{in.}^3 \approx 5 \text{ in.}^3$$

$$V_{\text{sphere}} = \frac{4}{3}\pi r^3 = \frac{4}{3}\pi(1 \text{ in.})^3 = \frac{4\pi}{3}\text{in.}^3 \approx 4 \text{ in.}^3$$

The volume of the spherical scoop of ice cream is less than the volume of the cone, so there will be no overflow.

Figure 6.57

(1 in.) (5 in.)

✓ **Check Point 6** A basketball has a radius of 4.5 inches. If the ball is filled with 350 cubic inches of air, is this enough air to fill it completely? no

---

Surface Area

② Compute the surface area of a three-dimensional figure.

In addition to volume, we can also measure the area of the outer surface of a three-dimensional object, called its **surface area**. Like area, surface area is measured in square units. For example, the surface area of the rectangular solid in **Figure 6.58** is the sum of the areas of the six outside rectangles of the solid.

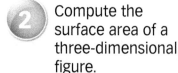

Figure 6.58

Surface Area =	lw + lw	+	lh + lh	+	wh + wh
	Areas of top and bottom rectangles		Areas of front and back rectangles		Areas of rectangles on left and right sides
=	$2lw$	+	$2lh$	+	$2wh$

Formulas for the surface area, abbreviated *SA*, of three-dimensional figures are given in **Table 6.4**.

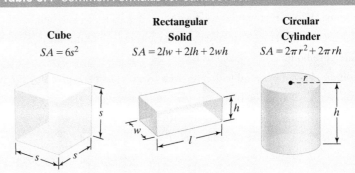

Table 6.4 Common Formulas for Surface Area

Cube	Rectangular Solid	Circular Cylinder
$SA = 6s^2$	$SA = 2lw + 2lh + 2wh$	$SA = 2\pi r^2 + 2\pi rh$

Example 7   Finding the Surface Area of a Solid

Find the surface area of the rectangular solid in **Figure 6.59**.

Solution

As shown in the figure, the length is 8 yards, the width is 5 yards, and the height is 3 yards. Thus, $l = 8$, $w = 5$, and $h = 3$.

$$SA = 2lw + 2lh + 2wh$$
$$= 2 \cdot 8 \text{ yd} \cdot 5 \text{ yd} + 2 \cdot 8 \text{ yd} \cdot 3 \text{ yd} + 2 \cdot 5 \text{ yd} \cdot 3 \text{ yd}$$
$$= 80 \text{ yd}^2 + 48 \text{ yd}^2 + 30 \text{ yd}^2 = 158 \text{ yd}^2$$

The surface area is 158 square yards.

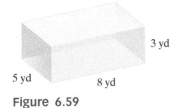

3 yd

5 yd   8 yd

Figure 6.59

Check Point 7   If the length, width, and height shown in **Figure 6.59** are each doubled, find the surface area of the resulting rectangular solid.   632 yd²

Achieving Success

FoxTrot

FOXTROT © 1997 Bill Amend. Reprinted with permission of UNIVERSAL UCLICK. All rights reserved.

A test-taking tip: Live for partial credit.

Always show your work. If worse come to worst, write *something* down, anything, even if it's a formula that you think might solve a problem or a possible idea or procedure for solving the problem. As Bill Amend's *FoxTrot* cartoon indicates, partial credit has salvaged more then a few test scores.

> **Concept and Vocabulary Check**   Exercises in the Concept and Vocabulary Check are intended for group and class discussions.

In Exercises 1–7, fill in each blank so that the resulting statement is true.

1. The volume, V, of a rectangular solid with length l, width w, and height h is given by the formula ____$V = lwh$____.

2. The volume, V, of a cube with an edge that measures s linear units is given by the formula ____$V = s^3$____.

3. A solid figure bounded by polygons is called a ____polyhedron____.

4. The volume, V, of a pyramid with base area B and height h is given by the formula ____$V = \frac{1}{3}Bh$____.

5. The volume, V, of a right circular cylinder with height h and radius r is given by the formula ____$V = \pi r^2 h$____.

6. The volume, V, of a right circular cone with height h and radius r is given by the formula ____$V = \frac{1}{3}\pi r^2 h$____.

7. The volume, V, of a sphere of radius r is given by the formula ____$V = \frac{4}{3}\pi r^3$____.

In Exercises 8–14, determine whether each statement is true or false. If the statement is false, make the necessary change(s) to produce a true statement.   Changes to false statements will vary.

8. A cube is a rectangular solid with the same length, width, and height.   true

9. A cube is an example of a polyhedron.   true

10. The volume of a pyramid is $\frac{1}{2}$ the volume of a rectangular solid having the same base and the same height.   false

11. Some three-dimensional figures are not polyhedrons.   true

12. A sphere is the set of points in space equally distant from its center.   true

13. Surface area refers to the area of the outer surface of a three-dimensional object.   true

14. The surface area, SA, of a rectangular solid with length l, width w, and height h is given by the formula $SA = lw + lh + wh$.   false

Respond to Exercises 15–16 using verbal or written explanations.   **15–16.** Answers will vary.

15. Explain the following analogy:

In terms of formulas used to compute volume, a pyramid is to a rectangular solid just as a cone is to a cylinder.

16. Explain why a cylinder is not a polyhedron.

Exercise Set 6.5

Practice Exercises

In Exercises 1–20, find the volume of each figure. If necessary, express answers in terms of π and then round to the nearest whole number.

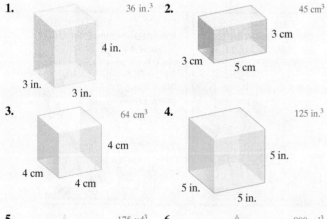

1. 36 in.³

2. 45 cm³

3. 64 cm³

4. 125 in.³

5. 175 yd³ $h = 15$ yd

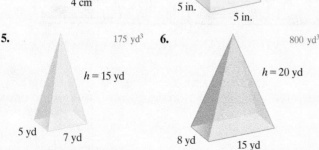

6. 800 yd³ $h = 20$ yd

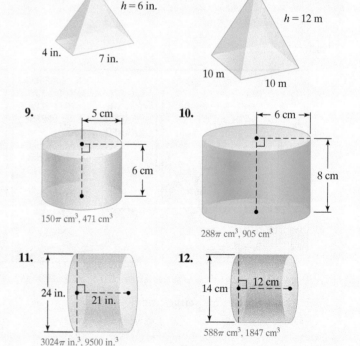

7. 56 in.³ $h = 6$ in.

8. 400 m³ $h = 12$ m

9. 150π cm³, 471 cm³

10. 288π cm³, 905 cm³

11. 3024π in.³, 9500 in.³

12. 588π cm³, 1847 cm³

13.

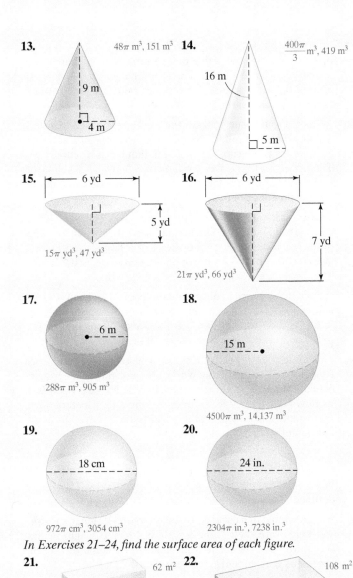

9 m

4 m

48π m³, 151 m³

14.

16 m

5 m

$\dfrac{400\pi}{3}$ m³, 419 m³

15.

6 yd

5 yd

15π yd³, 47 yd³

16.

6 yd

7 yd

21π yd³, 66 yd³

17.

6 m

288π m³, 905 m³

18.

15 m

4500π m³, 14,137 m³

19.

18 cm

972π cm³, 3054 cm³

20.

24 in.

2304π in.³, 7238 in.³

In Exercises 21–24, find the surface area of each figure.

21.

62 m²

3 m

2 m

5 m

22.

108 m²

3 m

4 m

6 m

23.

96 ft²

4 ft

4 ft

4 ft

24.

216 ft²

6 ft

6 ft

6 ft

In Exercises 25–30, use two formulas for volume to find the volume of each figure. Express answers in terms of π and then round to the nearest whole number.

25.

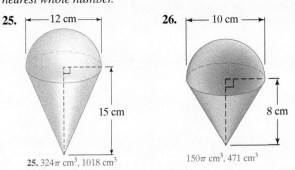

12 cm

15 cm

25. 324π cm³, 1018 cm³

26.

10 cm

8 cm

150π cm³, 471 cm³

27.

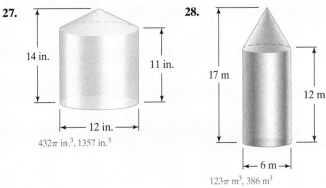

14 in.

11 in.

12 in.

432π in.³, 1357 in.³

28.

17 m

12 m

6 m

123π m³, 386 m³

29.

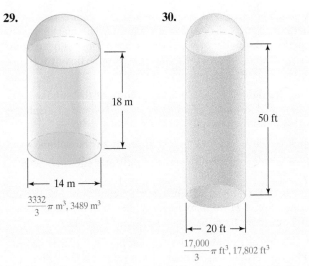

18 m

14 m

$\dfrac{3332}{3}\pi$ m³, 3489 m³

30.

50 ft

20 ft

$\dfrac{17,000}{3}\pi$ ft³, 17,802 ft³

Practice Plus

31. Find the surface area and the volume of the figure shown.

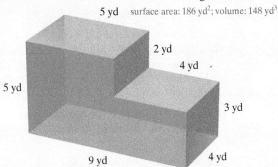

5 yd   surface area: 186 yd²; volume: 148 yd³

2 yd

4 yd

5 yd

3 yd

9 yd

4 yd

32. Find the surface area and the volume of the cement block in the figure shown.   surface area: 976 in.²; volume: 448 in.³

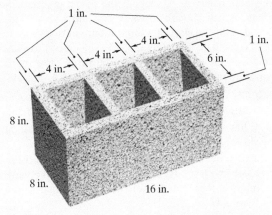

1 in.

4 in.

4 in.

1 in.

4 in.

4 in.

6 in.

8 in.

8 in.

16 in.

33. Find the surface area of the figure shown. 666 yd²

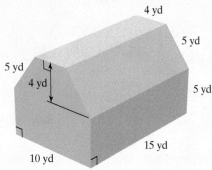

5 yd
4 yd
5 yd
5 yd
4 yd
5 yd
15 yd
10 yd

34. A machine produces open boxes using square sheets of metal measuring 12 inches on each side. The machine cuts equal-sized squares whose sides measure 2 inches from each corner. Then it shapes the metal into an open box by turning up the sides. Find the volume of the box. 128 in.³

35. Find the ratio, reduced to lowest terms, of the volume of a sphere with a radius of 3 inches to the volume of a sphere with a radius of 6 inches. $\frac{1}{8}$

36. Find the ratio, reduced to lowest terms, of the volume of a sphere with a radius of 3 inches to the volume of a sphere with a radius of 9 inches. $\frac{1}{27}$

37. A cylinder with radius 3 inches and height 4 inches has its radius tripled. How many times greater is the volume of the larger cylinder than the smaller cylinder? 9 times

38. A cylinder with radius 2 inches and height 3 inches has its radius quadrupled. How many times greater is the volume of the larger cylinder than the smaller cylinder? 16 times

Application Exercises

39. A building contractor is to dig a foundation 12 feet long, 9 feet wide, and 6 feet deep for a toll booth. The contractor pays $85 per load for trucks to remove the dirt. Each truck holds 6 cubic yards. What is the cost to the contractor to have all the dirt hauled away? $340

40. What is the cost of concrete for a walkway that is 15 feet long, 8 feet wide, and 9 inches deep if the concrete costs $30 per cubic yard? $100

41. A furnace is designed to heat 10,000 cubic feet. Will this furnace be adequate for a 1400-square-foot house with a 9-foot ceiling? no

42. A water reservoir is shaped like a rectangular solid with a base that is 50 yards by 30 yards, and a vertical height of 20 yards. At the start of a three-month period of no rain, the reservoir was completely full. At the end of this period, the height of the water was down to 6 yards. How much water was used in the three-month period? 21,000 yd³

43. The Great Pyramid outside Cairo, Egypt, has a square base measuring 756 feet on a side and a height of 480 feet.
   a. What is the volume of the Great Pyramid, in cubic yards? 3,386,880 yd³
   b. The stones used to build the Great Pyramid were limestone blocks with an average volume of 1.5 cubic yards. How many of these blocks were needed to construct the Great Pyramid? 2,257,920 blocks

44. Although the Eiffel Tower in Paris is not a solid pyramid, its shape approximates that of a pyramid with a square base measuring 120 feet on a side and a height of 980 feet. If it were a solid pyramid, what would be the Eiffel Tower's volume, in cubic yards? ≈ 174,222 yd³

45. You are about to sue your contractor who promised to install a water tank that holds 500 gallons of water. You know that 500 gallons is the capacity of a tank that holds 67 cubic feet. The cylindrical tank has a radius of 3 feet and a height of 2 feet 4 inches. Does the evidence indicate you can win the case against the contractor if it goes to court? yes

46. Two cylindrical cans of soup sell for the same price. One can has a diameter of 6 inches and a height of 5 inches. The other has a diameter of 5 inches and a height of 6 inches. Which can contains more soup and, therefore, is the better buy? the can with a diameter of 6 in. and a height of 5 in.

47. A circular backyard pool has a diameter of 24 feet and is 4 feet deep. One cubic foot of water has a capacity of approximately 7.48 gallons. If water costs $2 per thousand gallons, how much, to the nearest dollar, will it cost to fill the pool? $27

48. The tunnel under the English Channel that connects England and France is the world's longest tunnel. There are actually three separate tunnels built side by side. Each is a half-cylinder that is 50,000 meters long and 4 meters high. How many cubic meters of dirt had to be removed to build the tunnel? ≈ 3,769,911 m³

Critical Thinking Exercises

49. What happens to the volume of a sphere if its radius is doubled? The volume is multiplied by 8.

50. A scale model of a car is constructed so that its length, width, and height are each $\frac{1}{10}$ the length, width, and height of the actual car. By how many times does the volume of the car exceed its scale model? 1000 times

In Exercises 51–52, find the volume of the darkly shaded region. If necessary, round to the nearest whole number.

51. 168 cm³   **52.** 251 in.³

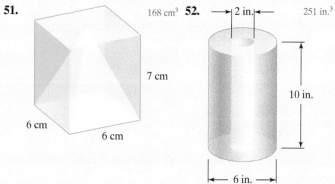

7 cm
6 cm
6 cm

2 in.
10 in.
6 in.

53. Find the surface area of the figure shown. 84 cm²

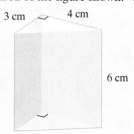

3 cm   4 cm
6 cm

Chapter 6 Summary

Definitions and Concepts	**Examples**

Section 6.1 Points, Lines, Planes, and Angles

Line AB ($\overleftrightarrow{AB}$ or $\overleftrightarrow{BA}$), half-line AB ($\overset{\circ}{\rightarrow}AB$), ray AB ($\overrightarrow{AB}$), and line segment AB ($\overline{AB}$ or $\overline{BA}$) are represented in the following figure:

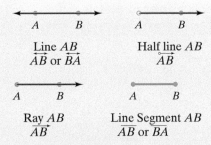

Line AB
$\overleftrightarrow{AB}$ or $\overleftrightarrow{BA}$

Half line AB
$\overset{\circ}{\rightarrow}AB$

Ray AB
$\overrightarrow{AB}$

Line Segment AB
$\overline{AB}$ or $\overline{BA}$

Angles are measured in degrees. A degree, 1°, is $\frac{1}{360}$ of a complete rotation. Acute angles measure less than 90°, right angles 90°, obtuse angles more than 90° but less than 180°, and straight angles 180°.

Complementary angles are two angles whose measures have a sum of 90°. The measure of an angle's complement is found by subtracting the angle's measure from 90°.

Supplementary angles are two angles whose measures have a sum of 180°. The measure of an angle's supplement is found by subtracting the angle's measure from 180°.

- An angle measures 4°.

  measure of its complement $= 90° - 4° = 86°$

  measure of its supplement $= 180° - 4° = 176°$

Additional Examples to Review

Example 1, page 363; Example 2, page 364; Example 3, page 365

Vertical angles have the same measure.

$$m\angle 1 = m\angle 3$$
$$m\angle 2 = m\angle 4$$

If parallel lines are intersected by a transversal, alternate interior angles, alternate exterior angles, and corresponding angles have the same measure.

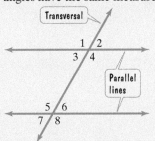

Transversal

Parallel lines

Alternate Interior Angles
$$m\angle 3 = m\angle 6$$
$$m\angle 4 = m\angle 5$$

Alternate Exterior Angles
$$m\angle 1 = m\angle 8$$
$$m\angle 2 = m\angle 7$$

Corresponding Angles
$$m\angle 1 = m\angle 5 \quad m\angle 3 = m\angle 7$$
$$m\angle 2 = m\angle 6 \quad m\angle 4 = m\angle 8$$

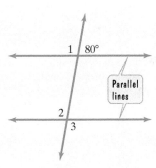

1   80°

Parallel lines

2

3

- $m\angle 1 = 180° - 80° = 100°$ ∡1 and the 80° angle are supplementary.

- $m\angle 2 = 100°$ ∡1 and ∡2 are corresponding angles with the same measure.

- $m\angle 3 = 100°$ ∡2 and ∡3 are vertical angles with the same measure.

Additional Examples to Review

Example 4, page 366; Example 5, page 367

Section 6.2 Triangles

The sum of the measures of the three angles of any triangle is 180°.

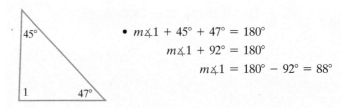

- $m\angle 1 + 45° + 47° = 180°$

$$m\angle 1 + 92° = 180°$$

$$m\angle 1 = 180° - 92° = 88°$$

Additional Examples to Review

Example 1, page 373; Example 2, page 373

Triangles can be classified by angles (acute, right, obtuse) or by sides (isosceles, equilateral, scalene).

Classification by Angles

Acute Triangle	**Right Triangle**	**Obtuse Triangle**
All angles are acute.	One angle measures 90°.	One angle is obtuse.

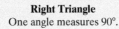

Classification by Sides

Isosceles Triangle	**Equilateral Triangle**	**Scalene Triangle**
Two sides have equal length. (Angles opposite these sides have the same measure.)	All sides have equal length. (Each angle measures 60°.)	No two sides are equal in length.

Similar triangles have the same shape, but not necessarily the same size. Corresponding angles have the same measure and corresponding sides are proportional. If the measures of two angles of one triangle are equal to those of two angles of a second triangle, then the two triangles are similar.

- The triangles are similar.

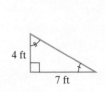
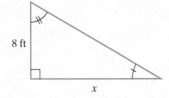

4 ft

8 ft

7 ft   x

$$\frac{4}{8} = \frac{7}{x}$$

$$4x = 8 \cdot 7$$

$$4x = 56$$

$$\frac{4x}{4} = \frac{56}{4}$$

$$x = 14$$

The missing length, x, is 14 feet.

Additional Examples to Review

Example 3, page 375; Example 4, page 376

The Pythagorean Theorem

The sum of the squares of the lengths of the legs of a right triangle equals the square of the length of the hypotenuse.

If the legs have lengths a and b and the hypotenuse has length c, then

$$a^2 + b^2 = c^2.$$

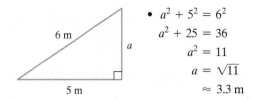

- $a^2 + 5^2 = 6^2$
  $a^2 + 25 = 36$
  $a^2 = 11$
  $a = \sqrt{11}$
  ≈ 3.3 m

Additional Examples to Review

Example 5, page 377; Example 6, page 377

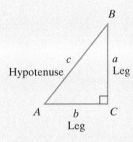

Section 6.3 Polygons, Perimeter, and Tessellations

A polygon is a closed geometric figure in a plane formed by three or more line segments. Names of some polygons include triangles (three sides), quadrilaterals (four sides), pentagons (five sides), hexagons (six sides), heptagons (seven sides), and octagons (eight sides). A regular polygon is one whose sides are all the same length and whose angles all have the same measure. The perimeter of a polygon is the sum of the lengths of its sides.

Types of Quadrilaterals

Name	Characteristics	Representation
Parallelogram	Quadrilateral in which both pairs of opposite sides are parallel and have the same measure. Opposite angles have the same measure.	
Rhombus	Parallelogram with all sides having equal length.	
Rectangle	Parallelogram with four right angles. Because a rectangle is a parallelogram, opposite sides are parallel and have the same measure.	
Square	A rectangle with all sides having equal length. Each angle measures 90°, and the square is a regular quadrilateral.	
Trapezoid	A quadrilateral with exactly one pair of parallel sides.	

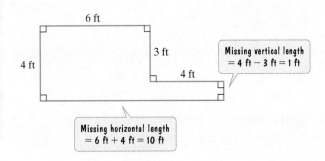

6 ft

4 ft

3 ft

4 ft

Missing vertical length
= 4 ft − 3 ft = 1 ft

Missing horizontal length
= 6 ft + 4 ft = 10 ft

- Perimeter

$$P = 6 + 3 + 4 + 1 + 10 + 4$$
$$= 28 \text{ ft}$$

Additional Example to Review

Example 1, page 384

The sum of the measures of the angles of an n-sided polygon is $(n - 2)180°$. If the n-sided polygon is a regular polygon, then each angle measures $\dfrac{(n - 2)180°}{n}$.

- The sum of the measures of the angles of a hexagon is

$$(n - 2)180° = (6 - 2)180°$$
$$= 4 \cdot 180°$$
$$= 720°.$$

- Each angle of a regular hexagon measures

$$\frac{720°}{6} = 120°.$$

Additional Example to Review

Example 2, page 385

A tessellation is a pattern consisting of the repeated use of the same geometric figures to completely cover a plane, leaving no gaps and having no overlaps. The angle requirement for the formation of a tessellation is that the sum of the measures of the angles at each vertex must be 360°.

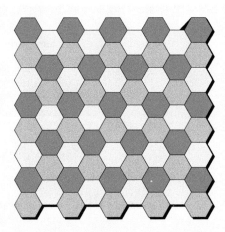

- Regular hexagons surround each vertex.
- Three angles come together at each vertex, measuring $120° + 120° + 120° = 360°$.
- The tessellation is possible because the sum of the measures of the angles at each vertex is 360°.

Additional Example to Review

Example 3, page 387

Section 6.4 Area and Circumference

Formulas for Area

Rectangle: $A = lw$

Square: $A = s^2$

Parallelogram: $A = bh$

Triangle: $A = \frac{1}{2}bh$

Trapezoid: $A = \frac{1}{2}h(a + b)$

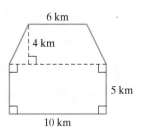

- Area of Trapezoid
  $$A = \tfrac{1}{2}h(a + b) = \tfrac{1}{2} \cdot 4 \text{ km} \cdot (6 \text{ km} + 10 \text{ km})$$
  $$= \tfrac{1}{2} \cdot 4 \text{ km} \cdot 16 \text{ km} = 32 \text{ km}^2$$
- Area of Rectangle
  $$A = lw = 10 \text{ km} \cdot 5 \text{ km} = 50 \text{ km}^2$$
- Total Area $= 32 \text{ km}^2 + 50 \text{ km}^2 = 82 \text{ km}^2$

Additional Examples to Review

Example 1, page 392; Example 2, page 393; Example 3, page 394; Example 4, page 395; Example 5, page 395

A circle is a set of points in the plane equally distant from a given point, its center. The radius (plural: radii), r, is a line segment from the center to any point on the circle. The diameter, d, is a line segment through the center whose endpoints both lie on the circle. The length of the diameter is twice the length of the radius.

$$\text{Circumference: } C = 2\pi r \quad \text{or} \quad C = \pi d$$
$$\text{Area: } A = \pi r^2$$

A circle has a diameter of 20 meters.

- Circumference: $C = \pi d = \pi(20 \text{ m}) = 20\pi \text{ m} \approx 62.8 \text{ m}$
- Area: $A = \pi r^2 = \pi(10 \text{ m})^2 = 100\pi \text{ m}^2 \approx 314.2 \text{ m}^2$

> The diameter measures **20 m**, so the radius measures $\frac{1}{2} \cdot$ **20 m = 10 m.**

Additional Examples to Review

Example 6, page 396; Example 7, page 397; Example 8, page 397

Section 6.5 Volume

Formulas for Volume

Rectangular Solid: $V = lwh$

Cube: $V = s^3$

Pyramid: $V = \frac{1}{3}Bh$

Cylinder: $V = \pi r^2 h$

Cone: $V = \frac{1}{3}\pi r^2 h$

Sphere: $V = \frac{4}{3}\pi r^3$

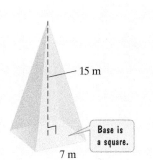

- Volume of the Rectangular Solid
  $$V = lwh = 14 \text{ cm} \cdot 8 \text{ cm} \cdot 5 \text{ cm}$$
  $$= 560 \text{ cm}^3$$

- Area of the Square Base
  $$B = 7 \text{ m} \cdot 7 \text{ m} = 49 \text{ m}^2$$

- Volume of the Pyramid
  $$V = \tfrac{1}{3}Bh = \tfrac{1}{3} \cdot 49 \text{ m}^2 \cdot 15 \text{ m}$$
  $$= \tfrac{1}{3} \cdot 49 \text{ m}^2 \cdot \overset{5}{\underset{1}{\cancel{15}}}\text{m}$$
  $$= 245 \text{ m}^3$$

- Volume of the Cylinder

$$V = \pi r^2 h = \pi(3 \text{ in.})^2 \cdot 5 \text{ in.}$$
$$= 45\pi \text{ in.}^3$$
$$\approx 141 \text{ in.}^3$$

Additional Examples to Review

Example 1, page 403; Example 2, page 403;
Example 3, page 404; Example 4, page 405;
Example 5, page 405; Example 6, page 406

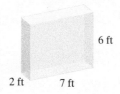

The measure of the area of the outer surface of a three-dimensional object is called its surface area and is measured in square units.

Common Formulas for Surface Area

	Rectangular	Circular
Cube	**Solid**	**Cylinder**
$SA = 6s^2$	$SA = 2lw + 2lh + 2wh$	$SA = 2\pi r^2 + 2\pi rh$

Surface Area of the Rectangular Solid

$SA = 2lw + 2lh + 2wh$

$= 2 \cdot 7 \text{ ft} \cdot 2 \text{ ft} + 2 \cdot 7 \text{ ft} \cdot 6 \text{ ft} + 2 \cdot 2 \text{ ft} \cdot 6 \text{ ft}$

$= 28 \text{ ft}^2 + 84 \text{ ft}^2 + 24 \text{ ft}^2 = 136 \text{ ft}^2$

Additional Example to Review

Example 7, page 407

Review Exercises

Section 6.1 Points, Lines, Planes, and Angles

In the figure shown, lines l and m are parallel. In Exercises 1–7, match each term with the numbered angle or angles in the figure.

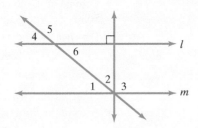

1. right angle   ∡3

2. obtuse angle   ∡5

3. vertical angles   ∡4 and ∡6

4. alternate interior angles   ∡1 and ∡6

5. corresponding angles   ∡1 and ∡4

6. the complement of ∡1   ∡2

7. the supplement of ∡6   ∡5

In Exercises 8–9, find the measure of the angle in which a question mark with a degree symbol appears.

8.

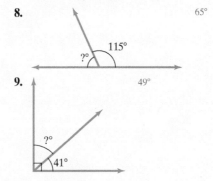

9.

10. If an angle measures 73°, find the measure of its complement.   17°

11. If an angle measures 46°, find the measure of its supplement.   134°

12. In the figure shown, find the measures of angles 1, 2, and 3.

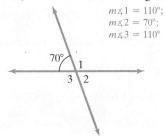

$m \angle 1 = 110°$;
$m \angle 2 = 70°$;
$m \angle 3 = 110°$

13. In the figure shown, two parallel lines are intersected by a transversal. One of the angle measures is given. Find the measure of each of the other seven angles.

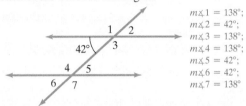

$m \angle 1 = 138°$;
$m \angle 2 = 42°$;
$m \angle 3 = 138°$;
$m \angle 4 = 138°$;
$m \angle 5 = 42°$;
$m \angle 6 = 42°$;
$m \angle 7 = 138°$

Section 6.2 Triangles

In Exercises 14–15, find the measure of angle A for the triangle shown.

14. **15.**

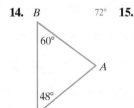

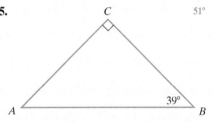

16. Find the measures of angles 1 through 5 in the figure shown.

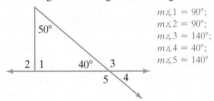

$m \angle 1 = 90°$;
$m \angle 2 = 90°$;
$m \angle 3 = 140°$;
$m \angle 4 = 40°$;
$m \angle 5 = 140°$

17. In the figure shown, lines *l* and *m* are parallel. Find the measure of each numbered angle.

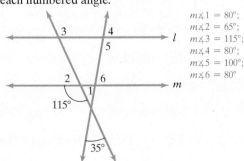

$m \angle 1 = 80°$;
$m \angle 2 = 65°$;
$m \angle 3 = 115°$;
$m \angle 4 = 80°$;
$m \angle 5 = 100°$;
$m \angle 6 = 80°$

In Exercises 18–19, use similar triangles and the fact that corresponding sides are proportional to find the length of each side marked with an x.

18.

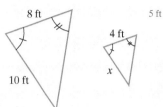

19.

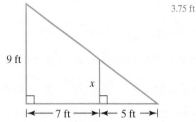

In Exercises 20–22, use the Pythagorean Theorem to find the missing length in each right triangle. Round, if necessary, to the nearest tenth.

20.

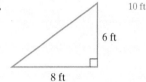

21. **22.**

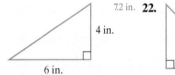
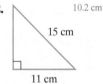

23. Find the height of the lamppost in the figure.   12.5 ft

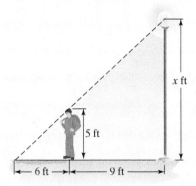

24. How far away from the building in the figure shown is the bottom of the ladder?   15 ft

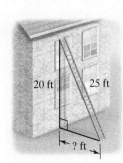

25. A vertical pole is to be supported by three wires. Each wire is 13 yards long and is anchored 5 yards from the base of the pole. How far up the pole will the wires be attached? 12 yd

Section 6.3 Polygons, Perimeter, and Tessellations

26. Write the names of all quadrilaterals that always have four right angles. rectangle, square

27. Write the names of all quadrilaterals with four sides always having the same measure. rhombus, square

28. Write the names of all quadrilaterals that do not always have four angles with the same measure.
parallelogram, rhombus, trapezoid

In Exercises 29–31, find the perimeter of the figure shown. Express the perimeter using the same unit of measure that appears in the figure.

29.

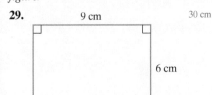

9 cm   30 cm

6 cm

30.

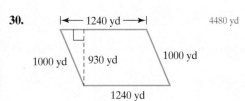

|← 1240 yd →| 4480 yd

1000 yd   930 yd   1000 yd

1240 yd

31.

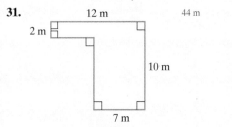

12 m   44 m

2 m

10 m

7 m

32. Find the sum of the measures of the angles of a 12-sided polygon. 1800°

33. Find the sum of the measures of the angles of an octagon. 1080°

34. The figure shown is a regular polygon. Find the measures of angle 1 and angle 2.

$m\angle 1 = 135°$;
$m\angle 2 = 45°$

2 1

35. A carpenter is installing a baseboard around a room that has a length of 35 feet and a width of 15 feet. The room has four doorways and each doorway is 3 feet wide. If no baseboard is to be put across the doorways and the cost of the baseboard is $1.50 per foot, what is the cost of installing the baseboard around the room? $132

36. Use the following tessellation to solve this exercise.

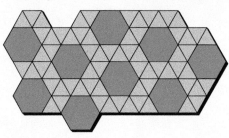

a. Name the types of regular polygons that surround each vertex. triangles, hexagons

b. Determine the number of angles that come together at each vertex, as well as the measures of these angles. 5 angles; 60°, 60°, 60°, 60°, 120°

c. Use the angle measures from part (b) to explain why the tessellation is possible. The sum is 360°.

37. Can a tessellation be created using only regular hexagons? Explain your answer. Yes, each angle measures 120° and 360° is a multiple of 120°.

Section 6.4 Area and Circumference

In Exercises 38–41, find the area of each figure.

38.

6.5 ft   32.5 ft²

5 ft

39.

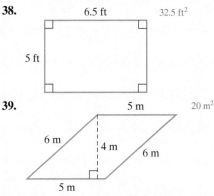

5 m   20 m²

6 m   4 m

6 m

5 m

40.

12 cm   10 cm
5 cm
|← 20 cm →|
50 cm²

41.

5 yd   135 yd²

12 yd   10 yd   15 yd

|← 22 yd →|

42. Find the circumference and the area of a circle with a diameter of 20 meters. Express answers in terms of π and then round to the nearest tenth.
$C = 20\pi$ m ≈ 62.8 m; $A = 100\pi$ m² ≈ 314.2 m²

In Exercises 43–44, find the area of each figure.

43.

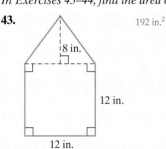

192 in.²

8 in.

12 in.

12 in.

44.

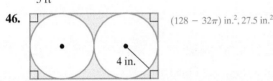

28 m²

In Exercises 45–46, find the area of each shaded region. Where necessary, express answers in terms of π and then round to the nearest tenth.

45.

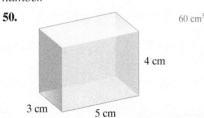

279.5 ft²

46.

(128 − 32π) in.², 27.5 in.²

47. What will it cost to carpet a rectangular floor measuring 15 feet by 21 feet if the carpet costs $22.50 per square yard? $787.50

48. What will it cost to cover a rectangular floor measuring 40 feet by 50 feet with square tiles that measure 2 feet on each side if a package of 10 tiles costs $13? $650

49. How much fencing, to the nearest whole yard, is needed to enclose a circular garden that measures 10 yards across?   31 yd

Section 6.5 Volume

In Exercises 50–54, find the volume of each figure. Where necessary, express answers in terms of π and then round to the nearest whole number.

50.

60 cm³

51.

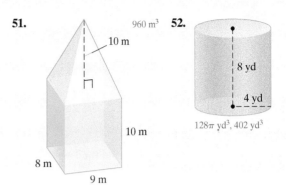

960 m³

52.

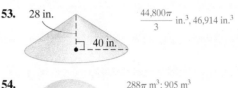

128π yd³, 402 yd³

53. 28 in.

$\frac{44,800\pi}{3}$ in.³, 46,914 in.³

40 in.

54.

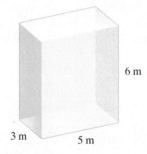

288π m³; 905 m³

6 m

55. Find the surface area of the figure shown.   126 m²

56. A train is being loaded with shipping boxes. Each box is 8 meters long, 4 meters wide, and 3 meters high. If there are 50 shipping boxes, how much space is needed?   4800 m³

57. An Egyptian pyramid has a square base measuring 145 meters on each side. If the height of the pyramid is 93 meters, find its volume.   651,775 m³

58. What is the cost of concrete for a walkway that is 27 feet long, 4 feet wide, and 6 inches deep if the concrete is $40 per cubic yard? $80

Chapter 6 Test

1. If an angle measures 54°, find the measure of its complement and supplement.   36°; 126°

In Exercises 2–4, use the figure shown to find the measure of angle 1.

2.

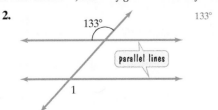

133°

3.

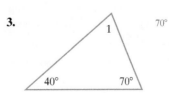

70°

4.

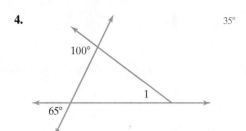

5. The triangles in the figure are similar. Find the length of the side marked with an x.   3.2 in.

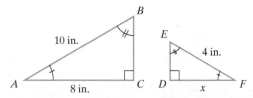

6. A vertical pole is to be supported by three wires. Each wire is 26 feet long and is anchored 24 feet from the base of the pole. How far up the pole should the wires be attached?   10 ft

7. Find the sum of the measures of the angles of a ten-sided polygon.   1440°

8. Find the perimeter of the figure shown.   40 cm

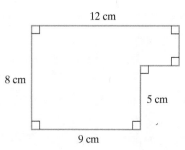

9. Which one of the following names a quadrilateral in which the sides that meet at each vertex have the same measure?   d

a. rectangle   **b.** parallelogram

c. trapezoid   **d.** rhombus

10. Use the following tessellation to solve this exercise.

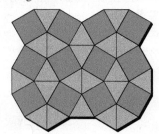

a. Name the types of regular polygons that surround each vertex.   triangles, squares   **b.** 5 angles; 60°, 60°, 60°, 90°, 90°

b. Determine the number of angles that come together at each vertex, as well as the measures of these angles.

c. Use the angle measures from part (b) to explain why the tessellation is possible.   The sum is 360°.

In Exercises 11–12, find the area of each figure.

11.

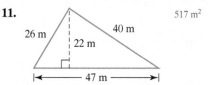

517 m²

12.

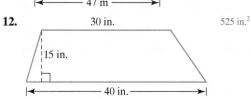

525 in.²

13. The right triangle shown has one leg of length 5 centimeters and a hypotenuse of length 13 centimeters.

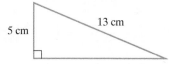

a. Find the length of the other leg.   12 cm

b. What is the perimeter of the triangle?   30 cm

c. What is the area of the triangle?   30 cm²

14. Find the circumference and area of a circle with a diameter of 40 meters. Express answers in terms of π and then round to the nearest tenth. $C = 40\pi$ m $\approx$ 125.7 m; $A = 400\pi$ m² $\approx$ 1256.6 m²

15. A rectangular floor measuring 8 feet by 6 feet is to be completely covered with square tiles measuring 8 inches on each side. How many tiles are needed to completely cover the floor?   108 tiles

In Exercises 16–18, find the volume of each figure. If necessary, express the answer in terms of π and then round to the nearest whole number.

16.

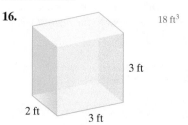

18 ft³

17.

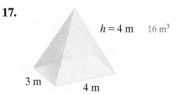

$h = 4$ m   16 m³

18.

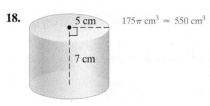

175π cm³ $\approx$ 550 cm³

Pathways to Probability: Counting Methods and Probability Theory

Two of America's best-loved presidents, Abraham Lincoln and John F. Kennedy, are linked by a bizarre series of coincidences:

- Lincoln was elected president in 1860. Kennedy was elected president in 1960.

- Lincoln's assassin, John Wilkes Booth, was born in 1839. Kennedy's assassin, Lee Harvey Oswald, was born in 1939.

- Lincoln's secretary, named Kennedy, warned him not to go to the theater on the night he was shot. Kennedy's secretary, named Lincoln, warned him not to go to Dallas on the day he was shot.

- Booth shot Lincoln in a theater and ran into a warehouse. Oswald shot Kennedy from a warehouse and ran into a theater.

- Both Lincoln and Kennedy were shot from behind, with their wives present.

- Andrew Johnson, who succeeded Lincoln, was born in 1808. Lyndon Johnson, who succeeded Kennedy, was born in 1908.

Source: Edward Burger and Michael Starbird, *Coincidences, Chaos, and All That Math Jazz,* W. W. Norton and Company, 2005.

Amazing coincidences? A cosmic conspiracy? Not really. In this chapter, you will see how the mathematics of uncertainty and risk, called probability theory, numerically describes situations in which to expect the unexpected. By assigning numbers to things that are extraordinarily unlikely, we can logically analyze coincidences without erroneous beliefs that strange and mystical events are occurring. We'll even see how wildly inaccurate our intuition can be about the likelihood of an event by examining an "amazing" coincidence that is nearly certain.

Here's where you'll find these applications:

Coincidences are discussed in the Blitzer Bonus on page 476. Coincidences that are nearly certain are developed in Exercise 73 of Exercise Set 7.7.

7.1 : The Fundamental Counting Principle

What am I supposed to learn?

After you have read this section, you should be able to:

 Use the Fundamental Counting Principle to determine the number of possible outcomes in a given situation.

Have you ever imagined what your life would be like if you won the lottery? What changes would you make? Before you fantasize about becoming a person of leisure with a staff of obedient elves, think about this: The probability of winning top prize in the lottery is about the same as the probability of being struck by lightning. There are millions of possible number combinations in lottery games and only one way of winning the grand prize. Determining the probability of winning involves calculating the chance of getting the winning combination from all possible outcomes. In this section, we begin preparing for the surprising world of probability by looking at methods for counting possible outcomes.

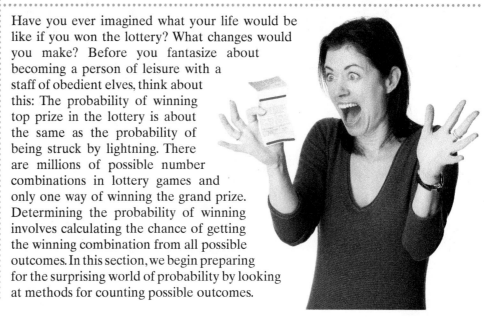

The Fundamental Counting Principle with Two Groups of Items

 Use the Fundamental Counting Principle to determine the number of possible outcomes in a given situation.

It's early morning, you're groggy, and you have to select something to wear for your 8 A.M. math class. Fortunately, your "school wardrobe" is rather limited—just two pairs of jeans to choose from (one blue, one black) and three T-shirts to choose from (one beige, one yellow, and one blue). Your early-morning dilemma is illustrated in **Figure 7.1**.

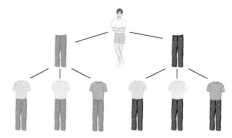

Figure 7.1 Selecting a wardrobe

The **tree diagram**, so named because of its branches, shows that you can form six different outfits from your two pairs of jeans and three T-shirts. Each pair of jeans can be combined with one of three T-shirts. Notice that the total number of outfits can be obtained by multiplying the number of choices for the jeans, 2, by the number of choices for the T-shirts, 3:

$$2 \cdot 3 = 6.$$

We can generalize this idea to any two groups of items—not just jeans and T-shirts—with the **Fundamental Counting Principle**.

The Fundamental Counting Principle

If you can choose one item from a group of M items and a second item from a group of N items, then the total number of two-item choices is $M \cdot N$.

Example 1 **Applying the Fundamental Counting Principle**

The Greasy Spoon Restaurant offers 6 appetizers and 14 main courses. In how many ways can a person order a two-course meal?

Solution

Choosing from one of 6 appetizers and one of 14 main courses, the total number of two-course meals is

$$6 \cdot 14 = 84.$$

A person can order a two-course meal in 84 different ways.

Check Point 1   A restaurant offers 10 appetizers and 15 main courses. In how many ways can you order a two-course meal?   150

Example 2 **Applying the Fundamental Counting Principle**

This is the semester that you will take your required psychology and social science courses. Because you decide to register early, there are 15 sections of psychology from which you can choose. Furthermore, there are 9 sections of social science that are available at times that do not conflict with those for psychology. In how many ways can you create a two-course schedule that satisfies the psychology–social science requirement?

Solution

The number of ways that you can satisfy the requirement is found by multiplying the number of choices for each course. You can choose your psychology course from 15 sections and your social science course from 9 sections. For both courses you have

$$15 \cdot 9, \text{ or } 135$$

choices. Thus, you can satisfy the psychology–social science requirement in 135 ways.

Check Point 2   Rework Example 2 given that the number of sections of psychology and nonconflicting sections of social science each decrease by 5.   40

The Fundamental Counting Principle with More Than Two Groups of Items

Whoops! You forgot something in choosing your lecture wardrobe—shoes! You have two pairs of sneakers to choose from—one black and one red, for that extra fashion flair! Your possible outfits including sneakers are shown in **Figure 7.2**.

The number of possible ways of playing the first four moves on each side in a game of chess is 318,979,564,000.

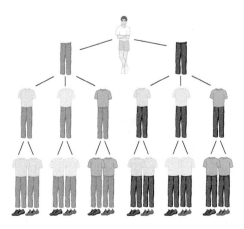

Figure 7.2
Increasing wardrobe selections

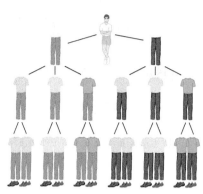

Figure 7.2 (repeated)

The tree diagram shows that you can form 12 outfits from your two pairs of jeans, three T-shirts, and two pairs of sneakers. Notice that the number of outfits can be obtained by multiplying the number of choices for jeans, 2, the number of choices for T-shirts, 3, and the number of choices for sneakers, 2:

$$2 \cdot 3 \cdot 2 = 12.$$

Unlike your earlier dilemma, you are now dealing with *three* groups of items. The Fundamental Counting Principle can be extended to determine the number of possible outcomes in situations in which there are three or more groups of items.

> **The Fundamental Counting Principle**
>
> The number of ways in which a series of successive things can occur is found by multiplying the number of ways in which each thing can occur.

For example, if you own 30 pairs of jeans, 20 T-shirts, and 12 pairs of sneakers, you have

$$30 \cdot 20 \cdot 12 = 7200$$

choices for your wardrobe.

Example 3   Options in Planning a College Course Schedule

Next semester you are planning to take three courses—math, English, and humanities. Based on time blocks and highly recommended professors, there are 8 sections of math, 5 of English, and 4 of humanities that you find suitable. Assuming no scheduling conflicts, how many different three-course schedules are possible?

Solution

This situation involves making choices with three groups of items.

Math	English	Humanities
8 choices	5 choices	4 choices

We use the Fundamental Counting Principle to find the number of three-course schedules. Multiply the number of choices for each of the three groups.

$$8 \cdot 5 \cdot 4 = 160$$

Thus, there are 160 different three-course schedules.

✔ **Check Point 3**   A pizza can be ordered with two choices of size (medium or large), three choices of crust (thin, thick, or regular), and five choices of toppings (ground beef, sausage, pepperoni, bacon, or mushrooms). How many different one-topping pizzas can be ordered?   30

Example 4   Car of the Future

Car manufacturers are now experimenting with lightweight three-wheel cars, designed for one person, and considered ideal for city driving. Intrigued? Suppose you could order such a car with a choice of 9 possible colors, with or without air conditioning, electric or gas powered, and with or without an onboard computer. In how many ways can this car be ordered with regard to these options?

Solution

This situation involves making choices with four groups of items.

Color	Air conditioning	Power	Computer
9 choices	2 choices: with or without	2 choices: electric or gas	2 choices: with or without

We use the Fundamental Counting Principle to find the number of ordering options. Multiply the number of choices for each of the four groups.

$$9 \cdot 2 \cdot 2 \cdot 2 = 72$$

Thus, the car can be ordered in 72 different ways.

✓ **Check Point 4**   The car in Example 4 is now available in 10 possible colors. The options involving air conditioning, power, and an onboard computer still apply. Furthermore, the car is available with or without a global positioning system (for pinpointing your location at every moment). In how many ways can this car be ordered in terms of these options?   160

Example 5   A Multiple-Choice Test

You are taking a multiple-choice test that has ten questions. Each of the questions has four answer choices, with one correct answer per question. If you select one of these four choices for each question and leave nothing blank, in how many ways can you answer the questions?

Solution

This situation involves making choices with ten questions.

Question 1	Question 2	Question 3	...	Question 9	Question 10
4 choices	4 choices	4 choices		4 choices	4 choices

We use the Fundamental Counting Principle to determine the number of ways that you can answer the questions on the test. Multiply the number of choices, 4, for each of the ten questions.

$$4 \cdot 4 \cdot 4 \cdot 4 \cdot 4 \cdot 4 \cdot 4 \cdot 4 \cdot 4 \cdot 4 = 4^{10} = 1,048,576 \quad \text{Use a calculator: } 4 \boxed{y^x} 10 \boxed{=}.$$

Thus, you can answer the questions in 1,048,576 different ways.

Are you surprised that there are over one million ways of answering a ten-question multiple-choice test? Of course, there is only one way to answer the test and receive a perfect score. The probability of guessing your way into a perfect score involves calculating the chance of getting a perfect score, just one way, from all 1,048,576 possible outcomes. In short, prepare for the test and do not rely on guessing!

✓ **Check Point 5**   You are taking a multiple-choice test that has six questions. Each of the questions has three answer choices, with one correct answer per question. If you select one of these three choices for each question and leave nothing blank, in how many ways can you answer the questions?   729

Example 6   Telephone Numbers in the United States

Telephone numbers in the United States begin with three-digit area codes followed by seven-digit local telephone numbers. Area codes and local telephone numbers cannot begin with 0 or 1. How many different telephone numbers are possible?

Solution

This situation involves making choices with ten groups of items.

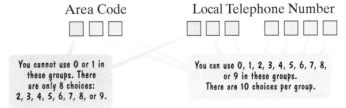

Area Code   Local Telephone Number

You cannot use 0 or 1 in these groups. There are only 8 choices: 2, 3, 4, 5, 6, 7, 8, or 9.

You can use 0, 1, 2, 3, 4, 5, 6, 7, 8, or 9 in these groups. There are 10 choices per group.

Blitzer Bonus
●●●●●●●●●●●●●●●●●

Running Out of Telephone Numbers

By the year 2020, portable telephones used for business and pleasure will all be videophones. At that time, U.S. population is expected to be 323 million. Faxes, beepers, cellphones, computer phone lines, and business lines may result in certain areas running out of phone numbers. Solution: Add more digits!

With or without extra digits, we expect that the 2020 videophone greeting will still be "hello," a word created by Thomas Edison in 1877. Phone inventor Alexander Graham Bell preferred "ahoy," but "hello" won out, appearing in the *Oxford English Dictionary* in 1883.

(*Source: New York Times*)

Here are the numbers of choices for each of the ten groups of items:

Area Code   Local Telephone Number
8 10 10   8 10 10   10 10 10 10

We use the Fundamental Counting Principle to determine the number of different telephone numbers that are possible. The total number of telephone numbers possible is

$$8 \cdot 10 \cdot 10 \cdot 8 \cdot 10 \cdot 10 \cdot 10 \cdot 10 \cdot 10 \cdot 10 = 6,400,000,000.$$

There are six billion four hundred million different telephone numbers that are possible.

Check Point 6   An electronic gate can be opened by entering five digits on a keypad containing the digits $0, 1, 2, 3, \ldots, 8, 9$. How many different keypad sequences are possible if the digit 0 cannot be used as the first digit?   90,000

Achieving Success

Read your lecture notes before starting your homework.

Often the homework problems, and later the test problems, are variations of the ones done by your professor in class.

Concept and Vocabulary Check

Exercises in the Concept and Vocabulary Check are intended for group and class discussions.

In Exercises 1–2, fill in each blank so that the resulting statement is true.

1. If you can choose one item from a group of M items and a second item from a group of N items, then the total number of two-item choices is _____ $M \cdot N$ _____.

2. The number of ways in which a series of successive things can occur is found by _____ multiplying _____ the number of ways in which each thing can occur. This is called the _Fundamental Counting_ Principle.

In Exercises 3–4, determine whether each statement is true or false. If the statement is false, make the necessary change(s) to produce a true statement.   Changes to false statements will vary.

3. If one item is chosen from M items, a second item is chosen from N items, and a third item is chosen from P items, the total number of three-item choices is $M + N + P$.   false

4. Regardless of the United States population, we will not run out of telephone numbers as long as we continue to add new digits.   true

Respond to Exercises 5–6 using verbal or written explanations.   5–6. Answers will vary.

5. Explain the Fundamental Counting Principle.

6. **Figure 7.2** on page 423 shows that a tree diagram can be used to find the total number of outfits. Describe one advantage of using the Fundamental Counting Principle rather than a tree diagram.

Exercise Set 7.1

Practice and Application Exercises

1. A restaurant offers 8 appetizers and 10 main courses. In how many ways can a person order a two-course meal?   80

2. The model of the car you are thinking of buying is available in nine different colors and three different styles (hatchback, sedan, or station wagon). In how many ways can you order the car?   27

3. A popular brand of pen is available in three colors (red, green, or blue) and four writing tips (bold, medium, fine, or micro). How many different choices of pens do you have with this brand? 12

4. In how many ways can a casting director choose a female lead and a male lead from five female actors and six male actors? 30

5. A family is planning a two-part trip. The first leg of the trip is from San Francisco to New York, and the second leg is from New York to Paris. From San Francisco to New York, travel options include airplane, train, or bus. From New York to Paris, the options are limited to airplane or ship. In how many ways can the two-part trip be made? 6

6. For a summer job, you are painting the parking spaces for a new shopping mall with a letter of the alphabet and a single digit from 1 to 9. The first parking space is A1 and the last parking space is Z9. How many parking spaces can you paint with distinct labels? 234

7. An ice cream store sells two drinks (sodas or milk shakes), in four sizes (small, medium, large, or jumbo), and five flavors (vanilla, strawberry, chocolate, coffee, or pistachio). In how many ways can a customer order a drink? 40

8. A pizza can be ordered with three choices of size (small, medium, or large), four choices of crust (thin, thick, crispy, or regular), and six choices of toppings (ground beef, sausage, pepperoni, bacon, mushrooms, or onions). How many one-topping pizzas can be ordered? 72

9. A restaurant offers the following limited lunch menu.

Main Course	Vegetables	Beverages	Desserts
Ham	Potatoes	Coffee	Cake
Chicken	Peas	Tea	Pie
Fish	Green beans	Milk	Ice cream
Beef		Soda	

If one item is selected from each of the four groups, in how many ways can a meal be ordered? Describe two such orders. 144; Answers will vary.

10. An apartment complex offers apartments with four different options, designated by A through D.

A	B	C	D
one bedroom	one bathroom	first floor	lake view
two bedrooms	two bathrooms	second floor	golf course view
three bedrooms			no special view

How many apartment options are available? Describe two such options. 36; Answers will vary.

11. Shoppers in a large shopping mall are categorized as male or female, over 30 or 30 and under, and cash or credit card shoppers. In how many ways can the shoppers be categorized? 8

12. There are three highways from city A to city B, two highways from city B to city C, and four highways from city C to city D. How many different highway routes are there from city A to city D? 24

13. A person can order a new car with a choice of six possible colors, with or without air conditioning, with or without automatic transmission, with or without power windows, and with or without a CD player. In how many different ways can a new car be ordered with regard to these options? 96

14. A car model comes in nine colors, with or without air conditioning, with or without a sun roof, with or without automatic transmission, and with or without antilock brakes. In how many ways can the car be ordered with regard to these options? 144

15. You are taking a multiple-choice test that has five questions. Each of the questions has three answer choices, with one correct answer per question. If you select one of these three choices for each question and leave nothing blank, in how many ways can you answer the questions? 243

16. You are taking a multiple-choice test that has eight questions. Each of the questions has three answer choices, with one correct answer per question. If you select one of these three choices for each question and leave nothing blank, in how many ways can you answer the questions? 6561

17. In the original plan for area codes in 1945, the first digit could be any number from 2 through 9, the second digit was either 0 or 1, and the third digit could be any number except 0. With this plan, how many different area codes are possible? 144

18. The local seven-digit telephone numbers in Inverness, California, have 669 as the first three digits. How many different telephone numbers are possible in Inverness? 10,000

19. License plates in a particular state display two letters followed by three numbers, such as AT-887 or BB-013. How many different license plates can be manufactured for this state? 676,000

20. How many different four-letter radio station call letters can be formed if the first letter must be W or K? 35,152

21. A stock can go up, go down, or stay unchanged. How many possibilities are there if you own seven stocks? 2187

22. A social security number contains nine digits, such as 074-66-7795. How many different social security numbers can be formed? 10^9 or 1,000,000,000

Critical Thinking Exercises

23. How many four-digit odd numbers are there? Assume that the digit on the left cannot be 0. 4500

24. In order to develop a more appealing hamburger, a franchise uses taste tests with 12 different buns, 30 sauces, 4 types of lettuce, and 3 types of tomatoes. If the taste tests were done at one restaurant by one tester who takes 10 minutes to eat each hamburger, approximately how long would it take the tester to eat all possible hamburgers? 720 hr

25. Write an original problem that can be solved using the Fundamental Counting Principle. Then solve the problem. Answers will vary.

7.2 : Permutations

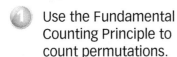

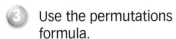
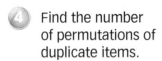

After you have read this section, you should be able to:

① Use the Fundamental Counting Principle to count permutations.

② Evaluate factorial expressions.

③ Use the permutations formula.

④ Find the number of permutations of duplicate items.

 Use the Fundamental Counting Principle to count permutations.

We open this section with six jokes about books. (Stay with us on this one.)

- *"Outside of a dog, a book is man's best friend. Inside of a dog, it's too dark to read."* —Groucho Marx

- *"I recently bought a book of free verse. For $12."* —George Carlin

- *"If a word in the dictionary was misspelled, how would we know?"* —Steven Wright

- *"I wrote a book under a pen name: Bic."* —Henny Youngman

- *"A bit of advice: Never read a pop-up book about giraffes."* —Jerry Seinfeld

- *"Fang just got out of the hospital. He was in a speed-reading contest. He hit a bookmark."* —Phyllis Diller

We can use the Fundamental Counting Principle to determine the number of ways these jokes can be delivered. You can choose any one of the six jokes as the first one told. Once this joke is delivered, you'll have five jokes to choose from for the second joke told. You'll then have four jokes left to choose from for the third delivery. Continuing in this manner, the situation can be shown as follows:

First Joke Delivered	Second Joke Delivered	Third Joke Delivered	Fourth Joke Delivered	Fifth Joke Delivered	Sixth Joke Delivered
6 choices	5 choices	4 choices	3 choices	2 choices	1 choice

Using the Fundamental Counting Principle, we multiply the choices:

$$6 \cdot 5 \cdot 4 \cdot 3 \cdot 2 \cdot 1 = 720.$$

Thus, there are 720 different ways to deliver the six jokes about books. One of the 720 possible arrangements is

Seinfeld joke—Youngman joke—Carlin joke—Marx joke—Wright joke—Diller joke.

Such an ordered arrangement is called a *permutation* of the six jokes.
   A **permutation** is an ordered arrangement of items that occurs when

- No item is used more than once. (Each joke is told exactly once.)

- The order of arrangement makes a difference. (The order in which these jokes are told makes a difference in terms of how they are received.)

Example 1   Counting Permutations

How many ways can the six jokes about books be delivered if George Carlin's joke is delivered first and Jerry Seinfeld's joke is told last?

Solution

The conditions of Carlin's joke first and Seinfeld's joke last can be shown as follows:

First Joke Delivered	Second Joke Delivered	Third Joke Delivered	Fourth Joke Delivered	Fifth Joke Delivered	Sixth Joke Delivered
1 choice: Carlin					1 choice: Seinfeld

Now let's fill in the number of choices for positions two through five. You can choose any one of the four remaining jokes for the second delivery. Once you've chosen this joke, you'll have three jokes to choose from for the third delivery. This leaves only two jokes to choose from for the fourth delivery. Once this choice is made, there is just one joke left to choose for the fifth delivery.

First Joke Delivered	Second Joke Delivered	Third Joke Delivered	Fourth Joke Delivered	Fifth Joke Delivered	Sixth Joke Delivered
1 choice: Carlin	4 choices	3 choices	2 choices	1 choice	1 choice: Seinfeld

We use the Fundamental Counting Principle to find the number of ways the six jokes can be delivered. Multiply the choices:

$$1 \cdot 4 \cdot 3 \cdot 2 \cdot 1 \cdot 1 = 24.$$

Thus, there are 24 different ways the jokes can be delivered if Carlin's joke is told first and Seinfeld's joke is told last.

☑ **Check Point 1**   How many ways can the six jokes about books be delivered if a man's joke is told first?   600

Blitzer Bonus
• • • • • • • • • • • • • • •

How to Pass the Time for $2\frac{1}{2}$ Million Years

If you were to arrange 15 different books on a shelf and it took you one minute for each permutation, the entire task would take 2,487,965 years.
Source: Isaac Asimov's Book of Facts

Example 2   Counting Permutations

You need to arrange seven of your favorite books along a small shelf. How many different ways can you arrange the books, assuming that the order of the books makes a difference to you?

Solution

You may choose any of the seven books for the first position on the shelf. This leaves six choices for second position. After the first two positions are filled, there are five books to choose from for third position, four choices left for the fourth position, three choices left for the fifth position, then two choices for the sixth position, and only one choice for the last position. This situation can be shown as follows:

First Shelf Position	Second Shelf Position	Third Shelf Position	Fourth Shelf Position	Fifth Shelf Position	Sixth Shelf Position	Seventh Shelf Position
7 choices	6 choices	5 choices	4 choices	3 choices	2 choices	1 choice

We use the Fundamental Counting Principle to find the number of ways you can arrange the seven books along the shelf. Multiply the choices:

$$7 \cdot 6 \cdot 5 \cdot 4 \cdot 3 \cdot 2 \cdot 1 = 5040.$$

Thus, you can arrange the books in 5040 ways. There are 5040 different possible permutations.

☑ **Check Point 2**   In how many ways can you arrange five books along a shelf, assuming that the order of the books makes a difference?   120

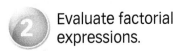

Evaluate factorial expressions.

Factorials: 0! Through 20!

0!	1
1!	1
2!	2
3!	6
4!	24
5!	120
6!	720
7!	5040
8!	40,320
9!	362,880
10!	3,628,800
11!	39,916,800
12!	479,001,600
13!	6,227,020,800
14!	87,178,291,200
15!	1,307,674,368,000
16!	20,922,789,888,000
17!	355,687,428,096,000
18!	6,402,373,705,728,000
19!	121,645,100,408,832,000
20!	2,432,902,008,176,640,000

Factorial Notation

The product in Example 2,

$$7 \cdot 6 \cdot 5 \cdot 4 \cdot 3 \cdot 2 \cdot 1$$

is given a special name and symbol. It is called 7 **factorial**, and written 7!. Thus,

$$7! = 7 \cdot 6 \cdot 5 \cdot 4 \cdot 3 \cdot 2 \cdot 1.$$

In general, if n is a positive integer, then $n!$ (*n factorial*) is the product of all positive integers from n down through 1. For example,

$$1! = 1$$
$$2! = 2 \cdot 1 = 2$$
$$3! = 3 \cdot 2 \cdot 1 = 6$$
$$4! = 4 \cdot 3 \cdot 2 \cdot 1 = 24$$
$$5! = 5 \cdot 4 \cdot 3 \cdot 2 \cdot 1 = 120$$
$$6! = 6 \cdot 5 \cdot 4 \cdot 3 \cdot 2 \cdot 1 = 720.$$

Factorial Notation

If n is a positive integer, the notation $n!$ (read "n factorial") is the product of all positive integers from n down through 1.

$$n! = n(n - 1)(n - 2)\cdots(3)(2)(1)$$

0! (zero factorial), by definition, is 1.

$$0! = 1$$

Example 3   Using Factorial Notation

Evaluate the following factorial expressions without using the factorial key on your calculator:

a. $\dfrac{8!}{5!}$ **b.** $\dfrac{26!}{21!}$ **c.** $\dfrac{500!}{499!}$.

Solution

a. We can evaluate the numerator and the denominator of $\frac{8!}{5!}$. However, it is easier to use the following simplification:

$$\frac{8!}{5!} = \frac{8 \cdot 7 \cdot 6 \cdot \boxed{5 \cdot 4 \cdot 3 \cdot 2 \cdot 1}}{\boxed{5 \cdot 4 \cdot 3 \cdot 2 \cdot 1}} = \frac{8 \cdot 7 \cdot 6 \cdot \boxed{5!}}{\boxed{5!}} = \frac{8 \cdot 7 \cdot 6 \cdot \cancel{5!}}{\cancel{5!}} = 8 \cdot 7 \cdot 6 = 336.$$

b. Rather than write out 26!, the numerator of $\frac{26!}{21!}$, as the product of all integers from 26 down to 1, we can express 26! as

$$26! = 26 \cdot 25 \cdot 24 \cdot 23 \cdot 22 \cdot 21!.$$

In this way, we can cancel 21! in the numerator and the denominator of the given expression.

$$\frac{26!}{21!} = \frac{26 \cdot 25 \cdot 24 \cdot 23 \cdot 22 \cdot 21!}{21!} = \frac{26 \cdot 25 \cdot 24 \cdot 23 \cdot 22 \cdot \cancel{21!}}{\cancel{21!}}$$

$$= 26 \cdot 25 \cdot 24 \cdot 23 \cdot 22 = 7,893,600$$

Technology

Most calculators have a key or menu item for calculating factorials. Here are the keystrokes for finding 9!:

Many Scientific Calculators:

9 $\boxed{x!}$ $\boxed{=}$

Many Graphing Calculators:

9 $\boxed{!}$ $\boxed{\text{ENTER}}$

On TI graphing calculators, this is selected using the MATH PRB menu.

Because $n!$ becomes quite large as n increases, your calculator will display these larger values in scientific notation.

c. In order to cancel identical factorials in the numerator and the denominator of $\frac{500!}{499!}$, we can express 500! as $500 \cdot 499!$.

$$\frac{500!}{499!} = \frac{500 \cdot 499!}{499!} = \frac{500 \cdot \cancel{499!}}{\cancel{499!}} = 500$$

 Check Point 3   Evaluate without using a calculator's factorial key:

a. $\dfrac{9!}{6!}$ 504   **b.** $\dfrac{16!}{11!}$ 524,160   **c.** $\dfrac{100!}{99!}$.   100

A Formula for Permutations

Suppose that you are the coach of a little league baseball team. There are 13 players on the team (and lots of parents hovering in the background, dreaming of stardom for their little "Albert Pujols"). You need to choose a batting order having 9 players. The order makes a difference, because, for instance, if bases are loaded and "Little Albert" is fourth or fifth at bat, his possible home run will drive in three additional runs. How many batting orders can you form?

You can choose any of 13 players for the first person at bat. Then you will have 12 players from which to choose the second batter, then 11 from which to choose the third batter, and so on. The situation can be shown as follows:

Batter 1	Batter 2	Batter 3	Batter 4	Batter 5	Batter 6	Batter 7	Batter 8	Batter 9
13 choices	12 choices	11 choices	10 choices	9 choices	8 choices	7 choices	6 choices	5 choices

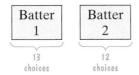 Use the permutations formula.

The total number of batting orders is

$$13 \cdot 12 \cdot 11 \cdot 10 \cdot 9 \cdot 8 \cdot 7 \cdot 6 \cdot 5 = 259{,}459{,}200.$$

Nearly 260 million batting orders are possible for your 13-player little league team. Each batting order is a permutation because the order of the batters makes a difference. The number of permutations of 13 players taken 9 at a time is 259,459,200.

We can obtain a formula for finding the number of permutations by rewriting our computation:

$$13 \cdot 12 \cdot 11 \cdot 10 \cdot 9 \cdot 8 \cdot 7 \cdot 6 \cdot 5$$

$$= \frac{13 \cdot 12 \cdot 11 \cdot 10 \cdot 9 \cdot 8 \cdot 7 \cdot 6 \cdot 5 \cdot \boxed{4 \cdot 3 \cdot 2 \cdot 1}}{\boxed{4 \cdot 3 \cdot 2 \cdot 1}} = \frac{13!}{4!} = \frac{13!}{(13-9)!}.$$

Thus, the number of permutations of 13 things taken 9 at a time is $\frac{13!}{(13-9)!}$. The special notation $_{13}P_9$ is used to replace the phrase "the number of permutations of 13 things taken 9 at a time." Using this new notation, we can write

$$_{13}P_9 = \frac{13!}{(13-9)!}.$$

The numerator of this expression is the factorial of the number of items, 13 team members: 13!. The denominator is also a factorial. It is the factorial of the difference between the number of items, 13, and the number of items in each permutation, 9 batters: $(13-9)!$.

Great Question!

Do I have to use the formula for $_nP_r$ to solve permutation problems?

No. Because all permutation problems are also Fundamental Counting problems, they can be solved using the formula for $_nP_r$ or using the Fundamental Counting Principle.

Technology

Graphing calculators have a menu item for calculating permutations, usually labeled $\boxed{_nP_r}$. For example, to find $_{20}P_3$, the keystrokes are

$$20\ \boxed{_nP_r}\ 3\ \boxed{\text{ENTER}}.$$

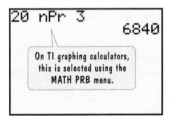

On TI graphing calculators, this is selected using the MATH PRB menu.

If you are using a scientific calculator, check your manual for the location of the menu item for calculating permutations and the required keystrokes.

The notation $_nP_r$ means the **number of permutations of n things taken r at a time**. We can generalize from the situation in which 9 batters were taken from 13 players. By generalizing, we obtain the following formula for the number of permutations if r items are taken from n items:

> ### Permutations of n Things Taken r at a Time
>
> The number of possible permutations if r items are taken from n items is
>
> $$_nP_r = \frac{n!}{(n-r)!}.$$

Example 4   Using the Formula for Permutations

Suppose that you and 19 of your friends have decided to form an Internet marketing consulting firm. The group needs to choose three officers—a CEO, an operating manager, and a treasurer. In how many ways can those offices be filled?

Solution

Your group is choosing $r = 3$ officers from a group of $n = 20$ people (you and 19 friends). The order in which the officers are chosen matters because the CEO, the operating manager, and the treasurer each have different responsibilities. Thus, we are looking for the number of permutations of 20 things taken 3 at a time. We use the formula

$$_nP_r = \frac{n!}{(n-r)!}$$

with $n = 20$ and $r = 3$.

$$_{20}P_3 = \frac{20!}{(20-3)!} = \frac{20!}{17!} = \frac{20 \cdot 19 \cdot 18 \cdot 17!}{17!} = \frac{20 \cdot 19 \cdot 18 \cdot \cancel{17!}}{\cancel{17!}} = 20 \cdot 19 \cdot 18 = 6840$$

Thus, there are 6840 different ways of filling the three offices.

✔ **Check Point 4**   A corporation has seven members on its board of directors. In how many different ways can it elect a president, vice-president, secretary, and treasurer?   840

Example 5   Using the Formula for Permutations

Suppose that you are working for The Sitcom Television Network. Your assignment is to help set up the television schedule for Monday evenings between 7 and 10 P.M. You need to schedule a show in each of six 30-minute time blocks, beginning with 7 to 7:30 and ending with 9:30 to 10:00. You can select from among the following situation comedies: *The Office, Seinfeld, That 70s Show, Cheers, The Big Bang Theory, Frasier, All in the Family, I Love Lucy, M*A*S*H, The Larry Sanders Show, Modern Family, Married . . . with Children,* and *Curb Your Enthusiasm.* How many different programming schedules can be arranged?

Solution

You are choosing $r = 6$ situation comedies from a collection of $n = 13$ classic sitcoms. The order in which the programs are aired matters. Family-oriented comedies have higher ratings when aired in earlier time blocks, such as 7 to 7:30. By contrast, comedies with adult themes do better in later time blocks. In short, we are looking for the number of permutations of 13 things taken 6 at a time. We use the formula

$$_nP_r = \frac{n!}{(n-r)!}$$

with $n = 13$ and $r = 6$.

$$_{13}P_6 = \frac{13!}{(13-6)!} = \frac{13!}{7!} = \frac{13 \cdot 12 \cdot 11 \cdot 10 \cdot 9 \cdot 8 \cdot \cancel{7!}}{\cancel{7!}} = 13 \cdot 12 \cdot 11 \cdot 10 \cdot 9 \cdot 8 = 1{,}235{,}520$$

There are 1,235,520 different programming schedules that can be arranged.

Check Point 5   How many different programming schedules can be arranged by choosing 5 situation comedies from a collection of 9 classic sitcoms?   15,120

Permutations of Duplicate Items

④ Find the number of permutations of duplicate items.

The number of permutations of the letters in the word SET is 3!, or 6. The six permutations are

SET,   STE,   EST,   ETS,   TES,   TSE.

Are there also six permutations of the letters in the name ANA? The answer is no. Unlike SET, with three distinct letters, ANA contains three letters, of which the two As are duplicates. If we rearrange the letters just as we did with SET, we obtain

ANA,   AAN,   NAA,   NAA,   ANA,   AAN.

Without the use of color to distinguish between the two As, there are only three distinct permutations: ANA, AAN, NAA.

There is a formula for finding the number of distinct permutations when duplicate items exist:

Permutations of Duplicate Items

The number of permutations of n items, where p items are identical, q items are identical, r items are identical, and so on, is given by

$$\frac{n!}{p!\, q!\, r! \dots}.$$

For example, ANA contains three letters ($n = 3$), where two of the letters are identical ($p = 2$). The number of distinct permutations is

$$\frac{n!}{p!} = \frac{3!}{2!} = \frac{3 \cdot \cancel{2!}}{\cancel{2!}} = 3.$$

We saw that the three distinct permutations are ANA, AAN, and NAA.

Example 6   Using the Formula for Permutations of Duplicate Items

In how many distinct ways can the letters of the word MISSISSIPPI be arranged?

Solution

The word contains 11 letters ($n = 11$), where four Is are identical ($p = 4$), four Ss are identical ($q = 4$), and 2 Ps are identical ($r = 2$). The number of distinct permutations is

$$\frac{n!}{p!\, q!\, r!} = \frac{11!}{4!\, 4!\, 2!} = \frac{11 \cdot 10 \cdot 9 \cdot 8 \cdot 7 \cdot 6 \cdot 5 \cdot \cancel{4!}}{\cancel{4!}\, 4 \cdot 3 \cdot 2 \cdot 1 \cdot 2 \cdot 1} = 34{,}650$$

There are 34,650 distinct ways the letters in the word MISSISSIPPI can be arranged.

Check Point 6   In how many ways can the letters of the word OSMOSIS be arranged?   420

Technology

Parentheses are necessary to enclose the factorials in the denominator when using a calculator to find

$$\frac{11!}{4!\, 4!\, 2!}.$$

```
11!/(4!4!2!)
            34650
```

Achieving Success

Avoid asking for help on a problem that you have not thought about. This is basically asking your professor to do the work for you. First try solving the problem on your own.

Concept and Vocabulary Check

Exercises in the Concept and Vocabulary Check are intended for group and class discussions.

In Exercises 1–3, fill in each blank so that the resulting statement is true.

1. 5!, called 5 _____factorial_____, is the product of all positive integers from _____5_____ down through _____1_____.
By definition, 0! = _____1_____.

2. The number of possible permutations if r objects are taken from n items is $_nP_r = $ _____$\frac{n!}{(n-r)!}$_____.

3. The number of permutations of n items, where p items are identical and q items are identical, is given by _____$\frac{n!}{p!q!}$_____.

In Exercises 4–7, determine whether each statement is true or false. If the statement is false, make the necessary change(s) to produce a true statement. Changes to false statements will vary.

4. A permutation occurs when the order of arrangement does not matter.   false

5. $8! = 8 + 7 + 6 + 5 + 4 + 3 + 2 + 1$ false

6. Because all permutation problems are also Fundamental Counting problems, they can be solved using the formula for $_nP_r$ or using the Fundamental Counting Principle.   true

7. Because the words BET and BEE both contain three letters, the number of permutations of the letters in each word is 3!, or 6.   false

Respond to Exercises 8–12 using verbal or written explanations.

8. What is a permutation?   8–12. Answers will vary.

9. Explain how to find $n!$, where n is a positive integer.

10. Explain the best way to evaluate $\frac{900!}{899!}$ without a calculator.

11. Describe what $_nP_r$ represents.

12. If 24 permutations can be formed using the letters in the word BAKE, why can't 24 permutations also be formed using the letters in the word BABE? How is the number of permutations in BABE determined?

Exercise Set 7.2

Practice and Application Exercises

Use the Fundamental Counting Principle to solve Exercises 1–12.

1. Six performers are to present their comedy acts on a weekend evening at a comedy club. How many different ways are there to schedule their appearances?   720

2. Five singers are to perform on a weekend evening at a night club. How many different ways are there to schedule their appearances?   120

3. In the *Cambridge Encyclopedia of Language* (Cambridge University Press, 1987), author David Crystal presents five sentences that make a reasonable paragraph regardless of their order. The sentences are as follows:
   - Mark had told him about the foxes.
   - John looked out of the window.
   - Could it be a fox?
   - However, nobody had seen one for months.
   - He thought he saw a shape in the bushes.

   In how many different orders can the five sentences be arranged?   120

4. In how many different ways can a police department arrange eight suspects in a police lineup if each lineup contains all eight people?   40,320

5. As in Exercise 1, six performers are to present their comedy acts on a weekend evening at a comedy club. One of the performers insists on being the last stand-up comic of the evening. If this performer's request is granted, how many different ways are there to schedule the appearances?   120

6. As in Exercise 2, five singers are to perform at a night club. One of the singers insists on being the last performer of the evening. If this singer's request is granted, how many different ways are there to schedule the appearances?   24

7. You need to arrange nine of your favorite books along a small shelf. How many different ways can you arrange the books, assuming that the order of the books makes a difference to you?   362,880

8. You need to arrange ten of your favorite photographs on the mantle above a fireplace. How many ways can you arrange the photographs, assuming that the order of the pictures makes a difference to you?   3,628,800

In Exercises 9–10, use the five sentences that are given in Exercise 3.

9. How many different five-sentence paragraphs can be formed if the paragraph begins with "He thought he saw a shape in the bushes" and ends with "John looked out of the window"? 6

10. How many different five-sentence paragraphs can be formed if the paragraph begins with "He thought he saw a shape in the bushes" followed by "Mark had told him about the foxes"? 6

11. A television programmer is arranging the order that five movies will be seen between the hours of 6 P.M. and 4 A.M. Two of the movies have a G rating, and they are to be shown in the first two time blocks. One of the movies is rated NC-17, and it is to be shown in the last of the time blocks, from 2 A.M. until 4 A.M. Given these restrictions, in how many ways can the five movies be arranged during the indicated time blocks? 4

12. A camp counselor and six campers are to be seated along a picnic bench. In how many ways can this be done if the counselor must be seated in the middle and a camper who has a tendency to engage in food fights must sit to the counselor's immediate left? 120

In Exercises 13–32, evaluate each factorial expression.

13. $\dfrac{9!}{6!}$ 504

14. $\dfrac{12!}{10!}$ 132

15. $\dfrac{29!}{25!}$ 570,024

16. $\dfrac{31!}{28!}$ 26,970

17. $\dfrac{19!}{11!}$ 3,047,466,240

18. $\dfrac{17!}{9!}$ 980,179,200

19. $\dfrac{600!}{599!}$ 600

20. $\dfrac{700!}{699!}$ 700

21. $\dfrac{104!}{102!}$ 10,712

22. $\dfrac{106!}{104!}$ 11,130

23. $7! - 3!$ 5034

24. $6! - 3!$ 714

25. $(7 - 3)!$ 24

26. $(6 - 3)!$ 6

27. $\left(\dfrac{12}{4}\right)!$ 6

28. $\left(\dfrac{45}{9}\right)!$ 120

29. $\dfrac{7!}{(7 - 2)!}$ 42

30. $\dfrac{8!}{(8 - 5)!}$ 6720

31. $\dfrac{13!}{(13 - 3)!}$ 1716

32. $\dfrac{17!}{(17 - 3)!}$ 4080

In Exercises 33–40, use the formula for $_nP_r$ to evaluate each expression.

33. $_9P_4$ 3024

34. $_7P_3$ 210

35. $_8P_5$ 6720

36. $_{10}P_4$ 5040

37. $_6P_6$ 720

38. $_9P_9$ 362,880

39. $_8P_0$ 1

40. $_6P_0$ 1

Use an appropriate permutations formula to solve Exercises 41–56.

41. A club with ten members is to choose three officers—president, vice-president, and secretary-treasurer. If each office is to be held by one person and no person can hold more than one office, in how many ways can those offices be filled? 720

42. A corporation has six members on its board of directors. In how many different ways can it elect a president, vice-president, secretary, and treasurer? 360

43. For a segment of a radio show, a disc jockey can play 7 records. If there are 13 records to select from, in how many ways can the program for this segment be arranged? 8,648,640

44. Suppose you are asked to list, in order of preference, the three best movies you have seen this year. If you saw 20 movies during the year, in how many ways can the three best be chosen and ranked? 6840

45. In a race in which six automobiles are entered and there are no ties, in how many ways can the first three finishers come in? 120

46. In a production of *West Side Story*, eight actors are considered for the male roles of Tony, Riff, and Bernardo. In how many ways can the director cast the male roles? 336

47. Nine bands have volunteered to perform at a benefit concert, but there is only enough time for five of the bands to play. How many lineups are possible? 15,120

48. How many arrangements can be made using four of the letters of the word COMBINE if no letter is to be used more than once? 840

49. In how many distinct ways can the letters of the word DALLAS be arranged? 180

50. In how many distinct ways can the letters of the word SCIENCE be arranged? 1260

51. How many distinct permutations can be formed using the letters of the word TALLAHASSEE? 831,600

52. How many distinct permutations can be formed using the letters of the word TENNESSEE? 3780

53. In how many ways can the digits in the number 5,446,666 be arranged? 105

54. In how many ways can the digits in the number 5,432,435 be arranged? 630

In Exercises 55–56, a signal can be formed by running different colored flags up a pole, one above the other.

55. Find the number of different signals consisting of eight flags that can be made using three white flags, four red flags, and one blue flag. 280

56. Find the number of different signals consisting of nine flags that can be made using three white flags, five red flags, and one blue flag. 504

Critical Thinking Exercises 57–58. Answers will vary.

57. Write a word problem that can be solved by evaluating $5!$.

58. Write a word problem that can be solved by evaluating $_7P_3$.

59. Ten people board an airplane that has 12 aisle seats. In how many ways can they be seated if they all select aisle seats? 239,500,800

60. Six horses are entered in a race. If two horses are tied for first place and there are no ties among the other four horses, in how many ways can the six horses cross the finish line? 360

61. Five men and five women line up at a checkout counter in a store. In how many ways can they line up if the first person in line is a woman, and the people in line alternate woman, man, woman, man, and so on? 14,400

62. How many four-digit odd numbers less than 6000 can be formed using the digits 2, 4, 6, 7, 8, and 9? 144

63. $\dfrac{n(n - 1) \cdots 3 \cdot 2 \cdot 1}{2} = n(n - 1) \cdots 3$

63. Express $_nP_{n-2}$ without using factorials.

7.3 | Combinations

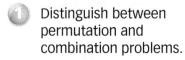

What am I supposed to learn?

After you have read this section, you should be able to:

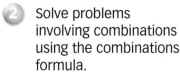

1. Distinguish between permutation and combination problems.

2. Solve problems involving combinations using the combinations formula.

Throughout the history of entertainment, performers have featured choreography in their acts. Singers who are known for their serious moves include Beyoncé, Lady Gaga, Shakira, Justin Timberlake, and Usher.

Imagine that you ask your friends the following question: "Of these five entertainers, which three would you select to be included in a documentary on singers and choreography?" You are not asking your friends to rank their three favorite artists in any kind of order—they should merely select the three to be included in the documentary.

One friend answers, "Beyoncé, Lady Gaga, and Usher." Another responds, "Usher, Lady Gaga, and Beyoncé." These two people have the same artists in their group of selections, even if they are named in a different order. We are interested *in which artists are named, not the order in which they are named,* for the documentary. Because the items are taken without regard to order, this is not a permutation problem. No ranking of any sort is involved.

Later on, you ask your sister which three artists she would select for the documentary. She names Justin Timberlake, Beyoncé, and Usher. Her selection is different from those of your two other friends because different entertainers are cited.

Mathematicians describe the group of artists given by your sister as a *combination.* A **combination** of items occurs when

- The items are selected from the same group (the five entertainers who are known for their choreography).
- No item is used more than once. (You may view Beyoncé as a phenomenal performer, but your three selections cannot be Beyoncé, Beyoncé, and Beyoncé.)
- The order of the items makes no difference. (Beyoncé, Lady Gaga, Usher is the same group in the documentary as Usher, Lady Gaga, Beyoncé.)

1 Distinguish between permutation and combination problems.

Do you see the difference between a permutation and a combination? A permutation is an ordered arrangement of a given group of items. A combination is a group of items taken without regard to their order. **Permutation** problems involve situations in which **order matters**. **Combination** problems involve situations in which the **order** of the items **makes no difference.**

Example 1 | Distinguishing between Permutations and Combinations

For each of the following problems, determine whether the problem is one involving permutations or combinations. (It is not necessary to solve the problem.)

a. Six students are running for student government president, vice-president, and treasurer. The student with the greatest number of votes becomes the president, the second highest vote-getter becomes vice-president, and the student who gets the third largest number of votes will be treasurer. How many different outcomes are possible for these three positions?

b. Six people are on the board of supervisors for your neighborhood park. A three-person committee is needed to study the possibility of expanding the park. How many different committees could be formed from the six people?

c. Baskin-Robbins offers 31 different flavors of ice cream. One of its items is a bowl consisting of three scoops of ice cream, each a different flavor. How many such bowls are possible?

Solution

a. Students are choosing three student government officers from six candidates. The order in which the officers are chosen makes a difference because each of the offices (president, vice-president, treasurer) is different. Order matters. This is a problem involving permutations.

b. A three-person committee is to be formed from the six-person board of supervisors. The order in which the three people are selected does not matter because they are not filling different roles on the committee. Because order makes no difference, this is a problem involving combinations.

c. A three-scoop bowl of three different flavors is to be formed from Baskin-Robbins's 31 flavors. The order in which the three scoops of ice cream are put into the bowl is irrelevant. A bowl with chocolate, vanilla, and strawberry is exactly the same as a bowl with vanilla, strawberry, and chocolate. Different orderings do not change things, and so this is a problem involving combinations.

✓ **Check Point 1**   For each of the following problems, determine whether the problem is one involving permutations or combinations. (It is not necessary to solve the problem.)

a. How many ways can you select 6 free DVDs from a list of 200 DVDs?   combinations

b. In a race in which there are 50 runners and no ties, in how many ways can the first three finishers come in?   permutations

A Formula for Combinations

② Solve problems involving combinations using the combinations formula.

We have seen that the notation $_nP_r$ means the number of permutations of n things taken r at a time. Similarly, the notation $_nC_r$ **means the number of combinations of n things taken r at a time**.

We can develop a formula for $_nC_r$ by comparing permutations and combinations. Consider the letters A, B, C, and D. The number of permutations of these four letters taken three at a time is

$$_4P_3 = \frac{4!}{(4-3)!} = \frac{4!}{1!} = \frac{4 \cdot 3 \cdot 2 \cdot 1}{1} = 24.$$

Here are the 24 permutations:

ABC,	ABD,	ACD,	BCD,
ACB,	ADB,	ADC,	BDC,
BAC,	BAD,	CAD,	CBD,
BCA,	BDA,	CDA,	CDB,
CAB,	DAB,	DAC,	DBC,
CBA,	DBA,	DCA,	DCB.

| This column contains only one combination, ABC. | This column contains only one combination, ABD. | This column contains only one combination, ACD. | This column contains only one combination, BCD. |

Because the order of items makes no difference in determining combinations, each column of six permutations represents one combination. There is a total of four combinations:

$$ABC, \quad ABD, \quad ACD, \quad BCD.$$

Thus, $_4C_3 = 4$: The number of combinations of 4 things taken 3 at a time is 4. With 24 permutations and only four combinations, there are 6, or 3!, times as many permutations as there are combinations.

In general, there are $r!$ times as many permutations of n things taken r at a time as there are combinations of n things taken r at a time. Thus, we find the number of combinations of n things taken r at a time by dividing the number of permutations of n things taken r at a time by $r!$.

$$_nC_r = \frac{_nP_r}{r!} = \frac{\dfrac{n!}{(n-r)!}}{r!} = \frac{n!}{(n-r)!\, r!}$$

Great Question!

Do I have to use the formula for $_nC_r$ to solve combination problems?

Yes. The number of combinations if r items are taken from n items cannot be found using the Fundamental Counting Principle and requires the use of the formula shown in the orange box.

> **Combinations of n Things Taken r at a Time**
>
> The number of possible combinations if r items are taken from n items is
>
> $$_nC_r = \frac{n!}{(n-r)!\, r!}.$$

Technology

Graphing calculators have a menu item for calculating combinations, usually labeled $_nC_r$. (On TI graphing calculators, $_nC_r$ is selected using the MATH PRB menu.) For example, to find $_8C_3$, the keystrokes on most graphing calculators are

$$8 \boxed{_nC_r} \; 3 \; \boxed{\text{ENTER}}.$$

If you are using a scientific calculator, check your manual to see whether there is a menu item for calculating combinations.

If you use your calculator's factorial key to find $\frac{8!}{5!3!}$, be sure to enclose the factorials in the denominator with parentheses

$$8\boxed{!} \div \boxed{(} 5\boxed{!} \times 3\boxed{!} \boxed{)}$$

pressing $\boxed{=}$ or $\boxed{\text{ENTER}}$ to obtain the answer.

Example 2 — Using the Formula for Combinations

A three-person committee is needed to study ways of improving public transportation. How many committees could be formed from the eight people on the board of supervisors?

Solution

The order in which the three people are selected does not matter. This is a problem of selecting $r = 3$ people from a group of $n = 8$ people. We are looking for the number of combinations of eight things taken three at a time. We use the formula

$$_nC_r = \frac{n!}{(n-r)!\, r!}$$

with $n = 8$ and $r = 3$.

$$_8C_3 = \frac{8!}{(8-3)!\, 3!} = \frac{8!}{5!\, 3!} = \frac{8 \cdot 7 \cdot 6 \cdot 5!}{5! \cdot 3 \cdot 2 \cdot 1} = \frac{8 \cdot 7 \cdot 6 \cdot \cancel{5!}}{\cancel{5!} \cdot 3 \cdot 2 \cdot 1} = 56$$

Thus, 56 committees of three people each can be formed from the eight people on the board of supervisors.

✓ **Check Point 2** You volunteer to pet-sit for your friend who has seven different animals. How many different pet combinations are possible if you take three of the seven pets? 35

Example 3 — Using the Formula for Combinations

In poker, a person is dealt 5 cards from a standard 52-card deck. The order in which the 5 cards are received does not matter. How many different 5-card poker hands are possible?

Solution

Because the order in which the 5 cards are received does not matter, this is a problem involving combinations. We are looking for the number of combinations of $n = 52$ cards drawn $r = 5$ at a time. We use the formula

$$_nC_r = \frac{n!}{(n-r)!\,r!}$$

with $n = 52$ and $r = 5$.

$$_{52}C_5 = \frac{52!}{(52-5)!\,5!} = \frac{52!}{47!\,5!} = \frac{52 \cdot 51 \cdot 50 \cdot 49 \cdot 48 \cdot \cancel{47!}}{\cancel{47!} \cdot 5 \cdot 4 \cdot 3 \cdot 2 \cdot 1} = 2{,}598{,}960$$

Thus, there are 2,598,960 different 5-card poker hands possible. It surprises many people that more than 2.5 million 5-card hands can be dealt from a mere 52 cards.

Figure 7.3 A royal flush

If you are a card player, it does not get any better than to be dealt the 5-card poker hand shown in **Figure 7.3**. This hand is called a *royal flush*. It consists of an ace, king, queen, jack, and 10, all of the same suit: all hearts, all diamonds, all clubs, or all spades. The probability of being dealt a royal flush involves calculating the number of ways of being dealt such a hand: just 4 of all 2,598,960 possible hands. In the next section, we move from counting possibilities to computing probabilities.

✓ **Check Point 3**   How many different 4-card hands can be dealt from a deck that has 16 different cards?   1820

Example 4   Using the Formula for Combinations and the Fundamental Counting Principle

In December 2011, the U.S. Senate consisted of 51 Democrats, 47 Republicans, and 2 Independents. How many distinct five-person committees can be formed if each committee must have 3 Democrats and 2 Republicans?

Solution

The order in which the members are selected does not matter. Thus, this is a problem involving combinations.

We begin with the number of ways of selecting 3 Democrats out of 51 Democrats without regard to order. We are looking for the number of combinations of $n = 51$ people taken $r = 3$ people at a time. We use the formula

$$_nC_r = \frac{n!}{(n-r)!\,r!}$$

with $n = 51$ and $r = 3$.

We are picking 3 Democrats out of 51 Democrats.

$$_{51}C_3 = \frac{51!}{(51-3)!\,3!} = \frac{51!}{48!\,3!} = \frac{51 \cdot 50 \cdot 49 \cdot \cancel{48!}}{\cancel{48!} \cdot 3 \cdot 2 \cdot 1} = \frac{51 \cdot 50 \cdot 49}{3 \cdot 2 \cdot 1} = 20{,}825$$

There are 20,825 ways to choose three Democrats for a committee.

Next, we find the number of ways of selecting 2 Republicans out of 47 Republicans without regard to order. We are looking for the number of combinations of $n = 47$ people taken $r = 2$ people at a time. Once again, we use the formula

$$_nC_r = \frac{n!}{(n-r)!\,r!}.$$

This time, $n = 47$ and $r = 2$.

We are picking 2 Republicans out of 47 Republicans.

$$_{47}C_2 = \frac{47!}{(47-2)!\,2!} = \frac{47!}{45!\,2!} = \frac{47 \cdot 46 \cdot 45!}{45! \cdot 2 \cdot 1} = \frac{47 \cdot 46}{2 \cdot 1} = 1081$$

There are 1081 ways to choose two Republicans for a committee.

We use the Fundamental Counting Principle to find the number of committees that can be formed:

$$_{51}C_3 \cdot {_{47}C_2} = 20{,}825 \cdot 1081 = 22{,}511{,}825$$

Thus, 22,511,825 distinct committees can be formed.

Check Point 4 A zoo has six male bears and seven female bears. Two male bears and three female bears will be selected for an animal exchange program with another zoo. How many five-bear collections are possible?   525

Achieving Success

Read ahead.

You might find it helpful to use some of your homework time to read (or skim) the section in the textbook that will be covered in your professor's next lecture. Having a clear idea of the new material that will be discussed will help you to understand the class a whole lot better.

Concept and Vocabulary Check

Exercises in the Concept and Vocabulary Check are intended for group and class discussions.

In Exercises 1–2, fill in each blank so that the resulting statement is true.

1. The number of possible combinations if r objects are taken from n items is $_nC_r = $ ___$\frac{n!}{(n-r)!\,r!}$___.

2. The formula for $_nC_r$ has the same numerator as the formula for $_nP_r$, but contains an extra factor of ___$r!$___ in the denominator.

In Exercises 3–4, determine whether each statement is true or false. If the statement is false, make the necessary change(s) to produce a true statement. Changes to false statements will vary.

3. Combination problems involve situations in which the order of the items makes a difference. false

4. Permutation problems involve situations in which the order of the items does not matter. false

Respond to Exercises 5–6 using verbal or written explanations. 5–6. Answers will vary.

5. What is a combination?

6. Explain how to distinguish between permutation and combination problems.

Exercise Set 7.3

Practice Exercises

In Exercises 1–4, does the problem involve permutations or combinations? Explain your answer. (It is not necessary to solve the problem.)

1. A medical researcher needs 6 people to test the effectiveness of an experimental drug. If 13 people have volunteered for the test, in how many ways can 6 people be selected?   combinations

2. Fifty people purchase raffle tickets. Three winning tickets are selected at random. If first prize is $1000, second prize is $500, and third prize is $100, in how many different ways can the prizes be awarded?   permutations

3. How many different four-letter passwords can be formed from the letters A, B, C, D, E, F, and G if no repetition of letters is allowed?   permutations

4. Fifty people purchase raffle tickets. Three winning tickets are selected at random. If each prize is $500, in how many different ways can the prizes be awarded?   combinations

In Exercises 5–20, use the formula for $_nC_r$ to evaluate each expression.

5. $_6C_5$ 6   **6.** $_8C_7$ 8   **7.** $_9C_5$ 126   **8.** $_{10}C_6$ 210

9. $_{11}C_4$ 330   **10.** $_{12}C_5$ 792 **11.** $_8C_1$ 8   **12.** $_7C_1$ 7

13. $_7C_7$ 1 **14.** $_4C_4$ 1 **15.** $_{30}C_3$ 4060 **16.** $_{25}C_4$ 12,650

17. $_5C_0$ 1 **18.** $_6C_0$ 1 **19.** $\dfrac{_7C_3}{_5C_4}$ 7 **20.** $\dfrac{_{10}C_3}{_6C_4}$ 8

Practice Plus

In Exercises 21–28, evaluate each expression.

21. $\dfrac{_7P_3}{3!} - {_7C_3}$ 0 **22.** $\dfrac{_{20}P_2}{2!} - {_{20}C_2}$ 0 **23.** $1 - \dfrac{_3P_2}{_4P_3}$ $\frac{3}{4}$

24. $1 - \dfrac{_5P_3}{_{10}P_4}$ $\frac{83}{84}$ **25.** $\dfrac{_7C_3}{_5C_4} - \dfrac{98!}{96!}$ -9499 **26.** $\dfrac{_{10}C_3}{_6C_4} - \dfrac{46!}{44!}$ -2062

27. $\dfrac{_4C_2 \cdot {_6C_1}}{_{18}C_3}$ $\frac{3}{68}$ **28.** $\dfrac{_5C_1 \cdot {_7C_2}}{_{12}C_3}$ $\frac{21}{44}$

Application Exercises

Use the formula for $_nC_r$ to solve Exercises 29–40.

29. An election ballot asks voters to select three city commissioners from a group of six candidates. In how many ways can this be done? 20

30. A four-person committee is to be elected from an organization's membership of 11 people. How many different committees are possible? 330

31. Of 12 possible books, you plan to take 4 with you on vacation. How many different collections of 4 books can you take? 495

32. There are 14 standbys who hope to get seats on a flight, but only 6 seats are available on the plane. How many different ways can the 6 people be selected? 3003

33. You volunteer to help drive children at a charity event to the zoo, but you can fit only 8 of the 17 children present in your van. How many different groups of 8 children can you drive? 24,310

34. Of the 100 people in the U.S. Senate, 18 serve on the Foreign Relations Committee. How many ways are there to select Senate members for this committee (assuming party affiliation is not a factor in the selection)? $\approx 3.07 \times 10^{19}$

35. To win at LOTTO in the state of Florida, one must correctly select 6 numbers from a collection of 53 numbers (1 through 53). The order in which the selection is made does not matter. How many different selections are possible? 22,957,480

36. To win in the New York State lottery, one must correctly select 6 numbers from 59 numbers. The order in which the selection is made does not matter. How many different selections are possible? 45,057,474

37. In how many ways can a committee of four men and five women be formed from a group of seven men and seven women? 735

38. How many different committees can be formed from 5 professors and 15 students if each committee is made up of 2 professors and 10 students? 30,030

39. The U.S. Senate of the 109th Congress consisted of 55 Republicans, 44 Democrats, and 1 Independent. How many committees can be formed if each committee must have 4 Republicans and 3 Democrats? 4,516,932,420

40. A mathematics exam consists of 10 multiple-choice questions and 5 open-ended problems in which all work must be shown. If an examinee must answer 8 of the multiple-choice questions and 3 of the open-ended problems, in how many ways can the questions and problems be chosen? 450

In Exercises 41–60, solve by the method of your choice.

41. In a race in which six automobiles are entered and there are no ties, in how many ways can the first four finishers come in? 360

42. A book club offers a choice of 8 books from a list of 40. In how many ways can a member make a selection? 76,904,685

43. A medical researcher needs 6 people to test the effectiveness of an experimental drug. If 13 people have volunteered for the test, in how many ways can 6 people be selected? 1716

44. Fifty people purchase raffle tickets. Three winning tickets are selected at random. If first prize is $1000, second prize is $500, and third prize is $100, in how many different ways can the prizes be awarded? 117,600

45. From a club of 20 people, in how many ways can a group of three members be selected to attend a conference? 1140

46. Fifty people purchase raffle tickets. Three winning tickets are selected at random. If each prize is $500, in how many different ways can the prizes be awarded? 19,600

47. How many different four-letter passwords can be formed from the letters A, B, C, D, E, F, and G if no repetition of letters is allowed? 840

48. Nine comedy acts will perform over two evenings. Five of the acts will perform on the first evening. How many ways can the schedule for the first evening be made? 15,120

49. Using 15 flavors of ice cream, how many cones with three different flavors can you create if it is important to you which flavor goes on the top, middle, and bottom? 2730

50. Baskin-Robbins offers 31 different flavors of ice cream. One of its items is a bowl consisting of three scoops of ice cream, each a different flavor. How many such bowls are possible? 4495

51. A restaurant lunch special allows the customer to choose two vegetables from this list: okra, corn, peas, carrots, and squash. How many outcomes are possible if the customer chooses two different vegetables? 10

52. There are six employees in the stock room at an appliance retail store. The manager will choose three of them to deliver a refrigerator. How many three-person groups are possible? 20

53. Suppose that you have three dress shirts, two ties, and two jackets. You need to select a dress shirt, a tie, and a jacket for work today. How many outcomes are possible? 12

54. Suppose that you have four flannel shirts. You are going to choose two of them to take on a camping trip. How many outcomes are possible? 6

55. A chef has five brands of hot sauce. Three of the brands will be chosen to mix into gumbo. How many outcomes are possible? 10

56. In the Mathematics Department, there are four female professors and six male professors. Three female professors will be chosen to serve as mentors for a special program designed to encourage female students to pursue careers in mathematics. In how many ways can the professors be chosen? 4

57. There are four Democrats and five Republicans on the county commission. From among their group they will choose a committee of two Democrats and two Republicans to examine a proposal to purchase land for a new county park. How many four-person groups are possible?   60

58. An office employs six customer service representatives. Each day, two of them are randomly selected and their customer interactions are monitored for the purposes of improving customer relations. In how many ways can the representatives be chosen?   15

59. A group of campers is going to occupy five campsites at a campground. There are 12 campsites from which to choose. In how many ways can the campsites be chosen?   792

60. Two teachers have driven to school in separate cars. When they arrive, there are seven empty spaces in the parking lot. They each choose a parking space. How many outcomes are possible?   42

(*Source* for Exercises 51–60: James Wooland, *CLAST Manual for Thinking Mathematically*)

Thousands of jokes have been told about marriage and divorce. Exercises 61–68 are based on the following observations:

- *"By all means, marry; if you get a good wife, you'll be happy. If you get a bad one, you'll become a philosopher."* —*Socrates*
- *"Marriage is a wonderful institution, but who wants to live in an institution?"* —*Groucho Marx*
- *"Whatever you may look like, marry a man your own age. As your beauty fades, so will his eyesight."* —*Phyllis Diller*
- *"Why do Jewish divorces cost so much? Because they're worth it."* —*Henny Youngman*
- *"I think men who have a pierced ear are better prepared for marriage. They've experienced pain and bought jewelry."* —*Rita Rudner*
- *"It wasn't actually a divorce. I was traded."* —*Tim Conway*

61. In how many ways can these six jokes be ranked from best to worst?   720

62. If Socrates's thoughts about marriage are excluded, in how many ways can the remaining five jokes be ranked from best to worst?   120

63. In how many ways can people select their three favorite jokes from these thoughts about marriage and divorce?   20

64. In how many ways can people select their two favorite jokes from these thoughts about marriage and divorce?   15

65. If the order in which these jokes are told makes a difference in terms of how they are received, how many ways can they be delivered if Socrates's comments are scheduled first and Marx's joke is told last?   24

66. If the order in which these jokes are told makes a difference in terms of how they are received, how many ways can they be delivered if a joke by a woman (Rudner or Diller) is told first?   240

67. In how many ways can people select their favorite joke told by a woman (Rudner or Diller) and their two favorite jokes told by a man?   12

68. In how many ways can people select their favorite joke told by a woman (Rudner or Diller) and their three favorite jokes told by a man?   8

Critical Thinking Exercises   69–70. Answers will vary.

69. Write a word problem that can be solved by evaluating $_7C_3$.

70. Write a word problem that can be solved by evaluating $_{10}C_3 \cdot _7C_2$.

71. A 6/53 lottery involves choosing 6 of the numbers from 1 through 53 and a 5/36 lottery involves choosing 5 of the numbers from 1 through 36. The order in which the numbers are chosen does not matter. Which lottery is easier to win? Explain your answer.   5/36 lottery; Answers will vary.

72. If the number of permutations of n objects taken r at a time is six times the number of combinations of n objects taken r at a time, determine the value of r. Is there enough information to determine the value of n? Why or why not?
$r = 3$; no; Answers will vary.

73. In a group of 20 people, how long will it take each person to shake hands with each of the other persons in the group, assuming that it takes three seconds for each shake and only 2 people can shake hands at a time? What if the group is increased to 40 people?   570 sec or 9.5 min; 2340 sec or 39 min

74. A sample of 4 telephones is selected from a shipment of 20 phones. There are 5 defective telephones in the shipment. How many of the possible samples of 4 phones do not include any of the defective ones?   1365

7.4 | Fundamentals of Probability

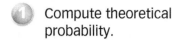
to learn?

After you have read this section, you should be able to:

1. Compute theoretical probability.

2. Compute empirical probability.

How many hours of sleep do you typically get each night? **Table 7.1** on the next page indicates that 75 million out of 300 million Americans are getting six hours of sleep on a typical night. The *probability* of an American getting six hours of sleep on a typical night is $\frac{75}{300}$. This fraction can be reduced to $\frac{1}{4}$, or expressed as 0.25, or 25%. Thus, 25% of Americans get six hours of sleep each night.

We find a probability by dividing one number by another. Probabilities are assigned to an *event*, such as getting six hours of sleep on a typical

Table 7.1 The Hours of Sleep Americans Get on a Typical Night	
Hours of Sleep	**Number of Americans, in millions**
4 or less	12
5	27
6	75
7	90
8	81
9	9
10 or more	6
	Total: 300

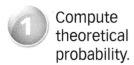

Compute theoretical probability.

night. Events that are certain to occur are assigned probabilities of 1, or 100%. For example, the probability that a given individual will eventually die is 1. Although Woody Allen whined, "I don't want to achieve immortality through my work. I want to achieve it through not dying," death (and taxes) are always certain. By contrast, if an event cannot occur, its probability is 0. Regrettably, the probability that Elvis will return and serenade us with one final reprise of "Don't Be Cruel" (and we hope we're not) is 0.

Probabilities of events are expressed as numbers ranging from 0 to 1, or 0% to 100%. The closer the probability of a given event is to 1, the more likely it is that the event will occur. The closer the probability of a given event is to 0, the less likely it is that the event will occur.

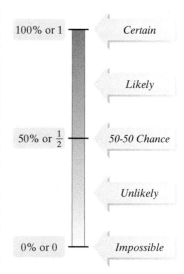

Possible Values for Probabilities

Theoretical Probability

You toss a coin. Although it is equally likely to land either heads up, denoted by H, or tails up, denoted by T, the actual outcome is uncertain. Any occurrence for which the outcome is uncertain is called an **experiment**. Thus, tossing a coin is an example of an experiment. The set of all possible outcomes of an experiment is the **sample space** of the experiment, denoted by S. The sample space for the coin-tossing experiment is

$$S = \{H, T\}.$$

Lands heads up   Lands tails up

An **event**, denoted by E, is any subset of a sample space. For example, the subset $E = \{T\}$ is the event of landing tails up when a coin is tossed.

Theoretical probability applies to situations like this, in which the sample space only contains equally likely outcomes, all of which are known. To calculate the theoretical probability of an event, we divide the number of outcomes resulting in the event by the total number of outcomes in the sample space.

Computing Theoretical Probability

If an event E has $n(E)$ equally likely outcomes and its sample space S has $n(S)$ equally likely outcomes, the **theoretical probability** of event E, denoted by $P(E)$, is

$$P(E) = \frac{\text{number of outcomes in event } E}{\text{total number of possible outcomes}} = \frac{n(E)}{n(S)}.$$

How can we use the formula $P(E) = \dfrac{n(E)}{n(S)}$ to compute the probability of a coin landing tails up? We use the following sets:

$$E = \{T\} \qquad S = \{H, T\}.$$

This is the event of landing tails up.   This is the sample space with all equally likely outcomes.

The probability of a coin landing tails up is

$$P(E) = \frac{\text{number of outcomes that result in tails up}}{\text{total number of possible outcomes}} = \frac{n(E)}{n(S)} = \frac{1}{2}.$$

Figure 7.4 Outcomes when a die is rolled

Theoretical probability applies to many games of chance, including dice rolling, lotteries, and card games. We begin with rolling a die. **Figure 7.4** illustrates that when a die is rolled, there are six equally likely possible outcomes. The sample space can be shown as

$$S = \{1, 2, 3, 4, 5, 6\}.$$

Example 1   **Computing Theoretical Probability**

A die is rolled once. Find the probability of rolling

a. a 3.   **b.** an even number.   **c.** a number less than 5.

d. a number less than 10.   **e.** a number greater than 6.

Solution

The sample space is $S = \{1, 2, 3, 4, 5, 6\}$ with $n(S) = 6$. We will use 6, the total number of possible outcomes, in the denominator of each probability fraction.

a. The phrase "rolling a 3" describes the event $E = \{3\}$. This event can occur in one way: $n(E) = 1$.

$$P(3) = \frac{\text{number of outcomes that result in 3}}{\text{total number of possible outcomes}} = \frac{n(E)}{n(S)} = \frac{1}{6}$$

The probability of rolling a 3 is $\frac{1}{6}$.

b. The phrase "rolling an even number" describes the event $E = \{2, 4, 6\}$. This event can occur in three ways: $n(E) = 3$.

$$P(\text{even number}) = \frac{\text{number of outcomes that result in an even number}}{\text{total number of possible outcomes}}$$

$$= \frac{n(E)}{n(S)} = \frac{3}{6} = \frac{1}{2}$$

The probability of rolling an even number is $\frac{1}{2}$.

c. The phrase "rolling a number less than 5" describes the event $E = \{1, 2, 3, 4\}$. This event can occur in four ways: $n(E) = 4$.

$$P(\text{less than 5}) = \frac{\text{number of outcomes that are less than 5}}{\text{total number of possible outcomes}} = \frac{n(E)}{n(S)} = \frac{4}{6} = \frac{2}{3}$$

The probability of rolling a number less than 5 is $\frac{2}{3}$.

d. The phrase "rolling a number less than 10" describes the event $E = \{1, 2, 3, 4, 5, 6\}$. This event can occur in six ways: $n(E) = 6$. Can you see that all of the possible outcomes are less than 10? This event is certain to occur.

$$P(\text{less than 10}) = \frac{\text{number of outcomes that are less than 10}}{\text{total number of possible outcomes}} = \frac{n(E)}{n(S)} = \frac{6}{6} = 1$$

The probability of any certain event is 1.

e. The phrase "rolling a number greater than 6" describes an event that cannot occur, or the empty set. Thus, $E = \varnothing$ and $n(E) = 0$.

$$P(\text{greater than 6}) = \frac{\text{number of outcomes that are greater than 6}}{\text{total number of possible outcomes}} = \frac{n(E)}{n(S)} = \frac{0}{6} = 0$$

The probability of an event that cannot occur is 0.

In Example 1, there are six possible outcomes, each with a probability of $\frac{1}{6}$:

$$P(1) = \frac{1}{6} \quad P(2) = \frac{1}{6} \quad P(3) = \frac{1}{6} \quad P(4) = \frac{1}{6} \quad P(5) = \frac{1}{6} \quad P(6) = \frac{1}{6}.$$

The sum of these probabilities is 1: $\frac{1}{6} + \frac{1}{6} + \frac{1}{6} + \frac{1}{6} + \frac{1}{6} + \frac{1}{6} = 1$. In general, **the sum of the theoretical probabilities of all possible outcomes in the sample space is 1**.

Check Point 1 A die is rolled once. Find the probability of rolling

a. a 2. $\frac{1}{6}$ **b.** a number less than 4. $\frac{1}{2}$

c. a number greater than 7. 0   **d.** a number less than 7. 1

Our next example involves a standard 52-card bridge deck, illustrated in **Figure 7.5**. The deck has four suits: Hearts and diamonds are red, and clubs and spades are black. Each suit has 13 different face values—A(ace), 2, 3, 4, 5, 6, 7, 8, 9, 10, J(jack), Q(queen), and K(king). Jacks, queens, and kings are called **picture cards** or **face cards**.

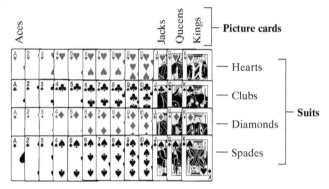

Figure 7.5 A standard 52-card bridge deck

Example 2   Probability and a Deck of 52 Cards

You are dealt one card from a standard 52-card deck. Find the probability of being dealt

a. a king.   **b.** a heart.   **c.** the king of hearts.

Solution

Because there are 52 cards in the deck, the total number of possible ways of being dealt a single card is 52. The number of outcomes in the sample space is 52: $n(S) = 52$. We use 52 as the denominator of each probability fraction.

a. Let E be the event of being dealt a king. Because there are four kings in the deck, this event can occur in four ways: $n(E) = 4$.

$$P(\text{king}) = \frac{\text{number of outcomes that result in a king}}{\text{total number of possible outcomes}} = \frac{n(E)}{n(S)} = \frac{4}{52} = \frac{1}{13}$$

The probability of being dealt a king is $\frac{1}{13}$.

b. Let E be the event of being dealt a heart. Because there are 13 hearts in the deck, this event can occur in 13 ways: $n(E) = 13$.

$$P(\text{heart}) = \frac{\text{number of outcomes that result in a heart}}{\text{total number of possible outcomes}} = \frac{n(E)}{n(S)} = \frac{13}{52} = \frac{1}{4}$$

The probability of being dealt a heart is $\frac{1}{4}$.

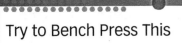

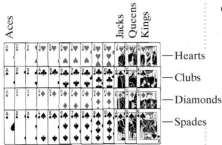

—Hearts
—Clubs
—Diamonds
—Spades

Figure 7.5 (repeated)
A standard 52-card bridge deck

Blitzer Bonus

Try to Bench Press This at the Gym

You have a deck of cards for every permutation of the 52 cards. If each deck weighed only as much as a single hydrogen atom, all the decks together would weigh a billion times as much as the Sun.

Source: Isaac Asimov's Book of Facts

c. Let E be the event of being dealt the king of hearts. Because there is only one card in the deck that is the king of hearts, this event can occur in just one way: $n(E) = 1$.

$$P(\text{king of hearts}) = \frac{\text{number of outcomes that result in the king of hearts}}{\text{total number of possible outcomes}}$$

$$= \frac{n(E)}{n(S)} = \frac{1}{52}$$

The probability of being dealt the king of hearts is $\frac{1}{52}$.

✓ **Check Point 2** You are dealt one card from a standard 52-card deck. Find the probability of being dealt

a. an ace. $\frac{1}{13}$ **b.** a red card. $\frac{1}{2}$ **c.** a red king. $\frac{1}{26}$

Probabilities play a valuable role in the science of genetics. Example 3 deals with cystic fibrosis, an inherited lung disease occurring in about 1 out of every 2000 births among Caucasians and in about 1 out of every 250,000 births among non-Caucasians.

Example 3   Probabilities in Genetics

Each person carries two genes that are related to the absence or presence of the disease cystic fibrosis. Most Americans have two normal genes for this trait and are unaffected by cystic fibrosis. However, 1 in 25 Americans carries one normal gene and one defective gene. If we use c to represent a defective gene and C a normal gene, such a carrier can be designated as Cc. Thus, CC is a person who neither carries nor has cystic fibrosis, Cc is a carrier who is not actually sick, and cc is a person sick with the disease. **Table 7.2** shows the four equally likely outcomes for a child's genetic inheritance from two parents who are both carrying one cystic fibrosis gene. One copy of each gene is passed on to the child from the parents.

Table 7.2 Cystic Fibrosis and Genetic Inheritance

		Second Parent	
		C	**c**
First	**C**	**CC**	**Cc**
Parent	**c**	**cC**	**cc**

Shown in the table are the four possibilities for a child whose parents each carry one cystic fibrosis gene.

If each parent carries one cystic fibrosis gene, what is the probability that their child will have cystic fibrosis?

Solution

Table 7.2 shows that there are four equally likely outcomes. The sample space is $S = \{CC, Cc, cC, cc\}$ and $n(S) = 4$. The phrase "will have cystic fibrosis" describes only the cc child. Thus, $E = \{cc\}$ and $n(E) = 1$.

$$P(\text{cystic fibrosis}) = \frac{\text{number of outcomes that result in cystic fibrosis}}{\text{total number of possible outcomes}} = \frac{n(E)}{n(S)} = \frac{1}{4}$$

If each parent carries one cystic fibrosis gene, the probability that their child will have cystic fibrosis is $\frac{1}{4}$.

✓ **Check Point 3** Use **Table 7.2** in Example 3 to solve this exercise. If each parent carries one cystic fibrosis gene, find the probability that their child will be a carrier of the disease who is not actually sick. $\frac{1}{2}$

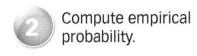

Compute empirical probability.

Empirical Probability

Theoretical probability is based on a set of equally likely outcomes and the number of elements in the set. By contrast, *empirical probability* applies to situations in which we observe how frequently an event occurs. We use the following formula to compute the empirical probability of an event:

Computing Empirical Probability

The empirical probability of event E is

$$P(E) = \frac{\text{observed number of times } E \text{ occurs}}{\text{total number of observed occurrences}}.$$

Example 4   Computing Empirical Probability

In 2010, there were approximately 242 million Americans ages 15 or older. **Table 7.3** shows the distribution, by marital status and gender, of this population.

Table 7.3 Marital Status of the U.S. Population, Ages 15 or Older, 2010, in Millions

	Married	Never Married	Divorced	Widowed	Total
Male	65	40	10	3	118
Female	65	34	14	11	124
Total	130	74	24	14	242

Total male:
65 + 40 + 10 + 3 = 118

Total female:
65 + 34 + 14 + 11 = 124

Total married:
65 + 65 = 130

Total never married:
40 + 34 = 74

Total divorced:
10 + 14 = 24

Total widowed:
3 + 11 = 14

Total adult population:
118 + 124 = 242

Source: U.S. Census Bureau

Great Question!

What do you mean by saying that one person is *randomly selected* from the population?

This means that every person in the population has an equal chance of being chosen. We'll have much more to say about random selections in Chapter 8, Statistical Pathways.

If one person is randomly selected from the population described in **Table 7.3**, find the probability, to the nearest hundreth, that the person

a. is divorced.   **b.** is female.

Solution

a. The probability of selecting a divorced person is the observed number of divorced people, 24 (million), divided by the total number in the population, 242 (million).

$$P(\text{selecting a divorced person from the population})$$
$$= \frac{\text{number of divorced people}}{\text{total number in the population}} = \frac{24}{242} \approx 0.10$$

The empirical probability of selecting a divorced person from the population in **Table 7.3** is approximately 0.10.

b. The probability of selecting a female is the observed number of females, 124 (million), divided by the total number in the population, 242 (million).

$$P(\text{selecting a female from the population})$$
$$= \frac{\text{number of females}}{\text{total number in the population}} = \frac{124}{242} \approx 0.51$$

The empirical probability of selecting a female from the population in **Table 7.3** is approximately 0.51.

✓ **Check Point 4** If one person is randomly selected from the population described in **Table 7.3**, find the probability, expressed as a decimal rounded to the nearest hundredth, that the person

a. has never been married. 0.31   **b.** is male. 0.49

In certain situations, we can establish a relationship between the two kinds of probability. Consider, for example, a coin that is equally likely to land heads or tails. Such a coin is called a **fair coin**. Empirical probability can be used to determine whether a coin is fair. Suppose we toss a coin 10, 50, 100, 1000, 10,000, and 100,000 times. We record the number of heads observed, shown in **Table 7.4**. For each of the six cases in the table, the empirical probability of heads is determined by dividing the number of heads observed by the number of tosses.

Table 7.4 Empirical Probabilities of Heads as the Number of Tosses Increases		
Number of Tosses	**Number of Heads Observed**	**Empirical Probability of Heads, or P(H)**
10	4	$P(H) = \frac{4}{10} = 0.4$
50	27	$P(H) = \frac{27}{50} = 0.54$
100	44	$P(H) = \frac{44}{100} = 0.44$
1000	530	$P(H) = \frac{530}{1000} = 0.53$
10,000	4851	$P(H) = \frac{4851}{10,000} = 0.4851$
100,000	49,880	$P(H) = \frac{49,880}{100,000} = 0.4988$

A pattern is exhibited by the empirical probabilities in the right-hand column of **Table 7.4**. As the number of tosses increases, the empirical probabilities tend to get closer to 0.5, the theoretical probability. These results give us no reason to suspect that the coin is not fair.

Table 7.4 illustrates an important principle when observing uncertain outcomes such as the event of a coin landing on heads. As an experiment is repeated more and more times, the empirical probablity of an event tends to get closer to the theoretical probability of that event. This principle is known as the **law of large numbers**.

Achieving Success

Do not expect to solve every word problem immediately. As you read each problem, underline the important parts. It's a good idea to read the problem at least twice. Be persistent, but use the **"Ten Minutes of Frustration" Rule**. If you have exhausted every possible means for solving a problem and you are still bogged down, stop after ten minutes. Put a question mark by the exercise and move on. When you return to class, ask your professor for assistance.

Concept and Vocabulary Check

Exercises in the Concept and Vocabulary Check are intended for group and class discussions.

In Exercises 1–4, fill in each blank so that the resulting statement is true.

1. The set of all possible outcomes of an experiment is called the _____sample space_____ of the experiment.

2. The theoretical probability of event E, denoted by _____$P(E)$_____, is the _number of outcomes in E_ divided by the _total number of possible outcomes_.

3. A standard bridge deck has _____52_____ cards with four suits: _____hearts_____ and _____diamonds_____ are red, and _____clubs_____ and _____spades_____ are black.

4. Probability that is based on situations in which we observe how frequently an event occurs is called _____empirical_____ probability.

In Exercises 5–8, determine whether each statement is true or false. If the statement is false, make the necessary change(s) to produce a true statement.   Changes to false statements will vary.

5. If an event is certain to occur, its probability is 1.   true

6. If an event cannot occur, its probability is −1.   false

7. The sum of the probabilities of all possible outcomes in an experiment is 1.   true

8. If an experiment is repeated more and more times, the theoretical probability of an event tends to get closer to the empirical probability of that event.   false

Respond to Exercises 9–14 using verbal or written explanations.   9–14. Answers will vary.

9. What is the sample space of an experiment? What is an event?

10. How is the theoretical probability of an event computed?

11. Describe the difference between theoretical probability and empirical probability.

12. Give an example of an event whose probability must be determined empirically rather than theoretically.

13. Use the definition of theoretical probability to explain why the probability of an event that cannot occur is 0.

14. Use the definition of theoretical probability to explain why the probability of an event that is certain to occur is 1.

Exercise Set 7.4

Practice and Application Exercises

In Exercises 1–54, express each probability as a fraction reduced to lowest terms.

In Exercises 1–10, a die is rolled. The set of equally likely outcomes is $\{1, 2, 3, 4, 5, 6\}$. Find the probability of rolling

1. a 4. $\frac{1}{6}$

2. a 5. $\frac{1}{6}$

3. an odd number. $\frac{1}{2}$

4. a number greater than 3. $\frac{1}{2}$

5. a number less than 3. $\frac{1}{3}$

6. a number greater than 4. $\frac{1}{3}$

7. a number less than 20. 1

8. a number less than 8. 1

9. a number greater than 20. 0

10. a number greater than 8. 0

In Exercises 11–20, you are dealt one card from a standard 52-card deck. Find the probability of being dealt

11. a queen. $\frac{1}{13}$

12. a jack. $\frac{1}{13}$

13. a club. $\frac{1}{4}$

14. a diamond. $\frac{1}{4}$

15. a picture card. $\frac{3}{13}$

16. a card greater than 3 and less than 7. $\frac{3}{13}$

17. the queen of spades. $\frac{1}{52}$

18. the ace of clubs. $\frac{1}{52}$

19. a diamond and a spade. 0

20. a card with a green heart. 0

In Exercises 21–26, a fair coin is tossed two times in succession. The set of equally likely outcomes is $\{HH, HT, TH, TT\}$. Find the probability of getting

21. two heads. $\frac{1}{4}$

22. two tails. $\frac{1}{4}$

23. the same outcome on each toss. $\frac{1}{2}$

24. different outcomes on each toss. $\frac{1}{2}$

25. a head on the second toss. $\frac{1}{2}$

26. at least one head. $\frac{3}{4}$

In Exercises 27–34, you select a family with three children. If M represents a male child and F a female child, the set of equally likely outcomes for the children's genders is $\{$MMM, MMF, MFM, MFF, FMM, FMF, FFM, FFF$\}$. Find the probability of selecting a family with   27. $\frac{3}{8}$

27. exactly one female child.

28. exactly one male child. $\frac{3}{8}$

29. exactly two male children. $\frac{3}{8}$

30. exactly two female children. $\frac{3}{8}$

31. at least one male child. $\frac{7}{8}$

32. at least two female children. $\frac{1}{2}$

33. four male children. 0

34. fewer than four female children. 1

In Exercises 35–40, a single die is rolled twice. The 36 equally likely outcomes are shown as follows:

	Second Roll					
	⚀	⚁	⚂	⚃	⚄	⚅
⚀	(1,1)	(1,2)	(1,3)	(1,4)	(1,5)	(1,6)
⚁	(2,1)	(2,2)	(2,3)	(2,4)	(2,5)	(2,6)
⚂	(3,1)	(3,2)	(3,3)	(3,4)	(3,5)	(3,6)
⚃	(4,1)	(4,2)	(4,3)	(4,4)	(4,5)	(4,6)
⚄	(5,1)	(5,2)	(5,3)	(5,4)	(5,5)	(5,6)
⚅	(6,1)	(6,2)	(6,3)	(6,4)	(6,5)	(6,6)

(First Roll labels the rows)

Find the probability of getting

35. two even numbers. $\frac{1}{4}$

36. two odd numbers. $\frac{1}{4}$

37. two numbers whose sum is 5. $\frac{1}{9}$

38. two numbers whose sum is 6. $\frac{5}{36}$

39. two numbers whose sum exceeds 12. 0

40. two numbers whose sum is less than 13. 1

Use the spinner shown to answer Exercises 41–48. Assume that it is equally probable that the pointer will land on any one of the ten colored regions. If the pointer lands on a borderline, spin again.

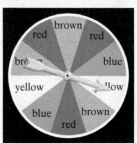

Find the probability that the spinner lands in

41. a red region. $\frac{3}{10}$

42. a yellow region. $\frac{1}{5}$

43. a blue region. $\frac{1}{5}$

44. a brown region. $\frac{3}{10}$

45. a region that is red or blue. $\frac{1}{2}$

46. a region that is yellow or brown. $\frac{1}{2}$

47. a region that is red and blue. 0

48. a region that is yellow and brown. 0

Exercises 49–54 deal with sickle cell anemia, an inherited disease in which red blood cells become distorted and deprived of oxygen. Approximately 1 in every 500 African-American infants is born with the disease; only 1 in 160,000 white infants has the disease. A person with two sickle cell genes will have the disease, but a person with only one sickle cell gene will have a mild, nonfatal anemia called sickle cell trait. (Approximately 8%–10% of the African-American population has this trait.)

		Second Parent	
		S	**s**
First	**S**	SS	Ss
Parent	**s**	sS	ss

If we use s to represent a sickle cell gene and S a healthy gene, the table above shows the four possibilities for the children of two Ss parents. Each parent has only one sickle cell gene, so each has the relatively mild sickle cell trait. Find the probability that these parents give birth to a child who

49. has sickle cell anemia. $\frac{1}{4}$

50. has sickle cell trait. $\frac{1}{2}$ **51.** is healthy. $\frac{1}{4}$

In Exercises 52–54, use the following table that shows the four possibilities for the children of one healthy, SS, parent, and one parent with sickle cell trait, Ss.

		Second Parent (with Sickle Cell Trait)	
		S	**s**
Healthy	**S**	SS	Ss
First Parent	**S**	SS	Ss

Find the probability that these parents give birth to a child who

52. has sickle cell anemia. 0

53. has sickle cell trait. $\frac{1}{2}$ **54.** is healthy. $\frac{1}{2}$

The table shows the distribution, by age and gender, of the 29.3 million Americans who live alone. Use the data in the table to solve Exercises 55–60.

Number of People in the United States Living Alone, in Millions

	Ages 15–24	Ages 25–34	Ages 35–44	Ages 45–64	Ages 65–74	Ages ≥75	Total
Male	0.7	2.2	2.6	4.3	1.3	1.4	12.5
Female	0.8	1.6	1.6	5.0	2.9	4.9	16.8
Total	1.5	3.8	4.2	9.3	4.2	6.3	29.3

Source: U.S. Census Bureau

Find the probability, expressed as a decimal rounded to the nearest hundredth, that a randomly selected American living alone is

55. male. 0.43   **56.** female. 0.57

57. in the 25–34 age range. 0.13

58. in the 35–44 age range. 0.14

59. a woman in the 15–24 age range. 0.03

60. a man in the 45–64 age range. 0.15

The table shows the number of Americans who moved in a recent year, categorized by where they moved and whether they were an owner or a renter. Use the data in the table, expressed in millions, to solve Exercises 61–66.

Number of People in the United States Who Moved, in Millions

	Moved to Same State	**Moved to Different State**	**Moved to Different Country**
Owner	11.7	2.8	0.3
Renter	18.7	4.5	1.0

Source: U.S. Census Bureau

Use the table to find the probability, expressed as a decimal rounded to the nearest hundredth, that a randomly selected American who moved was

61. an owner. 0.38   **62.** a renter. 0.62

63. a person who moved within the same state. 0.78

64. a person who moved to a different country. 0.03

65. a renter who moved to a different state. 0.12

66. an owner who moved to a different state. 0.07

The table shows the educational attainment of the U.S. population, ages 25 and over. Use the data in the table, expressed in millions, to solve Exercises 67–70.

Educational Attainment, in Millions, of the United States Population, Ages 25 and Over

	Less Than 4 Years High School	**4 Years High School Only**	**Some College (Less Than 4 years)**	**4 Years College (or More)**	**Total**
Male	14	25	20	23	82
Female	15	31	24	22	92
Total	29	56	44	45	174

Source: U.S. Census Bureau

Find the probability, expressed as a simplified fraction, that a randomly selected American, aged 25 or over,

67. had less than four years of high school. $\frac{1}{6}$

68. had four years of high school only. $\frac{28}{87}$

69. was a woman with four years of college or more. $\frac{11}{87}$

70. was a man with four years of college or more. $\frac{23}{174}$

Critical Thinking Exercises

71. Write a probability word problem whose answer is one of the following fractions: $\frac{1}{6}$ or $\frac{1}{4}$ or $\frac{1}{3}$. Answers will vary.

72. The target in the figure shown contains four squares. If a dart thrown at random hits the target, find the probability that it will land in an orange region. $\frac{3}{8} = 0.375$

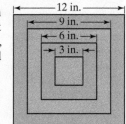

73. Some three-digit numbers, such as 101 and 313, read the same forward and backward. If you select a number from all three-digit numbers, find the probability that it will read the same forward and backward. $\frac{1}{10}$

7.5 Probability with the Fundamental Counting Principle, Permutations, and Combinations

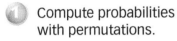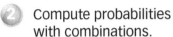
What am I supposed to learn?

After you have read this section, you should be able to:

① Compute probabilities with permutations.

② Compute probabilities with combinations.

According to actuarial tables, there is no year in which death is as likely as continued life, at least until the age of 115. Until that age, the probability of dying in any one year ranges from a low of 0.00009 for a girl at age 11 to a high of 0.465 for either gender at age 114. For a healthy 30-year-old, how does the probability of dying this year compare to the probability of winning the

top prize in a state lottery game? In this section, we provide the surprising answer to this question, as we study probability with the Fundamental Counting Principle, permutations, and combinations.

Probability of Dying at Any Given Age

Age	Probability of Male Death	Probability of Female Death	Age	Probability of Male Death	Probability of Female Death
10	0.00013	0.00010	70	0.03473	0.01764
20	0.00140	0.00050	80	0.07644	0.03966
30	0.00153	0.00050	90	0.15787	0.11250
40	0.00193	0.00095	100	0.26876	0.23969
50	0.00567	0.00305	110	0.39770	0.39043
60	0.01299	0.00792			

Source: George Shaffner, *The Arithmetic of Life and Death*

① Compute probabilities with permutations.

Probability with Permutations

Example 1   Probability and Permutations

We return to the six jokes about books by Groucho Marx, George Carlin, Steven Wright, Henny Youngman, Jerry Seinfeld, and Phyllis Diller that opened Section 7.2. Suppose that each joke is written on one of six cards. The cards are placed in a hat and then six cards are drawn, one at a time. The order in which the cards are drawn determines the order in which the jokes are delivered. What is the probability that a man's joke will be delivered first and a man's joke will be delivered last?

Solution

We begin by applying the definition of probability to this situation.

P(man's joke first, man's joke last)

$$= \frac{\text{number of permutations with man's joke first, man's joke last}}{\text{total number of possible permutations}}$$

Groucho Marx

George Carlin

Steven Wright

Henny Youngman

Jerry Seinfeld

Phyllis Diller

We can use the Fundamental Counting Principle to find the total number of possible permutations. This represents the number of ways the six jokes can be delivered.

First Joke Delivered	Second Joke Delivered	Third Joke Delivered	Fourth Joke Delivered	Fifth Joke Delivered	Sixth Joke Delivered
6 choices	5 choices	4 choices	3 choices	2 choices	1 choice

There are $6 \cdot 5 \cdot 4 \cdot 3 \cdot 2 \cdot 1$, or 720 possible permutations. Equivalently, there are 720 different ways to deliver the six jokes about books.

We can also use the Fundamental Counting Principle to find the number of permutations with a man's joke delivered first and a man's joke delivered last. These conditions can be shown as follows:

First Joke Delivered	Second Joke Delivered	Third Joke Delivered	Fourth Joke Delivered	Fifth Joke Delivered	Sixth Joke Delivered
5 choices					4 choices

Can be a joke by any one of the five men

Must be a joke by any one of the four remaining men

Now let's fill in the number of choices for positions two through five.

First Joke Delivered	Second Joke Delivered	Third Joke Delivered	Fourth Joke Delivered	Fifth Joke Delivered	Sixth Joke Delivered
5 choices	4 choices	3 choices	2 choices	1 choice	4 choices

Can be a joke by any one of the four remaining comics: three men or Phyllis Diller

Now 3 comics remain.

Now 2 comics remain.

Only 1 comic remains.

Thus, there are $5 \cdot 4 \cdot 3 \cdot 2 \cdot 1 \cdot 4$, or 480 possible permutations. Equivalently, there are 480 ways to deliver the jokes with a man's joke told first and a man's joke delivered last.

Now we can return to our probability fraction.

P(man's joke first, man's joke last)

$$= \frac{\text{number of permutations with man's joke first, man's joke last}}{\text{total number of possible permutations}}$$

$$= \frac{480}{720} = \frac{2}{3}$$

The probability of a man's joke delivered first and a man's joke told last is $\frac{2}{3}$.

✓ Check Point 1 Consider the six jokes about books by Groucho Marx, George Carlin, Steven Wright, Henny Youngman, Jerry Seinfeld, and Phyllis Diller. As in Example 1, each joke is written on one of six cards which are randomly drawn one card at a time. The order in which the cards are drawn determines the order in which the jokes are delivered. What is the probability that a joke by a comic whose first name begins with G is told first and a man's joke is delivered last? $\frac{4}{15}$

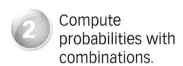 Compute probabilities with combinations.

Probability with Combinations

In 2012, Americans spent $65.5 billion on state and multi-state lotteries. With each lottery drawing, the probability that someone will win the jackpot is relatively high. If there is no winner, it is virtually certain that eventually someone will be graced with millions of dollars. So, why are you so unlucky compared to this undisclosed someone? In Example 2, we provide an answer to this question.

Example 2   Probability and Combinations: Powerball

Powerball is a multi-state lottery played in most U.S. states. It is the first lottery game to randomly draw numbers from two drums. The game is set up so that each player chooses five different numbers from 1 to 59 and one Powerball number from 1 to 35. Twice per week 5 white balls are drawn randomly from a drum with 59 white balls, numbered 1 to 59, and then one red Powerball is drawn randomly from a drum with 35 red balls, numbered 1 to 35. A player wins the jackpot by matching all five numbers drawn from the white balls in any order and matching the number on the red Powerball. With one $2 Powerball ticket, what is the probability of winning the jackpot?

Solution

Because the order of the five numbers shown on the white balls does not matter, this is a situation involving combinations. We begin with the formula for probability.

$$P(\text{winning the jackpot}) = \frac{\text{number of ways of winning the jackpot}}{\text{total number of possible combinations}}$$

We can use the combinations formula

$$_nC_r = \frac{n!}{(n-r)!\,r!}$$

to find the total number of possible combinations in the first part of the Powerball lottery. We are selecting $r = 5$ numbers from a collection of $n = 59$ numbers from the drum of white balls.

$$_{59}C_5 = \frac{59!}{(59-5)!\,5!} = \frac{59!}{54!\,5!} = \frac{59 \cdot 58 \cdot 57 \cdot 56 \cdot 55 \cdot 54!}{54! \cdot 5 \cdot 4 \cdot 3 \cdot 2 \cdot 1} = 5{,}006{,}386$$

There are 5,006,386 number combinations in the first part of Powerball.

Next, we must determine the number of ways of selecting the red Powerball. Because there are 35 red Powerballs in the second drum, there are 35 possible combinations of numbers.

We can use the Fundamental Counting Principle to find the total number of possible number combinations in Powerball.

$$_{59}C_5 \cdot 35 = 5{,}006{,}386 \cdot 35 = 175{,}223{,}510$$

| Combinations of white balls from the first drum | Combinations of red Powerballs from the second drum |

There are 175,223,510 number combinations in Powerball. If a person buys one $2 ticket, that person has selected only one combination of the numbers. With one Powerball ticket, there is only one way of winning the jackpot.

Now we can return to our probability fraction.

$$P(\text{winning the jackpot}) = \frac{\text{number of ways of winning the jackpot}}{\text{total number of possible combinations}}$$

$$= \frac{1}{175{,}223{,}510} \approx 5.707 \times 10^{-9} = 0.000000005707$$

The probability of winning the jackpot with one Powerball ticket is $\frac{1}{175,223,510}$, or about 1 in 175 million.

Blitzer Bonus

Comparing the Probability of Dying to the Probability of Winning Powerball's Jackpot

As a healthy nonsmoking 30-year-old, your probability of dying this year is approximately 0.001. Divide this probability by the probability of winning the Powerball jackpot with one ticket:

$$\frac{0.001}{0.000000005707} \approx 175{,}223.$$

A healthy 30-year-old is nearly 175,000 times more likely to die this year than to win the Powerball jackpot. It's no wonder that people who calculate probabilities often refer to lotteries as a "tax on stupidity."

Suppose that a person buys 5000 different Powerball tickets. Because that person has selected 5000 different combinations of the Powerball numbers, the probability of winning the jackpot is

$$\frac{5000}{175,223,510} \approx 2.85 \times 10^{-5} = 0.0000285.$$

The chances of winning the jackpot are about 285 in ten million. At $2 per Powerball ticket, it is highly probable that our Powerball player will be $10,000 poorer. Knowing a little probability helps a lotto.

✓ **Check Point 2**   Hitting the jackpot in Powerball is not the only way to win a monetary prize. For example, a minimum award of $10,000 is given to a player who correctly matches four of the five numbers drawn from the 59 white balls and the one number drawn from the 35 red Powerballs. Find the probability of winning this consolation prize. Express the answer as a fraction and as a decimal correct to nine places. ▸ $\frac{27}{17,522,351} \approx 1.541 \times 10^{-6} \approx 0.000001541$

Example 3   Probability and Combinations

A club consists of five men and seven women. Three members are selected at random to attend a conference. Find the probability that the selected group consists of

a. three men.   **b.** one man and two women.

Solution

The order in which the three people are selected does not matter, so this is a problem involving combinations.

a. We begin with the probability of selecting three men.

$$P(3 \text{ men}) = \frac{\text{number of ways of selecting 3 men}}{\text{total number of possible combinations}}$$

First, we consider the denominator of the probability fraction. We are selecting $r = 3$ people from a total group of $n = 12$ people (five men and seven women). The total number of possible combinations is

$$_{12}C_3 = \frac{12!}{(12-3)!\,3!} = \frac{12!}{9!\,3!} = \frac{12 \cdot 11 \cdot 10 \cdot 9!}{9! \cdot 3 \cdot 2 \cdot 1} = 220.$$

Thus, there are 220 possible three-person selections.

Next, we consider the numerator of the probability fraction. We are interested in the number of ways of selecting three men from five men. We are selecting $r = 3$ men from a total group of $n = 5$ men. The number of possible combinations of three men is

$$_5C_3 = \frac{5!}{(5-3)!\,3!} = \frac{5!}{2!\,3!} = \frac{5 \cdot 4 \cdot 3!}{2 \cdot 1 \cdot 3!} = 10.$$

Thus, there are 10 ways of selecting three men from five men. Now we can fill in the numbers in the numerator and the denominator of our probability fraction.

$$P(3 \text{ men}) = \frac{\text{number of ways of selecting 3 men}}{\text{total number of possible combinations}} = \frac{10}{220} = \frac{1}{22}$$

The probability that the group selected to attend the conference consists of three men is $\frac{1}{22}$.

12 Club Members

5 Men   7 Women

↓

Select 3

12 Club Members

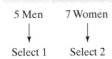

5 Men   7 Women
↓   ↓
Select 1   Select 2

b. We set up the fraction for the probability that the selected group consists of one man and two women.

$$P(1 \text{ man}, 2 \text{ women}) = \frac{\text{number of ways of selecting 1 man and 2 women}}{\text{total number of possible combinations}}$$

The denominator of this fraction is the same as the denominator in part (a). The total number of possible combinations is found by selecting $r = 3$ people from $n = 12$ people: $_{12}C_3 = 220$.

Next, we move to the numerator of the probability fraction. The number of ways of selecting $r = 1$ man from $n = 5$ men is

$$_5C_1 = \frac{5!}{(5-1)!\,1!} = \frac{5!}{4!\,1!} = \frac{5 \cdot 4!}{4! \cdot 1} = 5.$$

The number of ways of selecting $r = 2$ women from $n = 7$ women is

$$_7C_2 = \frac{7!}{(7-2)!\,2!} = \frac{7!}{5!\,2!} = \frac{7 \cdot 6 \cdot 5!}{5! \cdot 2 \cdot 1} = 21.$$

By the Fundamental Counting Principle, the number of ways of selecting 1 man and 2 women is

$$_5C_1 \cdot {_7C_2} = 5 \cdot 21 = 105.$$

Now we can fill in the numbers in the numerator and the denominator of our probability fraction.

$$P(1 \text{ man}, 2 \text{ women}) = \frac{\text{number of ways of selecting 1 man and 2 women}}{\text{total number of possible combinations}}$$

$$= \frac{_5C_1 \cdot {_7C_2}}{_{12}C_3} = \frac{105}{220} = \frac{21}{44}$$

The probability that the group selected to attend the conference consists of one man and two women is $\frac{21}{44}$.

✓ **Check Point 3**   A club consists of six men and four women. Three members are selected at random to attend a conference. Find the probability that the selected group consists of

a. three men. $\frac{1}{6}$

▶ **b.** two men and one woman. $\frac{1}{2}$

Achieving Success

Think about finding a tutor.

If you're attending all lectures, taking good class notes, reading the textbook, and doing all the assigned homework, but still having difficulty in a math course, you might want to find a tutor. Many on-campus learning centers and math labs have trained people available to help you. Sometimes a TA who has previously taught the course is available. **Make sure the tutor is both good at math and familiar with the particular course you're taking.** Bring your textbook, class notes, the problems you've done, and information about course policy and tests to each meeting with your tutor. That way he or she can be sure the tutoring sessions address your exact needs.

Concept and Vocabulary Check

Exercises in the Concept and Vocabulary Check
are intended for group and class discussions.

In Exercises 1–2, fill in each blank so that the resulting statement is true.

1. Six stand-up comics, A, B, C, D, E, and F, are to perform on a single evening at a comedy club. The order of performance is determined by random selection. The probability that comic E will perform first is the number of ___permutations___ with comic E performing first divided by _the total number of possible permutations_

2. The probability of winning a lottery with one lottery ticket is the number of ways of winning, which is precisely _____1_____, divided by the total number of possible ___combinations___.

In Exercises 3–4, determine whether each statement is true or false. If the statement is false, make the necessary change(s) to produce a true statement.   Changes to false statements will vary.

3. When working problems involving probability with permutations, the denominators of the probability fractions consist of the total number of possible permutations.   true

4. When working problems involving probability with combinations, the numerators of the probability fractions consist of the total number of possible combinations.   false

Respond to Exercises 5–6 using verbal or written explanations.   5–6. Answers will vary.

5. If you purchase ten lottery tickets, explain how you determine the probability of winning the lottery. How do you find the number in the numerator of the probability fraction and the number in the denominator of the probability fraction?

6. If people understood the mathematics involving probabilities and lotteries, as you now do, do you think they would continue to spend hundreds of dollars per year on lottery tickets? Explain your answer.

Exercise Set 7.5

Practice and Application Exercises

1. Martha, Lee, Nancy, Paul, and Armando have all been invited to a dinner party. They arrive randomly and each person arrives at a different time.
   a. In how many ways can they arrive?   120
   b. In how many ways can Martha arrive first and Armando last?   6
   c. Find the probability that Martha will arrive first and Armando last. $\frac{1}{20}$

2. Three men and three women line up at a checkout counter in a store.
   a. In how many ways can they line up?   720
   b. In how many ways can they line up if the first person in line is a woman, and then the line alternates by gender— that is a woman, a man, a woman, a man, and so on?   36
   c. Find the probability that the first person in line is a woman and the line alternates by gender. $\frac{1}{20}$

3. Six stand-up comics, A, B, C, D, E, and F, are to perform on a single evening at a comedy club. The order of performance is determined by random selection. Find the probability that
   a. Comic E will perform first. $\frac{1}{6}$
   b. Comic C will perform fifth and comic B will perform last. $\frac{1}{30}$
   c. The comedians will perform in the following order: D, E, C, A, B, F. $\frac{1}{720}$
   d. Comic A or comic B will perform first. $\frac{1}{3}$

4. Seven performers, A, B, C, D, E, F, and G, are to appear in a fundraiser. The order of performance is determined by random selection. Find the probability that
   a. D will perform first. $\frac{1}{7}$
   b. E will perform sixth and B will perform last. $\frac{1}{42}$
   c. They will perform in the following order: C, D, B, A, G, F, E. $\frac{1}{5040}$
   d. F or G will perform first. $\frac{2}{7}$

5. A group consists of four men and five women. Three people are selected to attend a conference.
   a. In how many ways can three people be selected from this group of nine?   84
   b. In how many ways can three women be selected from the five women?   10
   c. Find the probability that the selected group will consist of all women. $\frac{5}{42}$

6. A political discussion group consists of five Democrats and six Republicans. Four people are selected to attend a conference.
   a. In how many ways can four people be selected from this group of eleven?   330
   b. In how many ways can four Republicans be selected from the six Republicans?   15
   c. Find the probability that the selected group will consist of all Republicans. $\frac{1}{22}$

Mega Millions is a multi-state lottery played in most U.S. states. As of this writing, the top cash prize was $656 million, going to three lucky winners in three states. Players pick five different numbers from 1 to 56 and one number from 1 to 46. Use this information to solve Exercises 7–10. Express all probabilities as fractions.

7. A player wins the jackpot by matching all five numbers drawn from white balls (1 through 56) and matching the number on the gold Mega Ball® (1 through 46). What is the probability of winning the jackpot? $\frac{1}{175,711,536}$

8. A player wins a minimum award of $10,000 by correctly matching four numbers drawn from white balls (1 through 56) and matching the number on the gold Mega Ball® (1 through 46). What is the probability of winning this consolation prize? $\frac{85}{58,570,512}$

9. A player wins a minimum award of $150 by correctly matching three numbers drawn from white balls (1 through 56) and matching the number on the gold Mega Ball® (1 through 46). What is the probability of winning this consolation prize? $\frac{2125}{29,285,256}$

10. A player wins a minimum award of $10 by correctly matching two numbers drawn from white balls (1 through 56) and matching the number on the gold Mega Ball® (1 through 46). What is the probability of winning this consolation prize? $\frac{14,875}{12,550,824}$

11. A box contains 25 transistors, 6 of which are defective. If 6 are selected at random, find the probability that
    a. all are defective.   *   b. none are defective.   *

12. A committee of five people is to be formed from six lawyers and seven teachers. Find the probability that
    a. all are lawyers.   *   b. none are lawyers.   *

13. A city council consists of six Democrats and four Republicans. If a committee of three people is selected, find the probability of selecting one Democrat and two Republicans. $\frac{3}{10} = 0.3$

14. A parent-teacher committee consisting of four people is to be selected from fifteen parents and five teachers. Find the probability of selecting two parents and two teachers.   *

*Exercises 15–20 involve a deck of 52 cards. If necessary, refer to the picture of a deck of cards, **Figure 7.5** on page 445.*

15. A poker hand consists of five cards.
    a. Find the total number of possible five-card poker hands.   2,598,960
    b. A diamond flush is a five-card hand consisting of all diamonds. Find the number of possible diamond flushes.   *
    c. Find the probability of being dealt a diamond flush.   *

16. A poker hand consists of five cards.
    a. Find the total number of possible five-card poker hands.   2,598,960
    b. Find the number of ways in which four aces can be selected.   1
    c. Find the number of ways in which one king can be selected.   4
    d. Use the Fundamental Counting Principle and your answers from parts (b) and (c) to find the number of ways of getting four aces and one king.   4
    e. Find the probability of getting a poker hand consisting of four aces and one king. $\frac{4}{2,598,960} \approx 0.00000154$

17. If you are dealt 3 cards from a shuffled deck of 52 cards, find the probability that all 3 cards are picture cards.   *

18. If you are dealt 4 cards from a shuffled deck of 52 cards, find the probability that all 4 are hearts. $\frac{11}{4165} \approx 0.00264$

19. If you are dealt 4 cards from a shuffled deck of 52 cards, find the probability of getting two queens and two kings.   *

20. If you are dealt 4 cards from a shuffled deck of 52 cards, find the probability of getting three jacks and one queen.   *

Critical Thinking Exercises   21–22. Answers will vary.

21. Write and solve an original problem involving probability and permutations.

22. Write and solve an original problem involving probability and combinations whose solution requires $\frac{_{14}C_{10}}{_{20}C_{10}}$.

23. An apartment complex offers apartments with four different options, designated by A through D. There are an equal number of apartments with each combination of options.

A	B	C	D
one bedroom	one bathroom	first floor	lake view
two bedrooms	two bathrooms	second floor	golf course view
three bedrooms			no special view

If there is only one apartment left, what is the probability that it is precisely what a person is looking for, namely two bedrooms, two bathrooms, first floor, and a lake or golf course view? $\frac{1}{18}$

24. Suppose that it is a drawing in which the Powerball jackpot is promised to exceed $500 million. If a person purchases 175,223,510 tickets at $2 per ticket (all possible combinations), isn't this a guarantee of winning the jackpot? Because the probability in this situation is 1, what's wrong with doing this?   *

25. The digits 1, 2, 3, 4, and 5 are randomly arranged to form a three-digit number. (Digits are not repeated.) Find the probability that the number is even and greater than 500. $\frac{1}{10}$

26. In a five-card poker hand, what is the probability of being dealt exactly one ace and no picture cards?   *

Group Exercise

27. Research and present a group report on state and multi-state lotteries. Include answers to some or all of the following questions. As always, make the report interesting and informative. Which states do not have lotteries? Why not? How much is spent per capita on lotteries? What are some of the lottery games? What is the probability of winning top prize in these games? What income groups spend the greatest amount of money on lotteries? If your state has a lottery, what does it do with the money it makes? Is the way the money is spent what was promised when the lottery first began?   Answers will vary.

*See Answers to Selected Exercises.

7.6 Events Involving *Not* and *Or*; Odds

What am I supposed to learn?

After you have read this section, you should be able to:

① Find the probability that an event will not occur.

② Find the probability of one event or a second event occurring.

③ Understand and use odds.

What are you most afraid of? A shark attack? An airplane crash? The Harvard Center for Risk Analysis helps to put these fears in perspective. According to the Harvard Center, the odds in favor of a deadly shark attack are 1 to 280 million and the odds against a deadly airplane accident are 3 million to 1.

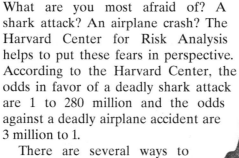

There are several ways to express the likelihood of an event. For example, we can determine the probability of a deadly shark attack or a deadly airplane accident. We can also determine the *odds in favor* and the *odds against* these events. In this section, we expand our knowledge of probability and explain the meaning of odds.

Probability of an Event Not Occurring

① Find the probability that an event will not occur.

If we know $P(E)$, the probability of an event E, we can determine the probability that the event will not occur, denoted by $P(\text{not } E)$. The event *not E* is the **complement** of E because it is the set of all outcomes in the sample space S that are not outcomes in the event E. In any experiment, an event must occur or its complement must occur. Thus, the sum of the probability that an event will occur and the probability that it will not occur is 1:

$$P(E) + P(\text{not } E) = 1.$$

Solving for $P(E)$ or for $P(\text{not } E)$, we obtain the following formulas:

Complement Rules of Probability

- The probability that an event E will not occur is equal to 1 minus the probability that it will occur.

$$P(\text{not } E) = 1 - P(E)$$

- The probability that an event E will occur is equal to 1 minus the probability that it will not occur.

$$P(E) = 1 - P(\text{not } E)$$

Using set notation, if E' is the complement of E, then $P(E') = 1 - P(E)$ and $P(E) = 1 - P(E')$.

Example 1 **The Probability of an Event Not Occurring**

If you are dealt one card from a standard 52-card deck, find the probability that you are not dealt a queen.

Great Question!

Do I have to use the formula for P(not E) to solve Example 1?

No. Here's how to work the example without using the formula:

P(not a queen)

$$= \frac{\text{number of ways a}}{\text{total number of outcomes}}$$

$$= \frac{48}{52} \quad \begin{array}{l}\text{With 4 queens,} \\ \text{52} - 4 = 48 \text{ cards} \\ \text{are not queens.}\end{array}$$

$$= \frac{4 \cdot 12}{4 \cdot 13} = \frac{12}{13}.$$

Solution

Because

$$P(\text{not } E) = 1 - P(E)$$

then

$$P(\text{not a queen}) = 1 - P(\text{queen}).$$

There are four queens in a deck of 52 cards. The probability of being dealt a queen is $\frac{4}{52} = \frac{1}{13}$. Thus,

$$P(\text{not a queen}) = 1 - P(\text{queen}) = 1 - \frac{1}{13} = \frac{13}{13} - \frac{1}{13} = \frac{12}{13}.$$

The probability that you are not dealt a queen is $\frac{12}{13}$.

✓ **Check Point 1** If you are dealt one card from a standard 52-card deck, find the probability that you are not dealt a diamond. $\frac{3}{4}$

Example 2 Using the Complement Rules

The circle graph in **Figure 7.6** shows the distribution, by age group, of the 191 million car drivers in the United States, with all numbers rounded to the nearest million. If one driver is randomly selected from this population, find the probability that the person

a. is not in the 20–29 age group. **b.** is less than 80 years old.

Express probabilities as simplified fractions.

Number of U.S. Car Drivers, by Age Group

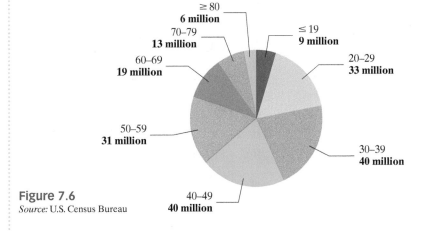

Figure 7.6
Source: U.S. Census Bureau

Solution

a. We begin with the probability that a randomly selected driver is not in the 20–29 age group.

$$P(\text{not in 20–29 age group}) = 1 - P(\text{in 20–29 age group})$$

$$= 1 - \frac{33}{191}$$

The graph shows 33 million drivers in the 20–29 age group.

This number, 191 million drivers, was given, but can be obtained by adding the numbers in the eight sectors.

$$= \frac{191}{191} - \frac{33}{191} = \frac{158}{191}$$

The probability that a randomly selected driver is not in the 20–29 age group is $\frac{158}{191}$.

Number of U.S. Car Drivers, by Age Group

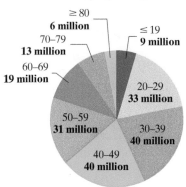

Figure 7.6 (repeated)

b. We could compute the probability that a randomly selected driver is less than 80 years old by adding the numbers in each of the seven sectors representing drivers less than 80 and dividing this sum by 191 (million). However, it is easier to use complements. The complement of selecting a driver less than 80 years old is selecting a driver 80 or older.

$$P(\text{less than 80 years old}) = 1 - P(\text{80 or older})$$

$$= 1 - \frac{6}{191} \quad \boxed{\text{The graph shows 6 million drivers 80 or older.}}$$

$$= \frac{191}{191} - \frac{6}{191} = \frac{185}{191}$$

The probability that a randomly selected driver is less than 80 years old is $\frac{185}{191}$.

✓ **Check Point 2** If one driver is randomly selected from the population represented in **Figure 7.6**, find the probability, expressed as a simplified fraction, that the person

a. is not in the 50–59 age group. $\quad \frac{160}{191}$ **b.** is at least 20 years old. $\quad \frac{182}{191}$

Find the probability of one event or a second event occurring.

Or Probabilities with Mutually Exclusive Events

Suppose that you randomly select one card from a deck of 52 cards. Let A be the event of selecting a king and B be the event of selecting a queen. Only one card is selected, so it is impossible to get both a king and a queen. The events of selecting a king and a queen cannot occur simultaneously. They are called *mutually exclusive events*.

> **Mutually Exclusive Events**
>
> If it is impossible for events A and B to occur simultaneously, the events are said to be **mutually exclusive**.

In general, if A and B are mutually exclusive events, the probability that either A or B will occur is determined by adding their individual probabilities.

> **Or Probabilities with Mutually Exclusive Events**
>
> If A and B are mutually exclusive events, then
>
> $$P(A \text{ or } B) = P(A) + P(B).$$
>
> Using set notation, $P(A \cup B) = P(A) + P(B)$.

Example 3 The Probability of Either of Two Mutually Exclusive Events Occurring

If one card is randomly selected from a deck of cards, what is the probability of selecting a king or a queen?

Solution

We find the probability that either of these mutually exclusive events will occur by adding their individual probabilities.

$$P(\text{king or queen}) = P(\text{king}) + P(\text{queen}) = \frac{4}{52} + \frac{4}{52} = \frac{8}{52} = \frac{2}{13}$$

The probability of selecting a king or a queen is $\frac{2}{13}$.

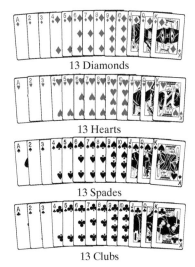

13 Diamonds

13 Hearts

13 Spades

13 Clubs

Figure 7.7 A deck of 52 cards

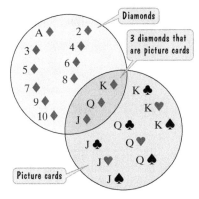

Figure 7.9

✔ **Check Point 3**   If you roll a single, six-sided die, what is the probability of getting either a 4 or a 5? $\frac{1}{3}$

Or Probabilities with Events That Are Not Mutually Exclusive

Consider the deck of 52 cards shown in **Figure 7.7**. Suppose that these cards are shuffled and you randomly select one card from the deck. What is the probability of selecting a diamond or a picture card (jack, queen, king)? Begin by adding their individual probabilities.

$$P(\text{diamond}) + P(\text{picture card}) = \frac{13}{52} + \frac{12}{52}$$

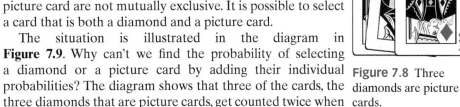

There are 13 diamonds in the deck of 52 cards.   There are 12 picture cards in the deck of 52 cards.

However, this sum is not the probability of selecting a diamond or a picture card. The problem is that there are three cards that are *simultaneously* diamonds and picture cards, shown in **Figure 7.8**. The events of selecting a diamond and selecting a picture card are not mutually exclusive. It is possible to select a card that is both a diamond and a picture card.

The situation is illustrated in the diagram in **Figure 7.9**. Why can't we find the probability of selecting a diamond or a picture card by adding their individual probabilities? The diagram shows that three of the cards, the three diamonds that are picture cards, get counted twice when we add the individual probabilities. First the three cards get counted as diamonds and then they get counted as picture cards. In order to avoid the error of counting the three cards twice, we need to subtract the probability of getting a diamond and a picture card, $\frac{3}{52}$, as follows:

Figure 7.8 Three diamonds are picture cards.

$P(\text{diamond or picture card})$

$$= P(\text{diamond}) + P(\text{picture card}) - P(\text{diamond and picture card})$$

$$= \frac{13}{52} + \frac{12}{52} - \frac{3}{52} = \frac{13 + 12 - 3}{52} = \frac{22}{52} = \frac{11}{26}.$$

Thus, the probability of selecting a diamond or a picture card is $\frac{11}{26}$.

In general, if A and B are events that are not mutually exclusive, the probability that A or B will occur is determined by adding their individual probabilities and then subtracting the probability that A and B occur simultaneously.

Or Probabilities with Events That Are Not Mutually Exclusive

If A and B are not mutually exclusive events, then

$$P(A \text{ or } B) = P(A) + P(B) - P(A \text{ and } B).$$

Using set notation,

$$P(A \cup B) = P(A) + P(B) - P(A \cap B).$$

Example 4   An *Or* Probability with Events That Are Not Mutually Exclusive

In a group of 25 baboons, 18 enjoy grooming their neighbors, 16 enjoy screeching wildly, while 10 enjoy grooming their neighbors and screeching wildly. If one baboon is selected at random from the group, find the probability that it enjoys grooming its neighbors or screeching wildly.

Solution

It is possible for a baboon in the group to enjoy both grooming its neighbors and screeching wildly. Ten of the brutes are given to engage in both activities. These events are not mutually exclusive.

$$P\left(\begin{array}{c}\text{grooming}\\\text{or screeching}\end{array}\right) = P(\text{grooming}) + P(\text{screeching}) - P\left(\begin{array}{c}\text{grooming}\\\text{and screeching}\end{array}\right)$$

$$= \frac{18}{25} + \frac{16}{25} - \frac{10}{25}$$

| 18 of the 25 baboons enjoy grooming. | 16 of the 25 baboons enjoy screeching. | 10 of the 25 baboons enjoy both. |

$$= \frac{18 + 16 - 10}{25} = \frac{24}{25}$$

The probability that a baboon in the group enjoys grooming its neighbors or screeching wildly is $\frac{24}{25}$.

✓ **Check Point 4** In a group of 50 students, 23 take math, 11 take psychology, and 7 take both math and psychology. If one student is selected at random, find the probability that the student takes math or psychology. $\frac{27}{50}$

Example 5 An *Or* Probability with Events That Are Not Mutually Exclusive

Figure 7.10 illustrates a spinner. It is equally probable that the pointer will land on any one of the eight regions, numbered 1 through 8. If the pointer lands on a borderline, spin again. Find the probability that the pointer will stop on an even number or on a number greater than 5.

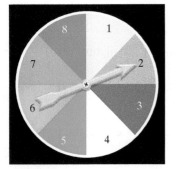

Figure 7.10 It is equally probable that the pointer will land on any one of the eight regions.

Solution

It is possible for the pointer to land on a number that is both even and greater than 5. Two of the numbers, 6 and 8, are even and greater than 5. These events are not mutually exclusive. The probability of landing on a number that is even or greater than 5 is calculated as follows:

$$P\left(\begin{array}{c}\text{even or}\\\text{greater than 5}\end{array}\right) = P(\text{even}) + P(\text{greater than 5}) - P\left(\begin{array}{c}\text{even and}\\\text{greater than 5}\end{array}\right)$$

$$= \frac{4}{8} + \frac{3}{8} - \frac{2}{8}$$

| Four of the eight numbers, 2, 4, 6, and 8, are even. | Three of the eight numbers, 6, 7, and 8, are greater than 5. | Two of the eight numbers, 6 and 8, are even and greater than 5. |

$$= \frac{4 + 3 - 2}{8} = \frac{5}{8}.$$

The probability that the pointer will stop on an even number or a number greater than 5 is $\frac{5}{8}$.

✓ **Check Point 5** Use **Figure 7.10** to find the probability that the pointer will stop on an odd number or a number less than 5. $\frac{3}{4}$

Example 6   *Or* **Probabilities with Real-World Data**

Table 7.5, first presented in Section 7.4, shows the marital status of the U.S. population in 2010. Numbers in the table are expressed in millions.

Table 7.5 Marital Status of the U.S. Population, Ages 15 or Older, 2010, in Millions					
	Married	**Never Married**	**Divorced**	**Widowed**	**Total**
Male	65	40	10	3	118
Female	65	34	14	11	124
Total	130	74	24	14	242

Source: U.S. Census Bureau

If one person is randomly selected from the population represented in **Table 7.5**, find the probability that

a. the person is divorced or male.

b. the person is married or divorced.

Express probabilities as simplified fractions and as decimals rounded to the nearest hundredth.

Solution

a. It is possible to select a person who is both divorced and male. Thus, these events are not mutually exclusive.

P(divorced or male)

$$= P(\text{divorced}) + P(\text{male}) - P(\text{divorced and male})$$

$$= \frac{24}{242} + \frac{118}{242} - \frac{10}{242}$$

Of the 242 million Americans, 24 million are divorced. | Of the 242 million Americans, 118 million are male. | Of the 242 million Americans, 10 million are divorced and male.

$$= \frac{24 + 118 - 10}{242} = \frac{132}{242} = \frac{22 \cdot 6}{22 \cdot 11} = \frac{6}{11} \approx 0.55$$

The probability of selecting a person who is divorced or male is $\frac{6}{11}$, or approximately 0.55.

b. It is impossible to select a person who is both married and divorced. These events are mutually exclusive.

P(married or divorced)

$$= P(\text{married}) + P(\text{divorced})$$

$$= \frac{130}{242} + \frac{24}{242}$$

Of the 242 million Americans, 130 million are married. | Of the 242 million Americans, 24 million are divorced.

$$= \frac{130 + 24}{242} = \frac{154}{242} = \frac{22 \cdot 7}{22 \cdot 11} = \frac{7}{11} \approx 0.64$$

The probability of selecting a person who is married or divorced is $\frac{7}{11}$, or approximately 0.64.

Check Point 6 If one person is randomly selected from the population represented in **Table 7.5** on page 463, find the probability that

a. the person is married or female. $\frac{189}{242} \approx 0.78$

b. the person is divorced or widowed. $\frac{19}{121} \approx 0.16$

Express probabilities as simplified fractions and as decimals rounded to the nearest hundredth.

Odds

3 Understand and use odds.

If we know the probability of an event E, we can also speak of the *odds in favor*, or the *odds against*, the event. The following definitions link together the concepts of odds and probabilities:

> **Probability to Odds**
>
> If $P(E)$ is the probability of an event E occurring, then
>
> **1.** The **odds in favor of E** are found by taking the probability that E will occur and dividing by the probability that E will not occur.
>
> $$\text{Odds in favor of } E = \frac{P(E)}{P(\text{not } E)}$$
>
> **2.** The **odds against E** are found by taking the probability that E will not occur and dividing by the probability that E will occur.
>
> $$\text{Odds against } E = \frac{P(\text{not } E)}{P(E)}$$
>
> The odds against E can also be found by reversing the ratio representing the odds in favor of E.

Example 7 From Probability to Odds

You roll a single, six-sided die.

a. Find the odds in favor of rolling a 2.

b. Find the odds against rolling a 2.

Solution

Let E represent the event of rolling a 2. In order to determine odds, we must first find the probability of E occurring and the probability of E not occurring. With $S = \{1, 2, 3, 4, 5, 6\}$ and $E = \{2\}$, we see that

$$P(E) = \frac{1}{6}$$

$$\text{and } P(\text{not } E) = 1 - \frac{1}{6} = \frac{6}{6} - \frac{1}{6} = \frac{5}{6}.$$

Now we are ready to construct the ratios for the odds in favor of E and the odds against E.

Great Question!

When you computed the odds in Example 7(a), the denominators of the two probabilities divided out. Will this always occur?

Yes.

a. Odds in favor of E (rolling a 2) $= \dfrac{P(E)}{P(\text{not } E)} = \dfrac{\frac{1}{6}}{\frac{5}{6}} = \dfrac{1}{6} \cdot \dfrac{6}{5} = \dfrac{1}{5}$

The odds in favor of rolling a 2 are $\frac{1}{5}$. The ratio $\frac{1}{5}$ is usually written 1:5 and is read "1 to 5." Thus, the odds in favor of rolling a 2 are 1 to 5.

b. Now that we have the odds in favor of rolling a 2, namely $\frac{1}{5}$ or 1:5, we can find the odds against rolling a 2 by reversing this ratio. Thus,

$$\text{Odds against } E \text{ (rolling a 2)} = \frac{5}{1} \text{ or } 5{:}1.$$

The odds against rolling a 2 are 5 to 1.

✓ **Check Point 7**   You are dealt one card from a 52-card deck.

a. Find the odds in favor of getting a red queen.   2:50 or 1:25

▶ **b.** Find the odds against getting a red queen.   50:2 or 25:1

Example 8   From Probability to Odds

The winner of a raffle will receive a new sports utility vehicle. If 500 raffle tickets were sold and you purchased ten tickets, what are the odds against your winning the car?

Solution

Let E represent the event of winning the SUV. Because you purchased ten tickets and 500 tickets were sold,

$$P(E) = \frac{10}{500} = \frac{1}{50} \quad \text{and} \quad P(\text{not } E) = 1 - \frac{1}{50} = \frac{50}{50} - \frac{1}{50} = \frac{49}{50}.$$

Now we are ready to construct the ratio for the odds against E (winning the SUV).

$$\text{Odds against } E = \frac{P(\text{not } E)}{P(E)} = \frac{\frac{49}{50}}{\frac{1}{50}} = \frac{49}{50} \cdot \frac{50}{1} = \frac{49}{1}$$

The odds against winning the SUV are 49 to 1, written 49:1.

✓ **Check Point 8**   The winner of a raffle will receive a two-year scholarship to the college of his or her choice. If 1000 raffle tickets were sold and you purchased five tickets, what are the odds against your winning the scholarship?   199:1

Now that we know how to convert from probability to odds, let's see how to convert from odds to probability. Suppose that the odds in favor of event E occurring are a to b. This means that

$$\frac{P(E)}{P(\text{not } E)} = \frac{a}{b} \quad \text{or} \quad \frac{P(E)}{1 - P(E)} = \frac{a}{b}.$$

By solving the equation on the right for $P(E)$, we obtain a formula for converting from odds to probability.

Odds to Probability

If the odds in favor of event E are a to b, then the probability of the event is given by

$$P(E) = \frac{a}{a + b}.$$

Example 9   From Odds to Probability

The odds in favor of a particular candidate winning an election are 2 to 5. What is the probability that this candidate will win the election?

Solution

Because odds in favor, a to b, means a probability of $\frac{a}{a + b}$, then odds in favor, 2 to 5, means a probability of

$$\frac{2}{2 + 5} = \frac{2}{7}.$$

The probability that this candidate will win the election is $\frac{2}{7}$.

✓ **Check Point 9**   The odds against a particular candidate winning an election are 15 to 1. Find the odds in favor of the candidate winning the election and the probability of the candidate winning the election. $1{:}15; \dfrac{1}{16}$

Blitzer Bonus

Big Fears and Their Odds

ODDS SCALE

The Fear	Odds in Favor
Fatal shark attack	1 to 280,000,000
Fatal airplane accident	1 to 3,000,000
Losing your job	1 to 252
Home burglary at night	1 to 181
Developing cancer	1 to 7
Catching a sexually transmitted disease	1 to 4
Developing heart disease	1 to 4
Dying from tobacco-related illnesses (smokers)	1 to 2

Source: David Ropeik, Harvard Center for Risk Analysis

Concept and Vocabulary Check

Exercises in the Concept and Vocabulary Check are intended for group and class discussions.

In Exercises 1–5, fill in each blank so that the resulting statement is true.

1. Because $P(E) + P(\text{not } E) = 1$, then $P(\text{not } E) = $ ____ $1 - P(E)$ ____ and $P(E) = $ ____ $1 - P(\text{not } E)$ ____.

2. If it is impossible for events A and B to occur simultaneously, the events are said to be __mutually exclusive__. For such events, $P(A \text{ or } B) = $ ____ $P(A) + P(B)$ ____.

3. If it is possible for events A and B to occur simultaneously, then $P(A \text{ or } B) = $ ____ $P(A) + P(B) - P(A \text{ and } B)$

4. The odds in favor of E can be found by taking the probability that ___*E will occur*___ and dividing by the probability that ___*E will not occur*___ .

5. If the odds in favor of event E are a to b, then the probability of the event, $P(E)$, is given by the formula ___$P(E) = \frac{a}{a+b}$___ .

In Exercises 6–9, determine whether each statement is true or false. If the statement is false, make the necessary change(s) to produce a true statement.   Changes to false statements will vary.

6. The probability that an event will not occur is equal to the probability that it will occur minus 1.   false

7. The probability of A or B can always be found by adding the probability of A and the probability of B.   false

8. The odds against E can always be found by reversing the ratio representing the odds in favor of E.   true

9. According to the National Center for Health Statistics, the lifetime odds in favor of dying from heart disease are 1 to 5, so the probability of dying from heart disease is $\frac{1}{5}$.   false

Respond to Exercises 10–16 using verbal or written explanations.   10–16. Answers will vary.

10. Explain how to find the probability of an event not occurring. Give an example.

11. What are mutually exclusive events? Give an example of two events that are mutually exclusive.

12. Explain how to find *or* probabilities with mutually exclusive events. Give an example.

13. Give an example of two events that are not mutually exclusive.

14. Explain how to find *or* probabilities with events that are not mutually exclusive. Give an example.

15. Explain how to find the odds in favor of an event if you know the probability that the event will occur.

16. Explain how to find the probability of an event if you know the odds in favor of that event.

Exercise Set 7.6

Practice and Application Exercises

In Exercises 1–6, you are dealt one card from a 52-card deck. Find the probability that you are not dealt

1. an ace. $\frac{12}{13}$

2. a 3. $\frac{12}{13}$

3. a heart. $\frac{3}{4}$

4. a club. $\frac{3}{4}$

5. a picture card. $\frac{10}{13}$

6. a red picture card. $\frac{23}{26}$

The graph shows the probability of cardiovascular disease, by age and gender. Use the information in the graph to solve Exercises 7–8. Express all probabilities as decimals, estimated to two decimal places.

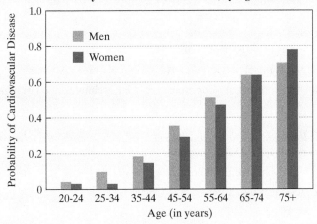

Probability of Cardiovascular Disease, by Age and Gender

Source: American Heart Association

7. **a.** What is the probability that a randomly selected man between the ages of 25 and 34 has cardiovascular disease? 0.10

   b. What is the probability that a randomly selected man between the ages of 25 and 34 does not have cardiovascular disease? 0.90

8. **a.** What is the probability that a randomly selected woman, 75 or older, has cardiovascular disease? 0.78

   b. What is the probability that a randomly selected woman, 75 or older, does not have cardiovascular disease? 0.22

The table shows the distribution, by age, of a random sample of 3000 American moviegoers ages 12 through 74. Use this distribution to solve Exercises 9–12.

Age Distribution of the U.S. Moviegoer Audience

Ages	Number
12–24	900
25–44	1080
45–64	840
65–74	180

Source: Nielsen survey of 3000 American moviegoers ages 12–74

If one moviegoer is randomly selected from this population, find the probability, expressed as a simplified fraction, that

9. the moviegoer is not in the 25–44 age range. $\frac{16}{25}$

10. the moviegoer is not in the 45–64 age range. $\frac{18}{25}$

11. the moviegoer's age is less than 65. $\frac{47}{50}$

12. the moviegoer's age is at least 25. $\frac{7}{10}$

In Exercises 13–18, you randomly select one card from a 52-card deck. Find the probability of selecting

13. a 2 or a 3. $\frac{2}{13}$

14. a 7 or an 8. $\frac{2}{13}$

15. a red 2 or a black 3. $\frac{1}{13}$

16. a red 7 or a black 8. $\frac{1}{13}$

17. the 2 of hearts or the 3 of spades. $\frac{1}{26}$

18. the 7 of hearts or the 8 of spades. $\frac{1}{26}$

19. The mathematics faculty at a college consists of 8 professors, 12 associate professors, 14 assistant professors, and 10 instructors. If one faculty member is randomly selected, find the probability of choosing a professor or an instructor.

20. A political discussion group consists of 30 Republicans, 25 Democrats, 8 Independents, and 4 members of the Green party. If one person is randomly selected from the group, find the probability of choosing an Independent or a member of the Green party. $\frac{12}{67}$ **19.** $\frac{9}{22}$

In Exercises 21–22, a single die is rolled. Find the probability of rolling

21. an even number or a number less than 5. $\frac{5}{6}$

22. an odd number or a number less than 4. $\frac{2}{3}$

In Exercises 23–26, you are dealt one card from a 52-card deck. Find the probability that you are dealt

23. a 7 or a red card. $\frac{7}{13}$

24. a 5 or a black card. $\frac{7}{13}$

25. a heart or a picture card. $\frac{11}{26}$

26. a card greater than 2 and less than 7, or a diamond. $\frac{25}{52}$

In Exercises 27–30, it is equally probable that the pointer on the spinner shown will land on any one of the eight regions, numbered 1 through 8. If the pointer lands on a borderline, spin again.

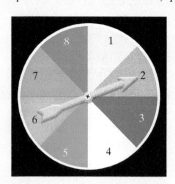

Find the probability that the pointer will stop on

27. an odd number or a number less than 6. $\frac{3}{4}$

28. an odd number or a number greater than 3. $\frac{7}{8}$

29. an even number or a number greater than 5. $\frac{5}{8}$

30. an even number or a number less than 4. $\frac{3}{4}$

Use this information to solve Exercises 31–34. The mathematics department of a college has 8 male professors, 11 female professors, 14 male teaching assistants, and 7 female teaching assistants. If a person is selected at random from the group, find the probability that the selected person is

31. a professor or a male. $\frac{33}{40}$

32. a professor or a female. $\frac{13}{20}$

33. a teaching assistant or a female. $\frac{4}{5}$

34. a teaching assistant or a male. $\frac{29}{40}$

35. In a class of 50 students, 29 are Democrats, 11 are business majors, and 5 of the business majors are Democrats. If one student is randomly selected from the class, find the probability of choosing a Democrat or a business major. $\frac{7}{10}$

36. A student is selected at random from a group of 200 students in which 135 take math, 85 take English, and 65 take both math and English. Find the probability that the selected student takes math or English. $\frac{31}{40}$

The table shows the educational attainment of the U.S. population, ages 25 and over, for a recent year. Use the data in the table, expressed in millions, to solve Exercises 37–44.

Educational Attainment of the U.S. Population, Ages 25 and Over, in Millions

	Less Than 4 Years High School	4 Years High School Only	Some College (Less than 4 years)	4 Years College (or More)	Total
Male	14	25	20	23	82
Female	15	31	24	22	92
Total	29	56	44	45	174

Source: U.S. Census Bureau

Find the probability, expressed as a simplified fraction, that a randomly selected American, aged 25 or over,

37. has not completed four years (or more) of college. $\frac{43}{58}$

38. has not completed four years of high school. $\frac{1}{6}$

39. has completed four years of high school only or less than four years of college. $\frac{50}{87}$

40. has completed less than four years of high school or four years of high school only. $\frac{85}{174}$

41. has completed four years of high school only or is a man. $\frac{113}{174}$

42. has completed four years of high school only or is a woman. $\frac{39}{58}$

Find the odds in favor and the odds against a randomly selected American, aged 25 and over, with

43. four years (or more) of college. 15:43; 43:15

44. less than four years of high school. 1:5; 5:1

The graph shows the distribution, by branch and gender, of the 1.42 million, or 1420 thousand, active-duty personnel in the U.S. military in 2011. Numbers are given in thousands and rounded to the nearest ten thousand. Use the data to solve Exercises 45–56.

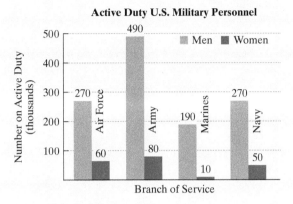

Active Duty U.S. Military Personnel

Source: U.S. Defense Department

If one person is randomly selected from the population represented in the bar graph, find the probability, expressed as a simplified fraction, that the person

45. is not in the Army. $\frac{85}{142}$ **46.** is not in the Marines. $\frac{61}{71}$

47. is in the Navy or is a man. **48.** is in the Army or is a woman.

49. is in the Air Force or the Marines. $\frac{53}{142}$ **47.** $\frac{127}{142}$ **48.** $\frac{69}{142}$

50. is in the Army or the Navy. $\frac{89}{142}$

Find the odds in favor and the odds against a randomly selected person from the population represented in the bar graph above being

51. in the Navy. 16:55; 55:16 **52.** in the Army. 57:85; 85:57

53. a woman in the Marines. **54.** a woman in the Air Force.

55. a man. 61:10; 10:61 **53.** 1:141; 141:1 **54.** 3:68; 68:3

56. a woman. 10:61; 61:10

In Exercises 57–60, a single die is rolled. Find the odds

57. in favor of rolling a number greater than 2. 2:1

58. in favor of rolling a number less than 5. 2:1

59. against rolling a number greater than 2. 1:2

60. against rolling a number less than 5. 1:2

The circle graphs show the percentage of children in the United States whose parents are college graduates in one-parent households and two-parent households. Use the information shown to solve Exercises 61–62.

**Percentage of U.S. Children
Whose Parents Are College Graduates**

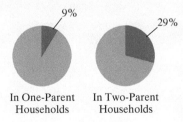

In One-Parent In Two-Parent
Households Households

Source: U.S. Census Bureau

61. a. What are the odds in favor of a child in a one-parent household having a parent who is a college graduate? 9:91

b. What are the odds against a child in a one-parent household having a parent who is a college graduate? 91:9

62. a. What are the odds in favor of a child in a two-parent household having parents who are college graduates? 29:71

b. What are the odds against a child in a two-parent household having parents who are college graduates? 71:29

In Exercises 63–72, one card is randomly selected from a deck of cards. Find the odds

63. in favor of drawing a heart. 1:3

64. in favor of drawing a picture card. 3:10

65. in favor of drawing a red card. 1:1

66. in favor of drawing a black card. 1:1

67. against drawing a 9. 12:1

68. against drawing a 5. 12:1

69. against drawing a black king. 25:1

70. against drawing a red jack. 25:1

71. against drawing a spade greater than 3 and less than 9. 47:5

72. against drawing a club greater than 4 and less than 10. 47:5

73. The winner of a raffle will receive a 21-foot outboard boat. If 1000 raffle tickets were sold and you purchased 20 tickets, what are the odds against your winning the boat? 49:1

74. The winner of a raffle will receive a 30-day all-expenses-paid trip throughout Europe. If 5000 raffle tickets were sold and you purchased 30 tickets, what are the odds against your winning the trip? 497:3

Of the 38 plays attributed to Shakespeare, 18 are comedies, 10 are tragedies, and 10 are histories. In Exercises 75–82, one play is randomly selected from Shakespeare's 38 plays. Find the odds

75. in favor of selecting a comedy. 9:10

76. in favor of selecting a tragedy. 5:14

77. against selecting a history. 14:5

78. against selecting a comedy. 10:9

79. in favor of selecting a comedy or a tragedy. 14:5

80. in favor of selecting a tragedy or a history. 10:9

81. against selecting a tragedy or a history. 9:10

82. against selecting a comedy or a history. 5:14

83. If you are given odds of 3 to 4 in favor of winning a bet, what is the probability of winning the bet? $\frac{3}{7}$

84. If you are given odds of 3 to 7 in favor of winning a bet, what is the probability of winning the bet? $\frac{3}{10}$

85. Based on his skills in basketball, it was computed that when Michael Jordan shot a free throw, the odds in favor of his making it were 21 to 4. Find the probability that when Michael Jordan shot a free throw, he missed it. Out of every 100 free throws he attempted, on the average how many did he make? $\frac{4}{25}$; 84

86. The odds in favor of a person who is alive at age 20 still being alive at age 70 are 193 to 270. Find the probability that a person who is alive at age 20 will still be alive at age 70. *

Exercises 87–88 give the odds against various flight risks. Use these odds to determine the probability of the underlined event for those in flight. (Source: Men's Health)

87. odds against <u>contracting an airborne disease</u>: 999 to 1 $\dfrac{1}{1000}$

88. odds against <u>deep-vein thrombosis</u> (blood clot in the leg): 28 to 1 $\dfrac{1}{29}$

Critical Thinking Exercises

89. In Exercise 35, find the probability of choosing
   a. a Democrat who is not a business major;
   b. a student who is neither a Democrat nor a business major.
   a. $\dfrac{12}{25}$ **b.** $\dfrac{3}{10}$

90. On New Year's Eve, the probability of a person driving while intoxicated or having a driving accident is 0.35. If the probability of driving while intoxicated is 0.32 and the probability of having a driving accident is 0.09, find the probability of a person having a driving accident while intoxicated.   0.06

91. The formula for converting from odds to probability is given in the box on page 466. Read the paragraph on the bottom of page 465 that precedes this box and derive the formula.   *

*See Answers to Selected Exercises.

7.7 Events Involving *And*; Conditional Probability

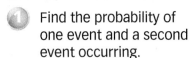

What am I supposed to learn?

After you have read this section, you should be able to:

 Find the probability of one event and a second event occurring.

② Compute conditional probabilities.

You are considering a job offer in South Florida. You were thrilled by images of Miami on MTV's *The Real World*. The job offer is just what you wanted and you are excited about living in the midst of Miami's tropical diversity. However, there is just one thing: the risk of hurricanes. You expect to stay in Miami ten years and buy a home. What is the probability that South Florida will be hit by a hurricane at least once in the next ten years?

In this section, we look at the probability that an event occurs at least once by expanding our discussion of probability to events involving *and*.

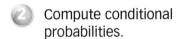 **Find the probability of one event and a second event occurring.**

And Probabilities with Independent Events

Consider tossing a fair coin two times in succession. The outcome of the first toss, heads or tails, does not affect what happens when you toss the coin a second time. For example, the occurrence of tails on the first toss does not make tails more likely or less likely to occur on the second toss. The repeated toss of a coin produces *independent events* because the outcome of one toss does not affect the outcome of others.

> ### Great Question!
>
> **What's the difference between *independent events* and *mutually exclusive events*?**
>
> Mutually exclusive events cannot occur at the same time. Independent events occur at different times, although they have no effect on each other.

> ### Independent Events
>
> Two events are **independent events** if the occurrence of either of them has no effect on the probability of the other.

When a fair coin is tossed two times in succession, the set of equally likely outcomes is

{ heads heads, heads tails, tails heads, tails tails } .

We can use this set to find the probability of getting heads on the first toss and heads on the second toss:

$$P(\text{heads and heads}) = \frac{\text{number of ways two heads can occur}}{\text{total number of possible outcomes}} = \frac{1}{4}.$$

We can also determine the probability of two heads, $\frac{1}{4}$, without having to list all the equally likely outcomes. The probability of heads on the first toss is $\frac{1}{2}$. The probability of heads on the second toss is also $\frac{1}{2}$. The product of these probabilities, $\frac{1}{2} \cdot \frac{1}{2}$, results in the probability of two heads, namely $\frac{1}{4}$. Thus,

$$P(\text{heads and heads}) = P(\text{heads}) \cdot P(\text{heads}).$$

In general, if two events are independent, we can calculate the probability of the first occurring and the second occurring by multiplying their probabilities.

> ### *And* Probabilities with Independent Events
>
> If A and B are independent events, then
>
> $$P(A \text{ and } B) = P(A) \cdot P(B).$$

Example 1   Independent Events on a Spinner

Figure 7.11 illustrates a spinner. It is equally probable that the pointer will land on any one of the eight regions, numbered 1 through 8. If the pointer lands on a borderline, spin again. Find the probability that the pointer will stop on a number less than 3 on two consecutive spins.

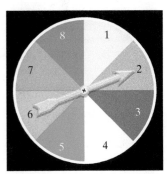

Figure 7.11 It is equally probable that the pointer will land on any one of the eight regions.

Solution

The spinner has eight equally likely outcomes and two of the outcomes, 1 and 2, are less than 3. Thus, the probability of landing on a number less than 3 on a spin is $\frac{2}{8}$, or $\frac{1}{4}$. The result that occurs on each spin is independent of all previous results. Thus,

$$P(\text{less than 3 and less than 3}) = P(\text{less than 3}) \cdot P(\text{less than 3}) = \frac{1}{4} \cdot \frac{1}{4} = \frac{1}{16}.$$

The probability that the pointer will stop on a number less than 3 on two consecutive spins is $\frac{1}{16}$.

Some people incorrectly believe that if a number less than 3 occurs on two consecutive spins, then a number greater than or equal to 3 is "due." Because the events are independent, the outcomes of previous spins have no effect on any other spins.

✓ **Check Point 1**   Use **Figure 7.11** to find the probability that the pointer will stop on a number greater than 5 on two consecutive spins. $\frac{9}{64}$

The *and* rule for independent events can be extended to cover three or more independent events. Thus, if A, B, and C are independent events, then

$$P(A \text{ and } B \text{ and } C) = P(A) \cdot P(B) \cdot P(C).$$

Example 2 Independent Events in a Family

The picture in the margin shows a family that had nine girls in a row. Find the probability of this occurrence.

Solution

If two or more events are independent, we can find the probability of them all occurring by multiplying their probabilities. The probability of a baby girl is $\frac{1}{2}$, so the probability of nine girls in a row is $\frac{1}{2}$ used as a factor nine times.

$$P(\text{nine girls in a row}) = \frac{1}{2} \cdot \frac{1}{2} \cdot \frac{1}{2} \cdot \frac{1}{2} \cdot \frac{1}{2} \cdot \frac{1}{2} \cdot \frac{1}{2} \cdot \frac{1}{2} \cdot \frac{1}{2}$$

$$= \left(\frac{1}{2}\right)^9 = \frac{1}{512}$$

The probability of a run of nine girls in a row is $\frac{1}{512}$. (If another child is born into the family, this event is independent of the other nine and the probability of a girl is still $\frac{1}{2}$.)

▶ ✔ **Check Point 2** Find the probability of a family having four boys in a row. $\frac{1}{16}$

Table 7.6 The Saffir/Simpson Hurricane Scale

Category	Winds (Miles per Hour)
1	74–95
2	96–110
3	111–130
4	131–155
5	> 155

Now let us return to the hurricane problem that opened this section. The Saffir/Simpson scale assigns numbers 1 through 5 to measure the disaster potential of a hurricane's winds. **Table 7.6** describes the scale. According to the National Hurricane Center, the probability that South Florida will be hit by a category 1 hurricane or higher in any single year is $\frac{5}{19}$, or approximately 0.26. In Example 3, we explore the risks of living in "Hurricane Alley."

Example 3 Hurricanes and Probabilities

If the probability that South Florida will be hit by a hurricane in any single year is $\frac{5}{19}$,

a. What is the probability that South Florida will be hit by a hurricane in three consecutive years?

b. What is the probability that South Florida will not be hit by a hurricane in the next ten years?

Solution

a. The probability that South Florida will be hit by a hurricane in three consecutive years is

$$P(\text{hurricane and hurricane and hurricane})$$

$$= P(\text{hurricane}) \cdot P(\text{hurricane}) \cdot P(\text{hurricane}) = \frac{5}{19} \cdot \frac{5}{19} \cdot \frac{5}{19} = \frac{125}{6859} \approx 0.018.$$

b. We will first find the probability that South Florida will not be hit by a hurricane in any single year.

$$P(\text{no hurricane}) = 1 - P(\text{hurricane}) = 1 - \frac{5}{19} = \frac{19}{19} - \frac{5}{19} = \frac{14}{19} \approx 0.737$$

The probability of not being hit by a hurricane in a single year is $\frac{14}{19}$. Therefore, the probability of not being hit by a hurricane ten years in a row is $\frac{14}{19}$ used as a factor ten times.

P(no hurricanes for ten years)

$$= P\binom{\text{no hurricane}}{\text{for year 1}} \cdot P\binom{\text{no hurricane}}{\text{for year 2}} \cdot P\binom{\text{no hurricane}}{\text{for year 3}} \cdot \ldots \cdot P\binom{\text{no hurricane}}{\text{for year 10}}$$

$$= \frac{14}{19} \quad \cdot \quad \frac{14}{19} \quad \cdot \quad \frac{14}{19} \quad \cdot \ldots \cdot \quad \frac{14}{19}$$

$$= \left(\frac{14}{19}\right)^{10} \approx (0.737)^{10} \approx 0.047$$

The probability that South Florida will not be hit by a hurricane in the next ten years is approximately 0.047.

Now we are ready to answer your question from the section opener:

What is the probability that South Florida will be hit by a hurricane at least once in the next ten years?

Because $P(\text{not } E) = 1 - P(E)$,

P(no hurricane for ten years) $= 1 - P$(at least one hurricane in ten years).

Equivalently,

P(at least one hurricane in ten years) $= 1 - P$(no hurricane for ten years)

$$= 1 - 0.047 = 0.953.$$

With a probability of 0.953, it is nearly certain that South Florida will be hit by a hurricane at least once in the next ten years.

> **The Probability of an Event Happening at Least Once**
>
> $P(\text{event happening at least once}) = 1 - P(\text{event does not happen})$

Check Point 3 If the probability that South Florida will be hit by a hurricane in any single year is $\frac{5}{19}$,

a. What is the probability that South Florida will be hit by a hurricane in four consecutive years? $\frac{625}{130,321} \approx 0.005$

b. What is the probability that South Florida will not be hit by a hurricane in the next four years? $\frac{38,416}{130,321} \approx 0.295$

c. What is the probability that South Florida will be hit by a hurricane at least once in the next four years? $\frac{91,905}{130,321} \approx 0.705$

Express all probabilities as fractions and as decimals rounded to three places.

And Probabilities with Dependent Events

Chocolate lovers, please help yourselves! There are 20 mouth-watering tidbits to select from. What's that? You want 2? And you prefer chocolate-covered cherries? The problem is that there are only 5 chocolate-covered cherries and it's impossible to tell what is inside each piece. They're all shaped exactly alike. At any rate, reach in, select a piece, enjoy, choose another piece, eat, and be well. There is nothing like savoring a good piece of chocolate in the midst of all this chit-chat about probability and hurricanes.

Great Question!

When solving probability problems, how do I decide whether to use the *or* formulas or the *and* formulas?

• *Or* problems usually have the word *or* in the statement of the problem. These problems involve only one selection.

Example:
If one person is selected, find the probability of selecting a man or a Canadian.

• *And* problems often do not have the word *and* in the statement of the problem. These problems involve more than one selection.

Example:
If two people are selected, find the probability that both are men.

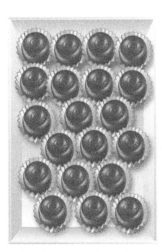

5 chocolate-covered cherries lie within the 20 pieces.

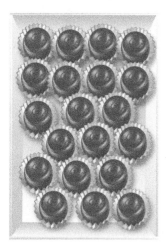

5 chocolate-covered cherries lie within the 20 pieces.

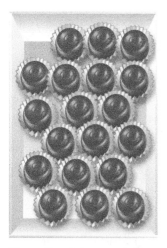

Once a chocolate-covered cherry is selected, only 4 chocolate-covered cherries lie within the remaining 19 pieces.

Another question? You want to know what your chances are of selecting 2 chocolate-covered cherries? Well, let's see. Five of the 20 pieces are chocolate-covered cherries, so the probability of getting one of them on your first selection is $\frac{5}{20}$, or $\frac{1}{4}$. Now, suppose that you did choose a chocolate-covered cherry on your first pick. Eat it slowly; there's no guarantee that you'll select your favorite on the second selection. There are now only 19 pieces of chocolate left. Only 4 are chocolate-covered cherries. The probability of getting a chocolate-covered cherry on your second try is 4 out of 19, or $\frac{4}{19}$. This is a different probability than the $\frac{1}{4}$ probability on your first selection. Selecting a chocolate-covered cherry the first time changes what is in the candy box. The probability of what you select the second time *is* affected by the outcome of the first event. For this reason, we say that these are *dependent events*.

Dependent Events

Two events are **dependent events** if the occurrence of one of them has an effect on the probability of the other.

The probability of selecting two chocolate-covered cherries in a row can be found by multiplying the $\frac{1}{4}$ probability on the first selection by the $\frac{4}{19}$ probability on the second selection:

P(chocolate-covered cherry and chocolate-covered cherry)

$$= P(\text{chocolate-covered cherry}) \cdot P\left(\begin{array}{c}\text{chocolate-covered cherry}\\ \text{given that one was selected}\end{array}\right)$$

$$= \frac{1}{4} \cdot \frac{4}{19} = \frac{1}{19}.$$

The probability of selecting two chocolate-covered cherries in a row is $\frac{1}{19}$. This is a special case of finding the probability that each of two dependent events occurs.

And Probabilities with Dependent Events

If A and B are dependent events, then

$$P(A \text{ and } B) = P(A) \cdot P(B \text{ given that } A \text{ has occurred}).$$

Example 4 An *And* Probability with Dependent Events

Good news: You won a free trip to Madrid and can take two people with you, all expenses paid. Bad news: Ten of your cousins have appeared out of nowhere and are begging you to take them. You write each cousin's name on a card, place the cards in a hat, and select one name. Then you select a second name without replacing the first card. If three of your ten cousins speak Spanish, find the probability of selecting two Spanish-speaking cousins.

Great Question!

You solved Example 4 using the *and* probability formula with dependent events. Because cousins are being *selected*, can I also solve the problem using the combinations formula?

Yes. Here's how it works:

P(two Spanish speakers)

$$= \frac{\text{number of ways of selecting 2 Spanish-speaking cousins}}{\text{number of ways of selecting 2 cousins}}$$

$$= \frac{{}_3C_2}{{}_{10}C_2}$$

> 2 Spanish speakers selected from 3 Spanish-speaking cousins

> 2 cousins selected from 10 cousins

$$= \frac{3}{45} = \frac{1}{15}$$

Solution

Because $P(A \text{ and } B) = P(A) \cdot P(B$ given that A has occurred), then

P(two Spanish-speaking cousins)

$= P$(speaks Spanish and speaks Spanish)

$= P(\text{speaks Spanish}) \cdot P\left(\begin{array}{c}\text{speaks Spanish given that a Spanish-speaking}\\ \text{cousin was selected first}\end{array}\right)$

$$= \frac{3}{10} \cdot \frac{2}{9}$$

> There are ten cousins, three of whom speak Spanish.

> After picking a Spanish-speaking cousin, there are nine cousins left, two of whom speak Spanish.

$$= \frac{6}{90} = \frac{1}{15} \approx 0.067.$$

The probability of selecting two Spanish-speaking cousins is $\frac{1}{15}$.

✓ **Check Point 4**   You are dealt two cards from a 52-card deck. Find the probability of getting two kings. $\frac{1}{221} \approx 0.00452$

The multiplication rule for dependent events can be extended to cover three or more dependent events. For example, in the case of three such events,

$P(A \text{ and } B \text{ and } C)$

$= P(A) \cdot P(B$ given that A occurred) $\cdot P(C$ given that A and B occurred).

Example 5   **An *And* Probability with Three Dependent Events**

Three people are randomly selected, one person at a time, from five freshmen, two sophomores, and four juniors. Find the probability that the first two people selected are freshmen and the third is a junior.

Solution

P(first two are freshmen and the third is a junior)

$= P(\text{freshman}) \cdot P\left(\begin{array}{c}\text{freshman given that a}\\ \text{freshman was selected first}\end{array}\right) \cdot P\left(\begin{array}{c}\text{junior given that a freshman was}\\ \text{selected first and a freshman was}\\ \text{selected second}\end{array}\right)$

$$= \frac{5}{11} \cdot \frac{4}{10} \cdot \frac{4}{9}$$

> There are 11 people, five of whom are freshmen.

> After picking a freshman, there are 10 people left, four of whom are freshmen.

> After the first two selections, 9 people are left, four of whom are juniors.

$$= \frac{8}{99}$$

The probability that the first two people selected are freshmen and the third is a junior is $\frac{8}{99}$.

Compute conditional probabilities.

Blitzer Bonus
••••••••••••••

Coincidences

The phone rings and it's the friend you were just thinking of. You're driving down the road and a song you were humming in your head comes on the radio. Although these coincidences seem strange, perhaps even mystical, they're not. Coincidences are bound to happen. Ours is a world in which there are a great many potential coincidences, each with a low probability of occurring. When these surprising coincidences happen, we are amazed and remember them. However, we pay little attention to the countless number of non-coincidences: How often do you think of your friend and she doesn't call, or how often does she call when you're not thinking about her? By noticing the hits and ignoring the misses, we incorrectly perceive that there is a relationship between the occurrence of two independent events.

   Another problem is that we often underestimate the probabilities of coincidences in certain situations, acting with more surprise than we should when they occur. For example, in a group of only 23 people, the probability that two individuals share a birthday (same month and day) is greater than $\frac{1}{2}$. Above 50 people, the probability of any two people sharing a birthday approaches certainty. You can verify the probabilities behind the coincidence of shared birthdays in relatively small groups by working Exercise 73 in Exercise Set 7.7.

Check Point 5   You are dealt three cards from a 52-card deck. Find the probability of getting three hearts. $\frac{11}{850} \approx 0.0129$

Conditional Probability

We have seen that for any two dependent events A and B,

$$P(A \text{ and } B) = P(A) \cdot P(B \text{ given that } A \text{ occurs}).$$

The probability of B given that A occurs is called *conditional probability,* denoted by $P(B \mid A)$.

> **Conditional Probability**
>
> The probability of event B, assuming that event A has already occurred, is called the **conditional probability** of B, given A. This probability is denoted by $P(B \mid A)$.

   It is helpful to think of the conditional probability $P(B \mid A)$ as the **probability that event B occurs if the sample space is restricted to the outcomes associated with event A**.

Example 6   Finding Conditional Probability

A letter is randomly selected from the letters of the English alphabet. Find the probability of selecting a vowel, given that the outcome is a letter that precedes h.

Solution

We are looking for

$$P(\text{vowel} \mid \text{letter precedes h}).$$

This is the probability of a vowel if the sample space is restricted to the set of letters that precede h. Thus, the sample space is given by

$$S = \{\, a, b, c, d, e, f, g \,\}.$$

There are seven possible outcomes in the sample space. We can select a vowel from this set in one of two ways: a or e. Therefore, the probability of selecting a vowel, given that the outcome is a letter that precedes h, is $\frac{2}{7}$.

$$P(\text{vowel} \mid \text{letter precedes h}) = \tfrac{2}{7}$$

Check Point 6   A letter is randomly selected from the letters of the English alphabet. Find the probability of selecting a letter that precedes h, given that the outcome is a vowel. (Do not include the letter y among the vowels.) $\frac{2}{5}$

Example 7   Finding Conditional Probability

You are dealt one card from a 52-card deck.

a. Find the probability of getting a heart, given that the card you were dealt is a red card.

b. Find the probability of getting a red card, given that the card you were dealt is a heart.

Solution

a. We begin with

$$P(\text{heart} \mid \text{red card}).$$

Probability of getting a heart if the sample space is restricted to the set of red cards

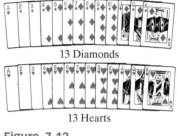

13 Diamonds

13 Hearts

Figure 7.12

The sample space is shown in **Figure 7.12**. There are 26 outcomes in the sample space. We can get a heart from this set in 13 ways. Thus,

$$P(\text{heart} \mid \text{red card}) = \frac{13}{26} = \frac{1}{2}.$$

b. We now find

$$P(\text{red card} \mid \text{heart}).$$

Probability of getting a red card if the sample space is restricted to the set of hearts

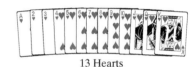

13 Hearts

Figure 7.13

The sample space is shown in **Figure 7.13**. There are 13 outcomes in the sample space. All of the outcomes are red. We can get a red card from this set in 13 ways. Thus,

$$P(\text{red card} \mid \text{heart}) = \frac{13}{13} = 1.$$

Example 7 illustrates that $P(\text{heart} \mid \text{red card})$ is not equal to $P(\text{red card} \mid \text{heart})$. In general, $P(B \mid A) \neq P(A \mid B)$.

Check Point 7 You are dealt one card from a 52-card deck.

a. Find the probability of getting a black card, given the card you were dealt is a spade. 1

b. Find the probability of getting a spade, given the card you were dealt is a black card. $\frac{1}{2}$

Example 8 Conditional Probabilities with Real-World Data

When women turn 40, their gynecologists typically remind them that it is time to undergo mammography screening for breast cancer. The data in **Table 7.7** are based on 100,000 U.S. women, ages 40 to 49, who participated in mammography screening.

Table 7.7 Mammography Screening on 100,000 U.S. Women, Ages 40 to 49			
	Breast Cancer	**No Breast Cancer**	**Total**
Positive Mammogram	720	6944	7664
Negative Mammogram	80	92,256	92,336
Total	800	99,200	100,000

Source: Gerd Gigerenzer, *Calculated Risks.* Simon and Schuster, 2002.

| | No Breast Cancer |
	Breast Cancer	Breast Cancer
Positive Mammogram	720	6944
Negative Mammogram	80	92,256

Table 7.7 (partly repeated)

Assuming that these numbers are representative of all U.S. women ages 40 to 49, find the probability that a woman in this age range

a. has a positive mammogram, given that she does not have breast cancer.

b. does not have breast cancer, given that she has a positive mammogram.

Solution

a. We begin with the probability that a U.S. woman aged 40 to 49 has a positive mammogram, given that she does not have breast cancer:

$$P(\text{positive mammogram} \mid \text{no breast cancer}).$$

This is the probability of a positive mammogram if the data are restricted to women without breast cancer:

	No Breast Cancer
Positive Mammogram	6944
Negative Mammogram	92,256
Total	99,200

Within the restricted data, there are 6944 women with positive mammograms and 6944 + 92,256, or 99,200 women without breast cancer. Thus,

$$P(\text{positive mammogram} \mid \text{no breast cancer}) = \frac{6944}{99,200} = 0.07.$$

Among women without breast cancer, the probability of a positive mammogram is 0.07.

b. Now, we find the probability that a U.S. woman aged 40 to 49 does not have breast cancer, given that she has a positive mammogram:

$$P(\text{no breast cancer} \mid \text{positive mammogram}).$$

This is the probability of not having breast cancer if the data are restricted to women with positive mammograms:

	Breast Cancer	No Breast Cancer	Total
Positive Mammogram	720	6944	7664

Within the restricted data, there are 6944 women without breast cancer and 720 + 6944, or 7664 women with positive mammograms. Thus,

$$P(\text{no breast cancer} \mid \text{positive mammogram}) = \frac{6944}{7664} \approx 0.906.$$

Among women with positive mammograms, the probability of not having breast cancer is $\frac{6944}{7664}$, or approximately 0.906.

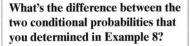

Great Question!

What's the difference between the two conditional probabilities that you determined in Example 8?

Example 8 shows that the probability of a positive mammogram among women without breast cancer, 0.07, is not the same as the probability of not having breast cancer among women with positive mammograms, approximately 0.9. We have seen that the conditional probability of B, given A, is not the same as the conditional probability of A, given B:

$$P(B \mid A) \neq P(A \mid B).$$

The conditional probability in Example 8(b) indicates a probability of approximately 0.9 that a woman aged 40 to 49 who has a positive mammogram is actually cancer-free. The likely probability of this false positive caused a rift between the U.S. Preventive Services Task Force and the American Cancer Society. The Task Force issued a report that indicated most women don't need to get mammograms until they reach 50. The American Cancer Society voiced its displeasure with this recommendation, stating that "the Task Force is essentially telling women that mammography at age 40 to 49 saves lives; just not enough of them."

Check Point 8 Use the data in **Table 7.7** at the bottom of page 477 to find the probability that a U.S. woman aged 40 to 49

a. $\frac{9}{10} = 0.9$

b. $\frac{45}{479} \approx 0.094$

a. has a positive mammogram, given that she has breast cancer.

b. has breast cancer, given that she has a positive mammogram.

Express probabilities as decimals and, if necessary, round to three decimal places.

We have seen that $P(B\,|A)$ is the probability that event B occurs if the sample space is restricted to event A. Thus,

$$P(B\,|A) = \frac{\text{number of outcomes of } B \text{ that are in the restricted sample space } A}{\text{number of outcomes in the restricted sample space } A}.$$

This can be stated in terms of the following formula:

A Formula for Conditional Probability

$$P(B\,|A) = \frac{n(B \cap A)}{n(A)} = \frac{\text{number of outcomes common to } B \text{ and } A}{\text{number of outcomes in } A}$$

Achieving Success

Be sure to use the Section Lecture Videos for each chapter. These interactive lectures highlight key examples from every section of the book.

Are you using any of the other textbook supplements for help and additional study? These include:

- The Student Solutions Manual. This contains fully worked solutions to the odd-numbered section exercises plus all Check Points, Concept and Vocabulary Checks, Chapter Reviews, and Chapter Tests.
- The MyMathLab course is a text-specific online homework and assessment system to help you master the material. Ask your instructor whether these are available to you.

Concept and Vocabulary Check

Exercises in the Concept and Vocabulary Check are intended for group and class discussions.

In Exercises 1–4, fill in each blank so that the resulting statement is true.

1. If the occurrence of one event has no effect on the probability of another event, the events are said to be ____independent____. For such events, $P(A \text{ and } B) = $ ____$P(A) \cdot P(B)$____.

2. The probability of an event occurring at least once is equal to 1 minus the probability that ___the event does not occur___

3. If the occurrence of one event has an effect on the probability of another event, the events are said to be ____dependent____. For such events, $P(A \text{ and } B) = $ ___$P(A) \cdot P(B$ given that A occurred)___

4. The probability of event B, assuming that event A has already occurred, is called the ____conditional____ probability of B, given A. This probability is denoted by ____$P(B\,|A)$____.

In Exercises 5–8, determine whether each statement is true or false. If the statement is false, make the necessary change(s) to produce a true statement. Changes to false statements will vary.

5. *And* probabilities can always be determined using the formula $P(A \text{ and } B) = P(A) \cdot P(B)$. false

6. Probability problems with the word *and* involve more than one selection. true

7. The probability that an event happens at least once can be found by subtracting the probability that the event does not happen from 1. true

8. $P(B\,|A)$ is the probability that event B occurs if the sample space is restricted to the outcomes associated with event A. true

Respond to Exercises 9–10 using verbal or written explanations. 9–10. Answers will vary.

9. Explain how to find *and* probabilities with independent events. Give an example.

10. Explain how to find *and* probabilities with dependent events. Give an example.

Exercise Set 7.7

Practice and Application Exercises

Exercises 1–26 involve probabilities with independent events.
Use the spinner shown to solve Exercises 1–10. It is equally probable that the pointer will land on any one of the six regions. If the pointer lands on a borderline, spin again. If the pointer is spun twice, find the probability it will land on

1. green and then red. $\frac{1}{6}$
2. yellow and then green. $\frac{1}{18}$
3. yellow and then yellow. $\frac{1}{36}$
4. red and then red. $\frac{1}{4}$
5. a color other than red each time. $\frac{1}{4}$
6. a color other than green each time. $\frac{4}{9}$

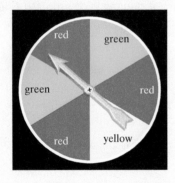

If the pointer shown above is spun three times, find the probability it will land on

7. green and then red and then yellow. $\frac{1}{36}$
8. red and then red and then green. $\frac{1}{12}$
9. red every time. $\frac{1}{8}$
10. green every time. $\frac{1}{27}$

In Exercises 11–14, a single die is rolled twice. Find the probability of rolling

11. a 2 the first time and a 3 the second time. $\frac{1}{36}$
12. a 5 the first time and a 1 the second time. $\frac{1}{36}$
13. an even number the first time and a number greater than 2 the second time. $\frac{1}{3}$
14. an odd number the first time and a number less than 3 the second time. $\frac{1}{6}$

In Exercises 15–20, you draw one card from a 52-card deck. Then the card is replaced in the deck, the deck is shuffled, and you draw again. Find the probability of drawing

15. a picture card the first time and a heart the second time. $\frac{3}{52}$
16. a jack the first time and a club the second time. $\frac{1}{52}$
17. a king each time. $\frac{1}{169}$
18. a 3 each time. $\frac{1}{169}$
19. a red card each time. $\frac{1}{4}$
20. a black card each time. $\frac{1}{4}$

*See Answers to Selected Exercises.

21. If you toss a fair coin six times, what is the probability of getting all heads? $\frac{1}{64}$
22. If you toss a fair coin seven times, what is the probability of getting all tails? $\frac{1}{128}$

In Exercises 23–24, a coin is tossed and a die is rolled. Find the probability of getting

23. a head and a number greater than 4. $\frac{1}{6}$
24. a tail and a number less than 5. $\frac{1}{3}$
25. The probability that South Florida will be hit by a major hurricane (category 4 or 5) in any single year is $\frac{1}{16}$.

    (*Source:* National Hurricane Center)

    a. What is the probability that South Florida will be hit by a major hurricane two years in a row? *
    b. What is the probability that South Florida will be hit by a major hurricane in three consecutive years? *
    c. What is the probability that South Florida will not be hit by a major hurricane in the next ten years? ≈ 0.524
    d. What is the probability that South Florida will be hit by a major hurricane at least once in the next ten years? *
26. The probability that a region prone to flooding will flood in any single year is $\frac{1}{10}$.

    a. What is the probability of a flood two years in a row?
    b. What is the probability of flooding in three consecutive years? $\frac{1}{1000} = 0.001$ **26. a.** $\frac{1}{100} = 0.01$
    c. What is the probability of no flooding for ten consecutive years? ≈ 0.349
    d. What is the probability of flooding at least once in the next ten years? ≈ 0.651

The graph shows that U.S. adults dependent on tobacco have a greater probability of suffering from some ailments than the general adult population. When making two or more selections from populations with large numbers, such as the U.S. adult population or the population dependent on tobacco, we assume that each selection is independent of every other selection. In Exercises 27–32, assume that the selections are independent events.

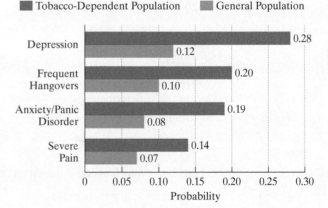

Probability That U.S. Adults Suffer from Various Ailments

Source: MARS 2005 OTC/DTC

27. If two adults are randomly selected from the general population, what is the probability that they both suffer from depression? 0.0144

28. It two adults are randomly selected from the population of cigarette smokers, what is the probability that they both suffer from depression? 0.0784

29. If three adults are randomly selected from the population of cigarette smokers, what is the probability that they all suffer from frequent hangovers? 0.008

30. If three adults are randomly selected from the general population, what is the probability that they all suffer from frequent hangovers? 0.001

31. If three adults are randomly selected from the population of cigarette smokers, what is the probability, expressed as a decimal correct to four places, that at least one person suffers from anxiety/panic disorder? 0.4686

32. If three adults are randomly selected from the population of cigarette smokers, what is the probability, expressed as a decimal correct to four places, that at least one person suffers from severe pain? 0.3639

Exercises 33–48 involve probabilities with dependent events.

In Exercises 33–36, we return to our box of chocolates. There are 30 chocolates in the box, all identically shaped. Five are filled with coconut, 10 with caramel, and 15 are solid chocolate. You randomly select one piece, eat it, and then select a second piece. Find the probability of selecting

33. two solid chocolates in a row. $\frac{7}{29}$

34. two caramel-filled chocolates in a row. $\frac{3}{29}$

35. a coconut-filled chocolate followed by a caramel-filled chocolate. $\frac{5}{87}$

36. a coconut-filled chocolate followed by a solid chocolate. $\frac{5}{58}$

In Exercises 37–42, consider a political discussion group consisting of 5 Democrats, 6 Republicans, and 4 Independents. Suppose that two group members are randomly selected, in succession, to attend a political convention. Find the probability of selecting

37. two Democrats. $\frac{2}{21}$

38. two Republicans. $\frac{1}{7}$

39. an Independent and then a Republican. $\frac{4}{35}$

40. an Independent and then a Democrat. $\frac{2}{21}$

41. no Independents. $\frac{11}{21}$

42. no Democrats. $\frac{3}{7}$

In Exercises 43–48, an ice chest contains six cans of apple juice, eight cans of grape juice, four cans of orange juice, and two cans of mango juice. Suppose that you reach into the container and randomly select three cans in succession. Find the probability of selecting

43. three cans of apple juice. $\frac{1}{57}$

44. three cans of grape juice. $\frac{14}{285}$

45. a can of grape juice, then a can of orange juice, then a can of mango juice. $\frac{8}{855}$

46. a can of apple juice, then a can of grape juice, then a can of orange juice. $\frac{8}{285}$

47. no grape juice. $\frac{11}{57}$

48. no apple juice. $\frac{91}{285}$

In Exercises 49–56, the numbered disks shown are placed in a box and one disk is selected at random.

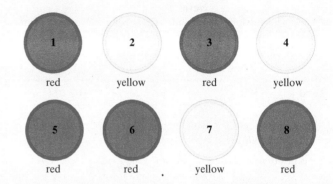

Find the probability of selecting

49. a 3, given that a red disk is selected. $\frac{1}{5}$

50. a 7, given that a yellow disk is selected. $\frac{1}{3}$

51. an even number, given that a yellow disk is selected. $\frac{2}{3}$

52. an odd number, given that a red disk is selected. $\frac{3}{5}$

53. a red disk, given that an odd number is selected. $\frac{3}{4}$

54. a yellow disk, given that an odd number is selected. $\frac{1}{4}$

55. a red disk, given that the number selected is at least 5. $\frac{3}{4}$

56. a yellow disk, given that the number selected is at most 3. $\frac{1}{3}$

The table shows the outcome of car accidents in Florida for a recent year by whether or not the driver wore a seat belt. Use the data to solve Exercises 57–60. Express probabilities as fractions and as decimals rounded to three places.

Car Accidents in Florida

	Wore Seat Belt	**No Seat Belt**	**Total**
Driver Survived	412,368	162,527	574,895
Driver Died	510	1601	2111
Total	412,878	164,128	577,006

Source: Alan Agresti and Christine Franklin, *Statistics*, Prentice Hall, 2007.

57. Find the probability of surviving a car accident, given that the driver wore a seat belt. *

58. Find the probability of not surviving a car accident, given that the driver did not wear a seat belt. *

59. Find the probability of wearing a seat belt, given that a driver survived a car accident. *

60. Find the probability of not wearing a seat belt, given that a driver did not survive a car accident. *

Shown again is the table indicating the marital status of the U.S. population in 2010. Numbers in the table are expressed in millions. Use the data in the table to solve Exercises 61–72. Express probabilities as simplified fractions and as decimals rounded to the nearest hundredth.

Marital Status of the U.S. Population, Ages 15 or Older, 2010, in Millions

	Married	Never Married	Divorced	Widowed	Total
Male	65	40	10	3	118
Female	65	34	14	11	124
Total	130	74	24	14	242

If one person is selected from the population described in the table, find the probability that the person

61. is not divorced. $\frac{109}{121} \approx 0.90$

62. is not widowed. $\frac{114}{121} \approx 0.94$

63. is widowed or divorced. $\frac{19}{121} \approx 0.16$

64. has never been married or is divorced. $\frac{49}{121} \approx 0.40$

65. is male or divorced. $\frac{6}{11} \approx 0.55$

66. is female or divorced. $\frac{67}{121} \approx 0.55$

67. is male, given that this person is divorced. $\frac{5}{12} \approx 0.42$

68. is female, given that this person is divorced. $\frac{7}{12} \approx 0.58$

69. is widowed, given that this person is a woman. $\frac{11}{124} \approx 0.09$

70. is divorced, given that this person is a man. $\frac{5}{59} \approx 0.08$

71. has never been married or is married, given that this person is a man. $\frac{105}{118} \approx 0.89$

72. has never been married or is married, given that this person is a woman. $\frac{99}{124} \approx 0.80$

73. Probabilities and Coincidence of Shared Birthdays

Use a calculator to solve this exercise. Round probabilities to three decimal places.

a. If two people are selected at random, the probability that they do not have the same birthday (day and month) is $\frac{365}{365} \cdot \frac{364}{365}$. Explain why this is so. (Ignore leap years and assume 365 days in a year.) Answers will vary.

b. If three people are selected at random, find the probability that they all have different birthdays. *

c. If three people are selected at random, find the probability that at least two of them have the same birthday. ≈ 0.008

d. If 20 people are selected at random, find the probability that at least 2 of them have the same birthday. 0.411

e. Show that if 23 people are selected at random, the probability that at least 2 of them have the same birthday is greater than $\frac{1}{2}$. *

Blitzer Bonus
● ● ● ● ● ● ● ● ● ● ● ● ● ● ● ●

More Probabilities of Shared Birthdays

- If 253 people are selected at random, the probability that at least 2 of them have the same birthday (day, month, *and year*) is approximately $\frac{1}{2}$.
- If 18 people are selected at random, the probability that at least 3 of them have the same birthday (day and month) is approximately $\frac{1}{2}$.
- If 14 people are selected at random, the probability that at least 2 of them are born within a day of each other is approximately $\frac{1}{2}$.
- If 7 people are selected at random, the probability that at least 2 of them are born within a week of each other is approximately $\frac{1}{2}$.

Source: Gina Kolata, *The New York Times Book of Mathematics,* Sterling Publishing, 2013

Critical Thinking Exercises

74. If the probability of being hospitalized during a year is 0.1, find the probability that no one in a family of five will be hospitalized in a year. 0.59049

75. If a single die is rolled five times, what is the probability it lands on 2 on the first, third, and fourth rolls, but not on either of the other rolls? $\frac{25}{7776} \approx 0.00322$

76. Nine cards numbered from 1 through 9 are placed into a box and two cards are selected without replacement. Find the probability that both numbers selected are odd, given that their sum is even. $\frac{5}{8}$

77. If a single die is rolled twice, find the probability of rolling an odd number and a number greater than 4 in either order. $\frac{11}{36}$

Group Exercises

78. Do you live in an area prone to catastrophes, such as earthquakes, fires, tornados, hurricanes, or floods? If so, research the probability of this catastrophe occurring in a single year. Group members should then use this probability to write and solve a problem similar to Exercise 25 in this Exercise Set. Answers will vary.

79. Group members should use the table for Exercises 61–72 to write and solve four probability problems different than those in the exercises. Two should involve *or* (one with events that are mutually exclusive and one with events that are not), one should involve *and*—that is, events in succession—and one should involve conditional probability. Answers will vary.

*See Answers to Selected Exercises.

Chapter 7 Summary

Definitions and Concepts	**Examples**

Section 7.1 The Fundamental Counting Principle

The Fundamental Counting Principle

The number of ways in which a series of successive things can occur is found by multiplying the number of ways in which each thing can occur.

- A car is available in 6 possible colors, with or without air conditioning, with or without a global positioning system, and electric, gas powered, or hybrid.

Color	Air conditioning	Global positioning system	Power
6 choices	2 choices: with or without	2 choices: with or without	3 choices: electric or gas or hybrid

The car can be ordered in $6 \cdot 2 \cdot 2 \cdot 3 = 72$ different ways.

Additional Examples to Review

Example 1, page 423; Example 2, page 423;
Example 3, page 424; Example 4, page 424;
Example 5, page 425; Example 6, page 425

Section 7.2 Permutations

A permutation from a group of items occurs when no item is used more than once and the order of arrangement makes a difference. The Fundamental Counting Principle can be used to determine the number of permutations possible.

- Seven acts (A, B, C, D, E, F, and G) perform in a variety show. B performs first, F performs next-to-last, and A performs last.

Act 1	Act 2	Act 3	Act 4	Act 5	Act 6	Act 7
1 choice: B	4 choices: C, D, E or G	3 choices	2 choices	1 choice	1 choice: F	1 choice: A

There are $1 \cdot 4 \cdot 3 \cdot 2 \cdot 1 \cdot 1 \cdot 1 = 24$ different ways to schedule their appearances.

Additional Examples to Review

Example 1, page 428; Example 2, page 429

Factorial Notation

$n! = n(n-1)(n-2) \cdots (3)(2)(1)$ and $0! = 1$

$$\frac{13!}{7!} = \frac{13 \cdot 12 \cdot 11 \cdot 10 \cdot 9 \cdot 8 \cdot 7!}{7!} = 13 \cdot 12 \cdot 11 \cdot 10 \cdot 9 \cdot 8$$
$$= 1{,}235{,}520$$

Additional Example to Review

Example 3, page 430

Permutations Formula

The number of permutations possible if r items are taken from n items is $_nP_r = \dfrac{n!}{(n-r)!}$.

If 16 equally qualified applicants fill five different positions, this can be done in

$$_{16}P_5 = \frac{16!}{(16-5)!} = \frac{16!}{11!} = \frac{16 \cdot 15 \cdot 14 \cdot 13 \cdot 12 \cdot 11!}{11!}$$
$$= 524{,}160 \text{ ways}$$

Additional Examples to Review

Example 4, page 432; Example 5, page 432

Permutations of Duplicate Items

The number of permutations of n items, where p items are identical, q items are identical, r items are identical, and so on, is

$$\frac{n!}{p!\,q!\,r!\,\ldots}.$$

- The number of distinct ways the letters of the word ERRATA can be arranged is

6 letters:
$n = 6$

2 identical Rs and 2 identical As

$$\frac{6!}{2!\,2!} = \frac{6 \cdot 5 \cdot 4 \cdot 3 \cdot 2 \cdot 1}{2 \cdot 1 \cdot 2 \cdot 1} = 180.$$

Additional Example to Review

Example 6, page 433

Section 7.3 Combinations

A combination from a group of items occurs when no item is used more than once and the order of items makes no difference.

Combinations Formula

The number of combinations possible if r items are taken from n items is $_nC_r = \dfrac{n!}{(n-r)!\,r!}$.

- The number of five-person committees that can be formed from nine people is

$$_9C_5 = \frac{9!}{(9-5)!5!} = \frac{9!}{4!5!} = \frac{9 \cdot 8 \cdot 7 \cdot 6 \cdot 5!}{4 \cdot 3 \cdot 2 \cdot 1 \cdot 5!} = 126.$$

Additional Examples to Review

Example 1, page 436; Example 2, page 438;
Example 3, page 438; Example 4, page 439

Section 7.4 Fundamentals of Probability

Theoretical probability applies to experiments in which the set of all equally likely outcomes, called the sample space, is known. An event is any subset of the sample space.

The theoretical probability of event E with sample space S is

$$P(E) = \frac{\text{number of outcomes in } E}{\text{total number of possible outcomes}} = \frac{n(E)}{n(S)}.$$

- 3 freshmen, 4 sophomores, 2 juniors, 1 senior

  Probability of selecting a junior:

$$P(\text{junior}) = \frac{\text{number of outcomes that result in a junior}}{\text{total number of possible outcomes}}$$

$$= \frac{2}{3+4+2+1} = \frac{2}{10} = \frac{1}{5}$$

Additional Examples to Review

Example 1, page 444; Example 2, page 445;
Example 3, page 446

Empirical probability applies to situations in which we observe the frequency of the occurrence of an event.

The empirical probability of event E is

$$P(E) = \frac{\text{observed number of times } E \text{ occurs}}{\text{total number of observed occurrences}}.$$

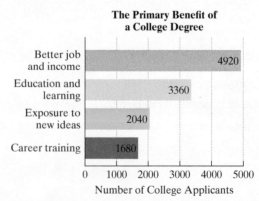

The Primary Benefit of a College Degree

Better job and income — 4920
Education and learning — 3360
Exposure to new ideas — 2040
Career training — 1680

0   1000   2000   3000   4000   5000
Number of College Applicants

Source: Princeton Review

- The bar graph is based on a survey of 12,000 college applicants and shows how the applicants responded to the question

  What is the primary benefit of a college degree?

P(selecting an applicant who believes the primary benefit is career training)

$$= \frac{\text{number who stated the primary benefit is career training}}{\text{total number of applicants}}$$

$$= \frac{1680}{12,000} = \frac{7}{50} = 0.14$$

Additional Example to Review

Example 4, page 447

Section 7.5 Probability with the Fundamental Counting Principle, Permutations, and Combinations

Probability of a permutation

$$= \frac{\text{the number of ways the permutation can occur}}{\text{total number of possible permutations}}$$

- Seven acts (A, B, C, D, E, F, and G) perform in a variety show. The order of performance is determined by random selection. What is the probability that C performs first and G performs last?

P(C performs first and G performs last)

$$= \frac{\text{number of permutations with C first and G last}}{\text{total number of possible permutations}}$$

$$= \frac{1 \; 5 \; 4 \; 3 \; 2 \; 1 \; 1}{7 \; 6 \; 5 \; 4 \; 3 \; 2 \; 1} = \frac{1 \cdot 5 \cdot 4 \cdot 3 \cdot 2 \cdot 1 \cdot 1}{7 \cdot 6 \cdot 5 \cdot 4 \cdot 3 \cdot 2 \cdot 1} = \frac{1}{7 \cdot 6} = \frac{1}{42}$$

Additional Example to Review

Example 1, page 451

Probability of a combination

$$= \frac{\text{the number of ways the combination can occur}}{\text{total number of possible combinations}}$$

A lottery game is set up so that each player chooses five different numbers from 1 to 20. If the five numbers match the five numbers drawn in the lottery, the player wins (or shares) top cash prize. The probability of winning with 100 lottery tickets is

$$\frac{\text{number of ways of winning}}{\substack{\text{total number of possible} \\ \text{combinations}}} = \frac{100}{_{20}C_5} = \frac{100}{15,504} \approx 0.0064$$

$$_{20}C_5 = \frac{20!}{(20-5)! \, 5!} = \frac{20!}{15! \, 5!} = \frac{20 \cdot 19 \cdot 18 \cdot 17 \cdot 16}{5 \cdot 4 \cdot 3 \cdot 2 \cdot 1} = 15,504$$

Additional Examples to Review

Example 2, page 453; Example 3, page 454

Section 7.6 Events Involving *Not* and *Or*; Odds

Complement Rules of Probability

$$P(\text{not } E) = 1 - P(E) \quad \text{and} \quad P(E) = 1 - P(\text{not } E)$$

- 3 freshmen, 4 sophomores, 2 juniors, 1 senior

Probability of not selecting a freshman:

$$P(\text{not freshman}) = 1 - P(\text{freshman}) = 1 - \frac{3}{10}$$

$$= \frac{10}{10} - \frac{3}{10} = \frac{7}{10}$$

Additional Examples to Review

Example 1, page 458; Example 2, page 459

If it is impossible for events A and B to occur simultaneously, the events are mutually exclusive.

If A and B are mutually exclusive events, then $P(A \text{ or } B) = P(A) + P(B)$.

If A and B are not mutually exclusive events, then

$$P(A \text{ or } B) = P(A) + P(B) - P(A \text{ and } B).$$

Probability to Odds

1. Odds in favor of $E = \dfrac{P(E)}{P(\text{not } E)}$

2. Odds against $E = \dfrac{P(\text{not } E)}{P(E)}$

Odds to Probability

If odds in favor of E are a to b then $P(E) = \dfrac{a}{a+b}$.

- 3 freshmen, 4 sophomores, 2 juniors, 1 senior

  Probability of selecting a sophomore or a senior:

  $$P(\text{sophomore or senior}) = P(\text{sophomore}) + P(\text{senior})$$
  $$= \frac{4}{10} + \frac{1}{10} = \frac{5}{10} = \frac{1}{2}$$

Additional Examples to Review

Example 3, page 460; Example 6(b), page 463

- 8 male juniors, 4 female juniors, 3 male seniors, 5 female seniors

  Probability of selecting a junior or a male:

  $$P(\text{junior or male}) = P(\text{junior}) + P(\text{male}) - P(\text{junior and male})$$

8 + 4 = 12 of the 20 students are juniors.	8 + 3 = 11 of the 20 students are male.	8 of the 20 students are male juniors.

  $$= \frac{12}{20} + \frac{11}{20} - \frac{8}{20} = \frac{15}{20} = \frac{3}{4}$$

Additional Examples to Review

Example 4, page 461; Example 5, page 462; Example 6(a), page 463

You roll a single, six-sided die. $S = \{1, 2, 3, 4, 5, 6\}$

- Probability of an outcome greater than 4:

  Two outcomes, 5 and 6, are greater than 4.

  $$P(E) = \frac{2}{6} = \frac{1}{3}$$

- Odds in favor of an outcome greater than 4:

  $$\frac{P(E)}{P(\text{not } E)} = \frac{\frac{1}{3}}{1 - \frac{1}{3}} = \frac{\frac{1}{3}}{\frac{2}{3}} = \frac{1}{3} \cdot \frac{3}{2} = \frac{1}{2} = 1 \text{ to } 2$$

- Odds against an outcome greater than 4:

  $$\frac{P(\text{not } E)}{P(E)} = \frac{1 - \frac{1}{3}}{\frac{1}{3}} = \frac{\frac{2}{3}}{\frac{1}{3}} = \frac{2}{3} \cdot \frac{3}{1} = \frac{2}{1} = 2 \text{ to } 1$$

  You can quickly obtain these odds by reversing the ratio for the odds in favor of E.

Additional Examples to Review

Example 7, page 464; Example 8, page 465

Odds against E are 7 to 10.

- Odds in favor of E are 10 to 7.

- $P(E) = \dfrac{10}{10 + 7} = \dfrac{10}{17}$

Additional Example to Review

Example 9, page 466

Section 7.7 Events Involving *And*; Conditional Probability

Two events are independent if the occurrence of either of them has no effect on the probability of the other.

If A and B are independent events,

$$P(A \text{ and } B) = P(A) \cdot P(B).$$

The probability of a succession of independent events is the product of each of their probabilities.

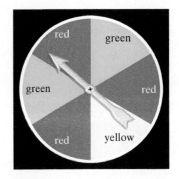

- The spinner is spun twice.

  $P(\text{red and then green}) = P(\text{red}) \cdot P(\text{green})$

  $$= \frac{3}{6} \cdot \frac{2}{6} = \frac{1}{2} \cdot \frac{1}{3} = \frac{1}{6}$$

- The spinner is spun three times.

  $P(\text{yellow on three consecutive spins})$

  $$= P(\text{yellow}) \cdot P(\text{yellow}) \cdot P(\text{yellow})$$

  $$= \frac{1}{6} \cdot \frac{1}{6} \cdot \frac{1}{6} = \frac{1}{216}$$

Additional Examples to Review

Example 1, page 471; Example 2, page 472; Example 3, page 472

Two events are dependent if the occurrence of one of them has an effect on the probability of the other.

If A and B are dependent events,

$$P(A \text{ and } B) = P(A) \cdot P(B \text{ given that } A \text{ has occurred}).$$

The multiplication rule for dependent events can be extended to cover three or more dependent events. In the case of three such events,

$P(A \text{ and } B \text{ and } C)$

$= P(A) \cdot P(B \text{ given } A \text{ occurred}) \cdot P(C \text{ given } A \text{ and } B \text{ occurred}).$

- Two balls are selected without replacement. Probability that both are red:

  $P(\text{red and red})$

  $$= P(\text{red}) \cdot P(\text{red given that a red was selected first})$$

  $$= \frac{4}{9} \cdot \frac{3}{8} = \frac{12}{72} = \frac{1}{6}$$

  > There are 9 balls, 4 of which are red.

  > After picking a red ball, there are 8 balls left, 3 of which are red.

Additional Examples to Review

Example 4, page 474; Example 5, page 475

The conditional probability of B, given A, written $P(B|A)$, is the probability of B if the sample space is restricted to A.

Numbers in the table represent the number of juniors and seniors ages 20, 21, and 22, by gender, in a group of 100 juniors and seniors.

	Age 20	Age 21	Age 22
Male	18	17	10
Female	20	23	12

- Probability that a randomly selected student is male, given that the student is 20 years old:

	Age 20
Male	18
Female	20

$P(\text{male}|\text{age is 20})$ ← Sample space is restricted to students who are 16.

$$= \frac{\text{number of males}}{\text{number of 20-year-olds}} = \frac{18}{18 + 20} = \frac{18}{38} = \frac{9}{19}$$

- Probability that a randomly selected student is 22, given that the student is female:

	Age 20	Age 21	Age 22
Female	20	23	12

$P(\text{age 22}|\text{female})$ ← Sample space is restricted to females.

$$= \frac{\text{number of 22-year-olds}}{\text{number of females}} = \frac{12}{20 + 23 + 12} = \frac{12}{55}$$

Additional Examples to Review

Example 6, page 476; Example 7, page 476; Example 8, page 477

Review Exercises

Section 7.1 The Fundamental Counting Principle

1. A restaurant offers 20 appetizers and 40 main courses. In how many ways can a person order a two-course meal? 800

2. A popular brand of pen comes in red, green, blue, or black ink. The writing tip can be chosen from extra bold, bold, regular, fine, or micro. How many different choices of pens do you have with this brand? 20

3. In how many ways can first and second prize be awarded in a contest with 100 people, assuming that each prize is awarded to a different person? 9900

4. You are answering three multiple-choice questions. Each question has five answer choices, with one correct answer per question. If you select one of these five choices for each question and leave nothing blank, in how many ways can you answer the questions? 125

5. A stock can go up, go down, or stay unchanged. How many possibilities are there if you own five stocks? 243

6. A person can purchase a condominium with a choice of five kinds of carpeting, with or without a pool, with or without a porch, and with one, two, or three bedrooms. How many different options are there for the condominium? 60

Section 7.2 Permutations and
Section 7.3 Combinations

In Exercises 7–10, evaluate each factorial expression.

7. $\dfrac{16!}{14!}$ 240

8. $\dfrac{800!}{799!}$ 800

9. $5! - 3!$ 114

10. $\dfrac{11!}{(11 - 3)!}$ 990

In Exercises 11–12, use the formula for $_nP_r$ to evaluate each expression.

11. $_{10}P_6$ 151,200

12. $_{100}P_2$ 9900

In Exercises 13–14, use the formula for $_nC_r$ to evaluate each expression.

13. $_{11}C_7$ 330

14. $_{14}C_5$ 2002

In Exercises 15–17, does the problem involve permutations or combinations? Explain your answer. (It is not necessary to solve the problem.)

15. How many different 4-card hands can be dealt from a 52-card deck? combinations

16. How many different ways can a director select from 20 male actors to cast the roles of Mark, Roger, Angel, and Collins in the musical *Rent*? permutations

17. How many different ways can a director select 4 actors from a group of 20 actors to attend a workshop on performing in rock musicals? combinations

In Exercises 18–28, solve each problem using an appropriate method.

18. Six acts are scheduled to perform in a variety show. How many different ways are there to schedule their appearances? 720

19. A club with 15 members is to choose four officers—president, vice-president, secretary, and treasurer. In how many ways can these offices be filled? 32,760

20. An election ballot asks voters to select four city commissioners from a group of ten candidates. In how many ways can this be done? 210

21. In how many distinct ways can the letters of the word TORONTO be arranged? 420

22. From the 20 CDs that you've bought during the past year, you plan to take 3 with you on vacation. How many different sets of three CDs can you take? 1140

23. You need to arrange seven of your favorite books along a small shelf. Although you are not arranging the books by height, the tallest of the books is to be placed at the left end and the shortest of the books at the right end. How many different ways can you arrange the books? 120

24. Suppose you are asked to list, in order of preference, the five favorite CDs you purchased in the past 12 months. If you bought 20 CDs over this time period, in how many ways can the five favorite be ranked? 1,860,480

25. In how many ways can five airplanes line up for departure on a runway? 120

26. How many different 5-card hands can be dealt from a deck that has only hearts (13 different cards)? 1287

27. A political discussion group consists of 12 Republicans and 8 Democrats. In how many ways can 5 Republicans and 4 Democrats be selected to attend a conference on politics and social issues? 55,440

28. In how many ways can the digits in the number 335,557 be arranged? 60

Section 7.4 Fundamentals of Probability

In Exercises 29–32, a die is rolled. Find the probability of rolling

29. a 6. $\frac{1}{6}$

30. a number less than 5. $\frac{2}{3}$

31. a number less than 7. 1

32. a number greater than 6. 0

In Exercises 33–37, you are dealt one card from a 52-card deck. Find the probability of being dealt

33. a 5. $\frac{1}{13}$

34. a picture card. $\frac{3}{13}$

35. a card greater than 4 and less than 8. $\frac{3}{13}$

36. a 4 of diamonds. $\frac{1}{52}$

37. a red ace. $\frac{1}{26}$

In Exercises 38–40, suppose that you reach into a bag and randomly select one piece of candy from 15 chocolates, 10 caramels, and 5 peppermints. Find the probability of selecting

38. a chocolate. $\frac{1}{2}$

39. a caramel. $\frac{1}{3}$

40. a peppermint. $\frac{1}{6}$

41. Tay-Sachs disease occurs in 1 of every 3600 births among Jewish people from central and eastern Europe, and in 1 in 600,000 births in other populations. The disease causes abnormal accumulation of certain fat compounds in the spinal cord and brain, resulting in paralysis, blindness, and mental impairment. Death generally occurs before the age of five. If we use *t* to represent a Tay-Sachs gene and *T* a healthy gene, the table at the top of the next column shows the four possibilities for the children of one healthy, *TT*, parent, and one parent who carries the disease, *Tt*, but is not sick.

		Second Parent	
		T	t
First	**T**	TT	Tt
Parent	**T**	TT	Tt

a. Find the probability that a child of these parents will be a carrier without the disease. $\frac{1}{2}$

b. Find the probability that a child of these parents will have the disease. 0

The table shows the employment status of the U.S. civilian labor force, ages 16 and over, by gender, in 2011. Use the data in the table, expressed in millions, to solve Exercises 42–44.

Employment Status of the U.S. Labor Force, Ages 16 or Older, 2011, in Millions

	Employed	Unemployed	Not in Labor Force	Total
Male	74	8	34	116
Female	66	6	52	124
Total	140	14	86	240

Source: U.S. Bureau of Labor Statistics

Find the probability, expressed as a simplified fraction, that a randomly selected person from the civilian labor force represented in the table

42. is employed. $\frac{7}{12}$

43. is female. $\frac{31}{60}$

44. is an unemployed male. $\frac{1}{30}$

Section 7.5 Probability with the Fundamental Counting Principle, Permutations, and Combinations

45. If cities A, B, C, and D are visited in random order, each city visited once, find the probability that city D will be visited first, city B second, city A third, and city C last. $\frac{1}{24}$

In Exercises 46–49, suppose that six singers are being lined up to perform at a charity fundraiser. Call the singers A, B, C, D, E, and F. The order of performance is determined by writing each singer's name on one of six cards, placing the cards in a hat, and then drawing one card at a time. The order in which the cards are drawn determines the order in which the singers perform. Find the probability that

46. singer C will perform last. $\frac{1}{6}$

47. singer B will perform first and singer A will perform last. $\frac{1}{30}$

48. the singers will perform in the following order: F, E, A, D, C, B. $\frac{1}{720}$

49. the performance will begin with singer A or C. $\frac{1}{3}$

50. A lottery game is set up so that each player chooses five different numbers from 1 to 20. If the five numbers match the five numbers drawn in the lottery, the player wins (or shares) the top cash prize. What is the probability of winning the prize

a. with one lottery ticket? *

b. with 100 different lottery tickets? *

51. A committee of four people is to be selected from six Democrats and four Republicans. Find the probability that

a. all are Democrats. $\frac{1}{14}$

b. two are Democrats and two are Republicans. $\frac{3}{7}$

52. If you are dealt 3 cards from a shuffled deck of red cards (26 different cards), find the probability of getting exactly 2 picture cards. $\frac{3}{26}$

Section 7.6 Events Involving *Not* and *Or*; Odds

In Exercises 53–57, a die is rolled. Find the probability of

53. not rolling a 5. $\frac{5}{6}$
54. not rolling a number less than 4. $\frac{1}{2}$
55. rolling a 3 or a 5. $\frac{1}{3}$
56. rolling a number less than 3 or greater than 4. $\frac{2}{3}$
57. rolling a number less than 5 or greater than 2. 1

In Exercises 58–63, you draw one card from a 52-card deck. Find the probability of

58. not drawing a picture card. $\frac{10}{13}$
59. not drawing a diamond. $\frac{3}{4}$
60. drawing an ace or a king. $\frac{2}{13}$
61. drawing a black 6 or a red 7. $\frac{1}{13}$
62. drawing a queen or a red card. $\frac{7}{13}$
63. drawing a club or a picture card. $\frac{11}{26}$

In Exercises 64–69, it is equally probable that the pointer on the spinner shown will land on any one of the six regions, numbered 1 through 6, and colored as shown. If the pointer lands on a borderline, spin again. Find the probability of

64. not stopping on 4. $\frac{5}{6}$
65. not stopping on yellow. $\frac{5}{6}$
66. not stopping on red. $\frac{1}{2}$
67. stopping on red or yellow. $\frac{2}{3}$
68. stopping on red or an even number. 1
69. stopping on red or a number greater than 3. $\frac{5}{6}$

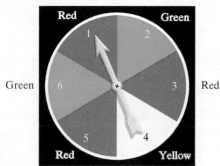

Use this information to solve Exercises 70–71. At a workshop on police work and the African-American community, there are 50 African-American male police officers, 20 African-American female police officers, 90 white male police officers, and 40 white female police officers. If one police officer is selected at random from the people at the workshop, find the probability that the selected person is

70. African American or male. $\frac{4}{5}$
71. female or white. $\frac{3}{4}$

Suppose that a survey of 350 college students is taken. Each student is asked the type of college attended (public or private) and the family's income level (low, middle, high). Use the data in the table to solve Exercises 72–75. Express probabilities as simplified fractions.

	Public	**Private**	**Total**
Low	120	20	140
Middle	110	50	160
High	22	28	50
Total	252	98	350

Find the probability that a randomly selected student in the survey

72. attends a public college. $\frac{18}{25}$
73. is not from a high-income family. $\frac{6}{7}$
74. is from a middle-income or a high-income family. $\frac{3}{5}$
75. attends a private college or is from a high-income family. $\frac{12}{35}$
76. One card is randomly selected from a deck of 52 cards. Find the odds in favor and the odds against getting a queen. in favor: 1:12; against: 12:1
77. The winner of a raffle will receive a two-year scholarship to any college of the winner's choice. If 2000 raffle tickets were sold and you purchased 20 tickets, what are the odds against your winning the scholarship? 99:1
78. The odds in favor of a candidate winning an election are given at 3 to 1. What is the probability that this candidate will win the election? $\frac{3}{4}$

Section 7.7 Events Involving *And*; Conditional Probability

Use the spinner shown to solve Exercises 79–83. It is equally likely that the pointer will land on any one of the six regions, numbered 1 through 6, and colored as shown. If the pointer lands on a borderline, spin again. If the pointer is spun twice, find the probability it will land on

79. yellow and then red. $\frac{2}{9}$ 80. 1 and then 3. $\frac{1}{36}$
81. yellow both times. $\frac{1}{9}$

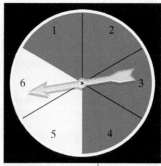

If the pointer is spun three times, find the probability it will land on

82. yellow and then 4 and then an odd number. $\frac{1}{36}$
83. red every time. $\frac{8}{27}$
84. What is the probability of a family having five boys born in a row? $\frac{1}{32}$

85. The probability of a flood in any given year in a region prone to flooding is 0.2. a. 0.04

  a. What is the probability of a flood two years in a row?

  b. What is the probability of a flood for three consecutive years? 0.008

  c. What is the probability of no flooding for four consecutive years? 0.4096

  d. What is the probability of a flood at least once in the next four years? 0.5904

In Exercises 86–87, two students are selected from a group of four psychology majors, three business majors, and two music majors. The two students are to meet with the campus cafeteria manager to voice the group's concerns about food prices and quality. One student is randomly selected and leaves for the cafeteria manager's office. Then, a second student is selected. Find the probability of selecting

86. a music major and then a psychology major. $\frac{1}{9}$

87. two business majors. $\frac{1}{12}$

88. A final visit to the box of chocolates: It's now grown to a box of 50, of which 30 are solid chocolate, 15 are filled with jelly, and 5 are filled with cherries. The story is still the same: They all look alike. You select a piece, eat it, select a second piece, eat it, and help yourself to a final sugar rush. Find the probability of selecting a solid chocolate followed by two cherry-filled chocolates. $\frac{1}{196}$

89. A single die is tossed. Find the probability that the tossed die shows 5, given that the outcome is an odd number. $\frac{1}{3}$

90. A letter is randomly selected from the letters of the English alphabet. Find the probability of selecting a vowel, given that the outcome is a letter that precedes k. $\frac{3}{10}$

91. The numbers shown at the top of next column are each written on a colored chip. The chips are placed into a bag and one chip is selected at random. Find the probability of selecting

  a. an odd number, given that a red chip is selected. $\frac{1}{2}$

  b. a yellow chip, given that the number selected is at least 3. $\frac{2}{7}$

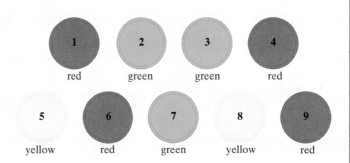

The data in the table are based on 145 Americans tested for tuberculosis. Use the data to solve Exercises 92–99. Express probabilities as simplified fractions.

	TB	No TB
Positive Screening Test	9	11
Negative Screening Test	1	124

Source: Deborah J. Bennett, *Randomness,* Harvard University Press, 1998.

Find the probability that a randomly selected person from this group

92. does not have TB. $\frac{27}{29}$

93. tests positive. $\frac{4}{29}$

94. does not have TB or tests positive. $\frac{144}{145}$

95. does not have TB, given a positive test. $\frac{11}{20}$

96. has a positive test, given no TB. $\frac{11}{135}$

97. has TB, given a negative test. $\frac{1}{125}$

Suppose that two people are randomly selected, in succession, from this group. Find the probability of selecting

98. two people with TB. $\frac{1}{232}$

99. two people with positive screening tests. $\frac{19}{1044}$

Exercises 100–106 involve a combination of topics from Section 7.4, Section 7.6, and Section 7.7.

The table shows the distribution, by age and gender, of the 31,593 deaths in the United States involving firearms in 2008.

Deaths in the United States Involving Firearms, 2008

	Under Age 5	Ages 5–14	Ages 15–19	Ages 20–24	Ages 25–44	Ages 45–64	Ages 65–74	Age ≥ 75	Total
Male	55	209	2331	3684	9591	7367	1930	2169	27,336
Female	33	79	259	411	1570	1434	261	210	4257
Total	88	288	2590	4095	11,161	8801	2191	2379	31,593

Source: National Safety Council

In Exercises 100–106, use the data in the table to find the probability, expressed as a fraction and as a decimal rounded to three places, that a firearm death in the United States

100. involved a male. *

101. involved a person in the 25–44 age range. *

102. involved a person less than 75 years old. *

103. involved a person in the 20–24 age range or in the 25–44 age range. *

104. involved a female or a person younger than 5. *

105. involved a person in the 20–24 age range, given that this person was a male. *

106. involved a male, given that this person was at least 75. *

*See Answers to Selected Exercises.

Chapter 7 Test

1. A person can purchase a particular model of a new car with a choice of ten colors, with or without automatic transmission, with or without four-wheel drive, with or without air conditioning, and with two, three, or four radio-CD speakers. How many different options are there for this model of the car?   240

2. Four acts are scheduled to perform in a variety show. How many different ways are there to schedule their appearances?   24

3. In how many ways can seven airplanes line up for a departure on a runway if the plane with the greatest number of passengers must depart first?   720

4. A human resources manager has 11 applicants to fill three different positions. Assuming that all applicants are equally qualified for any of the three positions, in how many ways can this be done?   990

5. From the ten books that you've recently bought but not read, you plan to take four with you on vacation. How many different sets of four books can you take?   210

6. In how many distinct ways can the letters of the word ATLANTA be arranged?   420

In Exercises 7–9, one student is selected at random from a group of 12 freshmen, 16 sophomores, 20 juniors, and 2 seniors. Find the probability that the person selected is

7. a freshman. $\frac{6}{25}$ 8. not a sophomore. $\frac{17}{25}$

9. a junior or a senior. $\frac{11}{25}$

10. If you are dealt one card from a 52-card deck, find the probability of being dealt a card greater than 4 and less than 10. $\frac{5}{13}$

11. Seven movies (A, B, C, D, E, F, and G) are being scheduled for showing. The order of showing is determined by random selection. Find the probability that film C will be shown first, film A next-to-last, and film E last. $\frac{1}{210}$

12. A lottery game is set up so that each player chooses six different numbers from 1 to 15. If the six numbers match the six numbers drawn in the lottery, the player wins (or shares) the top cash prize. What is the probability of winning the prize with 50 different lottery tickets? $\frac{10}{1001} \approx 0.00999$

In Exercises 13–14, it is equally probable that the pointer on the spinner shown will land on any one of the eight colored regions. If the pointer lands on a borderline, spin again.

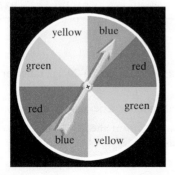

13. If the spinner is spun once, find the probability that the pointer will land on red or blue. $\frac{1}{2}$

14. If the spinner is spun twice, find the probability that the pointer lands on red on the first spin and blue on the second spin. $\frac{1}{16}$

15. A region is prone to flooding once every 20 years. The probability of flooding in any one year is $\frac{1}{20}$. What is the probability of flooding for three consecutive years?

16. One card is randomly selected from a deck of 52 cards. Find the probability of selecting a black card or a picture card. $\frac{8}{13}$

17. A group of students consists of 10 male freshmen, 15 female freshmen, 20 male sophomores, and 5 female sophomores. If one person is randomly selected from the group, find the probability of selecting a freshman or a female. $\frac{3}{5}$

18. A box contains five red balls, six green balls, and nine yellow balls. Suppose you select one ball at random from the box and do not replace it. Then you randomly select a second ball. Find the probability that both balls selected are red. $\frac{1}{19}$

19. A quiz consisting of four multiple-choice questions has four available options (a, b, c, or d) for each question. If a person guesses at every question, what is the probability of answering *all* questions correctly? $\frac{1}{256}$

20. A group is comprised of 20 men and 15 women. If one person is randomly selected from the group, find the odds against the person being a man.   3:4

21. The odds against a candidate winning an election are given at 1 to 4.
    a. What are the odds in favor of the candidate winning?   4:1
    b. What is the probability that the candidate will win the election? $\frac{4}{5}$

A class is collecting data on eye color and gender. They organize the data they collected into the table shown. Numbers in the table represent the number of students from the class that belong to each of the categories. Use the data to solve Exercises 22–26. Express probabilities as simplified fractions.

	Brown	Blue	Green
Male	22	18	10
Female	18	20	12

Find the probability that a randomly selected student from this class

22. does not have brown eyes. $\frac{3}{5}$

23. has brown eyes or blue eyes. $\frac{39}{50}$

24. is female or has green eyes. $\frac{3}{5}$

25. is male, given the student has blue eyes. $\frac{9}{19}$

26. If two people are randomly selected, in succession, from the students in this class, find the probability that they both have green eyes. $\frac{7}{150}$

15. $\frac{1}{8000} = 0.000125$

Statistical Pathways

Some random statistical factoids:

- 28% of liberals have insomnia, compared with 16% of conservatives. (*Mother Jones*)
- 17% of American workers would reveal company secrets for money, and 8% have done it already. (Monster.com)
- 31% of American adults find giving up their smartphone for a day more difficult than giving up their significant other. (Microsoft)
- Between the ages of 18 and 24, 73% of Americans have used text messages to send suggestive pictures, compared with 55% ages 25–29, 52% ages 30–34, 42% ages 35–44, 26% ages 45–54, 10% ages 55–64, and 7% ages 65 and older. (*Time*)
- 49% of Americans cite a "lot" of stress at age 22, compared with 45% at 42, 35% at 58, 29% at 62, and 20% by 70. (*Proceedings of the National Academy of Sciences*)
- 46% of Americans dread public speaking. (TNS Survey)

Statisticians collect numerical data from subgroups of populations to find out everything imaginable about the population as a whole, including whom they support in an election, what they watch on TV, how much money they make, or what worries them. Comedians and statisticians joke that 62.38% of all statistics are made up on the spot. Because statisticians both record and influence our behavior, it is important to distinguish between good and bad methods for collecting, presenting, and interpreting data.

Here's where you'll find these applications:

Throughout this chapter, you will gain an understanding of where data come from and how these numbers are used to make decisions. We'll return to activities U.S. adults dread in Example 7 of Section 8.4.

8.1 Sampling, Frequency Distributions, and Graphs

After you have read this section, you should be able to:

1. Describe the population whose properties are to be analyzed.

2. Select an appropriate sampling technique.

3. Organize and present data.

4. Identify deceptions in visual displays of data.

At the end of the twentieth century, there were 94 million households in the United States with television sets. The television program viewed by the greatest percentage of such households in that century was the final episode of *M*A*S*H*. Over 50 million American households watched this program.

Numerical information, such as the information about the top three TV shows of the twentieth century, shown in **Table 8.1**, is called **data**. The word **statistics** is often used when referring to data. However, statistics has a second meaning: Statistics is also a method for collecting, organizing, analyzing, and interpreting data, as well as drawing conclusions based on the data. This methodology divides statistics into two main areas. **Descriptive statistics** is concerned

*M*A*S*H* took place in the early 1950s, during the Korean War. By the final episode, the show had lasted four times as long as the Korean War.

with collecting, organizing, summarizing, and presenting data. **Inferential statistics** has to do with making generalizations about and drawing conclusions from the data collected.

Table 8.1 TV Programs with the Greatest U.S. Audience Viewing Percentage of the Twentieth Century

Program	Total Households	Viewing Percentage
1. *M*A*S*H* Feb. 28, 1983	50,150,000	60.2%
2. *Dallas* Nov. 21, 1980	41,470,000	53.3%
3. *Roots Part 8* Jan. 30, 1977	36,380,000	51.1%

Source: Nielsen Media Research

Populations and Samples

1. Describe the population whose properties are to be analyzed.

Consider the set of all American TV households. Such a set is called the *population*. In general, a **population** is the set containing all the people or objects whose properties are to be described and analyzed by the data collector.

The population of American TV households is huge. At the time of the *M*A*S*H* conclusion, there were nearly 84 million such households. Did over 50 million American TV households really watch the final episode of *M*A*S*H*? A friendly phone call to each household ("So, how are you? What's new? Watch any good television last night? If so, what?") is, of course, absurd. A **sample**, which is a subset or subgroup of the population, is needed. In this case, it would be appropriate to have a sample of a few thousand TV households to draw conclusions about the population of all TV households.

Blitzer Bonus

A Sampling Fiasco

In 1936, the *Literary Digest* mailed out over ten million ballots to voters throughout the country. The results poured in, and the magazine predicted a landslide victory for Republican Alf Landon over Democrat Franklin Roosevelt. However, the prediction of the *Literary Digest* was wrong. Why? The mailing lists the editors used included people from their own subscriber list, directories of automobile owners, and telephone books. As a result, its sample was anything but random. It excluded most of the poor, who were unlikely to subscribe to the *Literary Digest,* or to own a car or telephone in the heart of the Depression. Prosperous people in 1936 were more likely to be Republican than the poor. Thus, although the sample was massive, it included a higher percentage of affluent individuals than the population as a whole did. A victim of both the Depression and the 1936 sampling fiasco, the *Literary Digest* folded in 1937.

Example 1   Populations and Samples

The board of supervisors in a large city decides to conduct a survey among citizens of the city to determine their opinions about restricting land use in order to protect the environment.

a. Describe the population.

b. A supervisor suggests obtaining a sample by surveying people who hike along one of the city's nature trails. Each person will be asked to express his or her opinion on land use restriction to protect the environment. Does this seem like a good idea?

Solution

a. The population is the set containing all the citizens of the city.

b. Questioning people hiking along one of the city's nature trails is a terrible idea. The subset of hikers is likely to have a more positive attitude about environmental issues, thereby favoring land use restriction, than the population of all the city's citizens.

✓ **Check Point 1**   A city government wants to conduct a survey among the city's homeless to discover their opinions about required residence in city shelters from midnight until 6 A.M.

a. Describe the population.   the set containing all the city's homeless

b. A city commissioner suggests obtaining a sample by surveying all the homeless people at the city's largest shelter on a Sunday night. Does this seem like a good idea? Explain your answer.   no; People already in the shelters are probably less likely to be against mandatory residence in the shelters.

Random Sampling

There is a way to use a small sample to make generalizations about a large population: Guarantee that every member of the population has an equal chance to be selected for the sample. Surveying hikers along one of the city's nature trails does not provide this guarantee. Unless it can be established that all citizens of the city hike these trails, which seems unlikely, this sampling scheme does not permit each citizen an equal chance of selection.

> **Random Samples**
>
> A **random sample** is a sample obtained in such a way that every element in the population has an equal chance of being selected for the sample.

Suppose that you are elated with the quality of one of your courses. Although it's an auditorium section with 120 students, you feel that the professor is lecturing right to you. During a wonderful lecture, you look around the auditorium to see if any of the other students are sharing your enthusiasm. Based on body language, it's hard to tell. You really want to know the opinion of the population of 120 students taking this course. You think about asking students to grade the course on an A to F scale, anticipating a unanimous A. You cannot survey everyone. Eureka! Suddenly you have an idea on how to take a sample. Place cards numbered from 1 through 120, one number per card, in a box. Because the course has assigned seating by number, each numbered card corresponds to a student in the class. Reach in and randomly select six cards. Each card, and therefore each student, has an equal chance of being selected. Then use the opinions about the course from the six randomly selected students to generalize about the course opinion for the entire 120-student population.

Your idea is precisely how random samples are obtained. In random sampling, each element in the population must be identified and assigned a number. The numbers are generally assigned in order. The way to sample from the larger numbered population is to generate random numbers using a computer or calculator. Each numbered element from the population that corresponds to one of the generated random numbers is selected for the sample.

Call-in polls on radio and television are not reliable because those polled do not represent the larger population. A person who calls in is likely to have feelings about an issue that are consistent with the politics of the show's host. For a poll to be accurate, the sample must be chosen randomly from the larger population. The A. C. Nielsen Company uses a random sample of approximately 5000 TV households to measure the percentage of households tuned in to a television program.

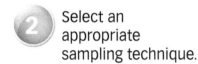

② Select an appropriate sampling technique.

Example 2 ┊ **Selecting an Appropriate Sampling Technique**

We return to the board of supervisors in the large city and their interest in how the city's citizens feel about restricting land use in order to protect the environment. Which of the following would be the most appropriate way to select a random sample?

a. Randomly survey construction workers who live in the city.

b. Survey the first 200 people whose names appear in the city's telephone directory.

c. Randomly select neighborhoods of the city and then randomly survey people within the selected neighborhoods.

Solution

Keep in mind that the population is the set containing all the city's citizens. A random sample must give each citizen an equal chance of being selected.

a. Randomly selecting construction workers who live in the city is not a good idea. The salaries of many construction workers depend on new construction, which would diminish with land use restriction to protect the environment. Furthermore, this sample does not give each citizen of the city an equal chance of being selected.

b. If the supervisors survey the first 200 people whose names appear in the city's telephone directory, all citizens do not have an equal chance of selection. For example, an individual whose last name begins with a letter toward the end of the alphabet has no chance of being selected.

c. Randomly selecting neighborhoods of the city and then randomly surveying people within the selected neighborhoods is an appropriate technique. Using this method, each citizen has an equal chance of being selected.

In summary, given the three options, the sampling technique in part (c) is the most appropriate.

Surveys and polls involve data from a sample of some population. Regardless of the sampling technique used, the sample should exhibit characteristics typical of those possessed by the target population. This type of sample is called a **representative sample**.

By selecting people from a shelter, homeless people who do not go to the shelters have no chance of being selected. An appropriate method would be to randomly select neighborhoods of the city and then randomly survey homeless people within the selected neighborhoods.

✓ **Check Point 2** Explain why the sampling technique described in Check Point 1(b) on page 495 is not a random sample. Then describe an appropriate ▶ way to select a random sample of the city's homeless.

Blitzer Bonus
●●●●●●●●●●●●●●●●

The United States Census

A census is a survey that attempts to include the entire population. The U.S. Constitution requires a census of the American population every ten years. When the Founding Fathers invented American democracy, they realized that if you are going to have government by the people, you need to know who and where they are. Nowadays about $400 billion per year in federal aid is distributed based on the Census numbers, for everything from jobs to bridges to schools. For every 100 people not counted, states and communities could lose as much as $130,000 annually, or $1300 per person each year, so this really matters.

Although the Census generates volumes of statistics, its main purpose is to give the government block-by-block population figures. The U.S. Census is not foolproof. The 1990 Census missed 1.6% of the American population, including an estimated 4.4% of the African-American population,

largely in inner cities. Only 67% of households responded to the 2000 Census, even after door-to-door canvassing. About 6.4 million people were missed and 3.1 million were counted twice. Although the 2010 Census was one of the shortest forms in history, counting each person was not an easy task, particularly with concerns about immigration status and privacy of data.

Of course, there would be more than $400 billion to spread around if it didn't cost so much to count us in the first place: about $15 billion for the 2010 Census. That included $338 million for ads in 28 languages, a Census-sponsored NASCAR entry, and $2.5 million for a Super Bowl ad. The ads were meant to boost the response rate, since any household that did not mail back its form got visited by a Census worker, another pricey item. In all, the cost of the 2010 Census worked out to approximately $49 per person.

Frequency Distributions

 3 Organize and present data.

After data have been collected from a sample of the population, the next task facing the statistician is to present the data in a condensed and manageable form. In this way, the data can be more easily interpreted.

Suppose, for example, that researchers are interested in determining the age at which adolescent males show the greatest rate of physical growth. A random sample of 35 ten-year-old boys is measured for height and then remeasured each year until they reach 18. The age of maximum yearly growth for each subject is as follows:

12, 14, 13, 14, 16, 14, 14, 17, 13, 10, 13, 18, 12, 15, 14, 15, 15, 14, 14, 13, 15, 16, 15, 12, 13, 16, 11, 15, 12, 13, 12, 11, 13, 14, 14.

A piece of data is called a **data item**. This list of data has 35 data items. Some of the data items are identical. Two of the data items are 11 and 11. Thus, we can say that the **data value** 11 occurs twice. Similarly, because five of the data items are 12, 12, 12, 12, and 12, the data value 12 occurs five times.

Collected data can be presented using a **frequency distribution**. Such a distribution consists of two columns. The data values are listed in one column. Numerical data are generally listed from smallest to largest. The adjacent column is labeled **frequency** and indicates the number of times each value occurs.

Table 8.2 A Frequency Distribution for a Boy's Age of Maximum Yearly Growth

Age of Maximum Growth	Number of Boys (Frequency)
10	1
11	2
12	5
13	7
14	9
15	6
16	3
17	1
18	1
Total:	$n = 35$

35 is the sum of the frequencies.

Grade	Frequency
A	3
B	5
C	9
D	2
F	1
	20

Example 3   Constructing a Frequency Distribution

Construct a frequency distribution for the data of the age of maximum yearly growth for 35 boys:

12, 14, 13, 14, 16, 14, 14, 17, 13, 10, 13, 18, 12, 15, 14, 15, 15, 14, 14, 13, 15, 16, 15, 12, 13, 16, 11, 15, 12, 13, 12, 11, 13, 14, 14.

Solution

It is difficult to determine trends in the data above in their current format. Perhaps we can make sense of the data by organizing them into a frequency distribution. Let us create two columns. One lists all possible data values, from smallest (10) to largest (18). The other column indicates the number of times the value occurs in the sample. The frequency distribution is shown in **Table 8.2**.

The frequency distribution indicates that one subject had maximum growth at age 10, two at age 11, five at age 12, seven at age 13, and so on. The maximum growth for most of the subjects occurred between the ages of 12 and 15. Nine boys experienced maximum growth at age 14, more than at any other age within the sample. The sum of the frequencies, 35, is equal to the original number of data items.

The trend shown by the frequency distribution in **Table 8.2** indicates that the number of boys who attain their maximum yearly growth at a given age increases until age 14 and decreases after that. This trend is not evident in the data in their original format.

✓ **Check Point 3**   Construct a frequency distribution for the data showing final course grades for students in a college math course, listed alphabetically by student name in a grade book:

F, A, B, B, C, C, B, C, A, A, C, C, D, C, B, D, C, C, B, C.

A frequency distribution that lists all possible data items can be quite cumbersome when there are many such items. For example, consider the following data items. These are statistics test scores for a class of 40 students.

82	47	75	64	57	82	63	93
76	68	84	54	88	77	79	80
94	92	94	80	94	66	81	67
75	73	66	87	76	45	43	56
57	74	50	78	71	84	59	76

It's difficult to determine how well the group did when the grades are displayed like this. Because there are so many data items, one way to organize these data so that the results are more meaningful is to arrange the grades into groups, or **classes**, based on something that interests us. Many grading systems assign an A to grades in the 90–100 class, B to grades in the 80–89 class, C to grades in the 70–79 class, and so on. These classes provide one way to organize the data.

Looking at the 40 statistics test scores, we see that they range from a low of 43 to a high of 94. We can use classes that run from 40 through 49, 50 through 59, 60 through 69, and so on up to 90 through 99, to organize the scores. In Example 4, we go through the data and tally each item into the appropriate class. This method for organizing data is called a **grouped frequency distribution**.

Example 4   Constructing a Grouped Frequency Distribution

Use the classes 40–49, 50–59, 60–69, 70–79, 80–89, and 90–99 to construct a grouped frequency distribution for the 40 test scores on the previous page.

Solution

We use the 40 given scores and tally the number of scores in each class.

Tallying Statistics Test Scores

Test Scores (Class)	Tally	Number of Students (Frequency)
40–49	|||	3
50–59	ℍℍ	6
60–69	ℍℍ	6
70–79	ℍℍ ℍℍ	11
80–89	ℍℍ ||||	9
90–99	ℍℍ	5

The second score in the list, 47, is shown as the first tally in this row.

The first score in the list, 82, is shown as the first tally in this row.

Table 8.3 A Grouped Frequency Distribution for Statistics Test Scores

Class	Frequency
40–49	3
50–59	6
60–69	6
70–79	11
80–89	9
90–99	5
Total:	$n = 40$

40, the sum of the frequencies, is the number of data items.

Omitting the tally column results in the grouped frequency distribution in **Table 8.3**. The distribution shows that the greatest frequency of students scored in the 70–79 class. The number of students decreases in classes that contain successively lower and higher scores. The sum of the frequencies, 40, is equal to the original number of data items.

The leftmost number in each class of a grouped frequency distribution is called the **lower class limit**. For example, in **Table 8.3**, the lower limit of the first class is 40 and the lower limit of the third class is 60. The rightmost number in each class is called the **upper class limit**. In **Table 8.3**, 49 and 69 are the upper limits for the first and third classes, respectively. Notice that if we take the difference between any two consecutive lower class limits, we get the same number:

$$50 - 40 = 10, \ 60 - 50 = 10, \ 70 - 60 = 10, \ 80 - 70 = 10, \ 90 - 80 = 10.$$

The number 10 is called the **class width**.

When setting up class limits, each class, with the possible exception of the first or last, should have the same width. Because each data item must fall into exactly one class, it is sometimes helpful to vary the width of the first or last class to allow for items that fall far above or below most of the data.

✓ **Check Point 4**   Use the classes in **Table 8.3** to construct a grouped frequency distribution for the following 37 exam scores:

Class	Frequency
40–49	1
50–59	5
60–69	4
70–79	15
80–89	5
90–99	7
	37

73	58	68	75	94	79	96	79
87	83	89	52	99	97	89	58
95	77	75	81	75	73	73	62
69	76	77	71	50	57	41	98
77	71	69	90	75.			

Table 8.2 (repeated) A Frequency Distribution for a Boy's Age of Maximum Yearly Growth	
Age of Maximum Growth	**Number of Boys (Frequency)**
10	1
11	2
12	5
13	7
14	9
15	6
16	3
17	1
18	1
Total:	$n = 35$

Histograms and Frequency Polygons

Take a second look at the frequency distribution for the age of a boy's maximum yearly growth, repeated in **Table 8.2**. A bar graph with bars that touch can be used to visually display the data. Such a graph is called a **histogram**. **Figure 8.1** illustrates a histogram that was constructed using the frequency distribution in **Table 8.2**. A series of rectangles whose heights represent the frequencies are placed next to each other. For example, the height of the bar for the data value 10, shown in **Figure 8.1**, is 1. This corresponds to the frequency for 10 given in **Table 8.2**. The higher the bar, the more frequent the age of maximum yearly growth. The break along the horizontal axis, symbolized by ᴧ, eliminates listing the ages 1 through 9.

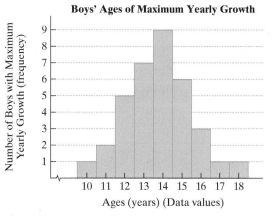

Figure 8.1 A histogram for a boy's age of maximum yearly growth

A line graph called a **frequency polygon** can also be used to visually convey the information shown in **Figure 8.1**. The axes are labeled just like those in a histogram. Thus, the horizontal axis shows data values and the vertical axis shows frequencies. Once a histogram has been constructed, it's fairly easy to draw a frequency polygon. **Figure 8.2** shows a histogram with a dot at the top of each rectangle at its midpoint. Connect each of these midpoints with a straight line. To complete the frequency polygon at both ends, the lines should be drawn down to touch the horizontal axis. The completed frequency polygon is shown in **Figure 8.3**.

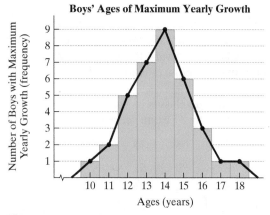

Figure 8.2 A histogram with a superimposed frequency polygon

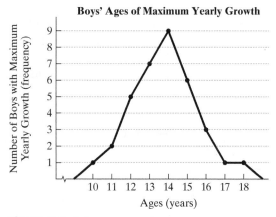

Figure 8.3 A frequency polygon

Stem-and-Leaf Plots

A unique way of displaying data uses a tool called a **stem-and-leaf plot**. Example 5 illustrates how we sort the data, revealing the same visual impression created by a histogram.

Example 5   Constructing a Stem-and-Leaf Plot

Use the data showing statistics test scores for 40 students to construct a stem-and-leaf plot:

82	47	75	64	57	82	63	93
76	68	84	54	88	77	79	80
94	92	94	80	94	66	81	67
75	73	66	87	76	45	43	56
57	74	50	78	71	84	59	76.

Solution

The plot is constructed by separating each data item into two parts. The first part is the *stem*. The **stem** consists of the tens digit. For example, the stem for the score of 82 is 8. The second part is the *leaf*. The **leaf** consists of the units digit for a given value. For the score of 82, the leaf is 2. The possible stems for the 40 scores are 4, 5, 6, 7, 8, and 9, entered in the left column of the plot.

Begin by entering each data item in the first row:

82	47	75	64	57	82	63	93.

Entering 82:

Stems	Leaves
4	
5	
6	
7	
8	2
9	

Adding 47:

Stems	Leaves
4	7
5	
6	
7	
8	2
9	

Adding 75:

Stems	Leaves
4	7
5	
6	
7	5
8	2
9	

Adding 64:

Stems	Leaves
4	7
5	
6	4
7	5
8	2
9	

Adding 57:

Stems	Leaves
4	7
5	7
6	4
7	5
8	2
9	

Adding 82:

Stems	Leaves
4	7
5	7
6	4
7	5
8	2 2
9	

Adding 63:

Stems	Leaves
4	7
5	7
6	4 3
7	5
8	2 2
9	

Adding 93:

Stems	Leaves
4	7
5	7
6	4 3
7	5
8	2 2
9	3

We continue in this manner and enter all the data items. **Figure 8.4** at the top of the next page shows the completed stem-and-leaf plot.

Stem	Leaves
4	1
5	82807
6	8299
7	359975533671715
8	73991
9	4697580

If you turn the page so that the left margin is on the bottom and facing you, the visual impression created by the enclosed leaves is the same as that created by a histogram. An advantage over the histogram is that the stem-and-leaf plot preserves exact data items. The enclosed leaves extend farthest to the right when the stem is 7. This shows that the greatest frequency of students scored in the 70s.

A Stem-and-Leaf Plot for 40 Test Scores

Stems	Leaves
4	7 5 3
5	7 4 6 7 0 9
6	4 3 8 6 7 6
7	5 6 7 9 5 3 6 4 8 1 6
8	2 2 4 8 0 0 1 7 4
9	3 4 2 4 4

Figure 8.4 A stem-and-leaf plot displaying 40 test scores

 Check Point 5 Construct a stem-and-leaf plot for the data in Check Point 4 on page 499.

Deceptions in Visual Displays of Data

4 Identify deceptions in visual displays of data.

Benjamin Disraeli, Queen Victoria's prime minister, stated that there are "lies, damned lies, and statistics." The problem is not that statistics lie, but rather that liars use statistics. Graphs can be used to distort the underlying data, making it difficult for the viewer to learn the truth. One potential source of misunderstanding is the scale on the vertical axis used to draw the graph. This scale is important because it lets a researcher "inflate" or "deflate" a trend. For example, both graphs in **Figure 8.5** present identical data for the percentage of people in the United States living below the poverty level from 2001 through 2005. The graph on the left stretches the scale on the vertical axis to create an overall impression of a poverty rate increasing rapidly over time. The graph on the right compresses the scale on the vertical axis to create an impression of a poverty rate that is slowly increasing, and beginning to level off, over time.

Table 8.4 U.S. Poverty Rate from 2001 to 2011

Year	Poverty Rate
2001	11.3%
2002	11.7%
2003	12.1%
2004	12.5%
2005	12.7%
2006	12.3%
2007	12.5%
2008	13.2%
2009	14.3%
2010	15.1%
2011	15.0%

Percentage of People in the United States Living below the Poverty Level, 2001–2005

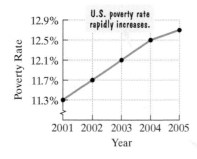

Year	Poverty Rate
2001	11.3%
2002	11.7%
2003	12.1%
2004	12.5%
2005	12.7%

The graph in both figures present this data.

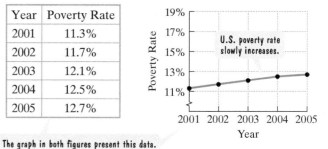

Figure 8.5
Source: U.S. Census Bureau

There is another problem with the data in **Figure 8.5**. Look at **Table 8.4**, which shows the poverty rate from 2001 through 2011. Depending on the time frame chosen, the data can be interpreted in various ways. Carefully choosing a time frame can help represent data trends in the most positive or negative light.

Things to Watch for in Visual Displays of Data

1. Is there a title that explains what is being displayed?
2. Are numbers lined up with tick marks on the vertical axis that clearly indicate the scale? Has the scale been varied to create a more or less dramatic impression than shown by the actual data?
3. Do too many design and cosmetic effects draw attention from or distort the data?
4. Has the wrong impression been created about how the data are changing because equally spaced time intervals are not used on the horizontal axis? Furthermore, has a time interval been chosen that allows the data to be interpreted in various ways?
5. Are bar sizes scaled proportionately in terms of the data they represent?
6. Is there a source that indicates where the data in the display came from? Do the data come from an entire population or a sample? Was a random sample used and, if so, are there possible differences between what is displayed in the graph and what is occurring in the entire population? (We'll discuss these *margins of error* in Section 8.4.) Who is presenting the visual display, and does that person have a special case to make for or against the trend shown by the graph?

Table 8.5 contains two examples of misleading visual displays.

Table 8.5 Examples of Misleading Visual Displays

Graphic Display

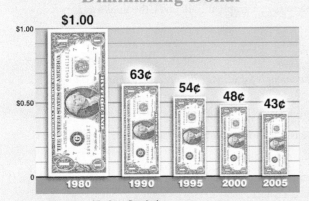

Source: Bureau of Labor Statistics

Presentation Problems

Although the length of each dollar bill is proportional to its spending power, the visual display varies both the length *and width* of the bills to show the diminishing power of the dollar over time. Because our eyes focus on the *areas* of the dollar-shaped bars, this creates the impression that the purchasing power of the dollar diminished even more than it really did. If the area of the dollar were drawn to reflect its purchasing power, the 2005 dollar would be approximately twice as large as the one shown in the graphic display.

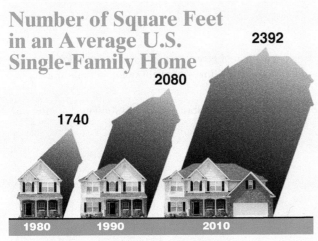

Source: National Association of Home Builders

Cosmetic effects of homes with equal heights, but different frontal additions and shadow lengths, make it impossible to tell if they proportionately depict the given areas. Time intervals on the horizontal axis are not uniform in size, making it appear that dwelling swelling has been linear from 1980 through 2010. The data indicate that this is not the case. There was a greater increase in area from 1980 through 1990, averaging 34 square feet per year, than from 1990 through 2010, averaging approximately 15.6 square feet per year.

Achieving Success

According to the Ebbinghaus retention model, you forget 50% of processed information within one hour of leaving the classroom. You lose 60% to 70% within 24 hours. After 30 days, 70% is gone. Reviewing and rewriting class notes is an effective way to counteract this phenomenon. At the very least, read your lecture notes at the end of each day. The more you engage with the material, the more you retain.

Concept and Vocabulary Check

Exercises in the Concept and Vocabulary Check are intended for group and class discussions.

In Exercises 1–6, fill in each blank so that the resulting statement is true.

1. A sample obtained in such a way that every element in the population has an equal chance of being selected is called a/an _____random_____ sample.

2. If data values are listed in one column and the adjacent column indicates the number of times each value occurs, the data presentation is called a/an ___frequency distribution___.

3. If the data presentation in Exercise 2 is varied by organizing the data into classes, the data presentation is called a/an grouped ___frequency distribution___. If one class in such a distribution is 80–89, the lower class limit is _____80_____ and the upper class limit is _____89_____.

4. Data can be displayed using a bar graph with bars that touch each other. This visual presentation of the data is called a/an _____histogram_____. The heights of the bars represent the _____frequencies_____ of the data values.

5. If the midpoints of the tops of the bars for the data presentation in Exercise 4 are connected with straight lines, the resulting line graph is a data presentation called a/an ___frequency polygon___. To complete such a graph at both ends, the lines are drawn down to touch the ___horizontal axis___.

6. A data presentation that separates each data item into two parts is called a/an ___stem-and-leaf plot___.

In Exercises 7–10, determine whether each statement is true or false. If the statement is false, make the necessary change(s) to produce a true statement.　Changes to false statements will vary.

7. A sample is the set of all the people or objects whose properties are to be described and analyzed by the data collector.　false

8. A call-in poll on radio or television is not reliable because the sample is not chosen randomly from a larger population.　true

9. One disadvantage of a stem-and-leaf plot is that it does not display the data items.　false

10. A deception in the visual display of data can result by stretching or compressing the scale on a graph's vertical axis.　true

Respond to Exercises 11–20 using verbal or written explanations.　11–20. Answers will vary.

11. What is a population? What is a sample?

12. Describe what is meant by a random sample.

13. Suppose you are interested in whether or not the students at your high school would favor a grading system in which students may receive final grades of A+, A, A−, B+, B, B−, C+, C, C−, and so on. Describe how you might obtain a random sample of 100 students from the entire student population.

14. For Exercise 13, would questioning every fifth student as he or she is leaving the school library until 100 students are interviewed be a good way to obtain a random sample? Explain your answer.

15. What is a frequency distribution?

16. What is a histogram?

17. What is a frequency polygon?

18. Describe how to construct a frequency polygon from a histogram.

19. Describe how to construct a stem-and-leaf plot from a set of data.

20. Describe two ways that graphs can be misleading.

Exercise Set 8.1

Practice and Application Exercises

1. The government of a large city needs to determine whether the city's residents will support the construction of a new jail. The government decides to conduct a survey of a sample of the city's residents. Which one of the following procedures would be most appropriate for obtaining a sample of the city's residents?   c
   a. Survey a random sample of the employees and inmates at the old jail.
   b. Survey every fifth person who walks into City Hall on a given day.
   c. Survey a random sample of persons within each geographic region of the city.
   d. Survey the first 200 people listed in the city's telephone directory.

2. The city council of a large city needs to know whether its residents will support the building of three new schools. The council decides to conduct a survey of a sample of the city's residents. Which procedure would be most appropriate for obtaining a sample of the city's residents?   c
   a. Survey a random sample of teachers who live in the city.
   b. Survey 100 individuals who are randomly selected from a list of all people living in the state in which the city in question is located.
   c. Survey a random sample of persons within each neighborhood of the city.
   d. Survey every tenth person who enters City Hall on a randomly selected day.

A questionnaire was given to college students in an introductory statistics class during the first week of the course. One question asked, "How stressed have you been in the last $2\frac{1}{2}$ weeks, on a scale of 0 to 10, with 0 being not at all stressed and 10 being as stressed as possible?" The students' responses are shown in the frequency distribution. Use this frequency distribution to solve Exercises 3–6.

Stress Rating	Frequency
0	2
1	1
2	3
3	12
4	16
5	18
6	13
7	31
8	26
9	15
10	14

Source: Journal of Personality and Social Psychology, 69, 1102–1112

*See Answers to Selected Exercises.

3. Which stress rating describes the greatest number of students? How many students responded with this rating?   7; 31

4. Which stress rating describes the least number of students? How many responded with this rating?   1; 1

5. How many students were involved in this study?   151

6. How many students had a stress rating of 8 or more?   55

7. A random sample of 30 college students is selected. Each student is asked how much time he or she spent on homework during the previous week. The following times (in hours) are obtained:

   16, 24, 18, 21, 18, 16, 18, 17, 15, 21, 19, 17, 17, 16, 19, 18, 15, 15, 20, 17, 15, 17, 24, 19, 16, 20, 16, 19, 18, 17.

   Construct a frequency distribution for the data.   *

8. A random sample of 30 male college students is selected. Each student is asked his height (to the nearest inch). The heights are as follows:

   72, 70, 68, 72, 71, 71, 71, 69, 73, 71, 73, 75, 66, 67, 75, 74, 73, 71, 72, 67, 72, 68, 67, 71, 73, 71, 72, 70, 73, 70.

   Construct a frequency distribution for the data.   *

A college professor had students keep a diary of their social interactions for a week. Excluding family and work situations, the number of social interactions of ten minutes or longer over the week is shown in the following grouped frequency distribution. Use this information to solve Exercises 9–16.

Number of Social Interactions	Frequency
0–4	12
5–9	16
10–14	16
15–19	16
20–24	10
25–29	11
30–34	4
35–39	3
40–44	3
45–49	3

Source: Society for Personality and Social Psychology

9. Identify the lower class limit for each class.   *

10. Identify the upper class limit for each class.   *

11. What is the class width?   5

12. How many students were involved in this study?   94

13. How many students had at least 30 social interactions for the week?   13

14. How many students had at most 14 social interactions for the week?   44

15. Among the classes with the greatest frequency, which class has the least number of social interactions?   5–9

16. Among the classes with the smallest frequency, which class has the least number of social interactions?   35–39

17. As of 2011, the following are the ages, in chronological order, at which U.S. presidents were inaugurated:

57, 61, 57, 57, 58, 57, 61, 54, 68, 51, 49, 64, 50, 48, 65, 52, 56, 46, 54, 49, 51, 47, 55, 55, 54, 42, 51, 56, 55, 51, 54, 51, 60, 62, 43, 55, 56, 61, 52, 69, 64, 46, 54, 47.

Source: Time Almanac

Construct a grouped frequency distribution for the data. Use 41–45 for the first class and use the same width for each subsequent class.   *

18. The IQ scores of 70 students enrolled in a liberal arts math course at a college are as follows:

102, 100, 103, 86, 120, 117, 111, 101, 93, 97, 99, 95, 95, 104, 104, 105, 106, 109, 109, 89, 94, 95, 99, 99, 103, 104, 105, 109, 110, 114, 124, 123, 118, 117, 116, 110, 114, 114, 96, 99, 103, 103, 104, 107, 107, 110, 111, 112, 113, 117, 115, 116, 100, 104, 102, 94, 93, 93, 96, 96, 111, 116, 107, 109, 105, 106, 97, 106, 107, 108.

Construct a grouped frequency distribution for the data. Use 85–89 for the first class and use the same width for each subsequent class.   *

19. Construct a histogram and a frequency polygon for the data involving stress ratings in Exercises 3–6.   *

20. Construct a histogram and a frequency polygon for the data in Exercise 7.   *

21. Construct a histogram and a frequency polygon for the data in Exercise 8.   *

The histogram shows the distribution of starting salaries (rounded to the nearest thousand dollars) for college graduates based on a random sample of recent graduates.

Starting Salaries of Recent College Graduates

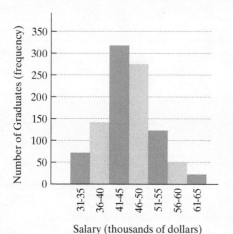

Salary (thousands of dollars)

In Exercises 22–25, determine whether each statement is true or false according to the graph.

22. The graph is based on a sample of approximately 500 recent college graduates.   false

23. More college graduates had starting salaries in the $51,000–$55,000 range than in the $36,000–$40,000 range.   false

24. If the sample is truly representative, then for a group of 400 college graduates, we can expect about 28 of them to have starting salaries in the $31,000–$35,000 range.   true

25. The percentage of starting salaries falling above those shown by any rectangular bar is equal to the percentage of starting salaries falling below that bar.   false

The frequency polygon shows a distribution of IQ scores.

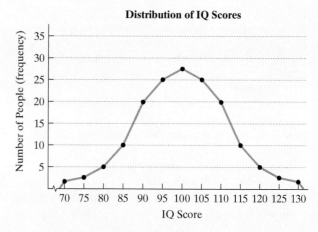

In Exercises 26–29, determine whether each statement is true or false according to the graph.

26. The graph is based on a sample of approximately 50 people.

27. More people had an IQ score of 100 than any other IQ score, and as the deviation from 100 increases or decreases, the scores fall off in a symmetrical manner.   true   **26.** false

28. More people had an IQ score of 110 than a score of 90.   false

29. The percentage of scores above any IQ score is equal to the percentage of scores below that score.   false

30. Construct a stem-and-leaf plot for the data in Exercise 17 showing the ages at which U.S. presidents were inaugurated.   *

31. A random sample of 40 high school teachers is selected from all high school teachers in a school district. The following list gives their ages:

63, 48, 42, 42, 38, 59, 41, 44, 45, 28, 54, 62, 51, 44, 63, 66, 59, 46, 51, 28, 37, 66, 42, 40, 30, 31, 48, 32, 29, 42, 63, 37, 36, 47, 25, 34, 49, 30, 35, 50.

Construct a stem-and-leaf plot for the data. What does the shape of the display reveal about the ages of the teachers?   *

32. In "Ages of Oscar-Winning Best Actors and Actresses" (*Mathematics Teacher* magazine) by Richard Brown and Gretchen Davis, the stem-and-leaf plots shown compare the ages of 30 actors and 30 actresses at the time they won the award.

Actors	Stems	Actresses
	2	146667
98753221	3	00113344455778
88776543322100	4	11129
6651	5	
210	6	011
6	7	4
	8	0

a. What is the age of the youngest actor to win an Oscar? 31

b. What is the age difference between the oldest and the youngest actress to win an Oscar? 59 years

c. What is the oldest age shared by two actors to win an Oscar? 56

d. What differences do you observe between the two stem-and-leaf plots? What explanations can you offer for these differences? Answers will vary.

In Exercises 33–37, describe what is misleading in each visual display of data.

33.
World Population, in Billions

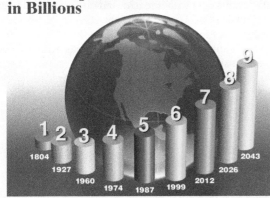

Source: U.S. Census Bureau

34.
Book Title Output in the United States

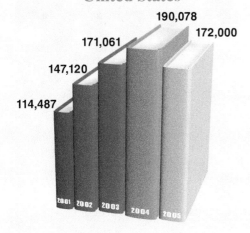

190,078
172,000
171,061
147,120
114,487

Source: R. R. Bowker

35.
Percentage of the World's Computers in Use, by Country

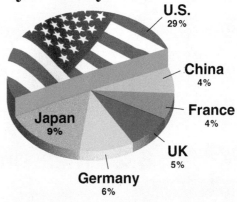

U.S.
29%

China
4%

France
4%

UK
5%

Germany
6%

Japan
9%

Source: Computer Industry Almanac

36.
Percentage of U.S. Households Watching ABC, CBS, and NBC in Prime Time

56%   51%   37%   22%   18%

1972–73   1982–83   1992–93 2002–03 2007–08

Source: Nielsen Media Research
Sizes of figures are not scaled proportionally in terms of the data they represent.

33. Time intervals on the horizontal axis do not represent equal amounts of time.
35. Percentages do not add up to 100%.

37.
Domestic Box-Office Receipts for Musical Films

Box-Office Receipts (millions of dollars)

$180
$160
$140
$120
$100
$80
$60
$40
$20

Chicago (2002) $170.7 | The Phantom of the Opera (2004) $51.3 | Rent (2005) $29.1 | The Producers (2005) $19.4 | Dreamgirls (2006) $103.4 | Hairspray (2007) $118.9 | Sweeney Todd (2007) $52.9 | Nine (2009) $19.7 | Rock of Ages (2012) $38.5

Source: Entertainment Weekly It is not clear whether the bars or the actors represent box-office receipts.

Critical Thinking Exercises

38. Create a graph that shows a more accurate way of presenting the data from one of the visual displays in Exercises 33–37.   Answers will vary.

39. Construct a grouped frequency distribution for the following data, showing the length, in miles, of the 25 longest rivers in the United States. Use five classes that have the same width.   *

2540	2340	1980	1900	1900
1460	1450	1420	1310	1290
1280	1240	1040	990	926
906	886	862	800	774
743	724	692	659	649

Source: U.S. Department of the Interior

*See Answers to Selected Exercises.

Group Exercises   40–41. Answers will vary.

40. The classic book on distortion using statistics is *How to Lie with Statistics* by Darrell Huff. This activity is designed for five people. Each person should select two chapters from Huff's book and then present to the class the common methods of statistical manipulation and distortion that Huff discusses.

41. Each group member should find one example of a graph that presents data with integrity and one example of a graph that is misleading. Use newspapers, magazines, the Internet, books, and so forth. Once graphs have been collected, each member should share his or her graphs with the entire group. Be sure to explain why one graph depicts data in a forthright manner and how the other graph misleads the viewer.

8.2 : Measures of Central Tendency

What am I supposed to learn?

After you have read this section, you should be able to:

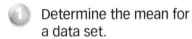 Determine the mean for a data set.

 Determine the median for a data set.

③ Determine the mode for a data set.

④ Determine the midrange for a data set.

During a lifetime, Americans average two weeks kissing. But wait, there's more:

- 130: The average number of "Friends" for a Facebook user
- 12: The average number of cars an American owns during a lifetime
- 300: The average number of times a 6-year old child laughs each day
- 550: The average number of hairs in the human eyebrow
- 28: The average number of years in the lifespan of a citizen during the Roman Empire
- 6,000,000: The average number of dust mites living in a US. bed.

Source: Listomania, Harper Design

These numbers represent what is "average" or "typical" in a variety of situations. In statistics, such values are known as **measures of central tendency** because they are generally located toward the center of a distribution. Four such measures are discussed in this section: the mean, the median, the mode, and the midrange. Each measure of central tendency is calculated in a different way. Thus, it is better to use a specific term (mean, median, mode, or midrange) than to use the generic descriptive term "average."

The Mean

 Determine the mean for a data set.

By far the most commonly used measure of central tendency is the *mean*. The **mean** is obtained by adding all the data items and then dividing the sum by the number of items. The Greek letter sigma, Σ, called a **symbol of summation**, is used to indicate the sum of data items. The notation Σx, read "the sum of x," means to add all the data items in a given data set. We can use this symbol to give a formula for calculating the mean.

Blitzer Bonus
· · · · · · · · · · · · · · · ·

Using Means to Compare How the U.S. Measures Up

- Mean Life Expectancy

78	82
U.S.	ITALY

- Mean Cost of an Angiogram

$798	$35
U.S.	CANADA

- Mean Number of Hours of TV Watched Daily

2.7	2.8
U.S.	ENGLAND

- Mean Number of Working Hours per Week

35.1	27.8
U.S.	GERMANY

- Mean Size of a Steak Served at Restaurants

13 ounces	8 ounces
U.S.	ENGLAND

Source: Time, USA Today

The Mean

The **mean** is the sum of the data items divided by the number of items.

$$\text{Mean} = \frac{\Sigma x}{n},$$

where Σx represents the sum of all the data items and n represents the number of items.

The mean of a sample is symbolized by $\bar{x}$ (read "x bar"), while the mean of an entire population is symbolized by μ (the lowercase Greek letter *mu*). Unless otherwise indicated, the data sets throughout this chapter represent samples, so we will use $\bar{x}$ for the mean: $\bar{x} = \frac{\Sigma x}{n}$.

Example 1   Calculating the Mean

Table 8.6 shows the ten highest-earning TV actors and the ten highest-earning TV actresses for the 2010–2011 television season. Find the mean earnings, in millions of dollars, for the ten highest-earning actors.

Table 8.6 Highest-Earning TV Actors and Actresses, 2010–2011

Actor	Earnings (millions of dollars)	Actress	Earnings (millions of dollars)
Charlie Sheen	$40	Eva Longoria	$13
Ray Romano	$20	Tina Fey	$13
Steve Carell	$15	Marcia Cross	$10
Mark Harmon	$13	Mariska Hargitay	$10
Jon Cryer	$11	Marg Helgenberger	$10
Laurence Fishburne	$11	Teri Hatcher	$9
Patrick Dempsey	$10	Felicity Huffman	$9
Simon Baker	$9	Courteney Cox	$7
Hugh Laurie	$9	Ellen Pompeo	$7
Chris Meloni	$9	Julianna Margulies	$7

Source: Forbes

Solution

We find the mean, $\bar{x}$, by adding the earnings for the actors and dividing this sum by 10, the number of data items.

$$\bar{x} = \frac{\Sigma x}{n} = \frac{40 + 20 + 15 + 13 + 11 + 11 + 10 + 9 + 9 + 9}{10} = \frac{147}{10} = 14.7$$

▷ The mean earnings of the ten highest-earning actors is $14.7 million.

One and only one mean can be calculated for any group of numerical data. The mean may or may not be one of the actual data items. In Example 1, the mean was 14.7, although no data item is 14.7.

✓ **Check Point 1**   Use **Table 8.6** to find the mean earnings, $\bar{x}$, in millions of dollars, for the ten highest-earning actresses. $9.5 million

In Example 1, some of the data items were identical. We can use multiplication when computing the sum for these identical items:

$$\bar{x} = \frac{40 + 20 + 15 + 13 + 11 + 11 + 10 + 9 + 9 + 9}{10}$$

$$= \frac{40 \cdot 1 + 20 \cdot 1 + 15 \cdot 1 + 13 \cdot 1 + 11 \cdot 2 + 10 \cdot 1 + 9 \cdot 3}{10}$$

> The data value 11 has a frequency of 2.
> The data value 9 has a frequency of 3.

When many data values occur more than once and a frequency distribution is used to organize the data, we can use the following formula to calculate the mean:

Table 8.7 Students' Stress-Level Ratings

Stress Rating x	Frequency f
0	2
1	1
2	3
3	12
4	16
5	18
6	13
7	31
8	26
9	15
10	14

Source: Journal of Personality and Social Psychology, 69, 1102–1112

Calculating the Mean for a Frequency Distribution

$$\text{Mean} = \bar{x} = \frac{\Sigma xf}{n},$$

where

x represents a data value.

f represents the frequency of that data value.

Σxf represents the sum of all the products obtained by multiplying each data value by its frequency.

n represents the *total frequency* of the distribution.

Example 2 ⁞ Calculating the Mean for a Frequency Distribution

In the previous Exercise Set, we mentioned a questionnaire given to students in an introductory statistics class during the first week of the course. One question asked, "How stressed have you been in the last $2\frac{1}{2}$ weeks, on a scale of 0 to 10, with 0 being not at all stressed and 10 being as stressed as possible?" **Table 8.7** shows the students' responses. Use this frequency distribution to find the mean of the stress-level ratings.

x	f	xf
0	2	$0 \cdot 2 = 0$
1	1	$1 \cdot 1 = 1$
2	3	$2 \cdot 3 = 6$
3	12	$3 \cdot 12 = 36$
4	16	$4 \cdot 16 = 64$
5	18	$5 \cdot 18 = 90$
6	13	$6 \cdot 13 = 78$
7	31	$7 \cdot 31 = 217$
8	26	$8 \cdot 26 = 208$
9	15	$9 \cdot 15 = 135$
10	14	$10 \cdot 14 = 140$

Totals: $n = 151$ $\quad\Sigma xf = 975$

> This value, the sum of the numbers in the second column, is the total frequency of the distribution.

Solution

We use the formula for the mean, $\bar{x}$:

$$\bar{x} = \frac{\Sigma xf}{n}.$$

First, we must find xf, obtained by multiplying each data value, x, by its frequency, f. Then, we need to find the sum of these products, Σxf. We can use the frequency distribution to organize these computations. Add a third column in which each data value is multiplied by its frequency. This column, shown on the left, is headed xf. Then, find the sum of the values, Σxf, in this column.

Now, substitute these values into the formula for the mean, $\bar{x}$. Remember that n is the *total frequency* of the distribution, or 151.

> Σxf is the sum of the numbers in the third column.

$$\bar{x} = \frac{\Sigma xf}{n} = \frac{975}{151} \approx 6.46$$

The mean of the 0 to 10 stress-level ratings is approximately 6.46. Notice that the mean is greater than 5, the middle of the 0 to 10 scale.

 **Check Point 2** Find the mean, $\bar{x}$, for the data items in the frequency distribution. (In order to save space, we've written the frequency distribution horizontally.) 36

Score, x	30	33	40	50
Frequency, f	3	4	4	1

② Determine the median for a data set.

The Median

The *median* age in the United States is 37.2. The oldest state by median age is Maine (42.7) and the youngest state is Utah (29.2). To find these values, researchers begin with appropriate random samples. The data items—that is, the ages—are arranged from youngest to oldest. The median age is the data item in the middle of each set of ranked, or ordered, data.

> ### The Median
>
> To find the **median** of a group of data items,
>
> 1. Arrange the data items in order, from smallest to largest.
> 2. If the number of data items is odd, the median is the data item in the middle of the list.
> 3. If the number of data items is even, the median is the mean of the two middle data items.

Example 3 ⁝⁝⁝ Finding the Median

Find the median for each of the following groups of data:

a. 84, 90, 98, 95, 88   **b.** 68, 74, 7, 13, 15, 25, 28, 59, 34, 47.

Solution

a. Arrange the data items in order, from smallest to largest. The number of data items in the list, five, is odd. Thus, the median is the middle number.

$$84, 88, 90, 95, 98$$

Middle data
item

The median is 90. Notice that two data items lie above 90 and two data items lie below 90.

b. Arrange the data items in order, from smallest to largest. The number of data items in the list, ten, is even. Thus, the median is the mean of the two middle data items.

$$7, 13, 15, 25, 28, 34, 47, 59, 68, 74$$

Middle data items
are 28 and 34.

$$\text{Median} = \frac{28 + 34}{2} = \frac{62}{2} = 31$$

The median is 31. Five data items lie above 31 and five data items lie below 31.

$$7 \quad 13 \quad 15 \quad 25 \quad 28 \mid 34 \quad 47 \quad 59 \quad 68 \quad 74$$

Five data items lie below 31.   Five data items lie above 31.

Median is 31.

Great Question!

What exactly does the median do with the data?

The median splits the data items down the middle, like the median strip in a road.

✓ **Check Point 3** Find the median for each of the following groups of data:

a. 28, 42, 40, 25, 35   35   **b.** 72, 61, 85, 93, 79, 87.   82

If a relatively long list of data items is arranged in order, it may be difficult to identify the item or items in the middle. In cases like this, the median can be found by determining its position in the list of items.

Great Question!

Does the formula
$$\frac{n + 1}{2}$$
give the value of the median?

No. The formula gives the *position* of the median, and not the actual value of the median. When finding the median, be sure to first arrange the data items in order from smallest to largest.

Position of the Median

If n data items are arranged in order, from smallest to largest, the median is the value in the

$$\frac{n + 1}{2}$$

position.

Example 4   Finding the Median Using the Position Formula

Table 8.8 gives the nine longest words in the English language. Find the median number of letters for the nine longest words.

Table 8.8 The Nine Longest Words in the English Language

Word	Number of Letters
Pneumonoultramicroscopicsilicovolcanoconiosis A lung disease caused by breathing in volcanic dust	45
Supercalifragilisticexpialidocious Meaning "wonderful", from song of this title in the movie *Mary Poppins*	34
Floccinaucinihilipilification Meaning "the action or habit of estimating as worthless"	29
Trinitrophenylmethylnitramine A chemical compound used as a detonator in shells	29
Antidisestablishmentarianism Meaning "opposition to the disestablishment of the Church of England"	28
Electroencephalographically Relating to brain waves	27
Microspectrophotometrically Relating to the measurement of light waves	27
Immunoelectrophoretically Relating to measurement of immunoglobulin	25
Spectroheliokinematograph A 1930s' device for monitoring and filming solar activity	25

Source: Chris Cole, rec.puzzles archive

Solution

We begin by listing the data items from smallest to largest.

$$25, 25, 27, 27, 28, 29, 29, 34, 45$$

There are nine data items, so $n = 9$. The median is the value in the

$$\frac{n + 1}{2} \text{ position} = \frac{9 + 1}{2} \text{ position} = \frac{10}{2} \text{ position} = \text{fifth position.}$$

We find the median by selecting the data item in the fifth position.

	Position 3	Position 4

$$25,\ 25,\ 27,\ 27,\ 28,\ 29,\ 29,\ 34,\ 45$$

Position 1	Position 2	Position 5

The median is 28. Notice that four data items lie above 28 and four data items lie below it. The median number of letters for the nine longest words in the English language is 28.

✓ **Check Point 4**   Find the median for the following group of data items:

$$1, 2, 2, 2, 3, 3, 3, 3, 3, 5, 6, 7, 7, 10, 11, 13, 19, 24, 26.\quad {}_5$$

Table 8.9 Time per Day, in Hours and Minutes, Spent Sleeping and Eating in Selected Countries

Country	Sleeping	Eating
France	8:50	2:15
U.S.	8:38	1:14
Spain	8:34	1:46
New Zealand	8:33	2:10
Australia	8:32	1:29
Turkey	8:32	1:29
Canada	8:29	1:09
Poland	8:28	1:34
Finland	8:27	1:21
Belgium	8:25	1:49
United Kingdom	8:23	1:25
Mexico	8:21	1:06
Italy	8:18	1:54
Germany	8:12	1:45
Sweden	8:06	1:34
Norway	8:03	1:22
Japan	7:50	1:57
S. Korea	7:49	1:36

Source: Organization for Economic Cooperation and Development

Example 5   Finding the Median Using the Position Formula

Table 8.9 gives the mean amount of time per day, in hours and minutes, spent sleeping and eating in 18 selected countries. Find the median amount of time per day, in hours and minutes, spent sleeping for these countries.

Solution

Reading from the bottom to the top of **Table 8.9**, the data items for sleeping appear from smallest to largest. There are 18 data items, so $n = 18$. The median is the value in the

$$\frac{n+1}{2}\ \text{position} = \frac{18+1}{2}\ \text{position} = \frac{19}{2}\ \text{position} = 9.5\ \text{position}.$$

This means that the median is the mean of the data items in positions 9 and 10.

	Position 3	Position 4		Position 7	Position 8

7:49, 7:50, 8:03, 8:06, 8:12, 8:18, 8:21, 8:23, 8:25, 8:27, 8:28, 8:29, 8:32, 8:32, 8:33, 8:34, 8:38, 8:50

Position 1	Position 2		Position 5	Position 6		Position 9	Position 10

$$\text{Median} = \frac{8\!:\!25 + 8\!:\!27}{2} = \frac{16\!:\!52}{2} = 8\!:\!26$$

The median amount of time per day spent sleeping for the 18 countries is 8 hours, 26 minutes.

✓ **Check Point 5**   Arrange the data items for eating in **Table 8.9** from smallest to largest. Then find the median amount of time per day, in hours and minutes, spent eating for the 18 countries.   1:06, 1:09, 1:14, 1:21, 1:22, 1:25, 1:29, 1:29, 1:34, 1:34, 1:36, 1:45, 1:46, 1:49, 1:54, 1:57, 2:10, 2:15; median: 1 hour, 34 minutes

When individual data items are listed from smallest to largest, you can find the median by identifying the item or items in the middle or by using the $\frac{n+1}{2}$ formula for its position. However, the formula for the position of the median is more useful when data items are organized in a frequency distribution.

Example 6 Finding the Median for a Frequency Distribution

The frequency distribution for the stress-level ratings of 151 students is repeated below using a horizontal format. Find the median stress-level rating.

Stress rating

x	0	1	2	3	4	5	6	7	8	9	10
f	2	1	3	12	16	18	13	31	26	15	14

Number of college students   Total: $n = 151$

Solution

There are 151 data items, so $n = 151$. The median is the value in the

$$\frac{n+1}{2} \text{ position} = \frac{151+1}{2} \text{ position} = \frac{152}{2} \text{ position} = 76\text{th position}.$$

We find the median by selecting the data item in the 76th position. The frequency distribution indicates that the data items begin with

$$0, 0, 1, 2, 2, 2, \ldots.$$

We can write the data items all out and then select the median, the 76th data item. A more efficient way to proceed is to count down the frequency column in the distribution until we identify the 76th data item:

x	f
0	2
1	1
2	3
3	12
4	16
5	18
6	13
7	31
8	26
9	15
10	14

We count down the frequency column.

1, 2

3

4, 5, 6

7, 8, 9, 10, 11, 12, 13, 14, 15, 16, 17, 18

19, 20, 21, 22, 23, 24, 25, 26, 27, 28, 29, 30, 31, 32, 33, 34

35, 36, 37, 38, 39, 40, 41, 42, 43, 44, 45, 46, 47, 48, 49, 50, 51, 52

53, 54, 55, 56, 57, 58, 59, 60, 61, 62, 63, 64, 65

66, 67, 68, 69, 70, 71, 72, 73, 74, 75, 76

Stop counting. We've reached the 76th data item.

The 76th data item is 7. The median stress-level rating is 7.

Check Point 6 Find the median for the following frequency distribution. 54.5

Age at presidential inauguration

x	42	43	46	51	52	54	55	56	60	61	64	69
f	1	1	1	3	1	2	2	2	1	2	1	1

Number of U.S. presidents assuming office in the 20th century with the given age

Statisticians generally use the median, rather than the mean, when reporting income. Why? Our next example will help to answer this question.

Example 7 — Comparing the Median and the Mean

Five employees in the assembly section of a television manufacturing company earn salaries of $19,700, $20,400, $21,500, $22,600, and $23,000 annually. The section manager has an annual salary of $95,000.

a. Find the median annual salary for the six people.

b. Find the mean annual salary for the six people.

Solution

a. To compute the median, first arrange the salaries in order:

$$\$19,700, \quad \$20,400, \quad \$21,500, \quad \$22,600, \quad \$23,000, \quad \$95,000.$$

Because the list contains an even number of data items, six, the median is the mean of the two middle items.

$$\text{Median} = \frac{\$21,500 + \$22,600}{2} = \frac{\$44,100}{2} = \$22,050$$

The median annual salary is $22,050.

b. We find the mean annual salary by adding the six annual salaries and dividing by 6.

$$\text{Mean} = \frac{\$19,700 + \$20,400 + \$21,500 + \$22,600 + \$23,000 + \$95,000}{6}$$

$$= \frac{\$202,200}{6} = \$33,700$$

The mean annual salary is $33,700.

In Example 7, the median annual salary is $22,050 and the mean annual salary is $33,700. Why such a big difference between these two measures of central tendency? The relatively high annual salary of the section manager, $95,000, pulls the mean salary to a value considerably higher than the median salary. When one or more data items are much greater than the other items, these extreme values can greatly influence the mean. In cases like this, the median is often more representative of the data.

This is why the median, rather than the mean, is used to summarize the incomes, by gender and race, shown in **Figure 8.6**. Because no one can earn less than $0, the distribution of income must come to an end at $0 for each of these eight groups. By contrast, there is no upper limit on income on the high side. In the United States, the wealthiest 20% of the population earn about 50% of the total income. The relatively few people with very high annual incomes tend to pull the mean income to a value considerably greater than the median income. Reporting mean incomes in **Figure 8.6** would inflate the numbers shown, making them nonrepresentative of the millions of workers in each of the eight groups.

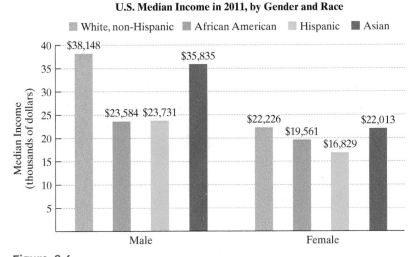

Figure 8.6
Source: U.S. Census Bureau

Check Point 7   **Table 8.10** shows the net worth, in millions of 2010 dollars, for ten U.S. presidents from Kennedy through Obama.

Table 8.10 Net Worth for Ten U.S. Presidents

President	Net Worth (millions of dollars)	President	Net Worth (millions of dollars)
Kennedy	$1000 (i.e. $1 billion)	Reagan	$13
Johnson	$98	Bush	$23
Nixon	$15	Clinton	$38
Ford	$7	Bush	$20
Carter	$7	Obama	$5

Source: Time

a. Find the mean net worth, in millions of dollars, for the ten presidents. <small>$122.6 million</small>

b. Find the median net worth, in millions of dollars, for the ten presidents. <small>$17.5 million</small>

c. Describe why one of the measures of central tendency is greater than the other. <small>Kennedy's net worth was much greater than the other presidents'.</small>

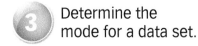

Determine the mode for a data set.

The Mode

Let's take one final look at the frequency distribution for the stress-level ratings of 151 college students.

Stress rating	x	0	1	2	3	4	5	6	7	8	9	10
Number of college students	f	2	1	3	12	16	18	13	31	26	15	14

> 7 is the stress rating with the greatest frequency.

The data value that occurs most often in this distribution is 7, the stress rating for 31 of the 151 students. We call 7 the *mode* of this distribution.

> ### The Mode
>
> The **mode** is the data value that occurs most often in a data set. If more than one data value has the highest frequency, then each of these data values is a mode. If there is no data value that occurs most often, then the data set has no mode.

Example 8   Finding the Mode

Find the mode for each of the following groups of data:

a. 7, 2, 4, 7, 8, 10   **b.** 2, 1, 4, 5, 3   **c.** 3, 3, 4, 5, 6, 6.

Solution

a. 7, 2, 4, 7, 8, 10

7 occurs most often.

The mode is 7.

b. 2, 1, 4, 5, 3

Each data item occurs the same number of times.

There is no mode.

c. 3, 3, 4, 5, 6, 6

Both 3 and 6 occur most often.

The modes are 3 and 6. The data set is said to be **bimodal**.

 Find the mode for each of the following groups of data:

a. 3, 8, 5, 8, 9, 10   8   **b.** 3, 8, 5, 8, 9, 3   3 and 8   **c.** 3, 8, 5, 6, 9, 10   no mode

 Determine the midrange for a data set.

The Midrange

Table 8.11 shows the ten hottest cities in the United States. Because temperature is constantly changing, you might wonder how the mean temperatures shown in the table are obtained.

First, we need to find a representative daily temperature. This is obtained by adding the lowest and highest temperatures for the day and then dividing this sum by 2. Next, we take the representative daily temperatures for all 365 days, add them, and divide the sum by 365. These are the mean temperatures that appear in **Table 8.11**.

Representative daily temperature,

$$\frac{\text{lowest daily temperature} + \text{highest daily temperature}}{2},$$

is an example of a measure of central tendency called the *midrange*.

Table 8.11 Ten Hottest U.S. Cities

City	Mean Temperature
Key West, FL	77.8°
Miami, FL	75.9°
West Palm Beach, FL	74.7°
Fort Myers, FL	74.4°
Yuma, AZ	74.2°
Brownsville, TX	73.8°
Phoenix, AZ	72.6°
Vero Beach, FL	72.4°
Orlando, FL	72.3°
Tampa, FL	72.3°

Source: National Oceanic and Atmospheric Administration

The Midrange

The **midrange** is found by adding the lowest and highest data values and dividing the sum by 2.

$$\text{Midrange} = \frac{\text{lowest data value} + \text{highest data value}}{2}$$

Example 9   Finding the Midrange

Newsweek magazine examined factors that affect women's lives, including justice, health, education, economics, and politics. Using these five factors, the magazine graded each of 165 countries on a scale from 0 to 100. The 12 best places to be a woman and the 12 worst places to be a woman are shown in **Table 8.12**.

Table 8.12 Women in the World

Best Places to Be a Woman		Worst Places to Be a Woman	
Country	Score	Country	Score
Iceland	100.0	Chad	0.0
Canada	99.6	Afghanistan	2.0
Sweden	99.2	Yemen	12.1
Denmark	95.3	Democratic Republic of the Congo	13.6
Finland	92.8	Mali	17.6
Switzerland	91.9	Solomon Islands	20.8
Norway	91.3	Niger	21.2
United States	89.8	Pakistan	21.4
Australia	88.2	Ethiopia	23.7
Netherlands	87.7	Sudan	26.1
New Zealand	87.2	Guinea	28.5
France	87.2	Sierra Leone	29.0

Source: Newsweek

Find the midrange score among the 12 best countries to be a woman.

Solution

Refer to **Table 8.12** on the previous page.

$$\text{Midrange} = \frac{\text{best place with the lowest score} + \text{best place with the highest score}}{2}$$

$$= \frac{87.2 + 100.0}{2} = \frac{187.2}{2} = 93.6$$

▶ The midrange score among the 12 best countries to be a woman is 93.6.

We can find the mean score among the 12 best countries to be a woman by adding up the 12 scores and then dividing the sum by 12. By doing so, we can determine that the mean score is approximately 92.5. It is much faster to calculate the midrange, which is often used as an estimate for the mean.

✓ **Check Point 9**   Use **Table 8.12** on the previous page to find the midrange score
▶ among the 12 worst countries to be a woman.   14.5

Example 10   Finding the Four Measures
of Central Tendency

Suppose your six exam grades in a class are

52, 69, 75, 86, 86, and 92.

Compute your final grade (90–100 = A, 80–89 = B, 70–79 = C, 60–69 = D, below 60 = F) using the

a. mean.   **b.** median.   **c.** mode.   **d.** midrange.

Solution

a. The mean is the sum of the data items divided by the number of items, 6.

$$\text{Mean} = \frac{52 + 69 + 75 + 86 + 86 + 92}{6} = \frac{460}{6} \approx 76.67$$

Using the mean, your final course grade is C.

b. The six data items, 52, 69, 75, 86, 86, and 92, are arranged in order. Because the number of data items is even, the median is the mean of the two middle items.

$$\text{Median} = \frac{75 + 86}{2} = \frac{161}{2} = 80.5$$

Using the median, your final course grade is B.

c. The mode is the data value that occurs most frequently. Because 86 occurs most often, the mode is 86. Using the mode, your final course grade is B.

d. The midrange is the mean of the lowest and highest data values.

$$\text{Midrange} = \frac{52 + 92}{2} = \frac{144}{2} = 72$$

Using the midrange, your final course grade is C.

✓ **Check Point 10**   *Consumer Reports* magazine gave the following data for the number of calories in a meat hot dog for each of 17 brands:

173, 191, 182, 190, 172, 147, 146, 138, 175, 136, 179, 153, 107, 195, 135, 140, 138.

Find the mean, median, mode, and midrange for the number of calories in a meat hot dog for the 17 brands. If necessary, round answers to the nearest tenth of a
▶ calorie.   mean: 158.6 cal; median: 153 cal; mode: 138 cal; midrange: 151 cal

Achieving Success

A recent government study cited in *Math: A Rich Heritage* (Globe Fearon Educational Publisher) found this simple fact: **The more college mathematics courses you take, the greater your earning potential will be.** Even jobs that do not require a college degree require mathematical thinking that involves attending to precision, making sense of complex problems, and persevering in solving them. No other discipline comes close to math in offering a more extensive set of tools for application and intellectual development. Take as much math as possible as you continue your pathways into higher education.

Concept and Vocabulary Check

Exercises in the Concept and Vocabulary Check are intended for group and class discussions.

In Exercises 1–5, fill in each blank so that the resulting statement is true.

1. $\frac{\Sigma x}{n}$, the sum of all the data items divided by the number of data items, is the measure of central tendency called the ___mean___.

2. The measure of central tendency that is the data item in the middle of ranked, or ordered, data is called the ___median___.

3. If n data items are arranged in order, from smallest to largest, the data item in the middle is the value in ___$\frac{n+1}{2}$___ position.

4. A data value that occurs most often in a data set is the measure of central tendency called the ___mode___.

5. The measure of central tendency that is found by adding the lowest and highest data values and dividing the sum by 2 is called the ___midrange___.

In Exercises 6–9, determine whether each statement is true or false. If the statement is false, make the necessary change(s) to produce a true statement.   Changes to false statements will vary.

6. Numbers representing what is average or typical about a data set are called measures of central tendency.   true

7. When finding the mean, it is necessary to arrange the data items in order.   false

8. If one or more data items are much greater than the other items, the mean, rather than the median, is more representative of the data.   false

9. A data set can contain more than one median, or no median at all.   false

Respond to Exercises 10–18 using verbal or written explanations.   10–18. Answers will vary.

10. What is the mean and how is it obtained?

11. What is the median and how is it obtained?

12. What is the mode and how is it obtained?

13. What is the midrange and how is it obtained?

14. The "average" income in the United States can be given by the mean or the median.

    a. Which measure would be used in anti-U.S. propaganda? Explain your answer.

    b. Which measure would be used in pro-U.S. propaganda? Explain your answer.

15. In a class of 40 students, 21 have examination scores of 77%. Which measure or measures of central tendency can you immediately determine? Explain your answer.

16. You read an article that states, "Of the 411 players in the National Basketball Association, only 138 make more than the average salary of $3.12 million." Is $3.12 million the mean or the median salary? Explain your answer.

17. A college student's parents promise to pay for next semester's tuition if an A average is earned in chemistry. With examination grades of 97%, 97%, 75%, 70%, and 55%, the student reports that an A average has been earned. Which measure of central tendency is the student reporting as the average? How is this student misrepresenting the course performance with statistics?

18. According to the National Oceanic and Atmospheric Administration, the coldest city in the United States is International Falls, Minnesota, with a mean Fahrenheit temperature of 36.8°. Explain how this mean is obtained.

Exercise Set 8.2

Practice Exercises

In Exercises 1–8, find the mean for each group of data items.

1. 7, 4, 3, 2, 8, 5, 1, 3   4.125

2. 11, 6, 4, 0, 2, 1, 12, 0, 0   4

3. 91, 95, 99, 97, 93, 95   95

4. 100, 100, 90, 30, 70, 100   ≈ 81.67

5. 100, 40, 70, 40, 60   62

6. 1, 3, 5, 10, 8, 5, 6, 8   5.75

7. 1.6, 3.8, 5.0, 2.7, 4.2, 4.2, 3.2, 4.7, 3.6, 2.5, 2.5   ≈ 3.45

8. 1.4, 2.1, 1.6, 3.0, 1.4, 2.2, 1.4, 9.0, 9.0, 1.8   3.29

In Exercises 9–12, find the mean for the data items in the given frequency distribution.

9.

Score x	Frequency f
1	1
2	3
3	4
4	4
5	6
6	5
7	3
8	2

* **10.**   ≈ 4.13

Score x	Frequency f
1	2
2	4
3	5
4	7
5	6
6	4
7	3

11.

Score x	Frequency f
1	1
2	1
3	2
4	5
5	7
6	9
7	8
8	6
9	4
10	3

* **12.**   *

Score x	Frequency f
1	3
2	4
3	6
4	8
5	9
6	7
7	5
8	2
9	1
10	1

In Exercises 13–20, find the median for each group of data items.

13. 7, 4, 3, 2, 8, 5, 1, 3   3.5

14. 11, 6, 4, 0, 2, 1, 12, 0, 0   2

15. 91, 95, 99, 97, 93, 95   95

16. 100, 100, 90, 30, 70, 100   95

17. 100, 40, 70, 40, 60   60

18. 1, 3, 5, 10, 8, 5, 6, 8   5.5

19. 1.6, 3.8, 5.0, 2.7, 4.2, 4.2, 3.2, 4.7, 3.6, 2.5, 2.5   3.6

20. 1.4, 2.1, 1.6, 3.0, 1.4, 2.2, 1.4, 9.0, 9.0, 1.8   1.95

Find the median for the data items in the frequency distribution in

21. Exercise 9.   5

22. Exercise 10.   4

23. Exercise 11.   6

24. Exercise 12.   5

In Exercises 25–32, find the mode for each group of data items. If there is no mode, so state.

25. 7, 4, 3, 2, 8, 5, 1, 3   3

26. 11, 6, 4, 0, 2, 1, 12, 0, 0   0

27. 91, 95, 99, 97, 93, 95   95

28. 100, 100, 90, 30, 70, 100   100

29. 100, 40, 70, 40, 60   40

30. 1, 3, 5, 10, 8, 5, 6, 8   5, 8

31. 1.6, 3.8, 5.0, 2.7, 4.2, 4.2, 3.2, 4.7, 3.6, 2.5, 2.5   2.5, 4.2

32. 1.4, 2.1, 1.6, 3.0, 1.4, 2.2, 1.4, 9.0, 9.0, 1.8   1.4

Find the mode for the data items in the frequency distribution in

33. Exercise 9.   5

34. Exercise 10.   4

35. Exercise 11.   6

36. Exercise 12.   5

In Exercises 37–44, find the midrange for each group of data items.

37. 7, 4, 3, 2, 8, 5, 1, 3   4.5

38. 11, 6, 4, 0, 2, 1, 12, 0, 0   6

39. 91, 95, 99, 97, 93, 95   95

40. 100, 100, 90, 30, 70, 100   65

41. 100, 40, 70, 40, 60   70

42. 1, 3, 5, 10, 8, 5, 6, 8   5.5

43. 1.6, 3.8, 5.0, 2.7, 4.2, 4.2, 3.2, 4.7, 3.6, 2.5, 2.5   3.3

44. 1.4, 2.1, 1.6, 3.0, 1.4, 2.2, 1.4, 9.0, 9.0, 1.8   5.2

Find the midrange for the data items in the frequency distribution in

45. Exercise 9.   4.5

46. Exercise 10.   4

47. Exercise 11.   5.5

48. Exercise 12.   5.5

Practice Plus

In Exercises 49–54, use each display of data items to find the mean, median, mode, and midrange.

49.

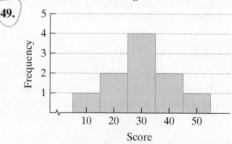

mean: 30;
median: 30;
mode: 30;
midrange: 30

50.

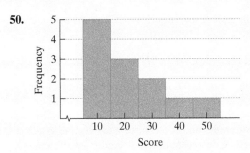

mean: ≈ 21.7;
median: 20;
mode: 10;
midrange: 30

51.

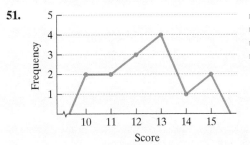

mean: ≈ 12.4;
median: 12.5;
mode: 13;
midrange: 12.5

52.

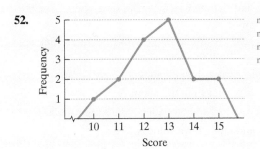

mean: ≈ 12.7;
median: 13;
mode: 13;
midrange: 12.5

53.

Stems	Leaves
2	1 4 5
3	0 1 1 3
4	2 5

mean: ≈ 31.3;
median: 31;
mode: 31;
midrange: 33

54.

Stems	Leaves
2	8
3	2 4 4 9
4	0 1 5 7

mean: ≈ 37.8;
median: 39;
mode: 34;
midrange: 37.5

Application Exercises

Exercises 55–57 present data on a variety of topics. For each data set described in boldface, find the

a. *mean.*

b. *median.*

c. *mode (or state that there is no mode).*

d. *midrange.*

55. Top Cities with New College Graduates *

Metro Area	Number of Recent College Graduates Who Moved to the Area from 2000–2011 (thousands)
New York	200
Chicago	97
Washington, D.C.	92
Los Angeles	92
San Francisco	64
Houston	51
Boston	51
Dallas-Fort Worth	50
Philadelphia	49
Denver	37
Seattle	34
Minneapolis-St. Paul	32
San Jose	27

Source: USA Today

56. Net Worth for the First 13 U.S. Presidents *

President	Net Worth (millions of 2010 dollars)
Washington	$525
Adams	$19
Jefferson	$212
Madison	$101
Monroe	$27
Adams	$21
Jackson	$119
Van Buren	$26
Harrison	$5
Tyler	$51
Polk	$10
Taylor	$6
Fillmore	$4

Source: Time

*See Answers to Selected Exercises.

57. Number of Social Interactions of College Students In Exercise Set 8.1, we presented a grouped frequency distribution showing the number of social interactions of ten minutes or longer over a one-week period for a group of college students. (These interactions excluded family and work situations.) Use the frequency distribution shown to solve this exercise. (This distribution was obtained by replacing the classes in the grouped frequency distribution previously shown with the midpoints of the classes.)   *

Social interactions in a week x	2	7	12	17	22	27	32	37	42	47
Number of college students f	12	16	16	16	10	11	4	3	3	3

The weights (to the nearest five pounds) of 40 randomly selected male high school seniors are organized in a histogram with a superimposed frequency polygon. Use the graph to answer Exercises 58–61.

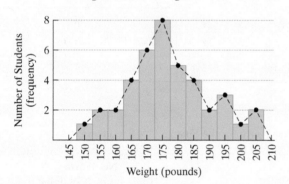

Weights of 40 Male High School Seniors

58. Find the mean weight.   176.875 lb

59. Find the median weight.   175 lb

60. Find the modal weight.   175 lb

61. Find the midrange weight.   177.5 lb

62. An advertisement for a speed-reading course claimed that the "average" reading speed for people completing the course was 1000 words per minute. Shown below are the actual data for the reading speeds per minute for a sample of 24 people who completed the course.

1000	900	800	1000	900	850
650	1000	1050	800	1000	850
700	750	800	850	900	950
600	1100	950	700	750	650

  a. Find the mean, median, mode, and midrange. (If you prefer, first organize the data in a frequency distribution.)   *

  b. Which measure of central tendency was given in the advertisement?   mode

  c. Which measure of central tendency is the best indicator of the "average" reading speed in this situation? Explain your answer.   Answers will vary.

63. In one common system for finding a grade-point average, or GPA,

$$A = 4, B = 3, C = 2, D = 1, F = 0.$$

The GPA is calculated by multiplying the number of credit hours for a course and the number assigned to each grade, and then adding these products. Then divide this sum by the total number of credit hours. Because each course grade is weighted according to the number of credits of the course, GPA is called a *weighted mean*. Calculate the GPA for this transcript:

  Sociology: 3 cr. A; Biology: 3.5 cr. C; Music: 1 cr. B; Math: 4 cr. B; English: 3 cr. C.   ≈2.76

Critical Thinking Exercises

64. Give an example of a set of six examination grades (from 0 to 100) with each of the following characteristics:

  a. The mean and the median have the same value, but the mode has a different value.   *

  b. The mean and the mode have the same value, but the median has a different value.   *

  c. The mean is greater than the median.   *

  d. The mode is greater than the mean.   *

  e. The mean, median, and mode have the same value.   *

  f. The mean and mode have values of 72.   *

65. On an examination given to 30 students, no student scored below the mean. Describe how this occurred.   All 30 students had the same grade.

Group Exercises   66–67. Answers will vary.

66. Select a characteristic, such as shoe size or height, for which each member of the group can provide a number. Choose a characteristic of genuine interest to the group. For this characteristic, organize the data collected into a frequency distribution and a graph. Compute the mean, median, mode, and midrange. Discuss any differences among these values. What happens if the group is divided (men and women, or people under a certain age and people over a certain age) and these measures of central tendency are computed for each of the subgroups? Attempt to use measures of central tendency to discover something interesting about the entire group or the subgroups.

67. A book on spotting bad statistics and learning to think critically about these influential numbers is *Damn Lies and Statistics* by Joel Best (University of California Press, 2001). This activity is designed for six people. Each person should select one chapter from Best's book. The group report should include examples of the use, misuse, and abuse of statistical information. Explain exactly how and why bad statistics emerge, spread, and come to shape policy debates. What specific ways does Best recommend to detect bad statistics?

*See Answers to Selected Exercises.

8.3 Measures of Dispersion

What am I supposed to learn?

After you have read this section, you should be able to:

 Determine the range for a data set.

 Determine the standard deviation for a data set.

When you think of Houston, Texas and Honolulu, Hawaii, do balmy temperatures come to mind? Both cities have a mean temperature of 75°. However, the mean temperature does not tell the whole story. The temperature in Houston differs seasonally from a low of about 40° in January to a high of close to 100° in July and August. By contrast, Honolulu's temperature varies less throughout the year, usually ranging between 60° and 90°.

Measures of dispersion are used to describe the spread of data items in a data set. Two of the most common measures of dispersion, the *range* and the *standard deviation*, are discussed in this section.

The Range

 Determine the range for a data set.

A quick but rough measure of dispersion is the **range**, the difference between the highest and lowest data values in a data set. For example, if Houston's hottest annual temperature is 103° and its coldest annual temperature is 33°, the range in temperature is

$$103° - 33°, \quad \text{or} \quad 70°.$$

If Honolulu's hottest day is 89° and its coldest day 61°, the range in temperature is

$$89° - 61°, \quad \text{or} \quad 28°.$$

> **The Range**
>
> The **range**, the difference between the highest and lowest data values in a data set, indicates the total spread of the data.
>
> $$\text{Range} = \text{highest data value} - \text{lowest data value}$$

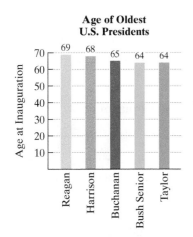

Age of Oldest U.S. Presidents

Figure 8.7

Source: Internet Public Library

Example 1   Computing the Range

Figure 8.7 shows the age of the five oldest U.S. presidents at the start of their first term. Find the age range for the five oldest presidents.

Solution

$$\text{Range} = \text{highest data value} - \text{lowest data value}$$
$$= 69 - 64 = 5$$

The range is 5 years.

 **Check Point 1**   Find the range for the following group of data items:

$$4, 2, 11, 7. \quad \text{\small 9}$$

The Standard Deviation

A second measure of dispersion, and one that is dependent on *all* of the data items, is called the **standard deviation**. The standard deviation is found by determining how much each data item differs from the mean.

In order to compute the standard deviation, it is necessary to find by how much each data item deviates from the mean. First compute the mean, $\bar{x}$. Then subtract the mean from each data item, $x - \bar{x}$. Example 2 shows how this is done. In Example 3, we will use this skill to actually find the standard deviation.

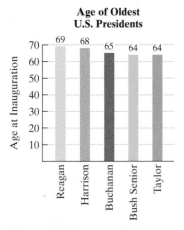

Age of Oldest U.S. Presidents

Figure 8.7 (repeated)

Table 8.13 Deviations from the Mean

Data Item x	Deviation: data item − mean $x - \bar{x}$
69	$69 - 66 =\;\; 3$
68	$68 - 66 =\;\; 2$
65	$65 - 66 = -1$
64	$64 - 66 = -2$
64	$64 - 66 = -2$

mean: 6

Data Item	Deviation
2	−4
4	−2
7	1
11	5

Example 2 **Preparing to Find the Standard Deviation; Finding Deviations from the Mean**

Find the deviations from the mean for the five data items 69, 68, 65, 64, and 64, shown in **Figure 8.7**.

Solution

First, calculate the mean, $\bar{x}$.

$$\bar{x} = \frac{\sum x}{n} = \frac{69 + 68 + 65 + 64 + 64}{5} = \frac{330}{5} = 66$$

The mean age for the five oldest U.S. presidents is 66 years. Now, let's find by how much each of the five data items in **Figure 8.7** differs from 66, the mean. For Reagan, who was 69 at the start of his first term, the computation is shown as follows:

$$\text{Deviation from mean} = \text{data item} - \text{mean}$$
$$= x - \bar{x}$$
$$= 69 - 66 = 3.$$

This indicates that Reagan's inaugural age exceeds the mean by three years.

The computation for Buchanan, who was 65 at the start of his first term, is given by

$$\text{Deviation from mean} = \text{data item} - \text{mean}$$
$$= x - \bar{x}$$
$$= 65 - 66 = -1.$$

This indicates that Buchanan's inaugural age is one year below the mean.

The deviations from the mean for each of the five given data items are shown in **Table 8.13**.

 **Check Point 2** Compute the mean for the following group of data items:

$$2, 4, 7, 11.$$

Then find the deviations from the mean for the four data items. Organize your work in table form just like **Table 8.13**. Keep track of these computations. You will be using them in Check Point 3.

The sum of the deviations from the mean for a set of data is always zero: $\Sigma(x - \bar{x}) = 0$. For the deviations from the mean shown in **Table 8.13**,

$$3 + 2 + (-1) + (-2) + (-2) = 5 + (-5) = 0.$$

This shows that we cannot find a measure of dispersion by finding the mean of the deviations, because this value is always zero. However, a kind of average of the deviations from the mean, called the **standard deviation**, can be computed. We do so by squaring each deviation and later introducing a square root in the computation. Here are the details on how to find the standard deviation for a set of data:

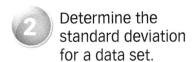

Determine the standard deviation for a data set.

Computing the Standard Deviation for a Data Set

1. Find the mean of the data items.

2. Find the deviation of each data item from the mean:

$$\text{data item} - \text{mean}.$$

3. Square each deviation:

$$(\text{data item} - \text{mean})^2.$$

4. Sum the squared deviations:

$$\Sigma(\text{data item} - \text{mean})^2.$$

5. Divide the sum in step 4 by $n - 1$, where n represents the number of data items:

$$\frac{\Sigma(\text{data item} - \text{mean})^2}{n - 1}.$$

6. Take the square root of the quotient in step 5. This value is the standard deviation for the data set.

$$\text{Standard deviation} = \sqrt{\frac{\Sigma(\text{data item} - \text{mean})^2}{n - 1}}$$

The standard deviation of a sample is symbolized by s, while the standard deviation of an entire population is symbolized by σ (the lowercase Greek letter *sigma*). Unless otherwise indicated, data sets represent samples, so we will use s for the standard deviation:

$$s = \sqrt{\frac{\Sigma(x - \bar{x})^2}{n - 1}}.$$

The computation of the standard deviation can be organized using a table with three columns:

Data item x	Deviation: $x - \bar{x}$ Data item − mean	(Deviation)^2: $(x - \bar{x})^2$ (Data item − mean)^2

In Example 2, we worked out the first two columns of such a table. Let's continue working with the data for the ages of the five oldest U.S. presidents and compute the standard deviation.

Example 3 **Computing the Standard Deviation**

Figure 8.7, giving the age of the five oldest U.S. presidents at the start of their first term, is shown on the previous page. Find the standard deviation for the ages of the five presidents.

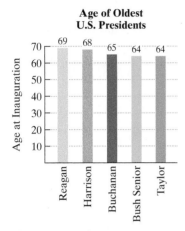

Age of Oldest U.S. Presidents

Figure 8.7 (repeated)

Solution

Step 1 Find the mean. From our work in Example 2, the mean is 66: $\bar{x} = 66$.

Step 2 Find the deviation of each data item from the mean: data item − mean or $x - \bar{x}$. This, too, was done in Example 2 for each of the five data items.

Step 3 Square each deviation: (data item − mean)² or $(x - \bar{x})^2$. We square each of the numbers in the (data item − mean) column, shown in **Table 8.14**. Notice that squaring the difference always results in a nonnegative number.

Table 8.14 Computing the Standard Deviation

Data item x	Deviation: data item − mean $x - \bar{x}$	(Deviation)²: (data item − mean)² $(x - \bar{x})^2$
69	$69 - 66 = 3$	$3^2 = 3 \cdot 3 = 9$
68	$68 - 66 = 2$	$2^2 = 2 \cdot 2 = 4$
65	$65 - 66 = -1$	$(-1)^2 = (-1) \cdot (-1) = 1$
64	$64 - 66 = -2$	$(-2)^2 = (-2) \cdot (-2) = 4$
64	$64 - 66 = -2$	$(-2)^2 = (-2) \cdot (-2) = 4$
Totals:	$\Sigma(x - \bar{x}) = 0$	$\Sigma(x - \bar{x})^2 = 22$

The sum of the deviations for a set of data is always zero.

Adding the five numbers in the third column gives the sum of the squared deviations: $\Sigma(\text{data item} - \text{mean})^2$.

Step 4 Sum the squared deviations: $\Sigma(\text{data item} - \text{mean})^2$. This step is shown in **Table 8.14**. The squares in the third column were added, resulting in a sum of 22: $\Sigma(x - \bar{x})^2 = 22$.

Step 5 Divide the sum in step 4 by $n - 1$, where n represents the number of data items. The number of data items is 5 so we divide by 4.

$$\frac{\Sigma(x - \bar{x})^2}{n - 1} = \frac{\Sigma(\text{data item} - \text{mean})^2}{n - 1} = \frac{22}{5 - 1} = \frac{22}{4} = 5.5$$

Step 6 The standard deviation, s, is the square root of the quotient in step 5.

$$s = \sqrt{\frac{\Sigma(x - \bar{x})^2}{n - 1}} = \sqrt{\frac{\Sigma(\text{data item} - \text{mean})^2}{n - 1}} = \sqrt{5.5} \approx 2.35$$

The standard deviation for the ages of the five oldest U.S. presidents is approximately 2.35 years.

Technology

Almost all scientific and graphing calculators compute the standard deviation of a set of data. Using the data items in Example 3,

69, 68, 65, 64, 64,

the keystrokes for obtaining the standard deviation on many scientific calculators are as follows:

69 $\boxed{\Sigma+}$ 68 $\boxed{\Sigma+}$ 65 $\boxed{\Sigma+}$

64 $\boxed{\Sigma+}$ 64 $\boxed{\Sigma+}$ $\boxed{\text{2nd}}$ $\boxed{\sigma n - 1}$.

Graphing calculators require that you specify if data items are from an entire population or a sample of the population.

✓ **Check Point 3** Find the standard deviation for the group of data items in Check Point 2 on page 524. Round to two decimal places. ≈ 3.92

Example 4 illustrates that as the spread of data items increases, the standard deviation gets larger.

Example 4 **Computing the Standard Deviation**

Find the standard deviation of the data items in each of the samples shown below.

Sample A	Sample B
17, 18, 19, 20, 21, 22, 23	5, 10, 15, 20, 25, 30, 35

Solution

Begin by finding the mean for each sample.
Sample A:

$$\text{Mean} = \frac{17 + 18 + 19 + 20 + 21 + 22 + 23}{7} = \frac{140}{7} = 20$$

Sample B:

$$\text{Mean} = \frac{5 + 10 + 15 + 20 + 25 + 30 + 35}{7} = \frac{140}{7} = 20$$

Although both samples have the same mean, the data items in sample B are more spread out. Thus, we would expect sample B to have the greater standard deviation. The computation of the standard deviation requires that we find $\Sigma(\text{data item} - \text{mean})^2$, shown in **Table 8.15**.

Table 8.15 Computing Standard Deviations for Two Samples

Sample A			Sample B		
Data Item x	Deviation: data item − mean $x - \bar{x}$	(Deviation)^2: (data item − mean)^2 $(x - \bar{x})^2$	Data Item x	Deviation: data item − mean $x - \bar{x}$	(Deviation)^2: (data item − mean)^2 $(x - \bar{x})^2$
17	$17 - 20 = -3$	$(-3)^2 = 9$	5	$5 - 20 = -15$	$(-15)^2 = 225$
18	$18 - 20 = -2$	$(-2)^2 = 4$	10	$10 - 20 = -10$	$(-10)^2 = 100$
19	$19 - 20 = -1$	$(-1)^2 = 1$	15	$15 - 20 = -5$	$(-5)^2 = 25$
20	$20 - 20 = 0$	$0^2 = 0$	20	$20 - 20 = 0$	$0^2 = 0$
21	$21 - 20 = 1$	$1^2 = 1$	25	$25 - 20 = 5$	$5^2 = 25$
22	$22 - 20 = 2$	$2^2 = 4$	30	$30 - 20 = 10$	$10^2 = 100$
23	$23 - 20 = 3$	$3^2 = 9$	35	$35 - 20 = 15$	$15^2 = 225$
Totals:		$\Sigma(x - \bar{x})^2 = 28$			$\Sigma(x - \bar{x})^2 = 700$

Each sample contains seven data items, so we compute the standard deviation by dividing the sums in **Table 8.15**, 28 and 700, by $7 - 1$, or 6. Then we take the square root of each quotient.

$$\text{Standard deviation} = \sqrt{\frac{\Sigma(x - \bar{x})^2}{n - 1}} = \sqrt{\frac{\Sigma(\text{data item} - \text{mean})^2}{n - 1}}$$

Sample A: Sample B:

$$\text{Standard deviation} = \sqrt{\frac{28}{6}} \approx 2.16 \qquad \text{Standard deviation} = \sqrt{\frac{700}{6}} \approx 10.80$$

Sample A has a standard deviation of approximately 2.16 and sample B has a standard deviation of approximately 10.80. The data in sample B are more spread out than those in sample A.

✓ **Check Point 4** Find the standard deviation of the data items in each of the samples shown below. Round to two decimal places.

Sample A: 73, 75, 77, 79, 81, 83   3.74

Sample B: 40, 44, 92, 94, 98, 100   28.06

Figure 8.8 illustrates four sets of data items organized in histograms. From left to right, the data items are

> **Figure 8.8(a):** 4, 4, 4, 4, 4, 4, 4
>
> **Figure 8.8(b):** 3, 3, 4, 4, 4, 5, 5
>
> **Figure 8.8(c):** 3, 3, 3, 4, 5, 5, 5
>
> **Figure 8.8(d):** 1, 1, 1, 4, 7, 7, 7.

Each data set has a mean of 4. However, as the spread of the data items increases, the standard deviation gets larger. Observe that when all the data items are the same, the standard deviation is 0.

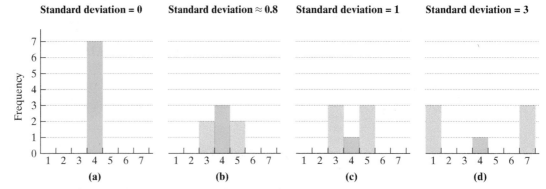

Figure 8.8 The standard deviation gets larger with increased dispersion among data items. In each case, the mean is 4.

Example 5　Interpreting Standard Deviation

Two fifth-grade classes have nearly identical mean scores on an aptitude test, but one class has a standard deviation three times that of the other. All other factors being equal, which class is easier to teach, and why?

Solution

The class with the smaller standard deviation is easier to teach because there is less variation among student aptitudes. Course work can be aimed at the average student without too much concern that the work will be too easy for some or too difficult for others. By contrast, the class with greater dispersion poses a greater challenge. By teaching to the average student, the students whose scores are significantly above the mean will be bored; students whose scores are significantly below the mean will be confused.

Check Point 5 Shown below are the means and standard deviations of the yearly returns on two investments from 1926 through 2004.

Investment	Mean Yearly Return	Standard Deviation
Small-Company Stocks	17.5%	33.3%
Large-Company Stocks	12.4%	20.4%

Source: Summary Statistics of Annual Total Returns 1926 to 2004 Yearbook, Ibbotson Associates, Chicago

a. Use the means to determine which investment provided the greater yearly return. small-company stocks

b. Use the standard deviations to determine which investment had the greater risk. Explain your answer. small-company stocks; Answers will vary.

Concept and Vocabulary Check

Exercises in the Concept and Vocabulary Check are intended for group and class discussions.

In Exercises 1–2, fill in each blank so that the resulting statement is true.

1. The difference between the highest and lowest data values in a data set is called the ___range___.

2. The formula

$$\sqrt{\frac{\Sigma(\text{data item} - \text{mean})^2}{n-1}}$$

gives the value of the ___standard deviation___ for a data set.

In Exercises 3–5, determine whether each statement is true or false. If the statement is false, make the necessary change(s) to produce a true statement. Changes to false statements will vary.

3. Measures of dispersion are used to describe the spread of data items in a data set.   true

4. The sum of the deviations from the mean for a data set is always zero.   true

5. Measures of dispersion get smaller as the spread of data items increases.   false

Respond to Exercises 6–12 using verbal or written explanations.   6–12. Answers will vary.

6. Describe how to find the range of a data set.

7. Describe why the range might not be the best measure of dispersion.

8. Describe how the standard deviation is computed.

9. Describe what the standard deviation reveals about a data set.

10. If a set of test scores has a standard deviation of zero, what does this mean about the scores?

11. Two classes took a statistics test. Both classes had a mean score of 73. The scores of class A had a standard deviation of 5 and those of class B had a standard deviation of 10. Discuss the difference between the two classes' performance on the test.

12. A sample of cereals indicates a mean potassium content per serving of 93 milligrams and a standard deviation of 2 milligrams. Write a description of what this means for a person who knows nothing about statistics.

Exercise Set 8.3

Practice Exercises

In Exercises 1–6, find the range for each group of data items.

1. $1, 2, 3, 4, 5$ 4
2. $16, 17, 18, 19, 20$ 4
3. $7, 9, 9, 15$ 8
4. $11, 13, 14, 15, 17$ 6
5. $3, 3, 4, 4, 5, 5$ 2
6. $3, 3, 3, 4, 5, 5, 5$ 2

In Exercises 7–10, a group of data items and their mean are given.

a. *Find the deviation from the mean for each of the data items.*

b. *Find the sum of the deviations in part (a).*   Always 0

7. $3, 5, 7, 12, 18, 27$; Mean $= 12$ *
8. $84, 88, 90, 95, 98$; Mean $= 91$ *
9. $29, 38, 48, 49, 53, 77$; Mean $= 49$ *
10. $60, 60, 62, 65, 65, 65, 66, 67, 70, 70$; Mean $= 65$ *

In Exercises 11–16, find a. *the mean;* b. *the deviation from the mean for each data item; and* c. *the sum of the deviations in part (b).*

11. $85, 95, 90, 85, 100$ *
12. $94, 62, 88, 85, 91$ *
13. $146, 153, 155, 160, 161$ *
14. $150, 132, 144, 122$ *
15. $2.25, 3.50, 2.75, 3.10, 1.90$ *
16. $0.35, 0.37, 0.41, 0.39, 0.43$ *   11–16. c. 0

*See Answers to Selected Exercises.

In Exercises 17–26, find the standard deviation for each group of data items. Round answers to two decimal places.

17. $1, 2, 3, 4, 5$ ≈ 1.58
18. $16, 17, 18, 19, 20$ ≈ 1.58
19. $7, 9, 9, 15$ ≈ 3.46
20. $11, 13, 14, 15, 17$ ≈ 2.24
21. $3, 3, 4, 4, 5, 5$ ≈ 0.89
22. $3, 3, 3, 4, 5, 5, 5$ 1
23. $1, 1, 1, 4, 7, 7, 7$ 3
24. $6, 6, 6, 6, 7, 7, 7, 4, 8, 3$ ≈ 1.49
25. $9, 5, 9, 5, 9, 5, 9, 5$ ≈ 2.14
26. $6, 10, 6, 10, 6, 10, 6, 10$ ≈ 2.14

In Exercises 27–28, compute the mean, range, and standard deviation for the data items in each of the three samples. Then describe one way in which the samples are alike and one way in which they are different.

27. Sample A: $6, 8, 10, 12, 14, 16, 18$ *
    Sample B: $6, 7, 8, 12, 16, 17, 18$
    Sample C: $6, 6, 6, 12, 18, 18, 18$

28. Sample A: $8, 10, 12, 14, 16, 18, 20$ *
    Sample B: $8, 9, 10, 14, 18, 19, 20$
    Sample C: $8, 8, 8, 14, 20, 20, 20$

Practice Plus

In Exercises 29–36, use each display of data items to find the standard deviation. Where necessary, round answers to two decimal places.

29.

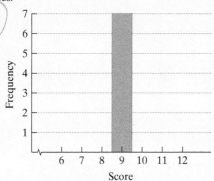

0

30.

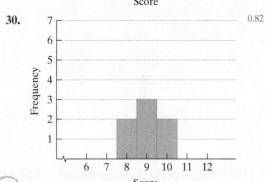

0.82

31.

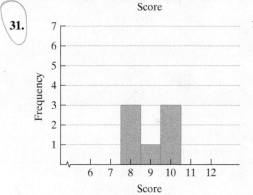

1

32.

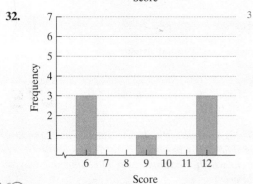

3

33.   7.91 **34.**   6.32

Stems	Leaves
0	5
1	0 5
2	0 5

Stems	Leaves
0	4 8
1	2 6
2	0

*See Answers to Selected Exercises.

35.   1.55 **36.**   4.27

Stems	Leaves
1	8 9 9 8 7 8
2	0 1 0 2

Stems	Leaves
1	3 5 3 8 3 4
2	3 0 0 4

Application Exercises

37. The data sets give the number of platinum albums for the five male artists and the five female artists in the United States with the most platinum albums. (Platinum albums sell one million units or more.)

Male Artists with the Most Platinum Albums

Artist	Platinum Albums
Garth Brooks	145
Elvis Presley	104
Billy Joel	80
Michael Jackson	71
Elton John	65

Female Artists with the Most Platinum Albums

Artist	Platinum Albums
Madonna	64
Barbra Streisand	63
Mariah Carey	61
Whitney Houston	54
Celine Dion	48

Source: RIAA

a. Without calculating, which data set has the greater mean number of platinum albums? Explain your answer. *
b. Verify your conjecture from part (a) by calculating the mean number of platinum albums for each data set. *
c. Without calculating, which data set has the greater standard deviation? Explain your answer. *
d. Verify your conjecture from part (c) by calculating the standard deviation for each data set. Round answers to two decimal places. *

38. The data sets give the ages of the first six U.S. presidents and the six most recent U.S. presidents (through Barack Obama).

Age of First Six U.S. Presidents at Inauguration

President	Age
Washington	57
J. Adams	61
Jefferson	57
Madison	57
Monroe	58
J. Q. Adams	57

Age of Six Most Recent U.S. Presidents at Inauguration

President	Age
Carter	52
Reagan	69
G. H. W. Bush	64
Clinton	46
G. W. Bush	54
Obama	47

Source: Time Almanac

a. Without calculating, which set has the greater standard deviation? Explain your answer.   most recent six *
b. Verify your conjecture from part (b) by calculating the standard deviation for each data set. Round answers to two decimal places.   first: 1.60; most recent: 9.29

Critical Thinking Exercises

39. Describe a situation in which a relatively large standard deviation is desirable.   Answers will vary.

40. If a set of test scores has a large range but a small standard deviation, describe what this means about students' performance on the test.   Answers will vary.

41. Use the data 1, 2, 3, 5, 6, 7. Without actually computing the standard deviation, which of the following best approximates the standard deviation?   a

    a. 2   **b.** 6   **c.** 10   **d.** 20

42. Use the data 0, 1, 3, 4, 4, 6. Add 2 to each of the numbers. How does this affect the mean? How does this affect the standard deviation?   *

*See Answers to Selected Exercises.

Group Exercises   43–44. Answers will vary.

43. As a follow-up to Group Exercise 66 on page 522, the group should reassemble and compute the standard deviation for each data set whose mean you previously determined. Does the standard deviation tell you anything new or interesting about the entire group or subgroups that you did not discover during the previous group activity?

44. Group members should consult a current almanac or the Internet and select intriguing data. The group's function is to use statistics to tell a story. Once "intriguing" data are identified, as a group

  a. Summarize the data. Use words, frequency distributions, and graphic displays.

  b. Compute measures of central tendency and dispersion, using these statistics to discuss the data.

8.4 : The Normal Distribution

What am I supposed to learn?

After you have read this section, you should be able to:

1 Recognize characteristics of normal distributions.

2 Understand the 68–95–99.7 Rule.

3 Find scores at a specified standard deviation from the mean.

4 Use the 68–95–99.7 Rule.

5 Convert a data item to a z-score.

6 Understand percentiles and quartiles.

7 Use and interpret margins of error.

8 Recognize distributions that are not normal.

Our heights are on the rise! In one million B.C., the mean height for men was 4 feet 6 inches. The mean height for women was 4 feet 2 inches. Because of improved diets and medical care, the mean height for men is now 5 feet 10 inches and for women it is 5 feet 5 inches. Mean adult heights are expected to plateau by 2050.

Mean Adult Heights

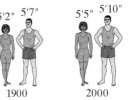

Source: National Center for Health Statistics

Recognize characteristics of normal distributions.

Suppose that a researcher selects a random sample of 100 adult men, measures their heights, and constructs a histogram for the data. The graph is shown in **Figure 8.9(a)**. **Figure 8.9(a)** and **(b)** illustrate what happens as the sample size increases. In **Figure 8.9(c)**, if you were to fold the graph down the middle, the left side would fit the right side. As we move out from the middle, the heights of the bars are the same to the left and right. Such a histogram is called **symmetric**. As the sample size increases, so does the graph's symmetry. If it were possible to measure the heights of all adult males, the entire population, the histogram would approach what is called the **normal distribution**, shown in **Figure 8.9(d)**. This distribution is also called the **bell curve** or the **Gaussian distribution**, named for the German mathematician Carl Friedrich Gauss (1777–1855).

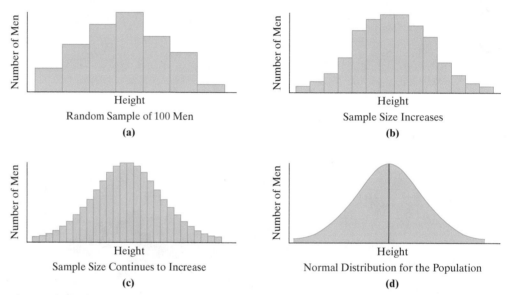

Figure 8.9 Heights of adult males

Figure 8.9(d) illustrates that the normal distribution is bell shaped and symmetric about a vertical line through its center. Furthermore, **the mean, median, and mode** of a normal distribution **are all equal** and located at the center of the distribution.

The shape of the normal distribution depends on the mean and the standard deviation. **Figure 8.10** illustrates three normal distributions with the same mean, but different standard deviations. As the standard deviation increases, the distribution becomes more dispersed, or spread out, but retains its symmetric bell shape.

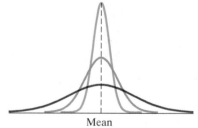

Figure 8.10

The normal distribution provides a wonderful model for all kinds of phenomena because many sets of data items closely resemble this population distribution. Examples include heights and weights of adult males, intelligence quotients, SAT scores, prices paid for a new car model, and life spans of light bulbs. In these distributions, the data items tend to cluster around the mean. The more an item differs from the mean, the less likely it is to occur.

The normal distribution is used to make predictions about an entire population using data from a sample. In this section, we focus on the characteristics and applications of the normal distribution.

2 Understand the 68–95–99.7 Rule.

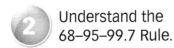
Blitzer Bonus
●●●●●●●●●●●●●●●

Well-Worn Steps and the Normal Distribution

These ancient steps each take on the shape of a normal distribution when the picture is viewed upside down. The center of each step is more worn than the outer edges. The greatest number of people have walked in the center, making this the mean, median, and mode for where people have walked.

The Standard Deviation and z-Scores in Normal Distributions

The standard deviation plays a crucial role in the normal distribution, summarized by the **68–95–99.7 Rule**. This rule is illustrated in **Figure 8.11**.

The 68–95–99.7 Rule for the Normal Distribution

1. Approximately 68% of the data items fall within 1 standard deviation of the mean (in both directions).
2. Approximately 95% of the data items fall within 2 standard deviations of the mean.
3. Approximately 99.7% of the data items fall within 3 standard deviations of the mean.

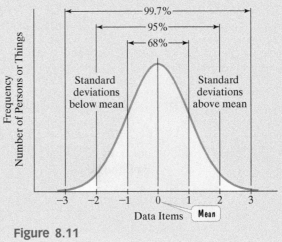

Figure 8.11

Figure 8.11 illustrates that a very small percentage of the data in a normal distribution lies more than 3 standard deviations above or below the mean. As we move from the mean, the curve falls rapidly, and then more and more gradually, toward the horizontal axis. The tails of the curve approach, but never touch, the horizontal axis, although they are quite close to the axis at 3 standard deviations from the mean. The range of the normal distribution is infinite. No matter how far out from the mean we move, there is always the probability (although very small) of a data item occurring even farther out.

3 Find scores at a specified standard deviation from the mean.

Example 1　Finding Scores at a Specified Standard Deviation from the Mean

Male adult heights in North America are approximately normally distributed with a mean of 70 inches and a standard deviation of 4 inches. Find the height that is

a. 2 standard deviations above the mean.
b. 3 standard deviations below the mean.

Solution

a. First, let us find the height that is 2 standard deviations above the mean.
$$\text{Height} = \text{mean} + 2 \cdot \text{standard deviation}$$
$$= 70 + 2 \cdot 4 = 70 + 8 = 78$$
A height of 78 inches is 2 standard deviations above the mean.

b. Next, let us find the height that is 3 standard deviations below the mean.
$$\text{Height} = \text{mean} - 3 \cdot \text{standard deviation}$$
$$= 70 - 3 \cdot 4 = 70 - 12 = 58$$
A height of 58 inches is 3 standard deviations below the mean.

The normal distribution of male adult heights in North America, with a mean of 70 inches and a standard deviation of 4 inches, is illustrated in **Figure 8.12**.

Normal Distribution of Male Adult Heights

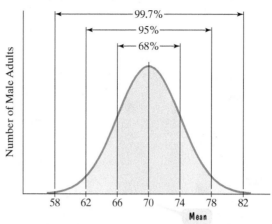

Male Adult Heights in North America

Figure 8.12

 Check Point 1   Female adult heights in North America are approximately normally distributed with a mean of 65 inches and a standard deviation of 3.5 inches. Find the height that is

a. 3 standard deviations above the mean.   75.5 in.

▶ **b.** 2 standard deviations below the mean.   58 in.

④ Use the 68–95–99.7 Rule.

Example 2   Using the 68–95–99.7 Rule

Use the distribution of male adult heights in **Figure 8.12** to find the percentage of men in North America with heights

a. between 66 inches and 74 inches.   **b.** between 70 inches and 74 inches.

c. above 78 inches.

Solution

a. The 68–95–99.7 Rule states that approximately 68% of the data items fall within 1 standard deviation, 4, of the mean, 70.

$$\text{mean} - 1 \cdot \text{standard deviation} = 70 - 1 \cdot 4 = 70 - 4 = 66$$
$$\text{mean} + 1 \cdot \text{standard deviation} = 70 + 1 \cdot 4 = 70 + 4 = 74$$

Figure 8.12 shows that 68% of male adults have heights between 66 inches and 74 inches.

b. The percentage of men with heights between 70 inches and 74 inches is not given directly in **Figure 8.12**. Because of the distribution's symmetry, the percentage with heights between 66 inches and 70 inches is the same as the percentage with heights between 70 and 74 inches. **Figure 8.13** indicates that 68% have heights between 66 inches and 74 inches. Thus, half of 68%, or 34%, of men have heights between 70 inches and 74 inches.

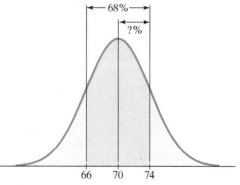

Figure 8.13 What percentage have heights between 70 inches and 74 inches?

c. The percentage of men with heights above 78 inches is not given directly in **Figure 8.12**. A height of 78 inches is 2 standard deviations, $2 \cdot 4$, or 8 inches, above the mean, 70 inches. The 68–95–99.7 Rule states that approximately 95% of the data items fall within 2 standard deviations of the mean. Thus, approximately $100\% - 95\%$, or 5%, of the data items are farther than 2 standard deviations from the mean. The 5% of the data items are represented by the two shaded green regions in **Figure 8.14**. Because of the distribution's symmetry, half of 5%, or 2.5%, of the data items are more than 2 standard deviations above the mean. This means that 2.5% of men have heights above 78 inches.

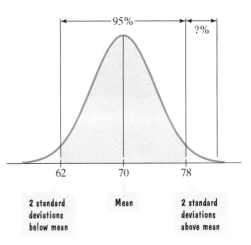

Figure 8.14 What percentage have heights above 78 inches?

✓ **Check Point 2**   Use the distribution of male adult heights in North America in **Figure 8.12** to find the percentage of men with heights

a. between 62 inches and 78 inches.   95%

b. between 70 inches and 78 inches.   47.5%

▶ **c.** above 74 inches.   16%

Because the normal distribution of male adult heights in North America has a mean of 70 inches and a standard deviation of 4 inches, a height of 78 inches lies 2 standard deviations above the mean. In a normal distribution, a **z-score** describes how many standard deviations a particular data item lies above or below the mean. Thus, the z-score for the data item 78 is 2.

The following formula can be used to express a data item in a normal distribution as a z-score:

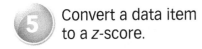 Convert a data item to a z-score.

Computing z-Scores

A z-score describes how many standard deviations a data item in a normal distribution lies above or below the mean. The z-score can be obtained using

$$z\text{-score} = \frac{\text{data item} - \text{mean}}{\text{standard deviation}}.$$

Data items above the mean have positive z-scores. Data items below the mean have negative z-scores. The z-score for the mean is 0.

Example 3   **Computing z-Scores**

The mean weight of newborn infants is 7 pounds and the standard deviation is 0.8 pound. The weights of newborn infants are normally distributed. Find the z-score for a weight of

a. 9 pounds.   **b.** 7 pounds.   **c.** 6 pounds.

Solution

We compute the z-score for each weight by using the z-score formula. The mean is 7 and the standard deviation is 0.8.

a. The z-score for a weight of 9 pounds, written z_9, is

$$z_9 = \frac{\text{data item} - \text{mean}}{\text{standard deviation}} = \frac{9 - 7}{0.8} = \frac{2}{0.8} = 2.5.$$

The z-score of a data item greater than the mean is always positive. A 9-pound infant is a chubby little tyke, with a weight that is 2.5 standard deviations above the mean.

b. The z-score for a weight of 7 pounds is

$$z_7 = \frac{\text{data item} - \text{mean}}{\text{standard deviation}} = \frac{7 - 7}{0.8} = \frac{0}{0.8} = 0.$$

The z-score for the mean is always 0. A 7-pound infant is right at the mean, deviating 0 pounds above or below it.

c. The z-score for a weight of 6 pounds is

$$z_6 = \frac{\text{data item} - \text{mean}}{\text{standard deviation}} = \frac{6 - 7}{0.8} = \frac{-1}{0.8} = -1.25.$$

The z-score of a data item less than the mean is always negative. A 6-pound infant's weight is 1.25 standard deviations below the mean.

Figure 8.15 shows the normal distribution of weights of newborn infants. The horizontal axis is labeled in terms of weights and z-scores.

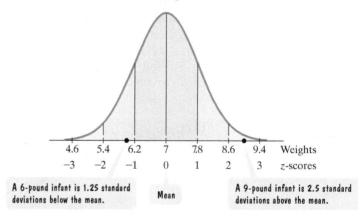

Normal Distribution of Weights of Newborn Infants

| 4.6 | 5.4 | 6.2 | 7 | 7.8 | 8.6 | 9.4 | Weights |
| −3 | −2 | −1 | 0 | 1 | 2 | 3 | z-scores |

A 6-pound infant is 1.25 standard deviations below the mean.   Mean   A 9-pound infant is 2.5 standard deviations above the mean.

Figure 8.15 Infants' weights are normally distributed.

✓ **Check Point 3**   The length of horse pregnancies from conception to birth is normally distributed with a mean of 336 days and a standard deviation of 3 days. Find the z-score for a horse pregnancy of

a. 342 days.   2   b. 336 days.   0   c. 333 days.   −1

In Example 4, we consider two normally distributed sets of test scores, in which a higher score generally indicates a better result. To compare scores on two different tests in relation to the mean on each test, we can use z-scores. The better score is the item with the greater z-score.

Example 4   Using and Interpreting z-Scores

A student scores 70 on an arithmetic test and 66 on a vocabulary test. The scores for both tests are normally distributed. The arithmetic test has a mean of 60 and a standard deviation of 20. The vocabulary test has a mean of 60 and a standard deviation of 2. On which test did the student have the better score?

Solution

To answer the question, we need to find the student's z-score on each test, using

$$z = \frac{\text{data item} - \text{mean}}{\text{standard deviation}}.$$

The arithmetic test has a mean of 60 and a standard deviation of 20.

$$z\text{-score for } 70 = z_{70} = \frac{70 - 60}{20} = \frac{10}{20} = 0.5$$

The vocabulary test has a mean of 60 and a standard deviation of 2.

$$z\text{-score for } 66 = z_{66} = \frac{66 - 60}{2} = \frac{6}{2} = 3$$

The arithmetic score, 70, is half a standard deviation above the mean, whereas the vocabulary score, 66, is 3 standard deviations above the mean. The student did much better than the mean on the vocabulary test.

 Check Point 4   The SAT (Scholastic Aptitude Test) has a mean of 500 and a standard deviation of 100. The ACT (American College Test) has a mean of 18 and a standard deviation of 6. Both tests measure the same kind of ability, with scores that are normally distributed. Suppose that you score 550 on the SAT and 24 on the ACT. On which test did you have the better score?   ACT

Example 5   Understanding z-Scores

Intelligence quotients (IQs) on the Stanford-Binet intelligence test are normally distributed with a mean of 100 and a standard deviation of 16.

a. What is the IQ corresponding to a z-score of −1.5?

b. Mensa is a group of people with high IQs whose members have z-scores of 2.05 or greater on the Stanford-Binet intelligence test. What is the IQ corresponding to a z-score of 2.05?

Solution

a. We begin with the IQ corresponding to a z-score of −1.5. The negative sign in −1.5 tells us that the IQ is $1\frac{1}{2}$ standard deviations below the mean.

$$IQ = \text{mean} - 1.5 \cdot \text{standard deviation}$$
$$= 100 - 1.5(16) = 100 - 24 = 76$$

The IQ corresponding to a z-score of −1.5 is 76.

b. Next, we find the IQ corresponding to a z-score of 2.05. The positive sign implied in 2.05 tells us that the IQ is 2.05 standard deviations above the mean.

$$IQ = \text{mean} + 2.05 \cdot \text{standard deviation}$$
$$= 100 + 2.05(16) = 100 + 32.8 = 132.8$$

The IQ corresponding to a z-score of 2.05 is 132.8. (An IQ score of at least 133 is required to join Mensa.)

Blitzer Bonus

The IQ Controversy

Is intelligence something we are born with or is it a quality that can be manipulated through education? Can it be measured accurately and is IQ the way to measure it? There are no clear answers to these questions.

In a study by Carolyn Bird (*Pygmalion in the Classroom*), a group of third-grade teachers was told that they had classes of students with IQs well above the mean. These classes made incredible progress throughout the year. In reality, these were not gifted kids, but, rather, a random sample of all third-graders. It was the teachers' expectations, and not the IQs of the students, that resulted in increased performance.

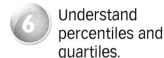

Check Point 5 Use the information in Example 5 on the previous page to find the IQ corresponding to a *z*-score of

▶ **a.** −2.25. 64 **b.** 1.75. 128

Percentiles and Quartiles

6 Understand percentiles and quartiles.

A *z*-score measures a data item's position in a normal distribution. Another measure of a data item's position is its **percentile**. Percentiles are often associated with scores on standardized tests. If a score is in the 45th percentile, this means that 45% of the scores are less than this score. If a score is in the 95th percentile, this indicates that 95% of the scores are less than this score.

> ### Percentiles
>
> If *n*% of the items in a distribution are less than a particular data item, we say that the data item is in the ***n*th percentile** of the distribution.

Example 6 Interpreting Percentile

The cutoff IQ score for Mensa membership, 132.8, is in the 98th percentile. What does this mean?

Solution

Because 132.8 is in the 98th percentile, this means that 98% of IQ scores fall below 132.8.

Caution: A score in the 98th percentile does *not* mean that 98% of the answers are correct. Nor does it mean that the score was 98%.

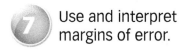

Check Point 6 A student scored in the 75th percentile on the SAT. What does this mean? 75% of the scores on the SAT are less than this student's score.

Three commonly encountered percentiles are the *quartiles*. **Quartiles** divide data sets into four equal parts. The 25th percentile is the **first quartile**: 25% of the data fall below the first quartile. The 50th percentile is the **second quartile**: 50% of the data fall below the second quartile, so the second quartile is equivalent to the median. The 75th percentile is the **third quartile**: 75% of the data fall below the third quartile. **Figure 8.16** illustrates the concept of quartiles for the normal distribution.

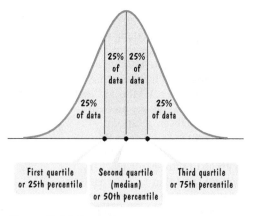

Figure 8.16 Quartiles

Polls and Margins of Error

7 Use and interpret margins of error.

What activities do you dread? Reading math textbooks with a turtle trying to catch a frisbee on the cover? (Be kind!) No, that's not America's most-dreaded activity. In a random sample of 1000 U.S. adults, 46% of those questioned responded, "Public speaking." The problem is that this is a single random sample. Do 46% of adults in the entire U.S. population dread public speaking?

Statisticians use properties of the normal distribution to estimate the probability that a result obtained from a single sample reflects what is truly happening in the population. If you look at the results of a poll like the one shown in **Figure 8.17**, you will observe that a *margin of error* is reported. Surveys and opinion polls often give a margin of error. Let's use our understanding of the normal distribution to see how to calculate and interpret margins of error.

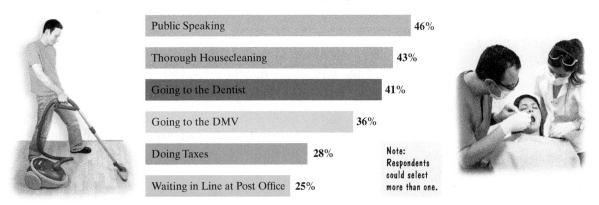

Activities U.S. Adults Say They Dread

Public Speaking	46%
Thorough Housecleaning	43%
Going to the Dentist	41%
Going to the DMV	36%
Doing Taxes	28%
Waiting in Line at Post Office	25%

Note: Respondents could select more than one.

Figure 8.17
Source: TNS survey of 1000 adults, March 2010.   Margin of error: ±3.2

Note the margin of error.

Suppose that $p\%$ of the population of U.S. adults dread public speaking. Instead of taking only one random sample of 1000 adults, we repeat the process of selecting a random sample of 1000 adults hundreds of times. Then, we calculate the percentage of adults for each sample who dread public speaking. With random sampling, we expect to find the percentage in many of the samples close to $p\%$, with relatively few samples having percentages far from $p\%$. **Figure 8.18** shows that the percentages of U.S. adults from the hundreds of samples can be modeled by a normal distribution. The mean of this distribution is the actual population percent, $p\%$, and is the most frequent result from the samples.

Mathematicians have shown that the standard deviation of a normal distribution of samples like the one in **Figure 8.18** is approximately $\frac{1}{2\sqrt{n}} \times 100\%$, where n is the sample size. Using the 68-95-99.7 Rule, approximately 95% of the samples have a percentage within 2 standard deviations of the true population percentage, $p\%$:

$$\text{2 standard deviations} = 2 \cdot \frac{1}{2\sqrt{n}} \times 100\% = \frac{1}{\sqrt{n}} \times 100\%.$$

If we use a single random sample of size n, there is a 95% probability that the percent obtained will lie within two standard deviations, or $\frac{1}{\sqrt{n}} \times 100$, of the true population percent. We can be 95% confident that the true population percent lies between

$$\text{the sample percent} - \frac{1}{\sqrt{n}} \times 100$$

and

$$\text{the sample percent} + \frac{1}{\sqrt{n}} \times 100$$

We call $\pm \frac{1}{\sqrt{n}} \times 100$ the **margin of error**.

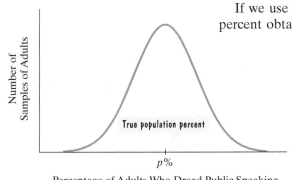

Percentage of Adults Who Dread Public Speaking

Figure 8.18

> **Margin of Error in a Survey**
>
> If a statistic is obtained from a random sample of size n, there is a 95% probability that it lies within $\dfrac{1}{\sqrt{n}} \times 100\%$ of the true population percent, where $\pm\dfrac{1}{\sqrt{n}} \times 100\%$ is called the **margin of error**.

Table 8.16 Activities U.S. Adults Dread

Activity	Percentage Who Dread the Activity
Public speaking	46%
Thorough house-cleaning	43%
Going to the dentist	41%
Going to the DMV	36%
Doing taxes	28%
Waiting in line at the post office	25%

Source: TNS survey of 1000 adults, March 2010.
Margin of error: ±3.2%

Example 7   Using and Interpreting Margin of Error

Table 8.16 shows that in a random sample of 1000 U.S. adults, 46% of those questioned said that they dread public speaking.

a. Verify the margin of error that was given for this survey.

b. Write a statement about the percentage of adults in the U.S. population who dread public speaking.

Solution

a. The sample size is $n = 1000$. The margin of error is

$$\pm\frac{1}{\sqrt{n}} \times 100\% = \pm\frac{1}{\sqrt{1000}} \times 100\% \approx \pm 0.032 \times 100\% = \pm 3.2\%.$$

b. There is a 95% probability that the true population percentage lies between

$$\text{the sample percent} - \frac{1}{\sqrt{n}} \times 100\% = 46\% - 3.2\% = 42.8\%$$

and

$$\text{the sample percent} + \frac{1}{\sqrt{n}} \times 100\% = 46\% + 3.2\% = 49.2\%.$$

We can be 95% confident that between 42.8% and 49.2% of all U.S. adults dread public speaking.

Blitzer Bonus

A Caveat Giving a True Picture of a Poll's Accuracy

Unlike the precise calculation of a poll's margin of error, certain polling imperfections cannot be determined exactly. One problem is that people do not always respond to polls honestly and accurately. Some people are embarrassed to say "undecided," so they make up an answer. Other people may try to respond to questions in the way they think will make the pollster happy, just to be "nice." Perhaps the following caveat, applied to the poll in Example 7, would give the public a truer picture of its accuracy:

The poll results are 42.8% to 49.2% at the 95% confidence level, but it's only under ideal conditions that we can be 95% confident that the true numbers are within 3.2% of the poll's results. The true error span is probably greater than 3.2% due to limitations that are inherent in this and every poll, but, unfortunately, this additional error amount cannot be calculated precisely. Warning: Five percent of the time—that's one time out of 20—the error will be greater than 3.2%. We remind readers of the poll that things occurring "only" 5% of the time do, indeed, happen.

We suspect that the public would tire of hearing this.

Number of Books U.S. Adults Read per Year

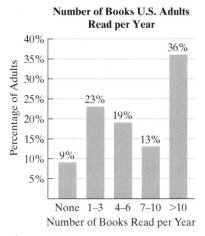

Figure 8.19
Source: Harris Poll of 2513 U.S. adults ages 18 and older conducted March 11 and 18, 2008

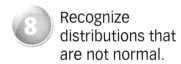

Recognize distributions that are not normal.

 **Check Point 7** A Harris Poll of 2513 U.S. adults ages 18 and older asked the question

How many books do you typically read in a year?

The results of the poll are shown in **Figure 8.19**.   a. ±2.0%

a. Find the margin of error for this survey. Round to the nearest tenth of a percent.

b. Write a statement about the percentage of U.S. adults who read more than ten books per year.   *

c. Why might some people not respond honestly and accurately to the question in this poll?   *

Other Kinds of Distributions

Although the normal distribution is the most important of all distributions in terms of analyzing data, not all data can be approximated by this symmetric distribution with its mean, median, and mode all having the same value.

In our discussion of measures of central tendency, we mentioned that the median, rather than the mean, is used to summarize income. **Figure 8.20** illustrates the population distribution of weekly earnings in the United States. There is no upper limit on weekly earnings. The relatively few people with very high weekly incomes tend to pull the mean income to a value greater than the median. The most frequent income, the mode, occurs toward the low end of the data items. The mean, median, and mode do not have the same value, and a normal distribution is not an appropriate model for describing weekly earnings in the United States.

The distribution in **Figure 8.20** is called a *skewed distribution*. A distribution of data is **skewed** if a large number of data items are piled up at one end or the other, with a "tail" at the opposite end. In the distribution of weekly earnings in **Figure 8.20**, the tail is to the right. Such a distribution is said to be **skewed to the right**.

In contrast to the distribution of weekly earnings, the distribution in **Figure 8.21** has more data items at the high end of the scale than at the low end. The tail of this distribution is to the left. The distribution is said to be **skewed to the left**. An example of a distribution skewed to the left is based on the student ratings of faculty teaching performance in many colleges. Most professors are given rather high ratings, while only a few are rated as terrible. These low ratings pull the value of the mean lower than the median.

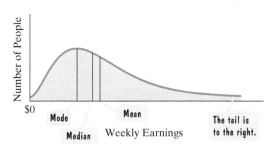

Figure 8.20 Skewed to the right

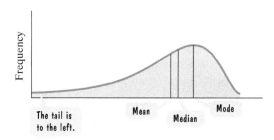

Figure 8.21 Skewed to the left

Great Question!

What's the bottom line on the relationship between the mean and the median for skewed distributions?

If the data are skewed to the right, the mean is greater than the median. If the data are skewed to the left, the mean is less than the median.

Achieving Success

Your academic pathways will soon be leading you to college math courses. In each course, check your performance by answering the following questions:

- Are you attending all lectures?
- For each hour of class time, are you spending at least two hours outside of class completing all homework assignments, checking answers, correcting errors, and using all resources to get the help that you need?
- Are you reviewing for quizzes and tests?
- Are you reading the textbook? Remember that in all college courses, you are responsible for the information in the text, whether or not it is covered in class.
- Are you keeping an organized notebook? Does each page have the appropriate section from the text on top? Do the pages contain examples your professor works during lecture and other relevant class notes? Have you included your worked-out homework exercises? Do you keep a special section for graded exams?
- Are you analyzing your mistakes and learning from your errors?
- Are there ways you can improve how you are doing in the course?

Copy this list of questions. Use them from time to time to assess how you are doing in your forthcoming college math classes.

Concept and Vocabulary Check

Exercises in the Concept and Vocabulary Check are intended for group and class discussions.

In Exercises 1–4, fill in each blank so that the resulting statement is true.

1. In a normal distribution, approximately ___68___% of the data items fall within 1 standard deviation of the mean, approximately ___95___% of the data items fall within 2 standard deviations of the mean, and approximately ___99.7___% of the data items fall within 3 standard deviations of the mean.

2. A z-score describes how many standard deviations a data item in a normal distribution lies above or below the ___mean___.

3. If n% of the items in a distribution are less than a particular data item, we say that the data item is in the nth ___percentile___ of the distribution.

4. If a statistic is obtained from a random sample of size n, there is a 95% probability that it lies within $\frac{1}{\sqrt{n}} \times 100\%$ of the true population percent, where $\pm \frac{1}{\sqrt{n}} \times 100\%$ is called the ___margin of error___.

In Exercises 5–8, determine whether each statement is true or false. If the statement is false, make the necessary change(s) to produce a true statement. Changes to false statements will vary.

5. The mean, median, and mode of a normal distribution are all equal. true

6. In a normal distribution, the z-score for the mean is 0. true

7. The z-score for a data item in a normal distribution is obtained using
$$z\text{-score} = \frac{\text{data item} - \text{standard deviation}}{\text{mean}}.$$ false

8. A score in the 50th percentile on a standardized test is the median. true

Respond to Exercises 9–18 using verbal or written explanations. 9–18. Answers will vary.

9. What is a symmetric histogram?

10. Describe the normal distribution and discuss some of its properties.

11. Describe the 68–95–99.7 Rule.

12. Describe how to determine the z-score for a data item in a normal distribution.

13. What does a z-score measure?

14. Give an example of both a commonly occurring and an infrequently occurring z-score. Explain how you arrived at these examples.

15. Describe when a z-score is negative.

16. If you score in the 83rd percentile, what does this mean?

17. If your weight is in the third quartile, what does this mean?

18. Two students have scores with the same percentile, but for different administrations of the SAT. Does this mean that the students have the same score on the SAT? Explain your answer.

Exercise Set 8.4

Practice and Application Exercises

The scores on a test are normally distributed with a mean of 100 and a standard deviation of 20. In Exercises 1–10, find the score that is

1. 1 standard deviation above the mean. 120
2. 2 standard deviations above the mean. 140
3. 3 standard deviations above the mean. 160
4. $1\frac{1}{2}$ standard deviations above the mean. 130
5. $2\frac{1}{2}$ standard deviations above the mean. 150
6. 1 standard deviation below the mean. 80
7. 2 standard deviations below the mean. 60
8. 3 standard deviations below the mean. 40
9. one-half a standard deviation below the mean. 90
10. $2\frac{1}{2}$ standard deviations below the mean. 50

Not everyone pays the same price for the same model of a car. The figure illustrates a normal distribution for the prices paid for a particular model of a new car. The mean is $17,000 and the standard deviation is $500.

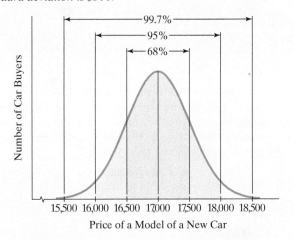

In Exercises 11–22, use the 68–95–99.7 Rule, illustrated in the figure, to find the percentage of buyers who paid

11. between $16,500 and $17,500. 68%
12. between $16,000 and $18,000. 95%
13. between $17,000 and $17,500. 34%
14. between $17,000 and $18,000. 47.5%
15. between $16,000 and $17,000. 47.5%
16. between $16,500 and $17,000. 34%
17. between $15,500 and $17,000. 49.85%
18. between $17,000 and $18,500. 49.85%
19. more than $17,500. 16%
20. more than $18,000. 2.5%
21. less than $16,000. 2.5%
22. less than $16,500. 16%

*See Answers to Selected Exercises.

Intelligence quotients (IQs) on the Stanford-Binet intelligence test are normally distributed with a mean of 100 and a standard deviation of 16. In Exercises 23–32, use the 68–95–99.7 Rule to find the percentage of people with IQs

23. between 68 and 132. 95%
24. between 84 and 116. 68%
25. between 68 and 100. 47.5%
26. between 84 and 100. 34%
27. above 116. 16%
28. above 132. 2.5%
29. below 68. 2.5%
30. below 84. 16%
31. above 148. 0.15%
32. below 52. 0.15%

A set of data items is normally distributed with a mean of 60 and a standard deviation of 8. In Exercises 33–48, convert each data item to a z-score.

33. 68 1
34. 76 2
35. 84 3
36. 92 4
37. 64 0.5
38. 72 1.5
39. 74 1.75
40. 78 2.25
41. 60 0
42. 100 5
43. 52 −1
44. 44 −2
45. 48 −1.5
46. 40 −2.5
47. 34 −3.25
48. 30 −3.75

Scores on a dental anxiety scale range from 0 (no anxiety) to 20 (extreme anxiety). The scores are normally distributed with a mean of 11 and a standard deviation of 4. In Exercises 49–56, find the z-score for the given score on this dental anxiety scale.

49. 17 1.5
50. 18 1.75
51. 20 2.25
52. 12 0.25
53. 6 −1.25
54. 8 −0.75
55. 5 −1.5
56. 1 −2.5

Intelligence quotients on the Stanford-Binet intelligence test are normally distributed with a mean of 100 and a standard deviation of 16. Intelligence quotients on the Wechsler intelligence test are normally distributed with a mean of 100 and a standard deviation of 15. Use this information to solve Exercises 57–58.

57. Use z-scores to determine which person has the higher IQ: an individual who scores 128 on the Stanford-Binet or an individual who scores 127 on the Wechsler. *

58. Use z-scores to determine which person has the higher IQ: an individual who scores 150 on the Stanford-Binet or an individual who scores 148 on the Wechsler. *

A set of data items is normally distributed with a mean of 400 and a standard deviation of 50. In Exercises 59–66, find the data item in this distribution that corresponds to the given z-score.

59. $z = 2$ 500

60. $z = 3$ 550

61. $z = 1.5$ 475

62. $z = 2.5$ 525

63. $z = -3$ 250

64. $z = -2$ 300

65. $z = -2.5$ 275

66. $z = -1.5$ 325

67. **Reducing Gun Violence** The data in the bar graph are from a random sample of 814 American adults. The graph shows four proposals to reduce gun violence in the United States and the percentage of surveyed adults who favored each of these proposals.

Proposals to Reduce Gun Violence in the United States

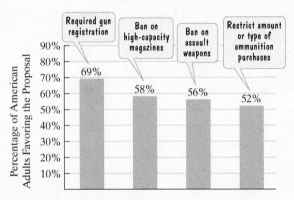

Source: Time/CNN poll using a sample of 814 American adults, January 14–15, 2013

a. Find the margin of error, to the nearest tenth of a percent, for this survey. ±3.5%

b. Write a statement about the percentage of adults in the U.S population who favor required gun registration to reduce gun violence. *

68. **How to Blow Your Job Interview** The data in the bar graph below are from a random sample of 1910 job interviewers. The graph shows the top interviewer turnoffs and the percentage of surveyed interviewers who were offended by each of these behaviors.

Top Interviewer Turnoffs

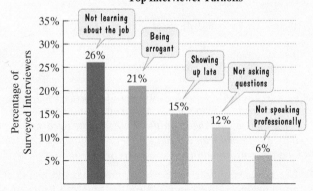

Source: Scott Erker, Ph.D., and Kelli Buczynski, "Are You Failing the Interview? 2009 Survey of Global Interviewing Practices and Perceptions." Development Dimensions International.

*See Answers to Selected Exercises.

a. Find the margin of error, to the nearest tenth of a percent, for this survey. ±2.3%

b. Write a statement about the percentage of interviewers in the population who are turned off by a job applicant being arrogant. *

69. Using a random sample of 4000 TV households, Nielsen Media Research found that 60.2% watched the final episode of $M*A*S*H$.

a. Find the margin of error in this percent. ±1.6%

b. Write a statement about the percentage of TV households in the population who tuned into the final episode of $M*A*S*H$. *

70. Using a random sample of 4000 TV households, Nielsen Media Research found that 51.1% watched *Roots, Part 8*.

a. Find the margin of error in this percent. ±1.6%

b. Write a statement about the percentage of TV households in the population who tuned into *Roots, Part 8*. *

71. In 1997, Nielsen Media Research increased its random sample to 5000 TV households. By how much, to the nearest tenth of a percent, did this improve the margin of error over that in Exercises 69 and 70? 0.2%

72. If Nielsen Media Research were to increase its random sample from 5000 to 10,000 TV households, by how much, to the nearest tenth of a percent, would this improve the margin of error? 0.4%

Critical Thinking Exercises

73. Give an example of a phenomenon that is normally distributed. Explain why. (Try to be creative and not use one of the distributions discussed in this section.) Estimate what the mean and the standard deviation might be and describe how you determined these estimates.

74. Give an example of a phenomenon that is not normally distributed and explain why. 73–74. Answers will vary.

Group Exercise

75. For this activity, group members will conduct interviews with a random sample of students on campus. Each student is to be asked. "What is the worst thing about being a college student?" One response should be recorded for each student. Answers will vary.

a. Each member should interview enough students so that there are at least 50 randomly selected students in the sample.

b. After all responses have been recorded, the group should organize the four most common answers. For each answer, compute the percentage of students in the sample who felt that this is the worst thing about being a student.

c. Find the margin of error for your survey.

d. For each of the four most common answers, write a statement about the percentage of all students on your campus who feel that this is the worst thing about being a student.

Chapter 8 Summary

Definitions and Concepts	Examples

Section 8.1 Sampling, Frequency Distributions, and Graphs

A population is the set containing all objects whose properties are to be described and analyzed. A sample is a subset of the population.

Random samples are obtained in such a way that each member of the population has an equal chance of being selected.

- A newspaper wants to find out which of its features are popular with its readers and decides to conduct a survey. Appropriate for obtaining a sample: Survey a random sample of readers from a list of all subscribers. Not appropriate for obtaining a sample: Survey a random sample of people from the telephone directory (This does not address the target population.); Survey the first 100 subscribers from an alphabetical list of all subscribers (This is not sufficiently random. A subscriber whose last name begins with a letter toward the end of the alphabet has no chance of being selected.)

Additional Examples to Review

Example 1, page 495; Example 2, page 496

Data can be organized and presented in frequency distributions (data values in one column with frequencies in the adjacent column), grouped frequency distributions (data values arranged into classes), histograms (bar graphs with data values on a horizontal axis and frequencies on a vertical axis), frequency polygons (line graphs with data values on a horizontal axis and frequencies on a vertical axis), and stem-and-leaf plots (data items separated into two parts).

Stretching or compressing the scale on a graph's vertical axis can create a more or less dramatic impression than shown by the actual data.

- 4, 5, 2, 2, 1, 3, 2, 3, 5, 5, 2, 1, 3, 3, 3

Frequency Distribution

Score	Frequency
1	2
2	4
3	5
4	1
5	3

Histogram

Frequency Polygon

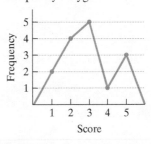

- 62, 63, 71, 74, 78, 79, 81, 81, 82, 83, 90, 92

Grouped Frequency Distribution

Class	Frequency
60–69	2
70–79	4
80–89	4
90–99	2

Stem-and-Leaf Plot

Stems	Leaves
6	2 3
7	1 4 8 9
8	1 1 2 3
9	0 2

Additional Examples and Graphs to Review

Example 3, page 498; Example 4, page 499; Figure 8.2, page 500; Figure 8.3, page 500; Example 5, page 501; Figure 8.5, page 502; Table 8.5, page 503

Section 8.2 Measures of Central Tendency

The mean, $\bar{x}$, is the sum of the data items divided by the number of items: $\bar{x} = \dfrac{\Sigma x}{n}$.

- 2, 7, 8, 8, 5

$$\bar{x} = \frac{\Sigma x}{n} = \frac{2 + 7 + 8 + 8 + 5}{5} = \frac{30}{5} = 6$$

Additional Example to Review

Example 1, page 509

The mean, $\bar{x}$, of a frequency distribution is computed using

$$\bar{x} = \frac{\Sigma xf}{n},$$

where x is a data value, f is its frequency, and n is the total frequency of the distribution.

-

Score, x	Frequency, f	xf
5	10	$5 \cdot 10 = 50$
6	15	$6 \cdot 15 = 90$
7	30	$7 \cdot 30 = 210$
8	35	$8 \cdot 35 = 280$
9	5	$9 \cdot 5 = 45$
10	5	$10 \cdot 5 = 50$
Totals: $n = 100$		$\Sigma xf = 725$

$$\bar{x} = \frac{\Sigma xf}{n} = \frac{725}{100} = 7.25$$

Additional Example to Review

Example 2, page 510

The median of ranked data is the item in the middle or the mean of the two middlemost items. The median is the value in the $\dfrac{n+1}{2}$ position in the list of ranked data.

- 5, 8, 1, 8, 4, 3, 6, 7

Numbers in order: 1, 3, 4, 5, 6, 7, 8, 8   Median $= \dfrac{5+6}{2} = \dfrac{11}{2} = 5.5$

Middle items are 5 and 6.

-

Score, x	Frequency, f
5	10
6	15
7	30
8	35
9	5
10	5
	$n = 100$

Median is in $\dfrac{n+1}{2}$ position $= \dfrac{100+1}{2} = \dfrac{101}{2} = 50.5$

The median is the mean of data items in position 50 (which is 7) and position 51 (which is also 7).

Median $= \dfrac{7+7}{2} = \dfrac{14}{2} = 7$

Additional Examples to Review

Example 3, page 511; Example 4, page 512;
Example 5, page 513; Example 6, page 514

When one or more data items are much greater than or much less than the other items, these extreme values greatly influence the mean, often making the median more representative of the data.

- $1, 2, 2, 5, 7, 8, 96$

$$\text{Mean} = \bar{x} = \frac{\Sigma x}{n} = \frac{1 + 2 + 2 + 5 + 7 + 8 + 96}{7}$$

$$= \frac{121}{7} \approx 17.3$$

Median (middlemost item) $= 5$

The median is more representative of the data than the mean.

Additional Example to Review

Example 7, page 515

The mode of a data set is the value that occurs most often. If there is no such value, there is no mode. If more than one data value has the highest frequency, then each of these data values is a mode.

- $4, 3, 7, 9, 3$
  Mode is 3.

- $5, 8, 4, 3, 4, 8$
  Modes are 4 and 8.

- $2, 5, 4, 4, 2, 5$
  No mode

-

Score, x	Frequency, f
5	10
6	15
7	30
8	35
9	5
10	5

The mode is 8 because it occurs most often (35 times).

Additional Example to Review

Example 8, page 516

The midrange is computed using

$$\frac{\text{lowest data value} + \text{highest data value}}{2}.$$

- $4, 3, 7, 9, 3$

$$\text{Midrange} = \frac{3 + 9}{2} = \frac{12}{2} = 6$$

- $2, 5, 4, 4, 2, 5$

$$\text{Midrange} = \frac{2 + 5}{2} = \frac{7}{2} \approx 3.5$$

Additional Examples to Review

Example 9, page 517; Example 10, page 518

Section 8.3 Measures of Dispersion

Range = highest data value − lowest data value

- $4, 3, 7, 9, 3$
  Range $= 9 - 3 = 6$

- $2, 5, 4, 4, 2, 5$
  Range $= 5 - 2 = 3$

Additional Example to Review

Example 1, page 523

$$\text{Standard deviation} = \sqrt{\frac{\Sigma(\text{data item} - \text{mean})^2}{n - 1}}$$

This is symbolized by $s = \sqrt{\dfrac{\Sigma(x - \bar{x})^2}{n - 1}}$.

As the spread of data items increases, the standard deviation gets larger.

- 5, 7, 8, 12

Find the standard deviation.

First find the mean: $\bar{x} = \dfrac{\Sigma x}{n} = \dfrac{5 + 7 + 8 + 12}{4} = \dfrac{32}{4} = 8.$

x	$x - \bar{x}$	$(x - \bar{x})^2$
5	$5 - 8 = -3$	$(-3)^2 = 9$
7	$7 - 8 = -1$	$(-1)^2 = 1$
8	$8 - 8 = 0$	$0^2 = 0$
12	$12 - 8 = 4$	$4^2 = 16$
		$\Sigma(x - \bar{x})^2 = 26$

Standard deviation:

$$s = \sqrt{\frac{\Sigma(x - \bar{x})^2}{n - 1}} = \sqrt{\frac{26}{4 - 1}} = \sqrt{\frac{26}{3}} \approx 2.94$$

Additional Examples to Review

Example 2, page 524; Example 3, page 525;
Example 4, page 526; Example 5, page 528

Section 8.4 The Normal Distribution

The normal distribution is a theoretical distribution for the entire population. The distribution is bell shaped and symmetric about a vertical line through its center, where the mean, median, and mode are located.

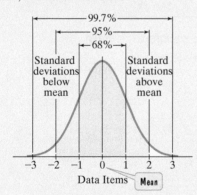

The 68–95–99.7 Rule

Approximately 68% of the data items fall within 1 standard deviation of the mean.

Approximately 95% of the data items fall within 2 standard deviations of the mean.

Approximately 99.7% of the data items fall within 3 standard deviations of the mean.

Cholesterol levels for U.S. men are normally distributed with a mean of 200 and a standard deviation of 15.

Normal Distribution of Male Cholesterol Levels

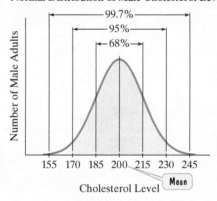

- mean $\pm$ 1 st. dev. = 200 $\pm$ 1 · 15 = 200 $\pm$ 15
  $\qquad\qquad\qquad\qquad\qquad\qquad$ = 185 or 215
- mean $\pm$ 2 st. dev. = 200 $\pm$ 2 · 15 = 200 $\pm$ 30
  $\qquad\qquad\qquad\qquad\qquad\qquad$ = 170 or 230
- mean $\pm$ 3 st. dev. = 200 $\pm$ 3 · 15 = 200 $\pm$ 45
  $\qquad\qquad\qquad\qquad\qquad\qquad$ = 155 or 245
- percentage of men with levels between 200 and 215:
  $\frac{1}{2} \cdot 68\% = 34\%$
- percentage of men with levels below 170:
  $\frac{1}{2}(100\% - 95\%) = \frac{1}{2}(5\%) = 2.5\%$

Additional Examples to Review

Example 1, page 533; Example 2, page 534

A z-score describes how many standard deviations a data item in a normal distribution lies above or below the mean.

$$z\text{-score} = \frac{\text{data item} - \text{mean}}{\text{standard deviation}}$$

If n% of the items in a distribution are less than a particular data item, that data item is in the nth percentile of the distribution. The 25th percentile is the first quartile, the 50th percentile, or the median, is the second quartile, and the 75th percentile is the third quartile.

If a statistic is obtained from a random sample of size n, there is a 95% probability that it lies within $\frac{1}{\sqrt{n}} \times 100\%$ of the true population statistic. $\pm \frac{1}{\sqrt{n}} \times 100\%$ is called the margin of error.

A distribution of data is skewed if a large number of data items are piled up at one end or the other, with a "tail" at the opposite end.

Scores on both tests are normally distributed.

- Test 1: Mean is 100 and standard deviation is 20. Your score is 140.

$$z\text{-score for }140 = \frac{140 - 100}{20} = \frac{40}{20} = 2$$

- Test 2: Mean is 95 and standard deviation is 16. Your score is 135.

$$z\text{-score for }135 = \frac{135 - 95}{16} = \frac{40}{16} = 2.5$$

- Your score on test 2 is farther above the mean (2.5 standard deviations) than your score on test 1 (2 standard deviations), so it is the better score.

Additional Examples to Review

Example 3, page 535; Example 4, page 537; Example 5, page 537

- A student scored in the 80th percentile on the SAT. This means that 80% of the scores on the SAT are less than this student's score.

Additional Example to Review

Example 6, page 538

- Using a random sample of 500 teens ages 12–17, a survey asked respondents to identify things they considered necessities.

Teens Ages 12–17 Who Consider These Necessities

Cellphone	47%
TV	42%
Microwave	27%
Google	22%
Video Games	22%

Source: Lemelson-MIT Invention Index Survey of 500 teens

Margin of error

$$= \pm \frac{1}{\sqrt{n}} \times 100\% = \pm \frac{1}{\sqrt{500}} \times 100\%$$
$$\approx \pm 0.045 \times 100\% = \pm 4.5\%$$

We can be 95% confident that between 47% − 4.5%, or 42.5%, and 47% + 4.5%, or 51.5%, of all teens ages 12–17 consider cellphones a necessity.

Additional Example to Review

Example 7, page 540

- A very easy exam has many high scores. The distribution of scores has a tail to the left and is said to be skewed to the left.

Additional Graphs to Review

Figure 8.20, page 541; Figure 8.21, page 541

Review Exercises

Section 8.1 Sampling, Frequency Distributions, and Graphs

1. The government of a large city wants to know if its citizens will support a three-year tax increase to provide additional support to the city's community college system. The government decides to conduct a survey of the city's residents before placing a tax increase initiative on the ballot. Which one of the following is most appropriate for obtaining a sample of the city's residents?   a

   a. Survey a random sample of persons within each geographic region of the city.

   b. Survey a random sample of community college professors living in the city.

   c. Survey every tenth person who walks into the city's government center on two randomly selected days of the week.

   d. Survey a random sample of persons within each geographic region of the state in which the city is located.

A random sample of ten college students is selected and each student is asked how much time he or she spent on homework during the previous weekend. The following times, in hours, are obtained:

$$8, 10, 9, 7, 9, 8, 7, 6, 8, 7.$$

Use these data items to solve Exercises 2–4.

2. Construct a frequency distribution for the data.   *

3. Construct a histogram for the data.   *

4. Construct a frequency polygon for the data.   *

The 50 grades on a chemistry test are shown. Use the data to solve Exercises 5–6.

44	24	54	81	18
34	39	63	67	60
72	36	91	47	75
57	74	87	49	86
59	14	26	41	90
13	29	13	31	68
63	35	29	70	22
95	17	50	42	27
73	11	42	31	69
56	40	31	45	51

5. Construct a grouped frequency distribution for the data. Use 0–39 for the first class, 40–49 for the second class, and make each subsequent class width the same as the second class.   *

6. Construct a stem-and-leaf plot for the data.   *

7. Describe what is misleading about the size of the barrels in the following visual display.   *

Average Daily Price per Barrel of Oil

Source: U.S. Department of Energy

Section 8.2 Measures of Central Tendency

In Exercises 8–9, find the mean for each group of data items.   9. 17

8. 84, 90, 95, 89, 98   91.2

9. 33, 27, 9, 10, 6, 7, 11, 23, 27

10. Find the mean for the data items in the given frequency distribution.   2.3

Score x	Frequency f
1	2
2	4
3	3
4	1

In Exercises 11–12, find the median for each group of data items.

11. 33, 27, 9, 10, 6, 7, 11, 23, 27   12. 28, 16, 22, 28, 34   28   11. 11

13. Find the median for the data items in the frequency distribution in Exercise 10.   2

In Exercises 14–15, find the mode for each group of data items. If there is no mode, so state.

14. 33, 27, 9, 10, 6, 7, 11, 23, 27   27

15. 582, 585, 583, 585, 587, 587, 589   585, 587

16. Find the mode for the data items in the frequency distribution in Exercise 10.   2

In Exercises 17–18, find the midrange for each group of data items.

17. 84, 90, 95, 88, 98   91   18. 33, 27, 9, 10, 6, 7, 11, 23, 27

19. Find the midrange for the data items in the frequency distribution in Exercise 10.   2.5   18. 19.5

20. A student took seven tests in a course, scoring between 90% and 95% on three of the tests, between 80% and 89% on three of the tests, and below 40% on one of the tests. In this distribution, is the mean or the median more representative of the student's overall performance in the course? Explain your answer.   Answers will vary.

21. The data items below are the ages of U.S. presidents at the time of their first inauguration.

57 61 57 57 58 57 61 54 68 51 49 64 50 48

65 52 56 46 54 49 51 47 55 55 54 42 51 56

55 51 54 51 60 62 43 55 56 61 52 69 64 46 54 47

   a. Organize the data in a frequency distribution. *

   b. Use the frequency distribution to find the mean age, median age, modal age, and midrange age of the presidents when they were inaugurated. *

Section 8.3 Measures of Dispersion

In Exercises 22–23, find the range for each group of data items.

22. 28, 34, 16, 22, 28   18

23. 312, 783, 219, 312, 426, 219   564

24. The mean for the data items 29, 9, 8, 22, 46, 51, 48, 42, 53, 42 is 35. Find **a.** the deviation from the mean for each data item and **b.** the sum of the deviations in part (a). *

25. Use the data items 36, 26, 24, 90, and 74 to find **a.** the mean, **b.** the deviation from the mean for each data item, and **c.** the sum of the deviations in part (b). *

In Exercises 26–27, find the standard deviation for each group of data items.

26. 3, 3, 5, 8, 10, 13   ≈ 4.05

27. 20, 27, 23, 26, 28, 32, 33, 35   ≈ 5.13

28. A test measuring anxiety levels is administered to a sample of ten college students with the following results. (High scores indicate high anxiety.)

        10, 30, 37, 40, 43, 44, 45, 69, 86, 86

Find the mean, range, and standard deviation for the data. *

29. Compute the mean and the standard deviation for each of the following data sets. Then, write a brief description of similarities and differences between the two sets based on each of your computations.

   Set A: 80, 80, 80, 80   Set B: 70, 70, 90, 90 *

30. Describe how you would determine

   a. which of the two groups, men or women, at your college has a higher mean grade point average.

   b. which of the groups is more consistently close to its mean grade point average.   Answers will vary.

Section 8.4 The Normal Distribution

The scores on a test are normally distributed with a mean of 70 and a standard deviation of 8. In Exercises 31–33, find the score that is

31. 2 standard deviations above the mean.   86

32. $3\frac{1}{2}$ standard deviations above the mean.   98

33. $1\frac{1}{4}$ standard deviations below the mean.   60

The ages of people living in a retirement community are normally distributed with a mean age of 68 years and a standard deviation of 4 years. In Exercises 34–40, use the 68–95–99.7 Rule to find the percentage of people in the community whose ages

34. are between 64 and 72.   **35.** are between 60 and 76.

36. are between 68 and 72.   **37.** are between 56 and 80.

38. exceed 72.   **39.** are less than 72.   **40.** exceed 76.

A set of data items is normally distributed with a mean of 50 and a standard deviation of 5. In Exercises 41–45, convert each data item to a z-score.

41. 50   0   **42.** 60   2   **43.** 58   1.6

44. 35   −3   **45.** 44   −1.2

46. A student scores 60 on a vocabulary test and 80 on a grammar test. The data items for both tests are normally distributed. The vocabulary test has a mean of 50 and a standard deviation of 5. The grammar test has a mean of 72 and a standard deviation of 6. On which test did the student have the better score? Explain why this is so. *

The number of miles that a particular brand of car tires lasts is normally distributed with a mean of 32,000 miles and a standard deviation of 4000 miles. In Exercises 47–49, find the data item in this distribution that corresponds to the given z-score.

47. $z = 1.5$ **48.** $z = 2.25$ **49.** $z = -2.5$

50. Using a random sample of 2281 American adults ages 18 and older, an Adecco survey asked respondents if they would be willing to sacrifice a percentage of their salary in order to work for an environmentally friendly company. The poll indicated that 31% of the respondents said "yes," 39% said "no," and 30% declined to answer.

   a. Find the margin of error, to the nearest tenth of a percent, for this survey.   ±2.1%

   b. Write a statement about the percentage of American adults who would be willing to sacrifice a percentage of their salary in order to work for an environmentally friendly company. *

51. The histogram indicates the frequencies of the number of syllables per word for 100 randomly selected words in Japanese.

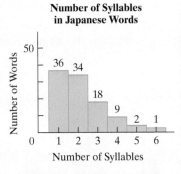

Number of Syllables in Japanese Words

   a. Is the shape of this distribution best classified as normal, skewed to the right, or skewed to the left? *

   b. Find the mean, median, and mode for the number of syllables in the sample of Japanese words. *

   c. Are the measures of central tendency from part (b) consistent with the shape of the distribution that you described in part (a)? Explain your answer. *

47. 38,000 miles   48. 41,000 miles   49. 22,000 miles

Chapter 8 Test

1. Politicians in the Florida Keys need to know if the residents of Key Largo think the amount of money charged for water is reasonable. The politicians decide to conduct a survey of a sample of Key Largo's residents. Which procedure would be most appropriate for a sample of Key Largo's residents? d

    a. Survey all water customers who pay their water bills at Key Largo City Hall on the third day of the month.

    b. Survey a random sample of executives who work for the water company in Key Largo.

    c. Survey 5000 individuals who are randomly selected from a list of all people living in Georgia and Florida.

    d. Survey a random sample of persons within each neighborhood of Key Largo.

Use these scores on a ten-point quiz to solve Exercises 2–4.

$$8, 5, 3, 6, 5, 10, 6, 9, 4, 5, 7, 9, 7, 4, 8, 8$$

2. Construct a frequency distribution for the data. *

3. Construct a histogram for the data. *

4. Construct a frequency polygon for the data. *

Use the 30 test scores listed below to solve Exercises 5–6.

79	51	67	50	78
62	89	83	73	80
88	48	60	71	79
89	63	55	93	71
41	81	46	50	61
59	50	90	75	61

5. Construct a grouped frequency distribution for the data. Use 40–49 for the first class and use the same width for each subsequent class. *

6. Construct a stem-and-leaf display for the data. *

7. The bar graph shows where teen women meet the guys they date. Describe what is misleading in this visual display of data. *

Where Teen Women Meet the Guys They Date

After-School Jobs 1%
After-School Activities 2%
Online 6%
Shopping Mall 9%
Through Friends 12%
School 70%

Source: J–14

Use the data items listed below to solve Exercises 8–11.

$$9, 4, 6, 4, 2$$

8. Find the mean. 5

9. Find the median. 4

10. Find the midrange. 5.5

11. Find the standard deviation. Round to two decimal places. 2.65

Use the frequency distribution shown to solve Exercises 12–14.

Score x	Frequency f
1	3
2	5
3	2
4	2

12. Find the mean. 2.25

13. Find the median. 2

14. Find the mode. 2

15. The annual salaries of four salespeople and the owner of a bookstore are   Answers will vary.

$17,500, $19,000, $22,000, $27,500, $98,500.

    Is the mean or the median more representative of the five annual salaries? Briefly explain your answer.

Monthly charges for cellphone plans in the United States are normally distributed with a mean of $62 and a standard deviation of $18. In Exercises 16–17, use the 68–95–99.7 Rule to find the percentage of plans that have monthly charges

16. between $62 and $80. 34%

17. that exceed $98. 2.5%

18. IQ scores are normally distributed in the population. Who has a higher IQ: a student with a 120 IQ on a scale where 100 is the mean and 10 is the standard deviation, or a teacher with a 128 IQ on a scale where 100 is the mean and 15 is the standard deviation? Briefly explain your answer. student

19. Using a random sample of 938 Americans ages 16 and older, a survey asked respondents to identify the word, phrase, or sentence they found most annoying in conversation. The bar graph summarizes the survey's results.

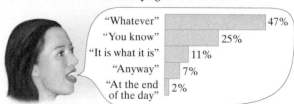

Which Is Most Annoying in Conversation?

"Whatever" 47%
"You know" 25%
"It is what it is" 11%
"Anyway" 7%
"At the end of the day" 2%

Source: Marist Poll Survey of 938 Americans ages 16 and older. August 3–6, 2010

    a. Find the margin of error, to the nearest tenth of a percent, for this survey. ±3.3%

    b. Write a statement about the percentage of Americans ages 16 and older who find "whatever" most annoying in conversation. *

*See Answers to Selected Exercises.

Appendix A

Basics of Percent

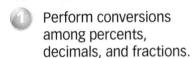

What am I supposed to learn?

After you have read this appendix, you should be able to:

1 Perform conversions among percents, decimals, and fractions.

2 Use the percent equation.

3 Solve applications involving the percent equation.

Percents are the result of expressing numbers as **part of 100**. The word *percent* means *per hundred*. For example, a poll showed that 55 out of every 100 college students prefer print textbooks. Thus,

$$\frac{55}{100} = 55\%,$$

indicating that 55% of students prefer print textbooks. The percent sign, %, is used to denote the number of parts out of 100.

Expressing Percents as Decimals

Because $100\% = \frac{100}{100} = 1$, multiplying or dividing by 100% is equivalent to multiplying or dividing by 1. This means that the value of a number does not change if it is multiplied or divided by 100%.

We can develop a procedure for writing a percent as a decimal by dividing the percent by 100%. Let's see how this works for 55% (the percent, or percentage, of students who prefer print textbooks).

$$55\% = \frac{55\%}{100\%} = \frac{55\%}{100\%} = \frac{55}{100} = 0.55$$

> Division by 100 moves the decimal point in the numerator two places to the left.

Generalizing from this example gives us a procedure for expressing a percent as a decimal.

 Perform conversions among percents, decimals, and fractions.

> **Expressing a Percent as a Decimal**
>
> 1. Move the decimal point in the percent two places to the left.
> 2. Remove the percent sign.

Example 1   Expressing Percents as Decimals

Express each percent as a decimal:

a. 19%   **b.** 180%   **c.** 0.23%   **d.** $\frac{3}{10}\%$.

Solution

Use the two steps in the box.

a. $19\% = 19.\% = 0.19\%$ The percent sign is removed.

> The decimal point starts at the far right.
>
> The decimal point is moved two places to the left.

Thus, $19\% = 0.19$.

b. $180\% = 180.\% = 1.80\,\%$ The percent sign is removed.

The decimal point starts at the far right.   The decimal point is moved two places to the left.

Thus, $180\% = 1.80$ or 1.8.

c. $0.23\% = 0.0023\%$ The percent sign is removed.

The decimal point is moved two places to the left.

Thus, $0.23\% = 0.0023$.

d. $\dfrac{3}{10}\% = 0.3\% = 0.003\%$

Thus, $0.3\% = 0.003$.

✓ **Check Point 1**   Express each percent as a decimal:

▶ **a.** 67% 0.67   **b.** 230% 2.30 or 2.3   **c.** 0.5% 0.005   **d.** $\dfrac{5}{8}\%$.   0.00625

Expressing Decimals as Percents

We can develop a procedure for writing a decimal as a percent by multiplying the decimal by 100%, or 1. Let's see how this works for a particular decimal, say, 0.43.

$$0.43 = 0.43(100\%) = (0.43 \cdot 100)\% = 43\%$$

Multiplication by 100 moves the decimal point two places to the right.

Generalizing from this example gives us a procedure for expressing a decimal as a percent.

Expressing a Decimal as a Percent

1. Move the decimal point two places to the right.

2. Attach a percent sign.

Example 2   Expressing Decimals as Percents

Express each decimal as a percent:

a. 0.27 **b.** 7.4 **c.** 0.0375 **d.** 0.002.

Solution

Use the two steps in the box.

Move decimal point two places to the right.   Move decimal point two places to the right.

a. $0.27\,\%$ Attach a percent sign.   **b.** $740.\%$ Attach a percent sign.

Thus, $0.27 = 27\%$.   Thus, $7.4 = 740\%$. Notice that a zero was attached as a placeholder.

Move decimal point two
places to the right.

c. $0\,03.75\%$ Attach a percent sign.

Thus, $0.0375 = 3.75\%$.

Move decimal point two
places to the right.

d. $0\,00.2\%$ Attach a percent sign.

Thus, $0.002 = 0.2\%$.

Check Point 2   Express each decimal as a percent:

a. 0.11 11%   **b.** 9.2 920%

c. 0.0625 6.25%   **d.** 0.003.   0.3%

Expressing Percents as Fractions

We have seen that percent means per hundred. This means that we can write a percent as a fraction by writing the number preceding the percent sign over 100. For example,

$$23\% = \frac{23}{100}.$$ Write 23, the number preceding
the percent sign, over 100.

Another way to think about this step is to divide the given percent by 100%, or 1. This does not affect the percent's value.

$$23\% = \frac{23\%}{100\%} = \frac{23\%}{100\%} = \frac{23}{100}$$

> ### Expressing a Percent as a Fraction
> Write the number preceding the percent sign over 100. Simplify, if possible.

Example 3   Expressing Percents as Fractions

Express each percent as a fraction or a mixed number:

a. 14% **b.** 175%

c. $33\frac{1}{3}\%$ **d.** $\frac{3}{4}\%$.

Solution

a. $14\% = \frac{14}{100}$ Write 14 over 100.

$= \frac{2 \cdot 7}{2 \cdot 50} = \frac{7}{50}$ Simplify.

b. $175\% = \frac{175}{100}$ Write 175 over 100.

$= \frac{25 \cdot 7}{25 \cdot 4} = \frac{7}{4}$ or $1\frac{3}{4}$ Simplify.

c. $33\frac{1}{3}\% = \dfrac{33\frac{1}{3}}{100}$ Write $33\frac{1}{3}$ over 100.

$= \dfrac{\frac{100}{3}}{1} \div 100$ Write $33\frac{1}{3}$ as an improper fraction. Replace the division bar by $\div$.

$= \dfrac{100}{3} \cdot \dfrac{1}{100}$ Express division as multiplication by a reciprocal.

$= \dfrac{\overset{1}{\cancel{100}}}{3} \cdot \dfrac{1}{\underset{1}{\cancel{100}}}$ Divide the numerator of the first fraction and the denominator of the second fraction by 100, the common factor.

$= \dfrac{1}{3}$ Multiply the remaining factors in the numerators and the denominators.

d. $\dfrac{3}{4}\% = \dfrac{\frac{3}{4}}{100}$ Write $\dfrac{3}{4}$ over 100.

$= \dfrac{3}{4} \div 100$ Replace the division bar by $\div$. (This step is optional.)

$= \dfrac{3}{4} \cdot \dfrac{1}{100}$ Express division as multiplication by a reciprocal.

$= \dfrac{3 \cdot 1}{4 \cdot 100}$ Multiply numerators and multiply denominators.

$= \dfrac{3}{400}$ Simplify.

✓ **Check Point 3** Express each percent as a fraction or a mixed number:

a. 26% $\frac{13}{50}$ **b.** 125% $\frac{5}{4}$ or $1\frac{1}{4}$

c. $66\frac{2}{3}\%$ $\frac{2}{3}$ **d.** $\dfrac{2}{5}\%$. $\frac{1}{250}$

We want whole numbers in numerators and denominators of simplified fractions. If fractions have numerators that are decimals, we can simplify by multiplying by 1 in the form $\dfrac{10}{10}$ or $\dfrac{100}{100}$ or $\dfrac{1000}{1000}$, and so on. This is illustrated in our next example.

Example 4 **Expressing Percents as Fractions**

Express each percent as a fraction:

a. 2.7% **b.** 0.35%.

Solution

a. $2.7\% = \dfrac{2.7}{100}$ Write 2.7 over 100.

$= \dfrac{2.7 \times 10}{100 \times 10}$ Multiply by 1 in the form $\frac{10}{10}$ to obtain a whole number in the numerator.

$= \dfrac{27}{1000}$ Simplify.

b. $0.35\% = \dfrac{0.35}{100}$ *Write 0.35 over 100.*

$\qquad = \dfrac{0.35 \times 100}{100 \times 100}$ *Multiply by 1 in the form $\frac{100}{100}$ to obtain a whole number in the numerator.*

$\qquad = \dfrac{35}{10,000}$ *Simplify.*

$\qquad = \dfrac{\cancel{5} \cdot 7}{\cancel{5} \cdot 2000} = \dfrac{7}{2000}$ *Express the fraction in simplest form.*

✓ **Check Point 4** Express each percent as a fraction:

a. 5.3% $\frac{53}{1000}$ **b.** 0.55%. $\frac{11}{2000}$

Expressing Fractions as Percents

In Section 1.3, we learned that any rational number $\frac{a}{b}$ can be expressed as a decimal by dividing the denominator, b, into the numerator, a. To write a fraction as a percent, first write the fraction as a decimal and then express the decimal as a percent.

> **Expressing a Fraction as a Percent**
> **1.** Write the fraction as a decimal: Divide the numerator by the denominator.
> **2.** Express the decimal as a percent: Move the decimal point two places to the right and attach a percent sign.

Example 5 Expressing Fractions as Percents

Express each fraction or mixed number as a percent:

a. $\dfrac{5}{8}$ **b.** $\dfrac{17}{25}$ **c.** $7\dfrac{3}{5}$ **d.** $\dfrac{19}{4}$ **e.** $\dfrac{2}{3}$.

Solution

a. Begin by writing $\dfrac{5}{8}$ as a decimal.

$$
\begin{array}{r}
0.625 \\
8\overline{)5.000} \\
\underline{4\,8} \\
20 \\
\underline{16} \\
40 \\
\underline{40} \\
0
\end{array}
$$

Thus, $\dfrac{5}{8} = 0.625 = 62.5\%$.   *Write 0.625 as a percent. Move the decimal point two places to the right and attach a percent sign.*

b. Begin by writing $\dfrac{17}{25}$ as a decimal.

$$
\begin{array}{r}
0.68 \\
25\overline{)17.00} \\
\underline{15\,0} \\
2\,00 \\
\underline{2\,00} \\
0
\end{array}
$$

Thus, $\dfrac{17}{25} = 0.68 = 68\%$.   *Write 0.68 as a percent. Move the decimal point two places to the right and attach a percent sign.*

c. To write the mixed number $7\frac{3}{5}$ as a percent, begin by writing $\frac{3}{5}$ as a decimal.

$$
\begin{array}{r}
0.6 \\
5)\overline{3.0} \\
\underline{3\ 0} \\
0
\end{array}
$$

Thus, $7\frac{3}{5} = 7.6 = 760\%$. *Write 7.6 as a percent. Move the decimal point two places to the right and attach a percent sign.*

d. Begin by writing $\frac{19}{4}$ as a decimal.

$$
\begin{array}{r}
4.75 \\
4)\overline{19.00} \\
\underline{16} \\
3\ 0 \\
\underline{2\ 8} \\
20 \\
\underline{20} \\
0
\end{array}
$$

Thus, $\frac{19}{4} = 4.75 = 475\%$. *Write 4.75 as a percent.*

e. Begin by writing $\frac{2}{3}$ as a decimal.

$$
\begin{array}{r}
0.66\ldots = 0.\overline{6} \\
3)\overline{2.00} \\
\underline{1\ 8} \\
20 \\
\underline{18} \\
2
\end{array}
$$

Now we write $0.\overline{6}$, or $0.666\ldots$, as a percent by moving the decimal point two places to the right and attaching a percent sign.

$$\frac{2}{3} = 0.\overline{6} = 0.666\ldots = 66.6\ldots\% = 66.\overline{6}\%.$$

A percent with a repeating decimal may be written as a mixed number. Because the remainder of our division is always 2, another way to write $\frac{2}{3}$ as a percent is

$$\frac{2}{3} = 66\frac{2}{3}\%.$$

Check Point 5 Express each fraction or mixed number as a percent:

a. $\frac{3}{8}$ *37.5%* **b.** $\frac{13}{25}$ *52%* **c.** $6\frac{4}{5}$ *680%* **d.** $\frac{17}{4}$ *425%* **e.** $\frac{1}{3}$. *33.$\overline{3}$% or $33\frac{1}{3}$%*

The Percent Equation

2 Use the percent equation.

In a survey of 10,400 students, 65% of those surveyed, or 6760 students, cited freedom of speech as our most important First Amendment right (Source: Knight Foundation survey). **Table A.1** shows the survey result in words and as a percent equation.

Table A.1 Most Important First Amendment Right

Survey Result	Percent Equation
65% of 10,400 students, or 6760 students, stated that freedom of speech is the most important First Amendment right.	$65\% \cdot 10,400 = 6760$ 65% of 10,400 is 6760.

The percent equation in **Table A.1** shows how percents are useful in comparing two numbers. In this statement, 65% compares the number 6760, called the **amount**, to the number 10,400, called the **base**.

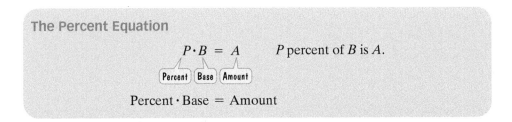

65% of 10,400 is 6760.

$$65\% \cdot 10,400 = 6760$$

Percent   Base   Amount

This is an example of the general **percent equation**.

The Percent Equation

$$P \cdot B = A \qquad P \text{ percent of } B \text{ is } A.$$

Percent Base Amount

Percent · Base = Amount

The percent equation contains three numbers. In a typical percent problem, two of the numbers are known and one is unknown. To solve the problem, we will let x (or any variable) represent the unknown number or the unknown percent. Then we'll write a percent equation that translates, or models, the conditions of the problem. Finally, we solve the equation and answer the problem's question.

Here are two important points to keep in mind when solving problems using the percent equation:

- The word "of" translates as multiplication.
- The percent must be expressed as a decimal in order to solve the equation.

Example 6   Solving a Percent Equation for the Amount

What number is 45% of 60?

Solution

Let x represent the unknown quantity.

Let x = the number.

Write an equation in x that models the conditions.

What number is 45% of 60?

$$x \quad = \quad 45\% \quad \cdot \quad 60$$

Solve the equation and answer the problem's question.

$x = 45\% \cdot 60$ This is the equation that models the problem's conditions.
$x = 0.45 \cdot 60$ Express 45% as a decimal: 45% = 0.45.
$x = 27$ Multiply:

         0.45
        × 60
        ———
         00
      27 00
      ———
      27.00.

Thus, 27 is 45% of 60.

 **Check Point 6**   What number is 25% of 80? 20

Example 7   Solving a Percent Equation for the Base

54% of what number is 81?

Solution

Let x represent the unknown quantity.

Let x = the number.

Write an equation in x that models the conditions.

54%   of   what number   is   81?

$$54\% \quad \cdot \quad x \quad = \quad 81$$

Solve the equation and answer the problem's question.

$54\% \cdot x = 81$ This is the equation that models the problem's conditions.

$0.54x = 81$ Express 54% as a decimal: 54% = 0.54.

$\dfrac{0.54x}{0.54} = \dfrac{81}{0.54}$ Isolate x by dividing both sides by 0.54.

$x = 150$ Simplify:

$$\begin{array}{r} 1\,50. \\ 0.54\overline{)81.00} \\ \underline{54}\downarrow \\ 270 \\ \underline{270}\downarrow \\ 00 \\ \underline{0} \\ 0 \end{array}$$

Thus, 54% of 150 is 81. Take a moment to check this statement by finding the product of 0.54 and 150. You should obtain 81.

▶ ✓ **Check Point 7**   12% of what number is 72?   600

If the unknown in the percent equation is the percent, your solution will be in decimal form. In order to answer the problem's question, convert this decimal to a percent in your final answer.

Example 8   Solving a Percent Equation for the Percent

What percent of 26 is 1.3?

Solution

Let x represent the unknown quantity.

Let x = the percent.

Write an equation in x that models the conditions.

What percent   of   26   is   1.3?

$$x \quad \cdot \quad 26 \quad = \quad 1.3$$

Solve the equation and answer the problem's question.

$$26x = 1.3$$ This is the equation that models the problem's conditions. We wrote x · 26 as 26x, although this is optional.

$$\frac{26x}{26} = \frac{1.3}{26}$$ Isolate x by dividing both sides by 26.

$$x = 0.05$$ Simplify:

$$\begin{array}{r} 0.05 \\ 26\overline{)1.30} \\ \underline{1\,30} \\ 0 \end{array}$$

$$x = 5\%$$ Express 0.05 as a percent: 0.05 = 5%.

Thus, 5% of 26 is 1.3. Check this statement by finding 5% of 26, or 0.05 · 26. You should obtain 1.3 for the product.

 **Check Point 8**   What percent of 96 is 2.4?   2.5%

Applications Involving the Percent Equation

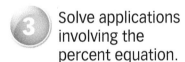 Solve applications involving the percent equation.

The percent equation can be used to determine the sales tax collected by states, counties, and cities on sales of items to customers. The sales tax is a percent of the cost of an item.

Example 9   Percent and Sales Tax

You purchase a computer for $1260.00. The local sales tax rate is 7.5% of the cost.

   a. How much sales tax is paid?
   b. What is the computer's total cost?

Solution

   a. We are looking for the sales tax amount.

   Let x = the amount paid in sales tax.

   We are told that the sales tax rate is 7.5% of the computer's cost.

Sales tax amount	is	7.5%	of	the computer's cost.
x	=	7.5%	·	1260.00

$$x = 7.5\% \cdot 1260.00$$ This is the equation that models the problem's conditions.
$$x = 0.075 \cdot 1260.00$$ Express 7.5% as a decimal: 7.5% = 0.075.
$$x = 94.50$$ Multiply:

$$\begin{array}{r} 1\,260 \\ \times\ 0.075 \\ \hline 6\,300 \\ 88\,200 \\ \hline 94.500 \end{array}$$

The sales tax is $94.50.

b. The computer's total cost is the purchase price plus the sales tax.

Computer's total cost	is	purchase price	plus	sales tax.

$$= \quad 1260.00 \quad + \quad 94.50$$

$$\begin{array}{r} 1 \\ 1260.00 \\ + \ 94.50 \\ \hline 1354.50 \end{array}$$

$$= \quad 1354.50$$

The computer's total cost is $1354.50.

✓ **Check Point 9**   You purchase a bicycle for $894.00. The local sales tax rate is 6% of the cost.

a. How much sales tax is paid? $53.64

▶ **b.** What is the bicycle's total cost? $947.64

None of us is thrilled about sales tax, but we do like buying things that are *on sale*, the theme of our next example.

Example 10   Buying at a Percent of the Regular Price

At a special sale, you buy a pair of running shoes for $94.80. The clerk informs you that this is 80% of the regular price. Without this special sale, what is the regular price for this model of running shoes?

Solution

Let x = the regular price of the shoes.
We are told that the sales price is 80% of the regular price.

Sales price	is	80%	of	the regular price.

$$94.80 \quad = \quad 80\% \quad \cdot \quad x$$

$94.80 = 80\% \cdot x$ This is the equation that models the problem's conditions.

$94.80 = 0.8x$ Express 80% as a decimal: $80\% = 0.80 = 0.8$.

$\dfrac{94.80}{0.8} = \dfrac{0.8x}{0.8}$ Isolate x by dividing both sides by 0.8.

$118.5 = x$ Simplify:

$$\begin{array}{r} 118.5 \\ 0.8\overline{)94.80} \\ \underline{8\downarrow}\quad \\ 14\quad \\ \underline{8\downarrow}\ \\ 68\ \\ \underline{64\downarrow} \\ 40 \\ \underline{40} \\ 0 \end{array}$$

and $\dfrac{0.8x}{0.8} = \dfrac{0.8}{0.8} \cdot x = 1x = x.$

The regular price for this model of running shoes is $118.50. Check this amount by taking 80% of 118.50, or $0.8 \cdot 118.5$. You should get $94.80, the sale price.

✓ **Check Point 10**   At a special sale, you buy an exercise machine for $602.00. The clerk informs you that this is 70% of the regular price. Without this special sale, ▶ what is the regular price of this exercise machine? $860

Example 11   Spending for the Average American Household

The circle graph in **Figure A.1** shows a breakdown of spending for the average U.S. household using 365 days worked as a basis of comparison. What percent of work time does the average U.S. household spend paying for federal taxes? Round to the nearest tenth of a percent.

Spending for the Average American Household, by 365 Days Worked

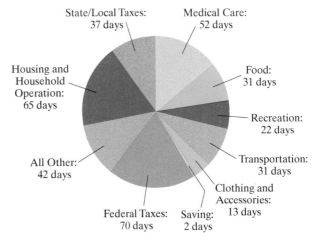

State/Local Taxes: 37 days

Medical Care: 52 days

Housing and Household Operation: 65 days

Food: 31 days

Recreation: 22 days

All Other: 42 days

Transportation: 31 days

Clothing and Accessories: 13 days

Federal Taxes: 70 days

Saving: 2 days

Figure A.1
Source: The Tax Foundation

Solution

Figure A.1 shows 70 days worked paying for federal taxes out of a total of 365 days worked. 70 is what percent of 365?

Let x = the percent.

70 is what percent of 365?

$$70 \quad = \quad x \quad \cdot \quad 365$$

$70 = 365x$ This is the equation that models the problem's conditions.

$\dfrac{70}{365} = \dfrac{365x}{365}$ Isolate x by dividing both sides by 365.

$0.192 \approx x$ Simplify:

$$
\begin{array}{r}
0.1917\ldots \\
365\overline{)70.0000} \\
\underline{36\ 5} \\
33\ 50 \\
\underline{32\ 85} \\
650 \\
\underline{365} \\
2850 \\
\underline{2555} \\
295
\end{array}
$$

Notice that to round to the nearest tenth of a percent, we need to include four decimal places in the decimal obtained from the division.

$19.2\% \approx x$ Express 0.192 as a percent: 0.192 = 19.2%.

The average American household spends approximately 19.2% of days worked paying for federal taxes.

A Brief Review • Rounding

- To round whole numbers, look at the digit to the right of the digit where rounding is to occur. If the digit to the right is 5 or greater, add 1 to the digit to be rounded. Replace all digits to the right with zeros. If the digit to the right is less than 5, do not change the digit to be rounded. Replace all digits to the right with zeros.
- To round decimals, drop the digits to the right of the rounding place rather than replacing these digits with zeros.

✓ **Check Point 11** Use **Figure A.1** on the previous page to solve this problem. What percent of work time does the average U.S. household spend paying for state and local taxes? Round to the nearest tenth of a percent. 10.1%

Appendix A Exercise Set

Practice Exercises

In Exercises 1–12, express each percent as a decimal.

1. 51% 0.51   2. 57% 0.57   3. 3% 0.03

4. 9% 0.09   5. 270% 2.7   6. 350% 3.5

7. 3.7% 0.037   8. 4.1% 0.041   9. 0.32% 0.0032

10. 0.27% 0.0027   11. $\frac{4}{10}$% 0.004   12. $\frac{7}{10}$% 0.007

In Exercises 13–22, express each decimal as a percent.

13. 0.89 89%   14. 0.73 73%

15. 5.1 510%   16. 8.3 830%

17. 0.0875 8.75%   18. 0.0125 1.25%

19. 0.8 80%   20. 0.1 10%

21. 0.008 0.8%   22. 0.001 0.1%

In Exercises 23–42, express each percent as a fraction or mixed number in simplest form.

23. 19% $\frac{19}{100}$ 24. 17% $\frac{17}{100}$

25. 16% $\frac{4}{25}$ 26. 24% $\frac{6}{25}$

27. 280% $\frac{14}{5}$ or $2\frac{4}{5}$ 28. 320% $\frac{16}{5}$ or $3\frac{1}{5}$

29. $5\frac{3}{4}$% $\frac{23}{400}$ 30. $7\frac{1}{4}$% $\frac{29}{400}$

31. $16\frac{2}{3}$% $\frac{1}{6}$ 32. $83\frac{1}{3}$% $\frac{5}{6}$

33. $\frac{3}{8}$% $\frac{3}{800}$ 34. $\frac{5}{8}$% $\frac{1}{160}$

35. 7.3% $\frac{73}{1000}$ 36. 3.7% $\frac{37}{1000}$

37. 12.5% $\frac{1}{8}$ 38. 37.5% $\frac{3}{8}$

39. 3.75% $\frac{3}{80}$ 40. 6.25% $\frac{1}{16}$

41. 0.45% $\frac{9}{2000}$ 42. 0.65% $\frac{13}{2000}$

In Exercises 43–52, express each fraction or mixed number as a percent.

43. $\frac{1}{4}$ 25%   44. $\frac{3}{4}$ 75%   45. $\frac{13}{50}$ 26%

46. $\frac{17}{50}$ 34%   47. $8\frac{2}{5}$ 840%   48. $9\frac{4}{5}$ 980%

49. $\frac{11}{5}$ 220%   50. $\frac{8}{5}$ 160%   51. $\frac{1}{6}$ 16.$\overline{6}$% or $16\frac{2}{3}$%

52. $\frac{5}{6}$ 83.$\overline{3}$% or $83\frac{1}{3}$%

In Exercises 53–72, perform conversions among decimals, percents, and fractions to complete the table.

	Decimal	Percent	Fraction
53.	0.46	46%	$\frac{23}{50}$
54.	0.35	35%	$\frac{7}{20}$
55.	0.125	12.5%	$\frac{1}{8}$
56.	0.375	37.5%	$\frac{3}{8}$
57.	0.14	14%	$\frac{7}{50}$
58.	0.225	22.5%	$\frac{9}{40}$
59.	1.6	160%	$\frac{8}{5}$ or $1\frac{3}{5}$
60.	2.4	240%	$\frac{12}{5}$ or $2\frac{2}{5}$
61.	8.26	826%	$\frac{413}{50}$ or $8\frac{13}{50}$
62.	6.42	642%	$\frac{321}{50}$ or $6\frac{21}{50}$

	Decimal	Percent	Fraction
63.	8.12	812%	$8\frac{3}{25}$
64.	9.28	928%	$9\frac{7}{25}$
65.	0.235	23.5%	$\frac{47}{200}$
66.	0.265	26.5%	$\frac{53}{200}$
67.	0.035	3.5%	$\frac{7}{200}$
68.	0.075	7.5%	$\frac{3}{40}$
69.	1.1$\overline{3}$	113.$\overline{3}$% or $113\frac{1}{3}$%	$\frac{17}{15}$
70.	1.4$\overline{6}$	146.$\overline{6}$% or $146\frac{2}{3}$%	$\frac{22}{15}$
71.	3.05	305%	$\frac{61}{20}$ or $3\frac{1}{20}$
72.	8.05	805%	$\frac{161}{20}$ or $8\frac{1}{20}$

In Exercises 73–100, write an equation that models the conditions. Then solve the equation and answer the problem's question.

73. What number is 15% of 40? $x = 0.15 \cdot 40; 6$

74. What number is 12% of 90? $x = 0.12 \cdot 90; 10.8$

75. $4\frac{1}{2}$% of 120 is what number? $0.045 \cdot 120 = x; 5.4$

76. $6\frac{1}{2}$% of 260 is what number? $0.065 \cdot 260 = x; 16.9$

77. 7% of what number is 602? $0.07x = 602; 8600$

78. 9% of what number is 828? $0.09x = 828; 9200$

79. 175 is 35% of what number? $175 = 0.35x; 500$

80. 104 is 26% of what number? $104 = 0.26x; 400$

81. What percent of 24 is 9? $x \cdot 24 = 9; 37.5\%$

82. What percent of 24 is 15? $x \cdot 24 = 15; 62.5\%$

83. 88 is what percent of 80? $88 = x \cdot 80; 110\%$

84. 66 is what percent of 40? $66 = x \cdot 40; 165\%$

85. 0.3 is 5% of what number? $0.3 = 0.05x; 6$

86. 0.8 is 4% of what number? $0.8 = 0.04x; 20$

87. 250% of 620 is what number? $2.5 \cdot 620 = x; 1550$

88. 150% of 260 is what number? $1.5 \cdot 260 = x; 390$

89. 3.2 is what percent of 50? $3.2 = x \cdot 50; 6.4\%$

90. 2.3 is what percent of 40? $2.3 = x \cdot 40; 5.75\%$

91. 120% of what number is 54? $1.2x = 54; 45$

92. 160% of what number is 88? $1.6x = 88; 55$

93. What percent of 800 is 2? $x \cdot 800 = 2; 0.25\%$

94. What percent of 55 is 6? $x \cdot 55 = 6; 10.9\overline{09}\%$ or $10\frac{10}{11}\%$

95. 800 is what percent of 2? $800 = x \cdot 2; 40,000\%$

96. 500 is what percent of 6? $500 = x \cdot 6; 8333.\overline{3}\%$ or $8333\frac{1}{3}\%$

97. $6\frac{1}{2}$% of what number is 39? $0.065x = 39; 600$

98. $7\frac{1}{2}$% of what number is 60? $0.075x = 60; 800$

99. What percent of 186 is 83.7? $x \cdot 186 = 83.7; 45\%$

100. What percent of 196 is 68.6? $x \cdot 196 = 68.6; 35\%$

Application Exercises

101. You purchase a new car for $32,800.00. The local sales tax rate is 6% of the cost.
  a. How much sales tax is paid? $1968
  b. What is the car's total cost? $34,768

102. You purchase a graphing calculator for $96.00. The local sales tax rate is 7% of the cost.
  a. How much sales tax is paid? $6.72
  b. What is the calculator's total cost? $102.72

103. Lunch at a restaurant cost $21.85. If you want to leave the server a tip that equals 20% of the cost of the lunch, how much of a tip should you leave? $4.37

104. Dinner at a restaurant cost $32.45. If you want to leave the server a tip that equals 20% of the cost of the dinner, how much of a tip should you leave? $6.49

105. A sofa is on sale for $600.00. If this is 80% of the regular price, what was the price before the sale? $750

106. A desk is on sale for $240.00. If this is 60% of the regular price, what was the price before the sale? $400

Views of Police. *According to a poll of 1500 adults conducted in 2014 by* USA Today/Pew Research Center, *a majority of Americans said police departments failed to do a good job at holding officers accountable. The circle graph shows a breakdown of the 1500 respondents who were surveyed. Use this information to solve Exercises 107–108.*

How Well Are Police Departments Holding Officers Accountable?

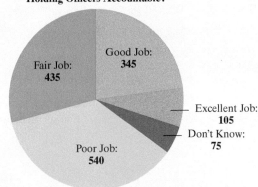

Source: USA Today/Pew Research Center

107. What percent of those surveyed said police departments did a poor job at holding officers accountable?   36%

108. What percent of those surveyed said police departments did an excellent job at holding officers accountable?   7%

Appendix B

Applications of Percent

What am I supposed to learn?

After you have read this appendix, you should be able to:

1. Solve commission problems.

2. Solve applications involving a percent increase.

3. Solve applications involving a percent decrease.

4. Calculate discount and sale price.

1 Solve commission problems.

One of the most common ways that we are given numerical information is with percents. In this appendix, we explore applications of the percent equation, $PB = A$. We open with an application in which percents play a significant role in determining how much money people who work in sales are paid.

Commission

People who work in sales frequently are paid on commission. A **commission** is a percent of an employee's sales. The percent is called the **commission rate**. By changing the terminology in the percent equation, we can obtain a formula for computing commission.

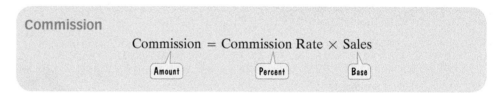

Commission

Commission = Commission Rate × Sales

Amount   Percent   Base

Example 1   Finding the Commission

A realtor sells a house for $425,000. If the realtor's commission rate is 1.5% of the selling price, find the commission.

Solution

Commission = Commission Rate × Sales

$$= \ 1.5\% \ \cdot \ \$425{,}000$$

$$= 0.015 \cdot \$425{,}000 \quad \text{Express 1.5\% as a decimal: 1.5\% = 0.015.}$$

$$= \$6375$$

$$\begin{array}{r} 425{,}000 \\ \times \quad .015 \\ \hline 2125\,000 \\ 4250\,000 \\ \hline 6375.000 \end{array}$$

The realtor's commission is $6375.

Check Point 1   The price of a car is $20,500. If the salesperson earns a 5% commission rate, find the commission. $1025

The commission equation contains three numbers:

Commission = Commission Rate × Sales.

If the commission rate and sales are known, their product gives the commission. This was illustrated in Example 1. If two different numbers are known, we will let x (or any variable of your choice) represent the unknown sales or the unknown commission rate. Then we'll write a commission equation that translates, or models, the conditions of the problem. Finally, we solve the equation and answer the problem's question.

Example 2   Finding the Sales

A salesperson earns a 9% commission rate on sales of electronics equipment. If she earned $450 in commission one week, what were her sales that week?

Solution

Let x represent the unknown quantity.

Let x = sales for the week.

Write an equation in x that models the conditions.

$$\text{Commission} = \text{Commission Rate} \times \text{Sales}$$
$$450 = 9\% \cdot x$$

Solve the equation and answer the problem's question.

$450 = 9\% \cdot x$ This is the equation that models the problem's conditions.

$450 = 0.09x$ Express 9% as a decimal: 9% = 0.09.

$\dfrac{450}{0.09} = \dfrac{0.09x}{0.09}$ Isolate x by dividing both sides by 0.09.

$5000 = x$ Simplify:

$$0.09\overline{)450.00}^{\,5000.}$$

For that week, sales were $5000.

✓ **Check Point 2** A salesperson earns a 7% commission rate on sales of furniture. If he earned $602 in commission one week, what were his sales that week? $8600

Example 3   Finding the Commission Rate

A salesperson earned $2125 in commission for selling $42,500 worth of appliances. What was the commission rate?

Solution

Let x represent the unknown quantity.

Let x = the commission rate.

Write an equation in x that models the conditions.

$$\text{Commission} = \text{Commission Rate} \times \text{Sales}$$
$$2125 = x \cdot 42,500$$

Solve the equation and answer the problem's question.

$2125 = 42,500x$ This is the equation that models the problem's conditions. We wrote x · 42,500 as 42,500x: It's more traditional to write the constant factor before the variable factor.

$\dfrac{2125}{42,500} = \dfrac{42,500x}{42,500}$ Isolate x by dividing both sides by 42,500.

$$42,500\overline{)2125.00}^{\,0.05}$$
$$\underline{2125\,00}$$
$$0$$

$0.05 = x$

$5\% = x$ Express 0.05 as a percent: 0.05 = 5%.

The commission rate is 5%.

✓ **Check Point 3** A salesperson earned $4500 in commission for selling $30,000 worth of sporting equipment. What was the commission rate?   15%

2 Solve applications involving a percent increase.

Percent Increase

Percents are used for comparing changes, such as increases or decreases in sales, population, prices, and production. If a quantity increases, its *percent increase* can be described using the percent equation:

$$\text{Percent Increase} \times \text{Original Amount} = \text{Amount of Increase}$$

Percent   Base   Amount

We can solve this equation for the percent increase: Divide both sides of the equation by the original amount. This gives us a formula for finding percent increase.

Finding Percent Increase

If a quantity increases, its percent increase can be found as follows:

1. Find the fraction for the percent increase:

$$\frac{\text{amount of increase}}{\text{original amount}}.$$

2. Find the percent increase by expressing the fraction in step 1 as a percent.

Example 4   Finding Percent Increase

Find the percent increase if:

a. 4 is increased to 6.

b. 4 is increased to 8.

c. 4 is increased to 16.

Solution

a. If 4 is increased to 6, the amount of increase is $6 - 4$, or 2.

$$\text{Percent increase} = \frac{\text{amount of increase}}{\text{original amount}}$$

This number must be the original amount and not the amount after the increase.

$$= \frac{6 - 4}{4} = \frac{2}{4} = \frac{1}{2} = 0.5 = 50\%$$

b. If 4 is increased to 8, the amount of increase is $8 - 4$, or 4.

$$\text{Percent increase} = \frac{\text{amount of increase}}{\text{original amount}}$$

$$= \frac{8 - 4}{4} = \frac{4}{4} = 1 = 100\%$$

c. If 4 is increased to 16, the amount of increase is $16 - 4$, or 12.

$$\text{Percent increase} = \frac{\text{amount of increase}}{\text{original amount}}$$

$$= \frac{16 - 4}{4} = \frac{12}{4} = 3 = 300\%$$

 Check Point 4   Find the percent increase if:

a. 10 is increased to 12.   20%

b. 10 is increased to 20.   100%

c. 10 is increased to 40.   300%

Example 5   Percent Increase in Life Expectancy

The bar graph in **Figure B.1** shows life expectancies at birth for men and women in four selected birth years.

Find the percent increase in life expectancy for a woman born in 1980 to a woman born in 2010. Round to the nearest tenth of a percent.

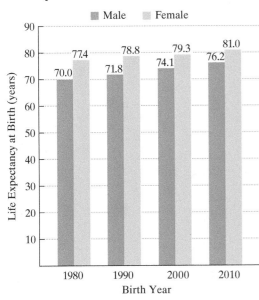

Figure B.1
Source: National Center for Health Statistics

Solution

The bar graph in **Figure B.1** shows a life expectancy of 77.4 years for women born in 1980 and a life expectancy of 81.0 years for women born in 2010. We find the amount of increase by subtracting the original 1980 life expectancy from the new 2010 life expectancy.

$$\text{Amount of increase} = 81.0 - 77.4$$
$$= 3.6$$

The amount of increase is 3.6 years.

$$\text{Percent increase} = \frac{\text{amount of increase}}{\text{original amount}}$$

$$= \frac{81.0 - 77.4}{77.4} = \frac{3.6}{77.4}$$

Now we need to express this fraction as a percent. We'll first find the fraction's equivalent decimal and then write the decimal as a percent. Divide the numerator, 3.6, by the denominator, 77.4.

```
          0.0465
  77.4)3.60000
       3 096
       ─────
        5040
        4644
        ─────
        3960
        3870
        ─────
          90
```

Include four decimal places so that we can round to the nearest tenth of a percent.

Thus,

$$\text{Percent increase} = \frac{3.6}{77.4} \approx 0.0465 = 4.65\% \approx 4.7\%.$$

The percent increase in life expectancy for a woman born in 1980 to a woman born in 2010 is approximately 4.7%.

✓ **Check Point 5**   Use **Figure B.1** to find the percent increase in life expectancy for a man born in 1980 to a man born in 2010. Round to the nearest tenth of a percent.   8.9%

Percent Decrease

③ Solve applications involving a percent decrease.

If a quantity decreases, its **percent decrease** can be found as follows:

Finding Percent Decrease

1. Find the fraction for the percent decrease:

$$\frac{\text{amount of decrease}}{\text{original amount}}.$$

2. Find the percent decrease by expressing the fraction in step 1 as a percent.

Example 6   Finding Percent Decrease

Find the percent decrease if:

a. 6 is decreased to 4.

b. 8 is decreased to 4.

c. 16 is decreased to 4.

Solution

a. If 6 is decreased to 4, the amount of decrease is 6 − 4, or 2.

$$\text{Percent decrease} = \frac{\text{amount of decrease}}{\text{original amount}}$$

This number must be the original amount and not the amount after the decrease. $= \dfrac{6-4}{6} = \dfrac{2}{6} = \dfrac{1}{3} = 0.33\dfrac{1}{3} = 33\dfrac{1}{3}\%$

b. If 8 is decreased to 4, the amount of decrease is 8 − 4, or 4.

$$\text{Percent decrease} = \frac{\text{amount of decrease}}{\text{original amount}}$$

$$= \frac{8-4}{8} = \frac{4}{8} = \frac{1}{2} = 0.5 = 50\%$$

c. If 16 is decreased to 4, the amount of decrease is 16 − 4, or 12.

$$\text{Percent decrease} = \frac{\text{amount of decrease}}{\text{original amount}}$$

$$= \frac{16-4}{16} = \frac{12}{16} = \frac{4 \cdot 3}{4 \cdot 4} = \frac{3}{4} = 0.75 = 75\%$$

Great Question!

In Example 4(c), we found that increasing 4 to 16 is a 300% increase. Shouldn't that mean that decreasing 16 to 4 is a 300% decrease?

No. Notice the difference between Example 4(c) and Example 6(c).

- 4 is increased to 16.

$$\text{Percent increase} = \frac{\text{amount of increase}}{\text{original amount}} = \frac{12}{4} = 3 = 300\%$$

- 16 is decreased to 4.

$$\text{Percent decrease} = \frac{\text{amount of decrease}}{\text{original amount}} = \frac{12}{16} = \frac{3}{4} = 75\%$$

Although an increase from 4 to 16 is a 300% increase, a decrease from 16 to 4 is *not* a 300% decrease. **A percent decrease involving nonnegative quantities can never exceed 100%.** When a quantity is decreased by 100%, it is reduced to zero.

✓ **Check Point 6**   Find the percent decrease if:

a. 10 is decreased to 6.   40%

b. 20 is decreased to 10.   50%

c. 40 is decreased to 10.   75%

Example 7   Sale Price and Percent Decrease

A jacket regularly sells for $135.00. The sale price is $60.75. Find the percent decrease of the sale price from the regular price.

Solution

First we find the amount of decrease by subtracting the sale price, $60.75, from the regular price, $135.00.

$$\text{Amount of decrease} = 135.00 - 60.75$$
$$= 74.25$$

The amount of decrease is $74.25.

$$\text{Percent decrease} = \frac{\text{amount of decrease}}{\text{original amount}}$$

$$= \frac{135.00 - 60.75}{135.00}$$

$$= \frac{74.25}{135.00}$$

$$= \frac{74.25}{135}$$

Now we need to express this fraction as a percent. We'll first find the fraction's equivalent decimal and then write the decimal as a percent. Divide the numerator, 74.25, by the denominator, 135.

$$
\begin{array}{r}
0.55 \\
135\overline{)74.25} \\
67\ 5\downarrow \\
\hline
6\ 75 \\
6\ 75 \\
\hline
0
\end{array}
$$

Thus,

$$\text{Percent decrease} = \frac{74.25}{135} = 0.55 = 55\%$$

The percent decrease of the sale price from the regular price is 55%. This means that the sale price of the jacket is 55% lower than the regular price.

> **Check Point 7**   A television regularly sells for $940. The sale price is $611. Find the percent decrease of the sale price from the regular price.   35%

Discount and Sale Price

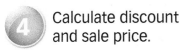

4 Calculate discount and sale price.

In Example 7, the sale price of the jacket is 55% (the percent decrease) lower than the original price. The original price, $135.00, is reduced, or discounted, by 55% of $135.00.

$$55\% \text{ of } \$135.00 = 0.55 \cdot \$135.00 = \$74.25$$

| Discount Rate | Original Price | | Discount (amount by which the price is decreased) |

The sale price of the jacket is

$$\$135.00 - \$74.25 = \$60.75.$$

| Original | Discount | Sale |
| Price | | Price |

Businesses lower prices using discount rates to attract customers and to reduce inventory. The discount rate is a percent of the original price.

Discount and Sale Price

Discount = discount rate × original price

Sale price = original price − discount

Example 8   Finding the Discount and Sale Price

An electronics store has a sign that reads "35% off all marked prices." The marked price on a fax machine is $320.00.

a. What is the discount?

b. What is the sale price?

Solution

a. We begin with the amount of the discount, or simply the discount.

Discount = Discount Rate x   Original Price

$$= \quad 35\% \quad \cdot \quad \$320.00$$

$$= 0.35 \cdot \$320.00 \qquad \text{Express 35\% as a decimal:}$$
$$\qquad\qquad\qquad\qquad 35\% = 0.35.$$

$$= \$112.00 \qquad\qquad\quad 320$$
$$\qquad\qquad\qquad\qquad\quad \times \;.35$$
$$\qquad\qquad\qquad\qquad\quad \overline{1600}$$
$$\qquad\qquad\qquad\qquad\quad 9600$$
$$\qquad\qquad\qquad\qquad\quad \overline{112.00}$$

The discount is $112.00.

b. Now that we've determined the discount, we can find the sale price of the fax machine.

Sale Price = Original Price — Discount

$$= \$320.00 - \$112.00$$

$$= \$208.00 \qquad\qquad \begin{array}{r} {\scriptstyle 110} \\ 32\cancel{0}.00 \\ - \;112.00 \\ \hline 208.00 \end{array}$$

The sale price of the fax machine is $208.00.

Great Question!

Let's say an item is on sale. What's the difference between the discount rate and the percent decrease in price?

They're the same. The percent decrease of the sale price from the original price *is* the discount rate. Consider Example 8, in which we are given that the discount rate on the fax machine is 35%. This also means that the percent decrease of the sale price ($208) from the regular price ($320) is 35%:

$$\text{Percent decrease} = \frac{\text{amount of decrease}}{\text{original amount}} = \frac{320 - 208}{320} = \frac{112}{320} = 0.35 = 35\%.$$

Check Point 8 A store specializing in sporting goods has a sign that reads "30% off the marked prices on all tents." The marked price on a tent is $160.00.

a. What is the discount? $48

b. What is the sale price? $112

Appendix B Exercise Set

Practice and Application Exercises

1. A real estate agent earns a commission rate of 3% on houses that he lists and sells. Find the agent's commission on a house that he listed and sold for $350,000. $10,500

2. A furniture salesperson is paid a commission rate of 4%. Last month, she sold $126,000 worth of furniture. Find the commission for that month. $5040

3. A salesperson earns a 2% commission rate on sales of camera equipment. Yesterday, she earned $116 in commission. What were her sales that day? $5800

4. A salesperson earns a 5% commission rate on sales at a mattress store. Yesterday, he earned $170 in commission. What were his sales that day? $3400

5. A salesperson earned $103 for selling $4120 worth of stereo equipment. What was the commission rate? 2.5%

6. A salesperson earned $221 for selling $2600 worth of lighting fixtures. What was the commission rate? 8.5%

In Exercises 7–12, complete the table.

	Sales	Commission Rate	Commission
7.	$42,500	5%	$2125
8.	$30,000	15%	$4500
9.	$120,000	3.5%	$4200
10.	$190,000	6.5%	$12,350

	Sales	Commission Rate	Commission
11.	$20,300	8%	$1624
12.	$32,600	9%	$2934

In Exercises 13–16, find the amount of increase and the percent increase.

	Original Amount	New Amount	Amount of Increase	Percent Increase
13.	40	70	30	75%
14.	50	75	25	50%
15.	12	30	18	150%
16.	16	56	40	250%

In Exercises 17–20, find the amount of decrease and the percent decrease.

	Original Amount	New Amount	Amount of Decrease	Percent Decrease
17.	60	48	12	20%
18.	80	60	20	25%
19.	200	30	170	85%
20.	300	15	285	95%

Although you want to choose a career that fits your interests and abilities, it is good to have an idea of what jobs pay when looking at career options. The bar graph shows the average starting salaries of U.S. college graduates with a bachelor's degree based on their college major. Use this information to solve Exercises 21–24. Round answers to the nearest percent.

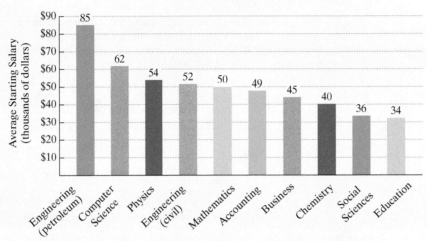

Average Starting Salaries of U.S. College Graduates, by College Major

Source: James Henslin, *Sociology Eleventh Edition*, Pearson, 2012.

21. Find the percent increase in the average starting salary from students majoring in education to students majoring in mathematics. 47%

22. Find the percent increase in the average starting salary from students majoring in education to students majoring in petroleum engineering. 150%

23. Find the percent decrease in the average starting salary from students majoring in business to students majoring in the social sciences. 20%

24. Find the percent decrease in the average starting salary from students majoring in physics to students majoring in chemistry. 26%

25. A sofa regularly sells for $840. The sale price is $714. Find the percent decrease of the sale price from the regular price. 15%

26. A fax machine regularly sells for $380. The sale price is $266. Find the percent decrease of the sale price from the regular price. 30%

In Exercises 27–32, find the discount and the sale price.

	Original Price	Discount Rate	Discount	Sale Price
27.	$80	40%	$32	$48
28.	$120	50%	$60	$60
29.	$62,800	15%	$9420	$53,380
30.	$134,600	18%	$24,228	$110,372
31.	$270.80	25%	$67.70	$203.10
32.	$360.40	45%	$162.18	$198.22

33. A sporting goods store going out of business has a sign that reads "85% off all marked prices." The marked price on a treadmill is $2370. What is the sale price? $355.50

34. A clothing store going out of business has a sign that reads "65% off all marked prices." The marked price on a jacket is $218.00. What is the sale price? $76.30

Appendix C

Bar, Line, and Circle Graphs

What am I supposed to learn?

After you have read this appendix, you should be able to read and interpret information from:

 bar graphs.

 line graphs.

 circle graphs.

 Read and interpret information from bar graphs.

Magazines, newspapers, and websites often display information using bar, line, and circle graphs. In this appendix, we review how to read and interpret data displayed by these graphs.

Bar Graphs

Bar graphs are convenient for comparing numbers of various items. The bars may be either horizontal or vertical. Their heights or lengths are used to show the amounts of different items, often expressed in whole numbers.

Figure C.1 is an example of a typical bar graph. It shows the number of marriages in the United States between whites and African Americans, in thousands, for five selected years from 1970 through 2010.

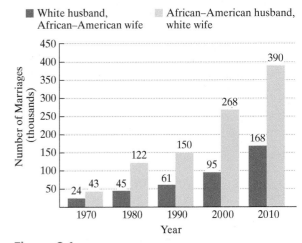

Number of Marriages in the United States between Whites and African Americans

Figure C.1
Source: James M. Henslin, *Sociology Eleventh Edition*, Pearson, 2012.

Example 1 **Reading and Interpreting Information from a Bar Graph**

A dramatic change in U.S. marriage patterns is the increase in marriages between African Americans and whites. **Figure C.1** illustrates this change. (In 1967, the U.S. Supreme Court struck down the state laws that prohibited such marriages.)

a. How many marriages between a white husband and an African–American wife were there in 2000?

b. In which year were there 122,000 marriages between an African–American husband and a white wife?

Solution

a. We begin with the number of marriages between a white husband and an African–American wife in 2000. Look at the bars labeled with the year 2000. The red bar to the left represents the number of marriages between a white husband and an African–American wife. The number above this bar is 95, representing 95 thousand. Thus, in 2000, there were 95,000 marriages between a white husband and an African–American wife.

Number of Marriages in the United States between Whites and African Americans

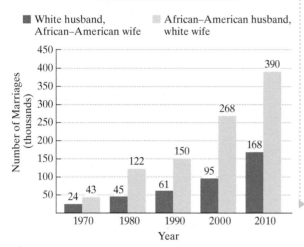

■ White husband,
  African-American wife

□ African-American husband,
  white wife

b. We are interested in which year there were 122,000 marriages between an African–American husband and a white wife. Look for the yellow bar labeled 122 (for 122 thousand, or 122,000). This is the bar to the right for the year labeled 1980. Thus, in 1980, there were 122,000 marriages between an African–American husband and a white wife.

 Check Point 1   Use the bar graph in **Figure C.1** to answer these questions:

a. How many marriages between an African–American husband and a white wife were there in 2010?   390,000

b. In which year were there 61,000 marriages between a white husband and an African–American wife?   1990

Figure C.1 (repeated)

Line Graphs

2 Read and interpret information from line graphs.

Line graphs are often used to illustrate trends over time. Some measure of time, such as months or years, frequently appears on the horizontal axis. Amounts are generally listed on the vertical axis. Points are drawn to represent the given information. The graph is formed by connecting the points with line segments.

Figure C.2 is an example of a typical line graph. The graph shows the average age at which women in the United States married for the first time from 1890 through 2010. The years are listed on the horizontal axis and the ages are listed on the vertical axis. The symbol ⌁ on the vertical axis shows that there is a break in values between 0 and 20. Thus, the first tick mark on the vertical axis represents an average age of 20.

Women's Average Age of First Marriage

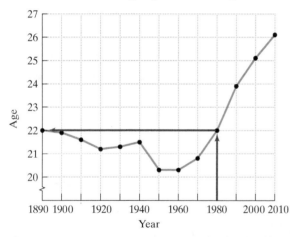

Figure C.2 shows how to find the average age at which women married for the first time in 1980.

Step 1   Locate 1980 on the horizontal axis.

Step 2   Locate the point on the line graph above 1980.

Step 3   Read across to the corresponding age on the vertical axis.

The age is 22. Thus, in 1980, women in the United States married for the first time at an average age of 22.

Figure C.2
Source: U.S. Census Bureau

Example 2   Using a Line Graph

The line graph in **Figure C.3** shows the percentage of U.S. college students who smoked cigarettes from 1982 through 2010.

a. Find an estimate for the percentage of college students who smoked cigarettes in 2010.

b. In which four-year period did the percentage of college students who smoked cigarettes decrease at the greatest rate?

c. In which year did 30% of college students smoke cigarettes?

Cigarette Use by U.S. College Students

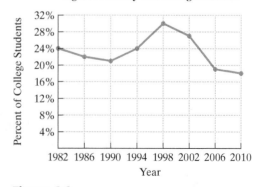

Figure C.3
Source: Rebecca Donatelle, *Health The Basics*, 10th Edition, Pearson, 2013.

Solution

a. Estimating the Percentage Smoking Cigarettes in 2010

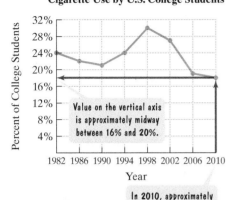

In 2010, approximately 18% of college students smoked cigarettes.

b. Identifying the Period of the Greatest Rate of Decreasing Cigarette Smoking

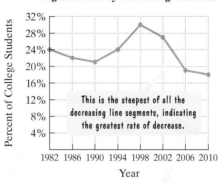

The percentage of college students decreased at the greatest rate in the four-year period from 2002 through 2006.

c. Identifying the Year when 30% of College Students Smoked Cigarettes

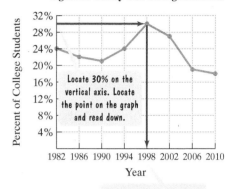

The year when 30% smoked cigarettes was 1998.

Check Point 2 Use the line graph in **Figure C.3** to solve this exercise.

a. Find an estimate for the percentage of college students who smoked cigarettes in 1986. 22%

b. In which four-year period did the percentage of college students who smoked cigarettes increase at the greatest rate? 1994 through 1998

c. In which years labeled on the horizontal axis did 24% of college students smoke cigarettes? 1982 and 1994

3 Read and interpret information from circle graphs.

Circle Graphs

Circle graphs, also called **pie charts**, show how a whole quantity is divided into parts. Circle graphs are divided into pieces, called **sectors**. **Figure C.4** shows a circle graph that indicates how the total cost of rearing a child in the United States to age 18 breaks down.

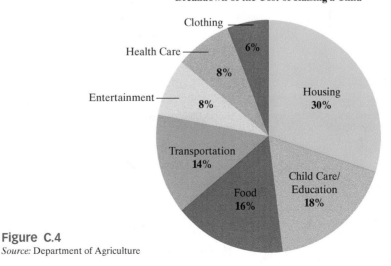

Figure C.4
Source: Department of Agriculture

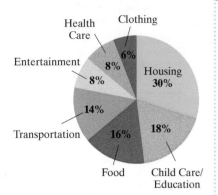

Breakdown of the Cost of Raising a Child

Health Care

Clothing

Entertainment

Housing 30%

6%

8%

8%

14%

16%   18%

Transportation

Food

Child Care/ Education

Figure C.4 (repeated)

Example 3   Baby on Board

Use the circle graph in **Figure C.4** to solve this problem.

a. What is the sum of the percents in **Figure C.4**? What does this represent?

b. What percent of raising a child includes food and transportation combined?

c. According to the U.S. Department of Agriculture, parents with a baby born in 2012 will spend somewhere from $217,000 to $500,000 to raise the child to age 18. If they spend $500,000, how much will be spent on food and transportation combined?

d. If parents spend $500,000 to raise a child, how much more will they spend on housing than on clothing?

Solution

a. Working clockwise from the housing sector, the sum of the percents displayed by the circle graph in **Figure C.4** is

$$30\% + 18\% + 16\% + 14\% + 8\% + 8\% + 6\% = 100\%.$$

The sum of the percents, 100%, represents the whole quantity, including all the categories involved in the cost of raising a child.

b. The percent spent on food and transportation combined is the percent spent on food plus the percent spent on transportation.

| Percent spent on food and transportation combined | is | the percent spent on food | plus | the percent spent on transportation. |

$$= 16\% + 14\%$$
$$= 30\%$$

The percent of raising a child that includes food and transportation combined is 30%.

c. The amount spent on food and transportation is 30% of the cost to raise the child.

| Amount spent on food and transportation | is | 30% | of | the cost to raise the child. |

$$= 0.3 \cdot \$500,000$$
$$= \$150,000$$

$$\begin{array}{r} 500,000 \\ \times\ .3 \\ \hline 150,000.0 \end{array}$$

The amount spent on food and transportation combined is $150,000.

d. To determine how much more is spent on housing than on clothing, we find the difference of these amounts.

| How much more is spent on housing than on clothing | is | the amount spent on housing | minus | the amount spent on clothing. |

$$= 30\% \text{ of the cost} - 6\% \text{ of the cost}$$
$$= 0.3 \cdot \$500,000 - 0.06 \cdot \$500,000$$
$$= \$150,000 - \$30,000$$
$$= \$120,000$$

If parents spend $500,000 to raise a child, they will spend $120,000 more on housing than on clothing.

 **Check Point 3**   Use the circle graph in **Figure C.4** to solve this problem.

a. What percent of raising a child includes entertainment and health care combined?   16%

b. If parents spend $300,000 to raise a child to 18, how much will be spent on entertainment and health care combined? $48,000

c. If parents spend $300,000 to raise a child, how much more will they spend on housing than on child care/education? $36,000

Appendix C Exercise Set

Practice and Application Exercises

The bar graph shows the average income in the United States, adjusted for inflation, from 2006 through 2012. Use the information displayed by the graph to solve Exercises 1–6.

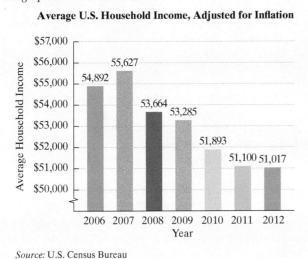

Average U.S. Household Income, Adjusted for Inflation

Source: U.S. Census Bureau

1. What was the average household income in 2008? $53,664
2. What was the average household income in 2009? $53,285
3. In which year was the average household income $51,100?   2011
4. In which year was the average household income $51,893?   2010
5. In which year was the average household income at a maximum? What was the average income for that year?   2007; $55,627
6. In which year was the average household income at a minimum? What was the average income for that year?   2012; $51,017

Seeing Storms. *The bar graph shows the number of predicted and actual hurricanes in the United States from 2000 through 2013. Use this information to solve Exercises 7–12.*

Number of Predicted and Actual Hurricanes in the United States

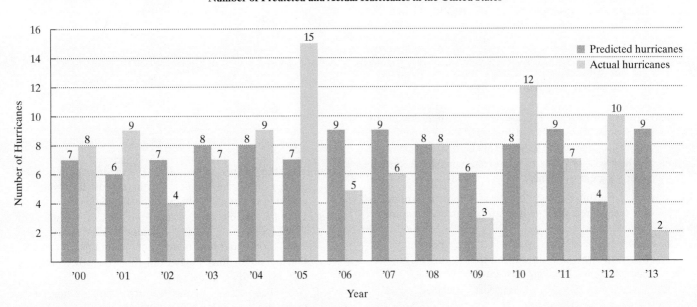

Source: Colorado State University

(In Exercises 7–12, be sure to refer to the bar graph at the bottom of the previous page.)

7. During which year was the number of actual hurricanes the greatest? How many hurricanes were there that year?   2005; 15 hurricanes

8. During which year was the number of predicted hurricanes the least? How many hurricanes were predicted that year?   2012; 4 hurricanes

9. By how much did the number of actual hurricanes increase from 2009 to 2010?   9 hurricanes

10. By how much did the actual number of hurricanes decrease from 2010 to 2011?   5 hurricanes

11. In which year did forecasters accurately predict the number of hurricanes?   2008

12. In which year did forecasters least accurately predict the number of hurricanes?   2005

According to a Gallup poll, in 2012 almost a third of Americans named Iran their country's greatest enemy. That number matched public sentiment in the United States toward Saddam Hussein's Iraq at the time of the 2003 invasion. The line graphs show the percentage of Americans who considered either Iraq or Iran their country's greatest enemy from 2001 through 2012. Use this information to solve Exercises 13–14.

Countries Americans Consider Their Greatest Enemy

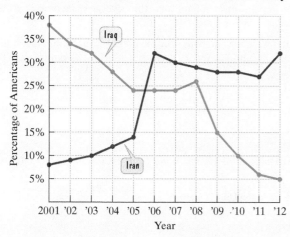

Source: Gallup

13. a. Find an estimate for the percentage of Americans who considered Iraq their country's greatest enemy in 2001.   ≈ 38%
   b. Between which two years did the percentage of Americans who considered Iraq their country's greatest enemy decrease at the greatest rate?   2008 and 2009
   c. In which year did 32% of Americans consider Iraq their country's greatest enemy?   2003

14. a. Find an estimate for the percentage of Americans who considered Iran their country's greatest enemy in 2001.   ≈ 8%
   b. Between which two years did the percentage of Americans who considered Iran their country's greatest enemy increase at the greatest rate?   2005 and 2006
   c. In which year did 12% of Americans consider Iran their country's greatest enemy?   2004

15. **School Absences Due to Bullying.** The line graph shows the percentage of high school students from 1993 through 2011 who missed at least one school day because they felt unsafe due to bullying.

Percentage of High School Students Who Missed at Least One Day Due to Bullying

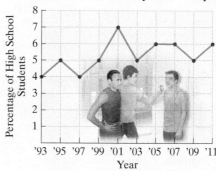

Source: Youth Risk Behavior Surveillance Survey

a. For the period shown, in which year did the percentage of high school students who missed school due to bullying reach a maximum? What percentage of students missed school for that year due to bullying?   2001; 7%
b. For the period shown, in which years did the percentage of high school students who missed school due to bullying reach a minimum? What percentage of students missed school for those years due to bullying?   1993 and 1997; 4%
c. In which two-year period did the percentage of high school students who missed school due to bullying decrease at the greatest rate?   2001–2003
d. In which years labeled on the horizontal axis did 6% of high school students miss school due to bullying?   2005, 2007, 2011

16. During a diagnostic evaluation, a 33-year-old woman experienced a panic attack a few minutes after she had been asked to relax her whole body. The graph shows the rapid increase in heart rate during the panic attack.

Heart Rate before and during a Panic Attack

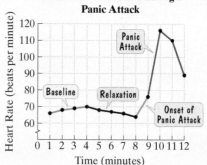

Source: Davis and Palladino, *Psychology*, Fifth Edition, Prentice Hall, 2007.

a. Use the graph to estimate the woman's maximum heart rate during the first 12 minutes of the diagnostic evaluation. After how many minutes did this occur?
b. Use the graph to estimate the woman's minimum heart rate during the first 12 minutes of the diagnostic evaluation. After how many minutes did this occur?
c. During which time period did the woman's heart rate increase at the greatest rate?   between 9 and 10 min
d. After how many minutes was the woman's heart rate approximately 75 beats per minute?   9 min

16. a. 115 beats per min; 10 min   **b.** 64 beats per min; 8 min

It may surprise you to see the data showing how little actual football there is in a televised National Football League (NFL) game. The circle graph shows the percent of time devoted to various aspects of an average 190-minute NFL television broadcast. Use the information displayed by the graph to solve Exercises 17–20.

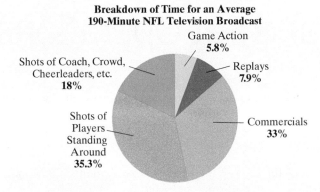

Breakdown of Time for an Average 190-Minute NFL Television Broadcast

Game Action 5.8%
Replays 7.9%
Shots of Coach, Crowd, Cheerleaders, etc. 18%
Shots of Players Standing Around 35.3%
Commercials 33%

Source: Wall Street Journal

17. What percent of broadcast time is not devoted to game action, or actual football? 94.2%

18. What percent of broadcast time is devoted to commercials and replays? 40.9%

19. In a 190-minute NFL broadcast, how much time is devoted to game action, or actual football? Round to the nearest minute. 11 min

20. In a 190-minute NFL broadcast, how much time is devoted to commercials? Round to the nearest minute. 63 min

High school students in the United States were asked what word came to mind when they saw or heard the word "computing." Teenage boys replied "video games," "design," "electronics," "solving problems," and "interesting." By contrast, the girls responded with "typing," "boredom," "math," and "nerd." These perceptions were reinforced when students were asked if they considered a career in computers a "good choice" for someone like them. The circle graphs show the percent of responses for teenage boys and teenage girls. Use the information displayed by the graphs to solve Exercises 21–24.

Percent of U.S. High School Students Who Consider a Career in Computers a "Good Choice"

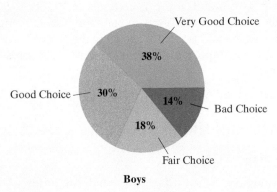

Very Good Choice 38%
Good Choice 30%
Bad Choice 14%
Fair Choice 18%

Boys

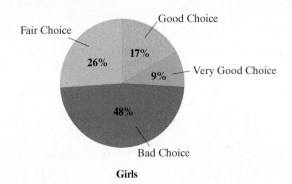

Good Choice 17%
Fair Choice 26%
Very Good Choice 9%
Bad Choice 48%

Girls

Source: Mother Jones, July/August 2014

21. How much greater was the percent of boys who considered a computer career a very good choice or a good choice than the percent of girls who considered a computer career a very good choice or a good choice? 42%

22. How much greater was the percent of girls who considered a computer career a fair choice or a bad choice than the percent of boys who considered a computer career a fair choice or a bad choice? 42%

23. In a sample of 500 boys and 500 girls, how many more boys considered a computer career a very good choice than girls who considered a computer career a very good choice? 145 more boys

24. In a sample of 500 boys and 500 girls, how many more girls considered a computer career a bad choice than boys who considered a computer career a bad choice? 170 more girls

Appendix D

Sets, Venn Diagrams, and Set Operations

What am I supposed to learn?

After you have read this appendix, you should be able to:

1. Use three methods to represent sets.

2. Define and recognize the empty set.

3. Understand the basic ideas of a Venn diagram.

4. Find the complement of a set.

5. Find the intersection of two sets.

6. Find the union of two sets.

A **set** is a collection of objects whose contents can be clearly determined. The objects in a set are called the **elements**, or **members**, of the set.

Methods for Representing Sets

An example of a set is the set of the days of the week, whose elements are Monday, Tuesday, Wednesday, Thursday, Friday, Saturday, and Sunday.

Capital letters are generally used to name sets. Let's use W to represent the set of the days of the week.

Three methods are commonly used to designate a set. One method is a **word description**. We can describe set W as the set of the days of the week. A second method is the **roster method**. This involves listing the elements of a set inside a pair of braces, { }. The braces at the beginning and end indicate that we are representing a set. The roster form uses commas to separate the elements of the set. Thus, we can designate the set W by listing its elements:

W = {Monday, Tuesday, Wednesday, Thursday, Friday, Saturday, Sunday}.

Grouping symbols such as parentheses, ( ), and square brackets, [ ], are not used to represent sets. Only commas are used to separate the elements of a set. Separators such as colons or semicolons are not used. Finally, the order in which the elements are listed in a set is not important. Thus, another way of expressing the set of the days of the week is

W = {Saturday, Sunday, Monday, Tuesday, Wednesday, Thursday, Friday}.

Example 1   Representing a Set Using a Description

Write a word description of the set

P = {Washington, Adams, Jefferson, Madison, Monroe}.

Solution

Set P is the set of the first five presidents of the United States.

> ✓ **Check Point 1**   Write a word description of the set
>
> $$L = \{a, b, c, d, e, f\}.$$ *L* is the set of the first six lowercase letters of the alphabet.

Example 2   Representing a Set Using the Roster Method

Set C is the set of U.S. coins with a value of less than a dollar. Express this set using the roster method.

Solution

$$C = \{\text{penny, nickel, dime, quarter, half-dollar}\}$$

> ✓ **Check Point 2**   Set M is the set of months beginning with the letter A. Express this set using the roster method.   *M* = {April, August}

The third method for representing a set is with **set-builder notation**. Using this method, the set of the days of the week can be expressed as

$$W = \{x \mid x \text{ is a day of the week}\}.$$

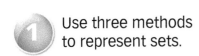

We read this notation as "Set W is the set of all elements x such that x is a day of the week." Before the vertical line is the variable x, which represents an element in general. After the vertical line is the condition x must meet in order to be an element of the set.

Table D.1 contains two examples of sets, each represented with a word description, the roster method, and set-builder notation.

Table D.1 Sets Using Three Designations		
Word Description	**Roster Method**	**Set-Builder Notation**
B is the set of members of the Beatles in 1963.	$B = \{$George Harrison, John Lennon, Paul McCartney, Ringo Starr$\}$	$B = \{x \mid x \text{ was a member of the Beatles in 1963}\}$
S is the set of U.S. states whose names begin with the letter A.	$S = \{$Alabama, Alaska, Arizona, Arkansas$\}$	$S = \{x \mid x \text{ is a U.S. state whose name begins with the letter A}\}$

The Empty Set

Consider the following sets:

$$\{x \mid x \text{ is a turtle that plays frisbee}\}$$
$$\{x \mid x \text{ is a number greater than 10 and less than 4}\}.$$

Can you see what these sets have in common? They both contain no elements. There are no turtles that play frisbee. There are no numbers that are both greater than 10 and also less than 4. Sets such as these that contain no elements are called the *empty set*, or the *null set*.

> **The Empty Set**
>
> The **empty set**, also called the **null set**, is the set that contains no elements. The empty set is represented by { } or ∅.

Notice that **{ } and ∅ have the same meaning**. However, **the empty set is not represented by {∅}**. This notation represents a set containing the element ∅.

Example 3   Recognizing the Empty Set

Which one of the following is the empty set?

a. {0}   b. 0

c. $\{x \mid x \text{ is a number less than 4 or greater than 10}\}$

d. $\{x \mid x \text{ is a square with exactly three sides}\}$

Solution

a. {0} is a set containing one element, 0. Because this set contains an element, it is not the empty set.

b. 0 is a number, not a set, so it cannot possibly be the empty set. It does, however, represent the number of members of the empty set.

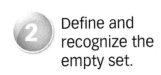

1 Use three methods to represent sets.

2 Define and recognize the empty set.

c. $\{x \mid x$ is a number less than 4 or greater than 10$\}$ contains all numbers that are either less than 4, such as 3, or greater than 10, such as 11. Because some elements belong to this set, it cannot be the empty set.

d. $\{x \mid x$ is a square with exactly three sides$\}$ contains no elements. There are no squares with exactly three sides. This set is the empty set.

 Check Point 3   Which one of the following is the empty set? b

a. $\{x \mid x$ is a number less than 3 or greater than 5 $\}$

b. $\{x \mid x$ is a number less than 3 and greater than 5 $\}$

c. nothing   **d.** $\{\varnothing\}$

Understand the basic ideas of a Venn diagram.

Great Question!

Is the size of the circle in a Venn diagram important?

No. The size of the circle representing set A in a Venn diagram has nothing to do with the number of elements in set A.

Universal Sets and Venn Diagrams

In discussing sets, it is convenient to refer to a general set that contains all elements under discussion. This general set is called the *universal set*. A **universal set**, symbolized by U, is a set that contains all the elements being considered in a given discussion.

We can obtain a more thorough understanding of sets and their relationship to a universal set by considering diagrams that allow visual analysis. **Venn diagrams**, named for the British logician John Venn (1834–1923), are used to show the visual relationship among sets.

Figure D.1 is a Venn diagram. The universal set is represented by a region inside a rectangle. Sets within the universal set are depicted by circles, or sometimes by ovals or other shapes. In this Venn diagram, set A is represented by the light blue region inside the circle.

The dark blue region in **Figure D.1** represents the set of elements in the universal set U that are not in set A. By combining the regions shown by the light blue shading and the dark blue shading, we obtain the universal set, U.

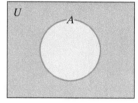

Figure D.1

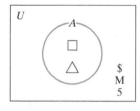

Figure D.2

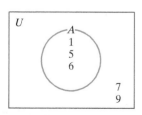

Figure D.3

Example 4   Determining Sets from a Venn Diagram

Use the Venn diagram in **Figure D.2** to determine each of the following sets:

a. U **b.** A **c.** the set of elements in U that are not in A.

Solution

a. Set U, the universal set, consists of all the elements within the rectangle. Thus, $U = \{\Box, \triangle, \$, M, 5\}$.

b. Set A consists of all the elements within the circle. Thus, $A = \{\Box, \triangle\}$.

c. The set of elements in U that are not in A, shown by the set of all the elements outside the circle, is $\{\$, M, 5\}$.

 Check Point 4   Use the Venn diagram in **Figure D.3** to determine each of the following sets:

a. U $\{1, 5, 6, 7, 9\}$ **b.** A $\{1, 5, 6\}$ **c.** the set of elements in U that are not in A. $\{7, 9\}$

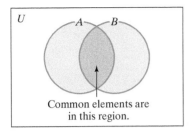

Common elements are
in this region.

Figure D.4

In mathematics, the word *some* means *there exists at least one*. If set A and set B have at least one element in common, then the circles representing the sets must overlap. This is illustrated in the Venn diagram in **Figure D.4**.

Example 5   Determining Sets from a Venn Diagram

Use the Venn diagram in **Figure D.5** to determine each of the following sets:

a. U

b. B

c. the set of elements in A but not B

d. the set of elements in U that are not in B

e. the set of elements in both A and B.

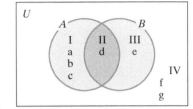

Figure D.5

The Roman numerals are labels for each region in the Venn diagram. They do *not* represent elements in the regions.

Solution

a. Set U, the universal set, consists of all elements within the rectangle. Taking the elements in regions I, II, III, and IV, we obtain $U = \{a, b, c, d, e, f, g\}$.

b. Set B consists of the elements in regions II and III. Thus, $B = \{d, e\}$.

c. The set of elements in A but not in B, found in region I, is $\{a, b, c\}$.

d. The set of elements in U that are not in B, found in regions I and IV, is $\{a, b, c, f, g\}$.

e. The set of elements in both A and B, found in region II, is $\{d\}$.

✓ **Check Point 5**   Use the Venn diagram in **Figure D.5** to determine each of the following sets:

a. A {a, b, c, d}

b. the set of elements in B but not A {e}

c. the set of elements in U that are not in A {e, f, g}

▶ **d.** the set of elements in U that are not in A or B.   {f, g}

The Complement of a Set

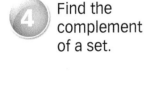

④ Find the complement of a set.

In arithmetic, we use operations such as addition and multiplication to combine numbers. We now turn to three set operations, called *complement, intersection*, and *union*. We begin by defining a set's complement.

> #### Definition of the Complement of a Set
> The **complement** of set A, symbolized by A', is the set of all elements in the universal set that are *not* in A.

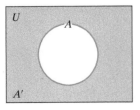

Figure D.6

The shaded region in **Figure D.6** represents the complement of set A, or A'. This region lies outside circle A, but within the rectangular universal set.

In order to find A', a universal set U must be given. A fast way to find A' is to cross out the elements in U that are given to be in set A. The set that remains is A'.

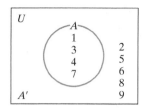

Figure D.7

Example 6 Finding a Set's Complement

Let $U = \{1, 2, 3, 4, 5, 6, 7, 8, 9\}$ and $A = \{1, 3, 4, 7\}$. Find A'.

Solution

Set A' contains all the elements of set U that are not in set A. Because set A contains the elements 1, 3, 4, and 7, these elements cannot be members of set A':

$$\{\cancel{1}, 2, \cancel{3}, \cancel{4}, 5, 6, \cancel{7}, 8, 9\}.$$

Thus, set A' contains 2, 5, 6, 8, and 9:

$$A' = \{2, 5, 6, 8, 9\}.$$

A Venn diagram illustrating A and A' is shown in **Figure D.7**.

 **Check Point 6** Let $U = \{a, b, c, d, e\}$ and $A = \{a, d\}$. Find A'. {b, c, e}

The Intersection of Sets

⑤ Find the intersection of two sets.

If A and B are sets, we can form a new set consisting of all elements that are in both A and B. This set is called the *intersection* of the two sets.

> **Definition of the Intersection of Sets**
>
> The **intersection** of sets A and B, written $A \cap B$, is the set of elements common to both set A and set B.

In Example 7, we are asked to find the intersection of two sets. This is done by listing the common elements of both sets. Because the intersection of two sets is also a set, we enclose these elements with braces.

Example 7 Finding the Intersection of Two Sets

Find each of the following intersections:

a. $\{7, 8, 9, 10, 11\} \cap \{6, 8, 10, 12\}$

b. $\{1, 3, 5, 7, 9\} \cap \{2, 4, 6, 8\}$

c. $\{1, 3, 5, 7, 9\} \cap \varnothing$.

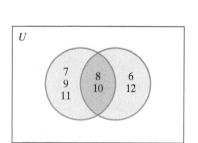

Figure D.8 The numbers 8 and 10 belong to both sets.

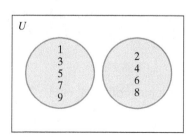

Figure D.9 These sets have no common elements.

Solution

a. The elements common to $\{7, 8, 9, 10, 11\}$ and $\{6, 8, 10, 12\}$ are 8 and 10. Thus,

$$\{7, 8, 9, 10, 11\} \cap \{6, 8, 10, 12\} = \{8, 10\}.$$

The Venn diagram in **Figure D.8** illustrates this situation.

b. The sets $\{1, 3, 5, 7, 9\}$ and $\{2, 4, 6, 8\}$ have no elements in common. Thus,

$$\{1, 3, 5, 7, 9\} \cap \{2, 4, 6, 8\} = \varnothing.$$

The Venn diagram in **Figure D.9** illustrates this situation.

c. There are no elements in $\varnothing$, the empty set. This means that there can be no elements belonging to both $\{1, 3, 5, 7, 9\}$ and $\varnothing$. Therefore,

$$\{1, 3, 5, 7, 9\} \cap \varnothing = \varnothing.$$

✓ **Check Point 7**   Find each of the following intersections:

a. $\{1, 3, 5, 7, 10\} \cap \{6, 7, 10, 11\}$ {7, 10}

b. $\{1, 2, 3\} \cap \{4, 5, 6, 7\}$ ∅

▸ c. $\{1, 2, 3\} \cap \varnothing.$ ∅

The Union of Sets

⑥ Find the union of two sets.

Another set that we can form from sets A and B consists of elements that are in A or B or in both sets. This set is called the *union* of the two sets.

> **Definition of the Union of Sets**
>
> The **union** of sets A and B, written $A \cup B$, is the set of elements that are members of set A or of set B or of both sets.

We can find the union of set A and set B by listing the elements of set A. Then, we include any elements of set B that have not already been listed. Enclose all elements that are listed with braces. This shows that the union of two sets is also a set.

Example 8   Finding the Union of Two Sets

Find each of the following unions:

a. $\{7, 8, 9, 10, 11\} \cup \{6, 8, 10, 12\}$

b. $\{1, 3, 5, 7, 9\} \cup \{2, 4, 6, 8\}$

c. $\{1, 3, 5, 7, 9\} \cup \varnothing.$

Solution

This example uses the same sets as in Example 7. However, this time we are finding the unions of the sets, rather than their intersections.

a. To find $\{7, 8, 9, 10, 11\} \cup \{6, 8, 10, 12\}$, start by listing all the elements from the first set, namely 7, 8, 9, 10, and 11. Now list all the elements from the second set that are not in the first set, namely 6 and 12. The union is the set consisting of all these elements. Thus,

$$\{7, 8, 9, 10, 11\} \cup \{6, 8, 10, 12\} = \{6, 7, 8, 9, 10, 11, 12\}.$$

b. To find $\{1, 3, 5, 7, 9\} \cup \{2, 4, 6, 8\}$, list the elements from the first set, namely 1, 3, 5, 7, and 9. Now add to the list the elements in the second set that are not in the first set. This includes every element in the second set, namely 2, 4, 6, and 8. The union is the set consisting of all these elements, so

$$\{1, 3, 5, 7, 9\} \cup \{2, 4, 6, 8\} = \{1, 2, 3, 4, 5, 6, 7, 8, 9\}.$$

c. To find $\{1, 3, 5, 7, 9\} \cup \varnothing$, list the elements from the first set, namely 1, 3, 5, 7, and 9. Because there are no elements in ∅, the empty set, there are no additional elements to add to the list. Thus,

$$\{1, 3, 5, 7, 9\} \cup \varnothing = \{1, 3, 5, 7, 9\}.$$

> **Great Question!**
>
> **When finding the union of two sets, what should I do if some elements appear in both sets?**
>
> List these common elements only once, *not twice*, in the union of the sets.

✓ **Check Point 8**   Find each of the following unions:

a. $\{1, 3, 5, 7, 10\} \cup \{6, 7, 10, 11\}$ {1, 3, 5, 6, 7, 10, 11}

b. $\{1, 2, 3\} \cup \{4, 5, 6, 7\}$ {1, 2, 3, 4, 5, 6, 7}

▸ c. $\{1, 2, 3\} \cup \varnothing.$ {1, 2, 3}

Appendix D Exercise Set

Practice Exercises

In Exercises 1–4, express each set using the roster method.

1. The set of the four seasons in a year   {winter, spring, summer, fall}

2. The set of months of the year that have exactly 30 days

3. $\{x \mid x$ is a month that ends with the letters b-e-r$\}$

4. $\{x \mid x$ is a lowercase letter of the alphabet that follows d and comes before j$\}$ {e, f, g, h, i}   **2.** {April, June, September, November}
   3. {September, October, November, December}

In Exercises 5–12, determine which sets are the empty set.

5. $\{\varnothing, 0\}$ not the empty set

6. $\{0, \varnothing\}$ not the empty set

7. $\{x \mid x$ is a woman who served as U.S. president before 2016$\}$ empty set

8. $\{x \mid x$ is a living U.S. president born before 1200$\}$ empty set

9. $\{x \mid x$ is the number of women who served as U.S. president before 2016$\}$ not the empty set

10. $\{x \mid x$ is the number of living U.S. presidents born before 1200$\}$ not the empty set

11. $\{x \mid x$ is a U.S. state whose name begins with the letter X$\}$ empty set

12. $\{x \mid x$ is a month of the year whose name begins with the letter X$\}$ empty set

In Exercises 13–16, let $U = \{a, b, c, d, e, f, g\}$, $A = \{a, b, f, g\}$, $B = \{c, d, e\}$, $C = \{a, g\}$, and $D = \{a, b, c, d, e, f\}$. Use the roster method to write each of the following sets.

13. A' {c, d, e}
14. B' {a, b, f, g}
15. C' {b, c, d, e, f}
16. D' {g}

In Exercises 17–20, let $U = \{1, 2, 3, 4, \ldots, 20\}$, $A = \{1, 2, 3, 4, 5\}$, $B = \{6, 7, 8, 9\}$, $C = \{1, 3, 5, 7, \ldots, 19\}$, and $D = \{2, 4, 6, 8, \ldots 20\}$. Use the roster method to write each of the following sets.

17. A' {6, 7, 8, 9, ... , 20}

18. B' {1, 2, 3, 4, 5, 10, 11, 12, 13, 14, 15, 16, 17, 18, 19, 20}

19. C' {2, 4, 6, 8, ... , 20}

20. D' {1, 3, 5, 7, ... , 19}

In Exercises 21–38, let $U = \{1, 2, 3, 4, 5, 6, 7\}$, $A = \{1, 3, 5, 7\}$, $B = \{1, 2, 3\}$, $C = \{2, 3, 4, 5, 6\}$. Find each of the following sets.

21. $A \cap B$ {1, 3}
22. $B \cap C$ {2, 3}
23. $A \cup B$ {1, 2, 3, 5, 7}
24. $B \cup C$ {1, 2, 3, 4, 5, 6}
25. A' {2, 4, 6}
26. B' {4, 5, 6, 7}
27. $A' \cap B'$ {4, 6}
28. $B' \cap C$ {4, 5, 6}
29. $A \cup C'$ {1, 3, 5, 7} or A
30. $B \cup C'$ {1, 2, 3, 7}
31. $A \cup \varnothing$ {1, 3, 5, 7} or A
32. $C \cup \varnothing$ {2, 3, 4, 5, 6} or C
33. $A \cap \varnothing$ $\varnothing$
34. $C \cap \varnothing$ $\varnothing$
35. $A \cup U$ {1, 2, 3, 4, 5, 6, 7} or U
36. $B \cup U$ {1, 2, 3, 4, 5, 6, 7} or U
37. $A \cap U$ {1, 3, 5, 7} or A
38. $B \cap U$ {1, 2, 3} or B

In Exercises 39–52, let $U = \{a, b, c, d, e, f, g, h\}$, $A = \{a, g, h\}$, $B = \{b, g, h\}$, $C = \{b, c, d, e, f\}$. Find each of the following sets.

39. $A \cap B$ {g, h}
40. $B \cap C$ {b}
41. $A \cup B$ {a, b, g, h}
42. $B \cup C$ {b, c, d, e, f, g, h}
43. A' {b, c, d, e, f} or C
44. B' {a, c, d, e, f}
45. $A' \cap B'$ {c, d, e, f}
46. $B' \cap C$ {c, d, e, f}
47. $A \cup C'$ {a, g, h} or A
48. $B \cup C'$ {a, b, g, h}
49. $A \cup \varnothing$ {a, g, h} or A
50. $C \cup \varnothing$ {b, c, d, e, f} or C
51. $A \cap \varnothing$ $\varnothing$
52. $C \cap \varnothing$ $\varnothing$

In Exercises 53–60, use the Venn diagram to represent each set in roster form.

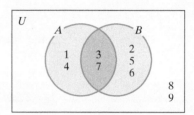

53. A {1, 3, 4, 7}
54. B {2, 3, 5, 6, 7}
55. U {1, 2, 3, 4, 5, 6, 7, 8, 9}
56. $A \cup B$ {1, 2, 3, 4, 5, 6, 7}
57. $A \cap B$ {3, 7}
58. A' {2, 5, 6, 8, 9}
59. B' {1, 4, 8, 9}
60. $(A \cap B)'$ {1, 2, 4, 5, 6, 8, 9}

Application Exercises

A math tutor working with a small group of students asked each student when he or she had studied for class the previous weekend. Their responses are shown in the Venn diagram.

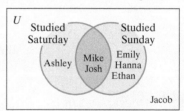

In Exercises 61–68, use the Venn diagram to list the elements of each set in roster form.   **62.** {Mike, Josh, Emily, Hanna, Ethan}

61. The set of students who studied Saturday   {Ashley, Mike, Josh}

62. The set of students who studied Sunday

63. The set of students who studied Saturday or Sunday

64. The set of students who studied Saturday and Sunday

65. The set of students who studied Saturday and not Sunday

66. The set of students who studied Sunday and not Saturday   {Emily, Hanna, Ethan}

67. The set of students who studied neither Saturday nor Sunday   {Jacob}

68. The set of students surveyed by the math tutor

63. {Ashley, Mike, Josh, Emily, Hanna, Ethan}   **64.** {Mike, Josh}   **65.** {Ashley}   **68.** {Ashley, Mike, Josh, Emily, Hanna, Ethan, Jacob}

The bar graph shows the percentage of Americans with gender preferences for various jobs.

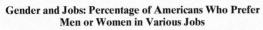

Gender and Jobs: Percentage of Americans Who Prefer Men or Women in Various Jobs

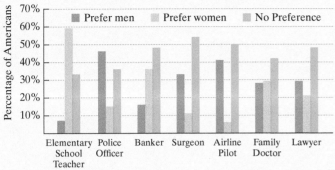

Source: Pew Research Center

In Exercises 69–74, use the information in the graph to place the indicated job in the correct region of the following Venn diagram.

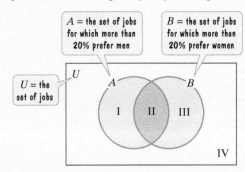

A = the set of jobs for which more than 20% prefer men

B = the set of jobs for which more than 20% prefer women

U = the set of jobs

69. elementary school teacher   III

70. police officer   I

71. surgeon   I

72. banker   III

73. family doctor   II

74. lawyer   II

Answers to Selected Exercises

CHAPTER 1

Section 1.1

Exercise Set 1.1

11. true; $5 + 9 + 5 + 8 = 27$ which is divisible by 3.　　**12.** true; $8 + 1 + 4 + 2 = 15$ which is divisible by 3.　　**13.** true; the last two digits of 10,612 form the number 12 and 12 is divisible by 4.　　**14.** true; the last two digits of 15,984 form the number 84 and 84 is divisible by 4.　　**17.** true; 104,538 is an even number so it is divisible by 2. $1 + 0 + 4 + 5 + 3 + 8 = 21$ which is divisible by 3 so it is divisible by 3. Any number divisible by 2 and 3 is divisible by 6.　　**18.** true; 163,944 is an even number so it is divisible by 2. $1 + 6 + 3 + 9 + 4 + 4 = 27$ which is divisible by 3 so it is divisible by 3. Any number divisible by 2 and 3 is divisible by 6.　　**19.** true; the last three digits of 20,104 form the number 104 and 104 is divisible by 8.　　**20.** true; the last three digits of 28,096 form the number 96 and 96 is divisible by 8.　　**23.** true; $5 + 1 + 7 + 8 + 7 + 2 = 30$ which is divisible by 3 so it is divisible by 3. The last two digits of 517,872 form the number 72 and 72 is divisible by 4 so it is divisible by 4. Any number divisible by 3 and 4 is divisible by 12.　　**24.** true; $7 + 8 + 5 + 1 + 7 + 2 = 30$ which is divisible by 3 so it is divisible by 3. The last two digits of 785,172 form the number 72 and 72 is divisible by 4 so it is divisible by 4. Any number divisible by 3 and 4 is divisible by 12.

Section 1.2

Check Point Exercises

1.

Exercise Set 1.2

1. 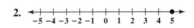　　**2.** 　　**3.**

4.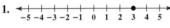

Section 1.3

Exercise Set 1.3

119. $\frac{1}{3}$ cup butter, $\frac{5}{2} = 2.5$ ounces unsweetened chocolate, $\frac{3}{4}$ cup sugar, 1 teaspoon vanilla, 1 egg, $\frac{1}{2}$ cup flour　　**120.** $\frac{1}{2}$ cup butter, $\frac{15}{4} = 3.75$ ounces unsweetened chocolate, $\frac{9}{8} = 1\frac{1}{8}$ cups sugar, $\frac{3}{2} = 1\frac{1}{2}$ teaspoons vanilla, $\frac{3}{2} = 1\frac{1}{2}$ eggs, $\frac{3}{4}$ cup flour　　**121.** $\frac{5}{6}$ cup butter, $\frac{25}{4} = 6.25$ ounces unsweetened chocolate, $\frac{15}{8} = 1\frac{7}{8}$ cups sugar, $\frac{5}{2} = 2\frac{1}{2}$ teaspoons vanilla, $\frac{5}{2} = 2\frac{1}{2}$ eggs, $\frac{5}{4} = 1\frac{1}{4}$ cups flour　　**122.** 1 cup butter, $\frac{15}{2} = 7.5$ ounces unsweetened chocolate, $\frac{9}{4} = 2\frac{1}{4}$ cups sugar, 3 teaspoons vanilla, 3 eggs, $\frac{3}{2} = 1\frac{1}{2}$ cups flour

137.

Section 1.5

Exercise Set 1.5

1. a. $\sqrt{100}$　**b.** $0, \sqrt{100}$　**c.** $-9, 0, \sqrt{100}$　**d.** $-9, -\frac{4}{5}, 0, 0.25, 9.2, \sqrt{100}$　**e.** $\sqrt{3}$　**f.** $-9, -\frac{4}{5}, 0, 0.25, \sqrt{3}, 9.2, \sqrt{100}$

2. a. $\sqrt{49}$　**b.** $0, \sqrt{49}$　**c.** $-7, 0, \sqrt{49}$　**d.** $-7, -0.\overline{6}, 0, \sqrt{49}$　**e.** $\sqrt{50}$　**f.** $-7, -0.\overline{6}, 0, \sqrt{49}, \sqrt{50}$　　**3. a.** $\sqrt{64}$　**b.** $0, \sqrt{64}$

c. $-11, 0, \sqrt{64}$　**d.** $-11, -\frac{5}{6}, 0, 0.75, \sqrt{64}$　**e.** $\sqrt{5}, \pi$　**f.** $-11, -\frac{5}{6}, 0, 0.75, \sqrt{5}, \pi, \sqrt{64}$　　**4. a.** $\sqrt{4}$　**b.** $0, \sqrt{4}$　**c.** $-5, 0, \sqrt{4}$

d. $-5, -0.\overline{3}, 0, \sqrt{4}$　**e.** $\sqrt{2}$　**f.** $-5, -0.\overline{3}, 0, \sqrt{2}, \sqrt{4}$

Section 1.6

Exercise Set 1.6

93. $(8.2 \times 10^7)(3.0 \times 10^9) = 2.46 \times 10^{17}$　　**94.** $(9.4 \times 10^7)(6.0 \times 10^9) = 5.64 \times 10^{17}$

Chapter 1 Review Exercises

79. a. $\sqrt{81}$　**b.** $0, \sqrt{81}$　**c.** $-17, 0, \sqrt{81}$　**d.** $-17, -\frac{9}{13}, 0, 0.75, \sqrt{81}$　**e.** $\sqrt{2}, \pi$　**f.** $-17, -\frac{9}{13}, 0, 0.75, \sqrt{2}, \pi, \sqrt{81}$

CHAPTER 2
Section 2.6

Check Point Exercises

1. a. 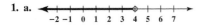   **b.** 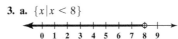   **c.**

2. $\{x \mid x \le 4\}$ **3. a.** $\{x \mid x < 8\}$ **b.** $\{x \mid x > -3\}$

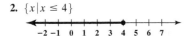   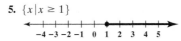   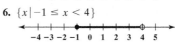

4. $\{x \mid x < -6\}$ **5.** $\{x \mid x \ge 1\}$ **6.** $\{x \mid -1 \le x < 4\}$

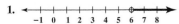   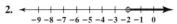   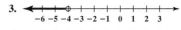

Exercise Set 2.6

1.   **2.**   **3.**

4.   **5.**   **6.**

7.   **8.**   **9.**

10.   **11.**   **12.**

13. $\{x \mid x > 5\}$ **14.** $\{x \mid x < 4\}$ **15.** $\{x \mid x \le 5\}$

16. $\{x \mid x \ge 6\}$ **17.** $\{x \mid x < 3\}$ **18.** $\{x \mid x \ge -4\}$

19. $\{x \mid x < 5\}$ **20.** $\{x \mid x \ge 3\}$ **21.** $\{x \mid x \ge -5\}$

22. $\{x \mid x < -3\}$ **23.** $\{x \mid x > 5\}$ **24.** $\{x \mid x \le 4\}$

25. $\{x \mid x < 5\}$ **26.** $\{x \mid x > 2\}$ **27.** $\{x \mid x < 8\}$

28. $\{x \mid x > 6\}$ **29.** $\{x \mid x > -6\}$ **30.** $\{x \mid x < -4\}$

31. $\{x \mid x > -5\}$ **32.** $\{x \mid x < -3\}$ **33.** $\{x \mid x \le 5\}$

34. $\{x \mid x \ge 3\}$ **35.** $\{x \mid x \le 3\}$ **36.** $\{x \mid x \le -7\}$

37. $\{x \mid x < 16\}$ **38.** $\{x \mid x < 19\}$ **39.** $\{x \mid x > -3\}$

40. $\{x \mid x < 3\}$ **41.** $\{x \mid x \ge -2\}$ **42.** $\{x \mid x \le -5\}$

43. $\{x \mid x > -4\}$ **44.** $\{x \mid x < 3\}$ **45.** $\{x \mid x \ge 4\}$

46. $\{x \mid x \le 6\}$ **47.** $\left\{x \mid x > \dfrac{11}{3}\right\}$ **48.** $\{x \mid x \ge 4\}$

   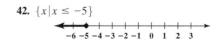      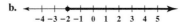

49. $\{x \mid x > 2\}$

50. $\{x \mid x < -1\}$

51. $\{x \mid x < 3\}$

52. $\{x \mid x \geq 0\}$

53. $\left\{x \mid x > \dfrac{5}{3}\right\}$

54. $\left\{x \mid x < -\dfrac{5}{4}\right\}$

55. $\{x \mid x \geq -10\}$

56. $\{x \mid x \geq -2\}$

57. $\{x \mid x < -6\}$

58. $\{x \mid x > 8\}$

59. $\{x \mid 3 < x < 5\}$

60. $\{x \mid 2 < x < 6\}$

61. $\{x \mid -1 \leq x < 3\}$

62. $\{x \mid -2 < x \leq 5\}$

63. $\{x \mid -5 < x \leq -2\}$

64. $\left\{x \mid \dfrac{3}{2} \leq x < \dfrac{11}{2}\right\}$

65. $\{x \mid 3 \leq x < 6\}$

66. $\{x \mid -4 \leq x < 2\}$

Chapter 2 Review Exercises

42. $\{x \mid x < 4\}$

43. $\{x \mid x > -8\}$

44. $\{x \mid x \geq -3\}$

45. $\{x \mid x > 6\}$

46. $\{x \mid x \geq 4\}$

47. $\{x \mid x \leq 2\}$

48. $\{x \mid -\frac{3}{4} < x \leq 1\}$

Chapter 2 Test

19. $\{x \mid x \leq -3\}$

20. $\{x \mid x > 7\}$

21. $\{x \mid -2 \leq x < \frac{5}{2}\}$

CHAPTER 3

Section 3.1

Check Point Exercises

1.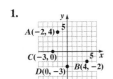

2.

3. $y = 2x$ $y = 10 + x$

x	(x, y)
0	$(0, 0)$
2	$(2, 4)$
4	$(4, 8)$
6	$(6, 12)$
8	$(8, 16)$
10	$(10, 20)$
12	$(12, 24)$

x	(x, y)
0	$(0, 10)$
2	$(2, 12)$
4	$(4, 14)$
6	$(6, 16)$
8	$(8, 18)$
10	$(10, 20)$
12	$(12, 22)$

b.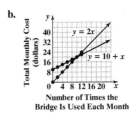

6.   The graph of g is the graph of f shifted down 3 units.

Exercise Set 3.1

1.

2. • (2, 5)

3. (−2, 3)

4. (−1, 4)

5. (−3, −5)

6. (−4, −2)

7.
(4, −1)

8. (3, −2)

9. (−4, 0)

10. (−5, 0)

11. (0, −3)

12. (0, −4)

13. (0, 0)

14. $\left(-3, -1\frac{1}{2}\right)$

15. $\left(-2, -3\frac{1}{2}\right)$

16. (−5, −2.5)

17. (3.5, 4.5)

18. (2.5, 3.5)

19. (1.25, −3.25)

20. (2.25, −4.25)

21.

22.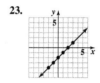

23.

24.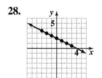

25.

26.

27.

28.

29.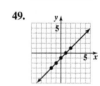

30.

31.

32.

47.

48.

49.

50.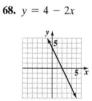

51.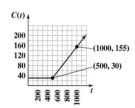

52.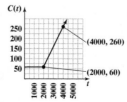

53.

54.

67. $y = 2x + 4$

68. $y = 4 - 2x$

69. $y = 3 - x^2$

70. $y = x^2 + 2$

79. $0.30t - 6$

(200, 54)
(120, 30)

80. $0.30t - 20$

(250, 55)
(200, 40)

81. $C(t) = \begin{cases} 30 & \text{if } 0 \le t \le 500 \\ 30 + 0.25(t - 500) & \text{if } t > 500 \end{cases}$

(1000, 155)
(500, 30)

82. $C(t) = \begin{cases} 60 & \text{if } 0 \le t \le 2000 \\ 60 + 0.10(t - 2000) & \text{if } t > 2000 \end{cases}$

(4000, 260)
(2000, 60)

Section 3.2

Check Point Exercises

1. **3.** **4.** **5.** **6.**

Exercise Set 3.2

1. **2.** **3.** **4.** **5.** **6.**

7. **8.** **21.** **22.** **23.** **24.**

25. **26.** **27.** **28.** **29.** **30.**

31. **32.**

33. a. $y = -3x$ or $y = -3x + 0$

    b. slope $= -3$; y-intercept $= 0$

    c.

34. a. $y = -2x$ or $y = -2x + 0$

    b. slope $= -2$; y-intercept $= 0$

    c.

35. a. $y = \dfrac{4}{3}x$ or $y = \dfrac{4}{3}x + 0$

    b. slope $= \dfrac{4}{3}$; y-intercept $= 0$

    c.

36. a. $y = \dfrac{5}{4}x$ or $y = \dfrac{5}{4}x + 0$

    b. slope $= \dfrac{5}{4}$; y-intercept $= 0$

    c.

37. a. $y = -2x + 3$

    b. slope $= -2$; y-intercept $= 3$

    c.

38. a. $y = -3x + 4$

    b. slope $= -3$; y-intercept $= 4$

    c.

39. a. $y = -\dfrac{7}{2}x + 7$

    b. slope $= -\dfrac{7}{2}$; y-intercept $= 7$

    c.

40. a. $y = -\dfrac{5}{3}x + 5$

    b. slope $= -\dfrac{5}{3}$; y-intercept $= 5$

    c.

41. **42.** **43.** **44.** **45.** **46.**

47. **48.**

67. Find the slope: $m = \dfrac{b - 0}{0 - a} = -\dfrac{b}{a}$. The y-intercept is b.

Write the slope-intercept equation: $y = -\dfrac{b}{a}x + b$.

Add $\dfrac{b}{a}x$ to each side: $\dfrac{b}{a}x + y = b$.

Divide each side by b: $\dfrac{x}{a} + \dfrac{y}{b} = 1$.

This is called the *intercept form* because the number dividing x is the x-intercept and the number dividing y is the y-intercept.

Section 3.3

Exercise Set 3.3

29.
There appears to be a positive correlation.

30.
There appears to be a positive correlation.

31.
There appears to be a negative correlation.

32.
There does not appear to be a correlation.

60. a.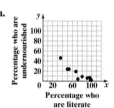

c.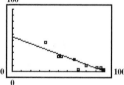

Section 3.4

Check Point Exercises

2. $\{(-3, 4)\}$

Exercise Set 3.4

5. $\{(4, 2)\}$ **6.** $\{(3, -1)\}$ **7.** $\{(3, 0)\}$ **8.** $\{(1, 0)\}$

9. $\{(-1, 4)\}$ **10.** $\{(1, 2)\}$ **11.** $\{(3, -4)\}$ **12.** $\{(1, -1)\}$

45. $\begin{cases} y = x - 4 \\ y = -\dfrac{1}{3}x + 4 \end{cases}$ **46.** $\begin{cases} y = \dfrac{1}{3}x + 2 \\ y = \dfrac{1}{3}x - 2 \end{cases}$ **57. a.** $C(x) = 18{,}000 + 20x$ **b.** $R(x) = 80x$

c. $(300, 24{,}000)$; When 300 canoes are produced and sold, both revenue and cost are $24,000.   **58. a.** $C(x) = 100{,}000 + 100x$ **b.** $R(x) = 300x$
c. $(500, 150{,}000)$; When 500 bicycles are produced and sold, both cost and revenue are $150,000.   **59. a.** $C(x) = 30{,}000 + 2500x$
b. $R(x) = 3125x$ **c.** $(48, 150{,}000)$; For 48 sold-out performances, both cost and revenue are $150,000.   **60. a.** $C(x) = 30{,}000 + 0.02x$
b. $R(x) = 0.5x$ **c.** $(62{,}500, 31{,}250)$; For 62,500 cards, both cost and revenue are $31,250.

Section 3.5

Check Point Exercises

1.   **2.** 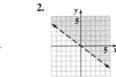   **3. a.** 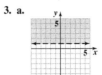   **b.**   **5.**   **6.**

Exercise Set 3.5

1.   **2.**   **3.**   **4.**   **5.**   **6.**

7.   **8.**   **9.**   **10.**   **11.**   **12.**

13.   **14.**   **15.**   **16.**   **17.**   **18.**

19.   **20.**   **21.**   **22.**   **23.** 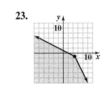   **24.**

25.   **26.**   **27.**   **28.** 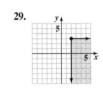   **29.**   **30.**

31.   **32.**   **33.**   **34.**   **35.**   **36.**

37.   **38.**   **39.** $y \geq -2x + 4$ **40.** $y \geq -3x + 2$ **41.** $\begin{cases} x + y \leq 4 \\ 3x + y \leq 6 \end{cases}$ **42.** $\begin{cases} x + y \leq 3 \\ 4x + y \leq 6 \end{cases}$

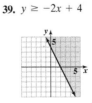   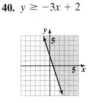

43. Point $A = (66, 160)$; $5.3(66) - 160 \geq 180$, or $189.8 \geq 180$, is true; $4.1(66) - 160 \leq 140$, or $110.6 \leq 140$, is true.
44. Point $B = (76, 220)$; $5.3(76) - 220 \geq 180$, or $182.8 \geq 180$, is true; $4.1(76) - 220 \leq 140$, or $91.6 \leq 140$, is true.

47. b.   **48. b.**   **51.** $\begin{cases} x \geq -2 \\ y > -1 \end{cases}$ **52.** $\begin{cases} y > x - 3 \\ y \leq x \end{cases}$

Chapter 3 Review Exercises

1.   **2.**   **3.**   **4.**   **5.**   **6.**

7.   **12.**   **13.**   **17.**   **18.**   **19.**

24.   **25.**   **26.**   **27.**   **28. a.** $y = -2x$
b. slope: -2; y-intercept: 0

c.

29. a. $y = \dfrac{5}{3}x$

b. slope: $\dfrac{5}{3}$; y-intercept: 0

30. a. $y = -\dfrac{3}{2}x + 2$

b. slope: $-\dfrac{3}{2}$; y-intercept: 2

31.   **32.**   **33.**

c.

c.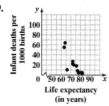

34. a. 254; If no women in a country are literate, the mortality rate of children under 5 is 254 per thousand.   **b.** -2.4; For each additional 1% of adult females who are literate, the mortality rate of children under 5 decreases by 2.4 per thousand.

38.

There appears to be a positive correlation.

39.

Life expectancy
(in years)

There appears to be a negative correlation.

60. c. $(240, 108{,}000)$; This means the company will break even if it produces and sells 240 desks. At this level, both revenue and cost will be $108,000.

62.   **63.**   **64.**   **65.**   **66.**

67.   **68.**   **69.**   **70.**   **71.**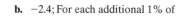

72.

73.

74.

Chapter 3 Test

1.

6.

8.

9.

13.   There appears to be a strong negative correlation.

21.

22.

23.

24.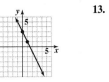

CHAPTER 4
Section 4.5

Check Point Exercises

1. b. $f(x) = x^2 - 6x + 8$
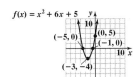
(6, 8), (5, 3), (4, 0), (3, −1), (2, 0), (1, 3), (0, 8), 10

5. $f(x) = x^2 + 6x + 5$
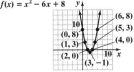
(−5, 0), (−1, 0), (0, 5), (−3, −4), 10

6. $f(x) = -x^2 - 2x + 5$

(−1, 6), (0, 5), (−3.4, 0), (1.4, 0), 8

7. c. $f(x) = -0.005x^2 + 2x + 5$
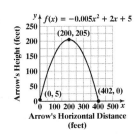
Arrow's Height (feet), (200, 205), (0, 5), (402, 0), Arrow's Horizontal Distance (feet)

Exercise Set 4.5

25. $f(x) = x^2 + 8x + 7$
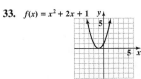

26. $f(x) = x^2 + 10x + 9$
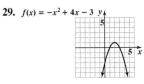

27. $f(x) = x^2 - 2x - 8$
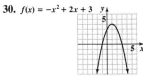

28. $f(x) = x^2 + 4x - 5$
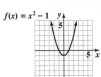

29. $f(x) = -x^2 + 4x - 3$
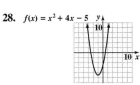

30. $f(x) = -x^2 + 2x + 3$
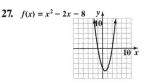

31. $f(x) = x^2 - 1$
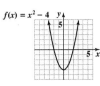

32. $f(x) = x^2 - 4$

33. $f(x) = x^2 + 2x + 1$
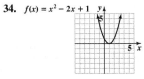

34. $f(x) = x^2 - 2x + 1$

35. $f(x) = -2x^2 + 4x + 5$

36. $f(x) = -3x^2 + 6x - 2$

47. c. $f(x) = -0.8x^2 + 2.4x + 6$
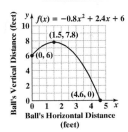
Ball's Vertical Distance (feet), (1.5, 7.8), (0, 6), (4.6, 0), Ball's Horizontal Distance (feet)

48. c. $f(x) = -0.8x^2 + 3.2x + 6$
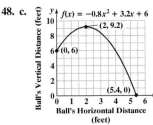
Ball's Vertical Distance (feet), (2, 9.2), (0, 6), (5.4, 0), Ball's Horizontal Distance (feet)

53.

$y = x^2 - 4x + 4$, $x + y = 2$, (2, 0), (1, 1)

Chapter 4 Review Exercises

81. $f(x) = x^2 - 6x - 7$

82. $f(x) = -x^2 - 2x + 3$

83. $f(x) = -3x^2 + 6x + 1$

84. $f(x) = x^2 - 4x$

86. d. 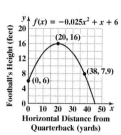 $f(x) = -0.025x^2 + x + 6$
(20, 16); (0, 6); (38, 7.9)
Football's Height (feet) / Horizontal Distance from Quarterback (yards)

Chapter 4 Test

24. $f(x) = x^2 + 2x - 8$
(−4, 0); (2, 0); (0, −8); (−1, −9)

25. 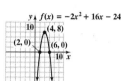 $f(x) = -2x^2 + 16x - 24$
(4, 8); (2, 0); (6, 0)

CHAPTER 5

Section 5.2

Exercise Set 5.2

33. a. 1900: 25.6 people per square mile; 2010: 87.4 people per square mile   **34. a.** 1800: 6.1 people per square mile; 2010: 87.4 people per square mile
37. Illinois: 222.2 people per square mile; Ohio: 257.6 people per square mile; Ohio has the greater population density by 35.4 people per square mile.
38. New York: 356.8 people per square mile; Rhode Island: 680.5 people per square mile; Rhode Island has the greater population density by 323.7 people per square mile.

Section 5.3

Exercise Set 5.3

51. a. $\frac{9}{5}$; Fahrenheit temperature increases by $\frac{9}{5}°$ for each 1° change in Celsius temperature.

Chapter 5 Review Exercises

22. 16,513.8 people per square mile; In Singapore there is an average of 16,513.8 people for each square mile.

CHAPTER 6

Section 6.1

Exercises Set 6.1

21. $x + x + 12° = 90°$; 39°, 51°   **22.** $x + x + 56° = 90°$; 17°, 73°   **29.** $m\angle 1 = 68°$; $m\angle 2 = 68°$; $m\angle 3 = 112°$; $m\angle 4 = 112°$; $m\angle 5 = 68°$; $m\angle 6 = 68°$; $m\angle 7 = 112°$ **30.** $m\angle 1 = 126°$; $m\angle 2 = 54°$; $m\angle 3 = 126°$; $m\angle 4 = 126°$; $m\angle 5 = 54°$; $m\angle 6 = 54°$; $m\angle 7 = 126°$
57. When two parallel lines are intersected by a transversal, corresponding angles have the same measure.

Section 6.2

Exercise Set 6.2

11. The three angles of the large triangle are given to have the same measures as the three angles of the small triangle.; 5 in.   **12.** The three angles of the large triangle are given to have the same measures as the three angles of the small triangle.; 8 in.   **13.** Two angles of the large triangle are given to have the same measures as two angles of the small triangle.; 6 m   **14.** One angle pair is given to have the same measure (right angles). The triangles also share a common angle.; 1.25 in.   **15.** One angle pair is given to have the same measure (right angles). Another angle pair consists of vertical angles with the same measure.; 16 in.   **16.** One angle pair is given to have the same measure. Another angle pair consists of vertical angles with the same measure.; 6 ft   **20.** Answers will vary; an example is: Because all of the corresponding angles have the same measure, the triangles must be similar.

Section 6.3

Exercise Set 6.3

5. a: square; b: rhombus; d: rectangle; e: parallelogram   **6.** a: square; b: rhombus   **8.** c: trapezoid; d: rectangle; e: parallelogram
45. If the polygons were all regular polygons, the sum would be 363°. The sum is not 360°.

46. If the polygons were all regular polygons, the sum would be $365\frac{1}{7}°$. The sum is not 360°.

c. $(300, 24,000)$; When 300 canoes are produced and sold, both revenue and cost are \$24,000.   **58. a.** $C(x) = 100,000 + 100x$ **b.** $R(x) = 300x$
c. $(500, 150,000)$; When 500 bicycles are produced and sold, both cost and revenue are \$150,000.   **59. a.** $C(x) = 30,000 + 2500x$
b. $R(x) = 3125x$ **c.** $(48, 150,000)$; For 48 sold-out performances, both cost and revenue are \$150,000.   **60. a.** $C(x) = 30,000 + 0.02x$
b. $R(x) = 0.5x$ **c.** $(62,500, 31,250)$; For 62,500 cards, both cost and revenue are \$31,250.

Section 3.5

Check Point Exercises

1. 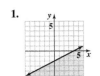 **2.**   **3. a.** 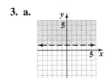 **b.**   **5.**   **6.**

Exercise Set 3.5

1.   **2.**   **3.**   **4.**   **5.**   **6.**

7.   **8.**   **9.**   **10.** 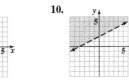 **11.** 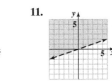 **12.**

13.   **14.**   **15.**   **16.**   **17.**   **18.**

19.   **20.**   **21.**   **22.** 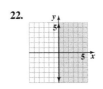 **23.**   **24.**

25.   **26.**   **27.**   **28.**   **29.** 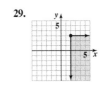 **30.**

31.   **32.**   **33.**   **34.**   **35.**   **36.**

37.   **38.**   **39.** $y \geq -2x + 4$ **40.** $y \geq -3x + 2$ **41.** $\begin{cases} x + y \leq 4 \\ 3x + y \leq 6 \end{cases}$ **42.** $\begin{cases} x + y \leq 3 \\ 4x + y \leq 6 \end{cases}$

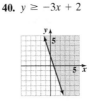

43. Point $A = (66, 160)$; $5.3(66) - 160 \geq 180$, or $189.8 \geq 180$, is true; $4.1(66) - 160 \leq 140$, or $110.6 \leq 140$, is true.
44. Point $B = (76, 220)$; $5.3(76) - 220 \geq 180$, or $182.8 \geq 180$, is true; $4.1(76) - 220 \leq 140$, or $91.6 \leq 140$, is true.

47. b.

48. b.

51. $\begin{cases} x \geq -2 \\ y > -1 \end{cases}$

52. $\begin{cases} y > x - 3 \\ y \leq x \end{cases}$

Chapter 3 Review Exercises

1.

2.

3.

4.

5.

6.

7.

12.

13.

17.

18.

19.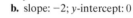

24.

25.

26.

27.

28. a. $y = -2x$
    b. slope: -2; y-intercept: 0
    c.

29. a. $y = \dfrac{5}{3}x$
    b. slope: $\dfrac{5}{3}$; y-intercept: 0

30. a. $y = -\dfrac{3}{2}x + 2$
    b. slope: $-\dfrac{3}{2}$; y-intercept: 2

31.

32.

33.

c.

c.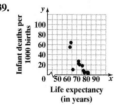

34. a. 254; If no women in a country are literate, the mortality rate of children under 5 is 254 per thousand.   **b.** -2.4; For each additional 1% of adult females who are literate, the mortality rate of children under 5 decreases by 2.4 per thousand.

38.

There appears to be a positive correlation.

39.

There appears to be a negative correlation.

60. c. $(240, 108{,}000)$; This means the company will break even if it produces and sells 240 desks. At this level, both revenue and cost will be $108,000.

62.

63.

64.

65.

66.

67.

68.

69.

70.

71.

CHAPTER 7

Section 7.5

Exercise Set 7.5

11. a. $\frac{1}{177,100} \approx 0.00000565$ **b.** $\frac{27,132}{177,100} \approx 0.153$ **12. a.** $\frac{2}{429} \approx 0.00466$ **b.** $\frac{7}{429} \approx 0.0163$ **14.** $\frac{70}{323} \approx 0.217$ **15. b.** 1287

c. $\frac{1287}{2,598,960} \approx 0.000495$ **17.** $\frac{11}{1105} \approx 0.00995$ **19.** $\frac{36}{270,725} \approx 0.000133$ **20.** $\frac{16}{270,725} \approx 0.0000591$ **24.** The prize is shared among all

winners. You are guaranteed to win but not to win $500 million.   **26.** $\frac{235,620}{2,598,960} \approx 0.0907$

Section 7.6

Exercise Set 7.6

86. $\frac{193}{463} \approx 0.417$

91.
$$\frac{P(E)}{1 - P(E)} = \frac{a}{b}$$
$$bP(E) = a(1 - P(E))$$
$$bP(E) = a - aP(E)$$
$$aP(E) + bP(E) = a$$
$$P(E)(a + b) = a$$
$$P(E) = \frac{a}{a + b}$$

Section 7.7

Exercise Set 7.7

25. a. $\frac{1}{256} \approx 0.00391$ **b.** $\frac{1}{4096} \approx 0.000244$ **d.** ≈ 0.476 **57.** $\frac{412,368}{412,878} \approx 0.999$ **58.** $\frac{1601}{164,128} \approx 0.010$ **59.** $\frac{412,368}{574,895} \approx 0.717$

60. $\frac{1601}{2111} \approx 0.758$ **73. b.** $\frac{365}{365} \cdot \frac{364}{365} \cdot \frac{363}{365} \approx 0.992$ **e.** $1 - \frac{365}{365} \cdot \frac{364}{365} \cdot \frac{363}{365} \cdot \ldots \cdot \frac{343}{365} \approx 0.507$

Chapter 7 Review Exercises

50. a. $\frac{1}{15,504} \approx 0.0000645$ **b.** $\frac{100}{15,504} \approx 0.00645$ **100.** $\frac{27,336}{31,593} \approx 0.865$ **101.** $\frac{11,161}{31,593} \approx 0.353$ **102.** $\frac{29,214}{31,593} \approx 0.925$ **103.** $\frac{15,256}{31,593} \approx 0.483$

104. $\frac{4312}{31,593} \approx 0.136$ **105.** $\frac{3684}{27,336} \approx 0.135$ **106.** $\frac{2169}{2379} \approx 0.912$

CHAPTER 8

Section 8.1

Exercise Set 8.1

7.

Time Spent on Homework (in hours)	Number of Students
15	4
16	5
17	6
18	5
19	4
20	2
21	2
22	0
23	0
24	2
	30

8.

Height (in inches)	Number of Students
66	1
67	3
68	2
69	1
70	3
71	7
72	5
73	5
74	1
75	2
	30

9. $0, 5, 10, \ldots, 40, 45$ **10.** $4, 9, 14, \ldots, 44, 49$

17.

Age at Inauguration	Number of Presidents
41–45	2
46–50	9
51–55	15
56–60	9
61–65	7
66–70	2
	44

18.

IQ Score	Number of Students
85–89	2
90–94	5
95–99	12
100–104	14
105–109	15
110–114	11
115–119	8
120–124	3
	70

19. a. 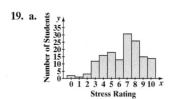 **b.**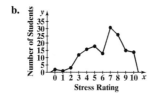

20.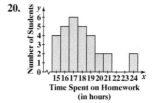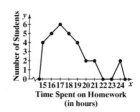

21.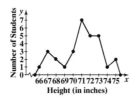

30.

Stem	Leaves
4	9 8 6 9 7 2 3 6 7
5	7 7 7 8 7 4 1 0 2 6 4 1 5 5 4 1 6 5 1 4 1 5 6 2 4
6	1 1 8 4 5 0 2 1 9 4

31.

Stem	Leaves
2	8 8 9 5
3	8 7 0 1 2 7 6 4 0 5
4	8 2 2 1 4 5 4 6 2 0 8 2 7 9
5	9 4 1 9 1 0
6	3 2 3 6 6 3

The greatest number of high school teachers are in their 40s.

34. Sizes of books are not scaled proportionally in terms of the data they represent.

39. Sample answer:

Length (miles)	Number of Rivers
501–1000	12
1001–1500	8
1501–2000	3
2001–2500	1
2501–3000	1
	25

Section 8.2

Exercise Set 8.2

9. ≈4.71 **11.** ≈6.26 **12.** ≈4.74 **55. a.** ≈67.4 thousand **b.** 51 thousand **c.** 51 thousand and 92 thousand **d.** 113.5 thousand
56. a. ≈86.6 million **b.** $26 million **c.** no mode **d.** $264.5 million **57. a.** ≈17.27 **b.** 17 **c.** 7, 12, 17 **d.** 24.5
62. a. mean: ≈854; median: 850; mode: 1000; midrange: 850 **64.** Sample answers: **a.** 75, 80, 80, 90, 91, 94 **b.** 50, 80, 80, 85, 90, 95
c. 70, 75, 80, 85, 90, 100 **d.** 75, 80, 85, 90, 95, 95 **e.** 75, 80, 85, 85, 90, 95 **f.** 68, 70, 72, 72, 74, 76

Section 8.3

Exercise Set 8.3

7. a.

Data item	Deviation
3	−9
5	−7
7	−5
12	0
18	6
27	15

8. a.

Data item	Deviation
84	−7
88	−3
90	−1
95	4
98	7

9. a.

Data item	Deviation
29	−20
38	−11
48	−1
49	0
53	4
77	28

10. a.

Data item	Deviation
60	−5
60	−5
62	−3
65	0
65	0
65	0
66	1
67	2
70	5
70	5

11. a. 91

b.

Data item	Deviation
85	−6
95	4
90	−1
85	−6
100	9

12. a. 84

b.

Data item	Deviation
94	10
62	−22
88	4
85	1
91	7

13. a. 155

b.

Data item	Deviation
146	−9
153	−2
155	0
160	5
161	6

14. a. 137

b.

Data item	Deviation
150	13
132	−5
144	7
122	−15

15. a. 2.70

b.

Data item	Deviation
2.25	−0.45
3.50	0.80
2.75	0.05
3.10	0.40
1.90	−0.80

16. a. 0.39

b.

Data item	Deviation
0.35	−0.04
0.37	−0.02
0.41	0.02
0.39	0
0.43	0.04

27. *Sample A*: mean: 12; range: 12; standard deviation: ≈ 4.32
Sample B: mean: 12; range: 12; standard deviation: ≈ 5.07
Sample C: mean: 12; range: 12; standard deviation: 6
The samples have the same mean and range, but different standard deviations.
28. *Sample A*: mean: 14; range: 12; standard deviation: ≈ 4.32
Sample B: mean: 14; range: 12; standard deviation: ≈ 5.07
Sample C: mean: 14; range: 12; standard deviation: 6
The samples have the same mean and range, but different standard deviations.
37. a. male artists; All of the data items for the men are greater than the greatest data item for the women.   **b.** male artists: 93; female artists: 58
c. male artists; There is greater spread in the data for the men.   **d.** male artists: 32.64; female artists: 6.82   **38. a.** most recent six presidents; The
ages for the last six presidents include ages both lower and higher than any of the ages for the first six presidents.   **42.** The mean is increased by 2.;
The standard deviation is unaffected.

Section 8.4

Check Point Exercises

7. b. We can be 95% confident that between 34% and 38% of Americans read more than ten books per year.   **c.** Sample answer: Some people may
be embarrassed to admit that they read few or no books in a year.

Exercise Set 8.4

57. The person who scores 127 on the Wechsler has the higher IQ.   **58.** The person who scores 148 on the Wechsler has the higher IQ.
67. b. We can be 95% confident that between 65.5% and 72.5% of the population favor required gun registration as a means to reduce gun violence.
68. b. We can be 95% confident that between 18.7% and 23.3% of interviewers are turned off by a job applicant being arrogant.   **69. b.** We can be
95% confident that between 58.6% and 61.8% of all TV households watched the final episode of *M*A*S*H*.   **70. b.** We can be 95% confident that
between 49.5% and 52.7% of all TV households watched *Roots, Part 8*.

Chapter 8 Review Exercises

2.

Time Spent on Homework (in hours)	Number of Students
6	1
7	3
8	3
9	2
10	1
	10

3.

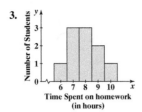

4.

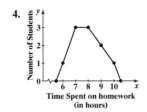

5.

Grades	Number of Students
0–39	19
40–49	8
50–59	6
60–69	6
70–79	5
80–89	3
90–99	3
	50

6.

Stem	Leaves
1	8 4 3 3 7 1
2	4 6 9 9 2 7
3	4 9 6 1 5 1 1
4	4 7 9 1 2 2 0 5
5	4 7 9 0 6 1
6	3 7 0 8 3 9
7	2 5 4 0 3
8	1 7 6
9	1 0 5

7. Sizes of barrels are not scaled proportionally in terms of the data they represent.

21. a.

Age at First Inauguration	Number of Presidents	Age at First Inauguration	Number of Presidents
42	1	57	4
43	1	58	1
44	0	59	0
45	0	60	1
46	2	61	3
47	2	62	1
48	1	63	0
49	2	64	2
50	1	65	1
51	5	66	0
52	2	67	0
53	0	68	1
54	5	69	1
55	4		Total: 44
56	3		

b. mean: ≈ 54.66 yr; median: 54.5 yr; mode: 51 yr, 54 yr (bimodal); midrange: 55.5 yr

24. a.

Data item	Deviation
29	−6
9	−26
8	−27
22	−13
46	11
51	16
48	13
42	7
53	18
42	7

b. 0

25. a. 50 **b.**

Data item	Deviation
36	−14
26	−24
24	−26
90	40
74	24

c. 0

28. mean: 49; range: 76; standard deviation: ≈ 24.32

29. Set A: mean: 80; standard deviation: 0; Set B: mean: 80; standard deviation: ≈ 11.55; Answers will vary. **46.** vocabulary test **50. b.** We can be 95% confident that between 28.9% and 33.1% of American adults would be willing to sacrifice a percentage of their salary to work for an environmentally friendly company. **51. a.** skewed to the right **b.** mean: 2.1 syllables; median: 2 syllables; mode: 1 syllable **c.** yes; The mean is greater than the median, which is consistent with a distribution skewed to the right.

Chapter 8 Test

2.

Score	Frequency
3	1
4	2
5	3
6	2
7	2
8	3
9	2
10	1
	16

3.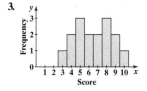

4.

5.

Class	Frequency
40–49	3
50–59	6
60–69	6
70–79	7
80–89	6
90–99	2
	30

6.

Stem	Leaves
4	8 1 6
5	1 0 5 0 9 0
6	7 2 0 3 1 1
7	9 8 3 1 9 1 5
8	9 3 0 8 9 1
9	3 0

7. The heights of the bars are not proportional to the percents they represent. **19. b.** We can be 95% confident that between 43.7% and 50.3% of Americans ages 16 and older find "whatever" most annoying in conversation.

Subject Index

Application Index

Absence from school, due to bullying, 580
Activities, most-dreaded, 539–540
Actors, casting combinations, 427, 435, 488
Advertising, sales and, 152
Age
　blood pressure and, 301–302
　body-mass index and, 239
　calculating, 12
　median age of the U.S. population, 208–209
　of Oscar winners, 506–507
　of presidents, 506, 530, 551
　stress level and, 204
Aging, near-light speed travel and, 52, 56
Ailments, tobacco and, 480–481
Alcohol, blood concentration of, 108, 349
Alligator, tail length of, 146
Angle(s)
　of snow on windows, 371
　on umbrellas, 370
Anxiety
　in college students, 551
　over dental work, 543
Apartments
　demand and supply for, 231
　option combinations, 427, 457
Aquarium
　volume of water in, 342–343, 345, 358
　weight of water in, 348
Architecture, golden rectangles in, 51
Area
　of islands, 345
　of kitchen floor tiling, 400
　to paint, 401
　of rectangular room, 401
　for shipping boxes, 419
Area codes, combinations of, 425–426, 427
Arrow, height of archer's, 313
Art
　golden rectangles in, 302
　painting and frame, area of, 322
Autism, brain development and, 55

Baboon grooming behavior, 461–462
Bachelor's degrees, gender differences in
　acquiring, 205
Ball(s). See also specific types of balls
　height and time in motion, 288–289, 315, 323
　height of bounces, 154
　random selection of colored, 492
Baseball, 335
　batting average, 147
　batting orders, 431
　distance from home plate to second base, 381
　height and time in motion, 323
　salaries in, 86
Baseboard installation, 391, 418
Basketball
　dimensions of court, 390

free throw odds in favor, 469
volume of, 406
Bicycle
　banking angle, 154
　front-to-rear sprocket ratio, 147
　hip angle of rider on, 371
　manufacturing, 230
　sales tax on, 562
Birthdays, probabilities and coincidence of
　shared, 482
Births
　in United States (2000-2009), 254, 262–263
　worldwide, 134
Blood, red blood cells in the body, 98
Blood alcohol concentration (BAC), 108, 349
Blood pressure, 301–302
　age and, 301–302
Blood volume, 148–149
Body-mass index (BMI), 155, 239
Book(s)
　arrangement of, 428–429, 434, 489
　book club selections, 441
　collections of, 441
　combinations of, 492
　number read a year, 541
Box(es)
　dimensions of, 294
　shipping, space needed by, 419
　volume of, 410
Breast cancer, mammography screening for,
　477–478
Budget deficit, federal, 97, 98
Bullying, school absences due to, 580
Buses, fare options, 135
Business
　ad costs, 251
　break-even point, 226–227, 230, 250
　CD sales, 146
　commission, 175
　customer service representatives, 442
　defective products, 442
　fractional ownership of franchise, 44
　hamburger restaurant, 427
　Internet marketing consultation, 432
　investment in, 230
　manufacturing costs, 111
　officers, 435
　profit, 167
　tree sales, 146

Caloric needs, 104–105
Camera, commission rate on sale of, 573
Campers, seating arrangements for, 435
Campsites, choosing, 442
Cancer, breast, 477–479
Cancer, cigarettes and, 213
Candy
　fat content of candy bars, 231

pH of human mouth and, 279
selection of, 489
Canoe manufacturing, 230
Car(s)
　depreciated value of, 134, 176
　gasoline consumed, 97
　option combinations, 424–425, 426, 427, 492
　outcome of accidents in, 481
　rental cost, 156, 158, 167
　sales commission on, 566
　sales tax on, 565
　skidding distance and speed of, 54
　stopping distance of, 183–184
　tires, durability of, 551
Cardiovascular disease, probability of, 467
Cards, probability of selecting, 445–446, 457,
　458–459, 460–461, 465, 467, 468, 469,
　474–475, 476–477, 480, 482, 488, 489, 490,
　492
Carpentry
　baseboard costs, 391, 418
　baseboard installation, 418
Carpet installation, cost of, 393, 400, 401, 419
Cellphones
　plans, 187–188, 191, 192
　subscription to, 167
Cereals, potassium content of, 529
Checkout line, gender combinations at, 435, 456
Chess tournament, round-robin, 293
Child mortality, literacy and, 248, 249
Children
　cost of raising, 577–579
　drug dosage for, 64
Chocolates, selection of, 473–474, 481, 489, 491
Cigarette smoking. See Smoking
City(ies)
　distance between, 335
　hottest, 517
　with new college graduates, 521
　visiting in random order, 489
Clock, movement around, in degrees, 363
Club, officers of, 489
Coin toss, 448, 480
College(s)
　attendance at, 490
　attitudes toward, 134
　final course grade, 163–164, 167, 175, 498, 499,
　518
College student(s)
　anxiety in, 551
　attitudes of, 127–128
　enrollment rates, 135
　heights of, 505
　IQ scores of, 506
　majors of, 491
　random selection of freshmen vs. other years,
　475–476, 492
　smoking among, 576–577

Application Index

Credits

Photos

Cover

cover Daniel R. Burch/Getty Images; **cover** Oleg Belov/Shutterstock; **cover** Oleg Belov/Shutterstock

FM

vii (right) Domen Colja/Alamy; **vii (left)** SHNS photo courtesy Comedy Central/Newscom; **vii (center)** Nicolas McComber/Getty Images; **viii (left)** Everett Collection Inc/Alamy; **viii (center)** Günay Mutlu/Getty Images; **viii (right)** Alex Bramwell/Getty Images; **xvii** Bob Blitzer

Chapter 1

1 Domen Colja/Alamy; 2 Image Source/Digital Vision/Getty Images; 16 (left) Alon Brik/Dreamstime; 16 (right) Shutterstock; 28 Burwell and Burwell Photography/E+/Getty Images; 41 CBS Television/votes, Robert/Album/Newscom; 46 Chitose Suzuki/AP Images; 47 Joshua Haviv/Fotolia; 51 Anastasios71/Shutterstock; 57 AF archive/Alamy; 58 Blitzer, Robert. F.; 67 U.S. Geological Survey; 72 Fotolia; 76 Sergey Galushko/iStock/Getty Images; 82 Jokerpro/Shutterstock

Chapter 2

101 Newscom; 102 Sergiy Bykhunenko/iStock/Getty Images; 104 (left) ImageSource/Age Fotostock; 104 (right) Tetra Images/Alamy; 108 Ikon Images/SuperStock; 112 SHNS photo courtesy Comedy Central/Newscom; 123 Island Effects/Vetta/Getty Images; 124 (top) (a) Stuart Ramson/AP Images; 124 (top) (b) Nancy Kaszerman/ZUMA Press, Inc./Alamy; 124 (top) (c) ZUMAPRESS, Inc./Alamy; 124 (top) (d) ZUMAPRESS, Inc./Alamy; 124 (bottom) (a) Alamy; 124 (bottom) (b) Michael N. Paras/Age Fotostock; 124 (bottom) (c) Image Source/Age Fotostock; 124 (bottom) (d) Jupiter Images;

127 Blitzer, Robert F.; 136 Doc White/Nature Picture Library/Alamy; 143 Chris Burt/Shutterstock; 147 (left) Alex Segre/Alamy; 147 (right) Moodboard/Alamy; 148 Stan Fellerman/Corbis; 149 Exactostock/SuperStock; 154 AP Images; 156 Steve Hamblin/Alamy; 164 Pearson Education, Inc; 175 (left) Paramount Picture/Everett Collection 175 (right) Photos 12/Alamy;

Chapter 3

177 Big Cheese Photo LLC/Alamy; 178 Dreamstime LLC; 193 Matthew Ward/Dorling Kindersley; 206 Paul Prescott/Shutterstock; 217 RubberBall/Alamy; 226 Sun/Newscom; 231 Terry J Alcorn/E+/Getty Images

Chapter 4

253 Pearson Education, Inc; 254 Günay Mutlu/Getty Images; 264 Robert F Blitzer; 267 Steve Shott/DkImages; 279 Christophe Testi/iStock/Getty Images; 283 Lise Gagne/iStock/Getty Images; 294 Brian Goodman/Shutterstock; 305 (top) Custom Life Science Images/Alamy; 305 (bottom) Any Chance Productions

Chapter 5

325 EcoPrint/Shutterstock; 326 Dorling Kindersley; 328 (a) U.S. Bureau of Engraving and Printing; 328 (b) U.S. Bureau of Engraving and Printing; 328 (c) U.S. Bureau of Engraving and Printing; 328 (d) U.S. Bureau of Engraving and Printing; 328 (e) Pearson Education, Inc; 328 (f) Pearson Education, Inc; 333 Dorling Kindersley; 336 Oksana Perkins/iStock/Getty Images; 338 Mtilghma/Fotolia; 346 (top) Vicki Reid/iStock/Getty Images; 346 (bottom) Ben Goode/Dreamstime; 351 Katvic/Fotolia

Chapter 6

361 Alla Shcherbak/iStock/Getty Images; 362 Hemis/Alamy; 371 Everett Collection

Inc/Alamy; 383 WDG Photo/Shutterstock; 384 Florin Tirlea/E+/Getty Images; 391 Iriana Shiyan/Fotolia; 402 Chas Metivier/KRT/Newscom; 404 Uyen Le/Photodisc/Getty Images

Chapter 7

421 (left) Historical/Corbis; 421 (right) Ted Spiegel/Historical Premium/Corbis; 422 Nicolas McComber/Getty Images; 423 Jaimie Duplass/Fotolia; 425 Oleksiy Mark/Shutterstock; 428 Craftvision/Getty Images; 436 (bottom left) Allstar Picture Library/Alamy; 436 (bottom right) Edd Westmacott/Alamy; 436 (center) JM5 Wenn Photos/Newscom; 436 (top left) United Archives GmbH/Alamy; 436 (top right) United Archives GmbH/Alamy; 442 Jupiterimages/Getty Images; 446 Pearson Education, Inc; 448 Ronstik/Getty Images; 451 (bottom) W. Breeze/Evening Standard/Getty Images; 451 (top) Monkey Business Images/Shutterstock; 452 (a) ZUMA Press, Inc./Alamy; 452 (b) Neal Preston/Historical Premium/Corbis; 452 (c) CBS Photo Archive/Getty Images; 452 (d) Bryan Bedder/Getty Images; 452 (e) Hulton Archive/Getty Images; 458 Mogens Trolle/Shutterstock; 462 Manoj Shah/Getty Images; 470 Tom Salyer/Alamy; 472 Bettmann/Corbis;

Chapter 8

493 Mstay/Getty Images; 494 Everett Collection; 495 Library of congress; 507 Pearson Education, Inc; 508 Alex Bramwell/Getty Images; 523 Epic StockMedia/Shutterstock; 531 Marmaduke St. John/Alamy; 533 Peter Barritt/Alamy; 535 Janice Richard/E+/Getty Images; 539 (left) Niels Laan/E+/Getty Images; 539 (right) Kemal Bas/iStock/Getty Images; 543 Image Source/Getty Images.

Text

Chapter 1

8 Excerpt from Coincidences, Chaos, and All That Math Jazz: Making Light of Weighy Ideas, 5e by Edward B. Burger, Michael P. Starbird. Copyright © 2005. Published by W.W. Norton.; **24** Quote by Cha Sa-soon from "Woman Passes 950th Driving Test." Published by Newsweek, © 2009.; **41** Dialogue excerpt from the television series "Numb3rs". Used by permission of CBS Television Studios.; **64** Excerpt from The Restaurant at the End of the Universe by Douglas Adams. Copyright © 1980. Published by Pan Books.

Chapter 2

121 Excerpt from Math Study Skills, 1st Edition, by Alan Bass. © 2008. Reprinted and Electronically reproduced by permission of Pearson Education, Inc., New Jersey, NJ.

Chapter 3

202 Excerpt from Math Study Skills, 1st Edition, by Alan Bass. © 2008. Published by Pearson Education Inc.

Chapter 6

387 Quote from The Man, The Legend, The Artist, Escher's Work.

Chapter 7

428 Excerpt from The Essential Groucho: Writings by, for, and about Groucho Marx edited by Stefan Kanfer. Copyright © 2008. Published by Penguin Books.; **428** A personal quote by Steve Wright.; **428** A personal quote by Henny Youngman.; **428** A personal quote by Jerry Seinfeld.; **428** A personal quote by Phyllis Diller.; **434** Excerpt from The Cambridge Encyclopedia Of Language by David Crystal. Copyright © 1987. Published by Cambridge University Press.; **442** A personal quote by Phyllis Diller.; **442** A personal quote by Rita Rudner.; **442** A personal quote by Henny Youngman.